Sustainable Global Resources of Seaweeds
Volume 2

Ambati Ranga Rao • Gokare A. Ravishankar
Editors

Sustainable Global Resources of Seaweeds Volume 2

Food, Pharmaceutical and Health Applications

Springer

Editors
Ambati Ranga Rao
Department of Biotechnology
Vignan's Foundation of Science,
Technology and Research
Deemed to be University
Vadlamudi, Guntur, Andhra Pradesh, India

Gokare A. Ravishankar
Dr. C. D. Sagar Centre for Life Sciences
Dayananda Sagar College of Engineering
Dayananda Sagar Institutions
Bengaluru, Karnataka, India

ISBN 978-3-030-92173-6 ISBN 978-3-030-92174-3 (eBook)
https://doi.org/10.1007/978-3-030-92174-3

This Springer imprint is published by the registered company Springer Nature Switzerland AG
The registered company address is: Gewerbestrasse 11, 6330 Cham, Switzerland

This volume is dedicated to Prof. Gopalakrishnan Subramanian, a doyen in the field of phycology. He is a remarkable scientist who is contributing immensely to the science of marine microalgae and has established a school of advanced studies in this area in India. He is fondly called as GS and his contributions will be remembered forever. His support to the research projects of the editors is gratefully acknowledged.

Born on January 8, 1943, in Kumbakonam, Tamil Nadu, India, he did his schooling in Town High School, Kumbakonam (1952–1958). He obtained his BSc and MSc in 1962 & 1964, respectively, from the University of Madras. He pursued his PhD degree in 1982 in Microbiology, from Madurai Kamaraj University.

His passion for research was the driving force to take up research in Cyanobacteria at Bharathidasan University, Tiruchirappalli, Tamil Nadu, India. He served this university as Professor and Head of microbiology for over 2 decades until his retirement in the year 2003. His contributions to marine cyanobacterial research were recognized by the Department of Biotechnology, Government of

India, resulting in the establishment of a National Facility for Marine Cyanobacteria (NFMC) at Bharathidasan University. This today houses the world's largest marine cyanobacterial germplasm collection. He has the unique distinction of serving as founder Director of NFMC from 1991 to 2003.

He played a pivotal role in serving as a Coordinator of the School of Life Sciences, Business Development Center, and also Bioinformatics Center at Bharathidasan University.

GS after retirement from the university became Research Advisor in a Pharma-Biotech company till 2010 and later served as Founder Director and Advisor of the Central Interdisciplinary Research Institute (CIDRF), Mahatma Gandhi Medical College & Research Institute Campus, Puducherry, from 2011 to 2018. Presently he occupies the Endowed Chair position at Sri Balaji Vidyapeeth Deemed University at Puducherry and also serves as Visiting Professor at Avinashilingam Institute of Home Science and Higher Education (Deemed to be University), Coimbatore.

His contributions include survey of the Indian coasts; collection; identification; culturing and supply of marine microalgae and cyanobacteria for researchers in India. He and his team were instrumental in developing technology for phycocyanin and carotenoid pigments for industrial use as well as development of effluent treatment and recycling processes in some industries.

He guided 15 postdoctoral, 14 PhD students, and 14 MPhil students. He has published over 110 research papers and has lectured extensively in India and abroad. His research projects were supported by the Department of Biotechnology, Department of Science and Technology, Department of Ocean Development, Indian Council of Agricultural Research, and by industries. He is actively sought after by these agencies as an expert member in various program advisory committees of the Government of India and also several universities in India.

He is recipient of a number of awards to name a few: innovators award—Exnora International; Prof M.O.P Iyyangar Memorial Lecture Award; State Level Dr.Radhakrishnab Award for Best Teacher; Rev.Balam Memorial Award; Lifetime achievement awards of the Krishnamurthy Institute of Algology (2009) and Association of Applied Microbiologists of India (2010).

He holds professional membership of Association of Microbiologists of India, Marine Biological Association of India, Phycological Society of India, American Association of Advancement of Science, American Society of Microbiologists, Seaweed Research & Utilization, and Association and Asia- Pacific Society for Applied Phycology.

It is a matter of privilege and honor for the editors to dedicate this volume to GS who is a devoted scientist, passionate teacher, an ardent supporter of excellence. Above all a great human being. We are sure that he would continue to motivate individuals who are sincere in pursuing research in algal technologies.

We wish him good health and active life ahead.

Preface

Upon realizing the need for a comprehensive treatise on seaweed cultivation and its utilization for food and nutritional security, we the editors present to you two volumes on seaweeds which are elegantly published by Springer Nature. The demand for increasing the food production for the growing world population, from nearly 8.0 billion presently, which is expected to reach 9.9 billion by 2050, has been a daunting task. Sustainable development goals of the UN for achieving zero hunger through enhanced food production and distribution also focus on dietary diversification of the world population to meet the food, nutritional, and nutraceutical needs. Moreover, the approaches to the enhancement in food production are also limited by the scarcity of water needed for agriculture. In this context utilization of seaweed biodiversity provides tremendous opportunities to employ them to produce large quantities of quality biomass for food applications, using marine water resources. Furthermore, they can be cultivated without the dependence on fertilizers and other agrochemical inputs which are otherwise needed in agricultural farming systems.

Seaweeds have been consumed, traditionally as food, in the Eastern countries for centuries. Their consumption is also rapidly expanding to other regions of the world. The seaweed recipes are gaining popularity owing to their culinary properties and health attributes. Seaweed constituents such as agar and carrageenan are used in food processing. Presently seaweeds are realizing applications in the health and wellness sectors. The growing global demand for seaweeds has resulted in the development of farming methods for large-scale production of biomass and international trade. Food safety and standards issues for seaweed products are gaining prominence for local and global needs. Research to unravel the hidden potentials of seaweed for health and therapeutic applications is in full swing.

Because of the utility value and economic implication of seaweeds and their products, for food, health, pharmaceutical, cosmeceutical, and a myriad of uses, their biodiversity is being increasingly explored for a wide variety of applications.

Here, we have made our best efforts to compile a vast body of knowledge by leveraging the experience of the experts in the field of seaweed science and technology for the benefit of all stakeholders. The material contained here will cater to a vast audience including students, teachers, scientists, food and health experts, technologists, policymakers, and environmentalists. It should also address the professionals in the food, agriculture, health, pharmaceutical, cosmeceutical, environmental, and several emerging technologies for sustainable utilization of seaweeds and their biodiversity for the benefit of the global population.

The above-mentioned aspects are presented in two volumes and the brief description of each of the volumes provides an insight into their broad contents.

Volumes I and II are contributed by 122 and 127 authors, respectively, from 21 countries.

Volume I deals with seaweed bioresources, cultivation, trade, and multifarious applications:

Seaweed farming in various parts of the world with a focus on Asia, Europe, Australia, and South America has been detailed by the authors from the respective parts of the world. They have also presented the state-of-the-art technologies and product development strategies, including the quality of the produce and market potentials. The ecological implications of seaweed cultivation concerning the industrial aquaculture scenario have been presented. Seaweeds as biofertilizers, aquaculture feeds, livestock feeds, and agricultural applications have been detailed. Industrial adoption of seaweeds for bioremediation, wastewater treatment, and bioenergy generation is dealt which have environmental applications with far-reaching implications. The biorefinery approach to valorization of the process with complete fractionation of constituents for value addition has been dealt with.

Volume II deals with seaweeds for food, pharmaceutical, and health applications:

The seaweed as food with nutritional advantages has been detailed. The use of several edible forms such as *Monostroma, Caulerpa, Palmaria, Gracilaria, Porphyra, Laminaria, Fucus, Undaria,* etc., used routinely in various recipes such as sea vegetables, salads, soups, and meat analogs has been dealt extensively. The use of seaweeds as sources of nutraceuticals has been explained in detail, and the products in the international market, as well as their trade, have been reviewed. The occurrence of pigments such as fucoxanthin, phlorotannins, phycocyanin, phycoerythrin, beta carotene, and several more is presented in detail. In addition to these, fatty acids, vitamins, and minerals have been described. These aforementioned seaweed constituents have innumerable health applications. The current understanding of their bioactive properties and pharmacological actions, including their bioefficacy, safety, and toxicity aspects, is described. Antidiabetic, antioxidant, antiobesity, cardioprotective, and prebiotic and probiotic properties with augmentation of gut-microbiome, antiviral effects, and many more direct benefits to health and disease management through seaweed-based products are described. The advancement of science and technology of seaweeds has brought to light the utility of seaweeds as a rich source of nutraceuticals and also as a source of micronutrients. Global trade of seaweed-based foods and quality considerations have been brought out elegantly which will govern the future expansion of the seaweed industry. Their use in cosmeceuticals is fast expanding with a high degree of value addition to the herbal formulations. The seaweed ingredients are also of value in extending the shelf life of food products. Biosynthetic and genomic studies to genetically modify seaweeds for enhanced productivity have been brought to light.

The collective efforts of scientists from all over the world represent global perspectives on the topics. The authors have put in their best in presenting you with their own research experience in handling various systems, thereby offering practical solutions to the industrial exploitation of seaweed resources.

The editors expect a wide range of readership of these volumes which fills in the demand for up-to-date knowledge in the utilization of seaweed biodiversity for commercial exploitation in a sustainable manner.

Guntur, Andhra Pradesh, India — Ambati Ranga Rao
Bengaluru, Karnataka, India — Gokare A. Ravishankar

Acknowledgments

We the editors of this volume wish to place on record the encouragement received from worldwide supporters interested in the science, technology, and trade of seaweeds. Most importantly, the spontaneity in acceptance to contribute chapters by the authors has resulted in timely publication as scheduled. We are grateful to the authors for their cooperation, despite the pandemic time, by keeping up the commitment to bringing to the table a vast body of knowledge and information on seaweeds for the benefit of all the stakeholders. We are highly indebted to the authors for their scholarly inputs provided to the international audience.

We wholeheartedly appreciate the efforts of Springer Nature Group who has done a marvelous job in collation, compilation, and publication with due diligence in bringing out these volumes. Our special thanks to Daniel Falatko and his team for their total dedication in bringing out this volume in such an elegant manner.

The editors are thankful to the support of their families during the pandemic for the cooperation and encouragement received to complete this task.

A. R. R. thanks his wife, Deepika; daughter, Jesvisree; parents, Venkateswaralu and Tulasidevi; brothers; sisters-in-law; sisters; and brothers-in-law.

G. A. R. thanks his wife, Shyla; son, Prashanth; daughter-in-law, Vasudha; and daughter,

Apoorva.

A. R. R. is indebted to Dr. L. Rathaiah, Chairman of Vignan Group of Institutions; Mr. L. Sri Krishnadevarayalu, Vice-Chairman (Member of Parliament); Prof. Dr. K. Ramamurthy Naidu, Chancellor; Dr. M. Y. S. Prasad, Vice-Chancellor; Dr. Kavi Kishor, Scientific Advisor; Dr. Madhusudhan Rao, Director, Engineering and Management; Dean Academics, Dean R&D; and Head, Biotechnology Department of Vignan's Foundation for Science, Technology and Research Deemed to be University for providing facility and support to fulfill this additional assignment.

G. A. R. expresses his gratitude to Dr. Premachandra Sagar, Vice Chairman, Dayananda Sagar Institutions, and Pro-Chancellor of Dayananda Sagar University, Bengaluru, for granting permission to take this additional responsibility and for his unstinted support.

G. A. R. offers his special thanks to the Council of Scientific and Industrial Research (CSIR) for including him as an expert in CSIR Mission on Seaweeds—a multi-institutional project on seaweeds cultivation and utilization as an economic activity supported by the Government of India under the "Atmanirbhar Bharath" program.

The editors are thankful to the institutions of the Government of India, Department of Biotechnology Department of Science and Technology, and Indian Council of Medical Research (ICMR) for their support of our research projects on algae-based foods at CSIR-Central Food Technological Research Institute (CFTRI), Mysuru, Karnataka, India.

Ambati Ranga Rao
Gokare A. Ravishankar

Contents

About the Editors

Ambati Ranga Rao is a senior scientist and associate professor in the Department of Biotechnology at Vignan's Foundation for Science, Technology, and Research (Deemed to be University), Andhra Pradesh, India. He is involved in both teaching cum research for graduate and undergraduate students. He is among the Top 2% World's Scientist as per the report by Stanford University, USA. Dr. Ranga Rao holds a bachelor's and a master's degree from Acharya Nagarjuna University, Andhra Pradesh, India, and a PhD degree from the University of Mysore. He started his research career as research assistant at the Department of Plant Cell Biotechnology, Council of Scientific and Industrial Research (CSIR)-Central Food Technological Research Institute (CFTRI), Mysuru, India, under the supervision of Dr. G. A. Ravishankar and Dr. R. Sarada. He was awarded Senior Research Fellow of Indian Council of Medical Research (ICMR), New Delhi. His PhD work at CFTRI focused on the production of astaxanthin from cultured green alga *Haematococcus pluvialis* and its biological activities. He worked extensively on process optimization of algal biomass production, mass culture of various algal species in raceway ponds and photobioreactors and downstream processing of algal metabolites and evaluation of their possible nutraceutical applications in *in vitro* and *in vivo* models. Furthermore, Dr. Ranga Rao was involved in a project on "Studies on field cultivation and harvesting of seaweeds-*Porphyra, Enteromorpha, Eucheuma* and their use in processed foods." Dr. Ranga Rao worked as lead scientist in Algal Technologies, Carot Labs Pvt. Ltd., India; Postdoctoral Research Associate in the Laboratory of Algal Research and Biotechnology, Arizona State University, USA, under the supervision of Prof. Milton Sommerfeld and Prof. Qiang Hu; Visiting Assistant Professor in Food Science and Technology Program, Beijing Normal University and Hong Kong Baptist University, United International College, China, under the supervision of Prof. Bo Lei; and Visiting Senior Research Fellow (Associate Professor Grade) in the Institute of Ocean and Earth Sciences, University of Malaya, Malaysia, under the guidance of Prof. Phang Siew-Moi. He is the author of 50 peer-reviewed publications, 60 international/national conferences/symposia/invited talks/FDPs/workshops/STC, and 25 chapters in books. His research citations exceed 3300 with h-index (19) and i10-index (26) as Google Scholar. He has delivered lectures at international/national conferences/symposia in the United States, Canada, Brazil, China, Malaysia, Indonesia, and Oman. He has edited three books (CRC Press and Academic Press, USA), as coeditor, namely, *Handbook of Algal Technologies and Phytochemicals: Volume-I Food, Health, and Nutraceutical Applications*; *Handbook of Algal Technologies and Phytochemicals: Volume II Phycoremediation, Biofuels, and Global Biomass Production*; and *Global Perspectives on Astaxanthin: from Industrial Production to Food, Health, and Pharmaceutical Application*. He was selected for the Junior Scientist of the Year Award by National Environmental Science Academy, New Delhi, India; honored TWAS-Young Affiliate by Regional Office of South East Asia and the Pacific Chinese Academy of Sciences (CAS), China; received Young Scientist Award at the World Food Science Congress by International Union of Food Science and Technology (IUFoST), Canada; and Carl Storm International Diversity Fellowship Award by Gordon Research Conferences, USA. He is a lifetime member of the Society of Applied Biotechnology, India; Association of Food Scientists and Technologists of India; National Environmental Science Academy, India; and Asia PGPR Society of Sustainable Agriculture, USA. He is an associate fellow of Andhra Pradesh Akademi of Sciences, Government of Andhra Pradesh, India, and also a fellow of the Society of Applied Biotechnology, India. He has received research grants and travel grant fellowships as both international and national awards, under Young Scientist schemes. He is also serving as an editorial board member, guest editor for special issues, and reviewer for reputed international and national journals.

Gokare A. Ravishankar Professor of biotechnology, is presently the vice president of R&D in Life Sciences and Biotechnology at Dayananda Sagar Institutions, Bengaluru, India. Earlier he had a distinguished research career of over 30 years at the CSIR-Central Food Technological Research Institute (CFTRI), Mysore, and in the institutions of Government of India. He served as chairman of the board of studies in biotechnology at the Visvesvaraya Technological University, Belgavi, and Academic Council Member of Dayananda Sagar University. He has also been a member of the boards of eight universities. He served as visiting professor to universities in Japan, Korea, Taiwan, and Russia. He is an internationally recognized expert in the areas of food science and technology, plant biotechnology, algal biotechnology, food biotechnology and postharvest technologies, plant secondary metabolites, functional foods, herbal products, genetic engineering, and biofuels. He is among the top 2% scientists in the world for research publications based on the report by Stanford University, USA. Dr. Ravishankar holds a master's and a PhD degree from Maharaja Sayajirao University of Baroda. He mentored over 40 PhD students, 62 master's students, 7 postdocs, and 8 international guest scientists and authored over 260 peer-reviewed research papers in international and national journals, 50 reviews, 55 patents in India and abroad, and edited 5 books, with a h-index of 70 and nearly 22,000 citations. He has presented over 220 lectures in various scientific meetings in India and abroad, including visits to about 30 countries. He has received international honors as a Fellow of the International Academy of Food Science and Technology (Canada), Institute of Food Technologists (USA), and is a Certified Food Scientist of USA. He is honored as fellow of several organizations in India, namely, the National Academy of Sciences, National Academy of Agricultural Sciences, Association of Microbiologists of India, Society of Agricultural Biochemists, Society of Applied Biotechnologists, Indian Botanical Society, Biotechnology Research Society of India, and the Association of Food Scientists and Technologists of India. He is also an elected member of the Plant Tissue Culture Association of India. Dr. Ravishankar received several coveted honors and awards as follows: Young Scientist award (Botany) by the then Prime Minister of India in 1992; National Technology Day Award of Government of India in 2003; Laljee Goodhoo Smarak Nidhi Award for food biotechnology R&D of industrial relevance; the prestigious, professor V. Subramanyan Food Industrial Achievement Award; Professor S.S. Katiyar Endowment Lecture Award in New Biology by Indian Science Congress; Professor Vyas Memorial Award of Association of Microbiologists of India; Professor V.N. Raja Rao Endowment Lecture Award in Applied Botany, University of Madras; Lifetime achievement award by the Society of Applied Biotechnologists; Dr. Diwaker Patel Memorial Award by Anand Agricultural University, Anand; Prof. C.S. Paulose Memorial Oration Award by Society for Biotechnologists of India; and Prof. Gadgil Memorial Lecture Award from Plant Tissue Culture Association of India. He has held honorary positions as President in the Society of Biological Chemists, Mysore Chapter, and President of Association of Microbiologists of India, Mysore and Bangalore Chapters. He is a lifetime member of the Nutrition Society of India and several biotechnology societies including the Society for Biotechnologists of India, Biotechnology Research Society of India, International Coffee Genome Network, American Society of Plant Biologists, and Global Harmonization Initiative. He is a consultant to World Bank projects in the domains of postharvest technologies, plant biotechnologies for value addition to crop plants, and food biotechnologies. He has also served as advisor and resource person in international conferences, seminars, workshops, and short courses; he has convened national and international seminars in biology, biotechnology, and food science and technology. He is an associate editor and reviewer of a large number of reputed research journals.

Contributors

Saied A. Aboushanab Institute of Chemical Engineering, Ural Federal University Named After the First President of Russia B. N. Yeltsin, Yekaterinburg, Russia

Lyutha Al-Subhi Department of Food Science and Nutrition, College of Agricultural and Marine Sciences, Sultan Qaboos University, Al-Khoud, Muscat, Oman

Mona Al-Zahrani School of Chemical Sciences, The University of Auckland, Auckland, New Zealand

Rudrappa Ambedkar Department of Microbiology and Biotechnology, Bangalore University, Bengaluru, Karnataka, India

Glacio Araújo Federal Institute of Education, Science and Technology of Ceará – IFCE, Aracati, Ceará, Brazil

Elena M. Balboa Department of Chemical Engineering, University of Vigo (Campus Ourense), Edificio Politécnico, Ourense, Spain

Revathy Baskaran Academy of Scientific and Innovative Research (AcSIR), Ghaziabad, Uttar Pradesh, India

Department of Fruit and Vegetable Technology, CSIR-Central Food Technological Research Institute, Mysore, Karnataka, India

Mauricio Alfredo Ondarza Beneitez Red de Glicociencia en Salud, Mexico, Hacienda de Cortez, Rincon de los Arcos, Irapuato, Guanajuato, Mexico

Neeru Bhatt Global Science Heritage, Toronto, ON, Canada

Tomohiro Bito Department of Agricultural, Life and Environmental Sciences, Faculty of Agriculture, Tottori University, Tottori, Japan

Maria do Carmo Barreto Faculty of Sciences and Technology, cE3c–Centre for Ecology, Evolution and Environmental Changes/Azorean Biodiversity Group and, University of Azores, Ponta Delgada, Portugal

Jaqueline Carneiro Biochemistry and Molecular Biology Department, Federal University of Paraná, PO, Curitiba, Brazil

Julio Cesar de Carvalho Bioprocess Engineering and Biotechnology Department, Federal University of Paraná, PO, Curitiba, Brazil

Kajal Chakraborty Bioprospecting Section of Marine Biotechnology Division, Central Marine Fisheries Research Institute, Cochin, Kerala, India

Manuel A. Coimbra LAQV-REQUIMTE, Department of Chemistry, University of Aveiro, Aveiro, Portugal

Guilherme Augusto Colusse Biochemistry and Molecular Biology Department, Federal University of Paraná, PO, Curitiba, Brazil

Bioprocess Engineering and Biotechnology Department, Federal University of Paraná, PO, Curitiba, Brazil

Joao Cotas Department of Life Sciences, MARE – Marine and Environmental Sciences Centre, University of Coimbra, Coimbra, Portugal

Wei Cui School of Medicine, Ningbo University, Ningbo, China

Irina G. Danilova Institute of Natural Sciences and Mathematics, Ural Federal University Named After the First President of Russia B. N. Yeltsin, Yekaterinburg, Russia

Institute of Immunology and Physiology, Ural Branch of the Russian Academy of Sciences, Yekaterinburg, Russia

Jesmi Debbarma ICAR-Central Institute of Fisheries Technology (ICAR-CIFT), Visakhapatnam Research Centre, Visakhapatnam, Andhra Pradesh, India

Charu Deepika Institute of Molecular Bioscience, The University of Queensland, Brisbane, QLD, Australia

Ponnuswamy Renuka Devi Department of Biotechnology, Anna University, Coimbatore, Tamil Nadu, India

Herminia Domínguez Department of Chemical Engineering, University of Vigo (Campus Ourense), Edificio Politécnico, Ourense, Spain

Maria Eugênia Rabello Duarte Biochemistry and Molecular Biology Department, Federal University of Paraná, PO, Curitiba, Brazil

Laurent Dufosse CHEMBIOPRO Chimie et Biotechnologie des Produits Naturels, ESIROI Département agroalimentaire, Université de la Réunion, Ile de La Réunion, Indian Ocean, France

Svetlana P. Ermakova G.B. Elyakov Pacific Institute of Bioorganic Chemistry, Far Eastern Branch of the Russian Academy of Sciences, Vladivostok, Russia Federation

Lydia Ferrara Department of Pharmacy, University of Naples Federico II, Naples, Italy

Artur Figueirinha Faculty of Pharmacy of the University of Coimbra, University of Coimbra, LAQV, REQUIMTE, Coimbra, Portugal

Faculty of Pharmacy of University of Coimbra, University of Coimbra, Coimbra, Portugal

Tejal K. Gajaria Academy of Scientific and Innovative Research, Ghaziabad, Uttar Pradesh, India

CSIR-Central Salt and Marine Chemicals Research Institute, Bhavnagar, Gujarat, India

Fengzheng Gao Bioprocess Engineering, AlgaePARC, Wageningen University, Wageningen, Netherlands

Sara García-Poza Department of Life Sciences, MARE – Marine and Environmental Sciences Centre, University of Coimbra, Coimbra, Portugal

Ana M. M. Gonçalves Department of Life Sciences, MARE – Marine and Environmental Sciences Centre, University of Coimbra, Coimbra, Portugal

Department of Biology and CESAM, University of Aveiro, Aveiro, Portugal

Fernando J. M. Gonçalves Department of Biology, University of Aveiro, Aveiro, Portugal

CESAM, University of Aveiro, Aveiro, Portugal

Shan He Key Laboratory of Marine Biotechnology, Ningbo University, Ningbo, China

Chai-Ling Ho Faculty of Biotechnology and Biomolecular Sciences, Universiti Putra Malaysia, UPM-Serdang, Selangor, Malaysia

Naveen Jayapala Department of Biochemistry, CSIR-Central Food Technological Research Institute, Mysore, Karńataka, India

Mudassar Anisoddin Kazi Applied Phycology and Biotechnology Division, CSIR-Central Salt and Marine Chemicals Research Institute, Bhavnagar, Gujarat, India

Sofia A. Khapchaeva Federal Research Centre "Fundamentals of Biotechnology" RAS, Moscow, Russia

Ayse Kose Faculty of Engineering, Department of Bioengineering, Ege University, Izmir, Turkey

Elena G. Kovaleva Institute of Chemical Engineering, Ural Federal University Named After the First President of Russia B. N. Yeltsin, Yekaterinburg, Russia

Mallikarjuna Rao Kovi Department of Plant Sciences, Faculty of Biosciences, Norwegian University of Life Sciences, Ås, Norway

Maushmi S. Kumar Shobhaben Pratapbhai Patel School of Pharmacy and Technology Management, SVKM'S NMIMS, Mumbai, Maharashtra, India

Rangaswamy Lakshminarayana Department of Microbiology and Biotechnology, Bangalore University, Bengaluru, Karnataka, India

Wei-Kang Lee Faculty of Biotechnology and Biomolecular Sciences, Universiti Putra Malaysia, UPM-Serdang, Selangor, Malaysia

Zhe Liang Biotechnology Research Institute, Chinese Academy of Agricultural Sciences, Beijing, China

Yi-Yi Lim Faculty of Biotechnology and Biomolecular Sciences, Universiti Putra Malaysia, UPM-Serdang, Selangor, Malaysia

Alexandr A. Lukyanov Lomonosov Moscow State University, Moscow, Russia

Olesya S. Malyarenko G.B. Elyakov Pacific Institute of Bioorganic Chemistry, Far Eastern Branch of the Russian Academy of Sciences, Vladivostok, Russia Federation

Vaibhav A. Mantri Applied Phycology and Biotechnology Division, CSIR-Central Salt and Marine Chemicals Research Institute, Bhavnagar, Gujarat, India

Academy of Scientific and Innovative Research (AcSIR), CSIR-Central Salt and Marine Chemicals Research Institute, Bhavnagar, Gujarat, India

Joao C. Marques Department of Life Sciences, MARE – Marine and Environmental Sciences Centre, University of Coimbra, Coimbra, Portugal

Pedro Monteiro Faculty of Pharmacy and Center for Neurosciences and Cell Biology, Health Sciences Campus, University of Coimbra, Coimbra, Portugal

Tiago Morais Lusalgae, Lda., Incubadora de Empresas da Figueira da Foz, Figueira da Foz, Portugal

Miguel Daniel Noseda Biochemistry and Molecular Biology Department, Federal University of Paraná, PO, Curitiba, Brazil

Cláudia Nunes CICECO, Department of Chemistry, University of Aveiro, Aveiro, Portugal

LAQV-REQUIMTE, Department of Chemistry, University of Aveiro, Aveiro, Portugal

Hilde-Gunn Opsahl-Sorteberg Department of Plant Sciences, Faculty of Biosciences, Norwegian University of Life Sciences, Ås, Norway

Diana Pacheco Department of Life Sciences, MARE – Marine and Environmental Sciences Centre, University of Coimbra, Coimbra, Portugal

Ratih Pangestuti Research and Development Divisions for Marine Bio-Industry, Indonesian Institute of Science (BBIL-LIPI), Lombok, West Nusa Tenggara, Republic of Indonesia

Leonel Pereira Department of Life Sciences, MARE – Marine and Environmental Sciences Centre, University of Coimbra, Coimbra, Portugal

Conrad O. Perera School of Chemical Sciences, The University of Auckland, Auckland, New Zealand

Madan Kumar Perumal Department of Biochemistry, CSIR-Central Food Technological Research Institute, Mysore, Karnataka, India

Academy of Scientific and Innovative Research (AcSIR), Ghaziabad, Uttar Pradesh, India

Diana C. G. A. Pinto LAQV-REQUIMTE, Department of Chemistry, University of Aveiro, Aveiro, Portugal

Tatiana V. Puchkova Federal Research Centre "Fundamentals of Biotechnology" RAS, Moscow, Russia

Stanislaus Mary Josephine Punitha Centre for Marine Science and Technology, Manonmaniam Sundaranar University, Kanyakumari, Tamil Nadu, India

Yanuariska Putra Research and Development Divisions for Marine Bio-Industry, Indonesian Institute of Science (BBIL-LIPI), Lombok, West Nusa Tenggara, Republic of Indonesia

Yimin Qin State Key Laboratory of Bioactive Seaweed Substances, Qingdao, China

Jiaxing College, Jiaxing, China

Chatragadda Ramesh Biological Oceanography Division, National Institute of Oceanography (CSIR-NIO), Dona Paula, Goa, India

Ambati Ranga Rao Department of Biotechnology, Vignan's Foundation of Science, Technology and Research Deemed to be University, Vadlamudi, Guntur, Andhra Pradesh, India

B. Madhusudana Rao ICAR-Central Institute of Fisheries Technology (ICAR-CIFT), Visakhapatnam Research Centre, Visakhapatnam, Andhra Pradesh, India

C. N. Ravishankar ICAR-Central Institute of Fisheries Technology (ICAR-CIFT), Cochin, Kerala, India

Gokare A. Ravishankar Dr. C. D. Sagar Centre for Life Sciences, Dayananda Sagar College of Engineering, Dayananda Sagar Institutions, Bengaluru, Karnataka, India

Gonçalo P. Rosa LAQV-REQUIMTE, Department of Chemistry, University of Aveiro, Aveiro, Portugal

Faculty of Sciences and Technology, cE3c–Centre for Ecology, Evolution and Environmental Changes/Azorean Biodiversity Group and, University of Azores, Ponta Delgada, Portugal

Kanchan Sambhwani Applied Phycology and Biotechnology Division, CSIR-Central Salt and Marine Chemicals Research Institute, Bhavnagar, Gujarat, India

Academy of Scientific and Innovative Research (AcSIR), CSIR-Central Salt and Marine Chemicals Research Institute, Bhavnagar, Gujarat, India

Sathianeson Satheesh Department of Marine Biology, Faculty of Marine Sciences, King Abdulaziz University, Jeddah, Saudi Arabia

Ana M. L. Seca LAQV-REQUIMTE, Department of Chemistry, University of Aveiro, Aveiro, Portugal

Faculty of Sciences and Technology, cE3c–Centre for Ecology, Evolution and Environmental Changes/Azorean Biodiversity Group and, University of Azores, Ponta Delgada, Portugal

Gabriela Jorge da Silva Faculty of Pharmacy and Center for Neurosciences and Cell Biology, Health Sciences Campus, University of Coimbra, Coimbra, Portugal

José W. A. Silva Federal Institute of Education, Science and Technology of Ceará – IFCE, Aracati, Ceará, Brazil

Shivdayal Singh Shobhaben Pratapbhai Patel School of Pharmacy and Technology Management, SVKM'S NMIMS, Mumbai, Maharashtra, India

Ksenia V. Sokolova Institute of Immunology and Physiology, Ural Branch of the Russian Academy of Sciences, Yekaterinburg, Russia

Alexei E. Solovchenko Lomonosov Moscow State University, Moscow, Russia

Derzhavin Tambov State University, Tambov, Russia

Eko Susanto Department of Fisheries Technology Product, Faculty of Fisheries and Marine Sciences, Diponegoro University, Semarang, Central Java, Republic of Indonesia

M. Cristina Taboada Department of Physiology, Faculty of Pharmacy, University of Santiago de Compostela, Santiago de Compostela, Spain

Ioannis Theodorou Department of Plant Sciences, Faculty of Biosciences, Norwegian University of Life Sciences, Ås, Norway

Morphogenesis of Macroalgae, LBI2M UMR8227, Station Biologique de Roscoff, CNRS-Sorbonne University, Paris, France

Marlene A. Trindade Department of Biology, University of Aveiro, Aveiro, Portugal

Roza V. Usoltseva G.B. Elyakov Pacific Institute of Bioorganic Chemistry, Far Eastern Branch of the Russian Academy of Sciences, Vladivostok, Russian Federation

Baskaran Vallikannan Department of Biochemistry, CSIR-Central Food Technological Research Institute, Mysore, Karnataka, India

Academy of Scientific and Innovative Research (AcSIR), Ghaziabad, Uttar Pradesh, India

Svetlana G. Vasilieva Lomonosov Moscow State University, Moscow, Russian

Chidambaram Kulandaisamy Venil Department of Biotechnology, Anna University, Coimbatore, Tamil Nadu, India

Kariyappa Vijay Department of Microbiology and Biotechnology, Bangalore University, Bengaluru, Karnataka, India

P. Viji ICAR-Central Institute of Fisheries Technology (ICAR-CIFT), Visakhapatnam Research Centre, Visakhapatnam, Andhra Pradesh, India

Nadarajan Viju Centre for Marine Science and Technology, Manonmaniam Sundaranar University, Kanyakumari, Tamil Nadu, India

Mostafa Waly Department of Food Science and Nutrition, College of Agricultural and Marine Sciences, Sultan Qaboos University, Al-Khoud, Muscat, Oman

Fumio Watanabe Department of Agricultural, Life and Environmental Sciences, Faculty of Agriculture, Tottori University, Tottori, Japan

Xiaojun Yan Key Laboratory of Marine Biotechnology, Ningbo University, Ningbo, China

Daina Yesuraj General Undergraduate Curriculum Requirements, University of Dubai, Dubai, UAE

Jinrong Zhang Key Laboratory of Marine Biotechnology, Ningbo University, Ningbo, China

Vasily S. Zotov Federal Research Centre "Fundamentals of Biotechnology" RAS, Moscow, Russia

Part I
Seaweeds and Its Useful Constituents Including Bioactive Compounds for Food and Health Applications

Chapter 1
Seaweeds: Potential Applications of the Aquatic Vegetables to Augment Nutritional Composition, Texture, and Health Benefits of Food and Food Products

Jesmi Debbarma, P. Viji, B. Madhusudana Rao, and C. N. Ravishankar

Abbreviations

AAE	Ascorbic acid equivalents
AAS	Amino acid score
Ala	Alanine
Arg	Arginine
Asp	Aspartic acid
CFU	Colony forming unit
Cl	Chlorine
Co	Cobalt
Cu	Copper
Cys	Cysteine
db	Decibel
DHA	Docosahexaenoic acid
DPPH	2,2-Diphenyl-1-picrylhydrazyl
dw	Dry weight
EAA	Essential amino acid
EAAI	Essential amino acid index
EPA	Eicosapentaenoic acid
EU	European Union
FAO	Food and Agriculture Organization
Fe	Iron

J. Debbarma (✉) · P. Viji · B. M. Rao
ICAR-Central Institute of Fisheries Technology (ICAR-CIFT), Visakhapatnam Research Centre, Visakhapatnam, Andhra Pradesh, India

C. N. Ravishankar
ICAR-Central Institute of Fisheries Technology (ICAR-CIFT), Cochin, Kerala, India

A. Ranga Rao, G. A. Ravishankar (eds.), *Sustainable Global Resources of Seaweeds Volume 2*, https://doi.org/10.1007/978-3-030-92174-3_1

Fr. wt.	Fresh weight
GAE	Gallic acid equivalents
Glu	Glutamic acid
Gly	Glycine
H_2O_2	Hydrogen peroxide
His	Histidine
I	Iodine
IDF	Insoluble dietary fibre
Ile	Isoleucine
K	Potassium
LDL	Low density lipoprotein cholesterol
Leu	Leucine
Lys	Lysine
Met	Methionine
Mg	Magnesium
Mn	Manganese
Mo	Molybdenum
MUFA	Monounsaturated fatty acids
Na	Sodium
nd	Not detected
nq	Not quantified
P	Phosphorus
Phe	Phenylalanine
ppm	Parts per million
Pro	Proline
PUFA	Polyunsaturated fatty acids
RSM	Response surface methodology
SDF	Soluble dietary fibre
Se	Selenium
Ser	Serine
SFA	Saturated fatty acids
Sr	Strontium
SRC	Solvent retention capacity
TDF	Total dietary fibre
Thr	Threonine
TMA-N	Trimethylamine nitrogen
Trp	Tryptophan
TS	Total solid
TVB-N	Total volatile base nitrogen
Tyr	Tyrosine
UNU	United Nations University
USA	United States of America
USD	United States dollar
Val	Valine
WHO	World Health Organization
Zn	Zinc

1 Introduction

Seaweeds, also known as macro algae, are heterogeneous plants that live in salt water, either in marine or brackish water environments, and grow abundantly on rocky solid substrates in waters up to 180 m depth. They are the crucial primary producers in the oceanic aquatic food web. Seaweeds contribute to a majority of the aquatic biomass representing 50% of the earth's primary productivity and has many industrial applications such as food, feed, fertilizer etc (Shama et al. 2019; Mohammad et al. 2019). Since ancient times, seaweeds are being utilized as an oriental diet in East Asian countries, especially in Chinese, Japanese, Korean and other South-East Asian cuisines. Historically, seaweeds have also met the food requirements of maritime communities across the world, mainly used as a nutritional ingredient in the diet of poor people living along the coastlines. Seaweeds are not commonly used as food in European countries and American countries. However, as nationals from these East Asian countries have migrated to other parts of the world, seaweed was introduced as food in those counties, for example, in some parts of North and South America. The demand for plant-based food or vegetarian food is increasing among the consumers as is the awareness of health issues and environmental sustainability concerning food choices. Edible seaweeds were widely consumed not only in East Asia, but also in other Asian counties e.g., Taiwan, Singapore, Thailand, Indonesia, Philippines, Malaysia, Cambodia and Vietnam, South Africa, Peru, Chile, the Canadian Maritimes, Scandinavia, South West England, Ireland, Wales, California, and Scotland. The polysaccharide from seaweeds such as agar, carrageenan and alginate have traditionally been used by Western counties as stabilizing, thickening and gelling agents in the food industry. Recently, France has approved the use of seaweed for human consumption as vegetables and condiments which had opened new opportunities in the food industry (Klnc et al. 2013). Today, the global seaweed industry is worth more than USD 6 billion per annum of which 85% comprises of food products for human consumption and seaweed derived extracts (carrageenan, agar and alginate) make up almost 40% of the world's hydrocolloids market.

Nearly, 221 commercially important seaweed species are available worldwide. About 10 species of seaweeds such as brown seaweed (*Saccharina japonica*, *Undaria pinnatifida* and *Sargasum fusiformies*), red seaweed (*Porphyra* spp., *Eucheuma* spp., *Kappaphycus alverezii* and *Gracilaria* spp.) and green seaweed (*Enteromorpha* spp., *Monostroma* spp. and *Caulerpa* spp.) are intensively cultivated for food, hydrocolloids extractions and other purposes (FAO 2016). Brown sea weeds are the most preferred one for human consumption (66.5%), followed by red seaweed (33%) and green seaweeds (5%) (Lorenzo et al. 2017). Among the top seven most cultivated seaweed taxa, three are exclusively grown for hydrocolloid extraction (*Eucheuma* spp. and *Kappaphycus alvarezii* for carrageenans; *Gracilaria* spp. for agar production). *Pyropia* spp., *Undaria pinnatifida, Saccharina japonica* and *Sargassum fusiforme* are most importantly used for human consumption. The five seaweed genera viz., *Saccharina, Undaria, Porphyra, Eucheuma, Kappaphycus* and *Gracilaria*—represent c. 98% of the world's cultivated seaweed production. However, only less than 1% of total seaweed production is still used for food and other uses i.e., other than for hydrocolloids globally. China, Indonesia, the republic of Korea and the Philippines were the top seaweed producing countries and were also those which cultivated the greatest diversity of seaweed species. China cultivates mostly kelp (i.e., *S. japonica* and *U. pinnatifida*) followed by red algae belonging to the genera *Gracilaria* and *Pyropia* for food purpose. On the other hand, *Kappaphycus* and *Eucheuma* coming under carrageenophytes are the mostly produced seaweed of Indonesia. China, Chile and Norway are the leaders in exploitation of the wild stocks of seaweeds, of which kelps are the most demanded group (FAO 2016).

Seaweeds are not a common weed but are valuable marine plant. Seaweeds are a rich source of minerals, most prominently iodine, calcium, iron and copper and. They are also rich in protein, fibre and vitamins, especially vitamin K and folic acid. Seaweeds are low in calories and fat. They are also known to be a good source of numerous bioactive compounds with diverse applications in different fields such as the pharmaceutical, agricultural, cosmeceutical and functional foods. Among health-conscious consumers, seaweed holds a reputation as a nutrient-rich-less caloric super food. Seaweeds and their extracts (derivatives and secondary metabolites) were reported to confer anti-microbial, anti-oxidant, anti-diabetic, anti-cancer, anti-inflammatory, anti-hyperlipidemic, anti-hypertensive and anti-obesity properties (Roohinejad et al. 2017; Kang et al. 2020).

Seaweeds provide a variety of metabolic compounds and other seaweed derived chemical compounds such as astaxanthin, carotene, superoxide dismutase, polypeptides, sterol and terpenoids with health benefits (Qin 2018). Seaweeds are the only sustainable resources for agar, alginate, carrageenan, fucoidan, laminaran, ulvan, mannitol, fucoxanthin, phlorotannins and other important food ingredients. Several studies have reported on the use of seaweeds and its derivatives for the development of functional food products which improve the functional, nutritional and textural qualities of food (López-López et al. 2009; Shitole et al. 2014; Vilar et al. 2019; Chen et al. 2020; Vieira et al. 2020; Lamont and McSweeney 2020; Enny Sholichah et al. 2021). Furthermore, several types of seaweeds and seaweeds derived natural food ingredients can complement the food products with the health benefits such as antioxidants, reduction of blood sugar, cholesterol level and blood pressure, improvement of memory, immunity, nutritional anemia and sleep, assisting weight loss, alleviating physical fatigue and visual fatigue etc. (Qin 2018). Additionally, regular seaweed consumption has been associated with a longer life expectancy (Willcox et al. 2009; Cardoso et al. 2015). The significant role of seaweeds in human nutrition has led to an increased interest in the consumption of seaweed and production of high-value seaweed enriched functional food all over the world. A wide range of seaweed enriched food products such as soup, noodles, cereals, meat-based products, surimi and surimi based restructured products, beverages, fermented and diary-based have been claimed to exert an important nutritional value and health benefits as well as the physicochemical and textural properties (Prabhasankar et al.

2009; Lane et al. 2014; Chauhan et al. 2015; Jayasinghe et al. 2016; Uchida et al. 2017; Jesmi et al. 2017; Marasabessy and Sudirjo 2017; Mamat et al. 2018; Chen et al. 2020; Gullón et al. 2020; Gopalakrishnan et al. 2020). Consequent to this, the global seaweed market is expected to grow to USD 22.1 billion by 2024 according to the Seafood Source report (Blank 2018). As seaweeds possesses a wide array of nutritional and diverse phycochemicals with therapeutic properties, they can be promoted as a perfect candidate as a supplement in the functional food industry or for the extraction of bioactive molecules (Catarino et al. 2017).

However, seaweeds are generally not preferred in in India for human consumption. But, recently there has been a growing interest to manufacture seaweed-based food products in the Indian functional food market and the seaweed as functional food ingredients create potential business opportunity to promote the diversification of exiting utilization of seaweed. Inspired by the growing interest in the utilization of seaweeds as potential functional food ingredient in development of novel food with significant health benefits, this review comprehends aspects related to nutritional composition, seaweed-based value added products, technological attributes, bioactive and its health benefits.

2 Classifications

Based on pigmentation, seaweeds can be categorized into three groups namely red (Rhodophyata), brown (Ochrophyta) and green (Chlorophyta) seaweeds (Fig. 1.1). The presence of red and blue pigments, phycocyanin and phycoerythrin in red seaweeds makes them distinct from brown and green seaweeds. Red seaweeds have a broad diversity of species as they exhibit a variety of life cycles and a diverse range of plant forms. The most common red seaweed species are *Poryphyra capensis*, *Aeodes orbitosa* and *Notogenia stiriata.* Size, species and overall morphology are the mostly varying parameters of brown seaweed species. The typical brown color of these species is attributed to the presence of fucoxanthene, a xanthophyll pigment. The abundantly grown species of brown seaweeds belong to kelp such as *Laminaria pallida*, *Fucus* sp. and *Zonaria* sp. The green algae represent the common seaweed grown in both fresh and saltwater. The green color of these seaweeds is given by chlorophyll a and b pigments. Common green seaweeds are *Chaetomorpha* and *Cladophora* sp. as well as sea lettuce which consists of the *Ulva* and *Monostroma* sp.

3 Seaweed for Human Food

Seaweeds have been included as a vital part of Asian cuisine for centuries. At present, Japan, the Republic of Korea and China are the largest consumers of seaweed as food (Table 1.1). The use of seaweed as human food has been traced back to the

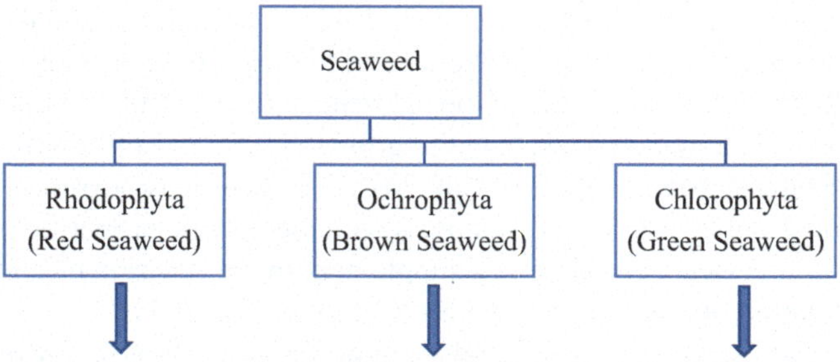

	Rhodophyta (Red Seaweed)	Ochrophyta (Brown Seaweed)	Chlorophyta (Green Seaweed)
Pigmentation ⟹	Phycocyanin, phycoerythrin and chlorophyll a	Xanthophyll pigment - fucaxanthin	Chlorophyll a and b, β-carotene
Habitats ⟹	Mostly occurring from low tide marks to 100 m depth, frequently found attached to coral reefs	Primarily found in shallow waters or on shoreline rocks and have very flexible stems that allow them to withstand the constant pounding of the waves	Common in the intertidal zone and mangrove swamps where light is abundant
Indian waters ⟹	431 species under 135 genera	Around 289 species under 37 genera	Approximately 213 species under 43 genera
Important edible seaweed genera ⟹	*Porphyra, Chondrus, Rhodymenia, Hypnea, Gigartina, Gracilaria, Laurencia, Iridaea, Phyllophora*	*Undaria, Durvillaea, Ecklonia, Sargassum, Turbinaria*	*Ulva, Enteromorpha, Monostroma, Caulerpa*

Gracilaria corticata *Padina sp.* *Caulerpa racemosa*

Fig. 1.1 Classification of seaweeds and its habitats with examples

Table 1.1 Traditional seaweed/ seaweed products commonly consumed as food

Products	Species	Country	Form	Use	Reference
Nori or Purple Laver	*Porphyra* spp., *Porphyra tenera*	Japan	Dried seaweed	Sushi for wrapping small rice ball and in soups	Mahadevan (2015), FAO (2016) and Sanjeewa et al. (2018)
Gim		Korea			
Aonori or green laver	*Monostroma latissimum, Enteromorpha prolifera* and *E. intestinalis*	Japan	Dried seaweed	As flavour enhancer in soups and foods The crushed pieces are used as a garnish	Mahadevan (2015)
Kombu	*Laminaria longissima, L. angustata, L. japonica, L. ochotensis* and *L. coriacea*	Japan	Dried seaweed	Served along with herring or sliced salmon, boiled with meat, soups, sauces and Kumbu tea. Nori and "Kombu" are also used in a special rice preparation called "onigiri" as a wrapping	Mahadevan (2015)
Haidai	*L. japonica*	China			
Wakame	*Undaria pinnatifida*	Japan	Dried seaweed	Used in salads, soups, noodles, pickles, etc	Mahadevan (2015)
Hiziki	*Hizikia fusiforme*	Japan	Dried seaweed	Stir fried with fried bean curd and vegetables such as carrot, or it may be simmered with other vegetables. It can also be incorporated into salads	Mahadevan (2015)
Arame	*Eisenia bicyclis*	Japan	Dried seaweeds or fresh	Appetizers, casseroles, muffins, pilafs, soups, toasted dishes etc	Mahadevan (2015)
Dulse	*Palmaria palmata*	Ireland and Maine	Flakes or dried powder	A snack food, seasoning and garnish	Yuan et al. (2005)
Mozuku	*Cladosiphon okamuranus*	Japan	Fresh and raw	Used as a fresh vegetable eaten along with soy sauce and salads	Mahadevan (2015)
Sea grapes or green caviar	*Caulerpa lentillifera, C. racemosa*	Japan	Fresh	Used in fresh salads	Mahadevan (2015)
Irish moss or carrageenan moss	*C. Crispus*			Thickening agent, puddings, seaweeds salads, sashimi garnishes and soup ingredient	Mahadevan (2015)

(continued)

Table 1.1 (continued)

Products	Species	Country	Form	Use	Reference
Winged kelp	*A. Esculenta*		Fresh, cooked, Raw	Cooked and Salad	Mahadevan (2015)
Ogo, ogonori or sea moss	*Gracilaria* spp.	Hawaii	Fresh	Salad vegetable	Mahadevan (2015)
Kelp		Japan, Korea	Whole/capsule/ powdered/granules	Soups, noodles, seasoning	Mahadevan (2015)

fourth century in Japan and the sixth century in China. Evidence is showing that approximately 21% of meals in Japan was served by seaweeds (Yoshinaga et al. 2001). "Nori" or "Purple Laver" (*Porphyra* spp.—a red seaweed), is a traditional Japanese food used for wrapping sushi and is also used in soups (FAO 2016). Nori is used primarily as a luxury food and it is often used in sushi, wrapped around a small portion of boiled rice with a slice of raw fish on the top. Nori is low in sodium and the characteristic taste of Nori is mainly contributed by three amino acids namely alanine, glycine and glutamic acid (Mahadevan 2015). Japan is the largest producer of nori, followed by the Republic of Korea and China. *Porphyra* spp. is one of the most nutritious seaweeds, which contain high protein (30–50% and about 75% of that is digestible) and Vitamin such as vitamins A, vitamin B group (B1, B2, B6, B12, niacin and folic acid) and vitamin C (Mahadevan 2015). It has 10 times as much vitamin A as that in spinach.

"Kombu", a Japanese dried seaweed product is derived from a mixture of *Laminaria spp.*, which include *Laminaria japonica*, *L. longissima*, *L. coriacea*, *L. angustata* and *L. ochotensis* (Mahadevan 2015). "Haidai" is the Chinese name for the dried *L. japonica*. In Japan, "Kombu" is also served with herring or sliced salmon, boiled with meat, soups, sauces and Kumbu tea. Nori and Kombu are also used for the wrapping of rice balls called "onigiri". *Laminaria* spp. are rich in protein (10%), fat (2%), minerals and vitamins especially vitamins B2, B12, and niacin. "Wakame" (*Undaria pinnatifida—a brown seaweed*) is used in salads, noodles, soups, pickles, etc. in Japan. Wakame has high dietary fibre, rich in ω-3 fatty acid, vitamin B group (especially niacin) and essential trace elements such as manganese, copper, cobalt, iron, nickel, and zinc (Mahadevan 2015). "Aonori" or green Laver is another Japanese seaweed product derived from green seaweeds *Monostroma* spp. and *Enteromorpha* spp. "Aonori," are mostly used to augment the taste of warm dishes like rice, soups, and salads. Another variety of Japanese food is "Kobumaki"—a traditional simmered food wrapped in kombu (often salmon or herring), generally prepared for the New Year holidays (Wells et al. 2017).

Sea grapes or green caviar is a common name for edible *Caulerpa* spp. such as *Caulerpa lentillifera* and *C. racemose*. Both have a grape-like appearance and are used in fresh salads preparation in Japan.

Seaweed consumption in Asia is traditional and consumed in varieties of dishes such as sushi wrapping, soup, salads, snacks, condiments, seasonings etc. While, consumption of seaweeds as food ingredients and food additives have also been found for many centuries in Europe and the Americas, in places such as Iceland, Scotland, Ireland, Maine (USA), Brittany (France), Nova Scotia (Canada), Peru, Wales and Chile (Michalak and Chojnacka 2018; Mouritsen et al. 2018). Unlike various Asian countries, consumption of seaweed is less prominent and non-traditional food in European and other Western countries like America despite occasional use as a conventional ingredient, especially in coastal areas (Bouga and Combet 2015). In recent decades seaweeds are being increasingly reinstated as part of more trendy and innovative cuisines like nutritional rich food and health food. Consumption of seaweeds as food has been reported in Azorean Islands, Portugal (Paiva et al. 2014), France (MacArtain et al. 2007) and Norway (Mouritsen et al.

2012; Chapman et al. 2015). *Pyropia sp., Fucus spiralis* and *Osmundea pinnatifida* were consumed in Portugal; *Himanthalia elongata, U. pinnatifida, Saccharina digitata, S. latissima, Ulva sp., Palmaria palmata, Gracilaria verrucosa*, and Chondrus crispus were consumed in France and *P. palmata, S. latissima, S. digitata* and *Alaria esculenta* were the seaweed species consumed in Norway. As mentioned earlier, France was the first European country who approved seaweeds for human consumption as vegetables and condiments.

4 Nutritional Composition of Seaweeds

Nutritional compositions of selected edible seaweeds are summarized in Table 1.2. Seaweeds are a good source of nutrients such as vitamins, minerals, protein, long-chain fatty acids (PUFA) and dietary fibre. Nutrient content among the different varieties of seaweeds varies widely depending on many factors such as species, locations, season and environmental conditions like water temperature, salinity, light and nutrients (Britton et al. 2020).

4.1 Protein and Amino Acid Profile of Seaweeds

Protein content of seaweeds varies from 5% to 47% of dry weight basis depending on the species, phylum, life cycle, seasonal and environmental conditions (Bocanegra et al. 2009; Černá 2011). Seaweed proteins are a good source of all the essential amino acids (EAAs). Several researchers estimated the amino acid content in different kinds of seaweeds and the amino acid composition of important edible seaweeds is summarized in Tables 1.3, 1.4 and 1.5. Comparison of amino acid content among the selected edible seaweeds was difficult because of the different units reported by researchers to express the amino acid content. Matanjun et al. (2008) reported presence of the 16 amino acids in *Euchema cottonii, Sargassum polycystum* and *Caulerpa lentillifera. E. cottonii* and *S. polycystum* have higher EAAs (60.59% and 61.66%, respectively) than in *C. lentillifera* (48.19%). Phenylalanine was the highest EAA found in all the three species. *S. polycystum* with the lowest protein content (5.4% dw) had the highest chemical score of 67.4% that was slightly lower than beef (69%) but relatively higher than casein (58%), oats (57%), rice (56%), soybeans (47%), wheat (43%) or peanuts (55%) (Brody 1999). The chemical score of *Porphyra columbina* (57%) was also better than and/or comparable with the above mention food items (Cian et al. 2013). The amino acid score of protein (AAS) evaluates the actual quantity of individual EAA in a food material and relates it to a reference protein (FAO/WHO/UNU 1985; FAO 1991). The amino acid score (AAS) and the essential amino acid index (EAAI) in red seaweeds was significantly higher than the brown seaweeds. AAS values ranged from 40% to 90% in red seaweeds and 20% to 70% in brown seaweeds (Dawczynski et al. 2007).

Cian et al. (2013) reported that the EAAs such as threonine, lysine, valine, leucine, methionine + cysteine and phenylalanine + tyrosine in *P. columbina* were higher than other EAAs like isoleucine, histidine and tryptophan. Brown seaweeds such as *Ascophyllum nodosum, Fucus vesiculosus and Bifurcaria bifurcate* possess all the EAAs ranging from 3075.28 mg/100 g dw to 5205.23 mg/100 g dw (except cysteine in the *A. nodosum* and *B. bifurcate*) as reported by Lorenzo et al. (2017). The glutamic acid was the most abundant amino acid present in all the three seaweeds ranging from 1874.47–1504.53 mg/100 dry matter, followed by aspartic acid (1677.01–800.84 mg/100 g dw) and alanine (985.40–655.73 mg/100 g dw). EAA/total AA ratio suggested that >40% of the amino acids were EAAs in *A. nodosum* and *F. vesiculosus*, 55% in *Gracelaria changii*, 36.87% in *Porphyra umbilicalis*, 42.72% in *Undaria pinnatifida* and 40.82% *Himanthalia elongata* (Cofrades et al. 2010; Chan and Matanjun 2017; Lorenzo et al. 2017). While, the ratio of EAA/NEAA in *Porphyra sp.* and *U. pinnatifida* was found to be higher than in *Laminaria sp.* and *Hizikia fusiforme* (Dawczynski et al. 2007).

4.2 Lipid and Fatty Acid Profile of Seaweeds

The fatty acid content of some of the selected important edible seaweeds was presented in Table 1.6. Seaweeds are considered as low caloric foods. The lipid contents in different varieties of seaweeds were as low as 0.29 to 6.54% dw and/or 2.3 to 20 mg/g fr. wt. for dried seaweeds and fresh seaweeds, respectively (Matanjun et al. 2008; Kumari et al. 2013; Lorenzo et al. 2017). The highest lipid content was found in *B. bifurcata* (6.54% dw) (Lorenzo et al. 2017). Kumari et al. (2013) studied the fatty acid profile of 33 species belonging to the orders Ulvales, Ulotrichales, Bryopsidales, Siphonocladales and Cladophorales and found that although low in lipid content, seaweeds have substantially high amounts of nutritionally important polyunsaturated fatty acids (PUFAs) such as linoleic acid, linolenic acid, arachidonic acid, eicosapentaenoic acid (EPA) and docosahexaenoic acid (DHA). PUFAs accounted for 10 to 70% of the total fatty acid content of seaweeds. Lorenzo et al. (2017) also reported that PUFAs were the most dominant fatty acids ranging from 43.47% to 48.19% in brown seaweeds, *A. nodosum*, *F. vesiculosus* and *B. bifurcate* followed by saturated fatty acids (SFAs) and monounsaturated fatty acids (MUFAs). The EPA (C20:5 ω-3) was detected in all three brown seaweeds (4.09 to 9.94% of the total fatty acids), while, DHA was only detected in *B. bifurcata* (11.10% of the total fatty acids). The presence of essential fatty acids such as linoleic acid (C18:2 ω6), linolenic acid (C18:3 ω3), arachidonic acid (C20:4 ω6) and EPA (C20:5 ω3) was reported in *E. cottonii* but the eicosanoid precursors arachidonic acid (C20:4 ω6) was not detected in green seaweed *C. lentillifera*. *E. cottonii* had the highest amount of ω-3 fatty acids accounting for 45.72% as compared to *C. lentillifera* (7.55%) (Matanjun et al. 2008).

SFAs content (Table 1.6) was more in the seaweeds such as *Caulerpa lentillifera*, *Ulva lactuca* and *Porphyra* spp. (Sánchez-Machado et al. 2004; Matanjun et al.

2008; Kumari et al. 2013). Except in *Gracilaria edulis*, all the selected important edible seaweed had nutritionally beneficial ω-6/ω3 ratio ranging from 0.1 to 2.62, which was below the WHO recommended ω6/ω3 ration 10 (Mahan and Escott-Stump 2000). Dietary PUFA in particular ω-3 fatty acids (EPA and DHA) can lower the low-density lipoprotein cholesterol (LDL) and reduced the risk of coronary heart disease.

4.3 Carbohydrates and Dietary Fibre

Seaweeds contain a great amount of carbohydrate and the amount of total carbohydrate content varies from 5.19 ± 0.21% to 51.81 ± 0.44%, 33.49 ± 1.70 to 45.9 ± 1.5% and 1.05 ± 0.031 to 55.77% in red seaweed, brown seaweed and green seaweed, respectively (Table 1.2) (Cian et al. 2013; Jesmi et al. 2016; Lorenzo et al. 2017; Ganesan et al. 2020). Although seaweeds have substantially high carbohydrate content, its greater portion is present as dietary fibre. That is beneficial for human health. Dietary fibre comprises a group of carbohydrate polymers, oligomers and lignin that cannot be digested in the human body. Based on the water solubility and fermentability, dietary fibre can be group into two categories, namely, insoluble dietary fibre (IDF) and soluble dietary fibre (SDF) (Dhingra et al. 2012). Dietary fibre content in seaweeds range from 25.05 ± 0.99% to 65.7 ± 0.9% depending on the species (Jiménez-Escrig and Sánchez-Muniz 2000; Santoso et al. 2006; Ratana-arporn and Chirapart 2006; Dawczynski et al. 2007; Matanjun et al. 2008; Ganesan et al. 2020). *Ulva reticulata* (65.7 ± 0.9%), *Pyropia orbicularis* (64.01 ± 0.88%) and *Gracilaria edulis* (63.175 ± 0.46%) have considerably higher dietary fibre content then than other seaweeds. The average fibre content of seaweeds was higher than that of terrestrial plants (MacArtain et al. 2007). Seaweeds are rich in soluble fractions but almost all the seaweeds contain more IDF than SDF. However, the content of SDF and IDF in seaweed vary widely depending on species and extraction method. Contrary to the result presented (Table 1.2), Gómez-Ordóñez et al. (2010) reported that 39.1–74.7% SDF content in the seaweed species namely *H. elongata*, *B. bifurcata*, *Laminaria saccharina*, *Mastocarpusstellatus* and *Gigartina pistillata*. Based on the recommended daily amount of dietary fibre of 24 g/day, a portion of 10 g TS of seaweeds provides around 15.4% of a person's daily fibre needs (MacArtain et al. 2007). Dietary fibre has a well-established health benefit when consumed in recommended quantity helps to reduce the risk of diabetes, prevents coronary heart disease, constipation and colon cancer (Braithwaite et al. 2014).

Though seaweed is a poor source of carbohydrate in terms of bioavailability, it is a rich source of polysaccharides dietary fibre. SDF in brown seaweeds is comprised of neutral glucans, laminarin and sulphated fucoidans (Lahaye and Kaeffer 1997). Glucose, galactose and fucose (which possibly corresponded to sulphated laminarin and fucoidan) were the major sugar components in brown seaweed and galactose in red seaweeds (Gómez-Ordóñez et al. 2010). The main polysaccharides present in

Table 1.2 Proximate composition and dietary fibre of Selected important edible seaweeds

Seaweeds	% of dry weight								Reference
	Moisture	Protein	Lipid	Ash	Carbohydrate	TDF	IDF	SDF	
Red Seaweed									
Porphyra colubina	12.79 ± 0.07	24.61 ± 0.21	0.25 ± 0.06	6.46 ± 0.09	–	48.02 ± 1.13	26.60 ± 2.20	21.42 ± 5.12	Cian et al. (2013)
Porphyra spp.	9.4 ± 0.056	43.99 ± 0.051	0.49 ± 0.040	10.30 ± 0.033	5.19 ± 0.21	31.63 ± 0.032	16.8	17.9	Admassu et al. (2018) and Jiménez-Escrig and Sánchez-Muniz (2000)
Eucheuma cottonii	10.55 ± 1.60	9.76 ± 1.33	1.10 ± 0.05	46.19 ± 0.42	26.49 ± 3.01	25.05 ± 0.99	6.80 ± 0.06	18.25 ± 0.93	Matanjun et al. (2008)
Pyropia orbicularis	12.05 ± 0.23	24.15 ± 0.11	0.54 ± 0.15	16.40 ± 0.45	51.81 ± 0.44	64.01 ± 0.88	53.98 ± 0.54	10.03 ± 1.41	Uribe et al. (2017)
Gracilaria acerosa	12.15 ± 0.85	13.45 ± 1.076	0.028 ± 0.14	0.103 ± 0.049	1.05 ± 0.031	13.45 ± 1.076	–	–	Syad et al. (2013)
Gracilaria edulis	87.14 ± 1.10	18.04 ± 0.03	0.72 ± 0.04	22.8 ± 0.04	24.80 ± 0.12	63.175 ± 0.46	–	–	Moisture content was in wet wt basis Jesmi et al. (2016) and Ganesan et al. (2020)
Brown seaweeds									
Ascophyllum nodosum	11.08 ± 0.53	8.70 ± 0.07	3.62 ± 0.17	30.89 ± 0.06	–	–	–	–	Lorenzo et al. (2017)
Fucus vesiculosus	11.23 ± 0.08	12.99 ± 0.04	3.75 ± 0.20	20.71 ± 0.04	–	–	–	–	Lorenzo et al. (2017)
Bifurcaria bifurcata	7.95 ± 0.06	8.92 ± 0.09	6.54 ± 0.27	31.68 ± 0.41	–	37.42 ± 0.78	22.79 ± 0.97	14.64 ± 0.68	Gómez-Ordóñez et al. (2010) and Lorenzo et al. (2017)
Sargassum polycystum	9.95 ± 0.55	5.40 ± 0.07	0.29 ± 0.01	42.40 ± 0.41	33.49 ± 1.70	39.67 ± 0.56	34.10 ± 0.28	5.57 ± 0.28	Matanjun et al. (2008)
Laminaria saccharina	6.64 ± 0.06	25.70 ± 0.11	0.79 ± 0.07	34.78 ± 0.08		30.23 ± 0.85	13.11 ± 0.56	17.12 ± 0.84	Gómez-Ordóñez et al. (2010)

(continued)

Table 1.2 (continued)

Seaweeds	% of dry weight								Reference
	Moisture	Protein	Lipid	Ash	Carbohydrate	TDF	IDF	SDF	
Undaria pinnatifida		19.8 ± 1.4	4.5 ± 0.7	–	45.9 ± 1.5	35.1	5.3	30	Jiménez-Escrig and Sánchez-Muniz (2000) and Dawczynski et al. (2007)
Green Seaweeds									
Caulerpa lentillifera	10.76 ± 0.80	10.41 ± 0.26	1.11 ± 0.05	37.15 ± 0.64	38.66 ± 0.96	32.99 ± 2.07	15.78 ± 1.20	17.21 ± 0.87	Matanjun et al. (2008)
Ulva lactuca	–	13.84 ± 3.55	0.86 ± 0.00	12.41 ± 0.32	43.19 ± 1.75	38.1	16.8	21.3	Jiménez-Escrig and Sánchez-Muniz (2000)
Ulva reticulata	22.51 ± 0.97	21.06 ± 0.42	4.84 ± 0.33	17.58 ± 2.0	55.77	65.7 ± 0.9	64.8 ± 1.8	0.9 ± 0.8	Santoso et al. (2006) and Ratana-arporn and Chirapart (2006)
Ulva fasciata	–	22.7 ± 0.22	0.89 ± 0.12	27.0 ± 0.024	32.0 ± 0.04	–	–	–	Ganesan et al. (2020)
Enteromorpha flexuosa	–	17.29 ± 1.24	0.76 ± 0.24	32.20 ± 0.92	30.10 ± 0.18	33.4	16.2	17.2	Jiménez-Escrig and Sánchez-Muniz (2000) and Ganesan et al. (2020)

TDF total dietary fibre, *IDF* insoluble dietary fibre, *SDF* soluble dietary fibre

Table 1.3 Amino acid composition of selected edible red seaweeds

Seaweeds	*Porphyra colubina*	*Porphyra spp.* (Dry laver)	*Eucheuma cottonii*	*Porphyria tenera* (Nori)	*Gracilaria corticata*	*Gracilaria edulis*
Country reported	Argentina	China	Malaysia	Czech Republic	India	India
Unit	g 100 g^{-1} protein	g 100 g^{-1} protein	mg g^{-1} dw	g 16 g^{-1} N	mg/g of seaweed dw	mg/g of seaweed dw
Asp	12.22 ± 0.20	4.21 ± 0.005	2.65 ± 0.15	10.06 ± 0.05	14.37 ± 0.78	12.67 ± 0.64
Glu	10.50 ± 0.56	5.52 ± 0.002	5.17 ± 0.13	10.34 ± 0.04	2.54 ± 0.06	2.77 ± 0.15
Ser	6.16 ± 0.09	1.20 ± 0.002	1.92 ± 0.04	4.61 ± 0.12	2.23 ± 0.18	2.73 ± 0.13
His	1.26 ± 0.08	0.49 ± 0.003	0.25 ± 0.10	1.94 ± 0.28	2.46 ± 0.27	0.18 ± 0.02
Gly	8.87 ± 0.14	2.35 ± 0.00	2.27 ± 0.32	5.53 ± 0.10	4.71 ± 0.18	3.42 ± 0.27
Thr	5.91 ± 0.13	1.77 ± 0.003	2.09 ± 0.01	4.86 ± 0.11	1.32 ± 0.09	20.57 ± 0.62
Arg	6.19 ± 0.16	2.10 ± 0.003	2.60 ± 0.14	7.24 ± 0.07	3.41 ± 0.30	3.33 ± 0.17
Ala	12.54 ± 0.29	4.54 ± 0.002	3.14 ± 0.11	6.71 ± 0.08	21.11 ± 0.54	1.46 ± 0.18
Pro	3.96 ± 0.41	1.43 ± 0.002	2.02 ± 0.09	3.60 ± 0.15	0.47 ± 0.20	0.46 ± 0.18
Tyr	2.55 ± 0.05	0.78 ± 0.002	1.01 ± 0.12	2.84 ± 0.19	1.25 ± 0.15	2.50 ± 0.24
Phe	3.70 ± 0.06	1.56 ± 0.005	19.07 ± 2.48	4.20 ± 0.13	1.42 ± 0.17	2.20 ± 0.10
Val	5.85 ± 0.11	2.40 ± 0.003	2.61 ± 0.07	5.42 ± 0.10	0.16 ± 0.01	0.15 ± 0.02
Met	1.68 ± 0.07	0.633 ± 0.003	0.83 ± 0.17	3.09 ± 0.17	8.73 ± 0.31	4.98 ± 0.48
Cys	1.89 ± 0.03	0.028 ± 0.002	nd	2.80 ± 0.19	1.49 ± 0.30	1.27 ± 0.06
Ile	2.71 ± 0.05	1.62 ± 0.00	2.41 ± 0.04	3.37 ± 0.16	2.53 ± 0.16	1.22 ± 0.07
Trp	0.63 ± 0.01	nd	–	1.50 ± 0.10	1.00 ± 0.07	0.59 ± 0.16
Leu	7.38 ± 0.11	2.64 ± 0.003	3.37 ± 0.06	5.64 ± 0.09	1.58 ± 0.35	0.38 ± 0.02
Lys	6.01 ± 0.10	1.68 ± 0.002	1.45 ± 0.48	3.81 ± 0.14	2.37 ± 0.27	0.22 ± 0.03
EAA	41.32	14.412 ± 0.013	34.68	41.07	22.76 ± 1.81	35.55 ± 1.75
NEAA	58.69	20.564	18.18	46.49	36.14 ± 3.33	29.86 ± 1.83
Reference	Cian et al. (2013)	Admassu et al. (2018)	Matanjun et al. (2008)	Mišurcová et al. (2014)	Rosemary et al. (2019)	Rosemary et al. (2019)

nd not detected, *EAA* essential amino acid, *NEAA* nonessential amino acid

Table 1.4 Amino acid composition of selected edible brown seaweeds

Seaweeds	*Ascophyllum nodosum*	*Fucus vesiculosus*	*Bifurcaria bifurcata*	*Sargassum polycystum*	*Laminaria japonica* (Kombu)	*Undaria pinnatifida* (Wakama)	*Hizikia fusiformes* (Hijiky)
Country reported	Camariñas (A Coruña, Spain)	Camariñas (A Coruña, Spain)	Camariñas (A Coruña, Spain)	Malaysia	Czech Republic	Czech Republic	Czech Republic
Unit	mg/100 g dw	mg/100 g dw	mg/100 g dw	mg g^{-1} dw	g16 g^{-1} N	g16 g^{-1} N	g16 g^{-1} N
Asp	846.64 ± 38.87	1677.01 ± 156.39	800.84 ± 105.55	4.47 ± 0.87	8.45 ± 0.10	5.95 ± 0.09	10.53 ± 0.13
Glu	1714.55 ± 133.17	1974 ± 150.67	1504.53 ± 178.74	8.08 ± 1.08	15.38 ± 0.05	6.64 ± 0.08	13.71 ± 0.10
Ser	378.62 ± 13.57	630.54 ± 47.00	357.10 ± 36.87	2.58 ± 0.16	2.93 ± 0.28	2.45 ± 0.21	4.13 ± 0.34
His	126.46 ± 10.65	194.59 ± 8.73	138.76 ± 12.70	0.26 ± 0.11	1.23 ± 0.67	1.06 ± 0.48	1.97 ± 0.22
Gly	417.70 ± 12.89	651.24 ± 30.84	390.14 ± 29.42	3.19 ± 0.35	3.77 ± 0.22	3.37 ± 0.15	5.63 ± 0.25
Thr	363.22 ± 17.12	613.08 ± 33.62	360.27 ± 38.25	2.60 ± 0.16	3.53 ± 0.24	2.71 ± 0.19	4.91 ± 0.29
Arg	316.79 ± 14.05	557.87 ± 38.44	330.11 ± 42.41	2.88 ± 0.17	3.31 ± 0.25	3.23 ± 0.16	4.57 ± 0.31
Ala	655.73 ± 34.75	985.40 ± 69.50	846.65 ± 82.87	4.25 ± 0.15	6.09 ± 0.14	4.65 ± 0.11	7.08 ± 0.20
Pro	399.24 ± 11.70	575.19 ± 39.15	318.40 ± 40.96	2.55 ± 0.14	5.05 ± 0.16	2.34 ± 0.22	2.37 ± 0.59
Tyr	162.85 ± 24.50	327.01 ± 30.59	175.00 ± 30.90	1.26 ± 0.06	1.43 ± 0.58	1.39 ± 0.37	2.25 ± 0.63
Phe	340.13 ± 17.74	541.53 ± 25.72	330.05 ± 32.32	30.42 ± 4.43	2.79 ± 0.30	2.88 ± 0.18	4.95 ± 0.28
Val	353.89 ± 32.95	582.70 ± 36.73	372.82 ± 49.05	3.13 ± 0.14	3.76 ± 0.22	3.31 ± 0.15	5.84 ± 0.24
Met	147.57 ± 18.71	218.21 ± 20.20	178.41 ± 18.08	1.25 ± 0.04	1.98 ± 0.42	2.07 ± 0.25	3.73 ± 0.38
Cys	0.00 ± 0.00	205.23 ± 25.43	0.00 ± 0.00	–	2.13 ± 0.39	0.96 ± 0.53	2.18 ± 0.65
Ile	295.26 ± 25.73	507.82 ± 32.42	299.73 ± 37.74	2.94 ± 0.16	2.51 ± 0.33	2.60 ± 0.20	4.42 ± 0.32
Trp	nd	nd	nd	–	1.65 ± 0.11	0.71 ± 0.02	1.06 ± 0.10
Leu	537.37 ± 38.87	862.14 ± 57.02	524.59 ± 61.38	4.67 ± 0.25	4.41 ± 0.19	4.42 ± 0.12	7.17 ± 0.20
Lys	431.72 ± 38.40	800.28 ± 74.20	393.06 ± 56.57	2.11 ± 0.77	3.20 ± 0.26	3.00 ± 0.17	3.31 ± 0.43
EAA	2912.42 ± 204.93	4878.22 ± 304.12	2927.79 ± 346.84	50.26	28.35	25.99	41.93
NEAA	4575.33 ± 198.91	7026.10 ± 512.60	4392.67 ± 502.38	26.38	45.23	27.75	47.88
Reference	Lorenzo et al. (2017)	Lorenzo et al. (2017)	Lorenzo et al. (2017)	Matanjun et al. (2008)	Mišurcová et al. (2014)	Mišurcová et al. (2014)	Mišurcová et al. (2014)

nd not detected, *EAA* essential amino acid, *NEAA* nonessential amino acid

Table 1.5 Amino acid composition of selected edible green seaweeds

Seaweeds	*Caulerpa lentillifera*	*Ulva lactuca*	*Ulva reticulata*	*Ulva fasciata*	*Enteromorpha flexuosa*
Country reported	Malaysia	Hong Kong	India	India	India
Unit	mg g^{-1} dw	mg/g protein dw	mg/100 mg dw	mg/g protein dw	mg/g protein dw
Asp	8.33 ± 0.11	89.7	1.08	60.8	63.2
Glu	13.47 ± 0.23	87.3	1.12	72.6	68.2
Ser	5.49 ± 0.20	55.4	0.5	–	–
His	1.44 ± 0.13	4.82	0.5	14.3	7.2
Gly	5.14 ± 0.03	67.4	1.03	52.2	42.3
Thr	5.84 ± 0.22	50.6	0	38.8	42.8
Arg	5.71 ± 0.22	84.4	0.93	29.2	40.3
Ala	6.88 ± 0.19	73.9	0.5	45.4	32.4
Pro	4.29 ± 0.11	44.6	0.58	37.8	29
Tyr	3.33 ± 0.08	35	0	33.6	22
Phe	19.95 ± 1.41	35	0.82	42.2	33.6
Val	6.18 ± 0.02	55	0.57	46.7	40
Met	1.58 ± 0.08	15.7	0.48	12.8	10.2
Cys	–	13.3	0	–	–
Ile	5.06 ± 0.12	38.2	0.51	40.2	25
Trp	–	nd	nd	–	–
Leu	7.79 ± 0.19	67.1	0.89	68.4	48.4
Lys	1.22 ± 0.05	65.8	0.52	52.4	20.2
EAA	54.77	416.62	5.22	301.5	220.2
NEAA	46.93	466.6	5.08	308.1	304.6
Reference	Matanjun et al. (2008)	Wong and Cheung (2000)	Ishakani et al. (2017)	Ganesan et al. (2020)	Ganesan et al. (2020)

nd not detected, *EAA* essential amino acid, *NEAA* nonessential amino acid

Table 1.6 Fatty acid profile of selected edible seaweeds

Seaweeds	% of total fatty acid content					Reference
	SFAs	MUFAs	PUFAs	PUFA/SFA	ω6/ω3	
Red Seaweed						
Porphyra colubina	26.37	13.76	59.48	2.23	0.5	Cian et al. (2013)
Porphyra spp.	60.48 ± 2.58	10.67 ± 1.55	28.86 ± 3.94	0.48	0.13	Sánchez-Machado et al. (2004)
Eucheuma cottonii	25.17 ± 0.38	23.28 ± 0.47	51.55 ± 0.57	2.05	0.1	Matanjun et al. (2008)
Porphyria tenera	34.3 ± 1.0	5.4 ± 1.1	60.4 ± 2.3	1.76 ± 0.08	0.7 ± 0.1	Kumari et al. (2013)
Gracilaria corticata	31 ± 2.5	3.5 ± 0.1	65.6 ± 2.5	2.13 ± 0.25	∞	Kumari et al. (2013)
Gracilaria edulis	40.1 ± 3.6	11.7 ± 4.5	48.4 ± 3.4	1.21 ± 0.15	59.2 ± 29	Kumari et al. (2013)
Brown seaweeds						
Ascophyllum nodosum	25.14 ± 0.49	31.15 ± 0.23	43.47 ± 0.54	1.73	2.62 ± 0.01	Lorenzo et al. (2017)
Fucus vesiculosus	29.26 ± 0.34	22.33 ± 0.33	48.19 ± 0.62	1.65	1.72 ± 0.01	Lorenzo et al. (2017)
Bifurcaria bifurcata	27.62 ± 0.77	26.51 ± 0.48	46.91 ± 1.37	1.7	1.41 ± 0.07	Lorenzo et al. (2017)
Sargassum polycystum	51.30 ± 0.51	28.36 ± 0.48	20.34 ± 0.43	0.4	0.98	Matanjun et al. (2008)
Laminaria ochroleuca	33.82 ± 2.21	19.33 ± 1.99	46.94 ± 4.58	1.40	0.83	Sánchez-Machado et al. (2004)
Undaria pinnatifida	20.39 ± 1.73	10.5 ± 1.78	69.11 ± 9.1	3.40	0.49	Sánchez-Machado et al. (2004)
Green Seaweeds						
Caulerpa lentillifera	46.41 ± 0.56	36.83 ± 0.55	16.76 ± 0.27	0.36	1.07	Matanjun et al. (2008)
Ulva lactuca	59.9 ± 2.3	12.2 ± 1.8	28.0 ± 0.7	0.47 ± 0.03	0.3 ± 0.01	Kumari et al. (2013)
Ulva reticulata	48.8 ± 5.9	7.1 ± 0.7	44.8 ± 5.9	0.94 ± 0.22	0.3 ± 0.2	Kumari et al. (2013)
Ulva fasciata	29.6 ± 2.1	7.5 ± 0.9	63.1 ± 1.9	2.14 ± 0.2	0.2 ± 0.01	Kumari et al. (2013)

SFAs saturated fatty acids, *MUFA* monounsaturated fatty acids, *PUFA* polyunsaturated fatty acids

brown seaweeds—*H. elongate* and *S. polyschides* were the alginates (297 g/kg TS and 163 g/kg TS, respectively) (Jard et al. 2013). *H. elongata* contains the highest concentration of alginates followed by *S. polyschides* and *S. muticum* (136 g/kg TS). On the other hand, red seaweed *U. lactuca* and *Gracilaria verrucosa* are rich in glucose (75 g/kg TS and 77 g/kg TS respectively) and xylose was found in higher concentration in *Palmaria palmata* compared to the other algae (233 g/kg TS). The presence of xylose in the sugars composition in seaweed lends beneficial application in food (MacArtain et al. 2007).

4.4 Mineral Content in Seaweed

Seaweed contains high ash content in the range of 6.46 to 46.19% in red seaweeds, 20.71 to 42.40% in brown seaweeds and 12.41 to 37.15% in green seaweeds as presented in Table 1.7. Seaweeds are rich in mineral macronutrients viz., calcium, chloride, magnesium, phosphorus, potassium, sodium, sulfur and mineral micronutrients viz., iodine, iron, copper, cobalt, boron, zinc, molybdenum, manganese, nickel and the total mineral content of accounts for 36% of seaweed dry mass. The concentration of potassium and sodium was high in *G. acerosa* and *S. wightii*, respectively and *P. columbina* was reported to have high mineral content with good Na/K relationship (Syad et al. 2013; Cian et al. 2013). *A. nodosum, F. vesiculosus* and *B. bifurcatathe* were rich in macronutrients such as K (3781.35 to 9316.28 mg/100 g), Mn (from 8.28 to 1.96 mg/100 g), Na (1836.82 to 4575.71 mg/100 g) and Ca (984.73 to 1160.27 mg/100 g), while *G. edulis, U. lactuca* and *Sargassum* sp. have both mineral macronutrients and the micronutrients (Jesmi et al. 2016; Lorenzo et al. 2017). *G. edulis* had a high content of Na, P, Ca and Fe (423.33 ± 1.15 mg 100 g^{-1}, 282.5 ± 0.5 mg 100 g^{-1}, 223.33 ± 0.58 mg 100 g^{-1} and 65.28 ± 0.33 mg 100 g^{-1}, respectively), whereas *Sargassum* sp. reported highest Se content (49.82 ± 0.09 mg 100 g^{-1}). *Acanthophora spicifera, Gracilaria edulis, Padina gymnospora, Ulva fasciata* and *Enteromorpha flexuosa* are rich in iron, iodine and calcium which were present in the range of 14.8–72 mg/100 g, 38.8–72.2 mg/100 g, and 410–870 mg/100 g, respectively (Ganesan et al. 2020).

Calcium holds 4–7% dw of the total mineral contents present in seaweeds and is available in the form of calcium phosphate (Rajapakse and Kim 2011). Calcium phosphate in seaweed is more bioavailable than the calcium in milk, which is present as calcium carbonate. Seaweeds are the primary source of iodine and can be considered as the cheap and best source of food ingredients to fulfill the minimum iodine requirement (150 mg/day) of the human body (Mišurcová et al. 2011). Brown seaweeds have higher iodine content (1500–8000 ppm) than red and green seaweeds (Rajapakse and Kim 2011).

Table 1.7 Mineral composition of selected edible seaweeds

Seaweeds	Mineral content (mg $100g^{-1}$ dw)										
	Na	K	P	Ca	Mg	Fe	Zn	Cu	Mn	Se	Reference
Red seaweeds											
P. columbinaa	414.22 ± 8.96	1444.17 ± 56.30	379.90 ± 7.90	443.70 ± 6.64	491.53 ± 3.44	22.00 ± 0.40	1.46 ± 0.09	0.51 ± 0.05	nq	nq	Cian et al. (2013)
Porphyra spp.	348.75 ± 1.06	1395.00 ± 4.24	nq	525.00 ± 1.41	261.75 ± 1.06	12.28 ± 0.32	2.79 ± 0.1	1.38 ± 0.02	2.26 ± 0.04	nq	Admassu et al. (2018)
E. cottonii	1771.84 ± 0.01	13155.19 ± 1.14	nq	329.69 ± 0.33	271.33 ± 0.20	2.61 ± 0.00	4.30 ± 0.02	0.03 ± 0.00	nq	0.59 ± 0.00	Matanjun et al. (2008)
A. spicifera	36.08 ± 1.08	52.08 ± 1.08	210 ± 0.12	430 ± 0.14	480 ± 1.02	52 ± 0.24	4.08 ± 0.28	nq	nq	nq	Ganesan et al. (2020)
G. edulis	32.03 ± 0.28	52.12 ± 0.07	124 ± 0.08	410 ± 0.08	580 ± 0.98	72 ± 0.24	5.21 ± 0.24	nq	nq	nq	Ganesan et al. (2020)
Brown seaweeds											
A. nodosum	4575.71 ± 50.05	3781.35 ± 13.40	nq	984.73 ± 47.26	867.82 ± 12.01	13.34 ± 0.90	nq	nq	1.96 ± 0.69	nq	Lorenzo et al. (2017)
F. vesiculosus	2187.51 ± 36.90	3745.05 ± 36.01	193.57 ± 1.13	1160.27 ± 23.10	732.37 ± 5.35	18.99 ± 0.32	nq	nq	8.28 ± 1.07	nq	Lorenzo et al. (2017)
B. bifurcata	1836.82 ± 52.12	9316.28 ± 101.94	169.54 ± 1.41	996.42 ± 12.83	528.01 ± 8.25	nq	nq	nq	nq	nq	Lorenzo et al. (2017)
S. polycystum	1362.13 ± 0.00	8371.23 ± 0.00		3792.06 ± 0.51	487.81 ± 0.24	68.21 ± 0.03	2.15 ± 0.00	nq	nq	1.14 ± 0.03	Matanjun et al. (2008)
P. gymnospora	36.36 ± 0.18	30.02 ± 0.17	164 ± 0.28	820 ± 0.34	780 ± 0.08	14.8 ± 0.32	4.19 ± 0.08	nq	nq	nq	Ganesan et al. (2020)
Green seaweeds											
C. lentillifera	8917.46 ± 0.00	1142.68 ± 0.00	nq	1874.74 ± 0.20	1028.62 ± 0.58	21.37 ± 0.00	3.51 ± 0.00	0.11 ± 0.00	nq	1.07 ± 0.00	Matanjun et al. (2008)

Seaweeds	Mineral content (mg $100g^{-1}$ dw)										
	Na	K	P	Ca	Mg	Fe	Zn	Cu	Mn	Se	Reference
U. fasciata	20.12 ± 0.02	27.20 ± 1.02	142 ± 0.18	740 ± 0.28	47 ± 0.04	47 ± 0.04	2.34 ± 0.48	nq	nq	nq	Ganesan et al. (2020)
E. flexuosa	13.20 ± 0.8	22.32 ± 1.08	270 ± 0.02	712 ± 0.04	40 ± 0.28	40 ± 0.28	1.52 ± 0.81	nq	nq	nq	Ganesan et al. (2020)
U. lactuca	351.67 ± 1.53	209 ± 1.73	nq	180.67 ± 1.15	-	34.47 ± 1.10	1.78 ± 0.02	1.83 ± 0.005	4.8 ± 0.02	1.60 ± 0.04	Jesmi et al. (2016)

nq not quantified

4.5 Vitamins in Seaweed

Seaweeds are a good source of both water-soluble vitamins such as B1, B2, B12 and C and fat-soluble vitamins such as Vitamin A and Vitamin E. Seaweeds are generally rich in B group vitamins (B1, B2, B12) than any other group. Ortiz et al. (2006) suggested that the daily vitamin requirements of the human body could be met by consuming 100 g of seaweeds. Water-soluble vitamin C, the major antioxidant was abundant in both the seaweeds *G. acerosa* and *S. wightii*, which constitutes 5.07 ± 0.20 mg and 5.07 ± 0.40 mg/g of dw, respectively, while Vitamin B1 and B2 were detected only in trace amounts (Syad et al. 2013). Also, fat-soluble vitamin E (1.33 ± 0.07 mg/g of dw to 1.35 ± 0.08 mg/g of dw) and vitamin A (0.0034 ± 0.0002 mg to 0.0040 ± 0.00012 mg/g dw) were present in *G. acerosa* and *S. wightii*. Vitamin D2 (2.59 mg%) was relatively abundant in *G. edulis* followed by vitamin E (1.02 mg%) and vitamin K1 (0.71 mg%) (Jesmi et al. 2016). Vitamin D2, vitamin K1 and vitamin E were present in trace amounts in *U. lactuca*, while only vitamin E (0.49 mg%) was present in *Sargassum sp.* (Jesmi et al. 2016). Vitamin A (2.67–0.31 mg/g) and vitamin B9 contents (1.00–0.07 mg/g) were relatively higher in *G. corticata* than *G. edulis*, while, *G. edulis* had a significantly higher content of vitamin B2 (1.54–0.39 mg/g) and vitamin B6 (4.77–0.23 mg/g) than *G. corticata*. In another study, Riboflavin (Vitamin B2) and niacin (Vitamin B3) are present in sufficient amount in *U. fasciata* and *G. edulis* (Ganesan et al. 2020). The highest amount of Vitamin B2 was detected in green seaweed *U. fasciata* (0.32 mg) than in red seaweed *G. edulis* (0.12 mg).

5 Seaweed Based Food Products

Nutrition plays a vital role in health promotion and prevention of infectious diseases (Nova et al. 2020). In recent years, there is growing consumer awareness of a nutritious diet that is fueling the demand for healthy and nutritive food with functional properties (Granato et al. 2020). Seaweeds are rich in multifunctional bioactive compounds that are associated with antioxidant, anti-inflammatory, anti-coagulant, anti-hypertensive, cardiovascular protective, anti-diabetic and anti-proliferative activities. In addition, compounds in seaweed constitute excellent functional food ingredients that find application in value-added product development. Fortification of food products with low levels of nutrients to provide a healthy and balanced diet has become a growing trend in the food industry. There is an increasing trend to supplement food products with nutrients at low levels, aiming for the desirable objective of a balanced and healthy diet (Cencic and Chingwaru 2010). Seaweeds constitute an estimable source of PUFA, dietary fibre, vitamins, amino acids and minerals as well as bioactive compounds with beneficial functional activities (Ibanez and Cifuentes 2013; Lage-Yusty et al. 2014; Jimenez-Escrig et al. 2015; Rafiquzzaman et al. 2015). The seaweed industry in India is a cottage industry focused on the production of agar, carrageenan and alginate, and its production mainly depends on *Gelidiella sp., Gracilaria sp., Sargassum sp.* and *Tubineria* sp. Additionally, a small part of seaweed production is being utilized for the production of animal feed and fertilizers as plant stimulants and growth promoters. Diversification of the utilization of seaweed as value-added food products will help to increase the consumption of seaweed as a food ingredient among the Indian population. Seaweeds as a food and functional ingredients can be presented in different forms, such as, whole seaweed meals, grounded seaweeds, seaweeds powders, seaweed extraction/seaweed bioactive compounds, seaweed puree/seaweeds homogenates and seaweed fermentation/fermented seaweed products (Michalak and Chojnacka 2018).

5.1 Seaweed Enriched Soup

Seaweed soups are part of East Asian and South-East Asian culture and serve as an important part of everyday diets and during ceremonial functions. Seaweed soups are popular food products and have been widely accepted by people in other parts of the world due to their delicious taste and nutritional value. Given the nutritional richness of seaweeds, seaweed soups are rich in fibre, vitamins, minerals, essential fatty acids, essential amino acids and other bioactive compounds, which have antioxidant, anti-cancer, anti-microbial properties etc. Consuming soup enhances satiety and reduce the total energy intake. It can act as a convenient vehicle for the delivery of the nutrients for meeting the recommended daily intake, thus improving the overall wellbeing and healthy nutritional status. Due to high nutritional value and water content, soups can be recommended for all types of consumers like children, adults and elderly people. Seaweed based soup mixtures are reported to supply enough iodine to meet the requirements for thyroid functioning (Zava and Zava 2011; Jayasinghe et al. 2016) and seaweed soup can be recommended as a therapeutic food to cure the mineral deficiency. Besides the delicious taste and nutrients source, soup can also bring health benefits to the human body. Soup intake and dietary fibre intake were positively correlated with plasma leptin levels (Kuroda et al. 2010).

Although seaweed soup is being consumed for centuries in Japan, the Republic of Korea and China, to our best knowledge very limited literature is available. However, in recent decades, seaweed soups are gaining interest attention of nutritionists. Agar from *Gracilaria verrucose* (1–4%, w/w), carrageenan from *Kappaphycus alvarezii* (1–4%, w/w) and *Ulva lactuca* dried powder (2.5%) were used to prepare vegetable soup (Jayasinghe et al. 2016). According to the organoleptic evaluation, vegetables (80%), grain (10%), legumes (3.5%), dried *Ulva* powder (2.5%) with agar agar (3%) or carrageenan (2%) and preservatives were found to be the best soup formulation mixture. Agar and carrageenan with thickening, gelling and stabilizing properties greatly improved the viscosity of the soup. While the addition of dried *Ulva* powder helps to significantly improve the physical and nutritional value of the soup. For example, soup containing 3% agar and 2.5% *Ulva* have higher protein (11.3 ± 0.8%) and higher minerals such as sodium (250.56 ± 0.75 mg/100 g), magnesium (45.8 ± 0.98 mg/100 g), potassium (53.39 ± 0.07 mg/100 g) and iodine (0.35 mg/L) compared to the commercially available vegetable soup powder. The shelf life of seaweed soup mixture stored at -18°C was estimated to be more than 6 months, while soup mixture had a short shelf life of about 3 months when stored at ambient temperature.

In another study, carrageenan extract from *Kappaphycus alvarezii* was used to prepare fish soup using croaker fish (*Johnius dussumieri*) (Jeyakumari et al. 2016). Carrageenan addition at 5% did not affect the consistency and flavor characteristics of the soup and also improved the functional properties and mineral content of the soup. Fish soup powder fortified with 10 to 50% seaweed extract from *Ulva lactuca* and it was found that soup powder with 30% was more acceptable (Lekshmi et al. 2017).

5.2 Seaweed Fortified Cereals

Fortification is a process to enhance the quality of the food by the inclusion of substances with important nutrient values especially vitamins, PUFA, micronutrients, dietary fibre, etc. to the food. Diversification of processed-seaweed products into different flour-based products can be an effective approach to maximize seaweed consumption. By taking advantage of the nutritional richness of seaweed, seaweed flour can be used in various processed food products such as noodles, pasta, bread, cake, etc.

5.2.1 Snack

Breadsticks, a new bakery based functional product, was developed with dried brown seaweed (*Himanthalia elongate*) and white flour (Cox and Abu-Ghannam 2013b). Incorporation of *H. elongate* at 17.07% concentration enhanced the phytochemical constitution in breadsticks with maximum total phenolic content (138.25 mg GAE/100 g db), DPPH free radical scavenging activity (65.01%), higher dietary fibre (7.95%), in addition to imparting an appealing colour and texture. Brown seaweeds are known to have higher bioactive components than red seaweeds and green seaweeds and they have antioxidant properties (Seafoodplus 2008).

Seaweed fortification of Crispy Enbal, as local baked food made from cassava in Kei Islands, Indonesia was developed using 15% seaweed pulp and seaweed flour from *Eucheuma cottonii* (Marasabessy and Sudirjo 2017). Crispy Enbal fortified with seaweed pulp had higher fibre content (7.48%), higher textural properties and was preferred organoleptically by consumers than crispy enbal fortified with seaweed flour. Ningsih and Anggraeni (2021) used dried *Euchema cottonii* seaweed powder as a substitute to wheat flour at concentrations of 2.5%, 5% and 7% to developed millecreps cake. The addition of 5% *Euchema cottonii* seaweed powder to the cake resulted in the products having bright yellow colour, mild seaweed taste and softer texture. However, the organoleptic characteristics of 5% *E. cottonii* incorporated cake did not change significantly as compared with the wheat flour control cake. Hence, *E. cottonii* can be used as raw material to develop cakes and can also be used to develop dietary fibre enriched functional cake. Organoleptically acceptable and dietary fibre enriched muffins were developed using 2% *K. alvarezii* composite flour and the fortification of seaweed also showed significant influences s on the physicochemical properties of the final products (Mamat et al. 2018).

5.2.2 Extruded Products

Singh et al. (2016) reported that nutritional value and functional properties of extruded snacks were improved with fortification of 6–8% *Ulva lactuca*. In another study, brown seaweed (*Sargassum tenerrimum*) was used different concentrations of 2.5, 5, 7.5 and 10% with corn flour and rice flour in the ratio of 70:30 to develop extracted products (Singh et al. 2017). Extrusion parameters like feed moisture, Sargassum concentration and barrel temperature were optimized using response surface methodology (RSM) for maximum expansion ratio (16.45%, 4.33% and 123.08 °C respectively), and porosity (16.06%, 4.51% and 124.04 °C respectively). *S. tenerrium* (2.5–7.5%) fortified extruded products with 13.5–17.5% feed moisture were accepted organoleptically with desirable hardness, puffiness and crispness, which are the typical physical characteristics of an extruded snack. Corn extruded snack seasoning with 4% polysaccharides extract from seaweeds (*Sirophysalistrinodis* and *Polycladiamyrica*) were acceptable organoleptically with a very mild seaweed flavour and taste similar to control snack (Etemadian et al. 2018). Also, seasoned snacks had lower lipid spoilage values (lower TBA and PV) than those recommended by the health authorities for snacks consumption.

S. wightii incorporated rice/corn-based extruded snacks with optimal functional and physical properties such as higher expansion ratio, hardness and lower bulk density were developed and the extruded products were rich in dietary fibre, minerals and carbohydrate and had the high antioxidant capacity and total phenolic content compared to similar extruded snack products (Gopalakrishnan et al. 2020).

5.2.3 Noodles

Fortification of noodles with seaweed can be a promising technology for improving its nutritional value in terms of dietary fibre and mineral content as well as for promoting seaweed as a dietary component in places where seaweed consumption is not popular. Seaweed puree from two different red seaweed species such as *E. cottoniii* and *G. verucossa* were used separately and in combination at 30% level to prepare the dried noodles by replacing the same amount of wheat flour (Dewi 2011). The results showed that substitution of seaweeds puree resulted in increasing the moisture (10.08 ± 2.02% to 13.94 ± 0.84%), crude fibre (2.00 ± 0.4% to 2.25 ± 0.18%), iodine content (1.06 ± 2.80 μg/g to 1.43 ± 0.76 μg/g) and imparted softer and spongier textural intensities to dried noodles as compared to noodle without seaweed puree. Moreover, the substitution of seaweed puree did not significantly change the sensory characteristics in terms of taste and colour. Seaweed noodles were also prepared using edible green seaweed (*Ulva reticulata*) puree, wheat flour, corn flour, tapioca starch, salt, and fish mince (Jesmi et al. 2017). The incorporation of green seaweed puree and fish mince increased the crude protein, crude fibre content and water holding capacity and enhanced the flavour of the noodles. The combination of fish mince and seaweed further enhanced the nutritional value of noodles with minimal alteration of textural and cooking attributes. Fortification of seaweed noodles (*Euchema cottonii*) with nano-calcium from bone catfish (*Clarias batrachus)* was evaluated to improve the quality and nutritional value of noodles (Halimah et al. 2016). Noodles were prepared with 30% of seaweeds and different concentrations of 1%, 1.5% and 2% nanocalcium. Seaweed noodles with 1% nano calcium gave the best results with calcium (1.49%), protein (5.74%), carbohydrates (47.22%), yellowish in appearance, have specific seaweed flavour and taste. Kumoro et al. (2016) reported that instant fried wheat noodles fortified with microalgae flours from *Spirulina platensis* (5%), and a seaweed flour from *Eucheuma cottonii* (5%) gave the best culinary properties in terms of texture, colour, flavour and taste to noodles.

Agusman and Wahyuni (2020) developed wheat noodles incorporated with *Caulerpa* sp. seaweed filtrate (0, 5, 10, 15 and 20%) and observed that the β-carotene contents and total phenolic compounds of noodles increased with increasing concentration of *Caulerpa* sp. filtrate. The β-carotene content and total phenolic compounds increased to a maximum of 3.68 ± 0.40 mg/kg and 85.2 ± 1.82 mg GAE/100 g, respectively, with the addition of 15% *Caulerpa* sp. filtrate. Noodles fortified with up to 20% *Caulerpa sp.* filtrate had higher tensile strength, lower elasticity and also imparted greenish colour and lower brightness to the fortified noodles. However, the addition of seaweed filtrate did not affect the overall consumer acceptance of the fortified noodles.

5.2.4 Pasta

Suitability of edible Japanese seaweed, wakame (*U. pinnatifida*) as an ingredient in pasta was evaluated to replace semolina at different concentrations (semolina:wakame—100:0; 95:5.0; 90:10; 80:20 and 70:30, w/w) (Prabhasankar et al. 2009). Pasta fortified with up to 20% of wakame found sensory acceptance with better bio-functional properties in terms of total phenolic content (0.94 mg gallic acid equivalents (GAE)/g), total antioxidant activity (2.82 mg ascorbic acid equivalents (AAE)/g), 2,2-Diphenyl-1-picrylhydrazyl (DPPH. 8.66 ± 0.44%) and superoxide radical scavenging activities (34.31 ± 1.15%). However, pasta with 10% wakame gave better sensory acceptability than 20% level. Also, the organoleptically acceptable pasta had only a mild seaweed flavour having a taste similar to control pasta. The nutritional value of the pasta was improved due to the higher content of bioactive components such as fucoxanthin and fucosterol provided through the addition of seaweeds. For example, the ratio of ω-3/ω-6 fatty acid in seaweed fortified pasta was 1:3.4 as compared to 1:15.2 in the control. As analyzed by scanning electron microscopy (SEM) interaction between starch granules and protein matrix was enhanced by the addition of wakame up to 20% level which resulted in improved quality of pasta. However, beyond 20% level of seaweed resulted in the formation of a weaker network between starch granules owing to gumming activity of seaweed with gluten matrix. Firdaus et al. (2017) reported the fortification of *E. cottonii* flour enhances the nutrition value especially iodine content (3.71 ppm), crude fibre (8.02%), the total of dietary fibre (20.88%), soluble fibre (11.69%), insoluble fibre (9.19%) and glycemic index (44.45) of pasta. Enny Sholichah et al. (2021) developed the gluten-free pasta enriched with *Kappaphycus alvarezii*. The addition of 40% of *K. alvarezii* puree produced the best pasta rich in nutrients especially protein (5.45%), dietary fibre (7.54%) and calcium (274.72 mg/100g) and seaweed also helped to enhance the viscous-elasticity and cooking properties.

5.2.5 Cookies

Cookies are ready-to-eat, popular baked products available in different sizes and shapes that are priced at affordable costs and are relished by people of all age groups (Vijerathna et al. 2019). Cookies have low water activity which allows for longer shelf life (Chauhan et al. 2015; Usman et al. 2015). Cookies were prepared by baking the mixture of dried *Bifurcaria bifurcata* seaweed powder at 3%, 6% and 9% (w/w) and chestnut flour doughs with other hydrocolloids and salts (Arufe et al. 2019). The addition of *B. bifurcata* seaweed powder to chestnut flour significantly modified cookie properties as well as antioxidant properties of baked cookies. Salt free freeze dried seaweed flour four different seaweed namely *Hizikia fusiforme, Codium fragile, Sargassum fulvellum* and *Enteromorpha linza* were used for developing cookies (Oh et al. 2020). The addition of 5% seaweed flour significantly changed the solvent retention capacity (SRC) and influenced the baking quality of the cookies. Cookies fortified with 5% *H. fusiforme* powder were the most preferred cookies, while other seaweed significantly influenced the odour and flavour of the fortified cookies and had a fishy smell. Gopalakrishnan et al. (2020) suggested that *Caulerpa racemose* can be successfully added to cookies, providing an innovative opportunity to utilize these seaweeds as a functional food. Apart from rich protein and dietary fibre contents, these cookies have high polyphenolic contents and antioxidative potential. The previous studies indicate that diversification of seaweed as a healthy snack can be fruitfully achieved by seaweed cookies.

5.2.6 Bread

Potential applications of seaweed hydrocolloids in the baking industry have been demonstrated by many studies. Hydrocolloids like sodium alginate and κ-carrageenan can modify the starch gelatinization, influence the rheological and textural properties and extend the shelf life of stored bread (Dziezak 1991; Rojas et al. 1999; Rosell et al. 2001). Hydrocolloids have been employed as gluten substitutes in the formulation of gluten-free bread which is attributed to its polymeric structure (Ylimaki et al. 1998). Water absorption of the dough and firmness increased when *K. alvarezii* powder was added at up to 8% to the wheat dough to develop functional bread (Mamat et al. 2013). However, the addition of seaweed powder decreased stickiness properties. Arufe et al. (2017) added *Fucus vesiculosus* seaweed powder in wheat bread which resulted in significant modifications in dough and bread properties such as increased density from 0.23 to 0.40 g/cm^3 and crumb firmness from 18 to 45 kPa. Seaweed also imparted green colour to the bread crust; however, it did not influence consumers preference. Another study reported that *Ascophyllum nodosum*—a brown seaweed and *Chondrus crispus*—a red seaweed supplemented wheat bread were acceptable at 4% and 2% concentrations, respectively. Supplementation of seaweed also improved the fibre and protein content of bread (Lamont and McSweeney 2020).

5.3 *Seaweed and Seaweed Polysaccharides Fortified Meat Products*

Seaweeds are demonstrated as a functional food ingredient to the meat products to enrich nutrients, as a fat replacement and to develop low salt meat products (Lopez-Lopez et al. 2009; Gullón et al. 2020). Physicochemical, structural and quality characteristics of hamburger patties and gel/emulsion meat systems were enhanced with the addition (1–5 g/100 g seaweed powder) of different types of seaweed namely *Sagassum thunbergia, Gelidiumamansii, Himanthaliaelongata, Undaria pinnatifida and Porphyra umbilicalis* proving the usefulness of seaweed in formulating innovative healthier meat products and overcoming the technical constraints associated with low salt products. Manish Kumar and Sharma (2004) demonstrated that 0.5% carrageenan can be used to replace fat in the formulation of low fat pork patties without degrading its textural and sensory attributes. Carrageenan fortified low fat pork can be stored at 4 ± 1 °C for 21 days and 35 days in aerobic and vacuum packaging, respectively. The addition of seaweed *H. elongata* at 5.5% level produced low-sodium pork frankfurters with better Na/K ratio, dietary fibre and calcium (Lopez-Lopez et al. 2009). Choi et al. (2015) reported that *Undaria pinnatifida* and *Hizikia fusiforme* improved quality characteristics such as cooking loss, sensory attributes and emulsion stability of low-salt frankfurters. Besides improving the textural and nutritional qualities, the addition of seaweed polysaccharides such as laminarin and fucoidan can enhance lipid stability in functional meat products (Moroney et al. 2013, 2015).

5.4 Seaweed and Seaweed Hydrocolloids Enriched Surimi and Fish Products

Surimi is a concentrated myofibrillar protein obtained by repeated washing of fish mince to remove the sarcoplasmic proteins, lipids, blood, and other impurities. Application of edible seaweeds and sulfated polysaccharides in surimi and surimi based products have been reported by many researchers (Alipour et al. 2018; Chen et al. 2020; Jannat-Alipour et al. 2019a, b). Hydrocolloids are the most commonly used additives in surimi to improve its textural and rheological properties. Seaweed powder, sulfated polysaccharides from *Ulva intestinalis,* κ-carrageenan were employed as potential new hydrocolloids in surimi formulation from silver carp (Alipour et al. 2018; Chen et al. 2020; Jannat-Alipour et al. 2019a). Results suggested that the incorporation of sulfated polysaccharides up to 0.25 g/100 g, seaweed powder up to 2.8 g $(100\ g)^{-1}$ and κ-carrageenan up to 5 g kg^{-1} into surimi gel significantly promote get strength, water holding capacity and textural properties and remarkably increased the whiteness of surimi gel. Also, SEM analysis showed surimi gels had a finer and denser network structure due to the presence of seaweed polysaccharides. Furthermore, edible green seaweed, *U. intestinalis* powder (2.77 g kg^{-1}) and its sulphated polysaccharide (0.5 g kg^{-1}) can be used as natural marine ingredients to extend the shelf life of surimi based restructured products up to 6 months at −18 °C (Jannat-Alipour et al. 2019b). All these studies illustrates that the nutritional and technological benefits of edible seaweeds and its derivatives can be utilized for the development of surimi based restructured products.

Taking the advantage of antioxidant properties of bioactive compounds from seaweeds, it has long been used as a natural potential antioxidant in fish products. Phlorotannin, the active component of *Fucus vesiculosus* inhibited the haemoglobin mediated lipid oxidation in cod protein isolates and washed cod muscle system during iced storage (Wang et al. 2010). The power of phlorotannin subfraction LH-2 at 300 mg/kg level was analogous to that of 100 mg/kg PG, one of the most potent antioxidants in muscle food systems. Babakhani et al. (2015) demonstrated the antioxidant effect of 50% ethanolic extracts of *Polysiphonia fucoides* on minced Atlantic mackerel (*Scomber scombrus*) during chilled storage. Mackerel mince added with 0.5 g/kg of extracts protected against the loss of α-tocopherol and tryptophan residues, which was accomplished by retarding the lipid and protein oxidation during storage.

Besides the use of seaweeds as a natural hydrocolloid and antioxidant, they have also been incorporated in fish products to supplement the nutrients which are lacking in fish meat and to improve their functional properties. Edible seaweed (*Fucus spiralis* and *Chondrus crispus*) incorporated canned chub mackerel was found to be organoleptically acceptable and the seaweed addition had enhanced levels of trace elements, namely I, Cu, Fe, Mn, Mg, Cl, Se, Co, Sr, Mo and Na which are often lacking or below recommended levels in regular diets (Vieira et al. 2020). This effect was more distinct when both seaweeds were used as a salt replacer in the brining step. Candra et al. (2020) and Widiyanti et al. (2021) reported that *E. cottonii* (up to 20%) was a potential filler in snakehead fish nugget and kekian from *Arius thalassinus* (a Chinese food made from fish mince with seasoning) gives better sensory properties such as colour, aroma, and texture. On the other hand, a higher concentration of *E. cottonii* in fish products increase the dietary fiber, moisture content and WHC but reduced the sensory acceptance, protein content and gel strength. Fish sausage fortified with dietary fibre extracted from the *G. edulis*, *U. reticulata* and *S. wightii* had better quality and textural properties (Gopalakrishnan et al. 2020).

5.5 Seaweed Enriched Dairy Products

There is a paucity of information on seaweed supplemented dairy products (Lopez-Lopez et al. 2009; Cofrades et al. 2011). The development of functional dairy foods enriched with seaweeds and their derivatives is a promising approach that benefits both the consumers and the dairy industry. Yoghurt and fermented milk products have significant nutritional properties and offer a favourable environment for the survival and growth of health beneficial probiotic bacteria. Cottage cheese fortified with kombu and wakame and yogurt fortified with an omega-3-rich algal oil has been investigated (Lalic and Berkovic 2005; Lane et al. 2014). The effect of seaweed supplementation on the sensory characteristic of yoghurt and quark was studied (Nuñez and Picon 2016). It was reported that supplementation of seaweed such *as U. pinnatifida, Ulva lactuca, Himanthalia elongata, Saccharina latissimi* and *Porphyra umbilicalis* at 0.5% level greatly influenced the sensory and textural attributes of the yoghurt and quark. *U. lactuca*, *U. pinnatifida* and *P. umbilicalis* were responsible for imparting seaweed odour and flavour to the dairy products, while masking the characteristic yoghurt odour, associated with acetaldehyde production by lactic acid bacteria. However, dairy products supplemented with *S. latissima* showed the lowest seaweed odour and flavour and provided the highest flavour quality to dairy products. Hence, this seaweed species can be the most suitable one for dairy product supplementation based on the sensory score. Use of ulvan polysaccharide from *Ulva lactuca*at 1% and 2% level along with 3% of probiotic starter culture containing of *Streptococcus thermophilus* TH-4, *Bifidobacterium sp.* Bb-12 and *Lactobacillus acidophilus* LA-5 and in synbiotic yoghurt resulted in the products having good chemical and physical properties, as well as stimulated the growth and activity of probiotic bacteria (Samah Shalaby and Amin 2019). While, yoghurt supplemented with higher concentration of ulvan polysaccharide (4%) delayed fermentation, altered yoghurt characteristic flavour and weakened the texture and structure. Yoghurt enriched with Fucoidan, fucose rich sulphated polysaccharides extracted from brown algae *Sargassum wightii,* exhibited good antioxidant potential and phenolic content (Gopalakrishnan et al. 2020). Moreover, fucoidan enrichment had only a minimal effect on the sensory attributes of yoghurt. Therefore, seaweeds and their bioactive compounds can be supplemented to enrich yoghurt and other dairy products to enhance its health benefit and also to improve the textural characteristic as well as fermentation. Yoghurt acts as an efficient delivery system of physiological benefits seaweeds through diet in the human body.

Del Olmo et al. (2018) reported that supplementation with edible seaweeds namely *H. elongata, L. ochroleuca, P. umbilicalis, U. lactuca*and *U. pinnatifida* (10 g of each dehydrated seaweed per kg of curd) to manufacture the cheese. While, the addition of seaweed did not significantly affect cheese textural attributes, cheese microbiota and enzymatic activities of microbial origin but influenced the cheese colour parameters. *H. elongata* was found to be better among all the five seaweeds in terms of imparting enhanced antioxidant activity and sensory attributes to the cheese. Furthermore, the total free amino acids in cheese were increased with the addition of 1% dehydrated seaweed to curd and also cheeses supplemented with *U. lactuca, U. pinnatifida* and *H. elongate* exhibited a higher free ω6/ω3 ratio (Del Olmo et al. 2019a).

Fermented milks, a popular product, are produced by inoculating a mixed culture containing *Bifidobacteria* and *L. acidophilus* and selected diary culture under well controlled hygienic process (Holzapfel 2002; Walstra et al. 2006). Recently, few studies were conducted to improve the textural and nutritional quality of fermented milks by supplementing with seaweed and its extract. The aqueous extract of *G. domingensis* had modified the texture of the fermented milk with comparable texture attributes to the products containing only gelatin (Uchida et al. 2017). This result might be due to the presence of protein and soluble polysaccharides in seaweed aqueous solution. Fermented milk supplemented with aqueous seaweed extract displayed at least 10% higher firmness, consistency, cohesiveness and viscosity index than control. Therefore, the aqueous extract of *G. domingensis* appears to be a promising alternative to gelatin as a texture modifier in fermented milks and related dairy products. In another study, it was reported that probiotic strains were stimulated in most cases when grown in UHT milk supplemented with seaweed extracts from *Chondrus crispus, H. elongata, L. ochroleuca, P. palmata, P. umbilicalis, U. lactuca and U. pinnatifida*, reaching counts above 10^8 cfu/ml at the end of fermentation (Del Olmo et al. 2019b). It can be suggested that seaweed extracts can be as used as a promising functional ingredient for the development of novel fermented milks enriched with seaweed nutrients

Fortification of milk chocolate bar with 5% seaweed flour (*Eucheuma cottonii*) not only improved the dietary fibre (8.65%) of the chocolate bar but also increased the water content and improved the appearance, aroma, texture as well as taste (Herdiani 2003; Stefani et al. 2019).

5.6 Fermented Products

Food fermentation is a process involving enzymatic conversion of major and minor food components at a controlled microbial growth. (Marco et al. 2017). The first report of Lactic acid fermentation of seaweeds was in 2004 on a cellulase-treated culture of *Ulva* spp. (Uchida and Murata 2004). Lactic acid fermentation of *U. pinnatifida* and laver has been demonstrated by Uchida and Murata 2004; Uchida et al. 2004, 2007; Tsuchiya et al. 2007). Further, conversion of carbohydrate extracts from seaweeds *U. pinnatifida* into ethanol production using *Escherichia coli* strains has also been reported (Uchida and Murata 2004; Lee et al. 2011; Jones and Mayfield 2012). Increased health consciousness of consumers has led to the development of novel functional healthy food based on plant-based diets and fermented food. Fermented seaweed sauces using nori (*Pyropia yezoensis*) was first reported by Uchida et al. (2017). The fermented nori sauces were rich in total nitrogen compounds (1.5 g N/100 mL) and potassium (880 mg/100 g). The unique taste of fermented nori sauce is attributed to its free amino acid composition that is rich in taurine (617 mg/100 g). Nori sauce was safe for human consumption without inorganic arsenic and allergy-causing substances including wheat, soy beans and crustaceans. On the other hand, fermented sauce from fresh nori and low quality dried nori from *P. yezoensis* was low in nitrogen compounds (0.20 g/100 ml) and free amino acids leading to the sourness in taste (Uchida et al. 2018). These findings recommend nori sauce as a novel nutritional source for humans. Novel functional seaweed-based products such as sauerkraut-style products were prepared from kelp by fermentation with *Lactobacillus plantarum* (approximately 10^6 CFU/g) and *Leuconostoc mesenteroides* (approximately 10^1 CFU/g) (Skonberg et al. 2021). Though higher percentage of kelp increased the total phenolic content and antioxidant activities, the fermentation process of sauerkraut was greatly affected with increasing concentrations.

Another seaweed fermented product was a type of fermented plant beverage. The fermented plant beverages are non-alcoholic beverages manufactured from cereals, fruits and vegetables and such products claimed to have health benefits such as anticancer and anti-inflammatory properties (McClatchey 2002; McKoy et al. 2002; Kantachote et al. 2005). The fermented beverage (FSB) produced from red seaweed, *Gracilaria fisheri*, under partial preparation inoculated with the probiotic LAB (P-S) have the strongest antibacterial activity against foodborne pathogenic bacteria (*Vibrio parahaemolyticus* > *Bacillus cereus* > *Salmonella typhi* > *Staphylococcus aureus* > *E. coli*) (Prachyakij et al. 2008). Also, this beverage provided the highest level of probiotic LAB and a level of yeast below the maximum recommended allowance for fermented beverages. Hayisama-ae et al. (2014) also demonstrated beverages from fermented *G. fisheri* using *Lactobacillus plantarum* DW12 having strong antibacterial activity against food borne pathogen.

5.7 Seaweed Nutrient Enriched Beverages

Functional food beverages are the fastest growing segment driven by cosmeceutical claims such as anti-ageing, relaxing, beauty-enhancing, or energy-supplying effects (Gruenwald 2009). Enrichment of beverages with natural bioactive ingredients could be achieved either by direct addition of the ingredients or/and by addition of the ingredients in encapsulated form into the products. Ready-to-consume beverages are the fastest moving products in the functional food market owing to their attributes of convenience, freshly prepared and customization (Hardy 2009). Phytochemicals, probiotics and prebiotics, dietary fibres, PUFA, proteins and peptides, plant sterols, as well as minerals and vitamins are the most used bioactive natural food ingredients for the enrichment of beverages. Fortification of beverages is a challenging process that needs extensive standardization in terms of bioavailability, solubility, pH, temperature, light, stability, colour and flavor so that fortification does not adversely affect the consumer acceptability of the finished product (Kasapoğlu et al. 2019). Being a rich source of dietary fibre, vitamins, minerals and bioactive compounds with significant nutraceutical properties (Shama et al. 2019), seaweeds are used to fortify the beverages in recent years. Instant powdered *E. cottoni* as beverages are natural drinks rich in dietary fibre and can be used as a healthy drink with beneficial health effects (Mailoa et al. 2015). Most recently, Gopalakrishnan et al. (2020) developed a reconstitutable formulation, CIFTEQ® Seaweed Nutridrink from a blend of micro encapsulated *S. wightii* extract and grape juice. The drink is a rich source of bioactive compounds such as fucoidan, fucoxanthin and seaweed polyphenols, vitamins like riboflavin and pantothenic acid, taurine, free amino acid and nutritionally essential elements such as calcium and iron.

5.8 Seaweeds as Dietary Supplements

Fucoidan, a bioactive polysaccharide extracted from *S. wightii* has recently been recognized as a novel food ingredient in EU and USA (Gopalakrishnan et al. 2020). A freeze-dried dietary supplement, CIFTEQ® FucoidanEx, is a branded product developed from bioactive polysaccharide extracted from *S. wightii* which is rich in a potential nutraceutical Fucoidan, taurine and essential micronutrients. It can be used as an ingredient in various nutraceuticals and cosmetic products formulations CIFTEQ® FucoTeaEx is another natural dietary supplement possessing the benefits of fucoidan from *S. wightii* and green tea phenolics (Gopalakrishnan et al. 2020). To enhance the bioavailability of this rich source of antioxidants, vitamins and minerals, a novel microencapsulation technology was adopted during the process. The product showed promising results on reducing the effect of drug-induced myocardial infarctions in rats as revealed from the e preclinical study conducted in ICAR-Central Institute of Fisheries Technology (ICAR-CIFT), Kochi

6 Effect of Seaweeds as Functional Ingredients in Food

6.1 Changes in Physio-Chemical and Textural Properties

Texture is an important sensory characteristic of food, which may impact liking and preferences. It can also directly influence the mouth feel of food. Several studies have been conducted to enhance the textural properties of food through the application of seaweed in various form like dried, puree, extracts etc. (Table 1.8). For example, the addition of *Himanthalia elongata* at 10–40% w/w in the formulation of beef patties leads to the improvement of the textural properties such as tenderness by 50% more and increased dietary fibre (1.64 g per 100g fr. wt. in 40% seaweed patties) and enhance antioxidant properties and cooking quality (Cox and Abu-Ghannam 2013a). The incorporation of seaweeds reduced the cooking losses of patties, due to their hydrocolloid content. Patties with 40% seaweed had the most acceptable sensory quality. Similar results were observed by López-López et al. (2009), where, the effect of wakame (3%) addition on the textural properties of raw beef patties was explained due to the role played by seaweed principal components, mostly dietary fibre. Dietary fibres are used as bulking and texturing agents, predominantly in the production of low-calorie foods. However, these properties are solely dependent on the water holding capacity of fibre. Dietary fibres from diverse sources have been studied alone or in combination with other functional ingredients to formulate the new products with improve textural properties, water holding and fat binding capacities, cooking qualities and also as a fat replacement. Meat is known to be low in dietary fibre, hence, the dietary fibre content in meat products can be augmented by the incorporation of food ingredients containing a high amount of dietary fibre, for example, seaweeds. Intake of dietary fibre delivers many health benefits such as reducing the risk of coronary heart disease, stroke, high blood pressure, diabetes by improving the blood glucose level, improve serum lipid concentrations, prevent obesity by weight loss and certain gastrointestinal disorders (Anderson et al. 2009).

The edible seaweeds *U. pinnatifida, P. umbilicalis* and *H. elongata* and added at 2.5% and 5% dw on low-salt gel/emulsion meat systems have higher hardness and chewiness and lower springiness and cohesiveness (Cofrades et al. 2008). Seaweed addition also improved the water and fat binding properties of the final products. Hanjabam et al. (2016) studied that the effect of *S. wightii* incorporation at 3% and 5% level on the quality of tuna jerky. The addition of seaweed produced samples with less tensile strength and increased dietary fibre and antioxidant properties.

The changes in the texture of seaweed supplemented cake and muffins may be related to their density and the volume changes (Lu et al. 2010). The volume of these products decreased sharply with the addition of seaweed due to its high water holding capacity. Mamat et al. (2018) reported that *Kappaphycus alvarezii* (2–10%) influences the textural properties of the muffins by increasing the hardness and decreasing the springiness in seaweed enriched muffin. Chewiness increased when *Eucheuma* powder (containing dietary fibre of 69.33%), was applied at 10% and

Table 1.8 Physicochemical, textural and sensory changes in seaweed enriched products

Products	Seaweeds	Form	Concentration	Physicochemical, textural and sensory changes	Reference
Beef patties	Himanthalia elongata	Rehydrated dried seaweeds	10–40% w/w	Positive improvement on texture and mouth feel Texture become tender Increased the dietary fibre content (1.64% in 40% seaweed enriched patties) Patties enriched with 40% seaweed shown the highest overall acceptability	Cox and Abu-Ghannam (2013a, b)
Sponge cake	*Euchema*	Dried powder	5–20% w/w	Increased chewiness and dietary fibre (8.07%) Up to 10% Eucheuma powder substitute to flour was the most acceptable sensorily	Huang and Yang (2019)
Tuna jerky	*Sargassum wightii*	Dried powder	0, 3, 5% w/w	Total fiber content increased from 0.91 to 2.49% in seaweed incorporated sample Lower tensile strength and L* value Addition of seaweed up to 3% did not influence the organoleptic properties	Hanjabam et al. (2016)
Muffins	*Kappaphycus alvarezii*	Dried powder	2–10% w/w	Improved the hardness and fibre content Decreased the springiness in muffins Lightness L* value of the muffins increased with the raising concentration of seaweed powder in the formulations	Mamat et al. (2018)
Low-salt gel/ emulsion meat systems	*U. pinnatifida, H. elongata* and *P. umbilicalis*	Dried powder	2.5%, 5% w/w	Improved water absorption and fat binding properties Higher hardness and chewiness and lower cohesiveness and springiness Type of seaweed markedly influence the colour changes in meat systems	Cofrades et al. (2008)
Wheat bread	*K. alvarezii*	Dried powder	2–8% w/w	Addition of seaweed powder up to 8% significantly impact the stickiness, firmness, volume and crumb color on dough and final product	Mamat et al. (2014)
Surimi gel	*Sargassum tenerrimum*	Aqueous extract	0.5–2.5% w/w	2% seaweed aqueous extract increased gel strength of surimi by 76.27% and lowered expressible moisture Slight decrease in whiteness value	Shitole et al. (2014)

(continued)

Table 1.8 (continued)

Products	Seaweeds	Form	Concentration	Physicochemical, textural and sensory changes	Reference
Surimi gel	*Ulva intestinalis*	Dried powder	1.4, 2.8 and 4.2 g $(100\ g)^{-1}$	Increased gel strength and water-holding capacity at > 2.8 g $(100\ g)^{-1}$ seaweed concentration Significant reduction in whiteness and pH values of surimi gels Acceptable sensorially with up to 2.8 g $(100\ g)^{-1}$ seaweed to surimi	Jannat-Alipour et al. (2019a)
Pasta	*Laminaria ochroleuca*	Puree of dehydrated seaweed powder	20% w/w	Comparable textural and mechanical characteristics to the control Increased fibre and mineral content	Fradinho et al. (2019)
Fish Cutlet	Euchema	dried powder	5, 7.5, 10, 12.5 and 15% w/w	Increased hardness Increased the force required for puncturing, increased dietary fibre	Senthil et al. (2005)
Pork patties	*Laminaria japonica*	Dried powder	1, 3, 5% w/w	Higher hardness, gumminess, chewiness and springiness	Choi et al. (2012)
Frankfurters	*P.umbilicalis, P. palmata, H.elongata* and *U. pinnatifida*	Dried powder	1% w/w	Darker in colour (lower L*, a* and b* value) Altered flavour and textural properties, mainly less hard and less chewy Frankfurters fortified with *H. elongata* was accepted organoleptically	Vilar et al. (2019)
Noodles	*Caulerpa* sp.	Aqueous filtrate	0, 5, 10, 15, and 20% w/w	*Caulerpa* sp. filtrate addition changed the tensile strength, elasticity, adhesiveness and appearance of the noodles without affecting the consumer acceptance Increased the phenol and β-carotene content Lightness L* decreased, while, greenness of the noodles increased	Agusman and Wahyuni (2020)
Cookies	*Bifurcaria bifurcata*	Dried seaweed	3, 6 and 9% w/w	Hardness of cookies increased by ≥3% seaweed addition Improved water retention capacity and gelatinization of starch	Arufe et al. (2019)
Yoghurt	*Ulva lactuca*	Ulvan polysaccharide (water and ethanol extract)	1%, 2% and 4% w/v	Ulvan polysaccharide at 1–2% provided synbiotic yoghurt with good physical and chemical properties Enhanced the flavour to yogurt At high concentration of 4%, there was delayed fermentation and yielded weaker texture	Samah Shalaby and Amin (2019)

20% in sponge cake formulation (Huang and Yang 2019). *Eucheuma* powder addition also increased the viscosity and viscoelasticity of the batters. *K. alvarezii* powder when added at 2–8% to bread formulation resulted in deceased stickiness properties and increased firmness of the final product and increased water absorption of the dough (Mamat et al. 2014).

Gel strength is a very important parameter to determine the quality and price of surimi. Gel strength is the most commonly used index to express the textural properties developed during the gelation of surimi (Benjakul et al. 2004). To improve the gel strength of surimi, a different range of food-grade ingredients including seaweed hydrocolloids and extracts have been studied. The addition of 2% water extract of *Sargassum tenerrimum* showed a significant increase in gel strength (by 76.27%) in lesser sardine surimi gel a (Shitole et al. 2014). One of the possible causes of this effect could be the cross-linking activity of phenolic compounds of seaweed extract that could induce the formation of both covalent and non-covalent bonds in a gel matrix (Prigent et al. 2003). On the other hand, the reduction in gel strength of surimi with a higher percentage of seaweed water extract (>2%) was associated with loss in the capability of protein cross-linking during gelation due to the self-aggregation of phenolic compounds. De Freitas and Mateus (2001) suggested that the existence of a higher number of phenolic compounds lowers its efficiency in interacting with protein molecules.

6.2 Changes in Organoleptic Characteristic

Though, functional properties such as nutritional, physicochemical and textural properties of restructured and/or processed food can be improved by reducing salt and fat, the addition of seaweed can hinder the organoleptic properties due to its unique flavour leading to less consumer acceptability of the final product. Several studies have been carried out on evaluating the effect of seaweed as a functional ingredient on textural and physicochemical properties, but very little attention has been paid concerning sensory implications, in particular to colour and flavour. Seaweed can change the appearance of the final products by integrating its natural pigments. However, the effect on colour changes in seaweed enriched products varied depending on the type, species and proportion of seaweed added. The pigment in the seaweed viz., chlorophyll, xanthophyll, phycophine, phycoerythrin and phycocyanin determines the colour of the seaweed and greatly influences the colour of the finished products. The presence of seaweeds caused a decrease of lightness (L*) and redness (a*) values, which was all the more pronounced the larger the concentration of seaweeds added in gel/emulsion meat system (Cofrades et al. 2008). The highest decrease in L* was observed by the addition of nori. Redness (a*) was increased with nori concentration, whereas, yellowness (b*) was more in the meat system containing brown algae such as wakame and sea spaghetti. Similarly, *L. japonica* powder addition lowered the lightness L*, yellowness and redness a* values of reduced-fat pork patties (Choi et al. 2012). This colour variation in the products is the result of the presence of pigments chlorophyll and phycoerythrin in red seaweeds and carotenoid, chlorophyll, phycophine, and xanthophyll in brown seaweeds (Kim et al. 2010). The crumb lightness L* decreased and yellowness b* values of bread increased significantly by the inclusion of seaweed *K. alvarazii* flour (Mamat et al. 2014). The colour in surimi gel is a very important parameter for the quality determination of the gel. Insignificant decreases in whiteness were observed in lesser sardine surimi gel with 2% seaweed (*S. tenerrimum*) water extract (Shitole et al. 2014). However, the incorporation of seaweed did not negatively affect the colour and odour likeness of surimi gel.

The overall acceptability, in particular, appearance and aroma of frankfurters containing *seaweeds* were primarily influenced by the seaweed flavour rather than by a reduction of salt and fat content (Vilar et al. 2019). The addition of seaweed in frankfurter had a negative impact on the appearance and aroma perceptions. Consumers perceived that frankfurters with seaweed were darker in colour in the order of *P. umbilicalis* > *U. pinnatifida* > *P. palmata* > *H. elongate*, which was confirmed by the ΔE* values. Colour intensity of the products strongly correlated to the liking of appearance, which can greatly influence the overall acceptability of the products. Similar findings were described by Fellendorf et al. (2016), where the impact of wakame as a functional ingredient was perceived negatively on white pudding flavour. On contrary, several studies are showing a positive influence on flavour when added at the right concentration. The overall acceptability, tenderness and juiciness of reduced-fat pork patties with *L. japonica* (1% and 3%, w/w) were significantly higher than the control (Choi et al. 2012). This was supported by other researchers Cho and Hong (2006) and Kim et al. (2008), where muffin and Korean type bread containing seaweed had better sensory characteristic for overall acceptability than control.

6.3 *Effect on the Shelf-Life Extension of Food During Storage*

Seaweeds are acknowledged for being a rich source of bioactive compounds such as carotenoids, sulfated polysaccharides and polyphenols which have remarkable antioxidant and antimicrobial activities. Hence, few studies have focused on the application of seaweeds as natural food preservatives to improve the quality and extend shelf life. Lipid oxidation and microbial growth in trout fillet during cold storage were delayed using ethanol extract of *Crassiphycus corneus* and *U. ohnoi* (Sáez et al. 2021). In addition to favorable antioxidant and antimicrobial properties, the seaweed extracts did not cause a significant impact on trout flesh colour, which is considered a crucial quality attribute for this species. This result was supported by the findings of shelf life extension of pangasius fillet, tiger shrimp and Indian mackerel by using aqueous extract of *Padina tetrastromatica*, ethanolic extract from *Hypneamusciformis* and *Acanthophoramuscoides* and methanolic extract of *Gracilaria verrucose*, respectively (Arulkumar et al. 2018, 2020; Deepitha et al. 2021). Research findings suggest that 2% aqueous extract of *P. tetrastromatica* showed a reduction in total volatile basic nitrogen, lipid oxidation and microbial count resulting in delaying the spoilage and meat discoloration of pangasius fillet stored at chilled condition (2°C) for 20 days (Deepitha et al. 2021). The shelf life of tiger shrimp during ice storage was increased by the ethanolic extracts from *H. musciformis* and *A. muscoides* which had acted as natural preservatives to reduce spoilage and maintain the quality (Arulkumar et al. 2020). Flake ice with methanolic extract of *G. verrucosa* was used a medium for preservation to increase the shelf life of Indian mackerel to 15 days (Arulkumar et al. 2018). This preservation had effectively inhibited microbial proliferation (both mesophilic and psychrophilic bacteria) and chemical spoilages of fish deterioration (TVB-N, TMA-N and biogenic amines). Seaweed extracts contain polyphenols and flavonoids which exhibits strong radical scavenging activity and H_2O_2 reducing power. These phytochemicals are responsible for their bioactivity and were efficient in improving the shelf life during chilled storage. The addition of *Himanthalia elongata* seaweed (10–40% w/w) as a source of antioxidants in beef patties has significantly lowered the microbiological counts and lipid oxidation. There was no microbial growth observed in the vacuum-packed patties with ≥20% seaweed over the 30th-day refrigerated storage at 4 °C (Cox and Abu-Ghannam 2013a). The microbial count in the control patties reached 3.05 log CFU g^{-1} and 5.41 log CFU g^{-1} at 21st and 30th day of storage, respectively and the microbial count was 1.09 log CFU g^{-1} in patties containing 10% seaweed. The lipid oxidation levels were also low (0.61 mg malondialdehyde kg^{-1} of the sample) as compared to that of the control sample. Hence, it can be concluded that seaweed addition in patties helps to extend the shelf life of the products at refrigerated storage.

Seaweeds and their polysaccharides based bioactive edible films and coating are considered as one of the most effective natural food preservatives to enhance the quality of freshness in food products such as fish, mushrooms etc. (Alotaibi and Tahergorabi 2018). Among the seaweed-based polysaccharides, sodium alginate and carrageenan were most studied as natural preservatives used in edible films and coating. Plenty of research has shown the effectiveness of these edible coating to improve shrimp and fish fillets preservation during chilled and refrigerated storage (Volpe et al. 2015; Bazargani-Gilani 2017; Rao et al. 2017; Balti et al. 2020). Red seaweed (*Gracilaria gracilis*) extract at varying concentrations (0.5, 1 and 1.5% (w/v)) were used to enrich active edible coatings from microalgal exopolysaccharides which could effectively improve the quality, retained all sensory attributes, inhibited bacterial growth and lipid oxidation and extended the shelf life of shrimp during refrigerated storage (4 ± 1°C) (Balti et al. 2020). Similarly, the effectiveness of edible active sodium alginate and carrageenan based coating in the preservation of fresh trout fillets from lipid oxidation and microbial growth during refrigerated storage were studied by Volpe et al. (2015) and Bazargani-Gilani (2017). Incorporation of natural preservatives in edible coating make the coatings antioxidant and antimicrobial leads to inhibition of bacterial proliferation and delayed lipid oxidation of the food on to which the coating is applied (Ojagh et al. 2010; Giatrakou and Savvaidis 2012).

7 Bioactive Compounds from Seaweeds and Their Health Benefits

Besides being rich in nutrients, seaweeds contain a wide range of novel bioactive compounds such as polysaccharides (agar, carrageenan, alginate, laminarin, ulvan etc.), polyphenols, pigments as well as proteins (lectins, phycobiliproteins, peptides, and amino acids), which are not found in terrestrial plants, vegetables and fruits (Brown et al. 2014). Bioactive compounds from seaweeds have well-acknowledged health benefits including antioxidant, antimicrobial, anti-cancer, anti-diabetic antiviral, antitumor, anticoagulant, immunomodulatory activities, anti-metastatic, anti-inflammatory, antiproliferative, anticoagulant activities, lower body weight and can help to prevent the risk of cardiovascular-associated disorders, such as hypertension, diabetes mellitus type 2 and metabolic syndrome etc. (Table 1.9) (Seo et al. 2012; Zheng et al. 2012; Woo et al. 2013; Kellogg et al. 2014; Kim et al. 2015; Kang et al. 2016; Lee et al. 2016; Rahnasto-Rilla et al. 2017; Tanna and Mishra 2018; Sørensen et al. 2019; Kang et al. 2020). Bioactive from seaweeds have long been used in industries including nutraceuticals, pharmaceuticals, food industry, biomedical materials, cosmetics as well as in fertilizer. Seaweeds are a cheap source of bioactive substances and besides being nutrient supplements; seaweeds can be also be used as a potential functional ingredient in the development of functional food in the prevention of diseases.

8 Opportunities and Challenges in Seaweed-Based Food Product Innovations and Commercialization

- Seaweeds are a rich source of nutrients, bioactive compounds, aid in enhancing the nutritional value and functionality of food products. Though a wide range of seaweed-based food products is available in the Asian food market, they are not available in India. The demand for seaweeds as an ingredient to develop new functional foods in the food industry opens up new opportunities in India.
- The growing awareness among consumers regarding sustainability and animal welfare concerns have increased the demand for vegan and/or plant-based food products. Seaweeds being plants have the potential to meet consumer demands and thus create opportunities for innovative product development.
- There is a huge demand for healthy functional products among urban consumers due to their health consciousness. The health beneficial properties of seaweeds and their derivatives can be utilized for the production of seaweed enriched food products as healthy food to prevent diseases thereby creating avenues for novel product development.
- Seaweeds consumption can be promoted among the Indian populations as a cheap alternative source of protein, dietary fibre and essential micronutrients.

Table 1.9 Seaweed bioactive and their possible health benefits

Bioactive	Seaweeds	Reported health benefits/bioactivities	Reference
Agar	*Gracilariopsis chorda, Gelidium amansii, Gracilaria verrucose, Gracilariopsis chorda, Gloiopeltis tenax, Grateloupia filicina*	Antioxidant, anti-diabetic and antimicrobial	Seo et al. (2012), Zheng et al. (2012), Woo et al. (2013), Mohibbullah et al. (2015) and Kang et al. (2016)
Alginate	*Sargassum sp., Padina sp.*	Anti-tumor, anti-oxidant	Tanna and Mishra (2018)
Carrageenan	*Acanthophora specifira, Hydroclathrus clathratus*	Antiviral, antitumor, anticoagulant and immunomodulatory activities	Gomaa and Elshoubaky (2015) and Tanna and Mishra (2018)
Fucoidan	*F. vesiculosus, F. evanescens, F. distichus, P. gymnospora, S. japonica, S. latissimi, U. pinnatifida, A. utricularis, H. Fusiform*	Anti-tumor, Anti-metastatic, anti-inflammatory, antiproliferative, anticoagulant activities and anti-diabetic	Ponce et al. (2003), Alekseyenko et al. (2007) and Rahnasto-Rilla et al. (2017)
Laminarin	*Laminaria sp., H. fusiform, U. pinnatifida*	Antitumor	Tanna and Mishra (2018)
Ulvan	*Ulva pertusa* and others Ulva species	Antitumor, anti-viral and anti-cancer	Tanna and Mishra (2018)
Ethanol Extract	*Sargassum thunbergii*	Anti-obesity effects Decreased body weight and fat accumulation reducing insulin	Kang et al. (2020)
Dried seaweeds incorporated into diet	*Alaria esculenta, Saccharina latissimi, Palmaria palmata*	Anti-diabetic, lower bodyweight, insulin levels, Increase high density lipoprotein (HDL) levels, control glycemic and ipid levels	Sørensen et al. (2019)
Fucoxanthin	*Ascophyllum nodosum, F. serratus, F. vesiculosus, H. fusiformis, Himanthalia elongate, L. digitata, L. saccharina, U. pinnatifida, S. horneri, C. lentillifera*	Dietary antioxidant and anti-diabetic effect	Plaza et al. (2010), Ma et al. (2014) and Sharma and Rhyu (2014)
Polyphenols	*Fucus sp., Haematococcus pluvialis, Laminaria sp., Porphyra sp., Spongiochloris spongiosa, Undaria sp., Sargassum muticum, Polysiphonia fucoids, Gelidium amansii*	Antioxidant, anti-cancer, prevents cardiovascular diseases, hypertension, diabetes mellitus type 2	Bocanegra et al. (2009), Rodríguez-Meizoso et al. (2010), Klejdus et al. (2009), Namvar et al. (2013), Sabeena Farvin and Jacobsen (2013), Kim et al. (2015) and Lee et al. (2016)
Phlorotannin	*A. nodosum, F. distichus*	Anti-diabetic effect, α-amylase inhibition and α-glucosidase inhibition	Nwosu et al. (2011) and Kellogg et al. (2014)

This not only promotes seaweed consumption but creates livelihood opportunities through seaweed farming for the coastal people.
- The marketing of seaweed-based products in India is a major challenge as seaweeds have not been a part of the regular diet in India. This issue can be overcome by studying consumer behaviour and developing appropriate market-driven products.
- The quality of seaweeds harvested from coastal waters with inflows from polluted drains poses a serious challenge to food safety. Monitoring the quality of seaweeds in terms of chemical and biological hazards is necessary for utilizing seaweeds as food or food ingredients

9 Conclusion

Seaweeds as a novel functional food ingredient is growing rapidly across the world in recent years. Seaweeds have been recognized. Though seaweeds have been a traditional food for centuries in Asia, seaweeds are becoming more widely recognized and accepted as wholesome, tasty and healthy food in western countries. Therefore, diversification of utilization of seaweed could supplement our existing food supply and food security challenges and also the development of functional food products enriched with seaweeds can meet the existing demand for healthy food.

References

Admassu H, Abera T, Abraha B, Yang R, Zhao W (2018) Proximate, Mineral and Amino acid Composition of Dried Laver (*Porphyra* spp.) seaweed. J Acad Ind Res 6(9):149–154

Agusman M, Wahyuni T (2020) The nutritional quality and preference of wheat noodles incorporated with *Caulerpa* sp. seaweed. Int Food Res J 27(3):445–453

Alekseyenko TV, Zhanayeva SY, Venediktova AA, Zvyagintseva TN, Kuznetsova TA, Besednova NN, Korolenko TA (2007) Antitumor and antimetastatic activity of fucoidan, a sulfated polysaccharide isolated from the Okhotsk Sea *Fucus evanescens* brown alga. Bull Exp Biol Med 143:730–732

Alipour HJ, Rezaei M, Shabanpour B, Tabarsa M (2018) Effects of sulfated polysaccharides from green alga *Ulva intestinalis* on physicochemical properties and microstructure of silver carp surimi. Food Hydrocoll 74:87–96

Alotaibi S, Tahergorabi R (2018) Development of a sweet potato starch-based coating and its effect on quality attributes of shrimp during refrigerated storage. LWT- Food Sci Technol 88:203–209

Anderson JW, Baird P, Davis RH Jr, Ferreri S, Knudtson M, Koraym A, Waters V, Williams CL (2009) Health benefits of dietary fibre. Nutr Rev 67:188–205

Arufe S, Della Valle G, Chiron H, Chenlo F, Sineiro J, Moreira R (2017) Effect of brown seaweed powder on physical and textural properties of wheat bread. Eur Food Res Technol 244(1):1–10

Arufe S, Chenlo F, Sineiro J, Moreira R (2019) Effect of brown seaweed addition and starch gelatinization on gluten-free chestnut flour doughs and cookies. J Food Measure Charact 13:2571–2580

Arulkumar A, Paramasivam S, Miranda JM (2018) Combined effect of icing medium and red alga *Gracilaria verrucosa* on shelflife extension of Indian Mackerel (*Rastrelliger kanagurta*). Food Bioproc Tech 11(10):1911–1922

Arulkumar A, Satheeshkumar K, Paramasivam S, Rameshthangam P, Miranda JM (2020) Chemical biopreservative effects of Red Seaweed on the Shelf Life of Black Tiger Shrimp (*Penaeus monodon*). Foods 9(5):634

Babakhani A, Farvin KHS, Jacobsen C (2015) Antioxidative effect of seaweed extracts in chilled storage of minced Atlantic Mackerel (*Scomber scombrus*): effect on lipid and protein oxidation. Food Bioproc Tech 9(2):352–364

Balti R, Ben Mansour M, Zayoud N, Le Balch R, Brodu N, Arhaliass A, Massé A (2020) Active exopolysaccharides based edible coatings enriched with red seaweed (*Gracilaria gracilis*) extract to improve shrimp preservation during refrigerated storage. Food Biosci 34:100522

Bazargani-Gilani B (2017) Activating sodium alginate-based edible coating using a dietary supplement for increasing the shelf life of rainbow trout fillet during refrigerated storage (4 ± 1 °C). J Food Saf 38(1):e12395

Benjakul S, Visessanguan W, Chantarasuwan C (2004) Effect of high temperature setting on gelling characteristics of surimi from some tropical fish. Int J Food Sci Technol 39:671–680

Blank C (2018) The rise of seaweed. https://www.seafoodsource.com/features/the-rise-ofseaweed. Accessed 22 May 2021

Bocanegra A, Bastida S, Benedí J, Ródenas S, Sánchez-Muniz FJ (2009) Characteristics and nutritional and cardiovascular-health properties of seaweeds. J Med Food 12:236–258

Bouga M, Combet E (2015) Emergence of seaweed and seaweed-containing foods in the UK: focus on labeling, iodine content, toxicity and nutrition. Foods 4:240–253

Braithwaite MC, Tyagi C, Tomar LK, Kumar P, Choonara YE, Pillay V (2014) Nutraceutical based therapeutics and formulation strategies augmenting their efficiency to complement modern medicine: an overview. J Funct Foods 6:82–99

Britton D, Schmid M, Revill AT, Virtue P, Nichols PD, Hurd CL, Mundy CN (2020) Seasonal and site-specific variation in the nutritional quality of temperate seaweed assemblages: implications for grazing invertebrates and the commercial exploitation of seaweeds. J Appl Phycol 33:603–616

Brody T (1999) Nutritional biochemistry, 2nd edn. Academic Press, London

Brown EM, Allsopp PJ, Magee PJ, Gill CI, Nitecki S, Strain CR, McSorley EM (2014) Seaweed and human health. Nutr Rev 72(3):205–216

Candra KP, Saputra H, Gunawan A, Saragih B, Syahrumsyah H, Agrointek Y (2020) The limit of red seaweed (*Eucheuma cottonii*) substitution in snakehead fish (*Channa striata*) nuggets based on sensory evaluation. Agrointek 14(2):339–346

Cardoso SM, Pereira OR, Seca AML, Pinto DCGA, Silva AMS (2015) Seaweeds as preventive agents for cardiovascular diseases: from nutrients to functional foods. Mar Drugs 13:6838–6865

Catarino MD, Silva AMS, Cardoso SM (2017) Fucaceae: a source of bioactive phlorotannins. Int J Mol Sci 18:1327

Cencic A, Chingwaru W (2010) The role of functional foods, nutraceuticals, and food supplements in intestinal health. Nutrients 2:611–625

Černá M (2011) Seaweed proteins and amino acids as nutraceuticals. In: Kim S-K (ed) Advances in food and nutrition research marine medicinal foods, vol 64, pp 297–312

Chan PT, Matanjun P (2017) Chemical composition and physicochemical properties of tropical red seaweed, *Gracilaria changii*. Food Chem 221:302–310

Chauhan A, Saxena DC, Singh S (2015) Total dietary fibre and antioxidant activity of gluten free cookies made from raw and germinated amaranth (*Amaranthus* spp.) flour. LWT- Food Sci Technol 63:939–945

Chapman AS, Stévant P, Larssen WE (2015) Food or fad? Challenges and opportunities for including seaweeds in a Nordic diet. Bot Mar 58(6):423–433

Chen J, Deng T, Wang C, Mi H, Yi S, Li X, Li J (2020) The effect of hydrocolloids on gel properties and protein secondary structure of silver carp surimi. J Sci Food Agric 100:2252–2260

Cho MS, Hong JS (2006) Quality characteristics of sulgidduk by the addition of sea tangle. Kor J Food Cook Sci 22(1):37–44

Choi Y-S, Choi JH, Han DJ, Kim HY, Kim HW, Lee MA, Chung HJ, Kim CJ (2012) Effects of Laminaria japonica on the physico-chemical and sensory characteristics of reduced-fat pork patties. Meat Sci 91(1):1–7

Choi YS, Kum JS, Jeon KH, Jong-Dae Park JD, Hyun-Wook Choi HW, Hwang KE, Jeong TJ, Kim YB, Kim CJ (2015) Effects of edible seaweed on physicochemical and sensory characteristics of reduced-salt Frankfurters. Korean J Food Sci An 35(6):748–756

Cian RE, Fajardo MA, Alaiz M, Vioque J, González RJ, Drago SR (2013) Chemical composition, nutritional and antioxidant properties of the red edible seaweed *Porphyra columbina*. Int J Food Sci Nutr 65(3):299–305

Cofrades S, López-López I, Solas MT, Bravo L, Jiménez-Colmenero F (2008) Influence of different types and proportions of added edible seaweeds on characteristics of low-salt gel/emulsion meat systems. Meat Sci 79:767–776

Cofrades S, López-Lopez I, Bravo L, Ruiz-Capillas C, Bastida S, Larrea MT, Jiménez-Colmenero F (2010) Nutritional and antioxidant properties of different brown and red Spanish edible seaweeds. Food Sci Technol Int 16:361–370

Cofrades S, Lopez-Lopez I, Ruiz-Capillas C, Triki M, Jimenez-Colmenero F (2011) Quality characteristics of low-salt restructured poultry with microbial transglutaminase and seaweed. Meat Sci 87:373–380

Cox S, Abu-Ghannam N (2013a) Enhancement of the phytochemical and fibre content of beef patties with *Himanthalia elongate* seaweed. Int J Food Sci Technol 48(11):2239–2249

Cox S, Abu-Ghannam N (2013b) Incorporation of *Himanthalia elongata* seaweed to enhance the phytochemical content of breadsticks using Response Surface Methodology (RSM). Int Food Res J 20(4):1537–1545

Dawczynski C, Schubert R, Jahreis G (2007) Amino acids, fatty acids, and dietary fibre in edible seaweed products. Food Chem 103(3):891–899

De Freitas V, Mateus N (2001) Structural features of procyanidin interactions with salivary proteins. J Agric Food Chem 49:940–945

Deepitha RP, Xavier KAM, Layana P, Nayak BB, Balange AK (2021) Quality improvement of pangasius fillets using aqueous seaweed (*Padina tetrastromatica*) extract. LWT- Food Sci Technol 137:110418

Del Olmo A, Picon A, Nuñez M (2018) Cheese supplementation with five species of edible seaweeds: effect on microbiota, antioxidant activity, colour, texture and sensory characteristics. Int Dairy J 84:36–45

Del Olmo A, López-Pérez O, Picon A, Gaya P, Nuñez M (2019a) Cheese supplementation with five species of edible seaweeds: effect on proteolysis, lipolysis and volatile compounds. Int Dairy J 90:104–113

Del Olmo A, Picon A, Nuñez M (2019b) Probiotic dynamics during the fermentation of milk supplemented with seaweed extracts: the effect of milk constituents. LWT- Food Sci Technol 107:249–255

Dewi EN (2011) Quality evaluation of dried noodle with seaweeds puree substitution. J Coast Dev 14(2):151–158

Dhingra D, Michael M, Rajput H, Patil RT (2012) Dietary fibre in foods: a review. J Food Sci Technol 49(3):255–266

Dziezak JD (1991) A focus on gums. Food Technol 45:115–132

Enny Sholichah E, Kumalasari R, Indrianti N, Ratnawati L, Restuti A, Munandar A (2021) Physicochemical, sensory, and cooking qualities of gluten-free pasta enriched with Indonesian edible Red Seaweed (*Kappaphycus alvarezii*). J Food Nutr Res 9(4):187–192

Etemadian Y, Shabanpour B, Ramzanpour Z, Shaviklo AR, Kordjazi M (2018) Production of the corn snack seasoned with brown seaweeds and their characteristics. J Food Measure Charact 12(3):2068–2079

FAO (1991) Protein quality evaluation. In: Report of the Joint FAO/WHO Expert Consultation. FAO Food and Nutrition Paper No. 51, Rome

FAO (2016) The State of World Fisheries and Aquaculture 2016. Contributing to food security and nutrition for all. FAO, Rome, p 200

FAO/WHO/UNU (1985) Energy and protein requirements. Report of a joint FAO/WHO/UNU Expert Consultation. WHO Technical Report Series No. 724. WHO, Geneva

Fellendorf S, O'Sullivan MG, Kerry JP (2016) Effect of using ingredient replacers on the physicochemical properties and sensory quality of low-salt and low-fat white puddings. Eur Food Res Technol 242(12):2105–2118

Firdaus M, Yahya RH, Nugraha G, Dwi Utari D (2017) Fortification of seaweed (*Eucheuma cottonii*) flour on nutrition, iodine, and glycemic index of pasta. IOP Conf Ser Earth Environ Sci 89:012011

Fradinho P, Raymundo A, Sousa I, Domínguez H, Torres MD (2019) Edible brown seaweed in gluten-free pasta: technological and nutritional evaluation. Foods 8(12):622

Ganesan AR, Subramani K, Shanmugam M, Seedevi P, Park S, Alfarhan AH, Rajagopal R, Balasubramanian B (2020) A comparison of nutritional value of underexploited edible seaweeds with recommended dietary allowances. J King Saud Univ Sci 32(1):1206–1211

Giatrakou V, Savvaidis I (2012) Bioactive packaging technologies with chitosan as a natural preservative agent for extended shelf-life food products natural preservative agent for extended shelf-life food products. In: Arvanitoyannis I (ed) Modified atmosphere and active packaging technologies. Taylor & Francis, Boca Raton, FL, pp 689–734

Gomaa HHA, Elshoubaky GA (2015) Antiviral activity of sulfated polysaccharides Carrageenan from some marine seaweeds. Int J Curr Pharmaceut Rev Res 7(1):34–42

Gómez-Ordóñez E, Jiménez-Escrig A, Rupérez P (2010) Dietary fibre and physicochemical properties of several edible seaweeds from the northwestern Spanish coast. Food Res Int 43(9):2289–2294

Gopalakrishnan A, Ravishankar CN, Pravin P, Jena JK (2020) ICAR technologies: high-value nutraceutical and nutritional products from seaweeds. Indian Council of Agricultural Research, New Delhi, p 22

Granato D, Barba FJ, Kovačević DB, Lorenzo JM, Cruz AG, Putnik P (2020) Functional foods: product development, technological trends, efficacy testing, and safety. Annu Rev Food Sci Technol 11:93–118

Gruenwald J (2009) Novel botanical ingredients for beverages. Clin Dermatol 27(2):210–216

Gullón P, Astray G, Gullón B, Franco D, Campagnol PCB, Lorenzo JM (2020) Inclusion of seaweeds as healthy approach to formulate new low-salt meat products. Curr Opin Food Sci 40:20–25

Halimah SN, Suryani RA, Wijayanti SW, Pangestu RA, Deni GD, Romadhon. (2016) Fortification seaweed noodles [*Euchema cottonii* (Weber-van Bosse, 1913)] with nano-calcium from bone catfish [*Clarias batrachus* (Linnaeus, 1758)]. Aquat Proc 7:221–225

Hanjabam MD, Zynudheen AA, Ninan G, Panda S (2016) Seaweed as an ingredient for nutritional improvement of Fish Jerky. J Food Process Preserv 41(2):e12845

Hardy N (2009) Global NPD launch patterns. In: Business insight: future innovations in food and drinks to 2015 NPD. Trend convergence and emerging growth opportunities. Business Insights, Woodbrook, Trinidad and Tobago, pp 113–114

Hayisama-ae W, Kantachote D, Bhongsuwan D, Nokkaew U, Chaiyasut C (2014) A potential synbiotic beverage from fermented red seaweed (*Gracilaria fisheri*) using Lactobacillus plantarum DW12. Int Food Res J 21(5):1789–1796

Herdiani F (2003) Utilization of seaweed (*Eucheuma cottoni*) to increase iodine content and food fiber in jam and dodol. Thesis. Institutu Agriculture, Bogor

Holzapfel WH (2002) Appropriate starter culture technologies for small-scale fermentation in developing countries. Int J Food Microbiol 75:197–212

Huang M, Yang H (2019) Eucheuma powder as a partial flour replacement and its effect on the properties of sponge cake. LWT- Food Sci Technol 110:262–268

Ibanez E, Cifuentes A (2013) Benefits of using algae as natural sources of functional ingredients. J Sci Food Agric 93:703–709

Ishakani AH, Vadher KH, Kadri RM, Patel MR (2017) Amino acid and fatty acid composition of seaweeds (*Ulva reticulata* and *Sargassum cinctum*): a novel natural source of nutrition. Int J Pure App Biosci 5(5):1210–1216

Jannat-Alipour H, Rezaei M, Shabanpour B, Tabarsa M (2019a) Edible green seaweed, *Ulva intestinalis* as an ingredient in surimi-based product: chemical composition and physicochemical properties. J Appl Phycol 31:2529–2539

Jannat-Alipour H, Rezaei M, Shabanpour B, Tabarsa M, Rafipour F (2019b) Addition of seaweed powder and sulphated polysaccharide on shelf_life extension of functional fish surimi restructured product. J Food Sci Technol 56(8):3777–3789

Jard G, Marfaing H, Carrère H, Delgenes JP, Steyer JP, Dumas C (2013) French Brittany macroalgae screening: Composition and methane potential for potential alternative sources of energy and products. Bioresour Technol 144:492–498

Jayasinghe PS, Pahalawattaarachchi V, Ranaweera KKDS (2016) Formulation of nutritionally superior and low cost seaweed based soup mix powder. J Food Process Technol 7(4):571

Jesmi D, Madhusudana Rao B, Murthy LN, Mathew S, Venkateshwarlu G, Ravishankar CN (2016) Nutritional profiling of the edible seaweeds *Gracilaria edulis*, *Ulva lactuca* and *Sargassum* sp. Indian J Fish 63(3):81–87

Jesmi D, Viji P, Rao BM, Prasad MM (2017) Nutritional and physical characteristics of noodles incorporated with green seaweed (*Ulva reticulata*) and fish (*Pangasianodon hypophthalmus*) mince. Indian J Fish 64(2): 90–95

Jeyakumari A, Joseph C, Zynudheen AA, Anandan R (2016) Quality evaluation of fish soup powder supplemented with carrageenan. Int J Sci Environ Technol 5(6):4362–4369

Jiménez-Escrig A, Sánchez-Muniz FJ (2000) Dietary fibre from edible seaweeds: chemical structure, physicochemical properties and effects on cholesterol metabolism. Nutr Res 20:585–598

Jimenez-Escrig A, Gomez-Ordonez E, Ruperez P (2015) Infrared characterisation, monosaccharide profile and antioxidant activity of chemical fractionated polysaccharides from the edible seaweed sugar Kombu (*Saccharina latissima*). Int J Food Sci Technol 50:340–346

Jones CS, Mayfield SP (2012) Algae biofuels: versatility for the future of bioenergy. Curr Opin Biotechnol 23:346–351

Kang MC, Kang N, Kim SY, Lima IS, Ko SC, Kim YT, Kim YB, Jeung HD, Choi KS, Jeon YJ (2016) Popular edible seaweed, *Gelidium amansii* prevents against diet induced obesity. Food Chem Toxicol 90:181–187

Kang MC, Lee HG, Kim HS, Song KM, Chun YG, Lee MH, Kim BK, Jeon YJ (2020) Anti-obesity effects of *Sargassum thunbergii* via downregulation of adipogenesis gene and upregulation of thermogenic genes in high-fat diet-induced obese mice. Nutrients 12(11):3325

Kantachote D, Ongsakol M, Charernjiratrakul W, Chaiyasut C, Poosaran N (2005) Fermented plant beverages and their application. In: Abstracts of proceedings of the 1st international conference on natural products for health and beauty from local wisdom to global marketplace. Maha Sarakham University, Thailand, pp 17–21

Kasapoğlu KN, Daşkaya-Dikmen C, Yavuz-Düzgün M, Karaça AC, Özçelik B (2019) Enrichment of beverages with health beneficial ingredients. In: Value-added ingredients and enrichments of beverages, pp 63–99

Kellogg J, Grace MH, Lila MA (2014) Phlorotannins from Alaskan seaweed inhibit carbolytic enzyme activity. Mar Drugs 12:5277–5294

Kim JH, Kim JH, Yoo SS (2008) Impacts of the proportion of sea-tangle on quality characteristics of muffin. Korean J Food and Cook Sci 24(5):565–572

Kim HW, Choi JH, Choi YS, Han DJ, Kim HY, Lee MA, Kim SY, Kim CJ (2010) Effects of sea tangle (*Lamina japonica*) powder on quality characteristics of breakfast sausages. J Kor Soc Food Sci Anim Resour 30(1):55–61

Kim SC, Lee JR, Park SJ (2015) *Porphyra tenera* induces apoptosis of oral cancer cells. Korea J Herbol 30:25–30

Klejdus B, Kopecký J, Benesová L, Vacek J (2009) Solid-phase/supercritical-fluid extraction for liquid chromatography of phenolic compounds in freshwater microalgae and selected cyanobacterial species. J Chromatogr A 1216(5):763–771

Klnc B, Cirik S, Turan G, Tekogul H, Koru E (2013) Seaweeds for food and industrial applications. In: Muzzalupo I (ed) Food industry, pp 735–748

Kumari P, Bijo AJ, Mantri VA, Reddy CRK, Jha B (2013) Fatty acid profiling of tropical marine macroalgae: an analysis from chemotaxonomic and nutritional perspectives. Phytochemistry 86:44–56

Kumoro AC, Johnny D, Alfilovita D (2016) Incorporation of microalgae and seaweed in instant fried wheat noodles manufacturing: nutrition and culinary properties study. Int Food Res J 23(2):715–722

Kuroda M, Ohta M, Okufuji T, Takigami C, Eguchi M, Hayabuchi H, Ikeda M (2010) Frequency of soup intake and amount of dietary fiber intake are inversely associated with plasma leptin concentrations in Japanese adults. Appetite 54:538–543

Lage-Yusty MA, Alvarado G, Ferraces-Casais P, Lopez-Hernandez J (2014) Modification of bioactive compounds in dried edible seaweeds. Int J Food Sci Technol 49:298–304

Lahaye M, Kaeffer B (1997) Seaweed dietary fibres: structure, physico-chemical and biological properties relevant to intestinal physiology. Sci Aliments 17(6):563–584

Lalic LM, Berkovic K (2005) The influence of algae addition on physicochemical properties of cottage cheese. Milchwissenschaft 60:151–154

Lamont T, McSweeney M (2020) Consumer acceptability and chemical composition of wholewheat breads incorporated with brown seaweed (*Ascophyllum nodosum*) or red seaweed (*Chondrus crispus*). J Sci Food Agric 101(4):1507–1514

Lane KE, Weili L, Smith C, Derbyshire E (2014) The bioavailability of an omega-3-rich algal oil is improved by nanoemulsion technology using yogurt as a food vehicle. Int J Food Sci Technol 49:1264–1271

Lee S, Oh Y, Kim D, Kwon D, Lee C, Lee J (2011) Converting carbohydrates extracted from marine algae into ethanol using various ethanolic *Escherichia coli* strains. Appl Biochem Biotechnol 164:878–888

Lee JH, Kim HH, Ko JY, Jang JH, Kim GH, Lee JS, Nah JW, Jeon YJ (2016) Rapid preparation of functional polysaccharides from *Pyropia yezoensis* by microwave-assistant rapid enzyme digest system. Carbohydr Polym 153:512–517

Lekshmi RGK, Tejpal CS, Mathew S (2017) Seaweeds—an untapped repository of biomolecules. In: National seminar on seaweeds a source of nutraceutical health care products and new materials—future perspective. SOFT and CIFT, pp 59–73

Lopez-Lopez I, Cofrades S, Ruiz-Capillas C, Jimenez-Colmenero F (2009) Design and nutritional properties of potential functional frankfurters based on lipid formulation, added seaweed and low salt content. Meat Sci 83:255–262

Lorenzo JM, Agregán R, Munekata PES, Franco D, Carballo J, Sahin S, Lacomba R, Barba FJ (2017) Proximate composition and nutritional value of three macroalgae: Ascophyllum nodosum, Fucus vesiculosus and Bifurcaria bifurcata. Mar Drugs 15:360

Lu TM, Lee CC, Mau JL, Lin SD (2010) Quality and antioxidant property of green tea sponge cake. Food Chem 119:1090–1095

Ma AC, Chen Z, Wang T, Song N, Yan Q, Fang YC, Guan HS, Liu HB (2014) Isolation of the molecular species of monogalactosyldiacylglycerols from brown edible seaweed *Sargassum horneri* and their inhibitory effects ontriglyceride accumulation in 3T3-L1 adipocytes. J Agric Food Chem 62(46):11157–11162

MacArtain P, Gill C, Brooks M, Campbell R, Rowland I (2007) Nutritional value of edible seaweeds. Nutr Rev 65:535–543

Mahadevan K (2015) Seaweeds: a sustainable food source. In: Seaweed sustainability—food and non-food applications, pp 347–364

Mahan KL, Escott-Stump S (2000) Nutricion y dietoterapia, de Krause. McGraw-Hill Interamericana, Madrid

Mailoa MN, Setha B, Febe FG (2015) Instant powdered *Eucheuma cottoni* as beverages rich in dietary fiber. Indian J Sci Technol 8(S9):154–157

Mamat H, Matanjun P, Ibrahim S, Amin MSF, Abdul HM, Rameli AS (2013) The effect of seaweed composite flour on the textural properties of dough and bread. J Appl Phycol 26(2):1057–1062

Mamat H, Matanjun P, Ibrahim S, Amin SFM, Hamid MA, Rameli AS (2014) The effect of seaweed composite flour on the textural properties of dough and bread. J Appl Phycol 26:1057–1062

Mamat H, Akanda JMH, Zainol MK, Ling YA (2018) The influence of seaweed composite flour on the physicochemical properties of Muffin. J Aquat Food Prod Technol 27(5):635–642

Manish Kumar M, Sharma BD (2004) The storage stability and textural, physico-chemical and sensory quality of low-fat ground pork patties with Carrageenan as fat replacer. Int J Food Sci Technol 39:31–42

Marasabessy I, Sudirjo F (2017) Seaweed fortification on crispy enbal as local food of Kei Islands. IOP Conf Ser Earth Environ Sci 89:1–5

Marco ML, Heeney D, Binda S, Cifelli CJ, Cotter PD, Foligne B, Gänzle M, Kort R, Pasin G, Pihlanto A, Smid EJ, Hutkins R (2017) Health benefits of fermented foods: microbiota and beyond. Curr Opin Biotechnol 44:94–102

Matanjun P, Mohamed S, Mustapha NM, Muhammad K (2008) Nutrient content of tropical edible seaweeds, *Eucheuma cottonii, Caulerpa lentillifera* and *Sargassum polycystum*. J Appl Phycol 21(1):75–80

McClatchey W (2002) From polyesian healers to health food stores: changing ethnopharmacology of Morinda. J Integr Cancer Ther 1:110–120

McKoy M, Thomas E, Simon O (2002) Preliminary investigation of the anti-inflammatory properties of an aqueous extract from *Morinda citrifolia* (noni). Proc West Pharmacol Soc 45:76–78

Michalak I, Chojnacka K (2018) Seaweeds as a component of the human diet. In: Chojnacka K, Wieczorek P, Schroeder G, Michalak I (eds) Algae biomass: characteristics and applications. Developments in applied phycology, vol 8. Springer, Cham, pp 57–71

Mišurcová L, Machů L, Orsavová J (2011) Seaweed minerals as nutraceuticals. In: Kim SK (ed) Marirne medical foods: advances in food and nutrition research, vol 64, pp 371–390

Mišurcová L, Buňka F, Vávra Ambrožová J, Machů L, Samek D, Kráčmar S (2014) Amino acid composition of algal products and its contribution to RDI. Food Chem 151:120–125

Mohammad JH, Ranga Rao A, Ravishankar GA (2019) Opportunities and challenges in seaweeds as feed stock for biofuel production. In: Ravishnkar GA, Rao AR (eds) Handbook of algal technologies and phytochemicals. Phycoremediation, biofuels and global biomass production, vol 2. CRC, Boca Raton, FL, pp 39–50

Mohibbullah M, Hannan MA, Choi JY, Bhuiyan MM, Hong YK, Choi JS, Choi IS, Moon IS (2015) The edible marine alga *Gracilariopsis chorda* alleviates hypoxia/reoxygenation-induced oxidative stress in cultured hippocampal neurons. J Med Food 18:960–971

Moroney NC, O'Grady MN, O'Doherty JV, Kerry JP (2013) Effect of a brown seaweed (*Laminaria digitata*) extract containing laminarin and fucoidan on the quality and shelf-life of fresh and cooked minced pork patties. Meat Sci 94(3):304–311

Moroney N, O'Grady M, Lordan S, Stanton C, Kerry J (2015) Seaweed polysaccharides (Laminarin and Fucoidan) as functional ingredients in pork meat: an evaluation of anti-oxidative potential, thermal stability and bioaccessibility. Mar Drugs 13(4):2447–2464

Mouritsen OG, Williams L, Bjerregaard R, Duelund L (2012) Seaweeds for umami flavour in the New Nordic cuisine. Flav 1:4

Mouritsen OG, Rhatigan P, Pérez-Lloréns JL (2018) World cuisine of seaweeds: science meets gastronomy. Int J Gastronomy Food Sci 14:55–65

Namvar F, Mohamad R, Baharara J, Zafar-Balanejad S, Fargahi F, Rahman HS (2013) Antioxidant, antiproliferative, and antiangiogenesis effects of polyphenol-rich seaweed (*Sargassum muticum*). Bio Med Res Int 2013:9

Ningsih SS, Anggraeni AA (2021) Sensory characteristics of mille crepes cake from seaweed powder. IOP Conf Ser Earth Environ Sci 672

Nova P, Martins AP, Teixeira C, Abreu H, Silva JG, Machado Silva AM, Freitas AC, Gomes AM (2020) Foods with microalgae and seaweeds fostering consumers health: a review on scientific and market innovations. J Appl Phycol 32:1789–1802

Nuñez M, Picon A (2016) Seaweeds in yogurt and quark supplementation: influence of five dehydrated edible seaweeds on sensory characteristics. Int J Food Sci Technol 52(2):431–438

Nwosu F, Morris J, Lund VA, Stewart D, Ross HA, McDougall GJ (2011) Anti-proliferative and potential anti-diabetic effects of phenolic-rich extracts from marine edible algae. Food Chem 126:1006–1012

Oh H, Lee P, Kim SY, Kim YS (2020) Preparation of cookies with various native seaweeds found on the Korean coast. J Aquat Food Prod Technol 29(2):1–8

Ojagh SM, Rezaei M, Razavi SH, Hosseini SMH (2010) Effect of chitosan coatings enriched with cinnamon oil on the quality of refrigerated rainbow trout. Food Chem 120:193–198

Ortiz J, Romero N, Robert P, Araya J, Hernandez JL, Bozzo C (2006) Dietary fiber, amino acid, fatty acid and tocopherol contents of the edible seaweeds *Ulva lactuca* and *Durvillaea antarctica*. Food Chem 99:98–104

Paiva L, Lima E, Patarra RF, Neto AI, Baptista J (2014) Edible Azorean macroalgae as source of rich nutrients with impact on human health. Food Chem 164:128–135

Plaza M, Santoyo S, Jaime L, García-Blairsy Reina G, Herrero M, Señoráns FJ, Ibáñez E (2010) Screening for bioactive compounds from algae. J Pharm Biomed Anal 51:450–455

Ponce NMA, Pujol CA, Damonte EB, Flores ML, Stoerz CA (2003) Fucoidans from the brown seaweed *Adenocystis utricularis*: extraction methods, antiviral activity and structural studies. Carbohydr Res 338:153–165

Prabhasankar P, Ganesan P, Bhaskar N, Hirose A, Stephen N, Gowda LR, Hosokawa M, Miyashita K (2009) Edible Japanese seaweed, wakame (*Undaria pinnatifida*) as an ingredient in pasta: chemical, functional and structural evaluation. Food Chem 115(2):501–508

Prachyakij P, Charernjiratrakul W, Kantachote D (2008) Improvement in the quality of a fermented seaweed beverage using an antiyeast starter of Lactobacillus plantarum DW3 and partial sterilization. World J Microbiol Biotechnol 24:1713–1720

Prigent SVE, Gruppen H, Visser AJ, Van Koningsveld GA, de Jong GAH, Voragen AGJ (2003) Effects of non-covalent interactions with 5-O (ortho)-caffeoylquinic acid (chlorogenic acid) on the heat denaturation and solubility of globular proteins. J Agric Food Chem 51:5088–5095

Qin Y (2018) Applications of bioactive seaweed substances in functional food products. Bioact Seaweeds Food Appl:111–134

Rafiquzzaman SM, Lee JM, Raju A, Lee J-H, Kim J-M, Kong I-S (2015) Characterization of the hypoglycaemic activity of glycoprotein purified from the edible brown seaweed, Undaria pinnatifida. Int J Food Sci Technol 50:143–150

Rahnasto-Rilla M, McLoughlin P, Kulikowicz T, Doyle M, Bohr V, Lahtela-Kakkonen M, Moaddel R (2017) The identification of a SIRT6 activator from brown algae *Fucus distichus*. Mar Drugs 15(6):190

Rajapakse N, Kim SK (2011) Nutritional and digestive health benefits of seaweed. In: Kim SK (ed) Marirne medical foods: advances in food and nutrition research, vol 64, pp 17–28

Rao BM, Jesmi D, Viji P (2017) Chilled storage of *Pangasianodon hypophthalmus* fillets coated with plant oil incorporated alginate gels: effect of clove leaf, clove bud, rosemary and thyme oils. J Aquat Food Prod Technol 26(6):744–755

Ratana-arporn P, Chirapart K (2006) Nutritional evaluation of tropical Green Seaweeds *Caulerpa lentillifera* and *Ulva reticulata*. Kasetsart J (Nat Sci) 40(Suppl):75–83

Rodríguez-Meizoso I, Jaime L, Santoyo S, Señoráns F, Cifuentes A, Ibáñez E (2010) Subcritical water extraction and characterization of bioactive compounds from *Haematococcus pluvialis* microalga. J Pharm Biomed Anal 51:456–463

Rojas JA, Rosell CM, Benedito de Barber C (1999) Pasting properties of different wheat flour-hydrocolloids systems. Food Hydrocoll 13:27–33

Roohinejad S, Koubaa M, Barba FJ, Saljoughian S, Amid M, Greiner R (2017) Application of seaweeds to develop new food products with enhanced shelf-life, quality and health-related beneficial properties. Food Res Int 99:1066–1083

Rosell CM, Rojas JA, Benedito de Barber C (2001) Influence of hydrocolloids on dough rheology and bread quality. Food Hydrocoll 15:75–81

Rosemary T, Arulkumar A, Paramasivam S, Mondragon-Portocarrero A, Miranda JM (2019) Biochemical, micronutrient and physicochemical properties of the dried red seaweeds *Gracilaria edulis* and *Gracilaria corticata*. Molecules 24(12):2225

Sabeena Farvin KH, Jacobsen C (2013) Phenolic compounds and antioxidant activities of selected species of seaweeds from Danish coast. Food Chem 138:1670–1681

Samah Shalaby MS, Amin HH (2019) Potential using of Ulvan polysaccharide from *Ulva lactuca* as a prebiotic in synbiotic yogurt production. J Prob Health 7:208

Sánchez-Machado DI, López-Cervantes J, López-Hernández J, Paseiro-Losada P (2004) Fatty acids, total lipid, protein and ash contents of processed edible seaweeds. Food Chem 85(3):439–444

Sanjeewa KKA, Lee W, Jeon YJ (2018) Nutrients and bioactive potentials of edible green and red seaweed in Korea. Fish Aquat Sci 21(1):19

Santoso J, Stark YY, Suzuki T (2006) Comparative contents of minerals and dietary fibers in several tropical seaweeds. Buletin Teknologi Hasil Perikanan 9(1):1–11

Sáez MI, Suárez MD, Alarcón FJ, Martínez TF (2021) Assessing the Potential of Algae Extracts for Extending the Shelf Life of Rainbow Trout (Oncorhynchus mykiss) Fillets. Foods 10(5):910

Seafoodplus (2008). www.seafoodplus.org/fileadmin/files/news/2004-01-22SFRTD1launchBrussels.pdf. Accessed 8 May 2010

Senthil MA, Mamatha BS, Mahadevaswamy M (2005) Effect of using seaweed (*Eucheuma*) powder on the quality of fish cutlet. Int J Food Sci Nutr 56(5):327–335

Seo MJ, Lee OH, Choi HS, Lee BY (2012) Extract from edible red seaweed (*Gelidium amansii*) inhibits lipid accumulation and ROS production during differentiation in 3T3-L1 cells. Prev Nutr Food Sci 17:129–135

Shama A, Joyce SG, Mari FD, Ranga Rao A, Ravishankar GA, Hudaa N (2019) Macroalgae and microalgae: novel sources of functional food and feed. In: Ravishankar GA, Rao AR (eds) Handbook of algal technologies and phytochemicals. Food, health and nutraceutical applications, vol 1. CRC, Boca Raton, FL, pp 207–219

Sharma BR, Rhyu DY (2014) Anti-diabetic effects of *Caulerpa lentillifera*: stimulation of insulin secretion in pancreatic β-cells and enhancement of glucose uptake in adipocytes. Asian Pac J Trop Biomed 4(7):575–580

Shitole SS, Balange AK, Gangan SS (2014) Use of Seaweed (*Sargassum tenerrimum*) extract as gel enhancer for lesser sardine (*Sardinella brachiosoma*) surimi. Int Aquat Res 6(1):55

Singh CB, Hassan MA, Gudipati V, Deshmukhe G, Xavier KAM, Balange AK (2016) Extruded snacks fortified with seaweed (*Ulva lactuca*): process optimization by response surface methodology. J Env Bio-Sci 30(2):289–297

Singh CB, Xavier KAM, Deshmukhe G, Gudipati V, Shitole SS, Balange AK (2017) Fortification of extruded product with brown seaweed (*Sargassum tenerrimum*) and its process optimization by response surface methodology. Waste Biomass Valor 9(5):755–764

Skonberg DI, Fader S, Perkins LB, Perry JJ (2021) Lactic acid fermentation in the development of a seaweed sauerkraut-style product: microbiological, physicochemical, and sensory evaluation. J Food Sci 86(1):334–342

Sørensen LE, Jeppesen PB, Christiansen CB, Hermansen K, Gregersen S (2019) Nordic seaweed and diabetes prevention: exploratory studies in KK-Ay mice. Nutrients 11(6):1435

Stefani S, Pratama RI, Rostini I, Afrianto E (2019) Seaweed flour fortification to the preference level of milk chocolate bar. Asian Food Sci J 12(1):1–10

Syad AN, Shunmugiah KP, Kasi PD (2013) Seaweeds as nutritional supplements: Analysis of nutritional profile, physicochemical properties and proximate composition of *G. acerosa* and *S. wightii*. Biomed Prev Nutr 3(2):139–144

Tanna B, Mishra A (2018) Metabolites unravel nutraceutical potential of edible seaweeds: an emerging source of functional food. Comprehens Rev Food Sci Food Saf 17(6):1613–1624

Tsuchiya K, Matsuda S, Hirakawa G, Shimada O, Horio R, Fujii T, Ishida A, Iwahara M (2007) GABA production from discolored laver by lactic acid fermentation and physiological function of fermented laver. Food Preserv Sci 33:121–125

Uchida M, Murata M (2004) Isolation of a lactic acid bacterium and yeast consortium from a fermented material of *Ulva* spp. (Chlorophyta). J Appl Microbiol 97:1297–1310

Uchida M, Amakasu H, Satoh Y, Murata M (2004) Combinations of lactic acid bacteria and yeast suitable for preparation of marine silage. Fish Sci 70:507–517

Uchida M, Murata M, Ishikawa F (2007) Lactic acid bacteria effective for regulating the growth of contaminant bacteria during the fermentation of *Undaria pinnatifida* (Phaeophyta). Fish Sci 73:694–704

Uchida M, Kurushima H, Ishihara K, Murata Y, Touhata K, Ishida N, Niwa K, Araki T (2017) Characterization of fermented seaweed sauce prepared from nori (*Pyropia yezoensis*). J Biosci Bioeng 123(3):327–332

Uchida M, Kurushima H, Hideshima N, Araki T, Ishihara K, Murata Y, Touhata, k. and Ishida, N. (2018) Preparation and characterization of fermented seaweed sauce manufactured from low-quality nori (dried and fresh fronds of *Pyropia yezoensis*). Fish Sci 84(3):589–596

Uribe E, Vega-Gálvez A, Heredia V, Pastén A, Di Scala K (2017) An edible red seaweed (*Pyropia orbicularis*): influence of vacuum drying on physicochemical composition, bioactive compounds, antioxidant capacity, and pigments. J Appl Phycol 30(1):673–683

Usman GO, Ameh UE, Alifa ON, Babatunde RM (2015) Proximate composition of biscuits produced from wheat flour and maize bran composite flour fortified with carrot extract. J Nutr Food Sci 5(5):1000395

Vieira EF, Soares C, Machado S, Oliva-Teles MT, Correia M, João Ramalhosa M, Carvalho A, Domingues VF, Antunes F, Morais S, Delerue-Matos C (2020) Development of new canned chub mackerel products incorporating edible seaweeds-influence on the minerals and trace elements composition. Molecules 25(5):1133

Vijerathna MPG, Wijesekara L, Perera R, Maralanda SMTA, Jayasinghe M, Wickramasinghe I (2019) Physico-chemical characterization of cookies supplemented with sugarcane bagasse fibres. Vidyodaya J Sci 22(1):29–39

Vilar EG, Ouyang H, O'Sullivan MG, Kerry JP, Hamill RM, O'Grady MN, Mohammed HO, Kilcawley KN (2019) Effect of salt reduction and inclusion of 1% edible seaweeds on the chemical, sensory and volatile component profile of reformulated frankfurters. Meat Sci 161:108001

Volpe MG, Siano F, Paolucci M, Sacco A, Sorrentino A, Malinconico M, Varricchio E (2015) Active edible coating effectiveness in shelf-life enhancement of trout (Oncorhynchus mykiss) fillets. LWT- Food Sci Technol 60(1):615–622

Walstra P, Wouters JTM, Geurts TJ (2006) In: Walstra P, Wouters JTM (eds) Fermented milks. Taylor & Francis Group, London

Wang T, Jónsdóttir R, Kristinsson HG, Thorkelsson G, Jacobsen C, Hamaguchi PY, Ólafsdóttir G (2010) Inhibition of haemoglobin-mediated lipid oxidation in washed cod muscle and cod protein isolates by Fucus vesiculosus extract and fractions. Food Chem 123(2):321–330

Wells ML, Potin P, Craigie JS, Raven JA, Merchant SS, Helliwell KE, Smith AG, Camire ME, Brawley SH (2017) Algae as nutritional and functional food sources: revisiting our understanding. J Appl Phycol 29:949–982

Widiyanti MA, Purnamayati L, Romadhon R (2021) Sensory and physicochemical characteristics of manyung (*Arius thalassinus*) kekian high fiber with the addition of *Eucheuma cottonii* seaweed. Caraka Tani J Sustain Agric 36(1):1–10

Willcox DC, Willcox BJ, Todoriki H, Suzuki M (2009) The Okinawan diet: Health implications of a low-calorie, nutrient-dense, antioxidant-rich dietary pattern low in glycemic load. J Am Coll Nutr 28:500S–516S

Wong KH, Cheung PCK (2000) Nutritional evaluation of some subtropical red and green seaweeds. Food Chem 71(4):475–482

Woo MS, Choi HS, Lee OH, Lee BY (2013) The edible red alga, Gracilaria verrucosa, inhibits lipid accumulation and ROS production, but improves glucose uptake in 3T3-L1 cells. Phytother Res 27:1102–1105

Ylimaki G, Hawrysh ZJ, Hardin RT, Thomson ABR (1998) Application of response-surface methodology to the development of rice flour yeast breads—objective measurements. J Food Sci 53:1800–1805

Yoshinaga J, Morita M, Yukawa M, Shiraishi K, Kawamura H (2001) Certified reference material for analytical quality assurance of minor and trace elements in food and related matrixes based on a typical Japanese diet: interlaboratory study. J AOAC Int 84:1202–1208

Yuan YV, Bone DE, Carrington MF (2005) Antioxidant activity of dulse (*Palmaria palmata*) extract evaluated in vitro. Food Chem 91(3):485–494

Zava TT, Zava DT (2011) Assessment of Japanese iodine intake based on seaweed consumption in Japan: a literature based analysis. Thyroid Res 4(1):14

Zheng J, Chen Y, Yao F, Chen W, Shi G (2012) Chemical composition and antioxidant/ antimicrobial activities in supercritical carbon dioxide fluid extract of *Gloiopeltis tenax*. Mar Drugs 10:2634–2647

Chapter 2
Gracilaria: An Emerging Source of Agar Feedstock—With Special Reference to Industrially Important Species

Kanchan Sambhwani, Mudassar Anisoddin Kazi, and Vaibhav A. Mantri

1 Introduction

The marine macrophytes, popularly termed seaweeds, are one of the naturally occurring renewable resources in coastal water bodies over the globe. They are distributed in the intertidal, subtidal, and a few of them even in deep waters as well. The earliest known use of seaweed resource can be date back to the Neolithic period through indirect evidence, but the preserved written records of about 1700 years ago are available in China. The coastal populations have been harvesting seaweeds, initially for industrial usage such as soda ash, potash, iodine, silica for glass making, for composting; but the applications in commodity products derived from seaweed hydrocolloids have emerged much later. The seaweed cultivation and processing sector have shown considerable growth despite economic rescission, due to several emerging applications including specialized products based on antioxidant, antibacterial, antiviral, anti-aging, anti-inflammatory, and anticancer rendering properties. It may be noted that, seaweeds are now increasingly cultivated than wild collection. The global production of seaweeds, has more than tripled, up from 10.6 million tons in 2000 to 32.4 million tons in 2018 (FAO 2018). The rapid growth registered in red seaweeds is the reason for this considerable expansion, including the farming of *Gracilaria* spp.

K. Sambhwani · V. A. Mantri (✉)
Applied Phycology and Biotechnology Division, CSIR-Central Salt and Marine Chemicals Research Institute, Bhavnagar, Gujarat, India

Academy of Scientific and Innovative Research (AcSIR), CSIR-Central Salt and Marine Chemicals Research Institute, Bhavnagar, Gujarat, India
e-mail: vaibhav@csmcri.res.in

M. A. Kazi
Applied Phycology and Biotechnology Division, CSIR-Central Salt and Marine Chemicals Research Institute, Bhavnagar, Gujarat, India

A. Ranga Rao, G. A. Ravishankar (eds.), *Sustainable Global Resources of Seaweeds Volume 2*, https://doi.org/10.1007/978-3-030-92174-3_2

The genus *Gracilaria* is the most diverse and largest genus in the Rhodophyta group of macroalga (Lyra et al. 2015), comprising of 287 species (Guiry and Guiry 2020). The species are distributed over temperate, subtropical, and tropical Pacific, Atlantic, and Indian Oceans (Gurgel et al. 2004). With its diversity this genus is important commercial commodity as source of agar to the Industry with annual sale of 246 million USD for producing 14,500 tons of agar (Porse and Rudolph 2017). This chapter appraises the insights of *Gracilaria* with specific case study of industrially important species of this genus.

2 Biology of *Gracilaria*

The genus *Gracilaria* comes under Rhodophyta or red algae group, understanding the systematics, morphology and life history of this gracilarioid algae is highly important because of their economic importance. The species consists of erect to prostrate thalli of primarily cylindrical in nature, but flattened are not uncommon. The thalli show the presence of holdfast or disc for the firm attachment to the substrate while branching of varying degrees is also reported depending on species. They typically occupy sheltered intertidal bays, back waters or estuaries. The subtidal muddy, rocky, or sandy bottoms are also preferred by some species. The typical life cycle is of isomorphic, tri-phasic which is commonly known as *Polysiphonia* type.

2.1 Traditional and Recent Classification

These algae are the part of class Florideophyceae. The ordinal classification in this class was mainly based on the morphology and development of female reproductive structures before and after fertilization. Gracilariaceae Nägeli the only family in the order Gracilariales Fredericq & Hommersand was earlier included in the order Gigartinales Schmitz. But the characteristic features of Gigartinales such as the presence of auxiliary cells and connecting cells before fertilization was absent in Gracilariaceae (Fredericq and Hommersand 1989b). Therefore, a new order Gracilariales was established to depict peculiar characteristics of a family Gracilariaceae by Fredericq and Hommersand (1989b). *Gracilaria* Greville is one of the 11 genera recognized in the family Gracilariaceae. Dawson (1949) segregated *Gracilariopsis* from *Gracilaria* on the absence of nutritive filaments connecting gonemoblast and pericarp. *Hydropuntia Montagne* (Polycavernosa by Chang 1963), was also then separated from *Gracilaria* based on spermatangial conceptales with numerous cavities and conspicuous, irregular tubular nutritive cells from the bottom of gonimoblast to the cystocarp floor (Gargiulo et al. 1992). Initially, four species were included in the genus *Gracilaria*, but without designated type species. *Gracilaria confervoides* (Linnaeus) Greville was then designated as lectotype of the genus by Schmitz et al. (1889). Later on, Papenfuss (1950) proposed a new combination of *Gracilaria verrucosa* (Hudson) Papenfuss for this type species. According to Papenfuss (1950) *Fucus confervoides described* by Linnaeus (1763) was a homonym to the *Fucus confervoides* Hudson (1762), and the species was first described as *Fucus verrucosus* by Hudson (1762). However, the name *G. verrucosa* is currently treated as rejected name and synonym of *Gracilariopsis longissima* (S. G. Gmelin) Steentoft, L. M. Irvine & Farnham (Guiry and Guiry 2020). The careful examination of the Linnean specimen of *F. confervoides* revealed its relatedness to *Gracilariopsis* Dawson rather than *Gracilaria* sensu Fredericq & Hommersand (Steentoft et al. 1991). Similarly, the description of specimens of Stackhouse also showed differences with *Gracilaria* sensu Dawson (1949) and Fredericq and Hommersand (1989a). Further to establish the type species for the genus *Gracilaria*, Steentoft et al. (1991) chose *G. compressa* (Agardh) Greville, the name included by Greville in his description. Currently *G. compressa* regarded as the synonym of *G. bursa-pastoris* (S. Gmelin) Silva (Steentoft et al. 1991). The comprehensive molecular analysis of Gracilariaceae by Gurgel and Fredericq (2004) revealed the presence of three major clades corresponding to, (1) *Curdiea* and *Melanthalia* (2) *Gracilariopsis* and (3) *Gracilaria* sensu lato including Hydropuntia. The study identified nine independent evolutionary lineages in *Gracilaria* sensu lato and advocated to assign generic rank to the two most basal lineages in this clade. The evolutionary lineages also showed biogeographic specificity in phylogenetic analysis representing common phenomena of ecological radiation and local speciation in *Gracilaria* (Gurgel and Fredericq 2004). The recent molecular analysis of the order Gracilariales based on *rbc*L DNA sequence identified two subfamilies, two tribes, two new genera and four subgenera (Gurgel et al. 2018). The genus *Gracilaria* now consists of two new subgenera, *Gracilaria* subgen. *Gracilaria* and *Gracilaria* subgen. *Corallopsis* Gurgel, J. N. Norris et Fredericq, stat. et. Currently174 species have been recognized as accepted species in the genus *Gracilaria*. Morphological studies contributed immensely to the taxonomic knowledge of the Genus *Gracilaria*. The major criteria for *Gracilaria* classification are: (1) tetrasporangia are scattered either in cortex or are produce in nemathecial sorri, (2) the spermantagia are either in deep conceptacles or in shallow pits, (3) the cystocarps were constricted at the base and rostrate or lacking constriction and non-rostrate. However, the species delimitation in *Gracilaria* using morphological features poses considerable limitations (Goff et al. 1994). Molecular studies using different molecular markers have helped in resolving issues in systematics, identifying new species and evolutionary history of this genus (Gurgel et al. 2004, 2008, 2018; Gargiulo et al. 2006; Kim et al. 2006; Sfriso et al. 2013; Lin et al. 2012; Soares et al. 2018). *rbc*L was seen as preferred marker in most of the studies due to its optimal resolution at both generic and species levels. In addition, other markers such as 18S, ITS, UPA, COI-5′ etc. were also applied in biodiversity studies of *Gracilaria*.

2.2 Life Cycle

The life cycle patterns in seaweeds are complex, and often represented by independent and free-living haploid as well as diploid stages alternating with each other. Understanding these patterns is essential due to their critical role in having control over reproduction necessary for successful seeding in farming operations. The life cycle stages due to their different ploidy level represent a differential response to environmental conditions and thus understanding their phenology play a vital role in resource management as well. Further, studies have also focused on their importance in evolutionary perspective. It may be noted that in red seaweeds life history patterns are less diverse and differ in the temporal difference in meiosis and syngamy, besides the degree of mitotic activity in both haploid and diploid phases also differ. In general sexual history pattern in the seaweeds constitutes rotation between haploid and diploid phases with meiosis mediating the transition from the diploid to the haploid state, while syngamy reconstituting a diploid state (Mantri 2010). The life cycle in *Gracilaria* is isomorphic, tri-phasic and diplobiontic type. It is also known as "*Polysiphonia* type" of life cycle pattern (Fig. 2.1). It may be noted that here there is single gametophytic phase, where the sporophytic phase is represented by two distinct phases qualifying it as triphasic nature. Further, the gametophytic phase is represented by two separate plants namely, male (bear spermatangium) as well as female (bear carpogonium) thalli respectively. Both male and female thalli are morphologically similar and indistinguishable in wild populations. The sperm (male) and female (egg) gamete both non-motile entities undergo fusion to form a

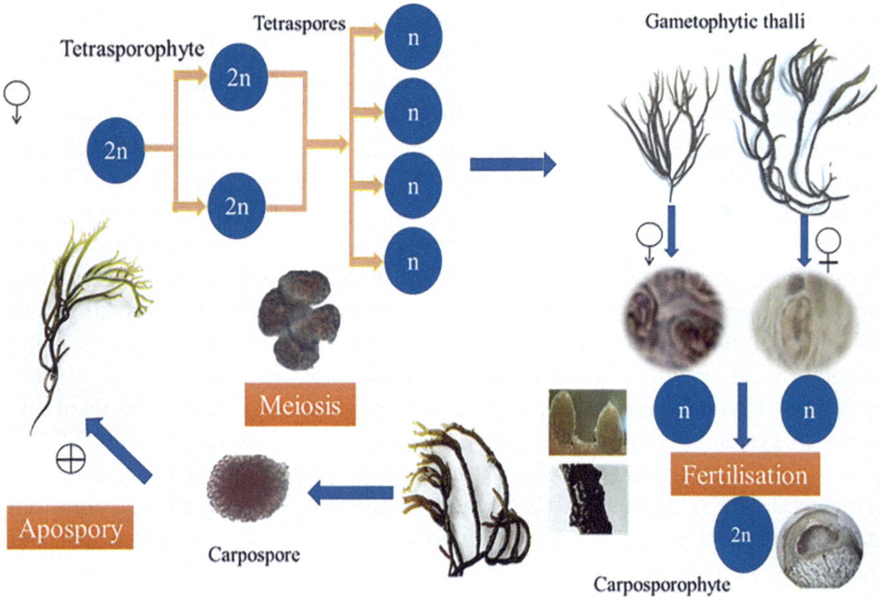

Fig. 2.1 Triphasic isomorphic life-cycle in genus *Gracilaria*

zygote. This diploid zygote germinates and develops into carpospophytic phase onto the female thalli. This phase is diploid and remains parasitic on female gametophyte tissue deriving nutrition to form a specialized structure called 'cystocarp'. The carpospores, which are formed via mitosis, are released in the water and get germinated to form another sporophyte called tetrasporophyte plant (morphologically similar to both male and female gametophyte). The tetrasporophyte cells get transformed and give rise to diploid tetrasporangia, which undergoes meiosis to form four terasporophytes in each tetrasporangium. The tetraspores now thus haploid get released into the water again forming separate male and female gametophytic thalli in equal proportion (two male and two females respectively). As stated before, it may be noted that, all the stages are similar in morphological attributes and the only stage that could be easily be distinguished with the naked eye in the wild, is the female gametophyte. It can be separated from the male and tetrasporophyte population that is growing in the vicinity due to the presence of protruding, hemispherical, cystocarps. Nevertheless, the formation of cystocarp takes place only after the fertilization event and practically no stage can be separated during the young stages of thallus development.

It may also be interesting to note that, several deviations to this typical *Polysiphonia* type life history pattern have been reported in the literature. The most common is the 'mixed phase' this is due to *in situ* germination of spores of one phase of the life cycle stage on to other. The development of a single type of individuals of gametophyte phase instead of both (male and female) has also been reported. Further, direct development of carposporophyte from cystocarps circumventing tetrasporophyte phase has also been reported. The parthenogenetic development has also been reported. The possible reasons for such deviations include, mutation, male or female repression and mitotic recombination.

2.3 Morphology and Ultrastructure

Members of *Gracilaria* consist of cylindrical, flattened or leaf-like thalli with decumbent to erect plants having variable branching. The plants are firmly attached to the substrata via a disc called holdfast. *Gracilaria* is a widespread and polymorphic genus, Sjöstedt (1926) was the first to recognise carpogonial branches in *Gracilaria*. The carpogonial branches develops from a transformed cortical cell that functions as the primordium (Yamamoto 1978). And the tetrasporongia develops from primary cortical cells scattered throughout the cortical layers of the thalli.

The thallus of *G. dura* is cylindrical and of 10 to 100 cm long with two to three orders of lateral branches of unilateral pattern and sometimes alternate. The plants are attached to the substratum by a rhizomatous base. The cortex is of a single layer having dark red-color pigmented cells. The medulla consists of large oval cells without intracellular space and decreasing in size toward the surface. The subcortical and medullary cells were filled with starch grains (Mantri et al. 2021).

Plants of *G. chilensis* are dark-brown to black with 20–21 cm thalli length with irregular branching up to three order. Cortex-to-medulla transition is gradual, with cells of increasing diameter towards the centre of the frond, cortex is two to three cells thick, outer cells anticlinally elongated (Byrne et al. 2002).

Plants of *G. lemaneiformis* are cylindrical and are up to 100 cm tall with irregular branching. Thalli are pseudoparenchymatous throughout with a sharp gradation from small-cells in outer cortical zone to a large-cells in thin-walled central medulla. Medullary cells are multinucleate and the cortex consists of one or two cell layers of pigmented, elongated outer cells that lack secondary pit-connections (Fredericq and Hommersand 1989b).

3 Economic Importance of *Gracilaria*

The genus *Gracilaria* is listed as the world's most important agarophyte resource (McLachlan and Bird 1983). Their economic importance is traced back to approx.14,000 years for food and medicine, some of its major sources are as described.

3.1 Agar Polysaccharide

Agar is the major commercial red seaweed extract and has immense application in the food, feed, textile, and research area. Agar, a water-soluble polysaccharide is a very important gelling biopolymer of plant origin found in cell matrix of red seaweeds especially in members of Gelidiaceae and Gracilariaceae families known as agarophytes (Madhavan 2015). Structurally agar is a linear chain of disaccharide agarobiose consisting of D-galactose and 3,6 L-galactose with sulphate ester, pyruvic acid and D-glucuronic acid in addition to agarobiose (Araki 1966). Agarose is a purified linear galactan. It forms the gel matrix in an aqueous medium that is ideal for the diffusion and electro-kinetic movement of biopolymers. This makes it suitable for applications in molecular biology, electrophoresis, and cell culture. Agarose is commonly prepared from superior quality agar or agar-bearing marine algae.

Basic extraction procedure of agar: The extraction procedure for the agar is multistep thus, slow and complicated (Kumar and Fotedar 2009). Generally, hot water as a solvent is used for extraction of the agar as it is water-soluble. After soaking of biomass, it is cooked at a high temperature. Cooking follows filtration and precipitation to remove phenolic dirt. Following the freeze-thaw procedure for gelling and finally, the gel formed is dried and ground to powder. But the phycocolloids extracted by this basic method are weak in terms of its rheological properties and are less preferable in industries (Manuhara et al. 2016).

To the basic extraction process studies have been reported for an increase in the agar quality with the ease of extraction procedure. Meena et al. (2007) have reported cost-effective method of quality increase of agar from *G. dura* by soaking the alga in different percentage of alkali and seen an increase of gel strength of the agar extracted and have reported that 10% aqueous alkali treatment gives the best result i.e. remarkable physiology of the polysaccharide and compared the treated polysaccharide with the commercially available agar. This step of pre-soaking the seaweed in alkali solution prior to polysaccharide extraction, converts the precursor, L/D-galactose-6-sulphate, moieties in the galactan backbone to 3,6-anhydrogalactose (Rees 1972), and results increase in gel strength of the extract (Praiboon et al. 2006). The galactan units with sulphate moiety serve as kinks in the polysaccharide which restrains the formation of double-helix network for the gelation process (Rees 1969). Depending on the extent of conversion of the precursor units to 3,6-anhydrogalactose, a spectrum of gel properties can be obtained when different levels of alkali treatment conditions are employed (Chirapart et al. 2006). Desulphating enzyme also removes the sulphate kinks from agar and increases its gel strength. Shukla et al. (2011) has reported sulphohydrolase from *G. dura* itself and its treatment increases quality as well as yield with it this greener approach also maintains the bioactivity of polysaccharide.

Meena et al., have reported the simplification of the downstream process by coagulating the alkali-treated agar extract via surfactant in place of tedious freeze-thaw cycles thus making it energy effective without compromising the yield and quality. This process of spontaneous coagulation of agarose from the extract after the addition of a nonionic surfactant is supposed the micellar aggregates of the surfactant molecules above the critical micelle concentration serving as a template for the linear galactan chains and in part replacing hydrogen bonding to water molecules with hydrogen bonding to the polar ethoxylate chain of the surfactant. This may have caused the breakdown of the gel network. The micellar aggregate may also have helped to promote hydrogen bonding between the agarose molecules themselves.

3.2 Pigments

Water soluble and nontoxic proteins, Phycobiliproteins are accessory pigments in, red algae (Rhodophyta). These proteins hold great fiscal importance in the field of food supplements, cosmetics and medical research such as fluorescent probes for the diagnostic assays and other diverse researches (Torres et al. 2019). In Rhodophyta, amongst the species of *Gracilaria*, phycoerythrin (R-phycoerythrin), phycocyanin (C phycocyanin), and allophycocyanin are present as pigments These pigments have been reported for their numerous bioactivities such as antioxidant activity, anti-inflammatory, and neuroprotective activity (Sonani 2016). R-phycoerythrin among the phycobiliproteins, has been singled out as an important tool in the field of medical diagnosis, and biomedical research, due to its excellent optical and spectroscopic properties, high absorption coefficient, and high fluorescence yield. This phycoerythrin varies, in amount, in the pigmental mutants of the genus *Gracilaria*.

In *Gracilaria lameneiformis* amino acid sequence was deduced for different pigmental mutants that interacted for secondary structure in different subunits of phycoerythrin.

Owing to the diverse applications of phycobiliproteins and in various fields, different methods have been reported for its maximum purity, high yield and cost-effective extraction method. In *Gracilaria crassa* various solvents such as phosphate buffer, distilled water, seawater has been checked for the better extraction of these phycobiliproteins (Sudhakar et al. 2015). Extraction using Ionic liquids is also performed in *Gracilaria* sp. resulting into a significant increase in the yield of phycobiliproteins extraction compared to the basic extraction procedure. Yield of 46.5% and their conformational structure or chromophore structural integrity was reported (Martins et al. 2016).

3.3 Bioactive Compounds

Seaweeds in general have proved as an extraordinary resource of bioactive compounds and *Gracilaria* is among such seaweed with a promising profile of such primary and secondary metabolites. The availability of large biomass produced through commercial cultivation also makes *Gracilaria* a suitable candidate for many studies in investigating bioactive potential. Many species of this genus have been evaluated for their antibacterial, antioxidant, antifungal, antiprotozoal, anti-inflammatory, antiviral, cytotoxic, antihypertensive, spermicidal and embryotoxic activities (de Almeida et al. 2011).

Gracilaria is one of the major sources of industrially important polysaccharide agar. Some of agar-oligosaccharides are known to show α-glucosidase inhibitor activity which can help treat diabetes (Saraswaty et al. 2015; Senthilkumar and Sudha 2012). Similarly, agar-type polysaccharide extracted by the process of cold-water extraction from *G. dominguensis* showed antitumor activity (Fernández et al. 1989). Neoagaro oligosaccharides (NAOSs) with different degrees of polymerization showed their potential to be used as antioxidants and prebiotic additives in food and feed industry (Zhang et al. 2019). Sulfated polysaccharides extracted from *G. birdiae* showed moderate antioxidant properties by inhibiting the formation of free radicals (Souza et al. 2012). The high molecular weight sulfated galactans of *G. corticata* were found effective in reducing the infection of herpes simplex virus types 1 and 2 (Chattopadhyay et al. 2008). Sulfated polysaccharides from *G. caudate* also showed anti-inflammatory and antinociceptive effects in mice (de Sousa et al. 2013).

Aqueous and alcoholic extracts from various *Gracilaria* species showed potent antitumor and antiproliferative activities. For example, methanolic extract from *G. salicornia* and *G. corticata* showed the cytotoxic effect against HT-29 cell line (Ghannadi et al. 2016). Similarly, methanolic extract from *G. tenuistipitata* exhibited apoptosis-based cytotoxicity against oral cancer cells was reported by inducing DNA damage, ROS induction, and mitochondrial depolarization (Yeh et al. 2012).

Methanolic extract from *G. edulis* also showed anticancer activity, antibacterial activity, and antioxidant activity (Hemasudha et al. 2019). The proliferation of Jurkat and molt- 4 human lymphoblastic leukemia cell lines was inhibited by aqueous extracts of *G. corticata* (Zandi et al. 2010).

Bioactive compounds cholest-5-en-3-ol and cholesteryl myristate isolated from three *Gracilaria* sp. of Malaysia demonstrated strong cytotoxic activity against HL-60 and MCF-7 cell lines (Andriani et al. 2016). Secondary metabolites such as 8-hydroxy-4 E,6 E -octadien-3-one and 3β-hydroxy-5α,6α-epoxy-7-megastigmen-9-one isolated from the seaweed *G. lemaneiformis* were found to show allelopathic effect on the growth of the red tide alga *Skeletonema costatum* (Lu et al. 2011).

4 Cultivation

Conventionally, to meet the raw material requirement seaweed industry has relied on gathering the biomass from wild resources, but now with the increase in demand for its products has led to the development of Gracilaria cultivation and with the advances in seaweed culture techniques, its pace is also accelerating.

4.1 Seedling Production

The cost and energy-effective extraction method of quality agar has led *Gracilaria* species industrially important. With the scarcity and limited distribution of biomass this genus is fast emerging candidate species for commercial cultivation. For establishing a commercial farm of the species, a year-round supply of adequate and viable seedling is crucial and environmental conditions play a vital role in the growth.

Epi and endophytes of *G.dura* are reported to contribute in the growth, health, and development of seaweeds by the direct production of the plant growth regulators (PGRs) and nitrogen fixation (Singh et al. 2011). They identified the isolates to species level by biochemical test, 16s RNA, and fatty acid profiling and concluded that on treating the seaweed with the protein from isolates bud regeneration occurs which depicts that protein is conjugated to IAA.

Using organogenetic capacity of this seaweed, the clonal propagation method has been approached for seedling production via apical tips of seaweed as they show high regeneration property. This technique is reported for several *Gracilaria* species till now such as *Gracilaria* chilensis (Collantes and Melo 1995), *Gracilaria textorii*, *Gracilaria acuminata* (Huang and Fujita 1997), *Gracilaria corticate* (Kumar et al. 2007), *Gracilaria changii* (Yeong et al. 2014) and *G.dura* (Saminathan et al. 2015). Further the environmental factors also influence this technique such as temperature, nutrient supply and irradiance.

4.2 Methods of Cultivation

Gracilaria has been harvested from the wild in a number of countries, including Argentina, Chile, China, India, the Philippines, and South Africa. However, due to over-exploitation of the biomass from natural beds and indiscriminate harvesting practice the farming became the sustainable means to support the industry. Santelices and Doty (1989) stated that the farming of *Gracilaria* was imitated in Taiwan, China, Chile, and Hawaii. The methods employed were mostly bottom planting (insertion or anchoring of thalli in the soft bottom) or floating techniques such as rope or nets hung horizontally or vertically. Besides land-based tank or pond cultivation methods are also popular. We under this section summarize only the most popular techniques used for farming of different *Gracilaria* species (Fig. 2.2).

4.2.1 Bottom Stocking

This is primarily aimed at enhancement or densification of natural stock under its known range of distribution or the conducive locations only. The individual thalli that are attached to small pebbles, stones, crustacean shells etc. are used taking their advantage of the already existing substratum. The method has been further

Fig. 2.2 Different methods of cultivation in *Gracilaria;* (**a**). Cultivation of *Gracilaria chilensis* by bottom stocking method in Chile; (**b** and **c**). Cultivation of *G. lemaneiformis* in China and *G. dura* in India by monoline method respectively; (**d**). Havering of *G. dura* cultivated by tube net method in India

improved to make use of those fragments that are not attached to their natural substratum. It may be noted that in this case the bunch of fragments are attached to small rocks using a rubber band or thread. The use of nylon mesh, sand-filled nylon tubes, wooden stakes has also been used to aid or enhance the anchoring support. These techniques stabilize the plantlets in the soft muddy bottom, where they naturally tend to develop an underground anchoring system and grow. The harvesting is attempted once the thalli reach a certain size keeping some part of the anchored thallus intact for re-growth. Three to four crops can be taken under these farming systems after which the new planting stock needs to be introduced to maintain growth rate profitable. The method although simple has labour intensive.

4.2.2 Floating Methods

There are several floating methods developed for the cultivation of various species of *Gracilaria* and adopted thorough out different farming areas in the world. However, most of these methods are divided into three techniques. The most common is 'tie-tie' method; as name indicated the thalli are first tied with the help of nylon rope and then tied to cultivation ropes. The ropes are then planted in the field singly (long line or monoline method) or in groups, raft. The second type is 'loop' is created by unwinding the rope and the group of thalli are inserted at regular interval. It may be noted, that in this avoid tying process, although there are chances of drifting of seed material if it is not fasten properly in the loop, it considerably saves on labour cost. These ropes are then planted in the field similarly in the first technique in long line or monoline or raft method. These both techniques are also used for the formation of nets using nylon ropes. The third type of technique is cultivating the thalli without typing or inserting them in the ropes but their bundles backed in secured structures such as pouches made from mosquito nets, fishnets or tubular nets. The advantage is not requiring manpower for tying but needs a high amount of initial seed material for seeding. The advantage of all these floating methods is that the cultivation structure can be relocated in the event of pest, infestation, or sudden fluctuation in salinity or other ambient parameters influencing growth adversely. The other advantage is one can adjust the depth at which the cultivation can be undertaken, unlike bottom planting. The seeding and harvesting operation can be attempted on the shore and the cultivation structures can be planted back in the deeper waters. However, there is lot of scope in further improving these techniques as per the need of operation and the species under consideration. We have very recently shown that triangular raft and vertical raft alignment can considerably improve the yield in industrially over-exploited *G. edulis* in Indian waters.

4.2.3 Pond or Tank Cultivation

Gracilaria is usually found in areas with adequate water motion and averse to growth in man-made impoundments. Nevertheless, few species are cultivated successfully in ponds and tanks on large scale for commercial farming. The ponds are constructed near the seashore or in-land where the salinity of the groundwater is sufficiently high to cultivate the marine species. They are smaller in size as seaweed biomass tends to concentrate at the bottom in larger ponds. The water depth is not kept constant and changed depending on ambient temperature, e.g., more water is maintained when the temperature is high, while less water is kept when the temperature is low. The periodic water exchange is highly essential to maintain an adequate supply of fresh nutrients to growing seedlings. The harvest is made depending on the growth pattern and some seed is always maintained from the previous crop to keep the cycle going. The use of abandoned ponds used for shrimp culture are also been used for the cultivation of *Gracilaria*. On the contrary use of tank cultivation is one of the most sophisticated methodologies adopted for biomass augmentation. The whole procedure is precisely controlled and mechanization and sophistication have been achieved. Although highest productivity per unit area has been achieved in this method, it remained most expensive form of seaweed aquaculture. Additionally, it has the potential to process the polluted water or use of effluents rich in the organic and inorganic matter for biomass production through green methodologies. This also forms excellent technique for rearing the spore based plant-let, or seedling production via clonal propagation and nursery set-up. The economic viability of this method can be drastically improved if *Gracilaria* cultivation should be looked beyond obtaining agar and production of value-added products such as pigments and secondary metabolites of nutraceutical and pharmaceutical applications.

4.3 Economics

Although considered important from a new investment perspective, economic analysis of farming or processing in seaweeds is seldom attempted. Valderrama et al. (2015) reported large differences in the economic performance of carrageenan seaweed farming from six countries reported. The comparison of large-scale versus family-sized system production of *Kappaphycus alvarezii* along the south-eastern coast of Brazil reported internal rates of return (IRR) of 38.17%, 70.73% and 87.81%, a payback period (PP) of 31, 17 and 14 months with a break-even Point (BP) at 78.41 tons of fresh weed. Nevertheless, highest cost in the commercial farming of *K. alvarezii* in the Philippines was attributed to labour (40%), followed by capital outlay (22%), materials (21%), seedlings (12%), and expedite towards interest (5%) (Hurtado et al. 2001). It has been also observed in several publications or reports that economic implications are projected an overly positive sketch of farming as well as down-stream processing including agarophytes and *Gracilaria*. But when it comes to energy applications such as biofuel production data on techno-economic analysis is elaborate (Dave et al. 2013; Fasahati et al. 2015; Krastina et al. 2017; Soleymani and Rosentrater 2017). We in this section give the overview of available data on economic of *Gracilaria* farming or processing. The economics of farming using 2000 bamboo rafts, occupying 1 ha sea front area for *G. edulis* and *G. dura* is available. The net income of USD 125.41 person^{-1} month^{-1} for *G. edulis* and USD 240.83 person^{-1} month^{-1} for *G. dura* has been reported from Mandapam, Tamil Nadu, India (Ganesan et al. 2017). Veeragurunathan et al. (2019) reported financial gains of USD 141 person^{-1} month^{-1} under commercial farming of *G. debilis* in six harvests along South-eastern coast of India. The estimated profit of USD 354.37 person^{-1} month^{-1} has been reported for *G. dura* using the tubular net method in India (Mantri et al. 2020). The bio-economic analysis of small-scale cultures of G. *chilensis* both in northern Chile is available (Zuniga-Jara and Contreras 2020). They reported a profit in *G. chilensis* after 8 years, which confirms the need of considerably more time for the income to recover the initial investments, which indicated that investors need to look at the farming of this alga in a long-term investment perspective. We look forward to having more techno-economic analysis studies not only in farming but also in the processing sector to present to encourage stakeholder investment in these species.

5 Industrially Important Species

The genus *Gracilaria* is the predominant genus in the family of Gracilariaceace (Rhodophyta) with over 150 described species distributed along different parts of tropical and sub-tropical coasts. Amongst them few are most economically important species due to their major contribution to the production from *Gracilaria*, based on 10-year documentation three of them are illustrated.

5.1 Gracilaria dura

G. dura from Indian waters has now a days attracted the industrial attention because of its cost effective and greener method of quality agar production.

Distribution: Members of the genus *Gracilaria* are found from tropical to temperate latitudes, with higher species richness in tropical regions. *G. dura* is widely distributed over the map such as Asia, Southeast Asia, South America, Atlantic island, and Africa. Detailed distribution with sources is illustrated in Table 2.1.

Polysaccharide: G. dura yields cost effective agarose and its content in this alga is as high as 20–25% of its dry weight. The gel strength of which is 1900 g cm^{-2} for 1% gel which is quite comparable to a commercial-grade agarose sample. Differences in isomorphic life stages of this alga is also reported on the agar content (Baghel et al. 2011), biochemical parameters (Sambhwani et al. 2020) as well as on

Table 2.1 Global distribution of three *Gracilaria* species

Gracilaria species	Distribution	Sources
G. dura	Atlantic Islands	Canary Island (Haroun et al. 2002, Afonso-Carrillo 2014), Lanzarote (Gil-Rodriguez and Afonso-Carrillo 1980)
	Europe	Adriatic Sea (Giaccone 1978), Atlantic France (Burel et al. 2019), Balearic Islands (Seoane-Camba 1975) Black Sea (Milchakova 2011), Britain (Bunker et al. 2017), Corsica (Coppejans 1979), Crimea (Sadogurskiy et al. 2019), France (Anon 2017), Greece (Athanasiadis 1987), Italy, Sardinia (Furnari et al. 2003), Spain (Gallardo et al. 2016)
	South America	Columbia (Pulido and Ruíz 2003)
	Africa	Eritrea (Lipkin and Silva 2002), Kenya (Oyieke and Gwande 2007), Mauritius, Tanzania (Silva et al. 1996), Morocco (Moussa et al. 2018)
	Middle- east	Egypt (Shabaka 2018), Israel (Einav and Isrrel 2008), Turkey (Taşkın et al. 2019)
	South-West Asia	Bangladesh (Ahmed et al. 2009), India (Rao and Gupta 2015), Lebanon (Lakkis 2013)
	Asia	China (Liu 2008), South China sea (Phang et al. 2016)
G. chilensis	South America	Chile (Bird et al. 1986, Kim et al. 2006, Guillemin et al. 2012), Temperate South America (Ramírez and Santelices 1991)
	Australia and New-Zealand	Chatham Islands, North Islands NZ, South Islands NZ (Nelson et al. 2013), New-Zealand (Byrne et al. 2002, Nelson and Dahlen 2014), South Australia, Rakiura Islands, Tasmania (Womersley 1996)
G. lameneiformis	North America	British Columbia, Gulf of California (Norris 1985), California (Abbott 1985)
	South America	Peru (Norris 1985), Venezuela (Ganesan 1990)
	South-East Asia	Philippines (Abbott 1994)
	Asia	China (Norris 1985, Liu 2008, Huan et al. 2013), Japan (Yoshida 1998)
	Pacific Islands	Hawaiian Islands (Abbott 1985)

its cultivation (Mantri et al. 2020). This has facilitated the segregation of phases for specific purposes as these phases are dis-proportionately abundant in nature.

Cultivation: To avoid the hindrance in the utilization of this economic resource by the industry due to its scarcity in nature, cultivation of this alga was studied and successfully reported by Mantri et al. (2009, 2019) and Veeragurunathan et al. (2015a, b). Further for feasibility in the cultivation, viable seedling production via clonal propagation method is also reported (Saminathan et al. 2015). To appertain, the cultivation of this alga using the tissue culture, protoplast fusion and development molecule markers for the identification of the life phase to process within its early stage have been adopted (Oza 1971; Kumar et al. 2007; Gupta et al. 2011).

5.2 Gracilaria chilensis

Since 1950 *Gracilaria chilensis* (now as *Agrophyton chilensis)* is one of the leading agarophyte in Chile. The generic name, *Agarophyton* is formed from Malayan and Greek words i.e., agar and *phyton,* thus the name means "agar-plant." The species, *A. chilensis* of this genus is currently the main sources of agar in the world (Zemke-White and Ohno 1999).

Distribution: The species is found over temperate regions such as at the coasts of South America, Australia and New-Zealand. The detailed distribution with sources is described in Table 2.1.

Polysaccharide: *G. chilensis* (Gracilariales, Rhodophyta) is an important economic macroalgae (Troell et al. 1997), due to high its agar yield and mineral content. It is considered high-quality raw material for extracting agar (Santelices and Doty 1989; Lobban and Harrison 1994). The extract from *G. chilensis* gave an overall yield of 31.1% of agar on its dry weight with high gel strength. This weed represents an excellent potential commercial resource of agar.

Cultivation: The species has been extensively cultivated in Chile. With the traditional method i.e., by the vegetative propagation of thalli, sporulation in polypropylene ropes is also successfully reported in the Raqui estuary of Chile. Considering the quality of *Gracilaria chilensis* as an efficient biofilter, Integrated Multi-Trophic Aquaculture (IMTA) systems is also studied for the species. The species in this system showed higher growth performance with mitigating the water environmental problems caused by several forms of fed aquaculture.

5.3 Gracilaria lemaneiformis

Due to high growth rate, fast adaptability and high agar yielding capacity *Gracilaria lemaneiformis* (as *Gracilariopsis lemaneiformis*) in China has now taken lead in the commercial mariculture.

Distribution: The species is found from tropical to warm-temperate oceans. It is recorded in coastal waters of North America, South America, South East Asia, Asia and Pacific islands. Detailed distribution with references is described in Table 2.1.

Polysaccharide: The average yield and gel strength of agar extracted from *G. lemaneiformis* are found to be 25.3 ± 1.0% of its dry weight and 1884 ± 105 g/cm^2 of 1.5% gel. An eco-friendly method of photobleaching extraction is also reported which results in quality enhancement by reducing the sulphate content and increasing the gel strength than by traditional alkali extraction method (Li et al. 2007). Recently a sustainable one step extraction process of agar is also reported for the species via bacteriolytic enzymes (Li et al. 2007).

Cultivation: This *G. lemaneiformis* has proved to be successor in industry-scale mariculture in China (Zhou et al. 2006). *G. lamaneiformis* cultivation has expanded from 50,536 tons to 196,778 tons of dry biomass in a decade using raft method at

Guangdong, Fujian, Shandong and other Chinese coastal waters thus leading this species as the largest contributor to Gracilaria production (Yang et al. 2015). The cultivation of *G. lemaneiformis* has proved a new approach for coastal environmental improvement also as it acts as biofilter and nutrient scrubber. This seaweed can also out compete with the harmful algal boom and decreases the phytoplankton densities, resulting in bioremediating mariculture system (Yang et al. 2015).

6 Conclusion and Future Prospects

Since outset of *Gracilaria*, the genus has become an important commercial commodity to the agar industries. Now a days, lamentably the accessibility of the species from its natural resource is not complementing to its demand. Considering the potential demand of these species in various industries such as pharmaceutical, food, hydrocolloid, and cosmetics, various government and non-government agencies are in contend since the last 3–4 years to promote agarophyte farming. The hydrocolloid from the farmed biomass has been recently validated by leading agar manufactures and they have shown interest in procuring the farmed biomass. In conclusion, the detailed insights of these economically important species of *Gracilaria* will lead to a better understanding of these species for viable benefits to the industries and farming management practices. Besides, the data will also be useful in bringing the other species of economic importance under the ambit of commercial farming and production of value-added compounds.

Acknowledgements Authors duly acknowledge the support and encouragement from the Director, CSIR-CSMCRI. KS duly acknowledge University Grant Commission (UGC-India) for the Senior Research Fellowship grant. We thank Dr. Marie-Laure Guillemin, Instituto de Ciencias Ambientales y Evolutivas, Chile for sharing the photo of cultivation of *Gracilaria chilensis* by bottom stocking method and Prof. Delin Duan, Chinese Academy of Sciences, China for sharing the photo of cultivation of *G. lemaneiformis* by monoline method. This manuscript has CSIR-CSMCRI PRIS registration number 52/2021.

References

Abbott IA (1985) New species of Gracilaria Grev. (Gracila riaceae, Rhodophyta) from California and Hawaii. pp. 115–121 in: Abbott, I.A. & Norris, J.N. (eds.), Taxonomy of Economic Seaweeds with Reference to Some Pacific and Caribbean Species, vol. 1. University of California, La Jolla

Abbott IA (1994) Taxonomy of Economic Seaweeds: With Reference to Some Pacific and Caribbean Species 4:111

Afonso-Carrillo J (2014) Lista actualizada de las algas marinas de las islas Canarias, 2014. Las Palmas: Elaborada para la Sociedad Española de Ficología (SEF): La Laguna, Tenerife, Canary Islands, Spain

Ahmed ZU, Khondker M, Begum ZNT, Hassan MA, Kabir SMH, Ahmad M (2009) Encyclopedia of Flora and Fauna of Bangladesh. Algae, Charophyta-Rhodophyta, (Achnanthaceae-Vaucheriaceae). Asiat Soc Bangladesh, Dhaka 4:543

Andriani Y, Syamsumir D, Yee T, Harisson F, Herng G, Oroscoa S, Orosco C, Ali AM, Latip J, Kikuzaki H, Mohamad H (2016) Biological activities of isolated compounds from three edible Malaysian red seaweeds, *Gracilaria changii*, G. manilaensis and *Gracilaria* sp. Nat Prod Commun 11:1117–1120

Anon (2017) Inventaire national du Patrimoine naturel. Paris: Muséum National d'Histoire Naturelle. Accessed September 2017

Araki C (1966) Some recent studies on the polysaccharides of agarophytes. In: Proceedings of the fifth international seaweed symposium, Halifax, August 25–28, 1965. Elsevier, Amsterdam, pp 3–17

Athanasiadis A (1987) A survey of the seaweeds of the Aegean Sea with taxonomic studies on species of the tribe Antithamnieae (Rhodophyta)

Baghel RS, Kumari P, Bijo AJ, Gupta V, Reddy CRK, Jha B (2011) Genetic analysis and marker assisted identification of life phases of red alga Gracilaria corticata (J. Agardh). Mol Biol Rep, 38(6):4211–4218

Bird CJ, McLachlan J, Oliveira ED (1986) Gracilaria chilensis sp. nov. (Rhodophyta, Gigartinales), from Pacific South America. Can J Bot 64(12):2928–2934

Bunker F, Brodie JA, Maggs CA, Bunker AR (2017) Seaweeds of Britain and Ireland. Princeton University Press

Burel T, Le Duff M, Gall EA (2019) Updated check-list of the seaweeds of the French coasts, Channel and Atlantic Ocean. An Aod-Les cahiers naturalistes de l'Observatoire marin, 7(1):1–38

Byrne K, Zuccarello GC, West J et al (2002) *Gracilaria* species (Gracilariaceae, Rhodophyta) from southeastern Australia, including a new species, *Gracilaria perplexa* sp. nov.: morphology, molecular relationships and agar content. Phycol Res 50:295–311. https://doi.org/10.1046/j.1440-1835.2002.00282.x

Chang CF (1963) *Polycavernosa*, a new genus of the Gracilariaceae. Stud Mar Sin 3:119–126

Chattopadhyay K, Ghosh T, Pujol C, Carlucci M, Damonte E, Ray B (2008) Polysaccharides from *Gracilaria corticata*: sulfation, chemical characterization and anti-HSV activities. Int J Biol Macromol 43:346–351

Chirapart A, Munkit J, Lewmanomont K (2006) Changes in yield and quality of agar from the Agarophytes, *Gracilaria fisheri* and *G. tenuistipitata* var. *liui* cultivated in earthen ponds. Nat Sci 40:529–540

Collantes G, Melo C (1995) Cultivo de tejidos y células de algas marinas. Alveal K. Ferrario ME, Oliveira EC, Sar E. (ed.), pp. 457–477

Coppejans E (1979) Végétation marine de la Corse (Méditerranée). III. Documents pour la flore des algues. Marine Vegetation of Corsica (Mediterranean Sea). III. Documents for the Algal Flora

Dave A, Huang Y, Rezvani S et al (2013) Techno-economic assessment of biofuel development by anaerobic digestion of European marine cold-water seaweeds. Bioresour Technol 135:120–127. https://doi.org/10.1016/j.biortech.2013.01.005

Dawson EY (1949) Studies of Northeast Pacific Gracilariaceae. Allan Hancock Found Publs Occ Pap 7:1–105

De Almeida CLF, de Falcão H, de Lima GR, de Montenegro C, Lira NS, de Athayde-Filho PF, Rodrigues LC, de Souza MFV, Barbosa-Filho JM, Batista LM (2011) Bioactivities from marine algae of the genus *Gracilaria*. Int J Mol Sci 12:4550–4573

de Sousa CL, Nicolau LA, Silva RO, Barros FC, Freitas AL, Aragao KS, Ribeiro Rde A, Souza MH, Barbosa AL, Medeiros JV (2013) Antiinflammatory and antinociceptive effects in mice of a sulfated polysaccharide fraction extracted from the marine red algae *Gracilaria caudata*. Immunopharmacol Immunotoxicol 35(1):93–100

Einav R, Israel A (2008) Checklist of seaweeds from the Israeli Mediterranean: Taxonomical and ecological approaches. Isr J Plant Sci 56(1–2):127–191

FAO (2018) The global status of seaweed production, trade and utilisation, vol 124. Goldfish Research Programme, Rome, p 120

Fasahati P, Woo HC, Liu JJ (2015) Industrial-scale bioethanol production from brown algae: effects of pre-treatment processes on plant economics. Appl Energy 139:175–187. https://doi.org/10.1016/j.apenergy.2014.11.032

Fernández LE, Valiente OG, Mainardi V, Bello JL (1989) Isolation and characterization of an antitumor active agar-type polysaccharide of *Gracilaria dominguensis*. Carbohydr Res 190(1):77–83

Fredericq S, Hommersand MH (1989a) Proposal of the Gracilariales *Ord. Nov.* (Rhodophyta) based on an analysis of the reproductive development of *Gracilaria verrucosa*. J Phycol 25(2):213–227

Fredericq S, Hommersand MH (1989b) Comparative morphology and taxonomic status of *Gracilariopsis* (Gracilariales, Rhodophyta). J Phycol 25:228–241. https://doi.org/10.1111/j.1529-8817.1989.tb00117.x

Furnari G, Giaccone G, Cormaci M, Alongi G, Serio D (2003) Biodiversità marina delle coste italiane: catalogo del macrofitobenthos. Biol Mar Mediterr 10(1):1–482

Gallardo B, Clavero M, Sánchez MI, Vilà M (2016) Global ecological impacts of invasive species in aquatic ecosystems. Glob Change Biol 22(1):151–163

Ganesan EK (1990) A catalog of benthic marine algae and seagrasses of Venezuela. pp. 1–237, 15 maps. Caracas: Fondo Editorial Conicit

Ganesan M, Eswaran K, Reddy CRK (2017) Farming of agarophytes in India – a long-time sustainability for the industry and preserving wild stocks. J Appl Phycol 29(5):2239–2248

Gargiulo GM, de Masi F, Tripodi G (1992) Morphology, reproduction and taxonomy of the Mediterranean species of *Gracilaria* (Gracilariales, Rhodophyta). Phycologia 3:53–80

Gargiulo GM, Morabito M, Genovese G, De Masi F (2006) Molecular systematics and phylogenetics of Gracilariacean species from the Mediterranean Sea. In: Eighteenth International Seaweed Symposium. Springer, Dordrecht, pp 271–278

Ghannadi A, Shabani L, Yegdaneh A (2016) Cytotoxic, antioxidant and phytochemical analysis of *Gracilaria* species from Persian gulf. Adv Biomed Res 5:139

Giaccone G (1978) Revisione della Flora Marina del Mare Adriatico (Revision of the marine flora of the Adriatic Sea). (WWF & Ann. Parco Mar. Miramare (Trieste) 6:1–118

Gil-Rodríguez MC, Afonso-Carrillo J (1980) Adiciones a la flora marina y catálogo ficológico para la isla de Lanzarote. Vieraea 10(1–2):59–70

Goff LJ, Moon DA, Coleman AW (1994) Molecular delineation of species and species relationships in the red algal agarophytes *Gracilariopsis* and *Gracilaria* (Gracilariales). J Phycol 30:521–537

Guillemin ML, Huanel OR, Martínez EA (2012) Characterization of genetic markers linked to sex determination in the haploid-diploid red alga gracilaria chilensis 1. J Phycol 48(2):365–372

Guiry MD, Guiry GM (2020) AlgaeBase. World-wide electronic publication, National University of Ireland, Galway. Available from: https://www.algaebase.org

Gupta V, Baghel RS, Kumar M, Kumari P, Mantri VA, Reddy CRK, Jha B (2011) Growth and agarose characteristics of isomorphic gametophyte (male and female) and sporophyte of *Gracilaria* dura and their marker assisted selection. Aquacult 318(3–4):389–396

Gurgel CFD, Fredericq S (2004) Systematics of the Gracilariaceae (Gracilariales, Rhodophyta): a critical assessment based on rbcL sequence analysis. J Phycol 40:154–159

Gurgel CFD, Fredericq S, Norris JN (2004) *Gracilaria apiculata* and *G. flabelliformis* (Gracilariaceae, Rhodophyta): restoring old names for common tropical western Atlantic species, including the recognition of three new subspecies, and a replacement name for *G. lacinulata*. Cryptogam Algol 40:154–159

Gurgel CFD, Fredericq S, Norris JN, Yoneshigue-Valentin Y (2008) Two new flat species of *Gracilaria* (Gracilariales, Rhodophyta) from Brazil: *G. abyssalis* sp. nov. and *G. brasiliensis* sp. nov. Phycologia 47(3):249–264

Gurgel CFD, Norris JN, Schmidt WE, Le HN, Fredericq S (2018) Systematics of the Gracilariales (Rhodophyta) including new subfamilies, tribes, subgenera, and two new genera, Agarophyton gen. nov. and Crassa gen. nov. Phytotaxa 374(1):1–23

Haroun RJ, Gil-Rodríguez MC, Castro JDD, Reine WPHV (2002) A checklist of the marine plants from the Canary Islands (central eastern Atlantic Ocean)

Hemasudha TS, Thiruchelvi R, Balashanmugam P (2019) Antioxidant, antibacterial, and anticancer activity from marine red algae *Gracilaria edulis*. Asian J Pharm Clin Res 12(2):276–279

Huan L, He L, Zhang B, Niu J, Lin A and Wang G (2013) AFLP and SCAR markers associated with the sex in *Gracilaria lemaneiformis* (Rhodophyta). J Phycol 49(4):728–732

Huang W, Fujita Y (1997) Callus induction and thallus regeneration in some species of red algae. Phycol Res 45:105–111. https://doi.org/10.1111/j.1440-1835.1997.tb00069.x

Hudson W (1762) Flora Anglica. J. Nourse, London, 506 pp

Hurtado AQ, Agbayani RF, Sanares R, De Castro-Mallare MTR (2001) The seasonality and economic feasibility of cultivating *Kappaphycus alvarezii* in Panagatan cays, Caluya, antique, Philippines. Aquaculture 199:295–310. https://doi.org/10.1016/S0044-8486(00)00553-6

Kim MS, Yang EC, Boo SM (2006) Taxonomy and phylogeny of flattened species of *Gracilaria* (Gracilariceae, Rhodophyta) from Korea based on morphology and protein-coding plastid rbc L and psb a sequences. Phycologia 45(5):520–528

Krastina J, Romagnoli F, Balina K (2017) SWOT analysis for a further LCCA-based techno-economic feasibility of a biogas system using seaweeds feedstock. In: Energy Procedia. Elsevier, Amsterdam, pp 491–496

Kumar V, Fotedar R (2009) Agar extraction process for *Gracilaria cliftonii*. Carbohydr Polym 78:813–819. https://doi.org/10.1016/j.carbpol.2009.07.001

Kumar GR, Reddy CRK, Jha B (2007) Callus induction and thallus regeneration from callus of phycocolloid yielding seaweeds from the Indian coast. J Appl Phycol 19(1):15–25

Lakkis S (2013) Flore et faune marines du Liban (Méditerranée orientale). Biologie, Biodiversité, Biogéographie 1:510

Li T, Wang C, Miao J (2007) Identification and quantification of indole-3-acetic acid in the kelp *Laminaria japonica* Areschoug and its effect on growth of marine microalgae. J Appl Phycol 19(5):479–484

Lin SM, Liu LC, Payri C (2012) Characterization of *Gracilaria vieillardii* (Gracilariaceae, Rhodophyta) and molecular phylogeny of foliose species from the western Pacific Ocean, including a description of *G. taiwanensis* sp. nov. Phycologia 51(4):421–431

Linnaeus C (1763) Species Plantarum, 2nd edn. Laurentii Salvii, Holmiae, pp 785–1684

Lipkin Y, Silva PC (2002) Marine algae and seagrasses of the Dahlak Archipelago, southern Red Sea. Nova Hedwigia 75(1/2):1–90

Liu Ruiyu [Liu, R.Y.] (Ed.) (2008) Checklist of biota of Chinese seas. pp. 1–1267. Beijing: Science Press, Academia Sinica. [in Chinese]

Lobban CS, Harrison PJ (1994) Seaweed ecology and physiology. Cambridge University Press

Lu H, Xie H, Gong Y, Wang Q, Yang Y (2011) Secondary metabolites from the seaweed *Gracilaria lemaneiformis* and their allelopathic effects on Skeletonema costatum. Biochem Syst Ecol 39(4–6):397–400

Lyra GDM, Gurgel CFD, Costa EDS, De Jesus PB, Caires TA, de Matos JCG, Oliveira MC, Oliveira EC, Nunes JDC (2015) A new tropical species of Gracilariaceae (Rhodophyta, Gracilariales): Gracilaria silviae sp. nov. Phytotaxa 222(3):199–210

Madhavan S (2015) A review on hydrocolloids-agar and alginate. J Pharm Sci 7:704–707

Mantri VA (2010) Studies on biology of *Gracilria dura* (C. Agardh) J. Agardh. PhD Thesis, Bhavnagar University, Bhavnagar, India. 119 pp

Mantri VA, Ganesan M, Gupta V, Krishnan P, Siddhanta AK (2019) An overview on agarophyte trade in India and need for policy interventions. J Appl Phycol 31:3011–3023

Mantri VA, Shah Y, Balar N et al (2021) Limited-scale field trial confirmed differences in growth and agarose characteristics in life-cycle stages of industrially important marine red alga *Gracilaria dura* (Gracilariales, Rhodophyta). J Appl Phycol 33:1059–1070. https://doi.org/10.1007/s10811-020-02356-1

Mantri VA, Shah Y, Thiruppathi S (2020) Feasibility of farming the agarose-yielding red alga *Gracilaria dura* using tube-net cultivation in the open sea along the Gujarat coast of NW India. J Appl Phycol 1:12–19

Mantri VA, Thakur MC, Kumar M, Reddy CRK, Jha B (2009) The carpospore culture of industrially important red alga *Gracilaria dura* (Gracilariales, Rhodophyta). Aquacult 297(1–4):85–90

Manuhara GJ, Praseptiangga D, Riyanto RA (2016) Extraction and characterization of refined K-carrageenan of red algae [*Kappaphycus Alvarezii* (Doty ex P.C. Silva, 1996)] originated from Karimun Jawa Islands. Aquat Procedia 7:106–111. https://doi.org/10.1016/j.aqpro.2016.07.014

Martins M, Vieira FA, Correia I et al (2016) Recovery of phycobiliproteins from the red macroalga: *Gracilaria* sp. using ionic liquid aqueous solutions. Green Chem 18:4287–4296. https://doi.org/10.1039/c6gc01059h

McLachlan J, Bird CJ (1983) Gracilaria in the seaweed market: a prospectus. In Symposium Internacional de Acuacultura, Coquimbo, Chile 133–157

Meena R, Siddhanta AK, Prasad K, Ramavat BK, Eswaran K, Thiruppathi S, Rao PS (2007) Preparation, characterization and benchmarking of agarose from *Gracilaria dura* of Indian waters. Carbohyd Polym 69(1):179–188

Milchakova NA (2011) Marine plants of the Black Sea. An Illustrated Field Guide

Moussa H, Hassoun M, Salhi G, Zbakh H, Riadi H (2018) Checklist of seaweeds of Al-Hoceima National Park of Morocco (Mediterranean marine protected area)

Nelson WA, Dalen J (2014) Taxonomic notes on the New Zealand macroalgal flora: synonymic checklist of the red algal order Gracilariales with selection of lectotypes and description of a new species. N Z J Bot 52(2):236–244

Nelson WA, Payri CE, Sutherland JE, Dalen J (2013) The genus *Melanthalia* (Gracilariales, Rhodophyta): new insights from New Caledonia and New Zealand. Phycologia 52:426–436. https://doi.org/10.2216/13-137.1

Norris JN (1985) Studies on Gracilaria Grev. (Gracilariaceae, Rhodophyta) from the Gulf of California, Mexico. Taxonomy of Economic Seaweeds. California Sea Grant College Program, California, I 123–135

Oyieke HA, Gwanda P (2007) The seaweeds of Kenya: checklist, history of seaweed study, coastal environment, and analysis of seaweed diversity and biogeography. S Afr J Bot 73:76–88

Oza RM (1971) Effect of IAA on the growth of fragment of *Gracilaria corticata*. J Ag Seaweed Res Util 1:48–49

Papenfuss GF (1950) Review of the genera of algae described by Stackhouse. Hydrobiologia 2(3):181–208

Phang SM, Yeong HY, Ganzon-Fortes ET, Lewmanomont K, Prathep A, Gerung GS, Tan KS (2016) Marine algae of the South China Sea bordered by Indonesia, Malaysia, Philippines, Singapore, Thailand and Vietnam. raffles Bulletin of Zoology

Porse H, Rudolph B (2017) The seaweed hydrocolloid industry: 2016 updates, requirements, and outlook. J Appl Phycol 29(5):2187–2200

Praiboon J, Chirapart A, Akakabe Y et al (2006) Physical and chemical characterization of agar polysaccharides extracted from the Thai and Japanese species of *Gracilaria*. Sci Asia 32:11–17. https://doi.org/10.2306/scienceasia1513-1874.2006.32(s1).011

Pulido GD, Ruíz MD (2003) Diversity of benthic marine algae of the Colombian Atlantic. Biota Colombiana 4(2):203–246

Ramírez ME, Santelices B (1991) Catálogo de algas bentónicas de las costas temperadas del Pacífico temperado de Sudamérica. Monografías Biológicas (Chile) 5:1–499

Rao PSN, Gupta RK (2015) Algae of India, A checklist of Indian Marine Algae (excluding Diatoms & Dinoflagellates). Botanical Survey of India, Kolkata 13:93

Rees DA (1969) Structure, conformation, and mechanism in the formation of polysaccharide gels and networks. Adv Carbohydr Chem Bi 24:267–332. https://doi.org/10.1016/S0065-2318(08)60352-2

Rees DA (1972) Shapely polysaccharides. The eighth Colworth medal lecture. Biochem J 126:257–273. https://doi.org/10.1042/bj1260257
Sadogurskiy SY, Belich TV, Sadogurskaya SA (2019) Macrophytes of the Marine Areas in the Nature Reserves of the Crimean Peninsula (the Black Sea and the Sea of Azov). Int J Algae 21(3)
Sambhwani K, Modi J, Singhala A, Bramhabatt H, Mishra A, Mantri VA (2020) Analysis of functional traits in female gametophytic and tetrasporophytic life-phases of industrially important red alga *Gracilaria dura* (Rhodophyta: Gracilariacae). J Appl Phycol 32(3):1962–1969. https://doi.org/10.1007/s10811-020-02116-1
Saminathan KR, Ashok KS, Veeragurunathan V, Mantri VA (2015) Seedling production in the industrially important agarophyte *Gracilaria dura* (Gracilariales, Rhodophyta). J Appl Phycol 27(4):1541–1548
Santelices B, Doty MS (1989) A review of Gracilaria farming. Aquacult 78(2):95–133
Saraswaty V, Mozefa T, Risdiana C, Rasyid A (2015) Bioactivity of polysaccharide from *Gracilaria verrucosa* as α-glucosidase inhibitor. Proc Chem 16:687–693
Schmitz FKJ, Falkenberg P, Hauptfleisch P, Engler A, Prantl KAE (1889) Systematische Übersicht der bisher bekannten Gattungen der Florideen. Universitats-Buchdruckerei, Bonn
Seoane-Camba J (1975) Algas bentonicas espanolas en los herbarios thuret-bornet y sauvageau del museum national d'histoire naturelle de paris. Ii. Algas de cataluna y baleares (excepto menorca)
Senthilkumar P, Sudha S (2012) Evaluation of alpha-amylase and alpha-glucosidase inhibitory properties of selected seaweeds from gulf of Mannar. Int Res J Pharm 3:128–130
Sfriso A, Wolf MA, Sciuto K, Morabito M, Andreoli C, Moro I (2013) *Gracilaria viridis* sp. nov. (Gracilariales, Rhodophyta): a new red algal species from the Mediterranean Sea. Phycologia 52(1):65–73
Shabaka SH (2018) Checklist of seaweeds and seagrasses of Egypt (Mediterranean Sea): A review. The Egypt J Aquat Res 44(3):203–212
Shukla MK, Kumar M, Prasad K et al (2011) Partial characterization of sulfohydrolase from *Gracilaria dura* and evaluation of its potential application in improvement of the agar quality. Carbohydr Polym 85:157–163. https://doi.org/10.1016/j.carbpol.2011.02.009
Silva PC, Basson PW, Moe RL (1996) Catalogue of the benthic marine algae of the Indian Ocean. Univ Calif Publ Bot 79:1–1259
Singh A, Nigam PS, Murphy JD (2011) Renewable fuels from algae: an answer to debatable land-based fuels. Bioresour Technol 102:10–16. https://doi.org/10.1016/j.biortech.2010.06.032
Sjöstedt LG (1926) Floridean studies, vol 22. CWK Gleerup, Aalborg
Soares LP, Gurgel CFD, Fujii MT (2018) *Gracilaria suzannae* sp. nov. (Gracilariales, Rhodophyta), a new flattened species from Northeast Brazil based on morphological and molecular evidence. Phycologia 57(3):345–353
Soleymani M, Rosentrater K (2017) Techno-economic analysis of biofuel production from macroalgae (seaweed). Bioengineering 4:92. https://doi.org/10.3390/bioengineering4040092
Sonani RR (2016) Recent advances in production, purification and applications of phycobiliproteins. World J Biol Chem 7:100. https://doi.org/10.4331/wjbc.v7.i1.100
Souza BWS, Cerqueira MA, Bourbon AI, Pinheiro AC, Martins JT, Teixeira JA, Coimbra MA, Vicente AA (2012) Chemical characterization and antioxidant activity of sulfated polysaccharide from the red seaweed *Gracilaria birdiae*. Food Hydrocoll 27:287–292
Steentoft M, Irvine LM, Bird CJ (1991) Proposal to conserve the type of *Gracilaria*, nom. Cons., as *G. compressa* and its lectotypification (Rhodophyta: Gracilariaceae). Taxon 40:663–666
Sudhakar MP, Jagatheesan A, Perumal K, Arunkumar K (2015) Methods of phycobiliprotein extraction from *Gracilaria crassa* and its applications in food colourants. Algal Res 8:115–120. https://doi.org/10.1016/j.algal.2015.01.011
Taşkın E, Çakır M, Akçalı B, Sungur Ö (2019) Benthic marine flora of the Marmara Sea (TURKEY). J Black Sea/Mediterr Environ 25(1):1–28

Torres P, Santos JP, Chow F, dos Santos DYAC (2019) A comprehensive review of traditional uses, bioactivity potential, and chemical diversity of the genus *Gracilaria* (Gracilariales, Rhodophyta). Algal Res 37:288–306

Troell M, Halling C, Nilsson A et al (1997) Integrated marine cultivation of *Gracilaria chilensis* (Gracilariales, Rhodophyta) and salmon cages for reduced environmental impact and increased economic output. Aquaculture 156:45–61. https://doi.org/10.1016/S0044-8486(97)00080-X

Valderrama D, Cai J, Hishamunda N (2015) The economics of Kappaphycus seaweed cultivation in developing countries: a comparative analysis of farming systems. Aquacult Econ Manag 19:251–277. https://doi.org/10.1080/13657305.2015.1024348

Veeragurunathan V, Eswaran K, Malarvizhi J, Gobalakrishnan M (2015a) Cultivation of *Gracilaria dura* in the open sea along the southeast coast of India. J Appl Phycol 27(6):2353–2365

Veeragurunathan V, Eswaran K, Saminathan KR, Mantri VA, Ajay G, Jha B (2015b) Feasibility of *Gracilaria dura* cultivation in the open sea on the south-eastern coast of India. Aquaculture 438:68–74

Veeragurunathan V, Prasad K, Malarvizhi J, Singh N, Meena R, Mantri VA (2019) *Gracilaria debilis* cultivation, agar characterization and economics: bringing new species in the ambit of commercial farming in India. J Appl Phycol 31:2609–2621

Womersley HBS (1996) The marine benthic flora of Southern Australia, Part IIIB. Canberra, ABRS. 392 p.

Yamamoto H (1978) Systematic and anatomical study of the genus Gracilaria in japan. 25(2):97–152. eprints.lib.hokudai.ac.jp

Yang Y, Liu Q, Chai Z, Tang Y (2015) Inhibition of marine coastal bloom-forming phytoplankton by commercially cultivated *Gracilaria lemaneiformis* (Rhodophyta). J Appl Phycol 27:2341–2352. https://doi.org/10.1007/s10811-014-0486-0

Yeh CC, Yang JI, Lee JC, Tseng CN, Chan YC, Hseu YC, Tang JY, Chuang LY, Huang HW, Chang FR (2012) Anti-proliferative effect of methanolic extract of *Gracilaria tenuistipitata* on oral cancer cells involves apoptosis, DNA damage, and oxidative stress. BMC Complement Altern Med 12(1):142

Yeong HY, Phang SM, Reddy CRK, Khalid N (2014) Production of clonal planting materials from *Gracilaria changii* and *Kappaphycus alvarezii* through tissue culture and culture of G. changii explants in airlift photobioreactors. J Appl Phycol 26:729–746. https://doi.org/10.1007/s10811-013-0122-4

Yoshida T (1998) Marine Algae of Japan. Uchida-roukakuho, Tokyo. pp. 1222

Zandi K, Tajbakhsh S, Nabipour I, Rastian Z, Yousefi F, Sharafian S, Sartavi K (2010) In vitro antitumor activity of *Gracilaria corticata* (a red alga) against Jurkat and molt-4 human cancer cell lines. Afr J Biotechnol 9(40):6787–6790

Zemke-White WL, Ohno M (1999) World seaweed utilisation: an end-of-century summary. J Appl Phycol 11:369–376. https://doi.org/10.1023/A:1008197610793

Zhang YH, Song XN, Lin Y (2019) Antioxidant capacity and prebiotic effects of *Gracilaria neoagaro* oligosaccharides prepared by agarase hydrolysis. Int J Biol Macromol 137:177–186. https://doi.org/10.1016/j.ijbiomac.2019.06.207

Zhou Y, Yang H, Hu H, Liu Y, Mao Y, Zhou H, Xu X, Zhang F (2006) Bioremediation potential of the macroalga *Gracilaria lemaneiformis* (Rhodophyta) integrated into fed fish culture in coastal waters of north China. Aquacult 252(2–4):264–276

Zuniga-Jara S, Contreras C (2020) An economic valuation of the commercial cultivation of Agarophyton chilensis in northern Chile. J Appl Phycol 32:3233–3242

Chapter 3
Seaweeds as Functional Food: A Comprehensive Review of Its Antioxidants and Therapeutic Merits Against Oxidative Stress-Mediated Chronic Diseases

Neeru Bhatt, Lyutha Al-Subhi, and Mostafa Waly

Abbreviations

AA	Arachidonic acid
C/EBPα	CCAAT-enhancer-binding proteins
CAT	Catalase
CVD	Cardiovascular diseases
DHA	Docosahexaenoic acid
EPA	Eicosapentaenoic acid
FXN	Fucoxanthin or FXN
HDL	High-density lipoprotein
LDL	Low-density lipoprotein
NK	Natural killer
PPARγ	Peroxisome proliferator-activated receptor gamma
PUFAs	Polyunsaturated fatty acids
RNS	Reactive nitrogen species
ROS	Reactive oxygen species
SOD	Superoxide dismutase
T2DM	Type 2 Diabetes mellitus
TC	Triglycerides
UCP1	Uncoupling protein 1
WHO	World Health Organization

N. Bhatt
Global Science Heritage, Toronto, ON, Canada

L. Al-Subhi · M. Waly (✉)
Department of Food Science and Nutrition, College of Agricultural and Marine Sciences, Sultan Qaboos University, Al-Khoud, Muscat, Oman

A. Ranga Rao, G. A. Ravishankar (eds.), *Sustainable Global Resources of Seaweeds Volume 2*, https://doi.org/10.1007/978-3-030-92174-3_3

1 Introduction

1.1 Basic Information About Seaweed

Phycology is an important branch of science dedicated to the study of algae. More than 30,000 species of algae are known for their one or more benefits. The microalgae are classed as phytoplankton, are the first trophic in the food chain, serving as nutrients to thousands of marine species. They are responsible for 30%–50% of the oxygen in the atmosphere (Sudhakar et al. 2018). Macroalgae are a varied group, their sizes ranging from a few centimeters to 100 m in length (Shama et al. 2019). Approximately 15,000 species of this group have been described. Since they are autotrophs, their habitat is limited to a certain depth, up to 60 m, always within the intertidal zone. Their growth is usually vertical, looking for sunlight (Ross et al. 2008). Unlike terrestrial plants they lack conductive tissues, rather they adsorb nutrients through their surface. They do not have roots, though some rhizoids or basal discs are present that allow them to adhere to rocks as a method of restraint, but not to nourish themselves. They form large underwater meadows and provide an ecosystem in which many different species of bacteria, corals, mollusks, fish, and other marine creatures coexist (Navarro-Barranco et al. 2018). The macroalgae are divided in to three main groups according to the pigments they contain; Chlorophyta (green algae), Rhodophyta (red algae), and Phaeophyta (brown algae) (Kuda et al. 2002).

2 Principal Bioactive Components of Seaweed

Seaweeds are known to have a wide variety of bioactive compounds with potential health benefits (Shama et al. 2019; Mohammad et al. 2019; Hessami et al. 2019). Some are exclusively present in seaweeds only (Rengasamy et al. 2020). These bioactive components have been exploited for their commercial applications (Kelman et al. 2012). Basic chemical composition and bioactive compounds in seaweeds are presented in Tables 3.1 and 3.2.

2.1 Phlorotannins

Phlorotannin (1,3,5-trihydroxy benzene), is a group of polyphenolic compounds which are commonly present in the brown algae (Fig. 3.1). Phlorotannins possess multiple physiological activities, with anti-carcinogenic, antibacterial, antiviral, anticancer, and anti-inflammatory properties (Brown et al. 2014). The antioxidant activity of phlorotannins is due to phenol rings, which act as electron traps to scavenge ROS. Phlorotannin is known to have an inhibitory effect on human salivary

Table 3.1 Chemical composition of seaweeds

Bioactive components	Concertation (%) On dry weight basis	References
Crude protein	10–30	FAO/WHO/UNU (1985), Wong and Cheung (2000) and Matanjun et al. (2009)
Lipids	1–6	Dawczynski et al. (2007) and Kumari et al. (2010)
Ash	8–40	Mabeau and Fleurence (1993)
Dietary fibre	33–50	Lahaye (1991) and Rupérez and Saura-Calixto (2001)
Polysaccharides	4–76	Paniagua-Michel et al. (2014)
Phlorotannin	0.5–20	Caro et al. (2012)
Phycoerythrin	Up to 30	Fleurence (1999)

Table 3.2 Bioactive components distributed in different seaweeds

Particulars	Red algae	Brown algae	Green algae	References
Carotenoids	β-Carotene, α-carotene, xanthophyll	Fucoxanthin, β-carotene, violaxanthin	β-Carotene, xanthophyll	MacArtain et al. (2007) and Burtin (2003)
Polysaccharides	Agars, arrageenans, xylans, floridean starch, galactan, porphyran	Alginic acid, fucoidan, laminarin (β-1, 3 glucan) sargassan	Sulphuric acid polysaccharides, sulphated galactans, xylans	Kumar et al. (2008) and Murata and Nakazoe (2001)
Phlorotannin		Fuhalols and phlorethols, fucols, fucophlorethols		Afonso et al. (2019)
Phycoerythrin	C-phycocyanin, R-phycocyanin, allophycocyanin or phycoerythrocyanin			Torres et al. (2019)

alpha-amylase, which may be useful as a natural nutraceutical to prevent diabetes (Kawamura-Konishi et al. 2012).

2.2 *Phycoerythrin*

Phycoerythrin (Fig. 3.2) is one of phycobiliproteins which has high economic value. Phycoerythrin are reported to involved in a number of biological activities, such as antiviral, antioxidant, anti-inflammatory, antidiabetic, antitumor, antihypertensive, immunosuppressive, and neuroprotective (Bungau et al. 2019; Caleja et al. 2017). It has high potential to be developed as natural dyes replacing synthetic dyes.

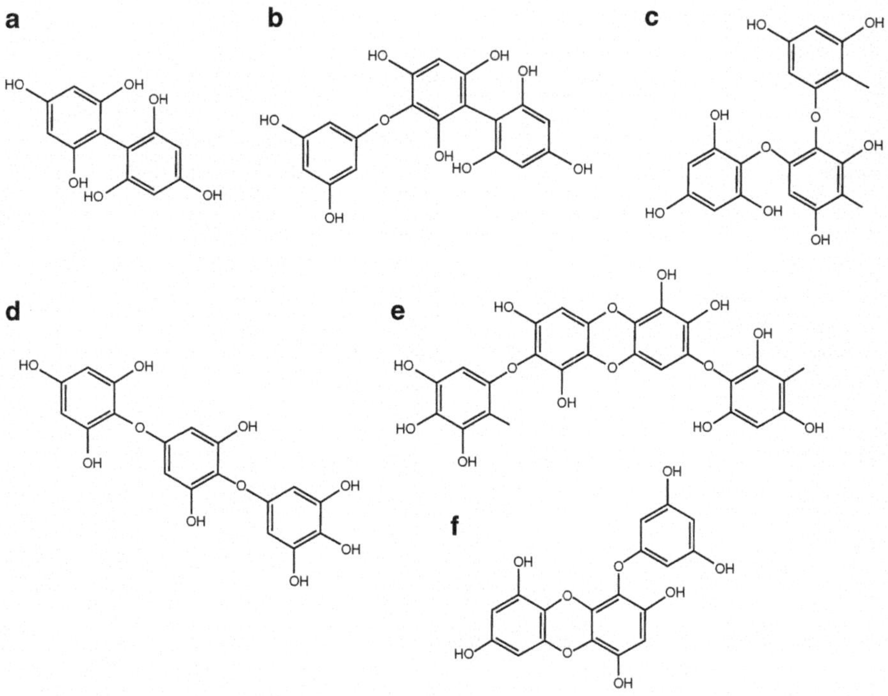

Fig. 3.1 Chemical structure of phlorotannins from brown seaweeds: (**a**) Fucol (**b**) Fucophlorethol (**c**) Phlorethol (**d**) Fuhalol (**e**) Carmalol; (**f**) Eckol

Fig. 3.2 Chemical structure of Phycoerythrin

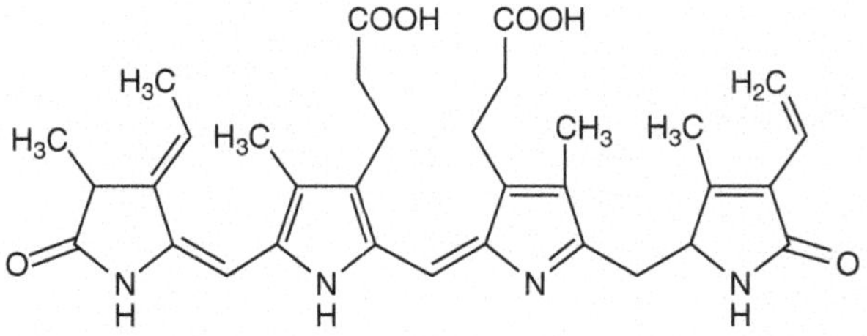

2.3 *Dietary Fiber*

Major soluble fibers include alginate from brown macroalgae, carrageenan and agar from red macroalgae, which overall can represent up to half of seaweed's dried weight (DW). Dietary fiber has been valued for their ability to reduce body weight by prolonging the gastric emptying rate, thus enhancing satiety leading to a reduction in food intake (Kristensen and Jensen 2011).

2.4 Sulfated Polysaccharides

Fucoidans are sulfated polysaccharides, which occur abundantly in seaweeds, especially in *Fucus vesiculosus.* They are composed of L-fucose and sulfate ester groups, and ulvans mainly composed of glucuronic acid and iduronic acid units together with rhamnose and xylose sulfates) (Cardoso et al. 2014). They possess antioxidants activity by either directly scavenging ROS, or induction of the activity of cellular endogenous antioxidant defenses, including SOD, CAT, glutathione transferase, and glucose-6-phosphate dehydrogenase (Rocha de Souza et al. 2007). In general, seaweed consumption has been related to a lower incidence of chronic diseases, such as diabetes, dyslipidemia, and coronary heart disease (Iso and Kubota 2007).

2.5 Carotenoids

Seaweed carotenoids are wonderful natural antioxidants which have been positively linked with the prevention of several chronic diseases such as cardiovascular, neurodegenerative diseases, age-related macular degeneration, obesity, and cancer (Meyers et al. 2014; Mikami et al. 2017). They are also known for their effects for longer life expectancy and a lesser risk of metabolic diseases in humans (Seca and Pinto 2018). The antioxidant activity of seaweed carotenoids is valuable against cellular inflammation, oxidative stress (Abdali et al. 2015; Aldini et al. 2011), radical scavenging ability and the quenching of singlet oxygen species (Stahl and Sies 2012) and coupled with other processes to have positive impacts on human health (Rodriguez-Concepcion et al. 2018). A regular dietary supply of carotenoids is needed as humans and other animals cannot synthesis carotenoids their own.

Additionally, the peptides, fatty acids and minerals, all together make seaweeds a perfect food supplement and therapeutic agent (Lordan et al., 2011). Algal population is also a good source of the long-chain omega-3 (n-3) polyunsaturated fatty acids, EPA and DHA (Cohen et al. 1995; Manerba et al. 2010) and can be used as alternative of fish oil.

The different bioactive components present in seaweeds positively influence physiological and metabolic functions of body in order to keep the human life safe and prevent pathogenesis. Inclusion of seaweed in the diet is a good option and should be encouraged. We will discuss here the preventive role of seaweeds against diseases cropped up due to oxidative stress.

3 Effect of Seaweeds on Chronic Diseases

3.1 Obesity

Obesity is a much common problem in present era and has emerged like epidemic in developed countries. Obesity concomitantly increase the chances of developing type 2 diabetes, hypertension, and dyslipidemia, and moderately increases the risk of developing osteoarthritis and CHD (Haslam and James 2005). Over the past decade, a new perspective on the biological function of seaweed carotenoids and their potential applications in the treatment of obesity and obesity-related diseases has emerged (Ojulari et al. 2019; Marseglia et al. 2015).

Seaweed carotenoid fucoxanthin (Fig. 3.3) has the capacity to lower down oxidative-inflammatory status related with obesity and is being applied in the treatment of the various diseases triggered by obesity (Beppu et al. 2012; Hosokawa et al. 2010). FXN is present in the chloroplasts of brown seaweed such as *Hizikia fusiforme, Fucus serratus, Laminaria, Alaria crassifolia, japMiyatonica, Fucus vesiculosus, Sargassum horneri,* and *Undaria pinnatifida* (Peng et al. 2011; Kumar and Brown 2013).The therapeutic property of FXN is attributed to its unique structure an allenic bond, which accounts for its distinctive mechanism of anti-adiposity action (Kim et al. 2013) mostly regulated on the nuclear receptor PPARγ and C/EBPα (Lee et al. 2019; Wang 2010), playing a very important role in the differentiation and function of mature adipocytes (Eeckhout et al. 2012). The activation of these nuclear receptors in adipocytes has been reported to enhance insulin resistance associated with obesity (Wang 2010; Schupp and Lazar 2010). PPARγ, when expressed in appreciable amounts in fat tissues, significantly induce adipogenesis (Eeckhout et al. 2012; Lee et al. 2019). Both PPARγ and C/EBPα bind most induced genes linked to adipogenesis and metabolism (Siersbæk et al. 2010). implying a coactive upregulation of adipogenic gene expression by these two key regulators (Yoshida et al. 2010).

Alginate, has been explored for its potential in weight management. In a pilot study, significantly lower energy intake in 12 healthy overweight and obese men was observed after consuming alginate rich bread (4% *Ascophyllum nodosum*) for a

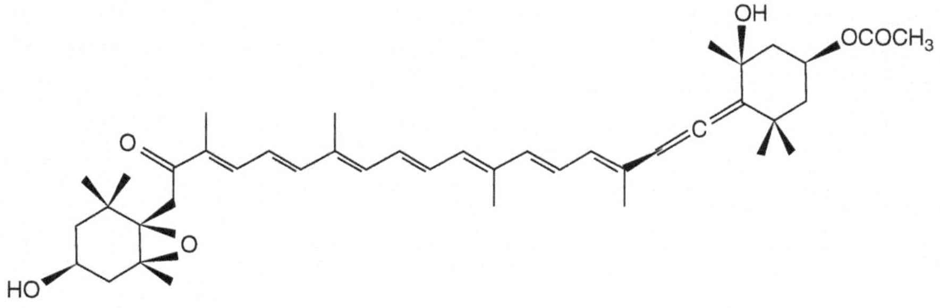

Fig. 3.3 Chemical structure of fucoxanthin

single breakfast meal was observed (Hall et al. 2012) probably due to the gastric stretching effects of alginate (1.15 g per serving). The alginate is responsible for colonic fermentation, resulted in the production of short-chain fatty acids, especially propionate. Propionate can alter cholesterol metabolism, (Wolever et al. 1995), glucose absorption and consequently weight loss (Paxman et al. 2008). An alginate-based drink was consumed by healthy obese, overweight, and normal-weight individuals over a period of 4-weeks. The results were comparable and significantly reduced energy intake was observed in healthy obese, overweight individuals but not in the normal-weight group. In a small pilot study alginate-based drinks affected both satiety and energy intake in 20 normal-weight subjects but had no effect on weight loss in 24 obese individuals following a calorie-restricted diet (Georg Jensen et al. 2012). The alginate may be used as an adjuvant to a calorie-controlled diet to facilitate weight loss. Fucoxanthin isolated from brown seaweed can reduce both body weight and the percentage of abdominal white adipose tissue (Georg Jensen et al. 2012).

3.2 Diabetes Mellitus

The mechanisms of anti-diabetic action of seaweed have been attributed to phlorotannins, fucoxanthin, polyphenolics, and polysaccharides (Murray et al. 2018). Cases of diabetes are continuously gearing up in developing and under developed countries (Selassie and Sinha 2011). Prolonged obesity and overweight often leads to T2DM along with impaired glucose tolerance, hypertension, and hyperlipidemia (Roberts 2010). Healthy diet and active life would counter the progression of the diseases (Ogurtsova et al. 2017; Galassi et al. 2006; Knowler et al. 2002). Inclusion of seaweed in the diet can positively influence risk of T2DM and its progression (Kim et al. 2008) may be due to the presence of numerous bioactive substances like polysaccharides, proteins, lipids, polyphenols, and pigments in seaweeds (Holdt and Kraan 2011). Different bioactive components of seaweed play different role, for instance alginate has shown positive effects on glucose metabolism (Torsdottir et al. 1991).

Supplementation of agar in a conventional diet along with routine exercise for 12 weeks significantly reduced body weight, body mass index and total cholesterol levels as compared with diet and exercise alone in a study of 76 individuals with impaired glucose tolerance and diabetes (Maeda et al. 2005). Another sulfated polysaccharide fucoidan, has shown a decreased effect of alpha-glucosidase activity in laboratory conditions (Lakshmana et al. 2019), blood glucose levels in db/db mice (Kim et al. 2012), while glycated hemoglobin and glucagon-like peptide-1 in type 2 diabetes patients (Sakai et al. 2019).

The dieckol inhibited alpha-glucosidase activity *in vitro* (Lee et al. 2010), and have shown protective effect against glucotoxicity-induced oxidative stress associated with diabetes (Kang et al. 2013; Lee et al. 2012), resulting a delay in T2DM development in a db/db rats (Lee et al. 2009). Dieckol was also reported to reduce postprandial hyperglycemia in prediabetic individuals (Lee and Jeon 2015). The antidiabetic effect of polyphenols isolated from brown algae was observed on alpha-glucosidase activity (Apostolidis and Lee 2010; Zhang et al. 2007). Quercetin flavonoid exerted positive effect in type 2 diabetic db/db mice (Hertog et al. 1993) also inhibited tumor necrosis factor alpha-induced insulin-resistance in skeletal muscle cells of mice (Jeong et al. 2012).

Fucoxanthin has the capacity to enhance insulin resistance, lower blood glucose levels, and cytokine production in adipose tissue. Numerous studies have reported that fucoxanthin induces UCP1 in abdominal white adipose tissue mitochondria, which lets oxidation of fatty acids and energy production (Maeda et al. 2005; Hosokawa et al. 2010; Nishikawa et al. 2012; Mikami et al. 2017). Besides, fucoxanthin also helpful in suppressing glycated hemoglobin in healthy subjects and the suppression was obvious in subjects with a certain UCP1 genotype that has been shown to be a predisposing factor for obesity (Mikami et al. 2017).

Ingestion of seaweed influences glycemic control, lowers blood lipids, and increases antioxidant enzyme activities, hence, can reduce risk factors for cardiovascular disease in patients with type 2 diabetes (Kim et al. 2008). In a recent study single ingestion of brown seaweed extract positively regulated insulin levels and sensitivity after a carbohydrate-rich meal but failed to alter postprandial glucose levels in healthy adults (Paradis et al. 2011).

3.3 Cancer

Cancer is a serious public health concern and its pathogenesis is very complex. According to WHO, in 2018, about 9.6 million cancer-related deaths and 18 million new cases were recorded (Global Cancer Observatory 2020). Natural products have been continuously investigated in order to achieve the successful cure of cancer. However, the work on marine plants is still in its infancy.

Various mechanisms have been proposed of anticancer activity of seaweeds. Fucoxanthin and other microalgae inhibits tumor growth by inducing G1 cell-cycle and restricting apoptosis (Satomi 2017). Additionally, fucoxanthin supports in modulating expression of various cellular molecules and cellular signal transduction pathways. The antitumor effect of sulfated polysaccharide isolated from *Champia feldmannii* against sarcoma 180 in mice was observed, indicating the immune stimulating properties of polysaccharides (Lins et al. 2008).

Another sulfated polysaccharide obtained from *Gracilaria lemaneiformis* showed notable anti-cancerous and immunomodulatory activities against transplanted H22 hepatoma cells in ICR mice. The anti-cancer activity of fucoidans has been investigated in several kinds of cancers for instance lung and gastric cancer (Han et al. 2008), breast cancer (Yamasaki-Miyamoto et al. 2009), and liver cancer HepG2 cell (Yan et al. 2010). Intravenous injection of radachlorin, derived from *S. platensis* suppressed partial or full tumor growth (Privalov et al. 2002). Furthermore, hot-water extract of *S. platensis* enhanced antitumor activity of natural killer (NK) cells in rats (Akao et al. 2009). In a recent study complex polysaccharide derived from Spirulina suppressed glioma cell growth by downregulating angiogenesis via partial regulation of interleukin-17 production (Kawanishi et al. 2013).

Red seaweeds of the genera *Hypnea Bryodies* and *Melanothamnus Somalensis* are rich in polyphenolic compounds that may suppress cancer through antioxidant properties, and recent research has been carried out documenting their anti-carcinogenic and antioxidant influence against oxidative stress in an experimental model of colon cancer. Both red seaweed extracts abolished the oxidative stress-associated colon carcinogenesis as evidenced by increasing the activity of antioxidant enzymes: catalase, glutathione peroxidase, glutathione S-transferase, glutathione reductase and superoxide dismutase (Waly et al. 2016).

It has been observed that cancer incidence is much lower among Asian populations who include seaweed in their diet, in comparison a Western-style diet (Ferlay et al. 2008). A direct relationship between high levels of seaweed consumption and lower incidences of cancer has been identified (Iso 2011). In a case-control study on 362 Korean premenopausal women with histologically confirmed breast cancer showed an inverse relation between seaweed consumption and breast cancer risk (Yang et al. 2010).

3.4 Cardiovascular Diseases

Among all the major health issues present in the modern world, cardiovascular diseases remain the main cause of death accounting for roughly 20% of all annual death worldwide (Gaziano 2005). CVDs include coronary heart disease, cerebrovascular disease, peripheral arterial disease, rheumatic heart disease, congenital heart disease and deep vein thrombosis and pulmonary embolism. Coronary heart disease and stroke are expected to remain the leading causes of death in the coming years. According to the WHO, about 23.6 million people will die from CVDs in 2030 (Alissa and Ferns 2011).

Seaweed consumption is inversely linked with CVDs. A collaborative cohort study investigating the effect of nutrition on mortality in Japan has associated seaweed consumption with lower incidences of CVD (Iso and Kubota 2007). An inverse association between seaweed intake and cardiovascular mortality among Japanese men and women, especially that from cerebral infarction was observed in a very recent cohort study (Kishida et al. 2020). The studies about the potential cardiovascular health properties of seaweed-are mostly conducted either *in vitro* or animal models, yet limited human data exists to substantiate the proposed anti-inflammatory, antiangiogenic, anticoagulant, and antiadhesive properties of seaweed components (Cumashi et al. 2007).

Tropical seaweed can reduce cholesterol, low-density lipoprotein, triglycerides, lipid peroxidation, and erythrocyte glutathione peroxidase (Matanjun et al. 2010). Carrageenan polysaccharide derived from *Chondrus crispus* may inhibit platelet aggregation and have anticoagulant properties (Liu et al. 2013). Fucoxanthin was found to reduce blood pressure and stroke risk in rats (Ikeda et al. 2003). Fucoidan extracted from brown seaweeds are valued for their anticoagulant properties (Berteau and Mulloy 2003) and may be useful in the prevention of heart disease. *Spirulina platensis* has shown antiatherogenic properties and reported to lower down cholesterol level in rabbits (Cheong et al. 2010), rats, (Nagaoka et al. 2005), and hamsters, (Riss et al. 2007) and to inhibit oxidative stress and apoptosis in cardiomyocytes (Khan et al. 2006). Supplementation of *Spirulina platensis* in patients with ischemic heart disease and atherogenic dyslipidemia showed antiatherosclerosis effects (Ionov and Basova 2003). The increased concentrations of DHA and EPA in serum and platelets and a lower ratio of LDL to HDL was reported by Conquer et al. in a double-blind, placebo-controlled study in 24 vegetarian subjects supplemented with 1.62 g/day DHA from algal-source (Conquer and Holub 1996). The concentrations of DHA and AA were increased significantly by the feeding of oils from *Sargassum horneri* and *Cystoseira hakodatensis* (Airanthi et al. 2011) in KK-Ay mouse due to the higher fucoxanthin content in the seaweeds. The significantly increased levels of TC, HDL-C and phospholipid in the test rats while those of hepatic cholesterol and triacylglycerol were decreased as compared with the control group. Fucoxanthin exerts its effects on cholesterol metabolism and in the transport system by down-regulation of the LDL receptor and the class B type 1 scavenger receptor, along with inducing sterol regulatory element binding protein expression (Beppu et al. 2012).

4 Conclusion

Damage to living cells by free radicals is linked to many chronic disorders. Although the endogenous defense mechanisms of humans can combat oxidation, an imbalance still exists if the diet is low in antioxidants. Seaweeds can appreciably combat such situations, as they are an excellent source of bioactive compounds, including complex polysaccharides, polyphenols, carotenoids, fatty acids, minerals and vitamins. Hence, inclusion of seaweeds in the diet is a good option and should be encouraged.

References

Abdali D, Samson SE et al (2015) How effective are antioxidant supplements in obesity and diabetes? Med Princ Pract 24:201–215

Afonso NC, Catarino MD et al (2019) Brown macroalgae as valuable food ingredients. Antioxidants (Basel) 8(9):365. https://doi.org/10.3390/antiox8090365

Airanthi MK, Sasaki N, Iwasaki S, Baba N, Abe M, Hosokawa M, Miyashita K (2011). Effect of brown seaweed lipids on fatty acid composition and lipid hydroperoxide levels of mouse liver.

J Agric Food Chem. 59(8):4156–4163. https://doi.org/10.1021/jf104643b. Epub 2011 Mar 30. PMID: 21405010

Akao Y, Ebihara T et al (2009) Enhancement of antitumor natural killer cell activation by orally administered *Spirulina* extract in mice. Cancer Sci 100(8):1494–1501

Aldini G, Yeum KJ et al (2011) Biomarkers for antioxidant defense and oxidative damage: principles and practical applications. Wiley, Hoboken, NJ

Alissa EM, Ferns GA (2011) Heavy metal poisoning and cardiovascular disease. J Toxicol 2011:870125. https://doi.org/10.1155/2011/870125

Apostolidis E, Lee CM (2010) In vitro potential of *Ascophyllum nodosum* phenolic antioxidant-mediated alpha-glucosidase and alpha-amylase inhibition. J Food Sci 75:97–102

Beppu F, Hosokawa M et al (2012) Effects of dietary fucoxanthin on cholesterol metabolism in diabetic/obese KK-A^y mice. Lipids Health Dis 11:112

Berteau O, Mulloy B (2003) Sulfated fucans, fresh perspectives: structures, functions, and biological properties of sulfated fucans and an overview of enzymes active toward this class of polysaccharide. Glycobiology 13:29R–40R

Brown EM, Allsopp PJ et al (2014) Seaweed and human health. Nutr Rev 72:205–216

Bungau S, Abdel-Daim MM et al (2019) Health benefits of polyphenols and carotenoids in age-related eye diseases. Cell Longev 12:9783429. https://doi.org/10.1155/2019/9783429

Burtin P (2003) Nutritional value of seaweeds. Electron J Environ Agric Food Chem 2(4):498–503

Caleja C, Barros L et al (2017) A comparative study between natural and synthetic antioxidants. Evaluation of their performance after incorporation into biscuits. Food Chem 216:342–346

Cardoso SM, Carvalho LG et al (2014) Bioproducts from seaweeds: a review with special focus on the Iberian Peninsula. Curr Org Chem 18:896–917

Caro Y, Anamal L et al (2012) Natural hydroxyanthraquinoid pigments as potent food grade colorant. Nat Prod Bioprosp 2(5):174–193

Cheong SH, Kim MY et al (2010) *Spirulina* prevents atherosclerosis by reducing hypercholesterolemia in rabbits fed a high-cholesterol diet. J Nutr Sci Vitaminol (Tokyo) 56:34–40

Cohen MC, McKenna C et al (1995) Requirements for controlled clinical trials of preoperative cardiovascular risk reduction. Control Clin Trials 16:89–95

Conquer JA, Holub BJ (1996) Supplementation with an algae source of docosahexaenoic acid increases (n-3) fatty acid status and alters selected risk factors for heart disease in vegetarian subjects. J Nutr 126:3032–3039

Cumashi A, Ushakova NA et al (2007) A comparative study of the anti-inflammatory, anticoagulant, antiangiogenic, and antiadhesive activities of nine different fucoidans from brown seaweeds. Glycobiology 17:541–552

Dawczynski C, Schubert R et al (2007) Amino acids, fatty acids, and dietary fibre in edible seaweed products. Food Chem 103:891–899

Eeckhout J, Oger F et al (2012) Coordinated regulation of PPARγ expression and activity through control of chromatin structure in adipogenesis and obesity. PPAR Res 2012:164140. https://doi.org/10.1155/2012/164140

FAO/WHO/UNU (1985) Energy and protein requirements. In: Report of a Joint FAO/WHO/UNU Expert Consultation. World Health Organization (WHO) Technical Report Series No. 724. WHO, Geneva

Ferlay J, Parkin DM et al (2008) Estimates of cancer incidence and mortality in Europe. Eur J Cancer 46:765–781

Fleurence J (1999) Seaweed proteins: biochemical, nutritional aspects and potential uses. Trends Food Sci Technol 103:25–28

Galassi A, Reynolds K et al (2006) Metabolic syndrome and risk of cardiovascular disease: a meta-analysis. Am J Med 119:812–819

Gaziano TA (2005) Cardiovascular disease in the developing world and its cost-effective management. Circulation 112(23):3547–3553

Georg Jensen M, Kristensen M et al (2012) Effect of alginate supplementation on weight loss in obese subjects completing a 12-wk energy-restricted diet: a randomized controlled trial. Am J Clin Nutr 96(1):5–13

Global Cancer Observatory. http://gco.iarc.fr/. Accessed 23 Oct 2020

Hall AC, Fairclough AC et al (2012) *Ascophyllum nodosum* enriched bread reduces subsequent energy intake with no effect on post-prandial glucose and cholesterol in healthy, overweight males. A pilot study. Appetite 58:379–386

Han JG, Syed AQ et al (2008) Antioxidant, immunomodulatory and anticancer activity of fucoidan isolated from *Fucus vesiculosus*. J Biotechnol 136:S571

Haslam DW, James WP (2005) Obesity. Lancet 366(9492):1197–1209

Hertog MG, Feskens EJ et al (1993) Dietary antioxidant flavonoids and risk of coronary heart disease: the Zutphen elderly study. Lancet 342:1007–1011

Hessami MJ, Cheng SF, Ranga Rao A, Yin YH, Phang SM (2019) Bioethanol production from agarophyte red seaweed, *Gelidium elegans* using a novel sample preparation method for analysing bioethanol content by gas chromatography. 3 Biotech 9(1):25

Holdt SL, Kraan S (2011) Bioactive compounds in seaweed: functional food applications and legislation. J Appl Phycol 23:543–597

Hosokawa M, Miyashita T et al (2010) Fucoxanthin regulates adipocytokine mRNA expression in white adipose tissue of diabetic/obese KK-ay mice. Arch Biochem Biophys 504:17–25

Ikeda K, Kitamura A et al (2003) Effect of *Undaria pinnatifida* (Wakame) on the development of cerebrovascular diseases in stroke-prone spontaneously hypertensive rats. Clin Exp Pharmacol Physiol 30:44–48

Ionov VA, Basova MM (2003) Use of blue-green micro-seaweed Spirulina platensis for the correction of lipid and hemostatic disturbances in patients with ischemic heart disease [in Russian]. Vopr Pitan 72:28–31

Iso H (2011) Lifestyle and cardiovascular disease in Japan. J Atheroscler Thromb 18:83–88

Iso H, Kubota Y (2007) Nutrition and disease in the Japan collaborative cohort study for evaluation of cancer (JACC). Asian Pac J Cancer Prev 8(Suppl):35–80. PMID 18260705

Jeong SM, Kang MJ et al (2012) Quercetin ameliorates hyperglycemia and dyslipidemia and improves antioxidant status in type 2 diabetic db/db mice. Nutr Res Pract 6:201–207

Kang MC, Wijesinghe WA et al (2013) Dieckol isolated from brown seaweed Ecklonia cava attenuates type II diabetes in db/db mouse model. Food Chem Toxicol 53:294–298

Kawamura-Konishi Y, Watanabe N et al (2012) Isolation of a new phlorotannin, a potent inhibitor of carbohydrate-hydrolyzing enzymes, from the brown alga Sargassum patens. J Agric Food Chem 60:5565–5570

Kawanishi Y, Tominaga A et al (2013) Regulatory effects of *Spirulina* complex polysaccharides on growth of murine RSV-M glioma cells through Toll-like receptor-4. Microbiol Immunol 57(1):63–73

Kelman D, Posner EK et al (2012) Antioxidant activity of Hawaiian marine algae. Drugs 10:403–416

Khan M, Varadharaj S et al (2006) C-phycocyanin ameliorates doxorubicin-induced oxidative stress and apoptosis in adult rat cardiomyocytes. J Cardiovasc Pharmacol 47:9–20

Kim MS, Kim JY et al (2008) Effects of seaweed supplementation on blood glucose concentration, lipid profile, and antioxidant enzyme activities in patients with type 2 diabetes mellitus. Nutr Res Pract 2:62–67

Kim KJ, Yoon KY, Lee BY (2012) Fucoidan regulate blood glucose homeostasis in C57BL/KSJ m+/+db and C57BL/KSJ db/db mice. Fitoterapia. 83(6):1105–1109. https://doi.org/10.1016/j.fitote.2012.04.027. Epub 2012 May 3. PMID: 22580164

Kim KN, Ahn G et al (2013) Inhibition of tumor growth in vitro and in vivo by fucoxanthin against melanoma B16F10 cells. Environ Toxicol Pharmacol 35:39–46

Kishida R, Yamagishi K et al (2020) Frequency of seaweed intake and its association with cardiovascular disease mortality: the JACC study. Atheroscler Thromb 27(12):1340–1347

Knowler WC, Barrett-Connor E et al (2002) Diabetes prevention program research group. Reduction in the incidence of type 2 diabetes with lifestyle intervention or metformin. N Engl J Med 346:393–403

Kristensen M, Jensen MG (2011) Dietary fibres in the regulation of appetite and food intake. Importance of viscosity. Appetite 56(1):65–70

Kuda T, Taniguchi E et al (2002) Fate of water-soluble polysaccharides in dried *Chorda filum* a brown alga during water washing. J Food Compos Anal 15:3–9

Kumar SA, Brown L (2013) Seaweeds as potential therapeutic interventions for the metabolic syndrome. Rev Endocr Metab Disord 14:299–308

Kumar CS, Ganesan P et al (2008) In vitro antioxidant activities of three selected brown seaweeds of India. Food Chem 107(2):707–713

Kumari P, Kumar M et al (2010) Tropical marine macroalgae as potential sources of nutritionally important PUFAs. Food Chem 120:749–757

Lahaye M (1991) Marine algae as sources of fibers: determination of soluble and insoluble dietary fiber contents in some, sea vegetables. J Sci Food Agric 54:587–594

Lakshmana S, Chandrasekaran R et al (2019) *In vitro* and *in silico* inhibition properties of fucoidan against α-amylase and α-D-glucosidase with relevance to type 2 diabetes mellitus. Carbohydr Polym 209:350–355

Lee SH, Jeon YJ (2015) Efficacy and safety of a dieckol-rich extract (AG-dieckol) of brown algae, *Ecklonia cava*, in pre-diabetic individuals: a double-blind, randomized, placebo-controlled clinical trial. Food Funct 6(3):853–858. https://doi.org/10.1039/c4fo00940a

Lee SH, Li Y et al (2009) α-Glucosidase and α-amylase inhibitory activities of phloroglucinol derivatives from edible marine brown alga, *Ecklonia cava*. J Sci Food Agric 89:1552–1558

Lee HJ, Kim HC et al (2010) Algae consumption and risk of type 2 diabetes: Korean national health and nutrition examination survey. J Nutr Sci Vitaminol 56:13–18

Lee SH, Park MH et al (2012) Dieckol isolated from *Ecklonia cava* protects against high-glucose induced damage to rat insulinoma cells by reducing oxidative stress and apoptosis. Biosci Biotechnol Biochem 76:1445–1451

Lee JE, Schmidt H et al (2019) Transcriptional and epigenomic regulation of adipogenesis. Mol Cell Biol 39:e00601–e00618

Lins KOAL, Bezerr DP et al (2008) Antitumor properties of a sulfated polysaccharide from the red seaweed *Champia feldmannii* (Diaz-Pifferer). J Appl Toxicol 29(1):20–26

Liu J, Hafting J et al (2013) Components of the cultivated red seaweed *Chondrus crispus* enhance the immune response of Caenorhabditis elegans to Pseudomonas aeruginosa through the pmk-1, daf-2/daf-16, and skn-1 pathways. Appl Environ Microbiol 79:7343–7350

Lordan S, Ross RP, Stanton C. Marine (2011) bioactives as functional food ingredients: potential to reduce the incidence of chronic diseases. Mar Drugs. 9(6):1056–100. https://doi.org/10.3390/md9061056. Epub 2011 Jun 14. PMID: 21747748; PMCID: PMC3131561

Mabeau S, Fleurence J (1993) Seaweed in food products: biochemical and nutritional aspects. Trends Food Sci Technol 4:103–107

MacArtain P, Gill CI et al (2007) Nutritional value of edible seaweeds. Nutr Rev 65(12):535–543

Maeda H, Hosokawa M et al (2005) Fucoxanthin from edible seaweed, *Undaria pinnatifida,* shows antiobesity effect through UCP1 expression in white adipose tissues. Biochem Biophys Res Commun 332:392–397

Manerba A, Vizzardi E et al (2010) N-3 PUFAs and cardiovascular disease prevention. Future Cardiol 6:343–350

Marseglia L, Mant S et al (2015) Oxidative stress in obesity: a critical component in human diseases. Int J Mol Sci 16:378–400

Matanjun P, Mohamed S et al (2009) Nutrient content of tropical edible seaweeds, *Eucheuma cottonii, Caulerpa lentillifera* and *Sargassum polycystum*. J Appl Phycol 21:75–80

Matanjun P, Mohamed S et al (2010) Comparison of cardiovascular protective effects of tropical seaweeds, *Kappaphycus alvarezii*, *Caulerpa lentillifera*, and *Sargassum polycystum*, on high-cholesterol/high-fat diet in rats. J Med Food 13:792–800

Meyers KJ, Mares JA et al (2014) Genetic evidence for role of carotenoids in age-related macular degeneration in the carotenoids in age-related eye disease study (CAREDS). Invest Ophthalmol Vis Sci 55:587–599

Mikami N, Hosokawa M et al (2017) Reduction of HbA1c levels by fucoxanthin-enriched akamoku oil possibly involves the thrifty allele of uncoupling protein 1 (UCP1): a randomized controlled trial in normal-weight and obese Japanese adults. J Nutr Sci 6:1–9

Mohammad JH, Ranga Rao A, Ravishankar GA (2019) Opportunities and challenges in seaweeds as feed stock for biofuel production. In: Ravishankar GA, Rao AR (eds) Handbook of algal technologies and phytochemicals. Phycoremediation, biofuels and global biomass production, vol 2. CRC, Boca Raton, FL, pp 39–50

Murata M, Nakazoe J (2001) Production and use of marine algae in Japan. Jpn Agric Res Q 35(4):281–290

Murray M, Dordevic A et al (2018) The impact of a single dose of a polyphenol-rich seaweed extract on postprandial glycaemic control in healthy adults: a randomized cross-over trial. Nutrients 10:270

Nagaoka S, Shimizu K et al (2005) A novel protein C-phycocyanin plays a crucial role in the hypocholesterolemic action of Spirulina platensis concentrate in rats. J Nutr 135:2425–2430

Navarro-Barranco C, Florido M et al (2018) Impoverished mobile epifaunal assemblages associated with the invasive macroalga *Asparagopsis taxiformis* in the Mediterranean Sea. Mar Environ Res 141:44–52

Nishikawa S, Hosokawa M et al (2012) Fucoxanthin promotes translocation and induction of glucose transporter 4 in skeletal muscles of diabetic/obese KK-A(y) mice. Phytomedicine 19:389–394

Ogurtsova K, da Rocha Fernandes JD et al (2017) IDF diabetes atlas: global estimates for the prevalence of diabetes for 2015 and 2040. Diabetes Res Clin Pract 128:40–50

Ojulari OV, Lee SG et al (2019) Beneficial effects of natural bioactive compounds from *Hibiscus sabdariffa* L. on obesity. Molecules 24:210

Paniagua-Michel JDJ, Olmos-Soto J et al (2014) Algal and microbial exopolysaccharides: new insights as biosurfactants and bioemulsifiers. In: Kim S-K (ed) Advances in food and nutrition research. Academic, San Diego, pp 221–257

Paradis ME, Couture P et al (2011) A randomised crossover placebo-controlled trial investigating the effect of brown seaweed (*Ascophyllum nodosum* and *Fucus vesiculosus*) on postchallenge plasma glucose and insulin levels in men and women. Appl Physiol Nutr Metab 36:913–919

Paxman JR, Richardson JC et al (2008) Daily ingestion of alginate reduces energy intake in free living subjects. Appetite 51:713–719

Peng J, Yuan J et al (2011) Fucoxanthin, a marine carotenoid present in seaweeds and diatoms: metabolism and bioactivities relevant to human health. Mar Drugs 9:1806–1828

Privalov VA, Lappa AV et al (2002) Clinical trials of a new chlorin photosensitizer for photodynamic therapy of malignant tumors. In: Dougherty TJ (ed) Optical methods for tumor treatment and detection: mechanisms and techniques in photodynamic therapy XI. 4612 of Proceedings of SPIE, pp 178–189

Rengasamy KR, Mahomoodally MF et al (2020) Bioactive compounds in seaweeds: an overview of their biological properties and safety. Food Chem Toxicol 135:111013

Riss J, Decorde K et al (2007) Phycobiliprotein C-phycocyanin from *Spirulina platensis* is powerfully responsible for reducing oxidative stress and NADPH oxidase expression induced by an atherogenic diet in hamsters. J Agric Food Chem 55:7962–7967

Roberts AW (2010) Cardiovascular risk and prevention in diabetes mellitus. Clin Med 10:495–499

Rocha de Souza MC, Marques CT et al (2007) Antioxidant activities of sulfated polysaccharides from brown and red seaweeds. J Appl Phycol 19:153–160

Rodriguez-Concepcion M, Avalos J et al (2018) A global perspective on carotenoids: metabolism, biotechnology, and benefits for nutrition and health. Prog Lipid Res 70:62–93

Ross AB, Jone JM et al (2008) Classification of macroalgae as fuel and its thermochemical behaviour. Bioresour Technol 99:6494–6504

Rupérez P, Saura-Calixto F (2001) Dietary fibre and physicochemical properties of edible Spanish seaweeds. Eur Food Res Technol 212:349–354
Sakai C, Abe S et al (2019) A randomized placebo-controlled trial of an Oral preparation of high molecular weight Fucoidan in patients with type 2 diabetes with evaluation of taste sensitivity. Yonago Acta Med 62:14–23
Satomi Y (2017) Antitumor and cancer-preventative function of fucoxanthin: a marine carotenoid. Anticancer Res 37:1557–1562
Schupp M, Lazar MA (2010) Endogenous ligands for nuclear receptors: digging deeper. J Biol Chem 285:40409–40415
Seca AM, Pinto DC (2018) Overview on the antihypertensive and anti-obesity effects of secondary metabolites from seaweeds. Mar Drugs 16:237
Selassie M, Sinha AC (2011) The epidemiology and ateology of obesity: a global challenge. Best Pract Res Clin Anaesthesiol 25:1–9
Shama A, Joyce SG, Mari FD, Ranga Rao A, Ravishankar GA, Hudaa N (2019) Macroalgae and microalgae: novel sources of functional food and feed. In: Ravishankar GA, Rao AR (eds) Handbook of algal technologies and phytochemicals. Food, health and nutraceutical applications, vol 1. CRC, Boca Raton, FL, pp 207–219
Siersbaek R, Nielsen R, Mandrup S (2010) PPARgamma in adipocyte differentiation and metabolism--novel insights from genome-wide studies. FEBS Lett. 584(15):3242–3249. https://doi.org/10.1016/j.febslet.2010.06.010. Epub 2010 Jun 11. PMID: 20542036
Stahl W, Sies H (2012) Photoprotection by dietary carotenoids: concept, mechanisms, evidence and future development. Mol Nutr Food Res 56:287–295
Sudhakar K, Mamat R et al (2018) An overview of marine macroalgae as bioresource. Renew Sustain Energy Rev 91:165–179
Torres MD, Florez-Feranadez N et al (2019) Integral utilization of red seaweed for bioactive production. Mar Drugs 17(6):314
Torsdottir I, Alpsten M et al (1991) A small dose of soluble alginate-fiber affects postprandial glycemia and gastric emptying in humans with diabetes. J Nutr 121:795–799
Waly MI, Al Alawi AA et al (2016) Red seaweed (*Hypnea Bryodies* and *Melanothamnus somalensis*) extracts counteracting Azoxymethane-induced hepatotoxicity in rats. Asian Pac J Cancer Prev 17(12):5071–5074
Wang YX (2010) PPARs: diverse regulators in energy metabolism and metabolic diseases. Cell Res 20:124–137
Wolever TM, Spadafora PJ et al (1995) Propionate inhibits incorporation of colonic [1,2-^{13}C] acetate into plasma lipids in humans. Am J Clin Nutr 61:1241–1247
Wong K, Cheung PC (2000) Nutritional evaluation of some subtropical red and green seaweeds. Part 1-proximate composition, amino acid profiles and some physico-chemical properties. Food Chem 7:475–482
Yamasaki-Miyamoto Y, Yamasaki M et al (2009) Fucoidan induces apoptosis through activation of Caspase-8 on human breast cancer MCF-7 cells. J Agric Food Chem 57:8677–8682
Yan MD, Li HY et al (2010) The anti-tumor activity of brown seaweed oligo-fucoidan via lncRNA expression modulation in HepG2 cells. Cytotechnology 71:363–374
Yang YJ, Nam SJ et al (2010) A case-control study on seaweed consumption and the risk of breast cancer. Br J Nutr 103:1345–1353
Yoshida H, Yanai H et al (2010) Administration of natural astaxanthin increases serum HDL-cholesterol and adiponectin in subjects with mild hyperlipidemia. Atherosclerosis 209:520–523
Zhang J, Tiller C et al (2007) Antidiabetic properties of polysaccharide- and polyphenolic-enriched fractions from the brown seaweed *Ascophyllum nodosum*. Can J Physiol Pharmacol 85:1116–1123

Chapter 4
Laminariaceae: Its Use in Food and Health Implications

Olesya S. Malyarenko, Roza V. Usoltseva, and Svetlana P. Ermakova

1 Introduction

Currently, there is an increase in the number of diseases of a non-infectious nature that are directly related to nutrition (diseases of the cardiovascular system, diabetes, obesity, osteoporosis, some malignant neoplasms, *etc.*), which represent a serious medical, social and economic problems. In this regard the search for new sources for obtaining food products of dietary therapeutics and dietary preventive nutrition, which have a functional effect on the metabolic and or biochemical functions of the body, is an effective solution to this problem nowadays. These criteria are fully met by algae-based food products. Brown algae are known to be a valuable source of phlorotannins and polysaccharides. Numerous studies have shown that their use in the diet contributes to the prevention of oncological, cardiovascular, and nervous systems diseases, strengthening the immune status due to the normalization of intestinal microbiocenosis.

The brown algae of family Laminariaceae are widespread and of nutritional and medicinal interest. Their algal fields (kelps) are a significant floristic component of the coastal area in the lower eulittoral and sublittoral zones of the temperate and polar seas. Representatives of the genera *Laminaria* and *Saccharina* are the best-studied macrophytes. Some of them form commercial reserves and are easy to cultivate; this is the key to organizing a raw material base for the production of useful substances.

Polysaccharides such as alginic acids, laminarans, and fucoidans are the main component of algal biomass; they are very diverse and vary significantly in their structural characteristics. Many of them have a pronounced biological effect, as

O. S. Malyarenko (✉) · R. V. Usoltseva · S. P. Ermakova
G.B. Elyakov Pacific Institute of Bioorganic Chemistry, Far Eastern Branch of the Russian Academy of Sciences, Vladivostok, Russian Federation

A. Ranga Rao, G. A. Ravishankar (eds.), *Sustainable Global Resources of Seaweeds Volume 2*, https://doi.org/10.1007/978-3-030-92174-3_4

well as unique physicochemical properties, which are closely interconnected with their structure (Fitton et al. 2015; Sanjeewa et al. 2017).

Phlorotannins are another class of biologically active compounds of brown algae. It is known that, in addition to antioxidant properties, polyphenols of brown algae have a whole range of biological activities, including hepatoprotective, proapoptotic, thrombolytic, fungicidal, and radioprotective activities. Their obvious promise as medicinal compounds is also of great interest (Thomas and Kim 2011; Pangestuti et al. 2018).

2 Composition and Structure of Biologically Active Compounds from Brown Algae of Family *Laminariaceae*

2.1 Polysaccharides

Alginic acids are found in all types of brown algae and are the main structural components of the cell wall and intercellular substance. They are most well-known of the polysaccharides of brown algae and have been successfully used for many years in the medical, pharmaceutical, food and cosmetic industries. By their chemical nature, these polysaccharides consist of 1,4-linked residues of β-D-mannuronic and α-L-guluronic acids. Alginic acids from different sources can differ in the ratio of mannuronic and guluronic acid residues in the polysaccharide molecule and in the distribution of monomeric units along the polymer chain (Usov 1999).

Laminarans are common water-soluble polysaccharides of brown algae, where they are reserve substances. These polysaccharides are built from residues of β-D-glucose, linked 1,3- and 1,6-glycosidic bonds. They differ in the ratio and manner of including 1,3- and 1,6-bonds in the main chain (Usoltseva et al. 2020). Structure of laminarans from *Saccharina cichorioides*, *Saccharina gurjanovae* and *Saccharina japonica* were studied. It was shown that these polysaccharides had most for laminarans structures and contains the main chain from 1,3-linked residues of β-D-glucose with the branches of single glucopyranose residues at the C6 (Zvyagintseva et al. 2003; Malyarenko et al. 2017; Shevchenko et al. 2007). The laminaran fraction with simplest structure was isolated from *S. gurjanovae.* It was practically linear 1,3-β-D-glucan with content of 1,6-linked glucose residues no more than 1–2% (Shevchenko et al. 2007). The ratio of bonds 1,3:1,6 in laminarans from *S. cichorioides* and *S. japonica* approximately equal 9:1 and 6:1, respectively (Malyarenko et al. 2017).

Fucoidans are found in the cell walls and extracellular matrices of brown algae. They are sulfated fucose-containing homo- and heteropolysaccharides. Residues of α-L-fucose are indispensable and often main component of the molecules of most of these polysaccharides. Fucoidans often contain other monosaccharides (galactose, mannose, xylose, ramnose, glucose, and uronic acids) and acetyl groups (Kusaykin et al. 2008; Usov and Bilan 2009). Data about fucoidan composition of some brown algae of family Laminariaceae are presented in Table 4.1.

Table 4.1 Composition of fucoidans from some brown algae of family *Laminariaceae*

Season and place of harvesting of algae		Yield, % [a]	SO_3Na^-	Monosaccharide composition of fucoidan
Laminaria digitata (Zhurishkina et al. 2015)				
Commercial preparation		–	25 [b]	Fuc/Gal/Man/Xyl/UA [c] 57.7/6.5/17.1/2.3/5.6
Laminaria hyperborea (Kopplin et al. 2018)				
Commercial preparation		–	Total sugar:SO_3Na^- [d] 1:1.7	Fuc/Gal [c] 97.8/2.2
Laminaria longipes (Usoltseva et al. 2019)				
Sea of Okhotsk, Russia July		0.35	32 [b]	Fuc [c] 100
Saccharina angustata (Teruya et al. 2010)				
Okinawa, Japan		1.4	Fuc:SO_3Na^- [d] 1:0.87	Fuc:Gal:Xyl:Glc:UA [d] 1:0.54:0.08:0.08:0.64
Saccharina cichorioides				
East Sea, Korea (Yoon et al. 2007)	F1 F2 F3	n.d.	Total sugar:SO_3Na^- [d] 1:0.09 1:2.19 1:1.38	Fuc:Gal:Man:Xyl:Glc [d] 0.12:0.54:0.28:0:0.06 0.6:0.36:0:0.03:0.01 0.15:0.22:0:0.01:0.63
Sea of Japan, Russia August (2 years old) May (1.5 years old) (Zvyagintseva et al. 2003)	F1 F2	7.2 6.5 3.5	Fuc:SO_3Na^- [d] 1:2.0 1:1.18 n.d.	Fuc/Gal/Man/Xyl/Glc [c] 72/8/8/7/5 100/0/0/0/0/ 85/7/5/0/3
Sea of Okhotsk, Russia July (Usoltseva et al. 2019)		4.1	36 [b]	Fuc/Gal [c] 98/2
Saccharina gurjanovae				
Sea of Okhotsk, Russia July (Shevchenko et al. 2007)	F1 F2 F3	1.8 1.3 0.4	Traces 36.2 [b] 38 [b]	Fuc/Gal/Man/Xyl [c] /ManA [b] 31.8/9.2/40.2/5.8/7.5 48.2/45.5/1.5/3.1/0 50.0/40.6/2.0/0/0
Sea of Okhotsk, Russia August (Shevchenko et al. 2015)	F1 F2	1.7 2.5	9.5 [b] 23.8 [b]	Fuc/Gal/Man/Xyl [c] 64.1/27.4/5.7/2.8 75.9/21.2/1.8/1.1
Saccharina japonica				
Sea of Japan, Russia 1-year cycle 2-year cycle August (Zvyagintseva et al. 2003)	F1 F2 F1 F2	1.1 2.4 1.3 2.7	Fuc:SO_3Na^- [d] n.d. 1:1.1 n.d. n.d.	Fuc/Gal/Man/Xyl/Rha/Glc [c] 54/25/6/7/3/3 82/12/0/1/2/1 82/12/0/1/2/1 86/8/0/1/0/0

(continued)

Table 4.1 (continued)

Season and place of harvesting of algae		Yield, % [a]	SO_3Na^-	Monosaccharide composition of fucoidan
Sea of Japan, Russia June (Vishchuk et al. 2011)	F1 F2	0.8 3.0	10 [b] 23 [b]	Fuc/Gal/Man/Xyl/Rha [c] 53/29/15/1.3/1.7 57/31/1/2/3
Rongcheng, Shandong, China March (Chen et al. 2017)	F1 F2	n.d.	13.4 [b] 29.6 [b]	Fuc/Gal/Man/Xyl/Glc/Rha [c]/UA [b] 46.4/20.5/8.7/3.6/6.5/3.2/2.8 56.2/12.4/9.8/2.3/3.2/7.7/1.9
Rongcheng County, Shandong, China March (Ke et al. 2020)	F1 F2	n.d.	7.5 [b] 21.5 [b]	Fuc/Gal/Man/Glc/Rha [c] 81.4/3.9/7.0/0.3/7.4 58.2/35.4/0.2/2.9/3.3
Rongcheng, Shandong, China May (Xue et al. 2001)		n.d.	28.3 [b]	Fuc/Gal/Man/Xyl/Rha/Glc/Ara [c]/UA [b] 52.8/27.4/6.1/1.9/3.3/3.4/3.0/18.4
Shazikou, Qingdao, China August (Wang et al. 2010a)	F1 F2 F3	n.d.	32.8 [b] 41.8 [b] 42 [b]	Fuc/Gal/Man/Rha/Glc/Ara [c]/UA [b] 42.7/31.9/5.7/4.2/12.4/3.2/ 6.0 81.1/15.3/1.4/0.2/1.4/n.d./0 76.4/17.6/1.9/0.7/2.8/n.d./0
Shazikou, Qingdao, China March (Wang et al. 2008)	F1 F2 F3	n.d.	23.3 [b] 36.4 [b] 36.7 [b]	Fuc/Gal/Man/Xyl/Rha/Glc [c] /UA [b] 20.4/43.3/11.9/4.6/7.8/11.9/ 11.8 76.6/20.5/2.9/n.d./n.d./n.d./1.6 21.5/78.5/n.d./n.d./n.d./n.d./1.2
Quangang, Fujian Province, China September (Ye et al. 2020)	F1 F2 F3 F4	0.14 0.28 0.44 0.56	12 [b] 15 [b] 23.8 [b] 37 [b]	Fuc/Gal/Man/Rha/Xyl [c] 39.1/32.8/24.2/1.6/2.3 35.3/16.6/17.7/1.8/28.6 39.4/22.8/11.0/2.0/24.8 24.2/41.5/17.9/n.d./16.4
Xiapu, Fujian, China (Ni et al. 2020)	F1 F2 F3 F4	0.8 3.2 1.2 2.2	7.3 [b] 7 [b] 7.7 [b] 30.7 [b]	Fuc/Gal/Man/Xyl/Glc [c] 35.5/ 27.2/13.1/10.2/12.5 45.4/ 9.6/19.8/12.2/10.2 38.8/ 23.4/13.4/7.6/13.9 79.5/16.8/1.8/1.1/n.d.
Saccharina latissima (Bilan et al. 2010)				
Ullapool, Scotland March	F1 F2 F3	0.13 0.4 1.08	3.5 [b] 16.2 [b] 36.8 [b]	Fuc/Gal/Man/Xyl/Glc/UA [b] 9.4/1.9/6.7/5.6/2.3/23.3 15.2/10.6/4.5/1.8/0.9/23.3 35.6/8.1/0.7/0.9/0.2/2.5
Saccharina longicruris (Rioux et al. 2010)				
Quebec, Canada May August November June		n.d.	17.6 [b] 20 [b] 19.1 [b] 13.1 [b]	Fuc/Gal/Man/Xyl/Glc/GalA/GlcA [b] 12.7/16.7/4.2/2.4/1.0/n.d./3.5 12.9/36.8/2.9/1.6/1.2/n.d./2.3 14.4/33.1/2.7/2.1/1.4/ 1.0/3.2 9.4/25.0/5.6/3.1/1/ 0.8/5.4
Saccharina sculpera (Ren et al. 2019)				

(continued)

Table 4.1 (continued)

Season and place of harvesting of algae		Yield, % [a]	SO_3Na^-	Monosaccharide composition of fucoidan
Lvshun Sea, Dalian, China June	F1	n.d.	27.1 [b]	Fuc/Gal/Man/Xyl/Glc/Rha/ GlcA [b]
	F2		20.6 [b]	16.7/31.9/6.4/2.2/2.5/1.5/6.8
	F3		34.7 [b]	23.6/11.8/11.2/5.8/0.8/1.5/12.0
				50.5/27.0/4.4/2.4/0.6/1.6/4.2

n.d. not determined, *Fuc* fucose, *Gal* galactose, *Man* mannose, *Xyl* xylose, *Glc* glucose, *Rha* rhamnose, *Ara* arabinose, *GalA* galacturonic acid, *ManA* mannuronic acid, *GlcA* glucuronic acid, *UA* uronic acid

[a]% of dry defatted algae weight

[b] Molar ratios

[c] mol. %

[d]% of sample weight

The detailed structures of some listed fucoidans were successfully established (Fig. 4.1).

So, commercial fucoidan from *L. hyperborea* contained main chain of 1,3-linked α-L-fucose residues with branches at C2 and C4 in form of both single fucose residues and short chains. Sulfate groups occupied 85% of all possible positions (Kopplin et al. 2018).

Unusual fucoidan from *L. longipes* was predominantly linear regular α-L-fucan, contained the repeating structural units →3)-α-L-Fuc*p*-(2SO_3^-)-(1 → 4)-α-L-Fuc*p*-(1 → 2)-α-L-Fuc*p*-(4SO_3^-)-(1 → with small amounts of disaccharide 1,4-linked fragments and 3-sulfated fucose residues (Usoltseva et al. 2019).

Sulfated polysaccharide from *S. cichorioides* was almost pure fucan, containing the main chain of 1,3-linked α-L-fucopyranose residues with a small degree of 1,4-linked α-L-fucopyranose residues. A small amount of single α-L-fucose residues were in the branches at the C2. Sulfate groups occupied position 2 and 4 of fucopyranose residues (Zvyagintseva et al. 2003).

S. japonica is a most popular alga of family Laminariaceae due to its known food use, and several groups of authors investigated fucoidans from this seaweed. Wang et al. (2010b) and Jin et al. (2013) showed that the main chain of galactofucan from *S. japonica* consisted of 1,3- and 1,4-linked α-L-fucopyranose residues(75 and 25%, respectively). Fucose residues of the main chain were substituted at C2 by single α-L-fucopyranose, and at C4 by disaccharide fragment β-D-Gal*p*-(1,6)-β-D-Gal*p*-(1- (65 and 35%, respectively). Every tetrasaccharide fragment had an average one branching point. Sulfate group were mainly located at C4 and less at C2 of fucose residue, some fucose residue was disulfated at C2 and C4. Galactose residues sulfated at C3 and/or C4. Two other polysaccharide fractions were obtained from this alga in work (Ke et al. 2020). It was shown that first fucoidan was composed of the main chain of 1,3-linked α-L-fucopyranose residues sulfated at C2 or C4 and three different monosaccharides (galactose, glucose, mannose) in branches at C2 and/or C4 of fucose residue. It was suggested that the second contained two backbones of alternating 1,3-linked galactopyranose and fucopyranose residues (galactofucan) and 1,6-linked galactose residues (galactan). Branches occupied position C4 of

A

SO3Na ↓ 2 (at first residue); SO3Na ↓ 4 (at second residue)

...→3)-α-L-Fuc*p*-(1→4)-α-L-Fuc*p*-(1→2)-α-L-Fuc*p*-(1→...

B

α-L-Fuc*p*-(1 ↓ 2) (branch at second residue)

...→3)-α-L-Fuc*p*-(1→3)-α-L-Fuc*p*-(1→3)-α-L-Fuc*p*-(1→3)-α-L-Fuc*p*-(1→...

SO_3Na at C4 and C2

C

R1 ↓ 2; **R2 ↓ 4**

...→3)-α-L-Fuc*p*-(1→3)-α-L-Fuc*p*-(1→4)-α-L-Fuc*p*-(1→...→3)-α-L-Fuc*p*-(1→....→3)-α-L-Fuc*p*-(1→...

R1 α-L-Fuc*p*-(1→
R2 β-D-Gal*p*-(1→6)-β-D-Gal*p*-(1→

SO_3Na at C4 and/or at C2 of Fuc*p*
at C3 and/or at C4 of Gal*p*

D

R ↓ 4

...→6)-β-D-Gal*p*-(1→6)-β-D-Gal*p*-(1→6)-β-D-Gal*p*-(1→...

R **β-D-Gal*p*-(1→**
or
α-L-Fuc*p*-(1→3)-β-D-Gal*p*-(1→
or
...→6)-β-D-Gal*p*-(1→6)-β-D-Gal*p*-(1→
4 ↑ β-D-Gal*p*-(1

SO_3Na at C4 of Fuc*p*
at C3 of Gal*p*

Fig. 4.1 Main structural fragments of fucan from *L. longipes* (**a**), fucan from *S. cichorioides* (**b**), galactofucan from *S. japonica* (**c**), and fucogalactan from *S. latissima* (**d**)

fucose residues and C2, C3, and/or C6 of galactose residues. Sulfate groups were mainly located at C2/C4 of fucose or galactose residues.

The main component of fucoidan from *S. latissima* was shown to be a fucan sulfate, containing a backbone of 1,3-linked α-L-fucopyranose residues sulfated at C4 and/or at C2 and branched at C2 by single sulfated a-L-fucopyranose residues (about one branching point per 5–6 backbone residues). Non-reducing terminal residues may be 2-mono-, 2,4-di-, and 2,3,4-trisulfated. The second component was determined as fucogalactan, having a backbone of 1,6-linked β-D-galactopyranose residues. About 20% of the galactan backbone residues were substituted at C4 by single β-D-galactopyranose, whereas about 15% had a disaccharide substituent α-L-Fuc*p*-(1,3)-β-D-Gal*p*-(1-. Some of side chains had a branched structure, containing 4,6-disubstituted β-D-galactopyranose residues, and occupied only about 5% of the backbone residues. It was shown that terminal fucopyranose residues in fucogalactan were sulfated at C4, whereas the majority of galactopyranose residues in these molecules were sulfated at C3 (Bilan et al. 2010).

Galactofucan from *S. gurjanovae* was sulfated and acetylated and had a main chain of a repeating 1,3-linked 2,4-disulfated α-L-fucose residues. Fucose chains could be sometimes terminated by 1,3-linked galactose residues. Short 1,4- and/or 1,6-linked sulfated galactose chains (DP = 1–5) were attached at positions C2, C3 of fucose residues. Sulfate groups occupied positions 2 and/or sometimes 3 of galactose residues, but a sulfation at C4 of the galactofucan could not be excluded (Shevchenko et al. 2007, 2015).

Galactofucan from *S. longicruris* contained predominantly 1,3-linked fucose and 1,6-linked galactose residues, and less 1,4-linked fucose, 1,3-,1,4- and 1,4,6-linked galactose residues. Sulfates were found at position 3 of galactose and position 4 of fucose residues. Also it was not excluded that fucose units might have been disulfated at position 2 and 4 (Rioux et al. 2010).

In summary, known fucoidans from algae of family Laminariaceae were polysaccharides characterized with a high structural diversity.

2.2 *Phlorotannins*

Phlorotannins are polymers of phloroglucinol (1,3,5-trihydroxybenzene) subunits, which are highly hydrophilic, contains a large number of hydroxyl groups, binds strongly to proteins, polysaccharides and other biopolymers, chelates 2-valent metals. The connection of phloroglucinol residues through C-C and/or C-O-C bonds leads to the formation of polymer molecules with various structures containing both phenyl and phenoxyl units. Based on the type of bond between monomers, phlorotannins can be classified into four groups: fuhalols and phloretols (ether bonds), fucols (phenyl bonds), fucophloroethols (ether and phenyl bonds), and eckols and carmalols (dibenzodioxin bonds). Within each groups, the binding of monomers to each other can occur at different positions of the phloroglucinol ring, which leads to the formation of structural isomers in addition to conformational ones. It is known that one type of alga often produces phlorotannins of various structures and degrees of polymerization (Li et al. 2011). Figure 4.2 shows structures of phlorotannin from brown algae.

There is only limited information about composition of phlorotannins of Laminariaceae brown algae. The phlorotannin composition from *L. digitata* was studied in work (Vissers et al. 2017). The purity of methanol extract was determined to be 60.1%, corresponding to a phlorotannin content of 4.3% of dry algae weight. The fucol-to-phlorethol linkage ratio was 1:26. The degree of polymerization of phlorotannins was determined to be up to 27. Structural isomers of phlorotannins up to a DP 18 were found; some of them have been shown to be branched.

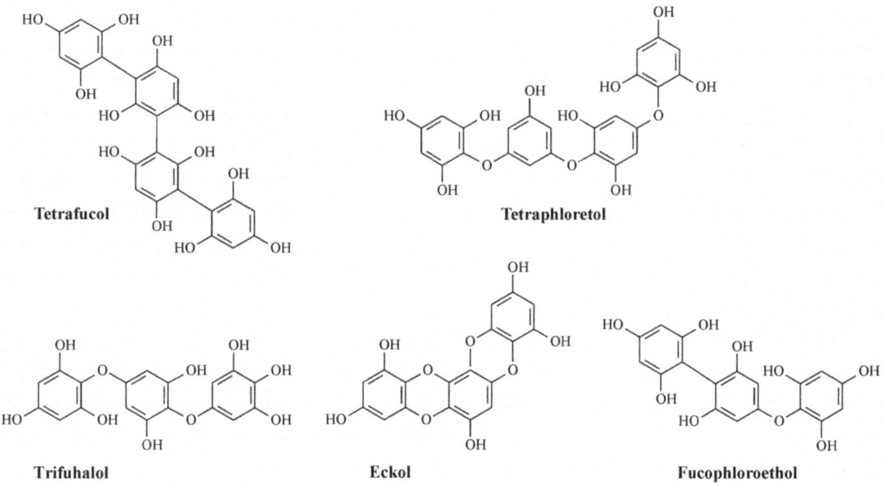

Fig. 4.2 Examples of main phlorotannin structures from brown algae

S. longicruris exhibited a phlorotannin profile with degrees of polymerization ranging from 3 to 39 dominated by DP of 17. Structural isomers also were detected (Steevensz et al. 2012).

3 Biological Activity

Brown algae have been proven to play an important role in the fight against chronic diseases. Brown algae are a rich source of polysaccharides and phlorotannins which are not found in terrestrial sources and can potentially be used both in food and medicine. Herein, we discussed the ability of bioactive components of brown algae of family Laminariaceae to improve the intestinal mucosa metabolism and prevent the carcinogenesis.

3.1 The Effect of Bioactive Components of Laminariaceae Brown Algae on Intestinal Mucosa Metabolism

It was shown that addition of *S. japonica* to basal diet of rats led to decreasing of the obesity-associated bacterial genera (*Allobaculum*, *Turicibacter*, *Coprobacillus*, *Mollicute*, and *Oscilibacter*), and the genera with pathogenic potentials (*Mollicute*, *Bacteroides*, *Clostridium*, *Escherichia*, and *Prevotella*) while leanness-associated genera (*Alistipes*, *Bacteroides*, and *Prevotella*), and lactic acid bacterial genera (*Subdoligranulum*, *Streptococcus*, *Lactobacillus*, *Enterococcus*, and *Bifidobacterium*) increased. Rats fed with seaweed also showed significantly higher serum IgG concentration, but had lower weight gain and serum triglyceride concentration and Fimicutes to Bacteroidetes ratio when compared with the control group. Thus, *S. japonica* has a great potential of as an effective prebiotic for promotion of host metabolism and reduction of obesity (Kim et al. 2018).

The polysaccharide-rich crude extract from *L. digitata* affected the composition and metabolism of the intestinal microbiota *in vitro*, altering the relative abundance of several families and genera, including Lachnospiraceae, *Streptococcus*, *Ruminococcus* and *Parabacteroides* of human fecal bacterial populations, as well as increasing the concentration of acetic acid, propionic acid, butyric acid and total short chain fatty acids (Strain et al. 2020). The polysaccharide extract from *L. digitata* has been shown to increase *Bifidobacteria* colony in the ileum of piglets (Mukhopadhya et al. 2012).

It was shown that alginate oligosaccharides from *Laminaria* sp. shift the intestinal microbiota profile of Atlantic salmon (*Salmo salar*); low inclusion of studied oligosaccharides can plausibly induce a prebiotic effect (Gupta et al. 2019).

The β-glucans (laminarans) from *L. hyperborea* and *L. digitata* reduced the Enterobacteriaceae population without influencing the lactobacilli and bifidobacteria populations in the ileum and colon of piglet (Sweeney et al. 2012). However, it was shown that *Bifidobacterium infantis* can be able to ferment laminaran from *L. digitata*. It is anticipated that β-glucans combined with bifidobacteria can be used as synbiotics for the functional food industry (Zhao and Cheung 2013).

The sulfated polysaccharides from *S. japonica* were fermented *in vitro* for 48 h by human fecal cultures. Wherein the pH in fecal cultures decreased from 6.5 to 5.1 and the levels of short chain fatty acids, such as acetic, butyric and lactic acids all significantly increased. Also beneficial strains (*Lactobacillus* and *Bifidobacterium*) were both significantly higher than those in control group. It was shown that consumption of sulfated polysaccharides from *S. japonica* is beneficial to the ecosystem of the intestinal tract by increasing the populations of probiotics and short chain fatty acids (Kong et al. 2016).

Prebiotics were proposed for stimulation of colonization/expansion of beneficial microflora in chickens. Extract of *Laminaria* spp. exerted positive effects on growth of broiler chickens, carcass and meat quality traits, carcass weight, carcass yield and breast muscle weight. Meat from prebiotic treated birds displayed a higher lipid oxidation levels compared to that from untreated ones (Tavaniello et al. 2018; Maiorano et al. 2017). Thus, polysaccharides of *Laminariaceae* algae are of a great interest for food and feed applications.

3.2 The Effect of Bioactive Components of Laminariaceae Brown Algae on Proliferation and Colony Formation of Cancer Cells

A fundamental feature of cancer cells is their ability to maintain uncontrolled cell division. In normal tissues, growth processes and the cell cycle are closely monitored to ensure the required number of cells and appropriate tissue architecture to support its function. In a normal cell, division is stimulated by growth factors that bind to a receptor on the cell surface that has an intracellular domain with tyrosine kinase activity. The activation of the tyrosine kinase domain leads to the activation of intracellular pathways that regulate the cell cycle, cell growth, and other biological processes of the cell, such as energy metabolism (Sever and Brugge 2015).

The anticancer activity of fucoidans and laminarans from brown algae is associated with the inactivation of membrane receptors, including receptor tyrosine kinases (RTK), epidermal growth factor receptor (EGFR), and platelet-derived growth factor receptor (PDGFR). Under the influence of various carcinogenic factors, cell transformation is initiated through the activation of intracellular signal transduction cascades (Bode and Dong 2003). The mitogen-activated protein kinase (MAPK) cascade is one of the most important signaling mechanisms activated by tumor promoters and involved in cell proliferation, differentiation and apoptosis (Sun et al. 2015). MAPK cascades include extracellular signal-regulated protein kinases (ERKs), c-Jun N-terminal kinases/stress-activated protein kinases (JNKs/SAPKs) and p38 kinases.

It has been shown that fucoidans from brown algae effectively prevent the transformation of tumor cells induced by tumor promoters by regulating MAPK activity. For example, fucoidans from *S. cichorioides* (Lee et al. 2008a) and *S. gurjanovae* (Lee et al. 2008b) at the concentration range of 1–100 μg/mL effectively inhibited the neoplastic transformation of the mouse epidermal cells JB6 Cl41, induced by EGF and TPA, respectively. The investigated polysaccharides were found to inhibit the phosphorylation of ERK, JNK and, therefore, the activity of AP-1. Fucoidan from *L. gurjanovae* also prevented EGF-induced activation of the EGFR receptor. The authors concluded that EGFR could be a potential target for fucoidan in inhibiting AP-1 activity and cell transformation (Lee et al. 2008b).

The laminaran (at concentration 5 mg/mL) from brown seaweed *L. digitata* (commercial preparation of “Sigma–Aldrich” company) was shown to inhibit the cell growth of colorectal cancer cells HT-29 by 40% with less cytotoxicity on IEC-6 intestinal epithelial cells. It was found that decreased proliferation of cancer cells depended on IGF-IR, which was associated with the down-regulation of MAPK/ERK cascade (Park et al. 2012). Moreover, the investigated laminaran was able to induce cell cycle arrest and inhibited the heregulin-stimulated phosphorylation of ErbB2, which activates c-Jun N-terminal kinase regulated a range of biological processes implicated in tumorigenesis and neurodegenerative disorders (Park et al. 2013).

Last decade, the influence of fucoidans and laminarans on the formation of colonies of various types of cancer cells has also been intensively studied. The soft agar method is often used to detect morphological transformation of cells caused by various carcinogenic factors. In soft agar, cancer cells divide independently of each other and form colonies, while normal cells cannot divide without contact with the extracellular matrix (Borowicz et al. 2014). Thus, fucoidans from brown algae *S. cichorioides* and *L. longipes* (Usoltseva et al. 2019). and *S. gurjanovae* (Shevchenko et al. 2015) significantly inhibited the formation of human colon carcinoma cells HT-29 and DLD-1, while fucoidans from *S. japonica* had an inhibitory effect on the growth of T-47D breast cancer cells and human melanoma cells

SK-MEL-28 in soft agar (Vishchuk et al. 2011). Also it was found that the laminarans from *S. cichorioides* and *S. japonica* as well as their sulfated inhibited the colony formation and migration of human colorectal adenocarcinoma, melanoma, and breast adenocarcinoma cells in variable degree (Malyarenko et al. 2017).

Laminaran from *L. digitata* (at dose 1.6 mg/mL) was reported to suppress the proliferation of colorectal cancer cells LOVO by 38.8% (Ji et al. 2013). Moreover, the sulfated laminaran was obtained and its antiproliferative activity was determined in human colon cancer LOVO cells. It was found that after sulfated modification the growth inhibitory activity of sulfated laminaran was enhanced and inhibition rate was 86% at the same dose. The molecular mechanism of antiproliferative effect of sulfated laminaran was associated with induction of apoptosis of LOVO cells. The laminaran was shown to up-regulate the expression levels of DR4, DR5, TRAIL, FADD, Bid, tBid and Bax, while the expression levels of pro-caspase-8, pro-caspase-3 and Bcl-2 were down-regulated.

It is known that reactive oxygen species are responsible for such cellular abnormalities as protein damage, deactivation of enzymes, DNA changes and lipid peroxidation, which, in turn, leads to pathological conditions—carcinogenesis, reperfusion injury, rheumatoid arthritis and diabetes (Barry and Gutteridge 2015). To maintain cell health, it is important to have a specific and effective antioxidant capable of scavenging free radicals. Several studies have demonstrated that phlorotannins and phlorotannin-enriched extracts from brown algae exhibit potent antioxidant activity (Tierney et al. 2013). Epidemiological, clinical and nutritional studies confirm that regular consumption of phlorotanin-containing foods or drugs is associated with a decrease in the risk of various chronic diseases, including cancer, metabolic and neurodegenerative disorders, cardiovascular diseases (Gómez-Guzmán et al. 2018).

The data on anticancer activity of phlorotannins of brown algae are very limited. Previously the phloroglucinol was found to suppress the metastatic ability of breast cancer cells. The molecular mechanism of its antimetastatic action was associated with the down-regulation of the transcription factor SLUG and regulation of PI3K/AKT/mTOR and Ras/Raf 1/ERK signaling pathways (Kim et al. 2015a). Another phlorotannin, diecol, was involved in the regulation of gene expression associated with metastases by the inhibition of the activity of matrix metalloproteinase-9 (MMP-9) and expression of vascular endothelial growth factor (VEGF) associated with migration of human breast cancer cells MCF-7. At the same time, diecol stimulated the expression of tissue inhibitors of metalloproteinase TIMP-1 and TIMP-2 (Kim et al. 2015b).

4 Conclusions and Future Perspectives

A growing body of scientific evidence indicates that phlorotannins, fucoidans, and laminarans from brown algae of family Laminariaceae have outstanding biologically active properties (antioxidant, anticancer, antimicrobial, antidiabetic, and anti-inflammatory) and might be used as active ingredients in functional foods. In turn, the bioactivity depends on their structural characteristics which are differing among the sources of isolation, methods of extraction and purification of these compounds. That is why the establishment of the complete chemical structure of phlorotannins, fucoidans, and laminarans used to study of their biological activity is an important task. In the future, comprehensive studies of the properties of phlorotannins, fucoidans, and laminarans, their bioavailability and metabolism in living organisms is considered to contribute to the development of a stable platform for the creation of functional food or drugs based on these unique algal components. This work was supported by the Scholarship of the President of the Russian Federation SP-1216.2021.4.

References

Barry H, Gutteridge JMC (2015) Free radicals in biology and medicine, 5th edn. Oxford University Press, Oxford

Bilan MI, Grachev AA, Shashkov AS, Kelly M, Sanderson CJ, Nifantiev NE, Usov AI (2010) Further studies on the composition and structure of a fucoidan preparation from the brown alga *Saccharina latissima*. Carbohydr Res 345:2038–2047

Bode AM, Dong Z (2003) Mitogen-activated protein kinase activation in UV-induced signal transduction. Sci Signal 167:RE2

Borowicz S, Van Scoyk M, Avasarala S, Karuppusamy Rathinam MK, Tauler J, Bikkavilli RK, Winn RA (2014) The soft agar colony formation assay. J Vis Exp 27(92):e51998

Chen A, Lan Y, Liu J, Zhang F, Zhang L, Li B, Zhao X (2017) The structure property and endothelial protective activity of fucoidan from *Laminaria japonica*. Int J Biol Macromol 105(Pt 2):1421–1429

Fitton JH, Stringer DN, Karpiniec SS (2015) Therapies from fucoidan: an update. Mar Drugs 13:5920–5946

Gómez-Guzmán M, Rodríguez-Nogales A, Algieri F, Gálvez J (2018) Potential role of seaweed polyphenols in cardiovascular-associated disorders. Mar Drugs 16(8):250

Gupta S, Lokesh J, Abdelhafiz Y, Siriyappagouder P, Pierre R, Sørensen M, Fernandes JMO, Kiron V (2019) Macroalga-derived alginate oligosaccharide alters intestinal bacteria of Atlantic salmon. Front Microbiol 10:2037

Ji CF, Ji YB, Meng DY (2013) Sulfated modification and anti-tumor activity of laminarin. Exp Ther Med 6(5):1259–1264

Jin W, Guo Z, Wang J, Zhang W, Zhang Q (2013) Structural analysis of sulfated fucan from *Saccharina japonica* by electrospray ionization tandem mass spectrometry. Carbohydr Polym 369:63–67

Ke S, Wie B, Qiu W, Zhou T, Wang S, Chen J, Chen J, Zhang H, Jin W, Wang H (2020) Structural characterization and α-glucosidase inhibitory and antioxidant of fucoidans extracted from *Saccharina japonica*. Chem Biodivers 17(7):e2000233

Kim EK, Tang Y, Kim YS, Hwang JW, Choi EJ, Lee JH, Lee SH, Jeon YJ, Park PJ (2015a) First evidence that Ecklonia cava-derived dieckol attenuates MCF-7 human breast carcinoma cell migration. Mar Drugs 13(4):1785–1797

Kim RK, Suh Y, Yoo KC, Cui YH, Hwang E, Kim HJ, Kang JS, Kim MJ, Lee YY, Lee SJ (2015b) Phloroglucinol suppresses metastatic ability of breast cancer cells by inhibition of epithelial-mesenchymal cell transition. Cancer Sci 106(1):94–101

Kim JY, Kwon YM, Kim IS, Kim JA, Yu DY, Adhikari B, Lee SS, Choi IS, Cho KK (2018) Effects of the brown seaweed *Laminaria japonica* supplementation on serum concentrations of IgG, triglycerides, and cholesterol, and intestinal microbiota composition in rats. Front Nutr 5:23

Kong Q, Dong S, Gao J, Jiang C (2016) In vitro fermentation of sulfated polysaccharides from *E. prolifera* and *L. japonica* by human fecal microbiota. Int J Biol Macromol 91:867–871

Kopplin G, Rokstad AM, Mélida H, Bulone V, Skjåk-Bræk G, Aachmann FL (2018) Structural characterization of fucoidan from *Laminaria hyperborea*: assessment of coagulation and inflammatory properties and their structure–function relationship. ACS Appl Bio Mater 1:1880–1892

Kusaykin MI, Bakunina IY, Sova VV, Ermakova SP, Kuznetsova TS, Besednova NN, Zaporozhets TS, Zvyagintseva TN (2008) Structure, biological activity, and enzymatic transformation of fucoidans from the brown seaweeds. Biotechnol J 3:904–915

Lee NY, Ermakova SP, Choi HK, Kusaykin MI, Shevchenko NM, Zvyagintseva TN, Choi HS (2008a) Fucoidan from *Laminaria cichorioides* inhibits AP-1 transactivation and cell transformation in the mouse epidermal JB6 cells. Mol Carcinog 47(8):629–637

Lee NY, Ermakova SP, Zvyagintseva TN, Kang KW, Dong Z, Choi HS (2008b) Inhibitory effects of fucoidan on activation of epidermal growth factor receptor and cell transformation in JB6 Cl41 cells. Food Chem Toxicol 46(5):1793–1800

Li YX, Wijeseker I, Li Y, Kim SK (2011) Phlorotannins as bioactive agents from brown algae. Process Biochem 46:2219–2224

Maiorano G, Stadnicka K, Tavaniello S, Abiuso C, Bogucka J, Bednarczyk M (2017) In ovo validation model to assess the efficacy of commercial prebiotics on broiler performance and oxidative stability of meat. Poult Sci 96:511–518

Malyarenko OS, Usoltseva RV, Shevchenko NM, Isakov VV, Zvyagintseva TN, Ermakova SP (2017) *In vitro* anticancer activity of the laminarans from far eastern brown seaweeds and their sulfated derivatives. J Appl Phycol 29(1):543–553

Mukhopadhya A, O'Doherty JV, Smith A, Bahar B, Sweeney T (2012) The microbiological and immunomodulatory effects of spray-dried versus wet dietary supplementation of seaweed extract in the pig gastrointestinal tract. J Anim Sci 90:28–30

Ni L, Wang L, Fu X, Duand D, Jeon YJ, Xu J, Gao X (2020) *In vitro* and *in vivo* anti-inflammatory activities of a fucose-rich fucoidan isolated from *Saccharina japonica*. Int J Biol Macromol 156:717–729

Pangestuti R, Siahaan EA, Kim SK (2018) Photoprotective substances derived from marine algae. Mar Drugs 16(11):399

Park HK, Kim IH, Kim J, Nam TJ (2012) Induction of apoptosis by laminarin, regulating the insulin-like growth factor-IR signaling pathways in HT-29 human colon cells. Int J Mol Med 30(4):734–738

Park HK, Kim IH, Kim J, Nam TJ (2013) Induction of apoptosis and the regulation of ErbB signaling by laminarin in HT-29 human colon cancer cells. Int J Mol Med 32(2):291–295

Ren D, Wang Q, Yang Y, Hu Y, Song Y, He Y, Liu S, Wu L (2019) Hypolipidemic effects of fucoidan fractions from *Saccharina sculpera* (Laminariales, Phaeophyceae). Int J Biol Macromol 140:188–195

Rioux LE, Turgeon SL, Beaulieu M (2010) Structural characterization of laminaran and galactofucan extracted from the brown seaweed *Saccharina longicruris*. Phytochemistry 71:1586–1595

Sanjeewa KKA, Lee JS, Kim WS, Jeon YJ (2017) The potential of brown-algae polysaccharides for the development of anticancer agents: an update on anticancer effects reported for fucoidan and laminaran. Carbohydr Polym 177:451–459

Sever R, Brugge JS (2015) Signal transduction in cancer. Cold Spring Harb Perspect Med 5(4):a006098

Shevchenko NM, Anastyuk SD, Gerasimenko NI, Dmitrenok PS, Isakov VV, Zvyagintseva TN (2007) Polysaccharide and lipid composition of the brown seaweed *Laminaria gurjanovae*. Russ J Bioorg Chem 33:88–98

Shevchenko NM, Anastyuk SD, Menshova RV, Vishchuk OS, Isakov VV, Zadorozhny PA, Sikorskaya TV, Zvyagintseva TN (2015) Further studies on structure of fucoidan from brown alga *Saccharina gurjanovae*. Carbohydr Polym 121:207–216

Steevensz AJ, Mackinnon SL, Hankinson R, Craft C, Connan S, Stengel DB, Melanson JE (2012) Profiling phlorotannins in brown macroalgae by liquid chromatography-high resolution mass spectrometry. Phytochem Anal 23(5):547–553

Strain CR, Collins KC, Naughton V, McSorley EM, Stanton C, Smyth TJ, Soler-Vila A, Rea MC, Ross PR, Cherry P, Allsopp PJ (2020) Effects of a polysaccharide-rich extract derived from Irish-sourced *Laminaria digitata* on the composition and metabolic activity of the human gut microbiota using an in vitro colonic model. Eur J Nutr 59:309–325

Sun Y, Liu WZ, Liu T, Feng X, Yang N, Zhou HF (2015) Signaling pathway of MAPK/ERK in cell proliferation, differentiation, migration, senescence and apoptosis. J Receptors Signal Transduct 35(6):600–604

Sweeney T, Collins CB, Reilly P, Pierce KM, Ryan M, O'Doherty JV (2012) Effect of purified b-glucans derived from Laminaria digitata, Laminaria hyperborea and Saccharomyces cerevisiae on piglet performance, selected bacterial populations, volatile fatty acids and pro-inflammatory cytokines in the gastrointestinal tract of pigs. Br J Nutr 108:1226–1234

Tavaniello S, Maiorano G, Stadnicka K, Mucci R, Bogucka J, Bednarczyk M (2018) Prebiotics offered to broiler chicken exert positive effect on meat quality traits irrespective of delivery route. Poult Sci 97:2979–2987

Teruya T, Takeda S, Tamaki Y, Tako M (2010) Fucoidan isolated from *Laminaria angustata* var. longissima induced macrophage activation. Biosci Biotechnol Biochem 74:1960–1962

Thomas NV, Kim SK (2011) Potential pharmacological applications of polyphenolic derivatives from marine brown algae. Environ Toxicol Pharmacol 32(3):325–235

Tierney MS, Smyth TJ, Rai DK, Soler-Vila A, Croft AK, Brunton N (2013) Enrichment of polyphenol contents and antioxidant activities of Irish brown macroalgae using food-friendly techniques based on polarity and molecular size. Food Chem 139(1–4):75–761

Usoltseva RV, Shevchenko NM, Malyarenko OS, Anastyuk SD, Kasprik AE, Zvyagintsev NV, Ermakova SP (2019) Fucoidans from brown algae *Laminaria longipes* and *Saccharina cichorioides*: structural characteristics, anticancer and radiosensitizing activity *in vitro*. Carbohydr Polym 221:157–165

Usoltseva RV, Belik AA, Kusaykin MI, Malyarenko OS, Zvyagintseva TN, Ermakova SP (2020) Laminarans and 1,3-β-D-glucanases. Int J Biol Macromol 163:1010–1025

Usov AI (1999) Alginic acids and alginates: analytical methods used for their estimation and characterisation of composition and primary structure. Russ Chem Rev 68:957–966

Usov AI, Bilan MI (2009) Fucoidans—sulfated polysaccharides of brown algae. Russ Chem Rev 78:785–799

Vishchuk OS, Ermakova SP, Zvyagintseva TN (2011) Sulfated polysaccharides from brown seaweeds *Saccharina japonica* and *Undaria pinnatifida*: isolation, structural characteristics, and antitumor activity. Carbohydr Res 346:2769–2776

Vissers AM, Caligiani A, Sforza S, Vincken JP, Gruppen H (2017) Phlorotannin composition of *Laminaria digitata*. Phytochem Anal 28(6):487–495

Wang J, Zhang Q, Zhang Z, Li Z (2008) Antioxidant activity of sulfated polysaccharide fractions extracted from *Laminaria japonica*. Int J Biol Macromol 42:127–132

Wang J, Zhang Q, Zhang Z, Song H, Li P (2010a) Potential antioxidant and anticoagulant capacity of low molecular weight fucoidan fractions extracted from *Laminaria japonica*. Int J Biol Macromol 46:6–12

Wang J, Zhang Q, Zhang Z, Zhang H, Niu X (2010b) Structural studies on a novel fucogalactan sulfate extracted from the brown seaweed *Laminaria japonica*. Int J Biol Macromol 47:126–131

Xue CH, Fang Y, Lin H, Chen L, Li ZJ, Deng D, Lu CX (2001) Chemical characters and antioxidative properties of sulfated polysaccharides from *Laminaria japonica*. J Appl Phycol 13:67–70

Ye J, Chen D, Ye Z, Huang Y, Zhang N, Lui EMK, Xue C, Xiao M (2020) Fucoidan isolated from *Saccharina japonica* inhibits LPS-induced inflammation in macrophages via blocking NF-kB, MAPK and JAK-STAT pathways. Mar Drugs 18(6):328

Yoon SJ, Pyun YR, Hwang JK, Mourao PAS (2007) A sulfated fucan from the brown alga *Laminaria cichorioides* has mainly heparin cofactor II-dependent anticoagulant activity. Carbohydr Res 342:2326–2330

Zhao J, Cheung PCK (2013) Comparative proteome analysis of *Bifidobacterium longum* subsp. *infantis* grown on β-glucans from different sources and a model for their utilization. J Agric Food Chem 61:4360–4370

Zhurishkina EV, Lapina IM, Ivanen DR, Stepanov SI, Shvetsova SV, Shavarda AL, Giliano NY, Kulminskaya AA (2015) Effect of fucoidans isolated from seaweeds *Laminaria digitata* and *Fucus vesiculosus* on cell lines HELA G-63, ECV 304 and PC 12 [in Russian]. Tsitologiia 57(10):727–735

Zvyagintseva TN, Shevchenko NM, Nazarenko EL, Gorbach VI, Urvantseva AM, Kiseleva MI, Isakov VV (2003) Water-soluble polysaccharides of some far-eastern brown seaweeds. Distribution, structure, and their dependence on the developmental conditions. J Exp Mar Biol Ecol 294:1–13

Chapter 5
Sargassum Species: Its Use in Food and Health Implications

Elena M. Balboa, M. Cristina Taboada, and Herminia Domínguez

Abbreviations

ABTS	2,2′-azino-bis(3-ethylbenzothiazoline-6-sulfonic acid)
ACE	Angiotensin I-converting enzyme
COX	Cyclooxygenase
DPPH	2,2 Diphenyl 1 picrylhydrazyl
FRAP	Ferric reducing antioxidant power assay
IFN-γ	Interferon-γ
IL-6	Interleukin
IM	Immuno-modulatory
iNOS	Inducible NO synthase (iNOS)
LOX	Lipoxygenase
LPS	Lipopolysaccharide
MMP	Matrix metalloproteinases
NF	Nuclear factor
NO	Nitric oxide
$ONOO^-$	Peroxynitrite
PGE2	Prostaglandin E2
ROS	Reactive oxygen species

E. M. Balboa · H. Domínguez (✉)
Department of Chemical Engineering, University of Vigo (Campus Ourense), Edificio Politécnico, Ourense, Spain
e-mail: elenamba@uvigo.es; herminia@uvigo.es

M. C. Taboada
Department of Physiology, Faculty of Pharmacy, University of Santiago de Compostela, Santiago de Compostela, Spain
e-mail: mariacristina.taboada@usc.es

A. Ranga Rao, G. A. Ravishankar (eds.), *Sustainable Global Resources of Seaweeds Volume 2*, https://doi.org/10.1007/978-3-030-92174-3_5

TDF	Total dietary fiber
TNF-α	Tumor necrosis factor

1 Introduction: *Sargassum* sp.

Sargassum (Family. Sargassaceae, order Fucales, subclass Cyclosporeae, class Phaeophyceae), is a genus of brown algae important in Asian diets. Some species have been regarded as a source of vitamins, carotenoids, proteins, and minerals (Sanjeewa et al. 2018a, b; Rushdi et al. 2020). This genus is distributed mainly in tropical and subtropical areas, and is used as fertilizer, insect repellant, feed, food and traditional medicine (Liu et al. 2012; Shama et al. 2019; Rushdi et al. 2020).

Recent studies have confirmed that the components of *Sargassum* sp. possess a variety of biological properties, such as antioxidant, anti-inflammatory, antimicrobial, antiproliferative, anticoagulant, neuroprotective, hepatoprotective, among others (Flórez-Fernández et al., 2017; Sugiura et al., 2018; Kim et al., 2021; Josephine and Kumar 2011; Yende et al. 2014; Sanjeewa et al. 2018a, b; Herawati and Sumanik 2019; Rushdi et al. 2020). Figure 5.1 summarizes the major components of this

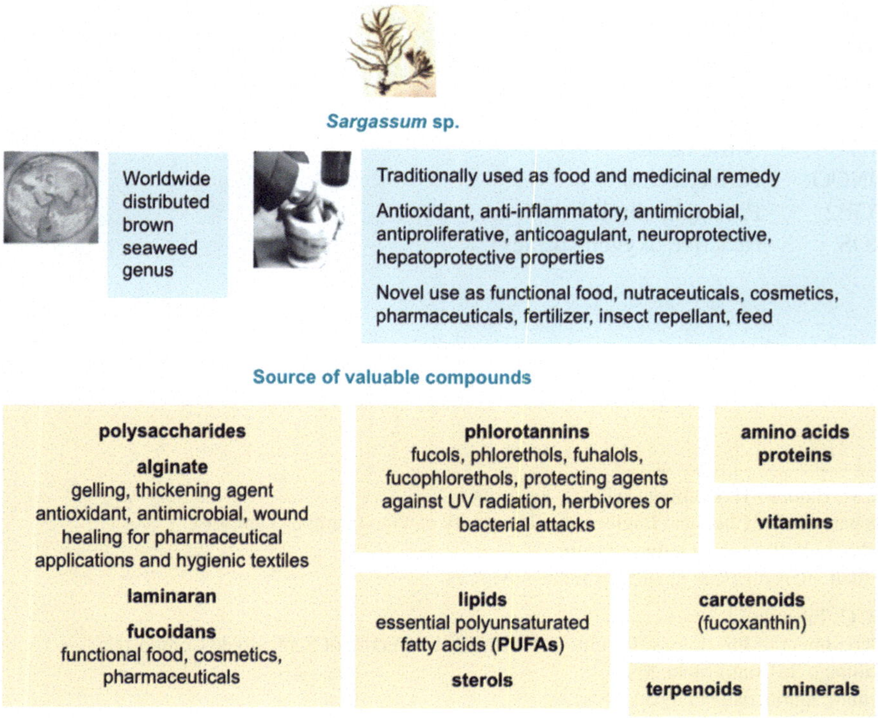

Fig. 5.1 Valuable compounds of *Sargassum* sp. and main uses

genus as well as traditional uses and potential applications. This chapter presents a summary of the major components in *Sargassum* species, as well as their potential feed and food applications and an updated overview on the biological properties that make this renewable resource a valuable ingredient for the development of functional foods, nutraceuticals, cosmetics, and pharmaceuticals.

2 Composition

Brown seaweeds contain 40–70% polysaccharides, the most relevant being alginate, laminaran, and fucoidans. Alginates, commercially used as a gelling and thickening agent, are found as the calcium, magnesium, and sodium salts of alginic acid, formed by linked β-(1,4)-D-mannuronic acid and α-(1,4)-L-guluronic acid. Their antioxidant and antimicrobial properties make them suitable to develop alginate film-coated textile fabrics for non implantable materials such as wound healing for pharmaceutical applications and also for hygienic textiles (Janarthanan and Senthil Kumar 2018). Laminarans are reserve polysaccharides, formed by β-1,1-glucan. Fucoidans are sulfated heteropolysaccharides exclusive from brown algae, composed of fucose, galactose, mannose, glucose, xylose, uronic acids and sulfate groups. Their complex composition and structure are highly dependent on the

Table 5.1 Proximal composition (g 100 g^{-1} db) of different species from *Sargassum* genus

Species	Protein	TDF	Ash	Crude lipid	Carbohydrate	Moisture	References
S. hemiphyllum	5.33	50.4	23.3	3.06	17.9	8.50	Wong and Cheung (2001)
S. henslowianum	11.33	61.1	21.3	4.56	1.74	8.50	Wong and Cheung (2001)
S. horneri	7.6	47.2	27.8	0.78	–	87.3 (wb)	Murakami et al. (2011)
S. ilicifolium	9.71	–	–	0.42	38.72 ± 0.96	–	Ganapathi et al. (2013)
S. muticum	6.92	60	26.04	1.6	13.5	–	Balboa et al. (2016)
S. myriocystum	13.21	–	–	0.17	40.21	–	Ganapathi et al. (2013)
S. patens	7.56	54.8	28.8	6.15	2.69	8.60	Wong and Cheung (2001)
S. plagiophyllum	12.30	–	–	0.16	33.38	–	Ganapathi et al. (2013)
S. polycystum	14.2	21.3	29	7.6	25.0	–	Perumal et al. (2019)
S. vulgare	7.69	22.59	27.09	0.56	34.18	7.89	Arguelles et al. (2019)

TDF total dietary fiber

species, parts of seaweeds, seasonal and geographic variations, and the extraction and purification processes. Fucoidans have been investigated in recent years to develop functional foods, cosmetics and pharmaceuticals, based on a variety of biological properties (Hwang et al. 2016; Josephine and Kumar 2011; Torres et al. 2020; Tian et al. 2020).

According to data in Table 5.1, summarizing the proximal composition of different species from *Sargassum*, other exclusive compounds found in brown seaweeds are phlorotannins. Phlorotannins, formed by phloroglucinol units, are usually classified as fucols, phlorethols, fuhalols and fucophlorethols, depending on the bonds present. They are synthesized as a protecting agent against UV radiation, herbivores or bacterial attacks and are gaining interest for their healthy properties (Mateos et al. 2020).

Lipids are minor constituents of seaweeds, but the high levels of essential polyunsaturated fatty acids (PUFAs) confer beneficial health effects, particularly due to the $\omega6{:}\omega3$ ratio close to 1. Other valuable components can be found in the lipidic fraction, among them carotenoids, such as the xanthophyll fucoxanthin, which is a free radical scavenger with anti-photoaging, anticarcinogenic, anti-inflammatory, anti-malaria and anti-diabetic properties (D'Orazio et al. 2012). Proteins, terpenoids, sterols, and vitamins are also found in *Sargassum* species (Magura et al. 2019; Sanjeewa et al. 2018a, b). Seaweeds are richer in minerals than terrestrial foodstuffs. Concentrations of essential and toxic mineral elements vary significantly with location. The health hazards due to consumption of some *Sargassum* species in humans should be considered, particularly in relation to the accumulation of inorganic arsenic (Yokoi and Konomi 2012; Cao et al. 2014; Magura et al. 2019; Tapia-Martinez et al. 2019).

Species variations, geographical and seasonal characteristics affect the chemical composition and structural features as well as the bioactivity of *Sargassum* components (Connan et al. 2006; Le Lann et al. 2008; Plouguerné et al. 2006; Balboa et al. 2016). The more constant composition of the specimens from tropical regions throughout the year would represent an advantage for the commercial exploitation of these species. Furthermore, the properties are also influenced by a number of factors related to the processing conditions (Tian et al. 2020).

3 Food and Feed Applications

Sargassum sp. are sources of bioactive compounds and functional constituents that can be used in food and feed applications both as the whole seaweed or as separate components. Traditionally, *Sargassum* sp. is boiled and consumed directly in many coastal countries. In India, Japan and Korea, it is consumed as salad, in soups, as rice dishes, and as a savory food ingredient (Kumar et al. 2019). When used dry, the type of technology can influence the properties. Charles et al. (2020) reported that oven-, vacuum-, and freeze-drying preserved color more efficiently than sun drying and oven-dried seaweed extracts exhibited higher levels of phenolic and antioxidant potential.

Recent studies have reported different health effects associated with the ingestion of the whole seaweed. A diet supplemented with 20% of *S. liebmannii* for 11 weeks in Sprague Dawley male rats showed an antiobesogenic effect and increased insulin sensitivity without toxicity signs, histological damage or alterations on hematological parameters, oxidative stress and cellular damage (Tapia-Martinez et al. 2019). The edible seaweed *Sargassum horneri* has been proposed as a functional food with an improvement effect on abnormal skin immune responses (Han et al. 2020). Diet supplementation can have beneficial effects concerning obesity and metabolic disorders, as recently reported by Murakami et al. (2021), who confirmed the effects of dried and powdered *S. horneri* at a weight ratio of 2% or 6% in C57BL/6J mice. They also observed improved insulin resistance and stimulated fecal excretion of triglyceride, as well as increased fecal polysaccharide content.

Alternatively, the administration of some components proved favourable and novel formulations have also been proposed. *Sargassum wightii* powder incorporated in the coffee brew to increase its sensory acceptability and overcome the fishy smell restricting its potential use. The phenolic levels increased more markedly when adding dried methanolic extracts (bromophenols, catechins, and tetraprenyltoluquinols) (Airanthi et al. 2011).

Yang et al. (2020) prepared a functional oil from the edible seaweed *Hizikia fusiformis*, with beneficial effects against cholinergic disorder, oxidative stress, and neuroinflammation, aspects playing important roles in Alzheimer's disease pathology. Jia et al. (2020a, b) have confirmed that *Sargassum fusiforme* polysaccharides can be suitable as antidiabetic and anti-digestive ingredients with strong alpha-glucosidase inhibitory activity. Their oral administration in diabetic rats can mitigate hyperglycemia, hyperinsulinemia, dyslipidemia, and oxidative stress and also promote glycogen synthesis in the liver and skeletal muscles. Furthermore, these polysaccharides could partially repair liver and muscle injuries caused by diabetes.

Different studies have confirmed the potential feed application of *Sargassum*, which could be illustrated with some recent examples. Al-Harthi and El-Deek (2012) used boiled or autoclaved *S. dentifebium*, with 8% protein, 48% minerals, a metabolizable energy of 1543 kcal/kg, as a feed resource for poultry. Choi et al. (2020) included *S. fusiforme* as an ingredient in ruminant diets for its fibre and mineral content with an arsenic content within allowable limits. Nazarudin et al. (2020) showed the prebiotic effects of the supplementation of *S. polycystum* up to 3% in diets for Asia sea bass fingerlings, also benefiting in survival, efficiency, and growth performance. Lee et al. (2020a, b) confirmed the immunomodulatory effects of *Sargassum horneri* hot-water extracts on the white shrimp *Litopenaeus vannamei*.

Table 5.2 Examples of biological properties of the different components in *Sargassum* especies

Species	Extract, fraction or compounds	Activity	References
S. angustifolium	Phenolics	AO (inhibition of linoleic acid peroxidation)	Rastian et al. (2007)
S. aquifolium	Petroleum ether extract	ABl (against human pathogenic bacteria)	Moni et al. (2019)
S. carpophyllum	Polysaccharides	AO ($O_2^{\cdot-}$, OH, DPPH and ABTS radicals) IS (promoting cytokine secretion (IL-2, TNF-α) of macrophages	Tian et al. (2020)
S. confusum	Extract	AOB (reduction of body fat and waist circumference, decrease of serum leptin levels)	Min et al. (2014)
S. cristaefolium	Fucoidan	AI	Wu et al. (2016)
S. duplicatum	Fucoidan	AT (HCT-116 cells)	Usoltseva et al. (2017)
	Fucoidan	AT (HT-29, HCT 116, DLD-1 cell lines)	Usoltseva et al. (2017)
	Phenols	AO (reducing, antiradical)	Johnson et al. (2019)
S. fulvellum	Fermented algae polysaccharides	AC	Zoysa et al. (2007)
	Fucoxanthin	AO, AOB, AT, ADb, anti- photoaging	D'Orazio et al. (2012)
	Polysaccharides	IM (increased NO and cytokine production (TNF-α, IL-6, IL-1β), increase of splenocyte proliferation and production of cytokines (IL-2, IFN-γ)	Byun (2015)
S. fusiforme	Polysaccharide	Stimulation of cytokine (IL-2, IL-6, IFN-γ, TNF-α) secretion of splenic lymphocytes immunosuppressed mice. Increase of phagocytic rates	Jin et al. (2017)
	Fucoidan	AC	Sun et al. (2017)
	Polysaccharide	UV protection, antiphotoaging in hairless mice. Enhancing SOD and CAT activity, reducing ROS and MDA, suppression of MMP-1 and -9	Ye et al. (2018)
	Polysaccharides	ADb (control of blood glucose levels, triglyceride, total cholesterol, amelioration of liver and kidney damage	Jia et al. (2020a, b)

(continued)

Table 5.2 (continued)

Species	Extract, fraction or compounds	Activity	References
S. horneri	Enzymatic digest (>30 kDa)	AC	Athukorala et al. (2007)
	Fucoidan	AI (inhibited NO production in RAW 264.7 cells)	Sanjeewa et al. (2019a)
	70% Ethanolic extracts	AI (inhibited NO and PGE2 production, downregulation of iNOS, COX-2 and proinflammatory cytokines (TNF-α, IL-6, IL-1β, NF-κB), suppression of phosphorylation of ERK1/2, JNK	Sanjeewa et al. (2019b)
	Fermented (phenolics and flavonoids)	Antiradical; AI (inhibition of inducible nitric oxide synthase (iNOS) and cyclooxygenase-2 (COX-2))	Kang et al. (2020)
S. filipendula	Sulfated fucoidans	IM (enhanced release of nitric oxide (NO) by murine macrophages (RAW 264.7), induce interleukin-6 (IL-6) and TNF-α release); AB (inhibited biofilm formation by *Klebsiella pneumoniae*, *Staphylococcus epidermidis*, *Trichomonas vaginalis*)	Telles et al. (2018)
S. fluitans	Fucoidans	AO (reducing ROS generation, increased the glutathione and catalase activity), Hepatoprotective	Chale-Dzul et al. (2017)
S. fulvellum	Polysaccharide	AIN (increases NO and cytokine (TNF-α, IL-6, IL-12p70) levels, and macrophage surface activation markers CD80 and CD86. RAW264.7 macrophage modulated MAP kinases and NF-κB signaling. In splenocytes, increased cell proliferation and Th1 cytokines (IFN-γ)	Sung et al. (2015)

(continued)

Table 5.2 (continued)

Species	Extract, fraction or compounds	Activity	References
S. fusiforme	Fucoxanthin-rich extract	Antiosteoporosis activity (suppressed osteoclast differentiation and accelerate osteoblast formation *in vitro* and *in vivo* in mice)	Koyama (2011)
	Polysaccharide	IM (increase of phagocytic rates and cytokine (IL-2, IL-6, TNF-α) secretion of peritoneal macrophages; lessening chemotherapy-induced immunosuppression)	Chen et al. (2012)
	Alginate	Anti-angiogenic effect on HMEC-1 cells AT (Bel7402, SMMC7721, HT-29 cell lines)	Cong et al. (2014)
	Phytosterols	Cholesterol-lowering, antiatherosclerotic	Chen et al. (2014)
	Fucoidan	AC	Sun et al. (2017)
	Polysaccharides	AO IM (overcome immunosuppression in mice; increased spleen index and antioxidant activities)	Wang et al. (2013)
	Fucoidan	ADb (decreased blood glucose, diet and water intake, improvement of liver function, suppression of oxidative stress in diabetic mice; decrease of diabetes-related intestinal bacteria)	Cheng et al. (2019)
	Polysaccharides	ADb; anti-digestive ingredient with strong alpha-glucosidase inhibitory activity	Jia et al. (2020a, b)
S. hemiphyllum	Fucoidan	AI (inhibition of IL-1β, IL-6, TNF-α, NO), reduction of mRNA expression levels of IL-β, iNOS, and COX-2 in LPS-stimulated RAW 264.7 cells	Hwang et al. (2011)

(continued)

Table 5.2 (continued)

Species	Extract, fraction or compounds	Activity	References
S. horneri	Enzymatic extracts	AO (DPPH, hydroxyl and alkyl radicals)	Park et al. (2004)
	Polysaccharide	AO (H_2O_2 -induced oxidative injury in RAW264.7 cells), decrease of intracellular ROS and NO, MDA levels, restoration of SOD and GSH-Px	Wen et al. (2014)
	Phenolics	AI (MMP-2, −9)	Karadeniz et al. (2018)
	Fucoidan	AC	Athukorala et al. (2007)
	Alginate	AI (reduction of COX-2, IL-6, TNF-α levels; inhibition of key molecular mediators of NF-κB and MAPK pathways, reduction of levels of fine dust-derived metal ions in keratinocytes)	Fernando et al. (2018)
	Ethanol extract	AI (down-regulating LPS stimulation of NO production in macrophages), inhibits the activity of pro-inflammatory cytokines (IL-1β, IL-6, TNF-α) including PGE2. iNOS, COX-2	Jayawardena et al. (2019)
	Ethanol extract	A-allergic in type I allergic responses on immunoglobulin E/bovine serum albumin-mediated activation and passive cutaneous anaphylaxis reaction in mice	Han et al. (2020)
	Ethanol extract (mojabanchromanol)	A-allergic (reduction of β-hexosaminidase release in immunoglobulin E (IgE)/bovine serum albumin (BSA)-stimulated bone marrow-derived cultured mast cells)	Kim et al. (2020)
	Ethanol extract	AO (DPPH, H_2O_2,$O_2^{\cdot-}$, $^{\cdot}OH$, RP, metal ion-chelating effect	Lee et al. (2020a, b)
	70% ethanol extract	AI (inhibition of particulate matter induced inflammatory response in lung macrophages)	Sanjeewa et al. (2020)
	Whole seaweed	AOB, amelioration of diet-induced metabolic diseases	Murakami et al. (2021)
	Chloroform fraction of methanol extracts (norisoprenoids)	AI (lipopolysaccharide (LPS)-induced RAW 264.7 macrophages)	Sanjeewa et al. (2021)

(continued)

Table 5.2 (continued)

Species	Extract, fraction or compounds	Activity	References
S. hystrix	Extracts	ADb (lower glucose, triglycerides and cholesterol in diabetic rats)	Gotama et al. (2018)
	Dried seaweed powder	Improvement of blood glucose, triacylglycerol, cholesterol, and cortisol blood levels AI (in *in vivo* stressed rats)	Husni et al. (2019)
S. ilicifolium	70% ethanolic extract and ethyl acetate fraction	Antiamnesia in rats	Sumithra and Arunachalam (2014)
S. integerrimum		Suppression of bone loss induced by ovariectomized rats and the associated hyperlipidaemia by activating Nrf2	Wu et al. (2019)
S. mcclurei	Fucoidan	AT (human colon cancer DLD-1 cells)	Thinh et al. (2013)
	Protein hydrolysates	AHy (inhibition of ACE)	Zheng et al. (2020)
S. macrocarpum	80% ethanol extracts	AI (inhibition of IL-12 p40, IL-6, and TNF-α production in CpG-stimulated bone marrow-derived macrophages (BMDMs) and against NF-κB activation.	Manzoor et al. (2014)
S. micracanthum	Sargachromenol	AI (inhibition of LPS-induced production of NO, PGE_2, iNOS and COX-2, inhibition of activation of the NF-B signaling pathway)	Yang et al. (2013a)

(continued)

Table 5.2 (continued)

Species	Extract, fraction or compounds	Activity	References
S. muticum	Solvent fractions of 80% ethanol extracts	AO (DPPH, superoxide radical, xanthine oxidase inhibitory activity) AMI (food spoilage strains)	Kim et al. (2007)
	CH_2Cl_2 fraction of 80% ethanol extract	AI (inhibition of NO and PGE2 production, decreased expression of iNOS, COX-2, IL-1β, and IL-6 mRNA)	Yoon et al. (2010)
	Norisoprenoid (apo-9′-fucoxanthinone)	AI (suppressed NO and PGE_2 production, downregulation of NF-κB activation in macrophages)	Yang et al. (2013b)
	Solvent fractions	AO (protection against oxidative stress on MCF-7 cells, inhibition of H_2O_2 production, inhibition of Caspase-9 activity)	Pinteus et al. (2017)
	Methanolic extracts	Neuroprotective (6-hydroxydopamine - induced neurotoxicity in human neuroblastoma cell line SH-SY5Y)	Silva et al. (2018)
	Fucoidan, Phlorotannin	AO (lower serum glucose decreased liver GPx and TBARS levels in rats)	Balboa et al. (2019)
		AI (lowered the level of IL-6, TNF-α, and IFN- γ in a collagen-induced arthritis in mice model)	Jeon et al. (2019)
S. myriocystum	Polysaccharide	AO (radical scavenging activity)	Badrinathan et al. (2012)
S. plagiophyllum	Fucoidan	AT (HepG2 and human lung cancer A549)	Suresh et al. (2013)
	Hexane fractions of methanol-ethyl acetate extracts	ADep in mice	Mesripour et al. (2019)
S. polycystum	Fucoidan	AT (breast cancer cells MCF7)	Palanisamy et al. (2017)
	Phenolics, steroids	AM (against 4 human pathogens)	Peruma et al. (2019)
	Phenols	AO (reducing power, antiradical)	Johnson et al. (2019)
	Saponins, steroids, cardiac glycosides	AMI (*Streptococcus pneumoniae, Klebsiella pneumoniae*)	Yu et al. (2019)
	Fucoidan	AP (HL-60 and MCF-7)	Fernando et al. (2020a)

(continued)

Table 5.2 (continued)

Species	Extract, fraction or compounds	Activity	References
S. sagamianum		Protection of pancreatic β-cells against high glucose-induced oxidative stress and apoptosis	Lee et al. (2019)
	80% ethanol	Inhibition of postprandial hyperglycemia by inhibiting α-glucosidase and α-amylase in streptozotocin-induced diabetic mice	Lee and Han (2018)
	Ethanolic extract	AI (inhibition of expression of NO, cytokines (IL-6, IL-1β, TNF-α), iNOS, and COX-2 in LPS-induced RAW 264.7 cells through suppression of the NF-κB p65 pathway); reduced edema in mouse ear	Kim et al. (2013)
S. serratifolium	Plastoquinones	ADb (inhibition of tyrosine phosphatase 1B, α-glucosidase, and $ONOO^-$-mediated albumin nitration)	Ali et al. (2017)
	Meroterpenoids	AO, AI, AOB, lipid-lowering effects	Kwon et al. (2018)
	Sargahydroquinoic acid	AOB (stimulation of lipid catabolic pathways and adipocyte browning)	Kwon et al. (2019)
S. siliquastrum	Fucoidan	UV protection; UVB-protective effects in human HaCaT keratinocyte	Fernando et al. (2020b)
S. swartzii	Methanolic extracts	An, AI (paw edemas and peritonitis in rats)	Hong et al. (2011)
S. tenerrimum	Phlorotannins	AA (active and passive cutaneous anaphylaxis in female BALB/c mice)	Haider et al. (2009)
S. thunbergii	Fucosterol	Cytotoxicity (HT-29, B16F10, HL-60 cell lines)	Kim et al. (2009)
	Sargaquinoic and sargahydroquinoic acid	AOB (reduced lipid accumulation and adipogenic differentiation) AOs: Attenuate bone loss in osteoporosis	Kim et al. (2016)
	Indole derivatives	AOB (inhibition adipocyte differentiation in 3T3-L1 cells)	Kang et al. (2017)
	Polysaccharide	Prebiotic	Fu et al. (2018)
	Indole-6-carboxaldehyde	AI (inhibitor of MMP-9)	Kim et al. (2019a, b)

(continued)

Table 5.2 (continued)

Species	Extract, fraction or compounds	Activity	References
S. vulgare	Phenolics	AO (inhibition of linoleic acid peroxidation)	Rastian et al. (2007)
	Fucoidan	AC, AT, AO	Dore et al. (2013)
	Phenolics	AO (antiradical); AM (*Staphylococcus aureus, Aeromonas hydrophila, Bacillus cereus*, Methicillin-resistant *S. aureus*)	Arguelles et al. (2019)
S. weizhouense	Polysaccharide	Inhibition of histone acetylation and inflammatory cytokines production, improving the resistance of host against PCV2 infection	Hai-lan et al. (2019)
S. wightii	Terpenoids	AAz: Anti-Alzheimer AO (DPPH, $OH^{\cdot}$, H_2O_2, reducing power) Cholinesterase inhibition	Syad et al. (2013)
	Ethanolic extracts: alkaloids, carbohydrates, glycosides, phenolics	Reduces glucose and cholesterol levels	Ramu et al. (2019)
	Solvent fractions	AO (ABTS, DPPH, chelating ability), AHY (angiotensin converting enzyme-I inhibitor), AI (anti-COX-1, -2, -5-LOX, DPP-4 inhibitory)	Maneesh et al. (2017)
	O-heterocyclic analogues	AO (DPPH), AH (ACEI)	Maneesh and Chakraborty (2018)
	Phlorotannins	AO (DPPH, ABTS, FRAP) ACE inhibition	Vijayan et al. (2018)
	Fucoidan	AO (antiradical, chelating activity)	Hanjabam et al. (2019)
	Sodium alginate	AO (DPPH), AB (*Staphylococcus aureus, E. coli*)	Janarthanan and Senthil Kumar (2018)
	Fucoxanthin	Inhibition of ACE	Raji et al. (2020)

(continued)

Table 5.2 (continued)

Species	Extract, fraction or compounds	Activity	References
Sargassum sp.	Fucoidan	AT (Lewis lung carcinoma, melanoma B16 cancer cells), stimulation of natural killer cell activity in rats	Ale et al. (2011)
	Ethanol extracts	AO (decreased ROS, oxidative stress and improves sperm motility)	Sobhani et al. (2015)
	Laminaran	Prebiotic	Chamidah (2018)
	Alkaloids, phenolics	ABI (*Staphylococcus aureus, E. coli, S. epidermidis*)	Setyati et al. (2018)
	Phenolic	ABI (pathogens causing periodontal disease)	Herawati and Sumanik (2019)
	Protein	Anticancer agent	Karim et al. (2019)

AA antiallergenic, *AAZ* antiAlzheimer, *ABl* antibacterial, *AC* anticoagulant, *ACE* angiotensin I-converting enzyme, *ACEI* angiotensin converting enzyme inhibition, *ADb* antidiabetic, *AH* antihypertensive, *AHY* antihypertensive, *AI* anti-inflammatory, *An* analgesic, *AO* antioxidant, *AOB* antiobesity, *AOs* antiosteoporosis, *AT* anti-thrombotic, *FRAP* ferric reducing antioxidant power assay, *IFN-γ* interferon-γ, *IL-6* interleukin, *IM* immuno-modulatory, *iNOS* inducible NO synthase (iNOS), *LPS* lipopolysaccharide, *NO* nitric oxide, *PGE2* prostaglandin E2, *TNF-α* tumor necrosis factor

4 Biological Properties

Recent studies have reported a number of health beneficial actions for crude extracts of *Sargassum*, but also for fractions and components obtained by solvent and/or chromatography. Both conventional solvent and innovative extraction technologies are valid to selectively recover bioactives, some of these properties and production techniques are summarized in Table 5.2.

Abundant recent information on the antioxidant properties, reported as reducing, chelating and radical scavenging capacity, as well as the ability to protect against oxidation in bulk oil and emulsions, is available both for the aqueous fractions and for organic solvent extracts. Usually, crude extracts obtained with organic solvents are evaluated, i.e. acetone, chloroform, and methanol for the extraction of phenolics with antiradical and reducing properties (Park et al. 2004; Johnson et al. 2019). Also, purer fractions have been prepared (Syad et al. 2013; Arguelles et al. 2019; Maneesh and Chakraborty 2018), i.e. the ethyl acetate: methanol fraction of *S. wightii*, which showed excellent radical scavenging capacity compared to α-tocopherol. The most active antioxidants are phlorotannins (Rastian et al. 2007; Vijayan et al. 2018), but also sulfated polysaccharides (Chale-Dzul et al. 2017; Hanjabam et al. 2019), carotenoids, and peptides could be responsible for these properties in the seaweed extracts. The antioxidant activity of polysaccharides has been enhanced by selenylation (Xiao et al. 2019) or by fermentation with marine lactic acid bacteria (Shobharani et al. 2013) or with *Bacillus amyloliquefaciens*, *Lysinibacillus xylanilyticus*, and *Lactobacillus casei* (Kang et al. 2020). Most studies refer to chemical tests for the evaluation of antioxidant properties, but also results from cell assays are reported, i.e. sulfated polysaccharides protected against hydrogen peroxide (H_2O_2)-induced oxidative injury in RAW264.7 cells (Wen et al. 2014). Serum glucose was significantly lowered in rats receiving a diet supplement with *S. muticum* fucoidan and phlorotannins, liver GPx and TBARS levels decreased, but no effect on SOD activity in either liver or erythrocytes was observed (Balboa et al. 2019). The antioxidant effects have also been confirmed on the sperm protection against oxidative reactions during cryopreservation to maintain the reproductive potential and motility of sperm (Sobhani et al. 2015).

Different *Sargassum* compounds and fractions have anti-inflammatory properties (Kim et al. 2013; Manzoor et al. 2014; Saraswati et al. 2019), with the ability to decrease the production of nitric oxide, prostaglandin E2, proinflammatory cytokines and inhibition of 5-lipoxygenase (LOX), COX-1, and COX-2 as well as matrix metalloproteinase (MMP)-2 and MMP-9 expression both *in vitro* and *in vivo*. Such actions have been observed for *Sargassum* crude extracts and solvent fractions. As a general trend, the most lipophilic compounds tended to be active anti-inflammatory agents, but this activity has also been reported for terpenoids, phlorotannin, fucoxanthin, and also for their combinations with fucoidans (Park et al. 2010; Yang et al. 2013a, b; Hwang et al. 2016; Sanjeewa et al. 2018a, b). The sulfate content and molecular weight of fucoidan are highly determining of these properties (Wu et al. 2016). Fermentation of *S. horneri*, causing an increase in the phenolic and flavonoid content, increased the anti-inflammatory properties, measured as inhibitory activity against inducible nitric oxide synthase (iNOS), and cyclooxygenase-2 (COX-2) expression compared to the control group (Kang et al. 2020).

The anticoagulant action was initially studied in fucoidans, to replace heparin. Fermentation of the whole seaweed with marine lactic acid bacteria as starter cultures enhanced this property (Shobharani et al. 2013). The cytotoxic activity on human cancer cell lines has been observed for solvent extracts (Kim et al. 2009) and also for fucoidans (Torres et al. 2020; Fernando et al. 2020b; Thinh et al. 2013).

Different components could be responsible for the antimicrobial properties, particularly phenolics, fatty acids, and sulfated polysaccharides (Kim et al. 2007; Setyati et al. 2018; Sudaryono et al. 2018; Arguelles et al. 2019; Herawati and Sumanik 2019), active against both food spoilage and pathogenic microorganisms. The antiviral activity of sulfated polysaccharides, closely dependent on the sulfate content and molecular weight, is well known and has also been reported for *Sargassum* fucoidans (Dinesh et al. 2016).

Immunomodulatory effects have been observed for the sulfated polysaccharides (Chen et al. 2012; Wang et al. 2013) and ethyl acetate fractions (Chandraraj et al. 2010) from this genus. Kim et al. (2019a, b) confirmed both *in vitro* and *in vivo* immunological activity on splenocyte proliferation and cytokine production *in vitro* of *S. horneri* extracts obtained by hot water extraction and supercritical fluid extraction. Optimal results could be obtained with a mixture of 10% of the aqueous extracts and 5% of the supercritical one.

Fucoxanthin-rich extracts have shown antiosteoporosis activity by suppressive effects against osteoclast differentiation and by accelerating osteoblast formation (Koyama 2011). Sargaquinoic and sargahydroquinoic acid were identified in extracts with the ability to attenuate bone loss in osteoporosis (Kim et al. 2016). *In vivo* studies confirmed that *S. integerrimum* prevented bone loss in rats presenting oestrogen deficiency with hyperlipidaemia. These compounds acted by upregulating nuclear factor (erythroid-derived 2)-like 2 (Nrf2) without side effects and have been proposed as a promising treatment option for osteoporosis induced by oestrogen deficiency and hyperlipidemia in postmenopausal women (Wu et al. 2019).

The antiallergic properties have been described both for the alginate fraction, which protected against inflammation caused by fine dust in keratinocytes (Fernando et al. 2018) and for ethanolic extracts (Kim et al. 2020), which also showed *in vivo* protection in mice against ovalbumin and shrimp allergens as effectively as the antiallergic drug disodium cromoglycate (Haider et al. 2009).

Hepatoprotective activity has been reported for fucoidan (Chale-Dzul et al. 2017), but also for *S. polycystum* alcoholic extracts, which improved antioxidant levels and prevented depletion of antioxidant liver mitochondrial enzymes in rats (Raghavendran et al. 2005).

Antihypertensive activities, measured as the angiotensin converting enzyme inhibitory potential, have been reported for phlorotannins (Vijayan et al. 2018) and for o-heterocyclic analogues, isolated from the ethylacetate:methanol fraction of *S. wightii* (Maneesh and Chakraborty 2018). Saringosterol, among other phytosterols, contributed to lower cholesterol (Chen et al. 2014). Anti-diabetic activity of different components has been reported, plastoquinones (sargahydroquinoic acid, sargachromenol and sargaquinoic acid) (Ali et al. (2017), 80% ethanol (Lee and Han 2018) and fucoidan (Kwon et al. 2019. Antiobesity has been confirmed in different assays. *Sargassum thunbergii* indole derivatives inhibitory effects on adipogenesis (Kang et al. 2017). Sargaquinoic and sargahydroquinoic acid and fucoxanthin were identified in extracts with antiobesity effects by reducing lipid accumulation and adipogenic differentiation (Kim et al. 2016; Flórez-Fernández et al. 2019). A meroterpenoid-rich fraction of an ethanolic extract was active on obesity and obesity-related hepatic steatosis (Kwon et al. 2018). Furthermore, Min et al. (2014) confirmed the potential of *S. confusum* extracts to reduce body fat in overweight women. Polysaccharides show prebiotic action, with potential for the formulation of functional food with beneficial effects on gut health (Fu et al. 2018; Chamidah 2018).

Lowered synthesis of acetylcholine is a key marker enzyme in Alzheimer's disease, and the inhibition of acetylcholinesterase (AChE) is an approach to symptomatic treatment. Solvent extracts from *Sargassum* have shown AChE inhibitory properties (Syad et al. 2013) and neuroprotective against Parkinson's disease (Silva et al. 2018), but also fucoidan was active.

Different properties in relation to skin care and protection are gaining relevance. Among them, the ability to increase the proliferation of rat dermal papilla cells, potentiating hair growth (Kang et al. 2016). Abundant studies have confirmed the protection against ultraviolet B-induced oxidative stress in human HaCaT keratinocytes (Piao et al. 2014), UVA light photodamage attenuation and protection against intracellular ROS generation. Also, low molecular weight fucoidans provide effective protection against UVB radiation, as an anti-photoaging effect (Ye et al. 2018; Fernando et al. 2020a).

5 Conclusions and Future Trends

Sargassum sp. are traditionally used for food and medicinal purposes in some coastal areas and the results from more recent research have confirmed the high content and variety of bioactive compounds with valuable health properties. Further studies in relation to their chemical and structural characterization, and aimed at the standardization of bioactives are needed to facilitate their future commercial utilization in food, nutraceuticals and pharmaceuticals.

References

Airanthi MW, Hosokawa M, Miyashita K 2011 Comparative Antioxidant Activity of Edible Japanese Brown Seaweeds. J Food Sci 76(1):C104–C111

Al-Harthi MA, El-Deek AA (2012) Nutrient profiles of brown marine algae (*Sargassum dentifebium*) as affected by different processing methods for chickens. J Food Agric Environ 10:475–480

Ale MT, Maruyama H, Tamauchi H, Mikkelsen JD, Meyer AS (2011) Fucose-containing sulfated polysaccharides from brown seaweeds inhibit proliferation of melanoma cells and induce apoptosis by activation of caspase-3 in vitro. Mar Drugs 9(12):2605–2621

Ali Y, Kim DH, Seong SH, Kim H-R, Jung HA, Choi JS (2017) α-Glucosidase and protein tyrosine phosphatase 1b inhibitory activity of plastoquinones from marine brown alga *Sargassum serratifolium*. Mar Drugs 15:368. https://doi.org/10.3390/md15120368

Arguelles EDLR, Monsalud RG, Sapin AB (2019) Chemical composition and *in vitro* antioxidant and antibacterial activities of *Sargassum vulgare* C. Agardh from lobo, Batangas, Philippines. J Int Soc Southeast Asian Agric Sci 25:112–122

Asanka Sanjeewa KK, Kim H-S, Lee H-G, JAyawardena TU, NAgahawatta DP, Yang HW, Udayanga D, Kim J-I, Jeon Y-J (2021) 3-hydroxy-5,6-epoxy-β-ionone isolated from invasive harmful brown seaweed sargassum horneri protects MH-S mouse lung cells from urban particulate matter-induced inflammation. Appl Sci 11(22):10929

Athukorala Y, Lee K-W, Kim S-K, Jeon Y-J (2007) Anticoagulant activity of marine green and brown algae collected from Jeju Island in Korea. Bioresour Technol 98(9):1711–1716.

Badrinathan S, Shiju TM, Christa ASS, Arya R, Pragasam V (2012) Purification and structural characterization of sulfated polysaccharide from *Sargassum myriocystum* and its efficacy in scavenging free radicals. Indian J Pharm Sci 74:549–555. https://doi.org/10.4103/0250-474X.110600

Balboa EM, Gallego-Fábrega C, Moure A, Domínguez H (2016) Study of the seasonal variation on proximate composition of oven-dried *Sargassum muticum* biomass collected in Vigo Ria, Spain. J Appl Phycol 28(3):1943–1953. https://doi.org/10.1007/s10811-015-0727-x

Balboa EM, Millán R, Domínguez H, Taboada C (2019) *Sargassum muticum* hydrothermal extract: effects on serum parameters and antioxidant activity in rats. Appl Sci 9:570. https://doi.org/10.3390/app9122570

Byun E-H (2015) Comparison study of immunomodulatory activity of polysaccharide and ethanol extracted from *Sargassum fulvellum*. J Kor Soc Food Sci Nutr 44:1621–1628. https://doi.org/10.3746/jkfn.2015.44.11.1621

Cao Y, Duan J, Guo J, Li W, Tao W (2014) Pharmacokinetic properties of arsenic species after oral administration of *Sargassum pallidum* extract in rats using an HPLC-HG-AFS method. J Pharm Biomed Anal 96:213–219. https://doi.org/10.1016/j.jpba.2014.03.045

Chale-Dzul J, Freile-Pelegrín Y, Robledo D, Moo-Puc R (2017) Protective effect of fucoidans from tropical seaweeds against oxidative stress in HepG2 cells. J Appl Phycol 29(5):2229–2238. https://doi.org/10.1007/s10811-017-1194-3

Chamidah A (2018) Prebiotic index evaluation of crude laminaran of *Sargassum* sp. using feces of wistar rats. In: IOP conference series: earth and environmental science, vol 139, p 012043. https://doi.org/10.1088/1755-1315/139/1/012043

Chandraraj S, Prakash B, Navanath K (2010) Immunomodulatory activities of ethyl acetate extracts of two marine sponges *Gelliodes fibrosa* and *Tedania anhelans* and brown algae *Sargassum ilicifolium* with reference to phagocytosis. Res J Pharm Biol Chem Sci 1:302–307

Charles AL, Sridhar K, Alamsjah MA (2020) Effect of drying techniques on color and bioactive potential of two commercial edible Indonesian seaweed cultivars. J Appl Phycol 32(1):563–572. https://doi.org/10.1007/s10811-019-01916-4

Chen X, Nie W, Fan S, Zhang J, Wang Y, Lu J, Jin L (2012) A polysaccharide from *Sargassum fusiforme* protects against immunosuppression in cyclophosphamide-treated mice. Carbohydr Polym 90:1114–1119. https://doi.org/10.1016/j.carbpol.2012.06.052

Chen Z, Liu J, Fu Z, Ye C, Zhang R, Song Y, Zhang Y, Li H, Ying H, Liu H (2014) 24(S)-saringosterol from edible marine seaweed *Sargassum fusiforme* is a novel selective LXRβ agonist. J Agric Food Chem 62:6130–6137. https://doi.org/10.1021/jf500083r

Cheng Y, Sibusiso L, Hou L, Jiang H, Chen P, Zhang X, Wu M, Tong H (2019) *Sargassum fusiforme* fucoidan modifies the gut microbiota during alleviation of streptozotocin-induced hyperglycemia in mice. Int J Biol Macromol 131:1162–1170. https://doi.org/10.1016/j.ijbiomac.2019.04.040

Choi YY, Lee SJ, Lee YJ, Kim HS, Eom JS, Kim SC, Kim ET, Lee SS (2020) New challenges for efficient usage of *Sargassum fusiforme* for ruminant production. Sci Rep 10(1):19655. https://doi.org/10.1038/s41598-020-76700-3

Cong Q, Xiao F, Liao W, Dong Q, Ding K (2014) Structure and biological activities of an alginate from *Sargassum fusiforme*, and its sulfated derivative. Int J Biol Macromol 69:252–259. https://doi.org/10.1016/j.ijbiomac.2014.05.056

Connan S, Delisle F, Deslandes E, & Ar Gall E (2006) Intra-thallus phlorotannin content and antioxidant activity in Phaeophyceae of temperate waters. Bot Mar 49:39–46.

De Zoysa M, Nikapitiya C, Jeon Y-J, Jee Y, Lee J (2008) Anticoagulant activity of sulfated polysaccharide isolated from fermented brown seaweed Sargassum fulvellum. J Appl Phycol 20(1):67–74.

Dinesh S, Menon T, Hanna LE, Suresh V, Sathuvan M, Manikannan M (2016) *In vitro* anti-HIV-1 activity of fucoidan from *Sargassum swartzii*. Int J Biol Macromol 82:83–88. https://doi.org/10.1016/j.ijbiomac.2015.09.078

Dore CMPG, Faustino Alves MGDC, Pofírio Will LSE, Costa TG, SAbry DA, De Souza Rego LAR, Accardo CM, Rocha HAO, Filgueira LGA, Leite EL (2013) A sulfated polysaccharide, fucans, isolated from brown algae Sargassum vulgare with anticoagulant, antithrombotic, antioxidant and anti-inflammatory effects. Carbohydr Polym, 91(1):467–475.

D'Orazio N, Gemello E, Gammone MA, De Girolamo M, Ficoneri C, Riccioni G (2012) Fucoxantin: a treasure from the sea. Mar Drugs 10:604–616. https://doi.org/10.3390/md10030604

Fernando IPS, Jayawardena TU, Sanjeewa KKA, Wang L, Jeon Y-J, Lee WW (2018) Anti-inflammatory potential of alginic acid from *Sargassum horneri* against urban aerosol-induced inflammatory responses in keratinocytes and macrophages. Ecotoxicol Environ Saf 160:24–31. https://doi.org/10.1016/j.ecoenv.2018.05.024

Fernando IPS, Sanjeewa KKA, Lee HG, Kim H-S, Prasanna Vaas APJ, de Silva HIC, Nanayakkara CM, Abeytunga DTU, Lee D-S, Lee J-S, Jeon Y-J (2020a) Fucoidan purified from *Sargassum polycystum* induces apoptosis through mitochondria-mediated pathway in HL-60 and MCF-7 cells. Mar Drugs 18(4):18040196. https://doi.org/10.3390/md18040196

Fernando IPS, Dias MKHM, Madusanka DMD, Han EJ, Kim MJ, Jeon Y-J, Ahn G (2020b) Step gradient alcohol precipitation for the purification of low molecular weight fucoidan from *Sargassum siliquastrum* and its UVB protective effects. Int J Biol Macromol 163:26–35. https://doi.org/10.1016/j.ijbiomac.2020.06.232

Flórez-Fernández N, González-Muñoz MJ, Ribeiro D, Fernandes E, Domínguez H, Freitas M (2017) Algae polysaccharides' chemical characterization and their role in the inflammatory process. Curr Med Chem 24:149–175

Flórez-Fernández N, Casas MP, González-Muñoz MJ, Domínguez H (2019) Microwave hydro-gravity pretreatment of *Sargassum muticum* before solvent extraction of antioxidant and anti-obesity compounds. J Chem Technol Biotechnol 94:256–264. https://doi.org/10.1002/jctb.5771

Fu X, Cao C, Ren B, Zhang B, Huang Q, Li C (2018) Structural characterization and *in vitro* fermentation of a novel polysaccharide from *Sargassum thunbergii* and its impact on gut microbiota. Carbohydr Polym 183:230–239. https://doi.org/10.1016/j.carbpol.2017.12.048

Ganapathi K, Subramanian V, Mathan S (2013) Bioactive potentials of brown seaweeds, *Sargassum myriocystum* J. Agardh, *S. plagiophyllum* C. Agardh, and *S. ilicifolium* (turner) J. Agardh. Int Res J Pharmaceut Appl Sci 3(5):105–111

Gotama TL, Husni A, Ustadi. (2018) Antidiabetic activity of *Sargassum hystrix* extracts in streptozotocin-induced diabetic rats. Prev Nutr Food Sci 23:189–195. https://doi.org/10.3746/pnf.2018.23.3.189

Haider S, Li Z, Lin H, Jamil K, Wang BP (2009) *In vivo* study of antiallergenicity of ethanol extracts from *Sargassum tenerrimum*, *Sargassum cervicorne* and *Sargassum graminifolium turn*. Eur Food Res Technol 229:435–441. https://doi.org/10.1007/s00217-009-1066-4

Hai-lan C, Hong-lian T, Jian Y, Manling S, Heyu F, Na K, Wenyue H, Si-yu C, Ying-yi W, Ting-jun H (2019) Inhibitory effect of polysaccharide of *Sargassum weizhouense* on PCV2 induced inflammation in mice by suppressing histone acetylation. Biomed Pharmacother 112:108741. https://doi.org/10.1016/j.biopha.2019.108741

Han EJ, Kim H-S, Sanjeewa KKA, Jung K, Jee Y, Jeon Y-J, Fernando IPS, Ahn G (2020) *Sargassum horneri* as a functional food ameliorated IgE/BSA-induced mast cell activation and passive cutaneous anaphylaxis in mice. Mar Drugs 18(12):594. https://doi.org/10.3390/md18120594

Hanjabam MD, Kumar A, Tejpal CS, Krishnamoorthy E, Kishore P, Ashok KK (2019) Isolation of crude fucoidan from *Sargassum wightii* using conventional and ultra-sonication extraction methods. Bioact Carbohydr Diet Fibre 20:100200. https://doi.org/10.1016/j.bcdf.2019.100200

Herawati D, Sumanik BC (2019) Role of brown algae (*Sargassum* sp) as antibacterial (*Porphyromonas gingivalis*) in periodontal diseases. Int J Appl Pharmaceut 11:12–15. https://doi.org/10.22159/ijap.2019.v11s4.35272

Hong DD, Hien HM, Anh HTL (2011) Studies on the analgesic and anti-inflammatory activities of *Sargassum swartzii* (turner) C. Agardh (Phaeophyta) and *Ulva reticulata* Forsskal (Chlorophyta) in experiment animal models. Afr J Biotechnol 10:2308–2314

Husni A, Lailatussifa R, Isnansetyo A (2019) *Sargassum hystrix* as a source of functional food to improve blood biochemistry profiles of rats under stress. Prev Nutr Food Sci 24:150–158. https://doi.org/10.3746/pnf.2019.24.2.150

Hwang PA, Phan NN, Lu WJ, Hieu BTN, Lin YC (2016) Low-molecular-weight fucoidan and high-stability fucoxanthin from brown seaweed exert prebiotics and anti-inflammatory activities in Caco-2 cells. Food Nutr Res 60:1–9. https://doi.org/10.3402/fnr.v60.32033

Hwang P-A, Chien S-Y, Chan Y-L, Lu MK, Wu CH, Kong Z-L, Wu C-J (2011) Inhibition of lipopolysaccharide (LPS)-induced inflammatory responses by Sargassum hemiphyllum sulfated polysaccharide extract in RAW 264.7 Macrophage Cells. J Agric Food Chem 59(5):2062–2068

Janarthanan M, Senthil Kumar M (2018) Extraction of alginate from brown seaweeds and evolution of bioactive alginate film coated textile fabrics for wound healing application. J Ind Text 49:328–351. https://doi.org/10.1177/1528083718783331

JayawardenaTU, Kim H-S, Sanjeewa KKA, Kim S-Y, Rho J-R, Jee Y, Ahn G, Jeon Y-J (2019) *Sargassum horneri* and isolated 6-hydroxy-4,4,7a-trimethyl- 5,6,7,7a-tetrahydrobenzofuran-2(4H)-one (HTT); LPS-induced inflammation attenuation via suppressing NF-κB, MAPK and oxidative stress through Nrf2/HO-1 pathways in RAW 264.7 macrophages. Algal Res 40:101513. https://doi.org/10.1016/j.algal.2019.101513

Jeon H, Yoon W-J, Ham Y-M, Yoon S-A, Kang SC (2019) Anti-arthritis effect through the anti-inflammatory effect of *Sargassum muticum* extract in collagen-induced arthritic (CIA) mice. Molecules 24:276. https://doi.org/10.3390/molecules24020276

Jia R-B, Li Z-R, Wu J, Ou Z-R, Sun B, Lin L, Zhao M (2020a) Antidiabetic effects and underlying mechanisms of anti-digestive dietary polysaccharides from: *Sargassum fusiforme* in rats. Food Funct 11(8):7023–7036. https://doi.org/10.1039/D0FO01166E

Jia R-B, Wu J, Li Z-R, Ou Z-R, Lin L, Sun B, Zhao M (2020b) Structural characterization of polysaccharides from three seaweed species and their hypoglycemic and hypolipidemic activities in type 2 diabetic rats. Int J Biol Macromol 155:1040–1049. https://doi.org/10.1016/j.ijbiomac.2019.11.068

Jin H-G, Zhou M, Jin Q-H, Liu B-Y, Guan L-P 2017 Antidepressant-like effects of saringosterol, a sterol from Sargassum fusiforme by performing in vivo behavioral tests Med Chem Res 26(5):909–915

Johnson M, Kanimozhi SA, Joy Jeba Malar TR, Shibila T, Freitas PR, Tintino SR, Menezes IRA, da Costa JGM, Coutinho HDM (2019) The antioxidative effects of bioactive products from *Sargassum polycystum* C. Agardh and *Sargassum duplicatum* J. Agardh against inflammation and other pathological issues. Complement Ther Med 46:19–23. https://doi.org/10.1016/j.ctim.2019.06.014

Josephine A, Kumar SA (2011) *Sargassum wightii*-a nature's gift from the ocean. Chapter 9. In: Pomin VH (ed) Seaweed: ecology, nutrient composition and medicinal uses. Nova Science, London, pp 185–207

Kang M-C, Ding Y, Kim E-A, Choi YK, De Araujo T, Heo S-J, Lee S-H (2017) Indole derivatives isolated from brown alga *Sargassum thunbergii* inhibit adipogenesis through AMPK activation in 3T3-L1 preadipocytes. Mar Drugs 15:119. https://doi.org/10.3390/md15040119

Kang SM, Lee C, Jeong DH, Kim J, Boo KH, Kim JH, Kim CS (2020) Evaluation of biological activities of *Sargassum horneri* fermented by microorganisms. J Kor Soc Food Sci Nutr 49(11):1194–1201

Karadeniz F, Lee S-G, Oh JH, Kim J-A, Kong C-S (2018) Inhibition of MMP-2 and MMP-9 activities by solvent-partitioned *Sargassum horneri* extracts. Fish Aquat Sci 21:16. https://doi.org/10.1186/s41240-018-0093-0

Karim H, Ahmad A, Natzir R, Massi MN, Arfah R, Asmi N, Karim A (2019) Isolation and identification of bioactive proteins from the brown algae *Sargassum* sp. and their potential as anticancer agents. J Phys Conf Ser 1341:32009. https://doi.org/10.1088/1742-6596/1341/3/032009

Kim J-Y, Lee J-A, Kim K-N, Yoon W-J, Lee WJ, Park S-Y (2007) Antioxidative and antimicrobial activities of *Sargassum muticum* extracts. J Kor Soc Food Sci Nutr 36:663–669. https://doi.org/10.3746/jkfn.2007.36.6.663

Kim KN, Ham YM, Moon JY, Kim MJ, Kim DS, Lee WJ, Lee NH, Hyun C-G (2009) *In vitro* cytotoxic activity of *Sargassum thunbergii* and *Dictyopteris divaricata* (Jeju seaweeds) on the HL-60 tumour cell line. Int J Pharm 5:298–306. https://doi.org/10.3923/ijp.2009.298.306

Kim M-J, Kim K-B, Jeong D-H, Ahn D-H (2013) Anti-inflammatory activity of ethanolic extract of *Sargassum sagamianum* in RAW 264.7 cells. Food Sci Biotechnol 22:1113–1120. https://doi.org/10.1007/s10068-013-0191-9

Kim J-A, Karadeniz F, Ahn B-N, Kwon MS, Mun O-J, Bae MJ, Seo Y, Kim M, Lee S-H, Kim YY, Mi-Soon J, Kong C-S (2016) Bioactive quinone derivatives from the marine brown alga

Sargassum thunbergii induce anti-adipogenic and pro-osteoblastogenic activities. J Sci Food Agric 96(3):783–790. https://doi.org/10.1002/jsfa.7148

Kim D-S, Sung N-Y, Han I-J, Lee B-S, Park S-Y, Nho EY, Eom J, Kim G, Kim K-A (2019a) Splenocyte-mediated immune enhancing activity of *Sargassum horneri* extracts. J Nutr Health 52(6):515–528. https://doi.org/10.4163/jnh.2019.52.6.515

Kim T-H, Heo S-J, Ko S-C, Park WS, Choi I-W, Yi M, Jung W-K (2019b) Indole-6-carboxaldehyde isolated from *Sargassum thunbergii* inhibits the expression and secretion of matrix metalloproteinase-9. Int J Mol Med 44:1979–1987. https://doi.org/10.3892/ijmm.2019.4319

Kim H-S, Han EJ, Fernando IPS, Sanjeewa KKA, Jayawardena TU, Kim H-J, Jee Y, Kang S-H, Jang J-H, Jang J-P, Herath KHINM, Jeon Y-J, Ahn G (2020) Anti-allergy effect of mojabanchromanol isolated from *Sargassum horneri* in bone marrow-derived cultured mast cells. Algal Res 48:101898. https://doi.org/10.1016/j.algal.2020.101898

Kim H-S, Fernando IPS, Lee S-H, Ko S-C, Kang MC, Ahn G, Je J-G, Sanjeewa KKA, Rho J-R, Shin HJ, Lee W, Lee D-S, Jeon Y-J (2021) Isolation and characterization of anti-inflammatory compounds from *Sargassum horneri* via high-performance centrifugal partition chromatography and high-performance liquid chromatography. Algal Res 54:102209. https://doi.org/10.1016/j.algal.2021.102209

Koyama T (2011) Extracts of marine algae show inhibitory activity against osteoclast differentiation. Adv Food Nutr Res 64:443–454. https://doi.org/10.1016/B978-0-12-387669-0.00034-X

Kumar Y, Tarafdar A, Kumar D, Badgujar PC (2019) Effect of Indian brown seaweed *Sargassum wightii* as a functional ingredient on the phytochemical content and antioxidant activity of coffee beverage. J Food Sci Technol 56:4516–4525. https://doi.org/10.1007/s13197-019-03943-y

Kwon M, Lim S-J, Joung E-J, Lee B, Oh C-W, Kim H-R (2018) Meroterpenoid-rich fraction of an ethanolic extract from *Sargassum serratifolium* alleviates obesity and non-alcoholic fatty liver disease in high fat-fed C57BL/6J mice. J Funct Foods 47:288–298. https://doi.org/10.1016/j.jff.2018.05.063

Kwon M, Lee B, Lim S-J, Choi JS, Kim H-R (2019) Sargahydroquinoic acid, a major compound in *Sargassum serratifolium* (C. Agardh), widely activates lipid catabolic pathways, contributing to the formation of beige-like adipocytes. J Funct Foods 58:355–366. https://doi.org/10.1016/j.jff.2019.04.045

Lee J-S, Han J-S (2018) *Sargassum sagamianum* extract alleviates postprandial hyperglycemia in diabetic mice. Prev Nutr Food Sci 23:122–126. https://doi.org/10.3746/pnf.2018.23.2.122

Lee J-S, Lee H-A, Han J-S (2019) *Sargassum sagamianum* extract protects INS-1 pancreatic β cells against high glucose-induced apoptosis. Cytotechnology 71:389–399. https://doi.org/10.1007/s10616-019-00295-5

Lee JH, Kim HJ, Jee Y, Jeon Y-J, Kim HJ (2020a) Antioxidant potential of *Sargassum horneri* extract against urban particulate matter-induced oxidation. Food Sci Biotechnol 29:855–865. https://doi.org/10.1007/s10068-019-00729-y

Lee P-T, Quan Tran HT, Huang H-T, Nan F-H, Lee M-C (2020b) *Sargassum horneri* extracts stimulate innate immunity, enhance growth performance, and upregulate immune genes in the white shrimp *Litopenaeus vannamei*. Fish Shellfish Immunol 102:276–285. https://doi.org/10.1016/j.fsi.2020.04.049

Le Lann K, Jégou C, & Stiger-Pouvreau V (2008) Effect of different conditioning treatments on total phenolic content and antioxidant activities in two Sargassacean species: Comparison of the frondose Sargassum muticum (Yendo) Fensholt and the cylindrical Bifurcaria bifurcata R Ross Phycol Res 56:238–245.

Liu L, Heinrich M, Myers S, Dworjanyn SA (2012) Towards a better understanding of medicinal uses of the brown seaweed *Sargassum* in traditional Chinese medicine: a phytochemical and pharmacological review. J Ethnopharmacol 142:591–619. https://doi.org/10.1016/j.jep.2012.05.046

Magura J, Moodley R, Jonnalagadda SB (2019) Toxic metals (as and Pb) in *Sargassum elegans* Suhr (1840) and its bioactive compounds. Int J Environ Health Res 29:266–275. https://doi.org/10.1080/09603123.2018.1537439

Maneesh A, Chakraborty K (2018) Previously undescribed antioxidative O-heterocyclic angiotensin converting enzyme inhibitors from the intertidal seaweed *Sargassum wightii* as potential antihypertensives. Food Res Int 13:474–486. https://doi.org/10.1016/j.foodres.2018.07.035

Maneesh A, Chakraborty K, Makkar F (2017) Pharmacological activities of brown seaweed *Sargassum wightii* (family Sargassaceae) using different *in vitro* models. Int J Food Prop 20:931–945. https://doi.org/10.1080/10942912.2016.1189434

Manzoor Z, Mathema VB, Chae D, Yoo E-S, Kang H-K, Hyun J-W, Lee NH, Ko M-H, Koh Y-S (2014) Extracts of the seaweed *Sargassum macrocarpum* inhibit the CpG-induced inflammatory response by attenuating the NF-κB pathway. Food Sci Biotechnol 23:293–297. https://doi.org/10.1007/s10068-014-0041-4

Mateos R, Pérez-Correa JR, Domínguez H (2020) Bioactive properties of marine phenolics. Mar Drugs 18(10):501. https://doi.org/10.3390/md18100501

Mesripour A, Rabian N, Yegdaneh A (2019) The effect of different partitions of seaweed *Sargassum plagyophylum* on depression behavior in mice model of despair. J Complement Integr Med 16:20180207. https://doi.org/10.1515/jcim-2018-0207

Min KS, Han D, Kwon S-O, Yeo K-M, Kim B-N, Ly SY (2014) The effect of *Sargassum confusum* on reduction of body fat in obese women. J Nutr Health 47:23–32. https://doi.org/10.4163/jnh.2014.47.1.23

Moni SS, Alam MF, Makeen HA, Alhazmi HA, Sultan M, Siddiqui R, Jabeen A, Sanobar S, Alam MS, Rehman ZU, Elmobark ME, Madkhali O, Haque A, Albratty M (2019) Solvent extraction, spectral analysis and antibacterial activity of the bioactive crystals of *Sargassum aquifolium* (turner) C.Agardh from Red Sea. Nat Prod Res:1–5. https://doi.org/10.1080/14786419.2019.1645659

Murakami K, Yamaguchi Y, Noda K, Fujii T, Shinohara N, Ushirokawa T, Sugawa-Katayama Y, Katayama M (2011) Seasonal variation in the chemical composition of a marine brown alga, *Sargassum horneri* (turner) C. Agardh. J Food Compos Anal 24(2):231–236. https://doi.org/10.1016/j.jfca.2010.08.004

Murakami S, Hirazawa C, Ohya T, Yoshikawa R, Mizutani T, Ma N, Moriyama M, Ito T, Matsuzaki C (2021) T*he edible brown seaweed Sargassum horneri* (turner) C. agardh ameliorates high-fat diet-induced obesity, diabetes, and hepatic steatosis in mice. Nutrients 13(2):551. https://doi.org/10.3390/nu13020551

Nazarudin MF, Yusoff F, Idrus ES, Aliyu-Paiko M (2020) Brown seaweed *Sargassum polycystum* as dietary supplement exhibits prebiotic potentials in Asian sea bass Lates calcarifer fingerlings. Aquacult Rep 18:100488. https://doi.org/10.1016/j.aqrep.2020.100488

Palanisamy S, Vinosha M, Marudhupandi T, Rajasekar P, Prabhu NM (2017) Isolation of fucoidan from Sargassum polycystum brown algae: Structural characterization, in vitro antioxidant and anticancer activity. Int J Biol Macromol, 102:405–412.

Park P-J, Shahidi F, Jeon Y-J (2004) Antioxidant activities of enzymatic extracts from an edible seaweed *Sargassum horneri* using ESR spectrometry. J Food Lipids 11:15–27. https://doi.org/10.1111/j.1745-4522.2004.tb00257.x

Park S-B, Chun K-R, Jung Y-M, Kim J-K, Suk K, Lee W-H (2010) The differential effect of high and low molecular weight fucoidans on the severity of collagen-induced arthritis in mice. Phytother Res 24:1384–1391. https://doi.org/10.1002/ptr.3140

Peruma B, Chitra R, Maruthupandian A, Viji M (2019) Nutritional assessment and bioactive potential of *Sargassum polycystum* C. Agardh (brown seaweed). Indian J Geo-Marine Sci 48:492–498

Piao MJ, Kim KC, Zheng J, Yao CW, Cha JW, Boo SJ, Yoon WJ, Kang HK, Yoo ES, Koh YS, Ko MH, Lee NH, Hyun JW (2014) The ethyl acetate fraction of Sargassum muticum attenuates ultraviolet B radiation-induced apoptotic cell death via regulation of MAPK- and caspase-dependent signaling pathways in human HaCaT keratinocytes. Pharm Biol 52(9):1110–1118.

Pinteus S, Lemos MFL, Silva J, Alves C, Neugebauer A, Freitas R, Duarte A, Pedrosa R (2017) An insight into *Sargassum muticum* cytoprotective mechanisms against oxidative stress on a human cell *in vitro* model. Mar Drugs 15:353. https://doi.org/10.3390/md15110353

Plouguerné E, Le Lann K, Connan S, Jechoux G, Deslandes E, & Stiger-Pouvreau V (2006) Spatial and seasonal variation in density, reproductive status, length and phenolic content of the invasive brown macroalga Sargassum muticum (Yendo) Fensholt along the coast of Western Brittany (France). Aquatic Bot 85:337–344

Raghavendran BH, Sathivel A, Devaki T (2005) Antioxidant effect of *Sargassum polycystum* (Phaeophyceae) against acetaminophen induced changes in hepatic mitochondrial enzymes during toxic hepatitis. Chemosphere 61:276–281. https://doi.org/10.1016/j.chemosphere.2005.01.049

Raji V, Loganathan C, Sadhasivam G, Kandasamy S, Poomani K, Thayumanavan P (2020) Purification of fucoxanthin from *Sargassum wightii* Greville and understanding the inhibition of angiotensin 1-converting enzyme: an *in vitro* and *in silico* studies. Int J Biol Macromol 148:696–703. https://doi.org/10.1016/j.ijbiomac.2020.01.140

Ramu S, Murali A, Jayaraman A (2019) Phytochemical screening and toxicological evaluation of *Sargassum wightii* Greville in Wistar rats. Turk J Pharmaceut Sci 16:466–475. https://doi.org/10.4274/tjps.galenos.2018.68442

Rastian Z, Mehranian M, Vahabzadeh F, Sartavi K (2007) Antioxidant activity of brown algae *Sargassum vulgar* and *Sargassum angustrifolum*. J Aquat Food Prod Technol 16:17–26. https://doi.org/10.1300/J030v16n02_03

Rushdi MI, Abdel-Rahman IAM, Saber H, Attia EZ, Abdelraheem WM, Madkour HA, Hassan HM, Elmaidomy AH, Abdelmohsen UR (2020) Pharmacological and natural products diversity of the brown algae genus: *Sargassum*. RSC Adv 10(42):24951–24972. https://doi.org/10.1039/d0ra03576a

Sanjeewa KKA, Fernando IPS, Kim SY, Kim HS, Ahn G, Jee Y, Jeon YJ (2018a) *In vitro* and *in vivo* anti-inflammatory activities of high molecular weight sulfated polysaccharide; containing fucose separated from *Sargassum horneri*. Short communication. Int J Biol Macromol 107:803–807. https://doi.org/10.1016/j.ijbiomac.2017.09.050

Sanjeewa KKA, Kang N, Ahn G, Jee Y, Kim Y-T, Jeon Y-J (2018b) Bioactive potentials of sulfated polysaccharides isolated from brown seaweed *Sargassum* spp in related to human health applications: a review. Food Hydrocoll 81:200–208. https://doi.org/10.1016/j.foodhyd.2018.02.040

Sanjeewa KKA, Jayawardena TU, Kim S-Y, Kim H-S, Ahn G, Kim J, Jeon Y-J (2019a) Fucoidan isolated from invasive *Sargassum horneri* inhibits LPS-induced inflammation via blocking NF-κB and MAPK pathways. Algal Res 41:101561. https://doi.org/10.1016/j.algal.2019.101561

Sanjeewa KKA, Jayawardena TU, Kim H-S, Kim S-Y, Ahn G, Kim H-J, Fu X, Jee Y, Jeon Y-J (2019b) Ethanol extract separated from *Sargassum horneri* (turner) abate LPS-induced inflammation in RAW 264.7 macrophages. Fish Aquat Sci 22:6. https://doi.org/10.1186/s41240-019-0121-8

Sanjeewa KKA, Jayawardena TU, Kim S-Y, Lee HG, Je J-G, Jee Y, Jeon Y-J (2020) *Sargassum horneri* (turner) inhibits urban particulate matter-induced inflammation in MH-S lung macrophages via blocking TLRs mediated NF-κB and MAPK activation. J Ethnopharmacol 249:112363. https://doi.org/10.1016/j.jep.2019.112363

Saraswati, Giriwono PE, Iskandriati D, Tan CP, Andarwulan N (2019) *Sargassum* seaweed as a source of anti-inflammatory substances and the potential insight of the tropical species: a review. Mar Drugs 17:590. https://doi.org/10.3390/md17100590

Setyati WA, Pramesti R, Zainuddin M, Puspita M, Renta PP (2018) Cytotoxicity and phytochemical profiling of *Sargassum* sp. extract as anti-MDR bacteria. In: IOP Conference Series: Earth and Environmental Science, vol 116, p 012024. https://doi.org/10.1088/1755-1315/116/1/012024

Shama A, Joyce SG, Mari FD, Ranga Rao A, Ravishankar GA, Hudaa N (2019) Macroalgae and microalgae: novel sources of functional food and feed. In: Ravishankar GA, Rao AR (eds) Handbook of algal technologies and phytochemicals. Food, health and nutraceutical applications, vol 1. CRC, Boca Raton, FL, pp 207–219

Shobharani P, Halami PM, Sachindra NM (2013) Potential of marine lactic acid bacteria to ferment *Sargassum* sp. for enhanced anticoagulant and antioxidant properties. J Appl Microbiol 114:96–107. https://doi.org/10.1111/jam.12023

Silva J, Alves C, Pinteus S, Mendes S, Pedrosa R (2018) Neuroprotective effects of seaweeds against 6-hydroxidopamine-induced cell death on an *in vitro* human neuroblastoma model. BMC Complement Altern Med 18:58. https://doi.org/10.1186/s12906-018-2103-2

Sobhani A, Eftekhaari TE, Shahrzad ME, Natami M, Fallahi S (2015) Antioxidant effects of brown algae *Sargassum* on sperm parameters: CONSORT-compliant article. Medicine 94:e1938. https://doi.org/10.1097/MD.0000000000001938

Sudaryono A, Chilmawati D, Susilowati T (2018) Oral administration of hot-water extract of tropical brown seaweed, *Sargassum cristaefolium*, to enhance immune response, stress tolerance, and resistance of white shrimp, *Litopenaeus vannamei*, to *Vibrio parahaemolyticus*. J World Aquacult Soc 4(5):877–888. https://doi.org/10.1111/jwas.12527

Sugiura Y, Usui M, Katsuzaki H, Imai K, Kakinuma M, Amano H, Miyata M (2018) Orally administered phlorotannins from Eisenia arborea suppress chemical mediator release and cyclooxygenase-2 signaling to alleviate mouse ear swelling. Mar Drugs 16(8):267. https://doi.org/10.3390/md16080267

Sumithra M, Arunachalam G (2014) Acute and subacute oral toxicity study of *Sargassum ilicifolium* turner C. Agardh in rodents. Int J PharmTech Res 6:821–828

Sun Y, Ding G-F, Xu Y-F (2017) Saringosterol and Fucosterol, Two Sterols from Sargassum fusiforme with Antibacterial and Antioxidant Properties. Oceanologia et Limnologia Sinica 48(3):640–646.

Sung N-Y, Kim H-M, Byun E-B, Park J-N, Park C, Byun M-W, Byun E-H (2015) Polysaccharide extracted from *Sargassum fulvellum* leads to macrophage activation and Th1 polarization in splenocytes. Fish Sci 81:777–785. https://doi.org/10.1007/s12562-015-0886-3

Suresh V, Anbazhagan C, Thangam R, Senthilkumar D, Senthilkumar NN, Kannan S, Rengasamy R, Palani P (2013) Stabilization of mitochondrial and microsomal function of fucoidan from Sargassum plagiophyllum in diethylnitrosamine induced hepatocarcinogenesis. Carbohydr Polym 92(2):1377–1385.

Syad AN, Shunmugiah KP, Kasi PD (2013) Antioxidant and anti-cholinesterase activity of *Sargassum wightii*. Pharm Biol 51:1401–1410. https://doi.org/10.3109/13880209.2013.793721

Tapia-Martinez J, Hernández-Cruz K, Franco-Colín M, Mateo-Cid LE, Mendoza-Gonzalez C, Blas-Valdivia V, Cano-Europa E (2019) Safety evaluation and antiobesogenic effect of *Sargassum liebmannii* J. Agardh (Fucales: Phaeophyceae) in rodents. J Appl Phycol 31:2597–2607. https://doi.org/10.1007/s10811-019-1752-y

Telles CBS, Mendes-Aguiar C, Fidelis GP, Frasson AP, Pereira WO, Scortecci KC, Camara RBG, Nobre LTDB, Costa LS, Tasca T, Rocha HAO (2018) Immunomodulatory effects and antimicrobial activity of heterofucans from *Sargassum filipendula*. J Appl Phycol 30(1):569–578. https://doi.org/10.1007/s10811-017-1218-z

Thinh PD, Menshova RV, Ermakova SP, Anastyuk SD, Ly BM, Zvyagintseva TN (2013) Structural characteristics and anticancer activity of fucoidan from the brown alga Sargassum mcclurei. Mar Drugs 11(5):1453–1476

Tian H, Liu H, Song W, Zhu L, Zhang T, Li R, Yin X (2020) Structure, antioxidant and immunostimulatory activities of the polysaccharides from *Sargassum carpophyllum*. Algal Res 49:101853. https://doi.org/10.1016/j.algal.2020.101853

Torres MD, Flórez-Fernández N, Simón-Vázquez N, Giménez-Abián JF, Díaz JF, González-Fernández A, Domínguez H (2020) Fucoidans: the importance of processing on their antitumoral properties. Algal Res 45:101748. https://doi.org/10.1016/j.algal.2019.101748

Usoltseva RV, Anastyuk SD, Shevchenko NM, Surits VV, Silchenko AS, Isakov VV, Zvyagintseva TN, Duc TP, Ermakova SP (2017) Polysaccharides from brown algae *Sargassum duplicatum*: the structure and anticancer activity *in vitro*. Carbohydr Polym 175:547–556. https://doi.org/10.1016/j.carbpol.2017.08.044

Vijayan R, Chitra L, Penislusshiyan S, Palvannan T (2018) Exploring bioactive fraction of *Sargassum wightii: in vitro* elucidation of angiotensin-i-converting enzyme inhibition and antioxidant potential. Int J Food Prop 21:674–684. https://doi.org/10.1080/10942912.2018.1454465

Wang W, Lu J-B, Wang C, Wang C-S, Zhang H-H, Li C-Y, Qian G-Y (2013) Effects of *Sargassum fusiforme* polysaccharides on antioxidant activities and intestinal functions in mice. Int J Biol Macromol 58:127–132. https://doi.org/10.1016/j.ijbiomac.2013.03.062

Wen Z-S, Liu L-J, OuYang X-K, Qu Y-L, Chen Y, Ding G-F (2014) Protective effect of polysaccharides from *Sargassum horneri* against oxidative stress in RAW264.7 cells. Int J Biol Macromol 68:98–106. https://doi.org/10.1016/j.ijbiomac.2014.04.037

Wong K, Cheung PC (2001) Influence of drying treatment on three *Sargassum* species. J Appl Phycol 13(1):43–50. https://doi.org/10.1023/A:1008149215156

Wu K, Gong Z, Zou L, Ye H, Wang C, Liu Y, Liang Y, Li Y, Ren J, Cui L, Liu Y (2019) *Sargassum integerrimum* inhibits oestrogen deficiency and hyperlipidaemia-induced bone loss by upregulating nuclear factor (erythroid-derived 2)-like 2 in female rats. J Orthopaed Trans 19:106–117. https://doi.org/10.1016/j.jot.2019.03.002

Wu G-J, Shiu S-M, Hsieh M-C, Tsai G-J (2016) Anti-inflammatory activity of a sulfated polysaccharide from the brown alga Sargassum cristaefolium. Food Hydrocoll 53:16–23

Xiao H, Chen C, Li C, Huang Q, Fu X (2019) Physicochemical characterization, antioxidant and hypoglycemic activities of selenized polysaccharides from *Sargassum pallidum*. Int J Biol Macromol 132:308–315. https://doi.org/10.1016/j.ijbiomac.2019.03.138

Yang E-J, Ham YM, Yang K-W, Lee NH, Hyun C-G (2013a) Sargachromenol from *Sargassum micracanthum* inhibits the lipopolysaccharide-induced production of inflammatory mediators in RAW 264.7 macrophages. Scientific World Journal 2013:712303. https://doi.org/10.1155/2013/712303

Yang E-J, Ham YM, Lee WJ, Lee NH, Hyun C-G (2013b) Anti-inflammatory effects of apo-9′-fucoxanthinone from the brown alga, *Sargassum muticum*. DARU 21:62. https://doi.org/10.1186/2008-2231-21-62

Yang W-C, Zhang Y-Y, Li Y-J, Nie Y-Y, Liang J-Y, Liu Y-Y, Liu J-S, Zhang Y-P, Song C, Qian Z-J, Zhang Y (2020) Chemical composition and aisease-related activities of a functional oil from the edible seaweed *Hizikia fusiforme*. Chem Biodivers 17(8):e2000055. https://doi.org/10.1002/cbdv.202000055

Ye Y, Ji D, You L, Zhou L, Zhao Z, Brennan C (2018) Structural properties and protective effect of *Sargassum fusiforme* polysaccharides against ultraviolet B radiation in hairless kun Ming mice. J Funct Foods 43:8–16. https://doi.org/10.1016/j.jff.2018.01.025

Yende S, Harle U, Chaugule B (2014) Therapeutic potential and health benefits of *Sargassum* species. Pharmacogn Rev 8:1–7. https://doi.org/10.4103/0973-7847.125514

Yokoi K, Konomi A (2012) Toxicity of so-called edible hijiki seaweed (*Sargassum fusiforme*) containing inorganic arsenic. Regul Toxicol Pharmacol 63:291–297. https://doi.org/10.1016/j.yrtph.2012.04.006

Yoon W-J, Ham YM, Lee WJ, Lee NH, Hyun C-G (2010) Brown alga *Sargassum muticum* inhibits proinflammatory cytokines, iNOS, and COX-2 expression in macrophage RAW 264.7 cells. Turk J Biol 34:25–34

Yu K-X, Norhisham SN, Ng CH (2019) Antimicrobial potential of padina australis and *Sargassum polycystum* against respiratory infections causing bacteria. Int J Med Toxicol Legal Med 22:138–141. https://doi.org/10.5958/0974-4614.2019.00030.5

Zheng Y, Zhang Y, San S (2020) Efficacy of a novel ACE-inhibitory peptide from *Sargassum maclurei* in hypertension and reduction of intracellular endothelin-1. Nutrients 12:653. https://doi.org/10.3390/nu12030653

Chapter 6
Food Applications and Health Benefits of The Genus *Gigartina* (Rhodophyta)

João Cotas, Sara García-Poza, Diana Pacheco, Glacio Araújo, José W. A. Silva, Ana M. M. Gonçalves, and Leonel Pereira

1 Introduction

Macroalgae are multicellular, plant-like organisms that normally live attached to rocks or other hard substrates in coastal areas. There are about 10,000 species of algae, of which 6500 are red algae (also called Rhodophyta), 2000 are brown algae (Phaeophyceae), and 1500 are green algae (Chlorophyta and Charophytes) (Guiry and Guiry 2021). These three taxonomic groups have different evolutionary histories showing specific ultrastructural and biochemical characteristics (Barbier et al. 2019). Algae are a source of novel bioactive compounds, such as phlorotannins and certain polysaccharides, which do not occur in terrestrial plants and can confer certain health-promoting properties (Brown et al. 2014). The different species of marine algae present differences in the composition and concentration of these bioactive compounds (Brown et al. 2014; Shama et al. 2019).

The use of seaweed by harvesting or gathering from the natural environment is an ancient human practice in several regions of the world, such as in China and Europe. Also, there is evidence of the use of seaweed for food and medicine for more than 14,000 years in southern Chile (Dillehay et al. 2008; Radulovich et al.

J. Cotas · S. García-Poza · D. Pacheco · L. Pereira (✉)
Department of Life Sciences, MARE–Marine and Environmental Sciences Centre, University of Coimbra, Coimbra, Portugal
e-mail: leonel.pereira@uc.pt

G. Araújo · J. W. A. Silva
Federal Institute of Education, Science and Technology of Ceará–IFCE, Aracati, Ceará, Brazil

A. M. M. Gonçalves
Department of Life Sciences, MARE–Marine and Environmental Sciences Centre, University of Coimbra, Coimbra, Portugal

Department of Biology and CESAM, University of Aveiro, Aveiro, Portugal

A. Ranga Rao, G. A. Ravishankar (eds.), *Sustainable Global Resources of Seaweeds Volume 2*, https://doi.org/10.1007/978-3-030-92174-3_6

2015). Today, algae are still part of the usual diet in many Asian countries. The popularity of seaweed consumption is growing in Western cultures, due both to the influx of Asian cuisine and the notional health benefits associated with their consumption. The addition of seaweed and seaweed isolates to foods has already shown potential to improve satiety and reduce postprandial glucose and lipid absorption rates in acute human feeding studies, highlighting their potential use in food development against obesity (Brownlee et al. 2011).

There is a direct correlation between the consumption of seaweed, with a lower incidence of chronic diseases such as cancer, hyperlipidemia and coronary heart disease (CD), above all based on epidemiological studies that compare Japanese and Western diets (Kim et al. 2009; Iso 2011). Although the consumption of seaweed in Western countries is growing, most of the harvested seaweed is used for the manufacture of hydrocolloids. For example, algae are used in the production of alginate, agar, and carrageenan, which are gelling agents used in the food industry (Smit 2004; Brown et al. 2014). The concept of using marine bioactive compounds from seaweed in therapeutic applications and to promote health, although common in Asian countries, it is an innovative approach in Western countries (Brown et al. 2014).

However, precise identification of specific seaweed species is vital for commercial exploitation (Falshaw and Furneaux 2009). The identification and taxonomy of the genus *Gigartina*, based only on morphology, is very difficult as there are similar species such as *Sarcothalia* sp. (Parsons et al. 1977) with close morphological traits. Thus, the new genomic tools that can sequence the ribulose-bisphosphate carboxylase gene (rbcL) in the seaweed chloroplast helped in the differentiation and clarification of the *Gigartina* species (Nelson and Broom 2008; Falshaw and Furneaux 2009). The *Gigartina* species are identical chemotaxonomic and different in terms of compounds quality from other identical red seaweeds species, such as the polysaccharide fraction which can be used as taxonomic marker as described by Falshaw and Furneaux (2009).

Actually, the *Gigartina* genus is considered important due to its content of carrageenan, which is currently exploited industrially (Falshaw and Furneaux 1998; Leandro et al. 2020; Guiry and Guiry 2021). Another commercial application is for direct food and feed usage (Cyrus et al. 2015; Avila-Peltroche and Padilla-Vallejos 2020).

This chapter focus on a relatively unexploited genus of red seaweeds, the *Gigartina* genus, highlighting their current ecological and economic relevance.

2 *Gigartina* Genus: Ecology

Environmentally, seaweeds play an important role as primary producers in aquatic ecosystems, being able to absorb and accumulate pollutants, contaminants and heavy metals, actually they are used as a biomonitoring organisms (Dawes et al. 1998), and shelter, nursery systems and food sources for diverse marine organisms (Prathep 2005).

Table 6.1 Species from the *Gigartina* genus (Guiry and Guiry 2021)

Gigartina ancistroclada Montagne 1845	*Gigartina angulata* J. Agardh 1876
Gigartina brachiata Harvey 1859	*Gigartina bracteata* (S. G. Gmelin) *Setchell &* N. L. Gardner 1933
Gigartina chondroides Bory 1828	*Gigartina clavifera* J. Agardh 1876
Gigartina cranwelliae Laing 1939	*Gigartina densa* Edyvane & Womersley 1994
Gigartina dilatata (J. D. Hooker & Harvey) N. M. Adams 1994	*Gigartina disticha* Sonder 1845
Gigartina divaricata J. D. Hooker & Harvey 1845	*Gigartina ewenii* W. A. Nelson & R. D'Archino 2014
Gigartina falshawiae D'Archino & W. A. Nelson 2019	*Gigartina fissa* (*Suhr*) J. Agardh 1876
Gigartina flabellata Kützing 1849	*Gigartina grandifida* J. Agardh 1876
Gigartina imperialis Papenfuss, nom. inval. 1976	*Gigartina insignis* (Endlicher & Diesing) F. Schmitz 1896
Gigartina kroneana Rabenhorst 1878	*Gigartina laciniata* J. Agardh 1876
Gigartina laingii Lindauer ex V. J. Chapman 1979	*Gigartina lanceata* var. *longifolia* (J. Agardh) V. J. Chapman 1979
Gigartina lessonii (Bory) J. Agardh 1851	*Gigartina macrocarpa* J. Agardh 1876
Gigartina minima Kylin 1938	*Gigartina minuta* V. J. Chapman 1979
Gigartina muelleriana Setchell & N. L. Gardner 1933	*Gigartina multidichotoma* E. Y. Dawson 1961
Gigartina nana (C. Agardh) J. Agardh	*Gigartina pachymenioides* Lindauer 1949
Gigartina paitensis W. R. Taylor 1947	*Gigartina paxillata* Papenfuss 1947
Gigartina pinnata J. Agardh 1851	*Gigartina pistillata* (S. G. Gmelin) Stackhouse 1809
Gigartina polycarpa (Kützing) Setchell & N. L. Gardner 1933	*Gigartina decipiens* var. *protea* (J. Agardh) V. J. Chapman 1979
Gigartina clavifera var. *pseudopistillata* Laing & Gourlay 1929	*Gigartina recurva* Edyvane & Womersley 1994
Gigartina rubens J. Agardh 1899	*Gigartina runcinata* Grunow *1868*
Gigartina sonderi Edyvane & Womersley 1994	*Gigartina tysonii* Reinbold 1912
Gigartina wehliae Sonder 1871	

Actually, the *Gigartina* genus is a red seaweed group comprising of 43 species (Table 6.1).

Thus, there is a general lack of bibliography about this genus. The most known seaweed of the genus is the *Gigartina pistillata*, which is the genus holotype (Guiry and Guiry 2021). This species is an edible red seaweed which is observed at both Northeast and Southeast Atlantic and Southeast Asia. Its morphology is defined as the model type of the genus *Gigartina*. Their thalli are erect, up to 20 cm tall, dark-red, or red-brown, cartilaginous, flexible, dichotomously branched, attached to the substratum through a small disk (Pereira 2016). *Gigartina* genus occupies mostly the intertidal rocky shores (D'Archino et al. 2020), and they have a higher growth

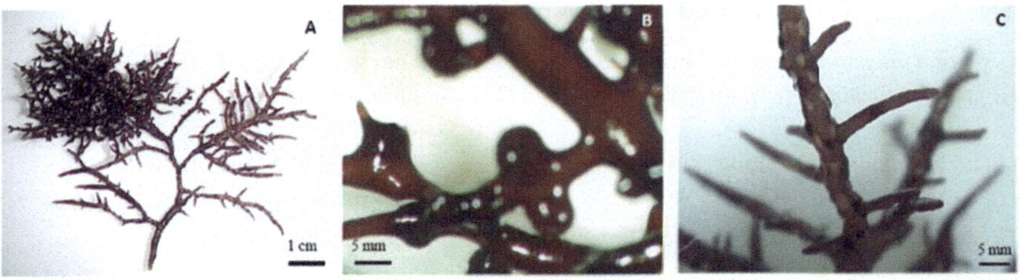

Fig. 6.1 *Gigartina pistillata* heterosporous thalli (**a**) showing cystocarps at the left side of the frond, and tetrasporangial sori at the right side of the frond; branch with cystocarps (**b**); and branch with tetrasporangial sori (**c**). (Adapted from Pereira (2004))

rate from the spring until the start of the summer, however, the chemical composition does not vary greatly between seasons (Amimi et al. 2007).

Gigartina genus can have a sporadic occurrence of the heterosporous thalli (i.e., having tetraspores and carpospores in the same thalli of one specimen), notwithstanding though having an isomorphic triphasic life cycle (Fig. 6.1) (Pereira et al. 2013; Pereira and Ribeiro-Claro 2016). In *Gigartina* genus, the seaweed life cycle phase has a high impact in the carrageenan chemical conformation. Gametophytic life cycle produces a kappa/iota-type carrageenan, while tetrasporophytic life cycle synthesizes a lambda-type carrageenan (Pereira and Mesquita 2003; Pereira et al. 2013).

3 *Gigartina* as a Food Source: Nutraceutical Potential?

Red seaweeds (Rhodophyta) constitute a phylum with a high biodiversity taxonomic group. Therefore, this taxonomic group represent a vast biotechnological potential, particularly for the food sector (Cotas et al. 2020b).

Nutraceutical food items are gaining popularity due to their health benefits and the increased societal awareness (Tanna and Mishra 2019).

Functional foods are nutritious food products that often provide medicinal and wellness benefits. Moreover, our life quality is greatly influenced by what we consume on a daily basis, so functional foods are now an essential part of our diet. Nutraceuticals are supplements that are applied to food to have extra nutritional and physiological benefits. Thus, nutraceutical food products are not considered food nor medication (Tibbetts et al. 2016; Tanna and Mishra 2018).

This historical seaweed utilization is corroborated by several research studies that point out *Gigartina* species nutraceutical potential (Gurgel et al. 2007; Rinaudo 2007; Mariya and Ravindran 2013; Michalak and Chojnacka 2015; Fredericq and Schmidt 2016; Tanna and Mishra 2018). These species contain a wide range of compounds with nutraceutical potential, such as natural pigments, a rich lipidic profile, proteins, minerals, and polysaccharides (Cotas et al. 2020b).

Furthermore, the lipidic profile of Gigartinales species is recognized for the high content of essential fatty acids, such as eicosapentaenoic acid (EPA, C20:5ω-3) and arachidonic acid (DHA, C20:4ω-6), exhibiting the presence in lower quantities of linoleic acid (LA, C18:2ω-6), α-linolenic acid (ALA, C18:3ω-3) and stearidonic acid (SDA, C18:4ω-3) (Galloway et al. 2012; Villanueva et al. 2014). Polyunsaturated fatty acids (PUFAs) are pivotal for the good functioning of the human organism, since some of them are not synthesized by humans, the essential fatty acids, so there is a need to acquire them through the food diet (Broadhurst et al. 2002; van Ginneken et al. 2011). Moreover, a study conducted by Villanueva et al. (2014) showed through assays performed in rats, that the intake of *G. pistillata* enhanced their lipidic profile, as well as the reduction of the total cholesterol.

Red seaweeds are also a rich protein source, exhibiting in some cases, higher contents than vegetables. For instance, research conducted by Gómez-Ordóñez et al. (2010) assessed that the protein content of 15.59%.dry weight of *G. pistillata* collected in the northwestern Spanish coast.

Since seaweeds are so rich in a variety of minerals that are important for human health, they represent a valuable resource for nutraceutical application (Tanna and Mishra 2019). Seaweeds typically have a comparable mineral concentration to seawater. The mineral composition of seaweed, on the other hand, varies in response to biotic and abiotic influences (Cotas et al. 2020b). A diet rich in several minerals, such as calcium, zinc, iron and manganese is pivotal for the good functioning of the human body, and for example, *G. pistillata* can be a good resource, since 34.56% of its dry weight is composed by minerals (Gómez-Ordóñez et al. 2010).

Gigartina spp. are also a rich source of dietary fibers. For example, *G. pistillata* was reported to contain 21.90% of soluble and 7.41% of insoluble dietary fibers (Gómez-Ordóñez et al. 2010). Seaweeds contain large amounts of polysaccharides, the majority of which are not digested by humans, due to the lack of catabolic enzymes in the gastrointestinal tract; thus, they can be considered dietary fibers (Jiménez-Escrig and Sánchez-Muniz 2000). Furthermore, *G. pistillata* is known by the production of an important sulfated polysaccharide for the food industry, the carrageenan (Rosenfeld et al. 2015). However, this species produces different types of carrageenan (kappa, iota, and lambda), with different rheological properties, according to the stage of their life cycle (Zinoun et al. 1993; Barahona et al. 2012; Mateos-Aparicio et al. 2018; Cotas et al. 2020c). Nevertheless, the physical-chemical properties of the sulfated carrageenans present in *Gigartina* showed that these polysaccharides could help with water binding, feces accumulation, and regulates intestinal transit, making them a good source of dietary fiber for human consumption (Gómez-Ordóñez et al. 2010). In fact, there is already a product in the market enriched with sulfated polysaccharides made with *G. pistillata*, aiming to strengthen the immune system (Vibrant Health 2021).

3.1 Gigartina *as Source of Food Industry Compounds*

Seaweeds are the major producers of vegetal biomass in the marine environment, having a high importance for its bioactive substances, such as polysaccharides, extracted mainly from the cell wall (Liu et al. 2020). These compounds have been used in several applications in the food, pharmaceutical, personal care products, and cosmetic industries (Rani et al. 2020). Among the algae used, those of the genus *Gigartina* stand out, as they have essential characteristics for biotechnological studies (Häder 2021). The commercial lambda-carrageenan can be extracted from species belonging to the *Gigartina* sp. (Cotas et al. 2020c). Thus, Amimi et al. (2007) reported that in Morocco, the *G. pistillata* needed for carrageenan industry is collected in summer whereas the biomass and carrageenan content is higher in other seasons. This carrageenan is normally used in the food industry as thickener or gelling agent (Tanna and Mishra 2019). Moreover, this species has low content of carotenoids, mainly lutein, during the winter; similarly the pigment phycoerythrin as reported by Lalegerie et al. (2019), Other seaweeds from the *Gigartina* genus are currently being exploited for food and lambda-carrageenan industrial extraction, such as *G. paitensis* in Peru (Avila-Peltroche and Padilla-Vallejos 2020).

4 *Gigartina* as Source of Pharmaceutical Compounds

Although there are many *Gigartina* species, there is a general lack of pharmaceutical and biotechnological studies involving this genus. Despite that, there are interesting results reporting the bioactive compounds of these species. The *G. pistillata* aqueous extract demonstrated a strong reduction power of ferric ion, due to its sulphate content and degree of sulfation (Jiménez-Escrig et al. 2012).

G. pistillata demonstrates interesting levels of mycosporine-like amino acids (MAAs), mainly Palythine, Shinorine, MAA_14 and MAA_22, (Lalegerie et al. 2019). This composition highlights the potential of *G. pistillata* ethanolic extract in human health, for UVB/A protection, wound-healing property or for anti-proliferative activity (Chrapusta et al. 2017; Cotas et al. 2020a). Orfanoudaki et al. (2019) reported that *G. macrocarpa* also contain Shinorine and Palythine. Products such as Asterina-330, Porphyra-334, are commercial sunscreens containing *G.pistillata* ingredients rich in the phytochemicals and biological activities mentioned above (Cotas et al. 2020a, b).

The carrageenan extracted from *G. pistillata* exhibited anti-tumor potential against several colorectal cancer cell lines, mainly the HT-29 cell line. The carrageenan extracted from the tetrasporophyte stage demonstrates a high anti-tumor activity which needs to be further investigated (Cotas et al. 2020c). Furthermore, the lambda-carrageenan demonstrates a strong antiviral activity in nature (Kalitnik et al. 2013; Jang et al. 2021).

5 Conclusions and Future Perspectives

The *Gigartina* genus is one of the largest genera in red seaweeds. Although genetic evaluation of the genus shows it is in a state of constant evolution, all the species are similar ecologically and morphologically. *Gigartina* species are considered edible, exhibiting a valuable nutritional content that promotes human health benefits at the gastrointestinal tract. Moreover, this genus demonstrates pharmaceutical activities that can be further exploited.

Acknowledgments This work is financed by national funds through FCT - Foundation for Science and Technology, I.P., within the scope of the projects UIDB/04292/2020 granted to MARE - Marine and Environmental Sciences Centre and UIDP/50017/2020 + UIDB/50017/2020 (by FCT/MTCES) granted to CESAM - Centre for Environmental and Marine Studies. João Cotas thanks to the European Regional Development Fund through the Interreg Atlantic Area Program, under the project NASPA (EAPA_451/2016). Sara García-Poza thanks to the project MENU - Marine Macroalgae: Alternative recipes for a daily nutritional diet (FA_05_2017_011) which co-financed this research, funded by the Blue Fund under Public Notice No. 5 - Blue Biotechnology. Diana Pacheco thanks to PTDC/BIA-CBI/31144/2017-POCI-01 project -0145-FEDER-031144-MARINE INVADERS, co-financed by the ERDF through POCI (Operational Program Competitiveness and Internationalization) and by the Foundation for Science and Technology (FCT, IP). Ana M. M. Gonçalves acknowledges University of Coimbra for the contract IT057-18-7253.

References

Amimi A, Mouradi A, Bennasser L, Givernaud T (2007) Seasonal variations in thalli and carrageenan composition of *Gigartina pistillata* (Gmelin) Stackhouse (Rhodophyta, Gigartinales) harvested along the Atlantic coast of Morocco. Phycol Res 55:143–149. https://doi.org/10.1111/j.1440-1835.2007.00457.x

Avila-Peltroche J, Padilla-Vallejos J (2020) The seaweed resources of Peru. Bot Mar 63:381–394. https://doi.org/10.1515/bot-2020-0026

Barahona T, Encinas MV, Mansilla A, Matsuhiro B, Zúñiga EA (2012) A sulfated galactan with antioxidant capacity from the green variant of tetrasporic *Gigartina skottsbergii* (Gigartinales, Rhodophyta). Carbohydr Res 347:114–120. https://doi.org/10.1016/j.carres.2011.11.014

Barbier M, Charrier B, Araujo R, Holdt SL, Jacquemin B, Rebours C (2019) PEGASUS—PHYCOMORPH European guidelines for a sustainable aquaculture of seaweeds. Roscoff, France. https://doi.org/10.21411/2c3w-yc73

Broadhurst CL, Wang Y, Crawford MA, Cunnane SC, Parkington JE, Schmidt WF (2002) Brain-specific lipids from marine, lacustrine, or terrestrial food resources: potential impact on early African *Homo sapiens*. Comp Biochem Physiol B Biochem Mol Biol 131:653–673. https://doi.org/10.1016/S1096-4959(02)00002-7

Brown EM, Allsopp PJ, Magee PJ, Gill CI, Nitecki S, Strain CR, Mcsorley EM (2014) Seaweed and human health. Nutr Rev 72:205–216. https://doi.org/10.1111/nure.12091

Brownlee IA, Fairclough AC, Hall AC, Paxman JR (2011) The potential health benefits of seaweed and seaweed extract. In: Seaweed: ecology, nutrient composition and medicinal uses, pp 119–136

Chrapusta E, Kaminski A, Duchnik K, Bober B, Adamski M, Bialczyk J (2017) Mycosporine-like amino acids: potential Health and beauty ingredients. Mar Drugs 15:326. https://doi.org/10.3390/md15100326

Cotas J, Leandro A, Monteiro P, Pacheco D, Figueirinha A, Gonçalves AMM, da Silva GJ, Pereira L (2020a) Seaweed Phenolics: from extraction to applications. Mar Drugs 18:384. https://doi.org/10.3390/md18080384

Cotas J, Leandro A, Pacheco D, Gonçalves AMM, Pereira L (2020b) A comprehensive review of the nutraceutical and therapeutic applications of red seaweeds (Rhodophyta). Life 10:19. https://doi.org/10.3390/life10030019

Cotas J, Marques V, Afonso MB, Rodrigues CMP, Pereira L (2020c) Antitumour potential of *Gigartina pistillata* Carrageenans against colorectal cancer stem cell-enriched Tumourspheres. Mar Drugs 18:50. https://doi.org/10.3390/md18010050

Cyrus MD, Bolton JJ, Scholtz R, Macey BM (2015) The advantages of *Ulva* (Chlorophyta) as an additive in sea urchin formulated feeds: effects on palatability, consumption and digestibility. Aquacult Nutr 21:578–591. https://doi.org/10.1111/anu.12182

D'Archino R, Nelson WA, Sutherland JE (2020) Unnamed for over 30 years: *Gigartina falshawiae* sp. nov. (Gigartinaceae, Rhodophyta) and its confusion with *Iridaea tuberculosa* in New Zealand. Phycologia 59:45–53. https://doi.org/10.1080/00318884.2019.1667187

Dawes CJ, Orduña-Rojas J, Robledo D (1998) Response of the tropical red seaweed *Gracilaria cornea* to temperature, salinity and irradiance. J Appl Phycol 10:419–425. https://doi.org/10.1023/A:1008021613399

Dillehay TD, Ramirez C, Pino M, Collins MB, Rossen J, Pino-Navarro JD (2008) Monte Verde: seaweed, food, medicine, and the peopling of South America. Science 320:784–786. https://doi.org/10.1126/science.1156533

Falshaw R, Furneaux RH (1998) Structural analysis of carrageenans from the tetrasporic stages of the red algae, *Gigartina lanceata* and *Gigartina chapmanii* (Gigartinaceae, Rhodophyta). Carbohydr Res 307:325–331. https://doi.org/10.1016/S0008-6215(98)00030-5

Falshaw R, Furneaux RH (2009) Chemotaxonomy of New Zealand red algae in the family Gigartinaceae (Rhodophyta) based on galactan structures from the tetrasporophyte life-stage. Carbohydr Res 344:210–216. https://doi.org/10.1016/j.carres.2008.10.020

Fredericq S, Schmidt WE (2016) Red Algae. In: eLS, pp 1–7. https://doi.org/10.1002/9780470015902.a0000335.pub2

Galloway AWE, Britton-Simmons KH, Duggins DO, Gabrielson PW, Brett MT (2012) Fatty acid signatures differentiate marine macrophytes at ordinal and family ranks. J Phycol 48:956–965. https://doi.org/10.1111/j.1529-8817.2012.01173.x

Gómez-Ordóñez E, Jiménez-Escrig A, Rupérez P (2010) Dietary fibre and physicochemical properties of several edible seaweeds from the northwestern Spanish coast. Food Res Int 43:2289–2294. https://doi.org/10.1016/j.foodres.2010.08.005

Guiry MD, Guiry GM (2021) AlgaeBase. World-Wide Electronic Publication, National University of Ireland, Galway

Gurgel C, Deluqui F, Lopez-Bautista J (2007) Red algae. In: Encyclopedia of life sciences. John Wiley, Chichester, pp 1–5. https://doi.org/10.1002/9780470015902.a0000335

Häder D-P (2021) Phycocolloids from macroalgae. In: Natural bioactive compounds. Elsevier, Amsterdam, pp 187–201. https://doi.org/10.1016/B978-0-12-820655-3.00009-4

Iso H (2011) Lifestyle and cardiovascular disease in Japan. J Atheroscler Thromb 18:83–88. https://doi.org/10.5551/jat.6866

Jang Y, Shin H, Lee MK, Kwon OS, Shin JS, Kim Y, Kim CW, Lee H-R, Kim M (2021) Antiviral activity of lambda-carrageenan against influenza viruses and severe acute respiratory syndrome coronavirus 2. Sci Rep 11:821. https://doi.org/10.1038/s41598-020-80896-9

Jiménez-Escrig A, Sánchez-Muniz FJ (2000) Dietary fibre from edible seaweeds: chemical structure, physicochemical properties and effects on cholesterol metabolism. Nutr Res 20:585–598. https://doi.org/10.1016/S0271-5317(00)00149-4

Jiménez-Escrig A, Gómez-Ordóñez E, Rupérez P (2012) Brown and red seaweeds as potential sources of antioxidant nutraceuticals. J Appl Phycol 24:1123–1132. https://doi.org/10.1007/s10811-011-9742-8

Kalitnik AA, Byankina Barabanova AO, Nagorskaya VP, Reunov AV, Glazunov VP, Soloveva TF, Yermak IM (2013) Low molecular weight derivatives of different carrageenan types and their antiviral activity. J Appl Phycol 25:65–72. https://doi.org/10.1007/s10811-012-9839-8

Kim J, Shin A, Lee J-S, Youn S, Yoo K-Y (2009) Dietary factors and breast cancer in Korea: an ecological study. Breast J 15:683–686. https://doi.org/10.1111/j.1524-4741.2009.00817.x

Lalegerie F, Lajili S, Bedoux G, Taupin L, Stiger-Pouvreau V, Connan S (2019) Photo-protective compounds in red macroalgae from Brittany: considerable diversity in mycosporine-like amino acids (MAAs). Mar Environ Res 147:37–48. https://doi.org/10.1016/j.marenvres.2019.04.001

Leandro A, Pacheco D, Cotas J, Marques JC, Pereira L, Gonçalves AMM (2020) Seaweed's bioactive candidate compounds to food industry and global food security. Life 10:140. https://doi.org/10.3390/life10080140

Liu Z, Niu F, Xie Y, Xie S, Liu Y, Yang Y, Zhou C, Wan X (2020) A review: natural polysaccharides from medicinal plants and microorganisms and their anti-herpetic mechanism. Biomed Pharmacother 129:110469. https://doi.org/10.1016/j.biopha.2020.110469

Mariya V, Ravindran VS (2013) Biomedical and pharmacological significance of marine macro algae-review. Indian J Mar Sci 42:527–537

Mateos-Aparicio I, Martera G, Goñi I, Villanueva-Suárez M-J, Redondo-Cuenca A (2018) Chemical structure and molecular weight influence the in vitro fermentability of polysaccharide extracts from the edible seaweeds *Himathalia elongata* and *Gigartina pistillata*. Food Hydrocoll 83:348–354. https://doi.org/10.1016/j.foodhyd.2018.05.016

Michalak I, Chojnacka K (2015) Algae as production systems of bioactive compounds. Eng Life Sci 15:160–176. https://doi.org/10.1002/elsc.201400191

Nelson WA, Broom JES (2008) New Zealand Gigartinaceae (Rhodophyta): resurrecting *Gigartina grandifida* endemic to the Chatham Islands. N Z J Bot 46:177–187. https://doi.org/10.1080/00288250809509761

Orfanoudaki M, Hartmann A, Karsten U, Ganzera M (2019) Chemical profiling of mycosporine-like amino acids in twenty-three red algal species. Edited by K. Müller. J Phycol 55:393–403. https://doi.org/10.1111/jpy.12827

Parsons MJ, Pickmere SE, Bailey RW (1977) Carrageenan composition in New Zealand species of *Gigartina* (Rhodophyta): geographic variation and interspecific differences. N Z J Bot 15:589–595. https://doi.org/10.1080/0028825X.1977.10429632

Pereira L (2004) Estudos em macroalgas carragenófitas (Gigartinales, Rhodophyceae) da costa portuguesa: aspectos ecológicos, bioquímicos e citológicos. PhD Thesis, Universidade de Coimbra, Coimbra, Portugal, 293 pp. (in Portuguese). https://estudogeral.sib.uc.pt/handle/10316/10017

Pereira L (2016) Edible seaweeds of the world. CRC, Boca Raton, FL. https://doi.org/10.1201/b19970

Pereira L, Mesquita JF (2003) Carrageenophytes of occidental Portuguese coast: 1-spectroscopic analysis in eight carrageenophytes from Buarcos bay. Biomol Eng 20:217–222. https://doi.org/10.1016/S1389-0344(03)00056-X

Pereira L, Ribeiro-Claro PJA (2016) Morphological, cytological and chemical analysis of the heterosporous *Gigartina pistillata* (chapter 7). In: Pereira L (ed) Carrageenans: sources and extraction methods, molecular structure, bioactive properties and health effects. Nova Science, New York, pp 148–159. ISBN: 978-1-63485-503-7

Pereira L, Gheda SF, Ribeiro-claro PJA (2013) Analysis by vibrational spectroscopy of seaweed polysaccharides with potential use in food, pharmaceutical, and cosmetic industries. Int J Carbohydr Chem 2013:1–7. https://doi.org/10.1155/2013/537202

Prathep A (2005) Spatial and temporal variations in diversity and percentage cover of macroalgae at Sirinart marine National Park, Phuket Province, Thailand. Sci Asia 31:225. https://doi.org/10.2306/scienceasia1513-1874.2005.31.225

Radulovich R, Neori A, Valderrama D, Reddy CRK, Cronin H, Forster J (2015) Farming of seaweeds. In: Tiwari B, Troy D (eds) Seaweed sustainability. Elsevier, Amsterdam, pp 27–59. https://doi.org/10.1016/B978-0-12-418697-2.00003-9

Rani V, Prabhu A, Venkatesan J, Kim S-K (2020) Seaweed polysaccharides: promising molecules for biotechnological applications. In: Reference module in chemistry, molecular sciences and chemical engineering. Elsevier, Amsterdam. https://doi.org/10.1016/B978-0-12-819475-1.00023-7

Rinaudo M (2007) Seaweed polysaccharides. In: Comprehensive glycoscience: from chemistry to systems biology, vol 2–4, pp 691–735. https://doi.org/10.1016/B978-044451967-2/00140-9

Rosenfeld S, Aldea C, Mansilla A, Marambio J, Ojeda J (2015) Richness, systematics, and distribution of molluscs associated with the macroalga *Gigartina skottsbergii* in the Strait of Magellan, Chile: a biogeographic affinity study. ZooKeys 519:49–100. https://doi.org/10.3897/zookeys.519.9676

Shama A, Joyce SG, Mari FD, Ranga Rao A, Ravishankar GA, Hudaa N (2019) Macroalgae and microalgae: novel sources of functional food and feed. In: Ravishankar GA, Rao AR (eds) Handbook of algal technologies and phytochemicals. Food, health and nutraceutical applications, vol 1. CRC, Boca Raton, FL, pp 207–219

Smit AJ (2004) Medicinal and pharmaceutical uses of seaweed natural products: a review. J Appl Phycol 16:245–262. https://doi.org/10.1023/B:JAPH.0000047783.36600.ef

Tanna B, Mishra A (2018) Metabolites unravel nutraceutical potential of edible seaweeds: an emerging source of functional food. In: Comprehensive reviews in food science and food safety. Blackwell, Hoboken, NJ. https://doi.org/10.1111/1541-4337.12396

Tanna B, Mishra A (2019) Nutraceutical potential of seaweed polysaccharides: structure, bioactivity, safety, and toxicity. Compr Rev Food Sci Food Saf 18:817–831. https://doi.org/10.1111/1541-4337.12441

Tibbetts SM, Milley JE, Lall SP (2016) Nutritional quality of some wild and cultivated seaweeds: nutrient composition, total phenolic content and in vitro digestibility. J Appl Phycol 28:3575–3585. https://doi.org/10.1007/s10811-016-0863-y

van Ginneken VJT, Helsper JPFG, de Visser W, van Keulen H, Brandenburg W (2011) Polyunsaturated fatty acids in various macroalgal species from North Atlantic and tropical seas. In: Lipids in health and disease, vol 10. BioMed Central Ltd, London, p 104. https://doi.org/10.1186/1476-511X-10-104

Vibrant Health (2021) Gigartina Red Marine Algae—natural immune support. Available online at: https://www.vibranthealthuk.com/vibrant-health-red-marine-algae-supplement.html (accessed on December 8, 2021)

Villanueva MJ, Morcillo M, Tenorio MD, Mateos-Aparicio I, Andrés V, Redondo-Cuenca A (2014) Health-promoting effects in the gut and influence on lipid metabolism of *Himanthalia elongata* and *Gigartina pistillata* in hypercholesterolaemic Wistar rats. Eur Food Res Technol 238:409–416. https://doi.org/10.1007/s00217-013-2116-5

Zinoun M, Cosson J, Deslandes E (1993) Influence of culture conditions on growth and physicochemical properties of Carrageenans in *Gigartina teedii* (Rhodophyceae—Gigartinales). Bot Mar 36. https://doi.org/10.1515/botm.1993.36.2.131

Chapter 7
Gracilaria as the Major Source of Agar for Food, Health and Biotechnology Applications

Wei-Kang Lee, Yi-Yi Lim, and Chai-Ling Ho

Abbreviations

BMPs Bimetallic magnetic nanoparticles

1 Introduction to *Gracilaria*

The *Gracilaria* was first described as a red algal genus by Greville in 1830. Being one of the largest genus in Rhodophyta, a total of 175 *Gracilaria* species have been identified to date (Guiry and Guiry 2020). *Gracilaria* species can live in sandy and muddy habitats or live as free-floating seaweeds in temperate and tropical regions (Fig. 7.1). They can tolerate high salinity (up to 60 part per thousand; Nyberg 2007) and water temperature (up to 35 °C; Raikar et al. 2001). Most *Gracilaria* species live in the lower intertidal region and upper subtidal region (Oliveira et al. 2000). These seaweeds contain gelatinous material in their cell walls known as agar, providing seaweeds with high flexibility to withstand strong ocean waves and currents, and resistance to pathogens (Ficko-Blean et al. 2015). Agar also helps these seaweeds to maintain ionic equilibrium in the cells, protect against extreme salinity, pH, temperatures, and desiccation (Ficko-Blean et al. 2015).

W.-K. Lee · Y.-Y. Lim · C.-L. Ho (✉)
Faculty of Biotechnology and Biomolecular Sciences, Universiti Putra Malaysia, UPM-Serdang, Selangor, Malaysia
e-mail: clho@upm.edu.my

A. Ranga Rao, G. A. Ravishankar (eds.), *Sustainable Global Resources of Seaweeds Volume 2*, https://doi.org/10.1007/978-3-030-92174-3_7

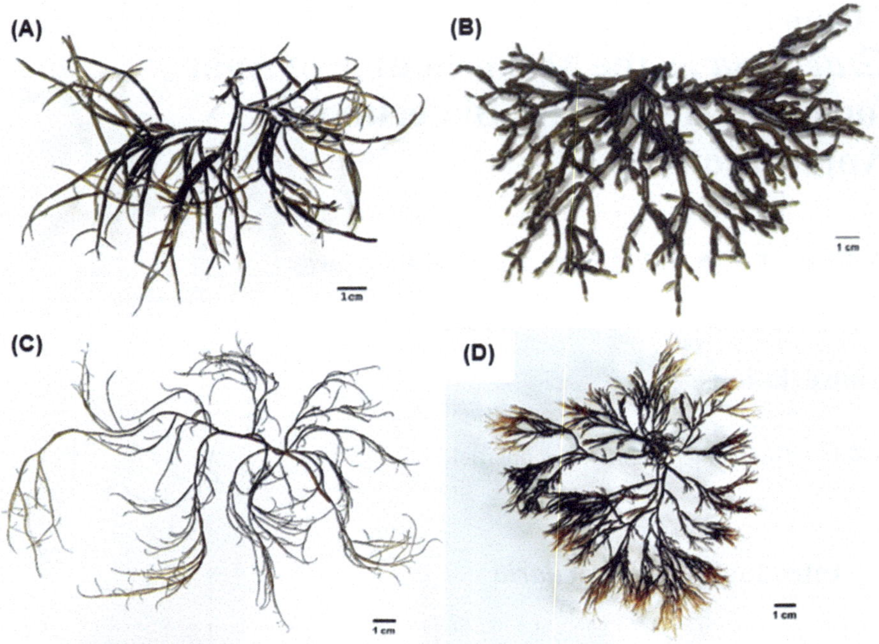

Fig. 7.1 *Gracilaria* species from Malaysia. (**a**) *G. changii*; (**b**) *G. salicornia* from Morib; (**c**) *G. manilaensis* from Kedah; (**d**) *G. edulis* from Port Dickson

2 Agar Polysaccharides from *Gracilaria*

2.1 Definition and History of Agar

Based on United States Pharmacopeia and the Food Chemicals Codex, agar is a hydrocolloid which is soluble in boiling water but not in cold water, has a clear aqueous solution at a concentration of 1.5% (w/v), forms gel between 32 and 39 °C and only melts when the temperature is higher than 85 °C. Agar was originally used to describe the gel-forming and thermal reversible polysaccharide which was extracted from *Gelidium* seaweeds. The depleted stocks of *Gelidium* agars has led to the exploitation of agar from *Gracilaria* species, which has comparable good agar gel strength when alkaline treatment was applied on the seaweeds before agar extraction (Wu 1990). To date, *Gracilaria* (91%) and *Gelidium* (9%) remain as the largest agar producers (Porse and Rudolph 2017).

2.2 Structure and Components of Agar

Agar consists of two components, agarose and agaropectin (Fig. 7.2). The solubility and gel-forming ability of agar are dependent on the relative hydrophobicity of the galactose disaccharide. Agarose is a neutral polysaccharide with repeated units of

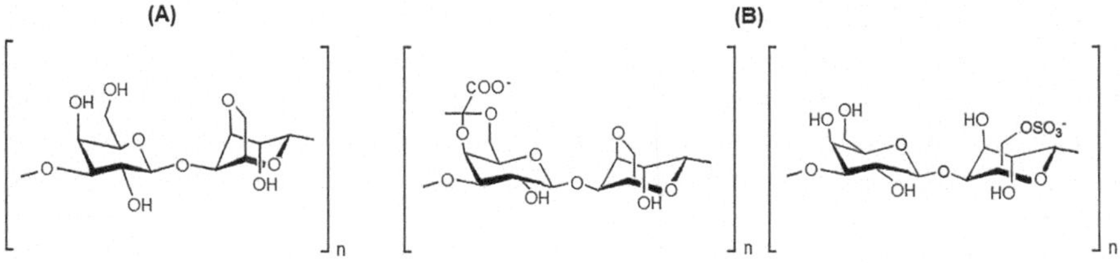

Fig. 7.2 Chemical structure of (**a**) agarose/agarobiose and (**b**) agaropectin in *Gracilaria*. OH group in agaropectin could be substituted by sulfate ester, methoxyl or pyruvic acid groups

agarobiose, consisting of β-1,3-linked D-galactose and α-1,4-linked 3,6-anhydro-L-galactose. Agarose can contribute up to 70% of agar polysaccharides. Agaropectin is a charged polysaccharide containing sulfate ester, methoxyl group and pyruvic acid attached to β-1,3-linked D-galactose and α-1,4-linked L-galactose (Araki 1966). The levels of side chain substitution are different according to the species of seaweed, for example, *Gracilaria* has a higher sulfate substitution compared to *Gelidium*.

2.3 Factors That Affect the Value of Agar

The average market value of agar was US$ 17 per kg in 2015 (Porse and Rudolph 2017). The sales of agar in 2015 was valued at US$ 246 million, with an average growth of 6% annually from 2009 (Porse and Rudolph 2017). The value of agar in the phycocolloid market is determined by the yield and the gelation characteristics of agar e.g. gel strength (the main indicator of gel quality), gel syneresis, viscosity, gelling and melting temperatures (Lee et al. 2017), that are affected by the quality of raw materials.

The economic return on the raw material largely depends on the agar yield of seaweed per dollar of production cost and agar yield per kg of seaweed (Cordover 2007). Most red seaweed species are poor producers of phycocolloids (Kaladharan and Kaliaperumal 1999). The genetic of seaweed species is the most important factor contributing to the variation in agar properties (Armisen and Galactas 1987). The yield and quality of agar can also be affected by many factors such as the physical environmental factors, nutrient availability, biotic factors, physiological state of seaweeds, and the agar extraction methods (Lee et al. 2017).

To date, *Gracilaria* has been successfully cultivated and farmed at commercial scale, with the biggest producers being China (70%; mainly *G. lemaneiformis*) and Indonesia (28%) (Yang et al. 2015; Porse and Rudolph 2017; FAO 2020). Cultivation of *Gracilaria* for agar production, provides a long term economic sustainability plan that prevents over-harvesting of wild *Gracilaria* species (FAO Fishery Statistics 2011) and overcomes the shortage of *Gracilaria* stocks from the natural beds (Mantri et al. 2020). *Gracilaria* species that have been farmed for agar production either in commercial scale or at trial stage include *G. chilensis*, *G. lemaneiformis*, *G. dura*, *G. verrucose*, *G. gracilis*, *G. edulis*, *G. tenuistipitata*, *G. lichenoides*,

G. blodgettii and *G. latifolium* (Soegiarto 1990; Rebello et al. 1997; Yang et al. 2015). Since these *Gracilaria* species are with varying agar yield and gel quality, selection of *Gracilaria* species with the best agar yield and gelling content is important to maximize business profitability. In recent years, molecular markers associated with good agar properties have been developed for *Gracilaria* species (Hu et al. 2019; Lim et al. 2019).

3 Extraction of Agar and Industrial Scenario

According to the Food and Agriculture Organization (FAO), the international quality standard of a food grade agar on an industrial scale include: a gel strength >750 g/cm^2 (as measured by the Nikan-Sui method), <18% moisture, <5% ash, <5 ppm lead and <3 ppm arsenic. The bacterial count should be less than 10,000 bacteria/g, with the absence of *Escherichia coli* and *Salmonella*, while the presence of other pathogenic bacteria specified. The same specifications are usually used for agar produced on an industrial scale (McHugh 1987). The leading manufacturing countries in agar production are Indonesia, Chile and China, where the Asia Pacific region contributes up to 69% of the total world agar production (Porse and Rudolph 2017). The major company players in the agar industry (62%) are Agarindo Bogatama (Indonesia), Green Fresh (Fujian) Foodstuff Co. (China), and Algas Marinas (Chile), Setexam (Morocco), MSC Co. (Republic of Korea), Java Biocolloid (Indonesia), and Puning Huey Shyang Seaweed Industrial Co. (China), and numerous smaller companies that made up the remaining 38% (Porse and Rudolph 2017).

Agar extraction from seaweeds involves multiple steps including cleaning and washing of seaweeds, pre-treatment, agar extraction, filtration, concentration and drying (Hernández-Carmona et al. 2013; McHugh 2003) (Fig. 7.3). Prior to agar extraction, seaweeds are cleaned and washed to remove epiphytes, sands, muds, salts and other impurities before being sun- or oven-dried. Pretreatment of dried *Gracilaria* with a 0.5% to 7% sodium hydroxide solution at 85–90 °C for 1–2 h (Hernández-Carmona et al. 2013) was able to improve the agar gel strength by decreasing the sulfate group from galactose-6-sulfate and increase the 3,6-anhydrogalactose content of agar (Vergara-Rodarte et al. 2010; Ahmad et al. 2011; Yarnpakdee et al. 2015). Despite alkali pre-treatment of seaweeds may produce agar with high gel strength (Arvizu-Higuera et al. 2008), the pre-treatment may also lower the agar yields. Degradation of agar may produce low molecular weight polymers that cannot be recovered by alcohol precipitation and filtration (Freile-Pelegrin and Robledo 1997; Kumar and Fotedar 2009).The concentration of sodium hydroxide, temperature and duration for desulfation of each *Gracilaria* species may need optimization (Freile-Pelegrin and Robledo 1997; Freile-Pelegrin and Murano 2005). Residual alkali can be removed by rinsing the seaweeds with water, or neutralized with weak acid.

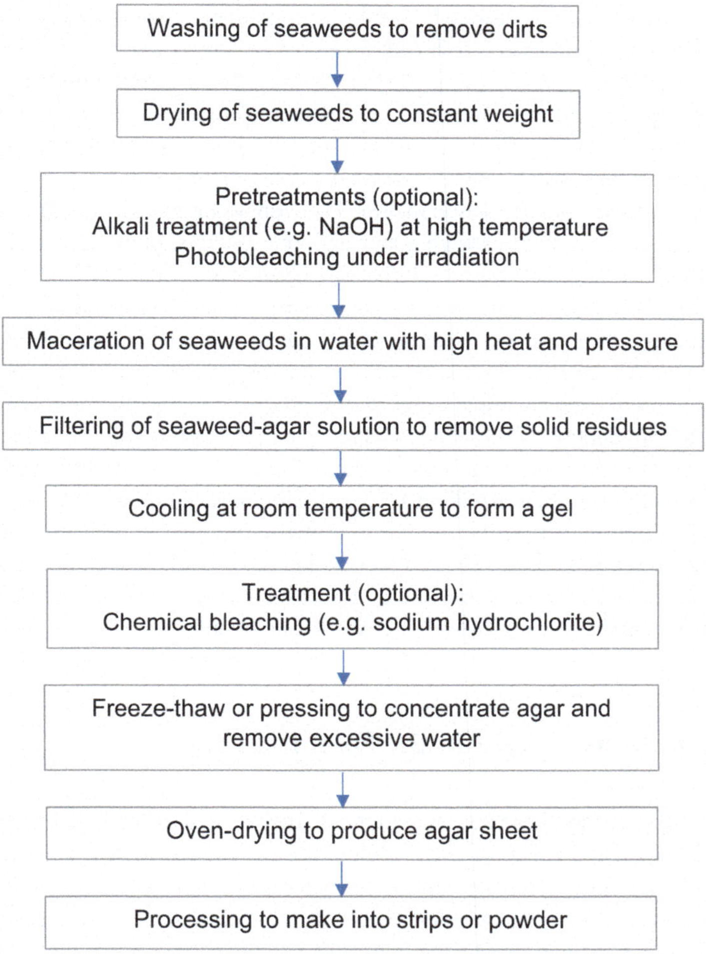

Fig. 7.3 General process of agar extraction from *Gracilaria* seaweeds

Traditionally, agar is extracted by maceration in water at high heat in an autoclave or waterbath. A combination of high heat and high pressure may reduce the processing time and increase the yield of agar, but the process may also potentially reduce the molecular weight of the agar polymers (Hernández-Carmona et al. 2013). Most *Gracilaria* species were reported to produce agar with a high gel strength when treated at high temperatures (80 °C to 90 °C) for a short duration (0.5 h to 3 h) (Arvizu-Higuera et al. 2008; Vergara-Rodarte et al. 2010). The water-soluble agar can be filtered under pressure to separate impurities and cooled at room temperature in a tray to form a gel. In order to produce pure white agar with a higher aesthetic property, chemical (sodium hypochlorite) or photobleaching are performed (Li et al. 2008). Diluted sulfuric acid could be added to adjust the pH. The alcohol precipitated gel formed upon bleaching has high water content (99%) which can be removed, either by a freeze-thaw process or by pressing. The diluted gel extract

(1–1.2% of agar), is concentrated tenfold after thawing. Alternatively, the water content of diluted agar can be reduced through a syneresis technique whereby the gel is squeezed in a hydraulic press. As the water being removed, impurities such as oligomers, organic and inorganic salts, and proteins are also removed (Hernández-Carmona et al. 2013). The final agar is oven-dried to constant weight and make into strips or powder depending on the specific downstream applications.

The conventional agar extraction method is not environment friendly and cost ineffective as the process requires high water and energy consumption, and discharges high volume of chemical waste (Hernández-Carmona et al. 2013). Eco-friendly extraction methods such as microwave-assisted extraction (MAE) (Sousa et al. 2013), photobleaching extraction (Li et al. 2008), hydrogen peroxide assisted extraction (Chen et al. 2020; An et al. 2021) and cold extraction (Maciel et al. 2008) have been developed to reduce the chemical usage and improve the extraction yield and quality. However, the application of these methods on agar extraction is still limited due to the expensive costs of high-tech equipment and lack of knowledge/skill in implementation of new methods (Abdul Khalil et al. 2018). Similarly, although sulfatase/sulfohydrolase (50 U) purified from *G. dura* was shown to be able to decrease the sulfate content and increase the agar gel strength (Shukla et al. 2011), the commercial feasibility is questionable, based on low yield of pure enzyme obtained per kg of seaweed and expensive production cost compared to alkaline treatment.

4 Applications of Agar in the Food Industry

The use of agar as food has been documented since its discovery in the seventeenth century (Armisen and Galactas 1987). About 90% of the world agar is used in food industry to alter food viscosity and texture (Kraan 2012). Agar has high gelling power whereby a low concentration (e.g. 1% w/v) of agar is suffice for gelation (Imeson 1997). Agar has an advantage over the other hydrocolloids that it does not require counter-ions such as calcium and potassium for gelation. The strong and unfavourable flavour of calcium and potassium might affect the taste of gel food products (Imeson 1997).

Agar is widely used as a food ingredient in many Asian countries. Since human cannot digest agar, it provides no nutritional value and does not add calorie to foods (Elleuch et al. 2011). Nonetheless, agar can act as a dietary fibre to soften stools, promote bowel movements and the growth of healthy gut bacteria, as well as to reduce cholesterol and glycemic response (Gómez-Ordóñez et al. 2010; O'Sullivan et al. 2010; Elleuch et al. 2011). Hence, agar is used as a safe and effective laxative (Imeson 2011). Agar has also been added to fruit pulps to create structured fruits (also known as fruit bars), as they are cheap, sustainable to produce and can prevent wastage by extending the consumption period while also preserving the nutritional and aesthetic characteristics of the fruits (Reis 2019; da Costa et al. 2020). The replacement of oil with agar micro-gels in the production of low oil mayonnaise has been reported by Kaneda and Shibata (2020). In a recent research, the D-Galactose molecule of agar was enzymatically converted to D-tagatose, a rare and low-calorie sweetener, making agar a potential source for commercial sweetener production in the food industry (Jeong et al. 2020).

Majority (>80%) of the commercial agar was used as food additives (McHugh 1987). Agar which is clear in colour, neutral and tasteless serves as the most suitable food additives (Ruperez and Saura-Calixto 2001). Desired texture and viscosity of foods can be achieved by changing the agar concentration, pH, temperature and salt, or by mixing with other hydrocolloids during food processing (Walstra 2003).

In addition, agar can be used as food thickener in canned food (Glicksman 1987), pie fillings, toppings for cupcakes and ice creams because it is able to form gel by retaining water molecules in their cross-linked polymer structures (Lim et al. 2018). Agar is also added into condiments such as sauces, syrups and ketchup to ensure the consistency of the texture. In the food industry, not only gelatin from the underutilized parts of fish (e.g. skin and bones) has been used to replace bovine and porcine gelation due to religious reason. Agar has also been used as a gelling agent to replace animal-origin gelatin in yoghurt production to suit the needs of vegetarian communities (Ganegama Arachchi et al. 2018).

Agar can retain a large quantity of sugar without affecting the gelling property (Nussinovitsch et al. 1991). This characteristic allows agar to be used in the preparation of confectioneries such as jams, marmalade and ice cream, toppings and fillings for pastry products, soft candies (jellies), and icings (Hansen 1993). The high gel hysteresis (i.e. the difference between gelling and melting temperatures) of agar can also prevent the agar gel cubes from melting and mixing with the fruits and sweet clack syrup during the heat sterilization of canned dessert (Imeson 1997).

5 Application of Agar in the Livestock Industry

In the livestock industry, agar is used to preserve the moisture content and to increase the survival rates of newly hatched chicks during their transportation to chicken farm (Olatunji 2020). The post-hatched chicks packed with aqua agar had a significantly higher feed intake and body weight gain compared to those without aqua agar, demonstrating the application of the agar-based aqueous system in reducing the negative impact of water deficiency and improving the growth performance (Incharoen et al. 2015).

6 Applications of Agars in the Healthcare Industry

6.1 Pharmacological Properties of Agar

Many sulfated polysaccharides extracted from red seaweeds possess anticoagulant properties. The anticoagulant effect of agaropectin (Qi et al. 2008) was comparable to that of heparin in inhibiting the coagulation process in rats. Hence, the sulfate groups in agar could be carefully tailored to replace heparin (Matsuhiro et al. 2014) which has limitations such as causing contamination by pathogenic agents, and side effects such as bleeding and platelet deficiency (Warkentin et al. 1995).

Antioxidant activity was reported from agaro-oligosaccharide which was derived from agarose in a concentration-dependent manner (Chen and Yan 2005; Chen et al. 2006). Agaro-oligosaccharide in neo-form (produced by hydrolysis of β-1, 4 linkage of agarose by β-agarase) was reported to have skin whitening effect (Jang et al. 2009). Both mono and oligo forms of agar (i.e. agarobiose, agarotetraose, neoagarotetraose, agarohexaose, agarooctaose, D-glucose, D-galactose, agarotriose) were found to have antioxidant, and antimelanogenic activities, and were able to reduce the nitric oxide production in macrophages (Enoki et al. 2012; Souza et al. 2012; Yun et al. 2013; Kim et al. 2017).

Although agar and agarose were not reported to have antiviral or antibacterial activities, the agarose nanoparticles were shown to have antimicrobial activity, associated with an increased chemical reactivity due to a higher surface area to volume ratio (Satar et al. 2016).

6.2 Commercial Usage of Agar

Agar is used as formative ingredients for oral tablets (Sharma et al. 2008), and as suspension reagent for barium sulfate radiological solutions (Miller 1965; Grabherr et al. 2008). Agar was shown to reduce cholesterol absorption by inhibiting the activity of human pancreatic cholesterol esterase, whereby the inhibition can be enhanced by increasing the degree of sulfation (Laurienzo 2010).

Agar or agarose was also used intensively as hydrogels in drugs and chemotherapeutics (Hou et al. 2018; Tan et al. 2021). Agar is a superabsorbent hydrogel due to its dense and compact interconnected polymer units. These interlinking polymer units form pores absorb water rapidly (Lyons et al. 2009). The water-retaining and mechanical properties of agar hydrogels can be improved by grafting with polyvinylpyrrolidone (Prasad et al. 2006). In addition, agar and agarose can be developed into antimicrobial wound dressings by the addition of a wide variety of materials such as locust bean gum, salep, tannic acid, zinc ions etc. (Ninan et al. 2016; Mao et al. 2017; Akkaya et al. 2020).

Agar has been used as bio-based natural materials for engineered drug delivery system due to its stability, high water content, biodegradability, low toxicity and biocompatibility due to structural similarity to human cell extracellular matrix (García-González et al. 2011; Rossi et al. 2011). The agarose drug delivery system has a high potential to be used for localized and effective delivery of drugs, antibiotics, chemotherapeutic agent, DNA/gene, protein/peptide, and cell (Grolman et al. 2019; Yazdi et al. 2020). Agarose-based films with different drugs and antibiotics retain their flexibility and thinness for external application (Felfel et al. 2018) as an assorted wound dressing and surgical materials (Basha et al. 2020). Cellular molecules (i.e. DNA, genes, RNA, plasmid, viral DNA, proteins, peptides and growth factors) packaged into agarose have prolonged stability and reduced degradation (Moribe et al. 2008) when tested in the treatment of various diseases (Setten et al. 2019).

7 Applications of Agars in the Biotechnology Industry

Agars are widely applied in biotechnology research. The agar with a lower gel strength is used as a solid support medium for microbial cultures e.g. bacteria and fungi (Peng et al. 2009) and plant tissue culture (Huang and Murashige 1977). The specifications for bacteriological-grade agar include controlled gelation and melting temperature, good gel clarity, low amount of electronegative groups and oligomers, and absence of contaminants such as thermophilic spores and hemolytic substances (Armisen 1991).

Agar, being stable and responsive to magnetic field, is used as a coating for magnetic beads for protein purification (Tong and Sun 2001). Agarose which has a higher gel strength and a low degree of side chain substitutions; is suitable to be used as a matrix for DNA, RNA and protein gel electrophoresis (Renn 1990). Besides, agarose is also applied in gel filtration chromatography (Freifelder 1982; Renn 1984). Some examples of commercially available agarose-based beads include Sepharose, WorkBeads 40 SEC, Superose and Superdex. In addition, agarose can be used as solid supportive material in immunodiffusion (Renn 1984).

The use of agar for enzyme immobilization has several advantages i.e. low cost and ease of preparation, good stability with no reactivity to protein (Mulagalapalli et al. 2007). Therefore, agar was used to immobilize enzymes such as maltase, manganese peroxidase, pullulanase, pectinase, amylase, chitosanase, pectinase and urease, bacteria and fungus (Lim et al. 2018 and references therein). Agar-based capsule has been developed to encapsulate probiotic strain *B. pseudocatenulatum* CECT 7765, to provide protection to the bacteria during the passage into the gastrointestinal tracks (Alehosseini et al. 2019). Besides, when immobilized with polyallylamine hydrochloride and alizarin red S, agar was used as non-enzymatic sensors to detect hydrogen peroxide (Soares et al. 2016).

8 *Gracilaria* Agars: Future Potentials and Developments

The applications of agar have been expanded into new and emerging field as biopolymer-based hydrogel film in medical, pharmaceutical, cosmetics and food packaging industries in recent years. Agar has been proposed to be non-toxic, biodegradable films and coatings from renewable source for food packaging industry, to replace plastic-based packaging materials that cause severe environmental problems (Mostafavi and Zaeim 2020).

The ternary blend agar/alginate/collagen (A/C/C) hydrogel film incorporated with silver nanoparticles and grapefruit seed extract was demonstrated to have good anti-fogging and strong antimicrobial properties against foodborne pathogenic bacteria, thus suitable as packaging films for highly respirating agricultural products (Rhim 2015). Besides that, agar films added with plant-based essential oils, zinc oxides, copper and titanium nanoparticles contain bioactive and phenolic agents that reduce oxidative degradation, spoilage, and have enhanced UV barrier and food colour in the food, horticultural and livestock products packaged in agar-based films (Choudhury et al. 2019).

Agar-based thin layer coatings of bimetallic magnetic nanoparticles (BMNPs) is another innovative usage of agar in the control of water pollution. The agar-based method offers a lower cost, more efficient and faster dye adsorption/removal method among the tested methods (Patra et al. 2016).

In cosmetics, agar has been used as a moisturizing agent in skin and hair products to absorb water, retain moisture and release water into skin/hair (Olatunji 2020). Agar-based composite films loaded with melanin nanoparticles were used in skincare products as thickener or packaging for oxidation sensitive foods to increase shelf life (Roy and Rhim 2019). In addition, agar is used as a natural and safe thickening agent in liquid bath soap to replace cocamide diethanolamine which has been labelled as 2B carcinogen (Dita and Sudarno 2020).

Agarose is suitable to be developed as materials for biomedical engineering due to its biocompatibility (less or non-immunogenic, non-toxic, nonabsorbable, nondegradable, high cell/matrix interaction and ability to mimic tissues with high accuracy), low production cost, easily available, stored and removed from the hosts. Agarose gels have been tested in rat models as dermal and cosmetic fillers (Fernández-Cossío et al. 2007; Karapantzou et al. 2020), proposed as a replacement for the cartilage tissue (Salati et al. 2020), and as the materials to produce phantom organs for medical training, diagnosis and imaging purposes (Rajeshkumar et al. 2020; Ahmad et al. 2020; Teixeira and Martins 2020).

9 Conclusions and Future Perspectives

Among the seaweed hydrocolloids, agar has the highest commodity price compared to carrageenan and alginates (Porse and Rudolph 2017). Cheap and renewable sources coupled with its bioactive and physiochemical properties, make agar an ideal raw material to be used in various industries (Fig. 7.4). New and innovative applications in the future may include the use of agar and agarose for cleaning and conservation of art surfaces (Sansonetti et al. 2020), glycerol-plasticized agarose separator in the lithium-metal battery to prevent the growth of lithium dendrites (Blin et al. 2020), and agarose-based structured optical fibre for *in vivo* imaging, monitoring and light delivery (Fujiwara et al. 2020).

With the growing global demand, there is a pressing need for genetic hybridization, improvement of existing strains and strain selection (e.g. disease resistance, fast growth and high amount of agar/agarose), consistent tank-grown seed stock, cost-effective mass culture system which is automated and optimized to overcome the shortage of raw materials and aquaculture issues such as fouling, inconsistent growth and agar quality (Callaway 2015; Lindell et al. 2015). Application of technologies in algal research such as genetic engineering, tissue culture, mathematical modelling, development of sensors and remote sensing would play major roles in the existing value chain of *Gracilaria* cultivation and agar production.

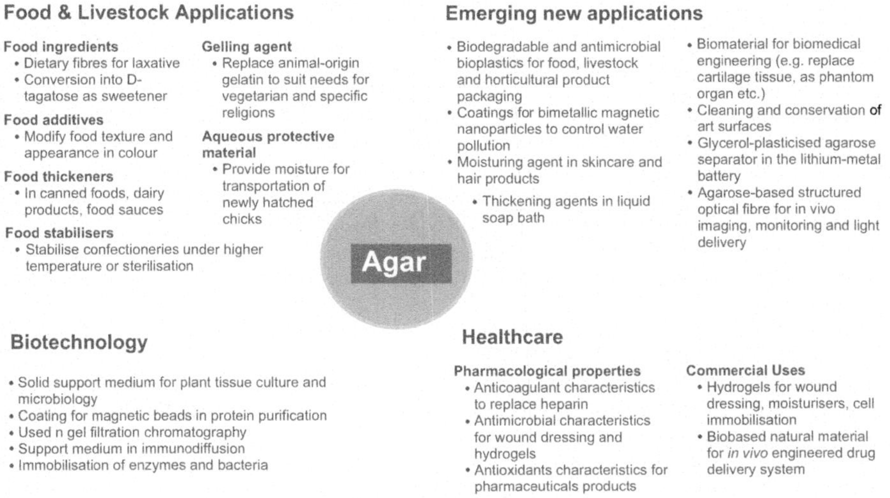

Fig. 7.4 Summary of agar applications in different fields

Continual optimization of hydrocolloid extraction and industrial processing, to minimize unusable biomass by-products, and to reduce the extraction time, massive usage of water in agar extraction, and the use of environmentally unfriendly chemicals is necessary (Khalil et al. 2018). The capital cost (e.g. equipment, operation and skilled technicians) of "greener" technologies such as ultrasound-assisted extraction, supercritical fluid extraction, pressurized solvent extraction for industrial scale is a challenge. Production of tailor-made agar hydrocolloid composition, specifically on appropriate side chain substitutions, would be useful to produce agar with rheology and bioactive properties specific to the desired commercial applications.

Acknowledgements We would like to acknowledge Ministry of Science, Technology and Innovation (MOSTI) of Malaysia and Universiti Putra Malaysia in supporting this work through eScience fund (Grant number: 02-01-04-SF1282) and Putra Graduate Initiative grant (GP-IPS/2016/9487700), respectively.

References

Abdul Khalil HPS, Lai TK, Tye YY, Rizal S, Chong EWN, Yap SW, Hamzah AA, Nurul Fazita MR, Paridah MT (2018) A review of extractions of seaweed hydrocolloids: properties and applications. *eXPRESS*. Polym Lett 12(4):296–317. https://doi.org/10.3144/expresspolymlett.2018.27

Ahmad R, Surif M, Ramli N, Yahya N, Nor ARM, Bekbayeva L (2011) A preliminary study on the agar content and agar gel strength of Gracilaria manilaensis using different agar extraction processes. World Appl Sci J 15(2):184–188

Ahmad MS, Suardi N, Shukri A, Mohammad H, Oglat AA, Alarab A, Makhamrah O (2020) Chemical characteristics, motivation and strategies in choice of materials used as liver phantom: a literature review. J Med Ultrasound 28:7–16. https://doi.org/10.4103/JMU.JMU_4_19

Akkaya NE, Ergun C, Saygun A, Yesilcubuk N, Akel-Sadoglu N, Kavakli IH, Turkmen HS, Catalgil-Giz H (2020) New biocompatible antibacterial wound dressing candidates; agar-locust

bean gum and agar-salep films. Int J Biol Macromol 155:430–438. https://doi.org/10.1016/j.ijbiomac.2020.03.214
Alehosseini A, del Pulgar EMG, Fabra MJ, Gómez-Mascaraque LG, Benítez-Páez A, Sarabi-Jamab M, Ghorani B, Lopez-Rubio A (2019) Agarose-based freeze-dried capsules prepared by the oil-induced biphasic hydrogel particle formation approach for the protection of sensitive probiotic bacteria. Food Hydrocoll 87:487–496. https://doi.org/10.1016/j.foodhyd.2018.08.032
An D, Xiao Q, Zhang C, Cai M, Zhang Y, Weng H, Chen F, Xiao A (2021) Preparation, characterization, and application of high-whiteness agar bleached with hydrogen peroxide. Food Hydrocoll 113:106520. https://doi.org/10.1016/j.foodhyd.2020.106520
Araki C (1966) Some recent studies on the polysaccharides of agarophytes. In: Young EG, McLachlan JL (eds) Proceeding of Fifth International seaweed symposium. Pergamon Press, Oxford, pp 3–17
Armisen R (1991) Agar and agarose biotechnological applications. In: International workshop on Gelidium. Springer, Dordrecht, pp 157–166
Armisen R, Galactas F (1987) Production, properties and uses of agar. In: McHugh DJ (ed) Production and utilisation of products from commercial seaweeds. FAO Fishery Technical Paper, Rome, pp 1–57
Arvizu-Higuera DL, Rodríguez-Montesinos YE, Murillo-Álvarez JI, Muñoz-Ochoa M, Hernández-Carmona G (2008) Effect of alkali treatment time and extraction time on agar from *Gracilaria vermiculophylla*. J Appl Phycol 20:515–519. https://doi.org/10.1007/978-1-4020-9619-8_9
Basha SI, Ghosh S, Vinothkumar K, Ramesh B, Mohan KM, Sukumar E (2020) Fumaric acid incorporated ag/agar-agar hybrid hydrogel: a multifunctional avenue to tackle wound healing. Mater Sci Eng C 111:110743. https://doi.org/10.1016/j.msec.2020.110743
Blin P, Boury B, Taguet A, Touja J, Monconduit L, Patra S (2020) Glycerol-plasticised agarose separator suppressing dendritic growth in Li metal battery. Carbohydr Polym 247:116697. https://doi.org/10.1016/j.carbpol.2020.116697
Callaway E (2015) Lab staple agar runs low: dwindling seaweed harvest imperils reagent essential for culturing microbes. Nature 528(7581):171–173. https://doi.org/10.1038/528171a
Chen HM, Yan XJ (2005) Antioxidant activities of Agaro-oligosaccharides with different degrees of polymerization in cell-based system. Biochim Biophys Acta Gen Subject 1722(1):103–111. https://doi.org/10.1016/j.bbagen.2004.11.016
Chen H, Yan X, Zhu P, Lin J (2006) Antioxidant activity and hepatoprotective potential of Agaro-oligosaccharides in vitro and in vivo. Nutr J 5(1):31. https://doi.org/10.1186/1475-2891-5-31
Chen H, Xiao Q, Weng H, Zhang Y, Yang Q, Xiao A (2020) Extraction of sulfated agar from *Gracilaria lemaneiformis* using hydrogen peroxide-assisted enzymatic method. Carbohydr Polym 232:115790. https://doi.org/10.1016/j.carbpol.2019.115790
Choudhury A, Jeelani PG, Biswal N, Chidambaram R (2019) Application of bionanocomposites on horticultural products to increase the shelf life. In: Gutierrez TJ (ed) Polymers for agri-food applications. Springer, Cham, pp 525–543
Cordover R (2007) Seaweed agronomy: cropping in inland saline groundwater evaporation basins: a report for the Rural Industries Research and Development Corporation. RIRDC, Barton, A. C. T
da Costa JN, Leal AR, do Nascimento LG, Rodrigues DC, Muniz CR, Figueiredo RW, Mata P, Noronha JP, de Sousa PHM (2020) Texture, microstructure and volatile profile of structured guava using agar and gellan gum. Int J Gastronomy Food Sci 20:100207. https://doi.org/10.1016/j.ijgfs.2020.100207
Dita LR, Sudarno J (2020) Utilisation of agar *Gracilaria* sp. as a natural thickener on liquid bath soap formulation. In: IOP Conference Series Earth and Environmental Science, vol 441, p 012021. https://doi.org/10.1088/1755-1315/441/1/012021
Elleuch M, Bedigian D, Roiseux O, Besbes S, Blecker C, Attia H (2011) Dietary fibre and fibre-rich by-products of food processing: characterisation, technological functionality and commercial applications: a review. Food Chem 124:411–421. https://doi.org/10.1016/j.foodchem.2010.06.077

Enoki T, Tominaga T, Takashima F, Ohnogi H, Sagawa H, Kato I (2012) Anti-tumor-promoting activities of Agaro-oligosaccharides on two-stage mouse skin carcinogenesis. Biol Pharm Bull 35(7):1145–1149. https://doi.org/10.1248/bpb.b12-00188

FAO (2020) The state of world fisheries and aquaculture 2020. FAO, Rome

FAO Fishery Statistics (2011) National Aquaculture Sector Overview (NASO). http://www.fao.org/fishery/naso/search/en. Accessed 16 Oct 2016

Felfel RM, Hossain KMZ, Kabir SF, Liew SY, Ahmed I, Grant DM (2018) Flexible and transparent films produced from cellulose nanowhisker reinforced agarose. Carbohydr Polym 194:328–338. https://doi.org/10.1016/j.carbpol.2018.04.005

Fernández-Cossío S, León-Mateos A, Sampedro FG, Oreja MTC (2007) Biocompatibility of agarose gel as a dermal filler: histologic evaluation of subcutaneous implants. Plast Reconstr Surg 120(5):1161–1169. https://doi.org/10.1097/01.prs.0000279475.99934.71

Ficko-Blean E, Hervé C, Michel G (2015) Sweet and sour sugars from the sea: the biosynthesis and remodeling of sulfated cell wall polysaccharides from marine macroalgae. Perspect Phycol 2:51–64. https://doi.org/10.1127/pip/2015/0028

Freifelder D (1982) Physical biochemistry. W.H. Freeman, San Francisco

Freile-Pelegrin Y, Murano E (2005) Agars from three species of *Gracilaria* (Rhodophyta) from Yucatán peninsula. Bioresour Technol 96:295–302. https://doi.org/10.1016/j.biortech.2004.04.010

Freile-Pelegrin Y, Robledo D (1997) Influence of alkali treatment on agar from *Gracilaria cornea* from Yucatán, México. J Appl Phycol 9:533–539. https://doi.org/10.1023/A:1007989931915

Fujiwara E, Cabral TD, Sato M, Oku H, Cordeiro CM (2020) Agarose-based structured optical fibre. Sci Rep 10(1):1–8. https://doi.org/10.1038/s41598-020-64103-3

Ganegama Arachchi GJ, Paththuwe Arachchi MJ, Wansapala MAJ, Jayarathna MPK (2018) Extraction of agar from locally grown *Gracilaria verrucosa* and development of gelatin free set-yoghurt product using agar. J Natl Aquat Resour Res Dev Agency:45–47

García-González CA, Alnaief M, Smirnova I (2011) Polysaccharide-based aerogels-promising biodegradable carriers for drug delivery systems. Carbohydr Polym 86(4):1425–1438. https://doi.org/10.1016/j.carbpol.2011.06.066

Glicksman M (1987) Utilisation of seaweed hydrocolloids in the food industry. Hydrobiologia 151(152):31–47. https://doi.org/10.1007/978-94-009-4057-4_3

Gómez-Ordóñez E, Jiménez-Escrig A, Rupérez P (2010) Dietary fibre and physicochemical properties of several edible seaweeds from the northwestern Spanish coast. Food Res Int 43:2289–2294. https://doi.org/10.1016/j.foodres.2010.08.005

Grabherr S, Dominietto M, Yu L, Djonov V, Müller B, Friess S (2008) Angiofil: a novel radio-contrast agent for post-mortem micro-angiography. In: Proceeding SPIE 7078, developments in X-ray tomography, vol 6, p 707810. https://doi.org/10.1016/10.1117/12.792077

Grolman JM, Singh M, Mooney DJ, Eriksson E, Nuutila K (2019) Antibiotic-containing agarose hydrogel for wound and burn care. J Burn Care Res 40(6):900–906. https://doi.org/10.1093/jbcr/irz113

Guiry MD, Guiry GM (2020) AlgaeBase. World-wide electronic publication. National University of Ireland, Galway

Hansen PMT (1993) Food hydrocolloids in the dairy industry. In: Nishinari K, Doi E (eds) Food hydrocolloids. Plenum Press, New York, pp 211–224

Hernández-Carmona G, Freile-Pelegrín Y, Hernández-Garibay E (2013) Conventional and alternative technologies for the extraction of algal polysaccharides. In: Dominguez H (ed) Functional ingredients from algae for foods and nutraceuticals. Woodhead, Philadelphia, pp 475–509

Hou M, Yang R, Zhang L, Zhang L, Liu G, Xu Z, Kang Y, Xue P (2018) Injectable and natural humic acid/agarose hybrid hydrogel for localised light-driven photothermal ablation and chemotherapy of cancer. ACS Biomater Sci Eng 4(12):4266–4277. https://doi.org/10.1021/acsbiomaterials.8b01147

Hu Y, Du Q, Mi P, Shang E, Sui Z (2019) Gene cloning and expression regulation in the pathway of agar and floridean starch synthesis of *Gracilariopsis lemaneiformis* (Rhodophyta). J Appl Phycol 31(3):1889–1896. https://doi.org/10.1007/s10811-018-1690-0

Huang LC, Murashige T (1977) Plant tissue culture media: major constitutents, their preparation and some applications. Methods Cell Sci 3:539–548. https://doi.org/10.1007/BF00918758

Imeson AP (1997) Thickening and gelling agents for food. Springer Science & Business Media, Berlin. https://doi.org/10.1007/978-1-4615-2197-6

Imeson A (2011) Food stabilisers, thickeners and gelling agents. Wiley, London. https://doi.org/10.1002/9781444314724

Incharoen T, Jomjanyouang W, Preecha N (2015) Effects of aqua agar as water replacement for posthatch chicks during transportation on residual yolk-sac and growth performance of young broiler chickens. Anim Nutr 1(4):310–312. https://doi.org/10.1016/j.aninu.2015.11.006

Jang MK, Lee DG, Kim NY, Yu KH, Jang HJ, Lee SW, Jang HJ, Lee YJ, Lee SH (2009) Purification and characterisation of neoagarotetraose from hydrolysed agar. J Microbiol Biotechnol 19(1):197–191. https://doi.org/10.4014/jmb.0906.06045

Jeong DW, Hyeon JE, Shin SK, Han SO (2020) Trienzymatic complex system for isomerisation of agar-derived D-galactose into D-tagatose as a low-calorie sweetener. J Agric Food Chem 68(10):3195–3202. https://doi.org/10.1021/acs.jafc.9b07536

Kaladharan P, Kaliaperumal N (1999) Seaweed industry in India. Naga 22:11–14

Kaneda I, Shibata S (2020) Rheological properties of low oil mayonnaise by replacing oil droplets with agar micro-gels. Nihon Reoroji Gakkaishi 48(2):113–120. https://doi.org/10.1678/rheology.48.113

Karapantzou C, Jakob M, Kinney B, Vandeputte J, Vale JP, Canis M (2020) The use of algeness in the face and neck: a safe, alternative filler for cosmetics and reconstruction. Ann Trans Med 8(6):362. https://doi.org/10.21037/atm.2020.02.52

Khalil HPS, Lai TK, Tye YY, Rizal S, Chong EWN, Yap SW, Hamzah AA, Fazita MR, Paridah MT (2018) A review of extractions of seaweed hydrocolloids: properties and applications. Express Polym Lett 12(4):296–317. https://doi.org/10.3144/expresspolymlett.2018.27

Kim JH, Yun EJ, Yu S, Kim KH, Kang NJ (2017) Different levels of skin whitening activity among 3, 6-anhydro-l-galactose, agarooligosaccharides, and neoagarooligosaccharides. Mar Drugs 15(10):321. https://doi.org/10.3390/md15100321

Kraan S (2012) Algal polysaccharides, novel applications and outlook, vol 22. INTECH Open Access, London, pp 489–524. https://doi.org/10.5772/51572

Kumar V, Fotedar R (2009) Agar extraction process for *Gracilaria cliftonii*. Carbohydr Polym 78:813–819. https://doi.org/10.1016/j.carbpol.2009.07.001

Laurienzo P (2010) Marine polysaccharides in pharmaceutical applications: an overview. Mar Drugs 8:2435–2465. https://doi.org/10.3390/md8092435

Lee WK, Lim YY, Leow ATC, Namasivayam P, Abdullah JO, Ho CL (2017) Factors affecting yield and gelling properties of agar. J Appl Phycol 29(3):1527–1540. https://doi.org/10.1007/s10811-016-1009-y

Li H, Yu X, Jin Y, Zhang W, Liu Y (2008) Development of an eco-friendly agar extraction technique from the red seaweed *Gracilaria lemaneiformis*. Bioresour Technol 99(8):3301–3305. https://doi.org/10.1016/j.biortech.2007.07.002

Lim YY, Lee WK, Leow ATC, Namasivayam P, Abdullah JO, Ho CL (2018) Sulfated galactans from red seaweeds and their potential applications. Pertanika J Scholar Res Rev 4(2):1–17

Lim YY, Lee WK, Lim PE, Phang SM, Leow ATC, Namasivayam P, Abdullah JO, Ho CL (2019) Expression analysis of potential transcript and protein markers that are related to agar yield and gel strength in *Gracilaria changii* (Rhodophyta). Algal Res 41:101532. https://doi.org/10.1016/j.algal.2019.101532

Lindell S, Green-Beach E, Bailey D, Beals M, Kim JK, Yarish C (2015) Multi-cropping seaweed *Gracilaria tikvahiae* with oysters for nutrient bioextraction and sea vegetables in Waquoit Bay, MA. In: National Shellfisheries Association 107th Annual Meeting. National Shellfisheries Association, Monterey, CA

Lyons JG, Geever LM, Nugent MJ, Kennedy JE, Higginbotham CL (2009) Development and characterisation of an agar-polyvinyl alcohol blend hydrogel. J Mech Behav Biomed Mater 2:485–493. https://doi.org/10.1016/j.jmbbm.2008.12.003

Maciel JS, Chaves LS, Souza BW, Teixeira DI, Freitas AL, Feitosa JP, de Paula RC (2008) Structural characterization of cold extracted fraction of soluble sulfated polysaccharide from red seaweed *Gracilaria birdiae*. Carbohydr Polym 71(4):559–565. https://doi.org/10.1016/j.carbpol.2007.06.026

Mantri VA, Shah Y, Thiruppathi S (2020) Feasibility of farming the agarose-yielding red alga *Gracilaria dura* using tube-net cultivation in the open sea along the Gujarat coast of NW India. Appl Phycol 1(1):12–19. https://doi.org/10.1080/26388081.2019.1648181

Mao C, Xiang Y, Liu X, Cui Z, Yang X, Yeung KWK, Pan H, Wang X, Chu PK, Wu S (2017) Photo-inspired antibacterial activity and wound healing acceleration by hydrogel embedded with ag/ag@ AgCl/ZnO nanostructures. ACS Nano 11(9):9010–9021. https://doi.org/10.1021/acsnano.7b03513

Matsuhiro B, Barahona T, Encinas MV, Mansilla A, Ortiz JA (2014) Sulfation of agarose from subantarctic *Ahnfeltia plicata* (Ahnfeltiales, Rhodophyta): studies of its antioxidant and anticoagulant properties in vitro and its copolymerisation with acrylamide. J Appl Phycol 26(5):2011–2019. https://doi.org/10.1007/s10811-014-0297-3

McHugh DJ (1987) Production and utilization of products from commercial seaweeds. FAO Fisheries Technical Paper No. 288, pp 1–189

McHugh DJ (2003) A guide to seaweed industry. FAO Fisheries and Aquaculture Department, Rome

Miller RE (1965) Barium sulfate suspensions 1. Radiology 84:241–251

Moribe K, Nomizu N, Izukura S, Yamamoto K, Tozuka Y, Sakurai M, Ishida A, Nishida H, Miyazaki M (2008) Physicochemical, morphological and therapeutic evaluation of agarose hydrogel particles as a reservoir for basic fibroblast growth factor. Pharm Dev Technol 13(6):541–547. https://doi.org/10.1080/10837450802309661

Mostafavi FS, Zaeim D (2020) Agar-based edible films for food packaging applications-a review. Int J Biol Macromol 159:1165–1176. https://doi.org/10.1016/j.ijbiomac.2020.05.123

Mulagalapalli S, Kumar S, Kalathur RCR, Kayastha AM (2007) Immobilisation of urease from pigeonpea (*Cajanus cajan*) on agar tablets and its application in urea assay. Appl Biochem Biotechnol 142:291–297. https://doi.org/10.1007/s12010-007-0022-7

Ninan N, Forget A, Shastri VP, Voelcker NH, Blencowe A (2016) Antibacterial and anti-inflammatory pH-responsive tannic acid-carboxylated agarose composite hydrogels for wound healing. ACS Appl Mater Interfaces 8(42):28511–28521. https://doi.org/10.1021/acsami.6b10491

Nussinovitsch A, Kopelman IJ, Mizrahi S (1991) Modelling the combined effect of fruit pulp, sugar and gum on some mechanical parameters of agar and alginate gels. Lebensmittel Wissenschaft Technologie 24:513–517

Nyberg CN (2007) Introduced marine macroalgae and habitat modifiers-their ecological role and significant attributes. https://gupea.ub.gu.se/bitstream/2077/4719/1/gupea_2077_4719_1.pdf. Accessed 15 Jun 2020

O'Sullivan L, Murphy B, McLoughlin P, Duggan P, Lawlor PG, Hughes H, Gardiner GE (2010) Prebiotics from marine macroalgae for human and animal health applications. Mar Drugs 8:2038–2064. https://doi.org/10.3390/md8072038

Olatunji O (2020) Agar. In: Aquatic biopolymers. Springer, Cham, pp 145–168

Oliveira EC, Alveal K, Anderson RJ (2000) Mariculture of the agar-producing Gracilarioid red algae. Rev Fish Sci 8:345–377. https://doi.org/10.1080/10408340308951116

Patra S, Roy E, Madhuri R, Sharma PK (2016) Agar based bimetallic nanoparticles as high-performance renewable adsorbent for removal and degradation of cationic organic dyes. J Ind Eng Chem 33:226–238. https://doi.org/10.1016/j.jiec.2015.10.008

Peng S, Zhou Q, Cai Z, Zhang Z (2009) Phytoremediation of petroleum contaminated soils by *mirabilis Jalapa L.* in a greenhouse plot experiment. J Hazard Mater 168:1490–1496. https://doi.org/10.1016/j.jhazmat.2009.03.036

Porse H, Rudolph B (2017) The seaweed hydrocolloid industry: 2016 updates, requirements, and outlook. J Appl Phycol 29:2187–2200. https://doi.org/10.1007/s10811-017-1144-0

Prasad K, Mehta G, Meena R, Siddhanta AK (2006) Hydrogel-forming agar-graft-PVP and κ-carrageenan-graft-PVP blends: rapid synthesis and characterisation. J Appl Polym Sci 102:3654–3663. https://doi.org/10.1002/app.24145

Qi H, Li D, Zhang J, Liu L, Zhang Q (2008) Study on extraction of agaropectin from *Gelidium amansii* and its anticoagulant activity. Chinese J Oceanol Limnol 26(2):186–189. https://doi.org/10.1007/s00343-008-0186-1

Raikar S, Lima M, Fujita Y (2001) Effect of temperature, salinity and light intensity on the growth of *Gracilaria* spp. (Gracilariales, Rhodophyta) from Japan, Malaysia and India. Indian J Marine Sci 30:98–104

Rajeshkumar G, Vishnupriyan R, Selvadeepak S (2020) Tissue mimicking material an idealised tissue model for clinical applications: a review. Mater Today Proc 22:2696–2703. https://doi.org/10.1016/j.matpr.2020.03.400

Rebello J, Ohno M, Ukeda H, Kusunose H, Sawamura M (1997) 3, 6 anhydrogalactose, sulfate and methoxyl contents of commercial agarophytes from different geographical origins. J Appl Phycol 9(4):367–370. https://doi.org/10.1023/A:1007954220257

Reis FR (2019) Reports on the processing of structured exotic fruits, dried exotic fruits, and other exotic fruit products. In: Reports on the processing of exotic fruits. Springer, Cham, pp 33–47

Renn DW (1984) Agar and agarose: indispensible partners in biotechnology. I EC Prod Res Dev 23:17–21. https://doi.org/10.1021/i300013a004

Renn DW (1990) Seaweeds and biotechnology: inseparable companions. Hydrobiologia 204(205):7–13. https://doi.org/10.1007/978-94-009-2049-1_2

Rossi F, Santoro M, Casalini T, Veglianese P, Masi M, Perale G (2011) Characterisation and degradation behavior of agar-carbomer based hydrogels for drug delivery applications: solute effect. Int J Mol Sci 12(6):3394–3408. https://doi.org/10.3390/ijms12063394

Roy S, Rhim JW (2019) Agar-based antioxidant composite films incorporated with melanin nanoparticles. Food Hydrocoll 94:391–398. https://doi.org/10.1016/j.foodhyd.2019.03.038

Ruperez P, Saura-Calixto F (2001) Dietary fibre and physicochemical properties of edible Spanish seaweeds. Eur Food Res Technol 212:349–354. https://doi.org/10.1007/s002170000264

Salati MA, Khazai J, Tahmuri AM, Samadi A, Taghizadeh A, Taghizadeh M, Zarrintaj P, Ramsey JD, Habibzadeh S, Seidi F, Saeb MR (2020) Agarose-based biomaterials: opportunities and challenges in cartilage tissue engineering. Polymers 12(5):1150. https://doi.org/10.3390/polym12051150

Sansonetti A, Bertasa M, Canevali C, Rabbolini A, Anzani M, Scalarone D (2020) A review in using agar gels for cleaning art surfaces. J Cult Herit 44:285–296. https://doi.org/10.1016/j.culher.2020.01.008

Satar R, Iizhar SA, Rasool M, Pushparaj PN, Ansari SA (2016) Investigating the antibacterial potential of agarose nanoparticles synthesised by nanoprecipitation technology. Pol J Chem Technol 18(2):9–12. https://doi.org/10.1515/pjct-2016-0022

Setten RL, Rossi JJ, Han SP (2019) The current state and future directions of RNAi-based therapeutics. Nat Rev Drug Discov 18(6):421–446. https://doi.org/10.1038/s41573-019-0017-4

Sharma V, Philip AK, Pathak K (2008) Modified polysaccharides as fast disintegrating excipients for orodispersible tablets of roxithromycin. AAPS PharmSciTech 9(1):87–94. https://doi.org/10.1208/s12249-007-9026-4

Shukla MK, Kumar M, Prasad K, Reddy CRK, Jha B (2011) Partial characterization of sulfohydrolase from *Gracilaria dura* and evaluation of its potential application in improvement of the agar quality. Carbohydr Polym 85(1):157–163. https://doi.org/10.1016/j.carbpol.2011.02.009

Soares MDFC, de Oliveira Farias EA, da Silva DA, Eiras C (2016) Development and characterisation of hybrid films based on agar and alizarin red S for applications as nonenzymatic sensors for hydrogen peroxide. J Mater Sci 51:7093–7107. https://doi.org/10.1007/s10853-016-9958-8

Soegiarto A (1990) Utilisation and farming of seaweeds in Indonesia. In: *Culture and use of algae in Southeast Asia. Proceedings of the Symposium on Culture and Utilisatin of Algae in Southeast Asia, December 8-11* 1981, *Tigbauan, Iloilo, Philippines*. Aquaculture Department, Southeast Asian Fisheries Development Center, pp 9–19

Sousa AM, Borges J, Silva AF, Gonçalves MP (2013) Influence of the extraction process on the rheological and structural properties of agars. Carbohydr Polym 96(1):163–171. https://doi.org/10.1016/j.carbpol.2013.03.070

Souza BW, Cerqueira MA, Bourbon AI, Pinheiro AC, Martins JT, Teixeira JA, Coimbra MA, Vicente AA (2012) Chemical characterisation and antioxidant activity of sulfated polysaccharide from the red seaweed *Gracilaria birdiae*. Food Hydrocoll 27(2):287–292. https://doi.org/10.1016/j.foodhyd.2011.10.005

Tan B, Huang L, Wu Y, Liao J (2021) Advances and trends of hydrogel therapy platform in localised tumor treatment: a review. J Biomed Mater Res 109(4):404–425. https://doi.org/10.1002/jbm.a.37062

Teixeira AM, Martins P (2020) Mechanical characterisation of an organic phantom candidate for breast tissue. J Biomater Appl 34(8):1163–1170. https://doi.org/10.1177/0885328219895738

Tong XD, Sun Y (2001) Agar-based magnetic affinity support for protein. Biotechnol Prog 17:738–743. https://doi.org/10.1021/bp010054s

Vergara-Rodarte MA, Hernández-Carmona G, Rodríguez-Montesinos YE, Arvizu-Higuera DL, Riosmena-Rodríguez R, Murillo-Álvarez JI (2010) Seasonal variation of agar from Gracilaria vermiculophylla, effect of alkali treatment time, and stability of its Colagar. J Appl Phycol 22(6):753–759. https://doi.org/10.1007/s10811-010-9516-8

Walstra P (2003) Physical chemistry of foods. Marcel Dekker, New York

Wang LF, Rhim JW (2015) Preparation and application of agar/alginate/collagen ternary blend functional food packaging films. Int J Biol Macromol 80:460–468. https://doi.org/10.1016/j.ijbiomac.2015.07.007

Warkentin TE, Levine MN, Hirsh J, Horsewood P, Roberts RS, Gent M, Kelton JG (1995) Heparin-induced thrombocytopenia in patients treated with low-molecular-weight heparin or unfractionated heparin. N Engl J Med 332:1330–1336. https://doi.org/10.1056/NEJM199505183322003

Wu C (1990) Training manual on Gracilaria culture and seaweed processing in China. FAO Fishery Technical Paper, China

Yang Y, Chai Z, Wang Q, Chen W, He Z, Jiang S (2015) Cultivation of seaweed *Gracilaria* in Chinese coastal waters and its contribution to environmental improvements. Algal Res 9:236–244. https://doi.org/10.1016/j.algal.2015.03.017

Yarnpakdee S, Benjakul S, Kingwascharapong P (2015) Physico-chemical and gel properties of agar from Gracilaria tenuistipitata from the lake of Songkhla, Thailand. Food Hydrocoll 51:217–226

Yazdi MK, Taghizadeh A, Taghizadeh M, Stadler FJ, Farokhi M, Mottaghitalab F, Zarrintaj P, Ramsey JD, Seidi F, Saeb MR, Mozafari M (2020) Agarose-based biomaterials for advanced drug delivery. J Control Release 326:523–543. https://doi.org/10.1016/j.jconrel.2020.07.028

Yun EJ, Lee S, Kim JH, Kim BB, Kim HT, Lee SH, Pelton JG, Kang NJ, Choi IG, Kim KH (2013) Enzymatic production of 3, 6-anhydro-L-galactose from agarose and its purification and *in vitro* skin whitening and anti-inflammatory activities. Appl Microbiol Biotechnol 97(7):2961–2970. https://doi.org/10.1007/s00253-012-4184-z

Chapter 8
Marine Algal Colorants for the Food Industry

Chidambaram Kulandaisamy Venil, Chatragadda Ramesh, Ponnuswamy Renuka Devi, and Laurent Dufossé

1 Introduction

Seaweeds are a commanding component of the marine plant life complex in the structure of coastal environmental biodiversity (Klinger 2015), and they are functioning as renewable living wherewithal to the global coastal communities (Rebours et al. 2014). They are the source materials approximately for 27 health benefits (Qin 2018), and a varied makeup applications (Wang et al. 2015). A range of chemical molecules existing in the seaweeds validate their prospective reach and their crucial biological properties (Rengasamy et al. 2020). Seaweeds are largely well known for their food items and as a reliable source of natural iodine.

Currently, coastal habitats are under tremendous pressure from various practices of development like tourism, shrimp farming, infrastructure construction, port facilities including dredging etc. In relationship to such activities, coastal habitats are being changed or degraded leading to water pollution, global warming, biodiversity loss etc. causing a main influence over the global food demand. Literature hints that the world's biodiversity is likely to see a sixth mass extinction induced by anthropogenic stimulates (Ceballos et al. 2015) and the future generation is to face the extinction of numerous species causing food problems, arising of new diseases etc.

C. K. Venil · P. R. Devi
Department of Biotechnology, Anna University, Coimbatore, Tamil Nadu, India

C. Ramesh
Biological Oceanography Division, National Institute of Oceanography (CSIR-NIO), Dona Paula, Goa, India
e-mail: chramesh@nio.org

L. Dufossé (✉)
CHEMBIOPRO Chimie et Biotechnologie des Produits Naturels, ESIROI Département agroalimentaire, Université de la Réunion, Ile de La Réunion, Indian Ocean, France
e-mail: laurent.dufosse@univ-reunion.fr

A. Ranga Rao, G. A. Ravishankar (eds.), *Sustainable Global Resources of Seaweeds Volume 2*, https://doi.org/10.1007/978-3-030-92174-3_8

Seaweeds are acknowledged macroalgae and they are amid the leading producers of food products for human consumption; and they act as biological engineers on rock-strewn coastlines of the global oceans. Being primary producers, they are sanctuary for nursery grounds and food sources for marine organisms. Seaweeds are natural sources possessing not only high biological value, but also having a grand economic value (Salehi et al. 2019; Shama et al. 2019). Latest delve into the bioactive molecules from the seaweeds has alerted on their potential new openings in the field of medicine (Baby et al. 2012; Jeeva et al. 2012). Furthermore, due to their distinct habitats and biology, seaweeds are rather easy to monitor, maneuver and assess. Consequently, they have been extensively exploited as typical organisms for analyzing biogeographic relationship and for trying various biological theories, both in intertidal and subtidal habitats (Bhagyaraj and Kunchithapatham 2016).

Seaweeds are pigment yielding marine algae and are categorized into three classes with diverse pigment structure namely brown algae (Ochrophyta), red algae (Rhodophyta) and green algae (Chlorophyta) encompassing nearly 1750, 6000 and 1200 species respectively (Silva et al. 2020). They are commonly found in the complex and dynamic ecosystems, highly influenced by temperature, salinity, light, pollutants and nutrients. Growth development and the pigment contents in some seaweeds emerge to vary according to the depth (Marinho-Soriano 2012), season, and geography (Schmid et al. 2017). They produce a wide range of secondary metabolites like phenolic compounds, halogenated compounds, sterols, terpenes, peptides, and other bioactive compounds and are deemed as potential organisms for providing biologically active compounds for the promotion of nutraceuticals, cosmetics and pharmaceutical products (Cotas et al. 2020). The pigments of seaweeds include chlorophylls, carotenoids [carotenes (β-carotene) and xanthophylls (fucoxanthin, violaxanthin, zeaxanthin, lutein etc.)] and phycobiliproteins. These pigments are bioactive compounds for antioxidant, anticancer, anti-inflammatory, antimicrobial activities etc. and serve as colorants in food/feed/nutraceutical ingredients as well (Aryee et al. 2018). Extracted seaweed stuffs are used as stabilizers and stiffeners in food industry and furthermore in cosmetics, pharmaceutical and biotechnology industries (Jeeva and Kiruba 2009; Wiencke and Bischof 2012). The existence of sulphated polysaccharides in the cell wall matrix crafts seaweeds as economically significant ones. The industry utilizes the high molecular weight polysaccharides (phycocolloids) in nutraceutical market.

The seaweed industry is a multi-billion-dollar opportunity owing to their added products like, hydrocolloids, fertilizers, animal feed, fuel biomass, bioactive compounds for pharmaceuticals, cosmetics, food grade pigments etc. (Gomez-Zavaglia et al. 2019). In the perspective of organic foods, seaweeds continue to be an under-exploited renewable natural resource in the vast global ocean terrain. Therefore, culturing and enhanced biomass productions of marine seaweeds have become the crucial characteristic of the contemporary study to accomplish the food and other demands arising there from in the days to come.

2 Biodiversity of Seaweeds

As an example of world's location, more than 7000 km of vast coastlines in India harbour a broad range of marine algae (Sahoo et al. 2003). Rao and Mantri (2006) have recounted that many rocky beaches, estuaries, coral reefs and lagoons along the Indian coast serve as the idyllic environment for the growth of seaweeds. Seaweeds are affluent in the eastern coasts, Gulf of Mannar, Tuticorin, Gulf of Kutch in western coast as well as Andaman and Nicobar islands and Lakshadweep in India. The most leading seaweeds amid green algae in India are: *Ulva* and *Caulerpa*, red algae: *Hypnea* and *Kappaphycus,* brown algae: *Sargassum* and *Turbinaria* (Sharma et al. 2019).

In Asian diets, seaweeds are consumed traditionally since ancient times and so have lately gained the attention in the European markets as well. Owing to the dietary benefits such as vitamins, minerals, nutritional fibre, they act as a complement to the latest diet (Ferraces-Casais et al. 2012). The seaweed industry is worth more than six billion USD per annum and about 85% are food products for human consumption. Seaweed derived extracts like carrageenan, agar and alginates constitute to 40% of the hydrocolloid food markets while the rest are sourced from microbes, plants and animals (Ganesan et al. 2019b).

In Asia, China is the largest producer of seaweeds chased by Indonesia, Japan, Korea, Malaysia and Philippines (Seth and Shanmugam 2016). Japanese coastal areas afford more than 90% of the demand for *Porphyra* sp., *Saccharina japonica* and *Undaria* sp. Though Japan is the leading importer of seaweeds, large industry is built there for agar, alginate and carrageenan from *Ecklonia* sp. and *Durvillea antarctica*, *Eucheuma* sp. and *Kappaphycus* sp. *Gracilaria* sp. is the most predominant species in Thailand followed by *Hypnea* sp., *Porphyra* sp., *Acanthophora* sp. and *Caulerpa* sp. In Korea, *Saccharina japonica* and *Undaria* sp. and *Porphyra* sp. comprise the seaweed production. *Gracilaria* sp. and *Sargassum* sp. are the main seaweeds in Vietnam (Gomez-Zavaglia et al. 2019).

3 Seasonal Variations and Environmental Threats

Seaweeds are closely linked to humans in many ways by providing valuable products ranging from food to medicine as well as storm protection (Ronnback et al. 2007). Seaweed growth and reproduction depends on various environmental factors such as temperature, salinity, nutrient supply and carbon-di-oxide concentrations (Harley et al. 2012). Seaweeds evolved physiological adaptations that enables them to optimize the performance with respect to temperatures they come across (Eggert 2012). Seaweeds are well adapted to thermal environment, and cause disruptive stress during environmental change. Seaweeds compete for nutrients, light and attachment space with other photo-autotrophs for resource availability. The elevated level of CO_2 affects the seaweeds depending on the carbon capture strategy (Hepburn et al. 2011).

Seaweeds perform a crucial biological role in the coastal habitats by supporting and providing oxygen and absorbing CO_2 and also performing as primary producers in the marine food chain. Having the capacity to eliminate heavy metals from the seawater, they act as bio-indicators of heavy metal pollution as well. Seaweeds can survive even in the difficult environmental stresses and in the adverse impact of climate changes (Harley et al. 2012). Nevertheless, seaweeds are destructed by natural disturbances like seismic activity, whirlwind, grazing, disease, human activities, sewage disposal, upland development and fishing practices.

4 Major Classes of Seaweed Pigments

Seaweeds contain three major classes of photosynthetic pigments namely, chlorophylls, carotenoids (carotenes and xanthophylls) and phycobiliproteins.

4.1 Chlorophylls

Chlorophylls from the marine algae are the copious pigments that demonstrate prospective health benefits such as antioxidant, neuroprotective, anti-inflammatory, anti-mutagenic effects etc. (Chen and Roca 2018). They are tetrapyrrolic pigments with utmost absorbance visible spectrum in the blue and red regions. These pigments have porphyrin ring with magnesium ion in the center with long hydrophobic chain. The pattern of chlorophyll varies by the side-chain substitutions and the most abundant chlorophyll pigment is chl a which is present in all the seaweeds. Seaweeds that contain chl a, chl b, lutein, beta carotene, neoxanthin, zeaxanthin have green colour and this is due to the presence of chl a and this type of seaweeds require sunlight for photosynthesis (Sathasivam and Ki 2018).

4.2 Carotenoids

Carotenoids are red, orange, or yellow natural terpenoid pigments that are dispersed broadly with exciting distinctiveness (Pangestuti et al. 2018). Seaweed carotenoids exist in the chloroplast, cytoplasm matrix and other macromolecules in the intracellular space. Takaichi (2011) have reported that seaweed carotenoids perform a significant role as a chemotaxonomic marker for taxonomic classification of seaweeds. Carotenoids being lipophilic molecules, protect against photooxidative process and ensue as efficient antioxidants. Carotenoids are derived from 40-carbon basal structure, which includes conjugated double bonds with central chain carrying cyclic end groups, and can be replaced with oxygen containing functional groups. Based on their composition, carotenoids are classified into two groups: carotenes (contain only carbon and H atoms) and xanthophylls (oxo-carotenoids carrying at least one oxygen atoms) (Mezzomo et al. 2016).

4.2.1 Carotenes

Carotenes are polyunsaturated hydrocarbons with 40 carbon atoms per molecule and varying hydrogen atoms. β-carotene, lipid soluble and produces orange—yellow pigment and serve as the precursor of vitamin A. β-carotene, the mainly leading seaweed carotene is a key source of vitamin A and is used in dietetic add-ons with peak amount of antioxidant properties (Boominathan and Mahesh 2015). Lutein and zeaxanthin shield the dye from oxidative stress and perform a crucial role in the prevention of cancer (Michaud et al. 2000). Lutein is the dihydroxy derivate of α-carotene while zeaxanthin is the dihydroxy derivative of β-carotene.

4.2.2 Xanthophylls

Xanthophylls are imitative of oxidized carotenes. Lutein is a major associate pigment in green and in some red algae while fucoxanthin is the leading carotenoid in brown algae and occurs in Light-Harvesting-Complexes (LHC) with chl a or chl c. Zeaxanthin is existing in green and red algae, while violaxanthin is existing in brown and green algae. In brown seaweeds, fucoxanthin is the dominant carotenoids (Kumar et al. 2013) with the characteristic allenic and conjugated carbonyl group in their polyene backbone. The allenic bond in carotenoids is responsible for the antioxidant potential (Sachindra et al. 2007). Fucoxanthin has been reported to possess strong anticancer activity among xanthophylls and carotenoids. Fucoxanthin, owing to its antioxidant activity, prevents liver cancer, skin cancer and through induction of apoptosis inhibits breast and prostate cancers (Pangestuti et al. 2018).

4.3 Phycobiliproteins

Phycobiliproteins are a group of water-soluble pigments found in the cytoplasm or chloroplasts of the marine algae. Based on their spectral properties, phycobiliproteins are divided into phycoerythrin (purple), phycocyanin (blue), allophycocyanin (bluish green) and phycoerythrocyanin (purple) (Stadnichuk and Tropin 2017). The commercially produced phycobiliproteins are phycocyanin and phycoerythrin. In red seaweeds, phycobiliprotein content is high, reach 50% of water-soluble proteins, and represent 20% of dry algal biomass. Red seaweeds appear to hold abounding pigments, which have prospective food and drug applications (Cotas et al. 2020). Exactly, phycobiliproteins, the light harvesting pigments in seaweeds have demonstrated their natural colorant, cosmetic, food, and drug applications (Cotas et al. 2020).

5 Process of Pigment Extraction from Seaweed Biomass

Pigments akin to chlorophylls and carotenoids are fat soluble and taken out using organic solvents (acetone, DMSO, methanol) whereas phycobilins are water-soluble pigments. The extraction of chlorophyll for industrial appliances comprises homogenization of algal biomass and extraction using natural solvents (Hosikian et al. 2010). The seaweed pigments can be extracted by means of other approaches like heat, light, oxygen and extreme pH triggers the ruin of pigment. The cell integrity is disrupted thereby removing the pigment molecules from the intrinsic proteins. Freeze thawing also breaks cellular membranes and liberate polysaccharides. The pigments are extracted using organic solvents and separated by TLC, column chromatography and HPLC.

Diverse methods have been applied to extract algal pigments (Aryee et al. 2018), and are being improved to achieve the zero-waste management from whole biomass (Fig. 8.1).

A simple aqueous and solvent extraction method was developed to utilize complete biomass for extracting pigments, lipids, agar, bioethanol, and minerals from *Gracilaria corticata* (Baghel et al. 2016; Mohammad et al. 2019; Hessami et al. 2019). Aqueous extraction method is more effective to obtain R-phycoerythrin from red algae (Dumay et al. 2015). Freeze-dried red alga *Grateloupia turuturu* samples resulted in more R-phycoerythrin than fresh samples in 80% ammonium sulfate precipitation method (Denis et al. 2009). Similarly, freeze drying method preserved more pigment substance (total carotenoids, 36.91 ± 2.18 µg/g dm) from *Durvillaea antarctica* than other drying methods; while convective drying resulted in high phytochemical and vitamin contents in *D. antarctica* (Uribe et al. 2020a). Freeze drying

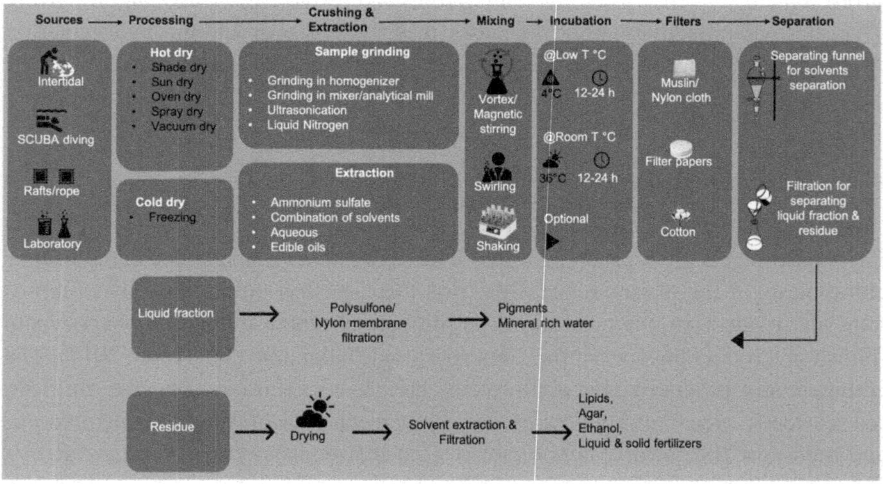

Fig. 8.1 An overview of pigment extraction process from seaweeds

method is evaluated to be more effective method to extract pigments from different seaweed species like *Gracilariopsis tenuifrons, Pterocladiella capillacea, Sargassum stenophyllum,* and *Ulva fasciata* (Amorim et al. 2020). In the case of *Pyropia orbicularis*, fresh samples alone yielded more pigment substances than other drying methods, while freeze drying method was effectual amid the other drying methods (Uribe et al. 2020b).

High yield of phycobiliprotein (0.28–0.50 mg g^{-1} ww) was achieved in distilled water than phosphate buffer and sea water extractions (Sudhakar et al. 2015). Pretreating the wet samples of *Sargassum horneri* by boiling and acid/alkali treatments, has improved the extraction rate of fucoxanthin from dried biomass (Teramukai et al. 2020). Also, fucoxanthin extraction from *S. horneri* was more effectual with edible oils such as short-chain and medium-chain triacylglycerols (Teramukai et al. 2020). Edible oil extraction method is more efficient to extract fucoxanthin from brown seaweeds than solvent extraction (Sudhakar et al. 2013).

Freeze drying and boiling methods are more effective practices employed to extract pigments and other phytochemical constituents correspondingly. Diverse nonionic surfactants have been utilized to extract carotenoids from *Sargassum* spp. and found that Tomadol (5.28 ± 2.01 mg g^{-1} ww) and Pluronic (1.86 ± 0.06 mg g^{-1} ww) yielded more pigments (Vieira et al. 2017). Ultrasonication method is also effective to extract phycoerythrin from *Kappaphycus alvarezii* at a frequency of 40 kHz (Uju et al. 2020), and carotenoids at 20 kHz from *Undaria pinnatifida* (Zhu et al. 2017). Encapsulation of unstable carotenoid pigment fucoxanthin extracted from brown seaweeds has been successful with carrier agents maltodextrin and Tween 80 (Indrawati et al. 2015).

6 Impact of Cooking on Seaweed Chemical Composition

Seaweed cooking evidently confirmed contiguous structure changes, which can be more effective than raw seaweeds. In the case of the edible red seaweed, *Chondrus crispus*, cooking resulted in not only increased levels of β-carotene and lutein, but improved the antioxidant activity (Pina et al. 2014). Microwaved seaweeds nori, sea lettuce, and kombu have yielded high chlorophyll derivatives than boiled seaweeds (Chena and Roca 2019).

Chlorophyll pigments from edible seaweeds have important biological properties like antioxidant, antimutagenic and anti-inflammatory effects and these pigments are bioavailable (Chena and Roca 2019). Chena and Roca (2019) have reported that seaweed pigments are easily absorbed by intestine and defiant to in vitro digestion. Different seaweeds have different chlorophyll content. The Rhodophyta and Phaeophyta have magnesium detached derivatives, phylum

Chlorophyta have magnesium containing chlorophyll derivatives (Amorim et al. 2020). The oxidized chlorophyll derivatives form part of photosynthetic tissue in edible seaweeds. The use of these algae for their direct food applications as a vegetable represents the world market for algae. In Asian countries, they are consumed as direct food, which includes *Porphyra* sp. *Monostroma* sp., *Ulva* sp., *Laminaria japonica*, *Undaria pinnatifida*, *Hizikia fusiforme*, *Cladosiphon okamuranus*, *Caulerpa lentillifera*, *Palmaria palmate*, *Chondrus crispus*, *Alaria esculenta*, *Gracilaria* sp. and *Callophyllis variegata* (Gomez-Zavaglia et al. 2019).

7 Pigments as Co-Products from the Polysaccharide Industry

Seaweeds are abundant in polysaccharides with 5–75% of dry weight. Phycocolloids, the macroalgal hydrocolloids are most important in the industrial commercialization representing half of the non-food macroalgal market products (Milledge et al. 2016). The phycocolloids are high molecular weight structural polysaccharides found in the cell wall of marine algae that form colloidal solutions, enabling to be used as thickeners, gelling agents and as stabilizers for various industrial sectors (Biotechnological, food, paint, textile, biomedical etc.). The hydrocolloid obtained from red algae includes sulfated galactans, agar and carrageenans while alginates are obtained from brown algae (Rhein-Knudsen et al. 2015). Of late, hydrocolloids are securing more market value in the food sector because of their functional food ingredients (Cardoso et al. 2014). The alginate and carrageenan have also been exploited for microencapsulation of probiotic cells thereby protecting the cells from acidic setting of the stomach and release the cell content into the intestine. Among all seaweed products, hydrocolloids accomplished the greatest commercial importance and manipulate the western markets.

Among the various polysaccharides, sulfated polysaccharides like fucoidan, fucan sulfate, ulvan and carrageenan have been more probed for their biological values (Tanna and Mishra 2019). The sulfated polysaccharides have a broad range of applications in nutraceuticals, pharma and cosmetics industries and they are in edible seaweeds like ulvan (chlorophyta), fucoidan (Phaeophyta) and carrageenan (Rhodophyta). These compounds demonstrate various biological activities like antioxidant, anticancer, anti-inflammatory, antidiabetic, anticoagulant, immunomodulatory activities etc. and this characteristic is credited to the association between polysaccharide and probiotic microbes to provide functional and medicinal property of sulfated polysaccharides (Ganesan et al. 2019a).

8 Applications of Seaweed Pigments

8.1 Biomedical Application of Seaweed Pigments

The use of seaweed pigments for therapeutic applications has a long history. Macroalgal polysaccharides are found to possess antiviral properties with sulphated polysaccharides towards human infectious viruses (Ahmadi et al. 2015). The dichloromethane from brown algae, *Sargassum paradoxum* exhibited antimicrobial activity and this is credited to meroditerpenoids (Brkljača and Urban 2014). Holdt and Kraan (2011) have reported that several compounds from macroalgae have antithrombic, anticoagulant activities. Chen et al. (2014) have reported that fucose containing compounds from seaweeds exhibited anticancer activity. The crude extract or partially purified polysaccharide mixtures from *S. muticum* showed anticancer activity (Namvar et al. 2013). *S. muticum* polysaccharides have been reported to produce gold, silver, zinc nanoparticles and have applications in pharmaceuticals, cosmetics and electronics industry (Azizi et al. 2014).

Aplysin, a bromosesquiterpene molecule, extracted from the red algae, *Laurencia tristicha,* confirmed its protective characteristics from alcoholic liver damage in rats (Ge et al. 2018). Caulerpine extracted from certain species of *Caulerpa* may play a role as growth hormone (Schwede et al. 1986) that has spasmolytic effect (Cavalcante-Silva et al. 2013).

Chlorophyll pigments extracted from frozen brown seaweed, *Undaria pinnatifida*, employed in dye sensitized solar cells proved their high fill factor and solar efficiency (Calogero et al. 2014). Promotion of seaweed mediated nanoparticles like silver nanoparticles have testified their prospective biological characteristics including antimicrobial, anticancer, and antifouling activities (Vijayan et al. 2016).

8.2 Nutritional Application of Seaweed Pigments

A range of seaweed species such as *Caulerpa* sp., *Durvillaea antarctica*, *Fucus vesiculosus*, *Gracilaria gracilis*, *Grateloupia turuturu*, *Mastocarpus stellatus*, *Pyropia yezoensis*, *Sargassum cristaefolium*, *Ulva lactuca*, etc. are globally utilized for food applications (Kolanjinathan et al. 2014). These seaweeds are a rich source of carotenoids, iodine, iron, magnesium, and zinc. Carotenoids are ubiquitous in biological systems, offering several ecological, biomedical, and food applications (Hari et al. 1994). Red seaweeds present many health benefits as they are rich source of food compounds like phycobiliproteins, carotenoids, carrageenin, agar, porphyran, vitamins, and minerals (Cotas et al. 2020). Nori is one of the famed edible red seaweed product obtained from *Porphyra* sp. and *Pyropia* sp. and used as a flavoring ingredient in cuisines like Sushi, in Southeast Asian countries such as Japan and Korea. Kelp is familiar seaweed highly utilized for multiple applications in food and drugs. Therefore, open water culture of kelp and other seaweed species are gaining added interest in the current research (Kim et al. 2019) (Table 8.1).

Blue evolution (www.blueevolution.com) industry from USA is developing a variety of seaweed food products as organic food products. Seaweed infused food products includes popcorn, pasta, bread, soups, oat bars, roasted chickpeas, Marsala Elk Bolognese, Kelp smoothie, kelp-based umami seasoning and so on.

9 Market Demand on Algal Pigments

Algal derived pigments are renewable resources that offer food and drug applications. Of late research of algal pigments is growing globally, and China is considered as a leading country in macroalgal research (Silva et al. 2020). A latest study on phycocyanin market by Meticulous Research group confirms that it is expected to reach \$245.5 million by 2027, with CAGR of 28.5% for the period of 2020 to 2027. β -carotene and astaxanthin are the highly demanded pigments in the global market with an expected market value of \$309 and \$225 million dollars by 2018, respectively (Bhosale and Bernstein 2005). Lutein, a xanthophyll pigment, is expected to gain a \$308 million market value by 2018 (Lin et al. 2015). The global market value of carotenoids is predicted to reach up to 2.0 billion by 2026 (Markets 2020). According to the global phycobiliproteins market research report, market demand for phycobiliproteins is expected to rise by 2026. Currently, phycobiliproteins (10 mg) price in Merck is ranged from \$200 to \$270.

10 Management of Macro-Algal Blooms for Pigments

Of late, unexpected events of several macroalgal species blossoming (Fletcher 1996) are occurring in the coastal waters owing to the factors like climate change, damaging aquaculture, tourism operations etc. in tropical and subtropical regions (Ye et al. 2011; Le Luherne et al. 2017). Directly or indirectly these macroalgal blooms damage the marine food web (Nelson et al. 2003) and the biodiversity in the coastal and estuarine waters (Wan et al. 2017), specially seagrass and coral reef ecosystems (Qiuying and Dongyan 2014), by causing smothering, light inhibition, space competition, and increasing hypoxia conditions via decomposition (Qiuying and Dongyan 2014). Considerably, the green seaweed, *Ulva* sp. blooms are frequently reported (Ye et al. 2011). Sometimes, these blooms are unmanageable in the coastal waters (Wan et al. 2017) leading to severe ecological and economic damages (Charlier et al. 2008; Leliaert et al. 2009; Ye et al. 2011). Similar systems are required to extract pigments from algal biomass washed ashore. Thus, identifying the macroalgal blooms in the coastal waters is an important concern to utilize the enormous biomass of macroalgae for food, pigments, and drug applications (Fig. 8.2).

Table 8.1 Pigments from important seaweed species for industrial applications

Seaweed	Pigment	Quantity	Extraction solvent	Industrial application	Reference
Caulerpa lentillifera	Carotenoids	1.88–2.21 ppm	–	Food and drug	Syamsuddin et al. (2019)
Chondrus crispus	Carotenoids	2.80 ± 0.42 mg kg^{-1} dw	Methanol:Hexane:Dichloromethane (50:25:25, v/v)	Food and drug	Pina et al. (2014)
Durvillaea antarctica	Carotenoids	36.91 ± 2.18 mg g^{-1} dw	Hexane:Acetone:Ethanol (2:1:1, v/v)	Food and drug	Uribe et al. (2020a)
Gracilaria crassa	Phycobiliproteins	0.28–0.50 mg g^{-1} ww	Distilled water	Food and cosmetics	Sudhakar et al. (2015)
Gracilaria gracilis	R-phycoerythrin	1.26 mg g^{-1} dw	Phosphate buffer (20 mM)	Food and cosmetics	Nguyen et al. (2020)
Gracilaria corticata	R-phycoerythrin	0.78 mg g^{-1} ww	Potassium phosphate buffer (0.1 M)	Food and cosmetics	Sudhakar et al. (2014)
Grateloupia turuturu	R-phycoerythrin	2.79 ± 0.20 μg g^{-1} dw	Distilled water and phosphate buffer (20 mM)	Food and cosmetics	Denis et al. (2009)
Kappaphycus alvarezii	β-Carotene	5.72 μg g^{-1} ww	100% methanol or acetone	Antioxidant	Brotosudarmo et al. (2018)
Kappaphycus alvarezii	R-phycoerythrin	1.91 μg mL^{-1}	0.1 M phosphate buffer	Food and cosmetics	Uju et al. (2020)
Mastocarpus stellatus	R-phycoerythrin	0.27 ± 0.01 mg g^{-1} dw	Phosphate buffer (20 mM)	Food and cosmetics	Nguyen et al. (2018)
Padina australis	Carotenoids	22.27 μg g^{-1} ww	100% methanol or acetone	Food and drug	Brotosudarmo et al. (2018)
Pyropia yezoensis	Carotenoids	3.46 ± 0.57 mg g^{-1} dw	Methanol:Dichloromethane (1:1, v/v)	Food and drug	Koizumi et al. (2018)
Pyropia orbicularis	Carotenoids	93.89 ± 6.39 μg g^{-1} dw	Hexane:Acetone:Ethanol (2:1:1, v/v)	Food and drug	Uribe et al. (2018)
Sargassum horneri	Fucoxanthin	1.3 mg g^{-1} oil	Edible oil tricaprylin	Food and drug	Teramukai et al. (2020)

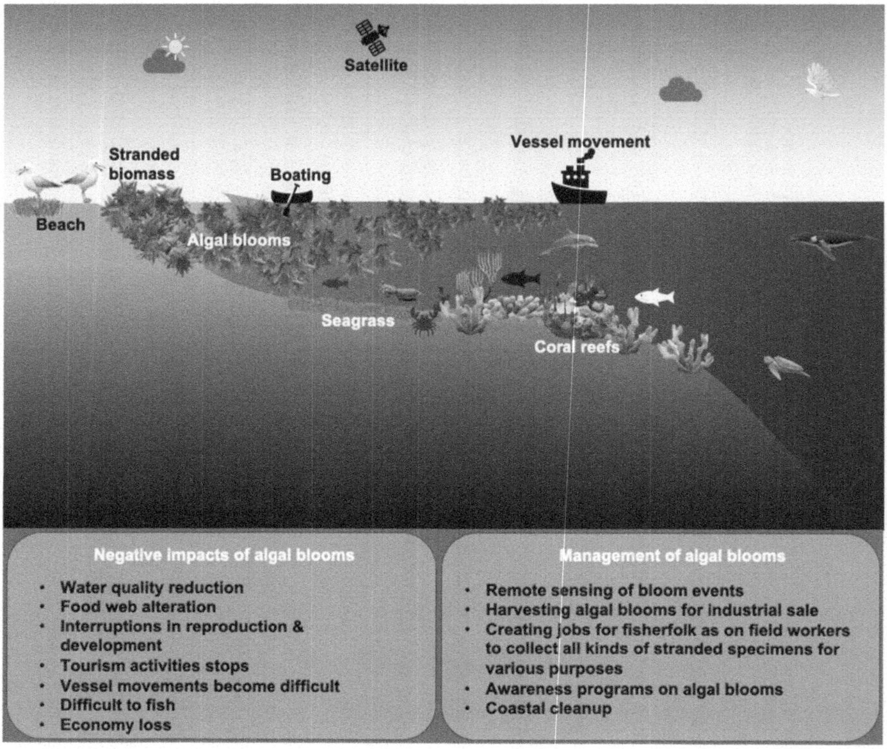

Fig. 8.2 Illustration showing the negative impacts of macroalgal blooms and their management

11 Future Perspective

Synthetic food colorants posed repeated side effects due to their short or long-term toxicity and eventually lead to cancer, hyperactivity in children and causes many diseases like asthma, organ damage as well as estrogen enhancers etc. (Zahra et al. 2017) so, issuing a strict edict to switch over to the natural foods and colorants from seaweeds may help to save many lives. On the other side, the increasing incidences of seaweed blooms in coral reefs around the world have become a major threat to newly recruiting coral polyps and existing corals, causing space competition and smothering, correspondingly. In this context, harvesting seaweed blooms from coastal waters and coral reef ecosystems would save coral reefs as well as benefit the food and drug industry. In the current global warming scenario, coastal marine biodiversity loss is evident from different geographical locations of the world. Thus, search for alternative food resources like sustainable seaweeds is underway in the current research. Culturing food grade seaweed species is one of the objectives of the blue-revolution to accomplish the growing global demand for natural sea foods. Many fisher folk and other laymen live along the global coastal waters are unaware about the market potential and the importance of seaweeds as a source of livelihood.

So, creating awareness programs and teaching seaweed culture practices may benefit the local people so as to culture seaweed for food and their livelihood.

Acknowledgements Dr. C. K. Venil thanks the UGC for awarding the Dr. D. S. Kothari Postdoctoral Fellowship (BL/17-18/0479). Also, the authors thank Anna University, Regional Campus—Coimbatore for providing necessary facilities to carry out the project work. Ramesh thanks the CSIR-NIO for the institutional support. This is NIO's contribution number: COLA-BXUGCU. Professor Laurent Dufossé deeply thanks the Conseil Régional de La Réunion, Réunion island, Indian Ocean, for continuous financial support of research activities dedicated to natural and biosourced pigments. Laurent Dufossé shows gratitude to Dr. Mireille Fouillaud and Dr. Yanis Caro for many years close relationship in microbial pigments research.

References

Ahmadi A, Zorofchian Moghadamtousi S, Abubakar S, Zandi K (2015) Antiviral potential of algae polysaccharides isolated from marine sources: a review. Biomed Res Int 2015:825203

Amorim AM, Nardelli AE, Chow F (2020) Effects of drying processes on antioxidant properties and chemical constituents of four tropical macroalgae suitable as functional bioproducts. J Appl Phycol 32:1495–1509

Aryee AN, Agyei D, Akanbi TO (2018) Recovery and utilization of seaweed pigments in food processing. Curr Opin Food Sci 19:113–119

Azizi S, Ahmad MB, Namvar F, Mohamad R (2014) Green biosynthesis and characterization of zinc oxide nanoparticles using brown marine macroalga *Sargassum muticum* aqueous extract. Mater Lett 116:275–277

Baby VA, Appavoo MR, Huxley VAJ (2012) Antibacterial effects of Ulva faciata extracts on nosocomial pathogen Pseudomonas aeruginosa. J Basic Appl Biol 5:204–208

Baghel RS, Trivedi N, Reddy CRK (2016) A simple process for recovery of a stream of products from marine macroalgal biomass. Bioresour Technol 203:160–165

Bhagyaraj I, Kunchithapatham VR (2016) Diversity and distribution of seaweeds in the shores and water lagoons of Chennai and Rameshwaram coastal areas, South-Eastern Coast of India. Biodivers J 7(4):923–934

Bhosale P, Bernstein PS (2005) Microbial xanthophylls. Appl Microbiol Biotechnol 68:445–455

Boominathan M, Mahesh A (2015) Seaweed carotenoids for cancer therapeutics. In: Kim SK (ed) Handbook of anticancer drugs from marine origin. Springer, Cham

Brkljača R, Urban S (2014) Chemical profiling (HPLC-NMR and HPLC-MS), isolation, and identification of bioactive meroditerpenoids from the Southern Australian marine brown alga *Sargassum paradoxum*. Mar Drugs 13:102–127

Brotosudarmo THP, Heriyanto S, Indriatmoko Y, Adhiwibawa MAS, Indrawati R et al (2018) Composition of the main dominant pigments from potential two edible seaweeds. Philipp J Sci 147:47–55

Calogero G, Citro I, Marco GD, Minicante SA, Morabito M, Genovese G (2014) Brown seaweed pigment as a dye source for photoelectrochemical solar cells. Spectrochim Acta Part A Mol Biomol Spectrosc 117:702–706

Cardoso SM, Carvalho LG, Silva PJ, Rodrigues MS, Pereira OR, Pereira L (2014) Bioproducts from seaweeds: a review with special focus on the Iberian Peninsula. Curr Org Chem 18(7):896–917

Cavalcante-Silva LHA, De Carvalho Correia AC, Barbosa-Filho JM, Da Silva BA, De Oliveira Santos BV, De Lira DP et al (2013) Spasmolytic effect of caulerpine involves blockade of Ca2+ influx on Guinea pig ileum. Mar Drugs 11:1553–1564

Ceballos G, Ehrlich PR, Barnosky AD, García A, Pringle RM, Palmer TM (2015) Accelerated modern human-induced species losses: entering the sixth mass extinction. Sci Adv 1:1400253

Charlier RH, Morand P, Finkl CW (2008) How Brittany and Florida coasts cope with green tides. Int J Environ Stud 65:191–208

Chen K, Roca M (2018) In vitro digestion of chlorophyll pigments from edible seaweeds. J Funct Foods 40:400–407

Chen SH, Zhao Y, Zhang Y, Zhang DH (2014) Fucoidan induces cancer cell apoptosis by modulating the endoplasmic reticulum stress cascades. PLoS One 9:e108157

Chena K, Roca M (2019) Cooking effects on bioaccessibility of chlorophyll pigments of the main edible seaweeds. Food Chem 295:101–109

Cotas J, Leandro A, Pacheco D, Gonçalves A, Pereira L (2020) A comprehensive review of the nutraceutical and therapeutic applications of red seaweeds (Rhodophyta). Life 10(3):19

Denis C, Ledorze C, Jaouen P, Fleurence J (2009) Comparison of different procedures for the extraction and partial purification of R-phycoerythrin from the red macroalga *Grateloupia turuturu*. Bot Mar 52:278–281

Dumay J, Morançais M, Nguyen HPT, Fleurence J (2015) Extraction and purification of r-phycoerythrin from marine red algae. In: Stengel DB, Connan S (eds) Natural products from marine algae: methods and protocols, methods in molecular biology, vol 1308. Springer Science+Business Media, New York, pp 109–117

Eggert A (2012) Seaweed responses to temperature. In: Wiencke C, Bischof K (eds) Seaweed biology. Springer, Berlin, pp 47–66

Ferraces-Casais P, Lage-Yusty MA, Rodríguez-Bernaldo de Quirós A et al (2012) Evaluation of bioactive compounds in fresh edible seaweeds. Food Anal Methods 5:828–834

Fletcher RL (1996) The occurrence of "green tides"—a review. Ecol Stud 123:7–43

Ganesan AR, Tiwari U, Rajauria G (2019a) Seaweed nutraceuticals and their therapeutic role in disease prevention. Food Sci Human Wellness 8(3):252–263

Ganesan M, Trivedi N, Gupta V, Madhav SV, Reddy CR, Levine IA (2019b) Seaweed resources in India—current status of diversity and cultivation: prospects and challenges. Bot Mar 62(5):463–482

Ge N, Liang H, Zhao Y, Liu Y, Gong A, Zhang W (2018) Aplysin protects against alcohol-induced liver injury via alleviating oxidative damage and modulating endogenous apoptosis-related genes expression in rats. J Food Sci 83:2612–2621

Gomez-Zavaglia A, Prieto Lage MA, Jimenez-Lopez C, Mejuto JC, Simal-Gandara J (2019) The potential of seaweeds as a source of functional ingredients of prebiotic and antioxidant value. Antioxidants 8(9):406. https://doi.org/10.3390/antiox8090406

Hari RK, Patel TR, Martin AM (1994) An overview of pigment production in biological systems: functions, biosynthesis, and applications in food industry. Food Rev Int 10:49–70

Harley CDG, Anderson KM, Demes KW, Jorve JP, Kordas RL, Coyle TA, Graham MH (2012) Effects of climate change on global seaweed communities. J Phycol 48(5):1064–1078

Hepburn CD, Pritchard DW, Cornwall CE, McLeod RJ, Beardall J, Raven JA, Hurd CL (2011) Diversity of carbon use strategies in a kelp forest community: implications for a high CO_2 ocean. Glob Chang Biol 17:2488–2497

Hessami MJ, Cheng SF, Ranga Rao A, Yin YH, Phang SM (2019) Bioethanol production from agarophyte red seaweed, *Gelidium elegans* using a novel sample preparation method for analysing bioethanol content by gas chromatography. 3 Biotech 9(1):25

Holdt S, Kraan S (2011) Bioactive compounds in seaweed: functional food applications and legislation. J Appl Phycol 23:543–597

Hosikian A, Lim S, Halim R, Danquah MK (2010) Chlorophyll extraction from microalgae: a review on the process engineering aspects. Int J Chem Eng 2010:391632

Indrawati R, Sukowijoyo H, Indriatmoko, Wijayanti RDE, Limantara L (2015) Encapsulation of brown seaweed pigment by freeze drying: characterization and its stability during storage. Proc Chem 14:353–360

Jeeva S, Kiruba S (2009) Bioremediating and biomediating potential of seaweeds. In: Abstracts of the National Seminar on marine resources: sustainable utilization and conservation, organized by Department of Plant Biology and Biotechnology. St. Mary's College, Thoothukudi, p 38

Jeeva S, Johnson M, Domettila C, Babu A, Mahesh M (2012) Preliminary phytochemical studies on some selected seaweeds from gulf of Mannar, India. Asian Pac J Trop Biomed 2(S1):S30–S33

Kim JK, Stekoll M, Yarish C (2019) Opportunities, challenges and future directions of open-water seaweed aquaculture in the United States. Phycologia 58:446–461

Klinger T (2015) The role of seaweeds in the modern ocean. Perspect Phycol 2:31–40

Koizumi J, Takatani N, Kobayashi N, Mikami K, Miyashita K, Yamano Y et al (2018) Carotenoid profiling of a red seaweed Pyropia yezoensis: insights into biosynthetic pathways in the order Bangiales. Mar Drugs 16:426

Kolanjinathan K, Ganesh P, Saranaj P (2014) Pharmacological importance of seaweeds: a review. World J Fish Mar Sci 6(1):1–15

Kumar SR, Hosokawa M, Miyashita K (2013) Fucoxanthin: a marine carotenoid exerting anti-cancer effects by affecting multiple mechanisms. Mar Drugs 11:5130–5147

Le Luherne E, Le Pape O, Murillo L, Randon M, Lebot C, Re Âveillac E (2017) Influence of green tides in coastal nursery grounds on the habitat selection and individual performance of juvenile fish. PLoS One 12:e0170110

Leliaert F, Zhang X, Ye N, Malta E, Engelen AH, Mineur F et al (2009) Identity of the Qingdao algal bloom. Phycol Res 57:147–151

Lin JH, Lee DJ, Chang JS (2015) Lutein production from biomass: marigold flowers versus micro-algae. Bioresour Technol 184:421–428

Marinho-Soriano E (2012) Effect of depth on growth and pigment contents of the macroalgae *Gracilaria bursapastoris*. Braz J Pharmacogn 22:730–735

Markets M (2020) Carotenoids market by type (Astaxanthin, beta-carotene, lutein, lycopene, canthaxanthin, and zeaxanthin), application (feed, food and beverages, dietary supplements, cosmetics, and pharmaceuticals), Source, Formulation, and Region—Global Forecast to 2026

Mezzomo N, Sandra R, Ferreira S (2016) Carotenoids functionality, sources, and processing by supercritical technology: a review. J Chem 2016:3164312

Michaud DS, Feskanich D, Rimm EB, Colditz GA, Speizer FE, Willett WC, Giovannucci E (2000) Intake of specific carotenoids and risk of lung cancer in 2 prospective US cohorts. Am J Clin Nutr 72(4):990–997

Milledge JJ, Nielsen BV, Bailey D (2016) High-value products from macroalgae: the potential uses of the invasive brown seaweed, *Sargassum muticum*. Rev Environ Sci Biotechnol 15:67–88

Mohammad JH, Ranga Rao A, Ravishankar GA (2019) Opportunities and challenges in seaweeds as feed stock for biofuel production. In: Ravishnkar GA, Rao AR (eds) Handbook of algal technologies and phytochemicals. Phycoremediation, biofuels and global biomass production, vol 2, CRC, Boca Raton, FL, pp 39–50

Namvar F, Mohamad R, Baharara J, Zafar-Balanejad S, Fargahi F, Rahman HS (2013) Antioxidant, antiproliferative, and antiangiogenesis effects of polyphenol-rich seaweed (*Sargassum muticum*). Biomed Res Int 2013:9

Nelson TA, Lee DJ, Smith BC (2003) Are "green tides" harmful algal blooms? Toxic properties of water-soluble extracts from two bloom-forming macroalgae, *Ulva fenestrata* and *Ulvaria obscura* (Ulvophyceae). J Phycol 39:874–879

Nguyen HPT, Morancais M, Fleurence J, Tran TNL, Dumay J (2018) Extracting and purifying pigment R-phycoerythrin from the red alga Mastocarpus Stellatus. In: Proceedings 2018 4th International Conference on Green Technology and Sustainable Development, GTSD 2018, pp 573–577

Nguyen HPT, Morançais M, Déléris P, Fleurence J, Nguyen-Le CT, Vo KH et al (2020) Purification of R-phycoerythrin from a marine macroalga *Gracilaria gracilis* by anion-exchange chromatography. J Appl Phycol 32:553–561

Pangestuti R, Siahaan EA, Kim SK (2018) Photoprotective substances derived from marine algae. Mar Drugs 16:399

Pina AL, Costa AR, Lage-Yusty MA, López-Hernández J (2014) An evaluation of edible red seaweed (*Chondrus crispus*) components and their modification during the cooking process. LWT- Food Sci Technol 56:175–180

Qin Y (2018) Applications of bioactive seaweed substances in functional food products. In: Qin Y (ed) Bioactive seaweeds for food applications: natural ingredients for healthy diet. Elsevier, London, pp 111–134

Qiuying H, Dongyan L (2014) Macroalgae blooms and their effects on seagrass ecosystems. J Ocean Univ China 13:791–798

Rao PVS, Mantri VA (2006) Indian seaweed resources and sustainable utilization: scenario at the dawn of a new century. Curr Sci 91:164–174

Rebours C, Marinho-Soriano E, Zertuche-González JA, Hayashi L, Vásquez JA, Kradolfer P et al (2014) Seaweeds: an opportunity for wealth and sustainable livelihood for coastal communities. J Appl Phycol 26:1939–1951

Rengasamy KRR, Mahomoodally MF, Aumeeruddy MZ, Zenginc G, Xiao J, Kim DH (2020) Bioactive compounds in seaweeds: an overview of their biological properties and safety. Food Chem Toxicol 135:111013

Rhein-Knudsen N, Ale MT, Meyer AS (2015) Seaweed hydrocolloid production: an update on enzyme assisted extraction and modification technologies. Mar Drugs 13(6):3340–3359

Ronnback P, Kautsky N, Pihl L, Troell M, So¨derqvist, T. and Wennhage, H. (2007) Ecosystem goods and services from Swedish coastal habitats: identification, valuation, and implications of ecosystem shifts. Ambio 36:534–544

Sachindra NM, Sato E, Maeda H, Hosokawa M, Niwano Y, Kohno M, Miyashita K (2007) Radical scavenging and singlet oxygen quenching activity of marine carotenoid fucoxanthin and its metabolites. J Agric Food Chem 55:8516–8522

Sahoo D, Sahu N, Sahoo D (2003) A critical survey of seaweed diversity of Chilika Lake, India. Algae 8:1–13

Salehi B, Sharifi-Rad J, Seca AML, Pinto DCGA, Michalak I, Trincone A, Mishra AP, Nigam M, Zam W, Martins N (2019) Current trends on seaweeds: looking at chemical composition, phytopharmacology, and cosmetic applications. Molecules 24(22):4182

Sathasivam R, Ki JS (2018) A review of the biological activities of microalgal carotenoids and their potential use in healthcare and cosmetic industries. Mar Drugs 16(1):26

Schmid M, Guihéneuf F, Stengel DB (2017) Ecological and commercial implications of temporal and spatial variability in the composition of pigments and fatty acids in five Irish macroalgae. Mar Biol 164:158

Schwede JG, Cardellina JH, Grode SH, James TR Jr, Blackman AJ (1986) Distribution of the pigment caulerpin in species of the green alga *Caulerpa*. Phytochemistry 26:155–158

Seth A, Shanmugam M (2016) Seaweeds as agricultural crops in India: new vistas. In: Dagar J, Sharma P, Sharma D, Singh A (eds) Innovative saline agriculture. Springer, New Delhi

Shama A, Joyce SG, Mari FD, Ranga Rao A, Ravishankar GA, Hudaa N (2019) Macroalgae and microalgae: novel sources of functional food and feed. In: Ravishankar GA, Rao AR (eds) Handbook of algal technologies and phytochemicals. Food, health and nutraceutical applications, vol 1. CRC, Boca Raton, FL, pp 207–219

Sharma A, Koneri R, Jha DK (2019) A review of pharmacological activity of marine algae in Indian coast. Int J Pharm Sci Res 10(8):3540–3549

Silva A, Silva SA, Carpena M, Garcia-Oliveira P, Gullón P, Barroso MF, Prieto M, Simal-Gandara J (2020) Macroalgae as a source of valuable antimicrobial compounds: extraction and applications. Antibiotics 9:642

Stadnichuk IN, Tropin IV (2017) Phycobiliproteins: structure, functions and biotechnological applications. Appl Biochem Microbiol 53:1–10

Sudhakar MP, Ananthalakshmi JS, Nair BB (2013) Extraction, purification and study on antioxidant properties of fucoxanthin from brown seaweeds. J Chem Pharm Res 5:169–175

Sudhakar MP, Saraswathi M, Nair BB (2014) Extraction, purification and application study of R-Phycoerythrin from *Gracilaria corticata* (J. Agardh) J. Agardh var. *corticata*. Indian J Nat Prod Resour 5:371–374

Sudhakar MP, Jagatheesan A, Perumal K, Arunkumar K (2015) Methods of phycobiliprotein extraction from *Gracilaria crassa* and its applications in food colorants. Algal Res 8:115–120

Syamsuddin R, Azis HY, Badraeni, Rustam (2019) Comparative study on the growth, carotenoid, fibre and mineral content of the seaweed *Caulerpa lentillifera* cultivated indoors and in the sea. IOP Conf Ser Earth Environ Sci 370:012019

Takaichi S (2011) Carotenoids in algae: distributions, biosyntheses and functions. Mar Drugs 9(6):1101–1118

Tanna B, Mishra A (2019) Nutraceutical potential of seaweed polysaccharides: structure, bioactivity, safety and toxicity. Comp Rev Food Sci Food Safety 18(3):817–831

Teramukai K, Kakui S, Beppu F, Hosokawa M, Miyashita K (2020) Effective extraction of carotenoids from brown seaweeds and vegetable T leaves with edible oils. Innov Food Sci Emerg Technol 60:102302

Uju, Dewi NPSUK, Santoso J, Setyaningsih I, Hardingtyas SD, Yopi (2020) Extraction of phycoerythrin from *Kappaphycus alvarezii* seaweed using ultrasonication. IOP Conf Ser Earth Environ Sci 414:012028

Uribe E, Vega-Gálvez A, Heredia V, Pastén A, Scala KD (2018) An edible red seaweed (*Pyropia orbicularis*): influence of vacuum drying on physicochemical composition, bioactive compounds, antioxidant capacity, and pigments. J Appl Phycol 30:673–683

Uribe E, Pardo-Orellana CM, Vega-Galvez A, Ah-Hen KS, Pasten A, Garcia V et al (2020a) Effect of drying methods on bioactive compounds, nutritional, antioxidant, and antidiabetic potential of brown alga *Durvillaea Antarctica*. Drying Technol 38:1915–1928

Uribe E, Vega-Gálvez A, García V, Pastén A, Rodríguez K, López J et al (2020b) Evaluation of physicochemical composition and bioactivity of a red seaweed (*Pyropia orbicularis*) as affected by different drying technologies. Drying Technol 38:1218–1230

Vieira FA, Guilherme RJR, Neves MC, Abreu H, Rodrigues ERO, Maraschin M et al (2017) Single-step extraction of carotenoids from brown macroalgae using non-ionic surfactants. Sep Purif Technol 172:268–276

Vijayan SR, Santhiyagu P, Ramasamy R, Arivalagan P, Kumar G, Ethiraj K et al (2016) Seaweeds: a resource for marine bionanotechnology. Enzyme Microb Technol 95:45–57

Wan AHL, Wilkes RJ, Heesch S, Bermejo R, Johnson MP, Morrison L (2017) Assessment and characterisation of Ireland's green tides (*Ulva* species). PLoS One 12:e0169049

Wang HD, Chen CC, Huynh P, Chang JS (2015) Exploring the potential of using algae in cosmetics. Bioresour Technol 184:355–362

Wiencke C, Bischof K (eds) (2012) Seaweed biology. Springer, Berlin

Ye N, Zhang X, Liang YMC, Xu D, Zhuang JZZ, Wang Q (2011) Green tides are overwhelming the coastline of our blue planet: taking the world's largest example. Ecol Res 26:477–485

Zahra N, Kalim I, Saeed K, Mumtaz Z, Amjad N, Nisa AU, Hina S, Masood S, Ahmed I, Ashraf M (2017) Effect of natural and synthetic dyes on human health. Int Res J Biol Sci 6(10):23–29

Zhu Z, Wu Q, Di X, Li S, Barba FJ, Koubaa M et al (2017) Multistage recovery process of seaweed pigments: investigation of ultrasound assisted extraction and ultra-filtration performances. Food Bioprod Process 104:40–47

Chapter 9
The New Products from Brown Seaweeds: Fucoxanthin and Phlorotannins

Xiaojun Yan, Jinrong Zhang, Shan He, Wei Cui, and Fengzheng Gao

Abbreviations

AD	Alzheimer disease
APP	Aβ precursor protein
BBB	Blood-brain barrier
ESI-Q-TOF MS	Electrospray ionization-quadrupole-time of flight mass spectrometry
HPLC	High-performance liquid chromatography
NMR	Nuclear magnetic resonance
OFAT	One-factor-at-a-time
UV	Ultra-violet

1 Introduction

In this article, we have given an introduction on the newly-developed process to stepwise isolate fucoxanthin and phlorotannins from crude alcohol extracts of brown seaweeds at commercial scale. Moreover, we have summarized the recent findings of fucoxanthin that could be developed as a promising dietary supplement in preventing Alzheimer disease, as well as the phlorotannins being bio-stimulants.

X. Yan (✉) · J. Zhang · S. He
Key Laboratory of Marine Biotechnology, Ningbo University, Ningbo, China
e-mail: yanxj@zjou.edu.cn; yanxiaojun@nbu.edu.cn

W. Cui
School of Medicine, Ningbo University, Ningbo, China

F. Gao
Bioprocess Engineering, AlgaePARC, Wageningen University, Wageningen, Netherlands

A. Ranga Rao, G. A. Ravishankar (eds.), *Sustainable Global Resources of Seaweeds Volume 2*, https://doi.org/10.1007/978-3-030-92174-3_9

1.1 Major Traditional Products in Brown Seaweed Industry: Alginate, Mannitol and Iodine

The brown seaweed industry has been established to produce the major products of brown seaweed, such as alginate, mannitol and iodine (Wang et al. 2018). Alginate has been the major product with wide applications in fabric, nutraceutical, cosmetic and pharmacological industries (Rehm and Moradali 2018). Mannitol has various beneficial applications, such as excipient, diuretic and sugar substitutes (Dexamethasone 2004). Another important commercial product from seaweeds is iodine, which has been used particularly in the pharmaceutical industry due to its beneficial activity for the prevention and control of iodine deficiency disorders (Zimmermann 2011).

However, the brown seaweed industry is facing certain challenges such as single industry structure, short industrial chain. Therefore, novel high-value products and technologies are greatly needed to enrich the product pipeline.

1.2 The Innovative Minor Bioactive Compounds in Brown Seaweed Industry: Fucoxanthin and Phlorotannins

Fucoxanthin is one of the most abundant and representative pigments presenting as a component of photosynthetic light-harvesting complex in brown seaweeds (Maoka et al. 2016; D'Orazio et al. 2012a). It has a unique structure featuring an allenic bond, a conjugated carbonyl, a 5,6-monoepoxide, and acetyl groups (Zarekarizi et al. 2019). Due to its structure characteristics, fucoxanthin is extremely prone to isomerization or degradation when heated or exposed to air and light (Zhao et al. 2014).

The functional activities of fucoxanthin includes antioxidant, anti-inflammatory, anti-obesity, anti-cancer, anti-diabetic, anti-malarial, anti-senile dementia and anti-angiogenic properties (Neumann et al. 2019; Guo et al. 2019a; Ferreira et al. 2018; Su et al. 2019; Lakey-Beitia et al. 2019). In addition, fucoxanthin has been approved to be safe as a food ingredient (Iio et al. 2011; Beppu et al. 2009a). These properties have attracted increasing attention of food, cosmetic and pharmaceutical industries. Fucoxanthin has already been used as a functional ingredient in various products such as pasta, biscuits, and dips by a number of food companies worldwide (Shannon and Abu-Ghannam 2017; Prabhasankar et al. 2009). Fucoxanthin supplements are generally recognized as safe by the European Food Safety Authority, Japanese Food for Specified Health Uses, and the US Food and Drug Administration (Shannon and Abu-Ghannam 2017). The broad health applications of fucoxanthin have emerged in recent years. World fucoxanthin production reached approximately 500 t in 2015 (Shannon and Abu-Ghannam 2017), with expected market size of $120 million by 2022 (Lourenco-Lopes et al. 2020). However, the fucoxanthin content in most of current products is very low.The application of fucoxanthin in food and pharmaceutical industries has remained underutilized possibly due to its instability and the high production costs of using inefficient preparation methods.

The production and commercialization of fucoxanthin are facing certain challenges due to its instability when exposed to oxygen or light (D'Orazio et al. 2012a). At present, fucoxanthin is commercially extracted from brown seaweeds at an industrial scale since its chemical synthesis is complex and inefficient (Zarekarizi et al. 2019). Diverse methods have been adopted in extraction and purification of fucoxanthin, including solvent extraction (Kim 2014), supercritical carbon dioxide extraction (Quitain et al. 2013), microwave-assisted extraction (Xiao et al. 2012), pressurized liquid extraction (Shang et al. 2011), enzyme-assisted extraction (Billakanti et al. 2013) and liquefied dimethyl ether extraction (Kanda et al. 2014), have been used to extract fucoxanthin from various macro- and microalgae. Moreover, chromatographic methods are often used to obtain fucoxanthin with high purity, including high-speed countercurrent chromatography (Xiao et al. 2012), silica gel column chromatography (Sudhakar et al. 2013), thin-layer chromatography (Piovan et al. 2014) and centrifugal partition chromatography (Kim et al. 2011). These chromatographic methods can provide fucoxanthin with satisfactory purity but have the disadvantages of low efficiency, low yield, and large solvent consumption. Therefore, the development of effective and environmentally friendly extraction and purification techniques is of great interest to support the commercialization of fucoxanthin, and to increase the commercial value of the final product (Lourenco-Lopes et al. 2020).

Phlorotannins, a complex mixture of polymers consisting of a different number of phloroglucinol (1,3,5-trihydroxybenzene), are the most abundant group of metabolites specific in brown algae. According to the different variants of chemical bonds between the monomers, phlorotannins are generally divided into several classes including fuhalols, phlorethols, fucols, fucophloroethols, and eckols. The physiological functions of phlorotannins include wound healing, chelation of heavy metal ions, bioadhesion, chemical defense against herbivores, *etc* (Cruces et al. 2017). Phlorotannins have a variety of bioactive properties including anti-cancer, anti-inflammation, anti-oxidant, anti-allergic, anti-wrinkling, UV radiation protection and hair growth promotion properties. Some available book chapters referencing phlorotannins provide an overview of occurrence, distribution, preparation, characterization, health beneficial activities, and applications of phlorotannins in pharmaceutical, food and cosmeceutical industries (Dominguez 2013). Moreover, the application of phlorotannins used as biostimulants, have profound influence on plant growth and protection viz. increase the number of roots, shoot elongation, seedling weight (Perez et al. 2016), photosynthetic pigments, proline, phytochemical and physiological enzymes (Rengasamy et al. 2016). Some brown algae extracts containing phlorotannins as natural fertilizers (including the seaweed concentrate Kelpak®) have been manufactured commercially since the middle of the twentieth century (Kim and Chojnacka 2015). The application of phenolic-containing algae extracts in agriculture have gained recognition among farmers and fertilizer manufacturers (Kim and Chojnacka 2015). At present, commercial extracts from different brown algae have gained wide acceptance in agriculture as plant biostimulants.

Brown algae extracts with various product types are now being used worldwide as biostimulants for a number of agricultural crops (Craigie 2011).

The structural diversity, preparation and chromatographic analysis of phlorotannins were reviewed in 2013 (Martinez and Castaneda 2013). The extraction, purification, and applications of phenolic compounds from seaweed were reviewed in 2020 (Cotas et al. 2020). At present, significant progress has been made in the extraction and isolation of phlorotannins with various methods. A number of extraction technologies, including solvent extraction (Obluchinskaya et al. 2019; Yoon et al. 2017; Leyton et al. 2016), microwave-assisted extraction (Amarante et al. 2020), ultrasound-assisted extraction (Vazquez-Rodriguez et al. 2020; Ummat et al. 2020), surfactant-mediated extraction, pressurized liquid extraction, and enzyme-assisted extraction (Yilmaz et al. 2019), have been developed to extract phlorotannins from various brown algae. Moreover, diverse methods are used to obtain phlorotannins-containing fractions, including macroporous resins chromatography (Leyton et al. 2017; Kim et al. 2014), High-speed counter-current chromatography (Zhou et al. 2019), centrifugal partition chromatography (Lee et al. 2014), liquid biphasic system (Chia et al. 2018), and liquid-liquid extraction (Li et al. 2017). At present, as high-value products, fractionated phlorotannins are used as ingredients of foods, pharmaceuticals, and cosmeceuticals, while phlorotannins-containing extracts are used for agricultural purposes (Chojnacka et al. 2018).

1.3 A New Process of Stepwise Isolation of Fucoxanthin and Phlorotannins from Crude Alcohol Extracts of Brown Seaweeds at Commercial Scale

Currently, fucoxanthin and phlorotannins are largely used as ingredients in the food processing industry due to their human beneficial properties, and as bioactive compounds in cosmetics or pharmaceuticals. Recently, we have developed a green extraction strategy to stepwise isolate minor bioactive compounds from brown seaweeds. Programs for seaweed industrial innovation have led to the successful establishment of a green process for the extraction and purification of fucoxanthin and phlorotannins at commercial scale. We have reported, for the first time, a systematic investigation of fucoxanthin extraction methods according to green chemistry principles, including safer extraction, products, auxiliaries, energy efficiency and environment-friendly process (Sanderson 2011). The limited availability of commercial-scale preparation of natural fucoxanthin from brown algae is mainly due to the poor purification efficiency. In our work, an efficient purification process has been developed, which enabled fucoxanthin with high purity to be obtained by stepwise precipitation during the reduced pressure distillation process. The flow diagram for extraction and purification of fucoxanthin and phlorotannins from brown seaweeds is shown in Fig. 9.1. The simple and green process could be

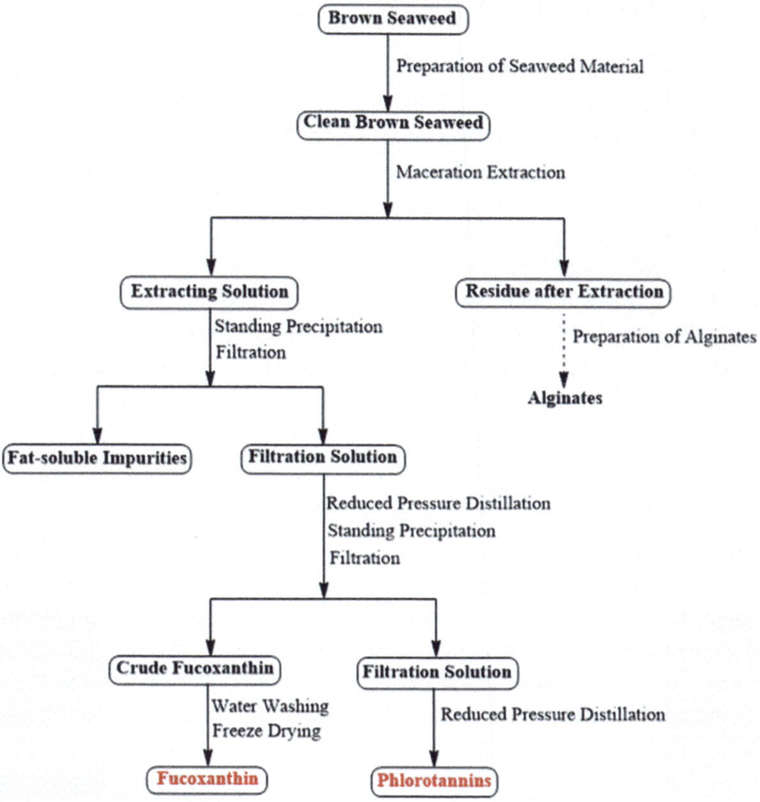

Fig. 9.1 Flow diagram for extraction of fucoxanthin and phlorotannins from brown seaweeds

demonstrated in detail when applied for the extraction and purification of fucoxanthin and phlorotannins from the brown seaweed *Sargussum horneri.*

2 Protocols for the New Processing Methods

2.1 The Extraction and Separation Procedures of Fucoxanthin of Phlorotannins from the Brown Seaweed **Sargussum horneri**

Based on a unique molecular structure with an unusual allenic bond, fucoxanthin is susceptible to degradation induced by oxygen, heat or light. Therefore, the extraction and separation procedures should be performed under low temperature and dimmed light conditions (Zhao et al. 2014).

2.1.1 Preparation of Seaweed

The brown seaweed *Sargassum horneri* was cultured and harvested from the regions near Gouqi Island (30°42′ N, 122°46′ E) in Zhejiang Province, the People's Republic of China (Fig. 9.2). The seaweed was frozen immediately after harvest and then stored at −20 °C before analysis. Frozen seaweed *S. horneri* was defrosted overnight in darkness at room temperature (25 °C), and was rinsed successively under running tap water at room temperature to remove sand, epiphytes and saline ions, and then centrifuged with a spin speed of 2000 rpm for 20 min to remove water from the surface. Prior to analysis, the wet seaweed was cut into 2-cm long strips using a stainless-steel knife to make easier extraction without fucoxanthin loss during extraction. The Sargassum species such as Sargassum fusiforme can be also used and treated as the same.

2.1.2 Maceration Extraction

One-factor-at-a-time (OFAT) method was used to determine the optimum condition for fucoxanthin extraction from wet brown seaweed *S. horneri*. Approximately 10.0 g clean *S. horneri* (water content was 83.95%) was extracted with 50 mL solvent in a centrifugation tube in an incubator with shaking at 100 rpm for 120 min in darkness at room temperature. After extraction, the mixture was filtered through a 0.45 μm membrane filter (Millipore, Billerica, MA, USA) and the filtrate was

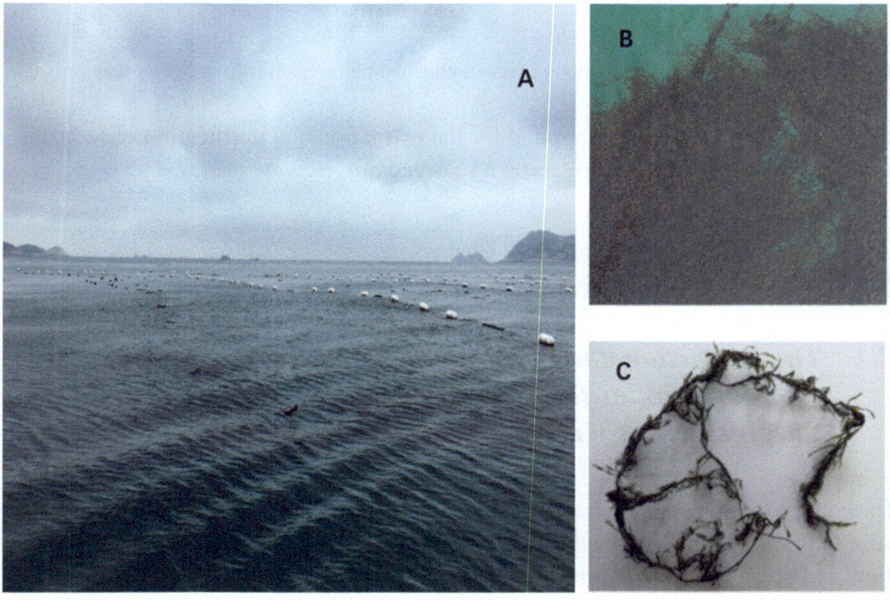

Fig. 9.2 The cultured *Sargassum horneri* and its biomass. (**a**) The mariculture of the seaweed, (**b**) the biomass in the sea, (**c**) the dried seaweed image

collected for further analysis. The influence of various conventional solvents (methanol, ethanol, acetone and ethyl acetate), ethanol-water solvent mixtures (ethanol concentration: 70%, 80%, 90% and 100%, V/V), solid-to-solvent ratio (2:1, 3:1, 4:1, 5:1 and 6:1 g/mL), duration of extraction (30, 60, 90, 120, 150 and 180 min), temperature of extraction (20, 25, 30 and 35 °C) and the number of extraction rounds on the yield of fucoxanthin were evaluated for optimization.

2.1.3 Removal of Fat-Soluble Impurities

The extraction solution was stored under dimmed light conditions at 10 °C for 48 h. After that, the fat-soluble impurities were precipitated out from the solution, and were removed by filtration through a 600-mesh screen.

2.1.4 Reduced Pressure Distillation

The filtered liquid was concentrated down to 1/4 of its original volume using a rotary evaporator (RV 10, IKA, Germany) under vacuum conditions of 0.003 MPa at certain temperatures. The extract was re-dissolved in 50 mL methanol and then filtered through a 0.45 μm membrane filter (Millipore, Billerica, MA, USA) before HPLC analysis. To assess the effects of the concentration temperature on the yield of fucoxanthin, various concentration temperatures (20, 25, 30, 35 and 40 °C) were tested.

2.1.5 Obtain Crude Fucoxanthin

The concentrated solution was stored under dimmed light conditions at 25 °C for 72 h. Then fucoxanthin was precipitated out from the solution, and was collected by filtration through a 600-mesh Screen.

2.1.6 Obtain Refined Fucoxanthin

Water was used to remove water-soluble impurities from the crude fucoxanthin. To 100 g wet crude fucoxanthin, 500 mL of water was added, stirred for 20 min and the solution was left to allow fucoxanthin to settle down, and was centrifuged with a spin speed of 8000 rpm for 20 min at 10 °C. The process was repeated three times to remove water-soluble impurities completely. Then the wet refined fucoxanthin was freeze-dried, and then sealed.

2.1.7 Obtain Phlorotannins

Reduced pressure distillation was used to remove the residual ethanol, and obtain concentrated solution of phlorotannins. In general, the filtered liquor was concentrated down to 1/30 of its original volume.

2.2 High-Performance Liquid Chromatography, Mass Spectrometry, and Nuclear Magnetic Resonance Spectrometer Analysis

Fucoxanthin quantification was carried out using HPLC and slightly modified (Mok et al. 2016). A Waters Alliance 2695 HPLC system was used for the analysis, equipped with a Waters model 2996 diode array detector and Waters Empower System (Waters Co., Milford, CT, USA). YMC C-30 carotenoid column (250 × 4.6 mm ID, 3 μm particle size, Waters, Ireland) was used for the separation. Methanol and water solvent system was used for mobile phase at a flow rate of 0.7 mL/min with a column temperature of 35 °C. The solvent gradient program was as follows: methanol/water ratio was increased from 90:10 to 100:0 over 20 min, and then, 100% methanol was held for the next 5 min. All determinations were performed in triplicate.

Standard fucoxanthin (Sigma-Aldrich, USA) in a concentration range of 1–200 μg/mL was used for calibration curve. The fucoxanthin content in the brown alga *S. horneri* was measured at 0.358 mg/g wet seaweed in average according to the HPLC quantification method (Mok et al. 2016).

The recovery during the extraction and purification procedure was calculated by the following equation:

$$\text{Recovery}(\%) = (W_1 \times P) \times 100 / (W_2 \times C)$$

where W_1 (mg) is the weight of the purified fucoxanthin, P (%) is the purity of the purified fucoxanthin, W_2 (g) is the weight of the brown seaweed used, and C (mg/g) is the fucoxanthin content in the brown alga *S. horneri*.

ESI-MS analyses were performed using an electrospray ionization-quadrupole-time of flight mass spectrometry (ESI-Q-TOF MS, Waters, Milford, CT, USA). ^{1}H-NMR spectra were recorded using an AVANCE 600 MHz NMR spectrometer with TMS as internal standard at 25 °C (Bruker, Fällanden, Switzerland).

3 Evaluation of the New Processing Method

3.1 *Optimization of Maceration Extraction Conditions*

One-factor-at-a-time (OFAT) method was used to determine the optimum condition for fucoxanthin extraction from wet brown seaweed *S. horneri*. The efficiency of the solvents for fucoxanthin extraction was as follows: methanol > ethanol> acetone > ethyl acetate. The result may be due to several reasons as follows. (1) The effectiveness of extraction mainly depended on the natural properties of the solute with solvents that can dissolve them based on the polarity (Warkoyo and Saati 2011). Polar carotenoids, such as fucoxanthin or epoxy carotenoids are solubilized in polar organism solvent, while non-polar carotenoids, such as carotenes show greater solubility in non-polar organism solvents (Rivera and Canela 2012). (2) The differences in the polarity of the extraction solvents have a major influence on fucoxanthin extraction (Reichardt 2010). The amount of the fucoxanthin obtained from solvents increased with solvent polarity. The most suitable solvents used to extract fucoxanthin from wet brown seaweed *S. horneri* were methanol and ethanol. They demonstrated a high yield of fucoxanthin due to solvent polarity. Hence, ethanol was chosen as the solvent for further optimization because it is safe, environmentally benign, naturally abundant, and cheap (Sanderson 2011).

3.1.1 Effects of Ethanol-Water Mixed Solvent on Fucoxanthin Yield

Ethanol and water are versatile and green solvents, and the ethanol-water system has been broadly used for many years, especially in the pharmaceutical industry. The yield of fucoxanthin extracted using different ethanol/water volume ratios was 100% ethanol >90% ethanol >80% ethanol >70% ethanol. Fucoxanthin yields increased with increasing volume fraction of ethanol (V/V) from 70% to 90% steeply, implying the importance of ethanol content in the extraction.

According to a previously reported study, extraction using absolute ethanol could result in the coextraction of total lipids and fucoxanthin from the microalga *Phaeodactylum tricornutum* (Kim et al. 2012a). In the present study, to reduce the extraction of total lipids and facilitate the following isolation of fucoxanthin, 80% ethanol was chosen as the solvent for further optimization.

3.1.2 Effects of Extraction Temperature on Fucoxanthin Yield

Fucoxanthin extracted from 80% ethanol at 30 °C showed the highest yield. The present findings are similar to the report, where fucoxanthin yield was maximum at 30 °C and decreased at higher temperatures (Kim et al. 2012a). Furthermore, fucoxanthin exists in algae, including brown seaweeds and microalgae, is extremely prone to isomerization or degradation when heated (Zhao et al. 2014). Hence, the optimum extraction temperature was room temperature due to the high product yield and energy efficiency.

3.1.3 Effects of Extraction Duration on Fucoxanthin Yield

The fucoxanthin yield increased insignificantly from 60 to 180 min of extraction time. In addition, approximately 80% of fucoxanthin was extracted from algal biomass within the first 10 min, and the maximum fucoxanthin concentration was obtained at approximately 60 min in a previously reported study (Kim et al. 2012b). To meet the complexity of large-scale industrial processes, 120 min has been chosen as the optimum duration for fucoxanthin extraction.

3.1.4 Effects of Solid to Solvent Ratio on Fucoxanthin Yield

Fucoxanthin yield increased as the solvent-to-solid ratio was increased from 2 to 6 mL/g. In general, the rate of extraction depends on the concentration gradient between raw material and solvent (Tan et al. 2011). Solvent-to-solid ratio of 3 mL/g was suitable for effective extraction of fucoxanthin from wet *S. horneri* due to the higher product yield, solvent saving and higher level of fucoxanthin content.

3.1.5 Effects of Extraction Rounds on Fucoxanthin Yield

The fucoxanthin yield increased with the increasing of extraction rounds. Approximately 90% of the total fucoxanthin was extracted in the first extraction round, indicating that one round of extraction was sufficient for effective fucoxanthin extraction from *S. horneri*. Hence, one round of extraction is recommended.

Hence, a simple and green method for natural fucoxanthin extraction from wet brown seaweed *S. horneri* by solvent extraction was developed with optimum extraction conditions: 80% ethanol concentration, extraction duration 120 min, extraction one time, solvent-to-solid ratio of 3 mL/g, and extraction temperature 25 °C.

3.1.6 Removal of Fat-Soluble Impurities

Extraction using ethanol results in the coextraction of lipids and fucoxanthin. Partial purification of fucoxanthin has been accomplished using a biphasic partitioning system. For example, lipids and fucoxanthin were coextracted from dry microalgae by absolute ethanol and then a hydroalcoholic phase was produced by adding water (40%; v/v). The addition of n-hexane to the hydroalcoholic phase results in a biphasic n-hexane/hydroalcoholic system, with the lipids partitioning extracted into the n-hexane and the fucoxanthin remained in the hydroalcoholic phase. The biphasic partitioning system was approved to be effective, and was used as a first step in the purification of fucoxanthin (Kim et al. 2012b).

Under the optimum extraction conditions, the ethanol proportion of the extraction solution is almost 65% (25 °C). Subsequently, the extracting solution was stored under dimmed light conditions at 10 °C for 48 h. The result demonstrated that most lipids precipitated, whereas fucoxanthin was still in the solution. As fat-soluble impurities, lipids were removed by filtration through a 600-mesh screen.

3.1.7 Selection of Concentration Temperature

After removal of fat-soluble impurities, the filtered solution was further concentrated to levels acceptable for the precipitation of fucoxanthin. Selection of suitable vacuum concentration equipment was one of the key factors for industrial-scale fucoxanthin production. Some equipments, such as cold evaporation concentrator (REDA Food Plants, Milan, Italy), can operate under vacuum conditions of 0.003 MPa at 25 °C for evaporating different solutions such as water and ethanol, and has been used in different industries including the food, cosmetics and pharmaceutical industries. So, considering the feasibility of equipment which meets the requirements in respect of large-scale preparation, the optimum concentration temperature for fucoxanthin preparation was determined to be 25 °C.

3.2 Purification of Fucoxanthin

There are many factors that can affect the precipitation, such as ethanol content, environment, equipment, standing, and stirring (Tai et al. 2020). According to our practical experiences, after the fat-soluble impurities were removed completely, the filtered liquor can be concentrated under vacuum conditions of 0.003 MPa at 25 °C. The concentration process was finished when the ethanol content reached almost 40% (25 °C). The concentrated solution was stored under dimmed light conditions at 25 °C for 72 h. The crude fucoxanthin was collected by filtration, and washed by water. The water-washing process was repeated for three times to remove water-soluble impurities completely. Then the wet refined fucoxanthin was freeze-dried, and then sealed (Fig. 9.3). Under the optimum condition in the present study, the refined fucoxanthin yield and purity were 69.5% and 95.8%, respectively.

3.3 Identification of Fucoxanthin

The *S. horneri* extract was analyzed by HPLC to detect fucoxanthin. The molecular mass of refined product was proposed as fucoxanthin based on the fragment pattern at m/z 659, and 581 corresponding to $[M + H]^+$, and $[M + Na]^+$, respectively. The

Fig. 9.3 Refine fucoxanthin extracted from *S. horneri*

refined product was subjected to 1D NMR spectroscopy, and was identification as all-trans-fucoxanthin according to the published literature (Mori et al. 2004), among the naturally occurring geometrical isomers of fucoxanthin (Haugan and Liaaen-Jensen 1994).

3.4 Obtain Phlorotannin

After fucoxanthin precipitation, the phlorotannins-containing filtered liquid was concentrated by reduced pressure distillation. In general, the filtered liquid was concentrated down to 1/30 of its original volume. Total phlorotannin content in the concentrated solution extracted from the alga *S. horneri* was 155.9 mg/L. The high concentrated solution will be used as plant biostimulant.

In this study, a new process to stepwise isolate fucoxanthin and phlorotannins from crude alcohol extracts of brown seaweeds has been successfully established at commercial scale. This mature technology has been achieved through three interrelated stages, including lab process, pilot magnification and full-scale manufacturing process. Currently, according to our green innovation technology, the first automatic production line of fucoxanthin with a high purity more than 50% in China has been put into operation in Shandong Province, with more than 500 kilogram produced per year.

The success of industrial-scale new bioactive compounds production involves many factors, such as the reasonable top-level design of the project, sufficient funding sources, support and cooperation from the government and cooperative enterprise, development of seaweed chemistry, physical-chemical properties of fucoxanthin and phlorotannins, suitable edible economic brown seaweeds, instruments and equipment in advanced pressure distillation.

All-trans-(6'R)-fucoxanthin is the naturally occurring stereoisomer of fucoxanthin in brown seaweeds, the yield and purity of refined fucoxanthin extracted from *S. horneri* were 69.5% and 95.8%, respectively. While the purity of refined fucoxanthin extracted from fresh *Laminaria Japonica* was no more than 30%. Our findings are similar to the previous report, where the extraction yield of fucoxanthin has been found to be very variable depending on the selected species (Lourenco-Lopes et al. 2020). And three high-value products, including fucoxanthin, phlorotannins and alginate, can be purified from the brown alga *S. horneri*.

Now, China Algae Industry Association (CAIA) has issued the social organization standard named Brown seaweed extract—Fucoxanthin (T/CAIAS 001–2021) in January 2021, based on our efforts, which will be of great significance to regulate the production and trade of fucoxanthin, as well as promote its application in food, cosmetics industries.

4 New Bioactivities of Fucoxanthin and Phlorotannins

4.1 Fucoxanthin as a Promising Dietary Supplement in Preventing Alzheimer Disease (AD)

Recently, many groups, including us, have reported that fucoxanthin has neuroprotective effects via acting on β-amyloid (Aβ) aggregation, oxidative stress, neuroinflammation, cholinergic dysregulation, and gut microbiota disorder, supporting the use of fucoxanthin as a food supplement for the prevention and/or the treatment of AD.

Fucoxanthin has been used as an ingredient in functional food for many years (D'Orazio et al. 2012b). Oral intake of fucoxanthin even at very high concentrations, *e.g.* 2000 mg/kg single administration or 100 mg/kg daily for one month, did not cause obvious toxic effects in animals (Beppu et al. 2009b, c). Furthermore, oral administration of fucoxanthin at the concentration of 3 mg per day for 1 month did not interfere physiological parameters in humans, indicating that fucoxanthin is safe (Hitoe and Shimoda 2017a, b). Recently, oral intake of fucoxanthin was reported to lead to detectable fucoxanthin and metabolites in the brain of rodents, suggesting that fucoxanthin could penetrate the blood-brain barrier (BBB), might be used to treat brain disorders (Zhang et al. 2015).

AD is a neurodegenerative disorder, and is characterized by the death of neurons and the impairments of cognition (Barnham et al. 2004). However, there is no effective therapy for AD at present. Therefore, it is needed to discover prevention or therapeutic medications to combat this disease (Lehman 2013; Wu et al. 2015). AD is a complex disorder with many pathological characteristics, such the misfolding of toxic proteins, the increase of neuroinflammation, the elevation of oxidative stress,

Table 9.1 Anti-AD targets and activities of fucoxanthin

Target	Anti-AD activity	Ref.
BACE-1	Fucoxanthin inhibits BACE with the IC_{50} at 5.3 μM *in vitro*	Jung et al. (2016)
AChE	Fucoxanthin directly inhibits AChE with the IC_{50} at 81.2 μM *in vitro*, and prevents scopolamine-induced cognitive impairments (50–200 mg/kg, *i.p.*) with similar efficacy as donepezil (3 mg/kg, *i.v.*) in mice	Lin et al. (2016)
Neuroinflammation	Fucoxanthin (5–50 μM) prevents Aβ-induced pro-inflammatory secretion in BV2 microglia	Pangestuti et al. (2013)
ROS	Fucoxanthin (1–3 μM) prevents H_2O_2-induced neurotoxicity in SH-SY5Y cells and primary cerebellar granule neurons.	Yu et al. (2017)
Aβ aggregation	Fucoxanthin (0.3–10 μM) directly inhibits Aβ aggregation *in vitro*, and prevents scopolamine-induced cognitive impairments and oxidative stress (50–200 mg/kg, *i.p.*) with similar efficacy as huperzine A (0.2 mg/kg, *i.v.*) in mice	Xiang et al. (2017)
	Fucoxanthin (1–3 μM) prevents Aβ oligomers-induced neurotoxicity and ROS production in SH-SY5Y cells	Lin et al. (2017)
Gut microbiota disorder	Fucoxanthin (125 mg/kg) reduces the ratio of *Firmicutes/Bacteroidetes* in the small intestine of mice	Guo et al. (2019b)

and the dysregulation of neurotransmitters in the brain (McCrimmon et al. 2012). Moreover, the abnormality of gut microbiota is recently recognized as another pathological characteristic of this disease. Due to the multiple pathological mechanisms, the multi-target agents might be useful to prevent or treat AD, because these agents might target these pathological events, simultaneously (Bostanciklioğlu 2019; Wyss-Coray and Mucke 2002).

Aβ, cleaved by Aβ precursor protein (APP), could aggregate into toxic Aβ aggregations, and induce the loss of functional neurons during AD (Reiss et al. 2018; Ferreira and Klein 2011). Fucoxanthin was reported to directly inhibit β-secretase (BACE-1), the main enzyme responsible for APP cleavage (Jung et al. 2016). Recently, we found that fucoxanthin could also inhibit Aβ oligomerization and fibrillization, possibly via interacting with Aβ from hydrophobic binding (Table 9.1) (Xiang et al. 2017). Furthermore, fucoxanthin could inhibit $A\beta_{1-42}$ oligomers-induced loss of SH-SY5Y cells (Lin et al. 2017). Intraperitoneal administration of fucoxanthin could prevent impairments of cognition in Aβ oligomer-treated mice (Xiang et al. 2017). All these results suggested that fucoxanthin might produce neuroprotective effects at least partially via the inhibition of Aβ production and aggregation.

The imbalance of oxidative stress and anti-oxidation presents in the central nervous system (CNS), and causes the redox potential during AD progress (Singh et al. 2019). Fucoxanthin could directly scavenge radical oxygen species (ROS) *in vitro* (Sachindra et al. 2007). Fucoxanthin could elevate the activities of intracellular antioxidant enzymes such as superoxide dismutase and catalase, and protect microglia cells and neurons against oxidative stress-induced neurotoxicity (Harvey et al.

2009). In primary cortical neurons, fucoxanthin could increase the nuclear transfer of the nuclear translocation of nuclear factor E2-related factor 2 (Nrf2), and elevate the transcription of Nrf2-related anti-oxidant enzymes, leading to the prevention of neurotoxicity (Hu et al. 2018). Moreover, fucoxanthin inhibited H_2O_2-induced increase of intracellular ROS in primary cerebellar granule neurons, suggesting that this agent could produce neuroprotection via reversing the imbalance between oxidation and anti-oxidation during AD progress (Table 9.1) (Yu et al. 2017).

Chronic over-activation microglia cells were found in the brain of AD patients, leading to the over-production of pro-inflammatory cytokines, such as interleukin-1β (IL-1β), interleukin-6 (IL-6), and tumor necrosis factor alpha (TNF-α). Such neuroinflammation was associated with functional neuronal loss and cognitive impairments in AD (Block et al. 2007). Therefore, neuroinflammation is another target for AD therapy. Fucoxanthin could reduce the production of IL-6 and TNF-α via the inhibition of the nuclear factor kappa light chain enhancer of activated B cells (NF-κB) signaling pathway in lipopolysacchatide-treated BV2 microglia cells, could prevent neurotoxins-related neuroinflammation (Table 9.1) (Zhao et al. 2017; Pangestuti et al. 2013).

Decreased activity of cholinergic transmission was regarded as the main cause of the impairments of learning and memory in AD patients (Toublet et al. 2019). Therefore, acetylcholinesterase (AChE) inhibitors, compounds used to inhibit the degradation of acetylcholine and elevate acetylcholine concentrations in the synapses, were used in the clinic for the treatment of AD (Hung and Fu 2017). We have found, for the first time, that fucoxanthin could directly inhibit AChE (Table 9.1) (Lin et al. 2016). By using molecular docking analysis, we further demonstrated that 5,6-monoepoxide motif within fucoxanthin may interact with the peripheral anion site of AChE, causing the non-competitive inhibition of this enzyme (Lin et al. 2016). Furthermore, fucoxanthin can significantly prevent the increased AChE activity, and increase the hippocampal acetylcholine concentrations in scopolamine-treated AD mice (Table 9.1) (Lin et al. 2016). These results support that fucoxanthin could increase cholinergic neurotransmission, and reduce the impairments of learning and memory in AD.

Recently, gut microbiota was found to play a very important role in regulating CNS activity. Therefore, gut microbiota disorder was positively associated with CNS dysfunction and cognitive abnormality (Bercik et al. 2011). Particularly, the increased ratio of *Firmicutes/Bacteroidetes* was associated with the impairments of tight junctions in the gastric intestinal epithelium, and might lead to neuroinflammation via the brain transfer of pro-inflammatory cytokines from vagus nerve (Hoffman et al. 2017). Fucoxanthin could reduce the ratio of *Firmicutes/Bacteroidetes* in the small intestine of mice, might produce neuroprotective effects via the inhibition of neuroinflammation via preventing epithelia tight junction impairments (Table 9.1) (Guo et al. 2019b).

In summary, fucoxanthin has potential anti-AD abilities, to directly inhibit the activity of BACE-1 and AChE, to prevent the fibrillization and oligomerization of Aβ, to reduce the activation of microglia cells, and to inhibit the disorder of gut

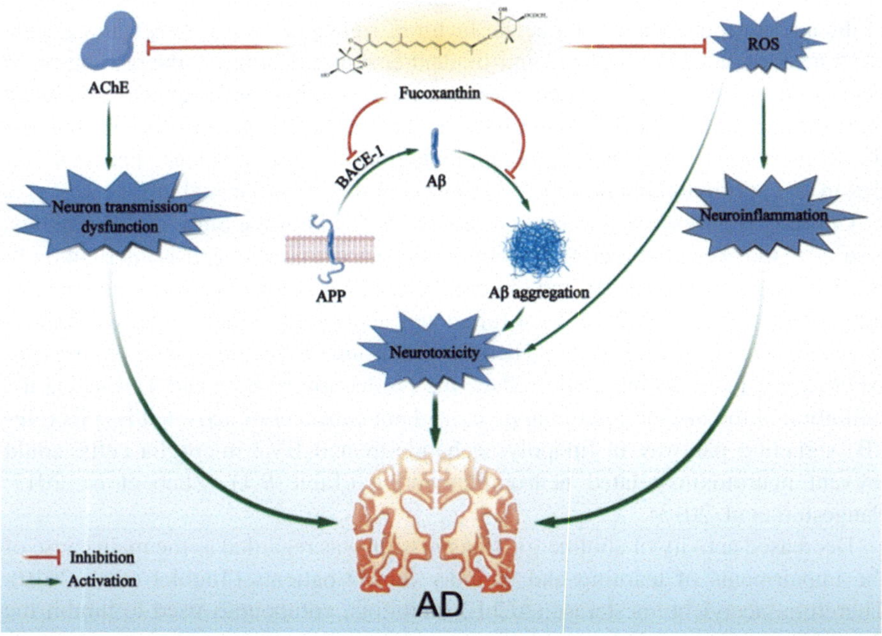

Fig. 9.4 Fucoxanthin produces anti-AD neuroprotective effects via the inhibition of the activities of AChE and BACE-1, the prevention of Aβ aggregation, and the reduction of intracellular ROS

microbiota (Fig. 9.4). Due to the complex pathogeneses of AD, fucoxanthin might be more useful for the treatment of AD.

4.2 *Phlorotannins as a Plant Biostimulant for Modern Agriculture*

Some brown algae extracts as natural fertilizers have been manufactured commercially since the middle of the twentieth century. The application of the fertilizers in agriculture production gained widely recognition, as natural alternative to artificial fertilizers. Currently, much attention has been focused on the considerable potential and promising applications of seaweed extracts on plant growth and development (Kim and Chojnacka 2015). Phlorotannins, as bioactive compounds found exclusively in brown algae, are well known from their antioxidant activity. The application of phlorotannins used as biostimulants, may increase the number of roots, shoot elongation, root elongation and rooting development of somatic embryo-derived papaya (*Carica papaya* L.) shoots, which result in a better survival rate state of papaya (96.5%) in ex vitro acclimatization conditions (Perez et al. 2016). Moreover, the foliar application of "*eckol*" a phlorotannin compound used as a plant biostimulant, may increase shoot and root length, shoot and root fresh and dry weight, leaf area and leaf number in the commercially cultivated cabbage. The levels of photosynthetic pigments, protein, proline and iridoid glycosides were significantly higher in cabbage leaves with "*eckol*" treatment. Meanwhile, the foliar application of "*eckol*" may enhance the myrosinase activity of the cabbage. These results demonstrate that the foliar application of "*eckol*" may enhance growth, phytochemical constituents and aphid resistance capacity in cabbage (Rengasamy et al. 2016). Currently, commercial phlorotannin-containing extracts from different brown algae have gained wide acceptance in agriculture as plant biostimulants. Utilization of algal biostimulants may improve seedling growth, shoot and root length and weight, chlorophyll content, or total protein content. Various extracts and formulations are now being using widely as biostimulants for a number of agricultural crops (Craigie 2011).

5 Conclusion

This chapter introduces the green processing technology for fucoxanthin and phlorotannins production from brown seaweeds and supports the planning of future R&D strategies and industrial applications.

The innovation of the processing introduced here is able to manufacture the minor bioactive compounds such as fucoxanthin and phlorotannins, which are demanded as new high value-added products for the brown seaweed industry. Currently, according to the principles of green chemistry, this new process to stepwise isolate fucoxanthin and phlorotannins from crude alcohol extracts of brown seaweeds has been successfully established at commercial scale. Most importantly, within the whole process, no organic solvents except for alcohol have been used, and the seaweed material after solvent extraction can be smoothly transferred into alginate manufacture or other biomass processing. The first automatic production line of fucoxanthin with a high purity more than 50% in China has been put into operation in Shandong Province, with the manufacture capacity reaching more than 500 kg produced per year. In addition, fucoxanthin has been reported as a promising dietary supplement in preventing Alzheimer disease, while phlorotannins can be used as plant biostimulant.

References

Amarante SJ, Catarino MD, Marcal C, Silva AMS, Ferreira R, Cardoso SM (2020) Microwave-assisted extraction of Phlorotannins from Fucus vesiculosus. Mar Drugs 18:559

Barnham KJ, Masters CL, Bush AI (2004) Neurodegenerative diseases and oxidative stress. Nat Rev Drug Discov 3:205–214

Beppu F, Niwano Y, Sato E, Kohno M, Tsukui T, Hosokawa M, Miyashita K (2009a) In vitro and in vivo evaluation of mutagenicity of fucoxanthin (FX) and its metabolite fucoxanthinol (FXOH). J Toxicol Sci 34:693–698

Beppu F, Niwano Y, Sato E, Kohno M, Tsukui T, Hosokawa M, Miyashita K (2009b) In vitro and in vivo evaluation of mutagenicity of fucoxanthin (FX) and its metabolite fucoxanthinol (FXOH). J Toxicol Sci 34:693–698

Beppu F, Niwano Y, Tsukui T, Hosokawa M, Miyashita K (2009c) Single and repeated oral dose toxicity study of fucoxanthin (FX), a marine carotenoid, in mice. J Toxicol Sci 34:501–510

Bercik P, Denou E, Collins J, Jackson W, Lu J, Jury J, Deng Y, Blennerhassett P, Macri J, McCoy KD, Verdu EF, Collins SM (2011) The intestinal microbiota affect central levels of brain-derived neurotropic factor and behavior in mice. Gastroenterology 141:599–609

Billakanti JM, Catchpole OJ, Fenton TA, Mitchell KA, MacKenzie AD (2013) Enzyme-assisted extraction of fucoxanthin and lipids containing polyunsaturated fatty acids from Undaria pinnatifida using dimethyl ether and ethanol. Process Biochem 48:1999–2008

Block ML, Zecca L, Hong J-S (2007) Microglia-mediated neurotoxicity: uncovering the molecular mechanisms. Nat Rev Neurosci 8:57–69

Bostanciklioğlu M (2019) The role of gut microbiota in pathogenesis of Alzheimer's disease. J Appl Microbiol 127:954–967

Chia SR, Show PL, Phang SM, Ling TC, Ong HC (2018) Sustainable approach in phlorotannin recovery from macroalgae. J Biosci Bioeng 126:220–225

Chojnacka K, Wieczorek PP, Schroeder G, Michalak I (2018) Algae as fertilizers, biostimulants, and regulators of plant growth, pp 115–122. https://doi.org/10.1007/978-3-319-74703-3

Cotas J, Leandro A, Monteiro P, Pacheco D, Figueirinha A, Goncalves AMM, da Silva GJ, Pereira L (2020) Seaweed phenolics: from extraction to applications. Mar Drugs 18:384

Craigie JS (2011) Seaweed extract stimuli in plant science and agriculture. J Appl Phycol 23:371–393

Cruces E, Rautenberger R, Rojas-Lillo Y, Cubillos VM, Arancibia-Miranda N, Ramirez-Kushel E, Gomez I (2017) Physiological acclimation of Lessonia spicata to diurnal changing PAR and UV radiation: differential regulation among down-regulation of photochemistry, ROS scavenging activity and phlorotannins as major photoprotective mechanisms. Photosynth Res 131:145–157

Dexamethasone (2004) A Medical dictionary, bibliography, and annotated research guide to internet references. Icon Group International

Dominguez H (2013) Functional ingredients from algae for foods and nutraceuticals. Elsevier Science, Burlington

D'Orazio N, Gemello E, Gammone MA, de Girolamo M, Ficoneri C, Riccioni G (2012a) Fucoxantin: a treasure from the sea. Mar Drugs 10:604–616

D'Orazio N, Gemello E, Gammone MA, de Girolamo M, Ficoneri C, Riccioni G (2012b) Fucoxantin: a treasure from the sea. Mar Drugs 10:604–616

Ferreira ST, Klein WL (2011) The Aβ oligomer hypothesis for synapse failure and memory loss in Alzheimer's disease. Neurobiol Learn Mem 96:529–543

Ferreira J, Ramos AA, Almeida T, Azqueta A, Rocha E (2018) Drug resistance in glioblastoma and cytotoxicity of seaweed compounds, alone and in combination with anticancer drugs: a mini review. Phytomedicine 48:84–93

Guo BB, Yang B, Pang XY, Chen TP, Chen F, Cheng K (2019a) Fucoxanthin modulates cecal and fecal microbiota differently based on diet. Food Funct 10:5644–5655

Guo B, Yang B, Pang X, Chen T, Chen F, Cheng KW (2019b) Fucoxanthin modulates cecal and fecal microbiota differently based on diet. Food Funct 10:5644–5655

Harvey CJ, Thimmulappa RK, Singh A, Blake DJ, Ling G, Wakabayashi N, Fujii J, Myers A, Biswal S (2009) Nrf2-regulated glutathione recycling independent of biosynthesis is critical for cell survival during oxidative stress. Free Radic Biol Med 46:443–453

Haugan JA, Liaaen-Jensen S (1994) Isolation and characterisation of four allenic (6′S)-isomers of fucoxanthin. Tetrahedron Lett 35:2101–2252

Hitoe S, Shimoda H (2017a) Seaweed Fucoxanthin supplementation improves obesity parameters in mild obese Japanese subjects. Funct Food Health Dis 7:246–262
Hitoe S, Shimoda H (2017b) Seaweed Fucoxanthin supplementation improves obesity parameters in mild obese Japanese subjects. Functional Foods in Health and Disease 7:246–262
Hoffman JD, Parikh I, Green SJ, Chlipala G, Mohney RP, Keaton M, Bauer B, Hartz AMS, Lin AL (2017) Age drives distortion of brain metabolic, vascular and cognitive functions, and the gut microbiome. Front Aging Neurosci 9:298
Hu L, Chen W, Tian F, Yuan C, Wang H, Yue H (2018) Neuroprotective role of fucoxanthin against cerebral ischemic/reperfusion injury through activation of Nrf2/HO-1 signaling. Biomed Pharmacother 106:1484–1489
Hung S-Y, Fu W-M (2017) Drug candidates in clinical trials for Alzheimer's disease. J Biomed Sci 24:47–47
Iio K, Okada Y, Ishikura M (2011) Single and 13-week oral toxicity study of Fucoxanthin oil from microalgae in rats. Food Hygiene and Safety Science 52:183–189
Jung HA, Ali MY, Choi RJ, Jeong HO, Chung HY, Choi JS (2016) Kinetics and molecular docking studies of fucosterol and fucoxanthin, BACE1 inhibitors from brown algae Undaria pinnatifida and Ecklonia stolonifera. Food Chem Toxicol 89:104–111
Kanda H, Kamo Y, Machmudah S, Wahyudiono, Goto M (2014) Extraction of Fucoxanthin from raw macroalgae excluding drying and cell wall disruption by liquefied dimethyl ether. Mar Drugs 12:2383–2396
Kim JC (2014) Solvent extraction of Fucoxanthin from Phaeodactylum tricornutum. Sep Sci Technol 49:410–415
Kim SK, Chojnacka K (2015) Marine algae extracts: processes, products, and applications. Wiley-VCH, Weinheim
Kim SM, Shang YF, Um BH (2011) A preparative method for isolation of Fucoxanthin from Eisenia bicyclis by centrifugal partition chromatography. Phytochem Anal 22:322–329
Kim SM, Jung YJ, Kwon ON, Cha KH, Um BH, Chung D, Pan CH (2012a) A potential commercial source of Fucoxanthin extracted from the microalga Phaeodactylum tricornutum. Appl Biochem Biotechnol 166:1843–1855
Kim SM, Kang SW, Kwon ON, Chung D, Pan CH (2012b) Fucoxanthin as a major carotenoid in Isochrysis aff. Galbana: characterization of extraction for commercial application. Journal of the Korean society for applied. Biol Chem 55:477–483
Kim J, Yoon M, Yang H, Jo J, Han D, Jeon YJ, Cho S (2014) Enrichment and purification of marine polyphenol phlorotannins using macroporous adsorption resins. Food Chem 162:135–142
Lakey-Beitia J, Kumar DJ, Hegde ML, Rao KS (2019) Carotenoids as novel therapeutic molecules against neurodegenerative disorders: chemistry and molecular docking analysis. Int J Mol Sci 20:5553
Lee JH, Ko JY, Oh JY, Kim CY, Lee HJ, Kim J, Jeon YJ (2014) Preparative isolation and purification of phlorotannins from Ecklonia cava using centrifugal partition chromatography by one-step. Food Chem 158:433–437
Lehman EJ (2013) Epidemiology of neurodegeneration in American-style professional football players. Alzheimers Res Ther 5:34
Leyton A, Pezoa-Conte R, Barriga A, Buschmann AH, Maki-Arvela P, Mikkola JP, Lienqueo ME (2016) Identification and efficient extraction method of phlorotannins from the brown seaweed Macrocystis pyrifera using an orthogonal experimental design. Algal Research-Biomass Biofuels and Bioproducts 16:201–208
Leyton A, Vergara-Salinas JR, Perez-Correa JR, Lienqueo ME (2017) Purification of phlorotannins from Macrocystis pyrifera using macroporous resins. Food Chem 237:312–319
Li YJ, Fu XT, Duan DL, Liu XY, Xu JC, Gao X (2017) Extraction and identification of Phlorotannins from the Brown alga, Sargassum fusiforme (Harvey) Setchell. Mar Drugs 15:49
Lin J, Huang L, Yu J, Xiang S, Wang J, Zhang J, Yan X, Cui W, He S, Wang Q (2016) Fucoxanthin, a marine carotenoid, reverses scopolamine-induced cognitive impairments in mice and inhibits acetylcholinesterase in vitro. Mar Drugs 14:67

Lin JJ, Yu J, Zhao JY, Zhang K, Zheng JC, Wang JL, Huang CH, Zhang JR, Yan XJ, Gerwick WH, Wang QW, Cui W, He S (2017) Fucoxanthin, a marine carotenoid, attenuates beta-amyloid oligomer-induced neurotoxicity possibly via regulating the PI3K/Akt and the ERK pathways in SH-SY5Y cells. Oxidative Med Cell Longev 2017:6792543

Lourenco-Lopes C, Garcia-Oliveira P, Carpena M, Fraga-Corral M, Jimenez-Lopez C, Pereira AG, Prieto MA, Simal-Gandara J (2020) Scientific approaches on extraction, purification and stability for the commercialization of Fucoxanthin recovered from Brown algae. Foods 9:1113

Maoka T, Nishino A, Yasui H, Yamano Y, Wada A (2016) Anti-oxidative activity of Mytiloxanthin, a metabolite of Fucoxanthin in shellfish and tunicates. Mar Drugs 14:93

Martinez JHI, Castaneda HGT (2013) Preparation and chromatographic analysis of Phlorotannins. J Chromatogr Sci 51:825–838

McCrimmon RJ, Ryan CM, Frier BM (2012) Diabetes and cognitive dysfunction. Lancet 379:2291–2299

Mok IK, Yoon JR, Pan CH, Kim SM (2016) Development, quantification, method validation, and stability study of a novel Fucoxanthin-fortified Milk. J Agric Food Chem 64:6196–6202

Mori K, Ooi T, Hiraoka M, Oka N, Hamada H, Tamura M, Kusumi T (2004) Fucoxanthin and its metabolites in edible Brown algae cultivated in deep seawater. Mar Drugs 2:63–72

Neumann U, Derwenskus F, Flister VF, Schmid-Staiger U, Hirth T, Bischoff SC (2019) Fucoxanthin, a carotenoid derived from Phaeodactylum tricornutum exerts antiproliferative and antioxidant activities in vitro. Antioxidants 8:183

Obluchinskaya ED, Daurtseva AV, Pozharitskaya ON, Flisyuk EV, Shikov AN (2019) Natural deep eutectic solvents as alternatives for extracting Phlorotannins from Brown algae. Pharm Chem J 53:243–247

Pangestuti R, Vo TS, Ngo DH, Kim SK (2013) Fucoxanthin ameliorates inflammation and oxidative reponses in microglia. J Agric Food Chem 61:3876–3883

Perez LP, Montesinos YP, Olmedo JG, Rodriguez RB, Sanchez RR, Montenegro ON, Escriba RCR, Daniels D, Gomez-Kosky R (2016) Effect of phloroglucinol on rooting and in vitro acclimatization of papaya (Carica papaya L. var. Maradol Roja). In Vitro Cellular & Developmental Biology-Plant 52:196–203

Piovan A, Filippini R, De Paoli M, Bresin B (2014) TLC densitometric method for the preliminary evaluation of fucoxanthin-based products. Nat Prod Res 28:1111–1115

Prabhasankar P, Ganesan P, Bhaskar N, Hirose A, Stephen N, Gowda LR, Hosokawa M, Miyashita K (2009) Edible Japanese seaweed, wakame (Undaria pinnatifida) as an ingredient in pasta: chemical, functional and structural evaluation. Food Chem 115:501–508

Quitain AT, Kai T, Sasaki M, Goto M (2013) Supercritical carbon dioxide extraction of Fucoxanthin from Undaria pinnatifida. J Agric Food Chem 61:5792–5797

Rehm BHA, Moradali MF (2018) Alginates and their biomedical applications. Springer, Singapore

Reichardt C (2010) Solvent and solvent effects in organic chemistry. Wiley-VCH, Weinheim

Reiss AB, Arain HA, Stecker MM, Siegart NM, Kasselman LJ (2018) Amyloid toxicity in Alzheimer's disease. Rev Neurosci 29:613–627

Rengasamy KRR, Kulkarni MG, Pendota SC, Van Staden J (2016) Enhancing growth, phytochemical constituents and aphid resistance capacity in cabbage with foliar application of eckol—a biologically active phenolic molecule from brown seaweed. New Biotechnol 33:273–279

Rivera S, Canela R (2012) Influence of sample processing on the analysis of carotenoids in maize. Molecules 17:11255–11268

Sachindra NM, Sato E, Maeda H, Hosokawa M, Niwano Y, Kohno M, Miyashita K (2007) Radical scavenging and singlet oxygen quenching activity of marine carotenoid fucoxanthin and its metabolites. J Agric Food Chem 55:8516–8522

Sanderson K (2011) Chemistry: it's not easy being green. Nature 469:18–20

Shang YF, Kim SM, Lee WJ, Um BH (2011) Pressurized liquid method for fucoxanthin extraction from Eisenia bicyclis (Kjellman) Setchell. J Biosci Bioeng 111:237–241

Shannon E, Abu-Ghannam N (2017) Optimisation of fucoxanthin extraction from Irish seaweeds by response surface methodology. J Appl Phycol 29:1027–1036

Singh A, Kukreti R, Saso L, Kukreti S (2019) Oxidative stress: a key modulator in neurodegenerative diseases. Molecules 24:1583

Su JQ, Guo K, Huang M, Liu YX, Zhang J, Sun LJ, Li DL, Pang KL, Wang GC, Chen L, Liu ZY, Chen YQ, Chen Q, Huang LQ (2019) Fucoxanthin, a marine xanthophyll isolated from Conticribra weissflogii ND-8: preventive anti-inflammatory effect in a mouse model of sepsis. Front Pharmacol 10:906

Sudhakar MP, Ananthalakshmi JS, Nair BB (2013) Extraction, purification and study on antioxidant properties of fucoxanthin from brown seaweeds. Journal of Chemical & Pharmaceutical Research 5:169–175

Tai YN, Shen JC, Luo Y, Qu HB, Gong XC (2020) Research progress on the ethanol precipitation process of traditional Chinese medicine. Chin Med 15:84

Tan PW, Tan CP, Ho CW (2011) Antioxidant properties: effects of solid-to-solvent ratio on antioxidant compounds and capacities of Pegaga (Centella asiatica). Int Food Res J 18:557–562

Toublet FX, Lecoutey C, Lalut J, Hatat B, Davis A, Since M, Corvaisier S, Freret T, de Oliveira Santos JS, Claeysen S, Boulouard M, Dallemagne P, Rochais C (2019) Inhibiting acetylcholinesterase to activate pleiotropic prodrugs with therapeutic interest in Alzheimer's disease. Molecules 24:2786

Ummat V, Tiwari BK, Jaiswal AK, Condon K, Garcia-Vaquero M, O'Doherty J, O'Donnell C, Rajauria G (2020) Optimisation of ultrasound frequency, extraction time and solvent for the recovery of polyphenols, Phlorotannins and associated antioxidant activity from Brown seaweeds. Mar Drugs 18:250

Vazquez-Rodriguez B, Gutierrez-Uribe JA, Antunes-Ricardo M, Santos-Zea L, Cruz-Suarez LE (2020) Ultrasound-assisted extraction of phlorotannins and polysaccharides from Silvetia compressa (Phaeophyceae). J Appl Phycol 32:1441–1453

Wang L, Park YJ, Jeon YJ, Ryu B (2018) Bioactivities of the edible brown seaweed, Undaria pinnatifida: a review. Aquaculture 495:873–880

Warkoyo W, Saati E (2011) The solvent effectiveness on extraction process of seaweed pigment. Makara J Technol 15:5–8

Wu H, Niu H, Shao A, Wu C, Dixon BJ, Zhang J, Yang S, Wang Y (2015) Astaxanthin as a potential neuroprotective agent for neurological diseases. Mar Drugs 13:5750–5766

Wyss-Coray T, Mucke L (2002) Inflammation in neurodegenerative disease—a double-edged sword. Neuron 35:419–432

Xiang S, Liu F, Lin J, Chen H, Huang C, Chen L, Zhou Y, Ye L, Zhang K, Jin J, Zhen J, Wang C, He S, Wang Q, Cui W, Zhang J (2017) Fucoxanthin inhibits β-amyloid assembly and attenuates β-amyloid oligomer-induced cognitive impairments. J Agric Food Chem 65:4092–4102

Xiao XH, Si XX, Yuan ZQ, Xu XF, Li GK (2012) Isolation of fucoxanthin from edible brown algae by microwave-assisted extraction coupled with high-speed countercurrent chromatography. J Sep Sci 35:2313–2317

Yilmaz GG, Pinchetti JLG, Cifuentes A, Herrero M, Ibanez E (2019) Comparison of extraction techniques and surfactants for the isolation of total polyphenols and Phlorotannins from the Brown algae Lobophora variegata. Anal Lett 52:2724–2740

Yoon M, Kim JS, Um MY, Yang H, Kim J, Kim YT, Lee C, Kim SB, Kwon S, Cho S (2017) Extraction optimization for Phlorotannin recovery from the edible Brown seaweed Ecklonia cava. Journal of Aquatic Food Product Technology 26:801–810

Yu J, Lin J-J, Yu R, He S, Wang Q-W, Cui W, Zhang J-R (2017) Fucoxanthin prevents H(2)O(2)-induced neuronal apoptosis via concurrently activating the PI3-K/Akt cascade and inhibiting the ERK pathway. Food Nutr Res 61:1304678

Zarekarizi A, Hoffmann L, Burritt D (2019) Approaches for the sustainable production of fucoxanthin, a xanthophyll with potential health benefits. J Appl Phycol 31:281–299

Zhang Y, Wu H, Wen H, Fang H, Hong Z, Yi R, Liu R (2015) Simultaneous determination of Fucoxanthin and its Deacetylated metabolite Fucoxanthinol in rat plasma by liquid chromatography-tandem mass spectrometry. Mar Drugs 13:6521–6536

Zhao D, Kim SM, Pan CH, Chung D (2014) Effects of heating, aerial exposure and illumination on stability of fucoxanthin in canola oil. Food Chem 145:505–513

Zhao D, Kwon SH, Chun YS, Gu MY, Yang HO (2017) Anti-neuroinflammatory effects of Fucoxanthin via inhibition of Akt/NF-κB and MAPKs/AP-1 pathways and activation of PKA/CREB pathway in lipopolysaccharide-activated BV-2 microglial cells. Neurochem Res 42:667–677

Zhou XZ, Yi MQ, Ding LJ, He S, Yan XJ (2019) Isolation and purification of a neuroprotective Phlorotannin from the marine algae Ecklonia maxima by size exclusion and high-speed counter-current chromatography. Mar Drugs 17:212

Zimmermann MB (2011) The role of iodine in human growth and development. Semin Cell Dev Biol 22:645–652

Chapter 10
Seaweed: Food Benefits in the Human Gut Microbiome Health

Mauricio Alfredo Ondarza Beneitez

1 Introduction

Recent research has focused on identifying the role of seaweeds in modulating the risk and development of chronic diseases such as cardiovascular disease (CVD) and cancer, using results mainly from cellular and animal studies to propose potential mechanisms behind the observed effects (Cardoso et al. 2015). Trials using Seaweed performed by Winberg from the University of Wollongong in Australia has shown that fiber supplement a new group of good bacteria in the colon (Winberg 2015). Moreover, it was also found that most of the bacterial groups that had been cultivated belong to those that produce short-chain fatty acids like butyrate, which in turn favors the growth of the mucous lining in the gut. Studies have revealed an exceptionally low short chain fatty acids presence, which contributes to inflammation in the colon as well as to the proliferation of an unhealthy and out of balance microbiome.

Nutrients in seaweed are excellent prebiotics, packed with bio-available vitamins and minerals as well as fiber and omega 3 fatty acids (Wells et al. 2017). Seaweed should be part of the foundation of health and must be placed at the core of our dietary intake since our bodies cannot manufacture them. Seaweed selection must be based on species free from harmful contaminants, as well as being non-allergenic (Fig. 10.1).

Ascophyllum species is potentially a powerful prebiotic to support the gut microbial population (Lopez-Santamarina et al. 2020). Additionally, these species have been also found to be among the most balanced nutritionally food materials. The link between our diets, gut-health/ bacteria, and the immune system has been focused of recent findings. It is well known that nearly 70% of our body's immune

M. A. O. Beneitez (✉)
Red de Glicociencia en Salud, Mexico, Hacienda de Cortez, Rincon de los Arcos, Irapuato, Guanajuato, Mexico

A. Ranga Rao, G. A. Ravishankar (eds.), *Sustainable Global Resources of Seaweeds Volume 2*, https://doi.org/10.1007/978-3-030-92174-3_10

Fig. 10.1 Open sea seaweed culture (Ondarza Beneitez 2012)

system resides in the digestive tract. Moreover, our gut plays an essential role in our mental and physical health (Fig. 10.2).

Our bodies suffer extra demands through food additives, processed foods, stress, and pollution. Research has revealed that a diet rich in whole and unprocessed foods will allow the growth of good bacteria in the gut, which will bolster a strong immune system (Fig. 10.3).

1.1 Does Seaweed Improve Our Gut Health?

Seaweed is high in fiber-it can make up about 25–75% of seaweed's dry weight (Penalver et al. 2020). This is higher than the fiber content of most fruits and vegetables. Additionally, sugars found in seaweed called sulphated polysaccharides have been shown to increase the growth of good gut bacteria. These polysaccharides can also increase the production of short chain fatty acids (SCFA), which provide support and nourishment to the cells lining your gut.

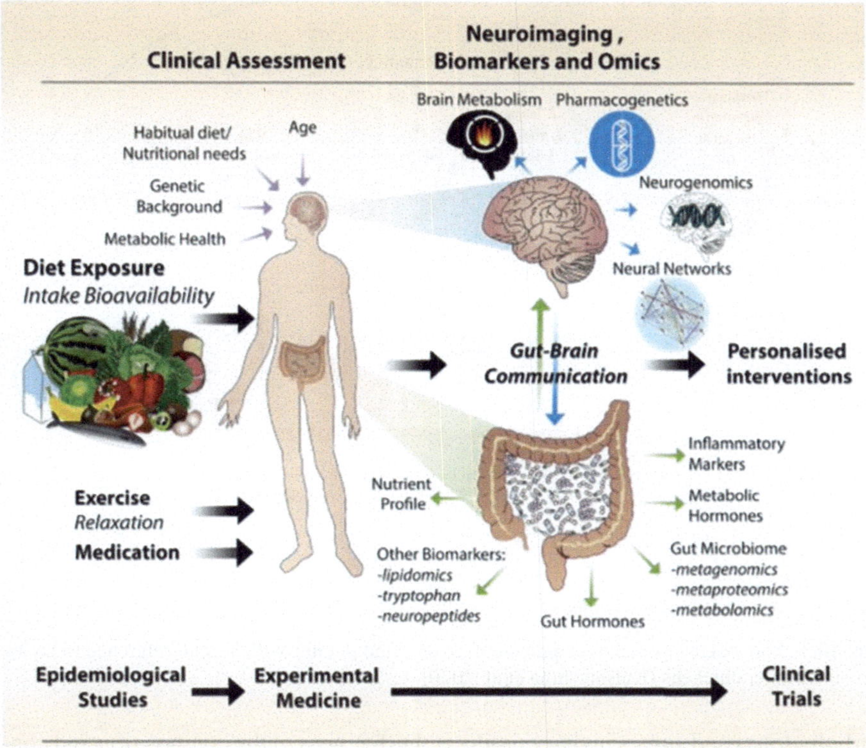

Fig. 10.2 Gut microbiota health (what we eat has an impact on mental health, Adan et al. 2019)

A specific carbohydrate food in nori seaweed could help drive changes to our gut bacteria by encouraging the growth of certain probiotic strains. Studies conducted at the Stanford University School of Medicine, show that consumption of seaweed could help establish a specific strain of Bacteroides in our gut-adding that a specific carbohydrate found in nori could selectively favor the growth (Fig. 10.4).

1.2 Bacteria for Our Gut Health

Acknowledging the need to improve gut health is gaining traction, with more studies on how the digestive system works and increasing numbers of consumer products to shape your microbiome. Probiotics (such as the yogurt drinks you can buy in store) contain live, "good bacteria" which have a positive impact when inhabiting your digestive system. Prebiotics on the other hand, act as food for these bacteria, enabling them to thrive and do their job. If your gut bacteria don't get enough food, (high fiber non-digestible) they feed on the mucus lining of the gut which can lead to irritable bowl. The bacteria help to digest and absorb foods such as indigestible

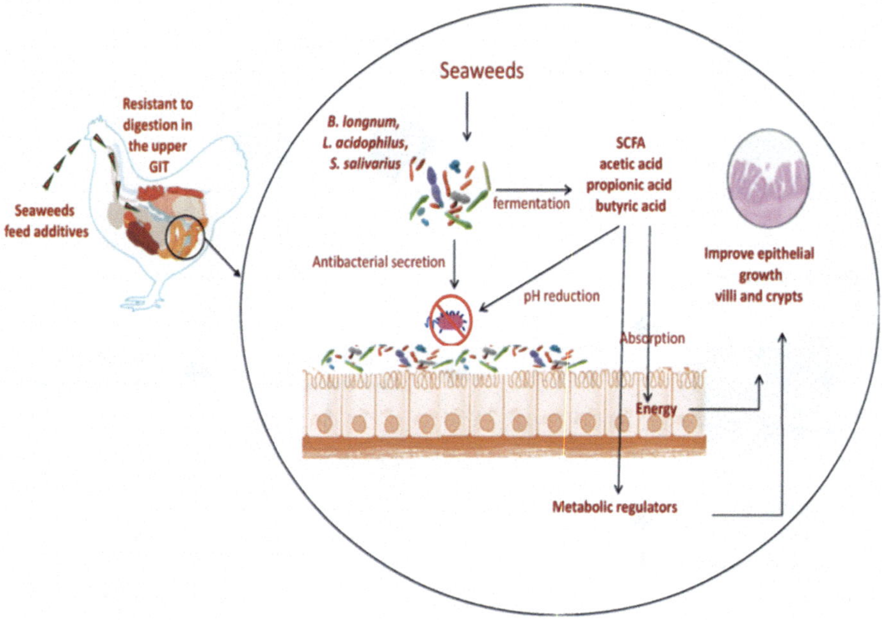

Fig. 10.3 Macroalgae as dietary supplements in selected poultry with special reference to laying hen and broiler chickens (Kulshreshtha et al. 2020)

carbohydrates and sugars in dairy products. Differences in the composition and concentration of bioactive compounds found in different species of seaweeds may be responsible for the potential health benefits (Liu et al. 2012) (Fig. 10.5).

2 Weight Management

Dietary fiber intake has been reported to aid weight loss (Kristensen and Jensen 2011). Consumption of seaweed-enriched bread (4% *Ascophyllum nodosum*) for a single breakfast meal resulted in significantly lower energy intake at a test lunch meal after 4 h and significantly reduced energy intake in the subsequent 24 h (−506 kcal) in 12 healthy overweight and obese men (Hall et al. 2012). These findings as well as those of An et al. (2013) who demonstrated the potentially beneficial effects of alginate through colonic fermentation that resulted in the production of short-chain fatty acids (most notably propionate), make alginate a seaweed component of interest for managing energy intake.

The microflora interacts with its host at both the local (intestinal mucosa) and systemic level, resulting in a broad range of immunological, physiological, and metabolic effects. Besides directly influencing the host, diet may play a role in modulating the effects of bacteria on the host, and this can become either beneficial or

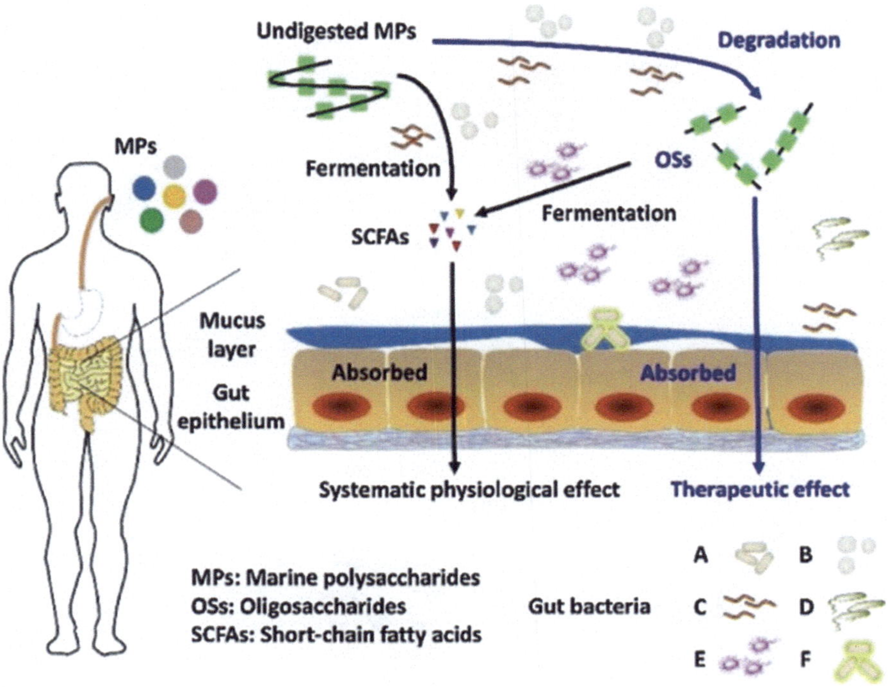

Fig. 10.4 Gut microbiota fermentation of marine polysaccharides and its effects on intestinal ecology (Shang et al. 2018)

detrimental to health (Rowland 1999; Rastall et al. 2005; Blaut and Clavel 2007; Cani et al. 2007).

There is recent evidence showing that seaweed-derived fibers can have positive effects on gut health (Vaugelade et al. 2000; Deville et al. 2004) and have also shown the potential prebiotic activity of low-molecular-weight polysaccharides (Deville et al. 2007; Ramnani et al. 2012). Marine-derived products are emerging as novel sources of prebiotic carbohydrates (O'Sullivan et al. 2010). Alginate, xanthan gum, and carrageenan gum seem to increase probiotic survival, providing live bacteria with a physical barrier against adverse digestive conditions (Ding and Shah 2009). A combination of alginate with chitosan is also emerging as an effective delivery system (Islam et al. 2010; Chavarri et al. 2010).

3 Antiviral Properties

It is the sulfated polysaccharides of seaweed species that appear to have antiviral properties, and in vitro and in vivo animal research has identified carrageenans, fucoidans, and sulfated rhamno galactans as having substantial antiviral activity against enveloped viruses such as herpes and HIV. Fucoidan has been shown to

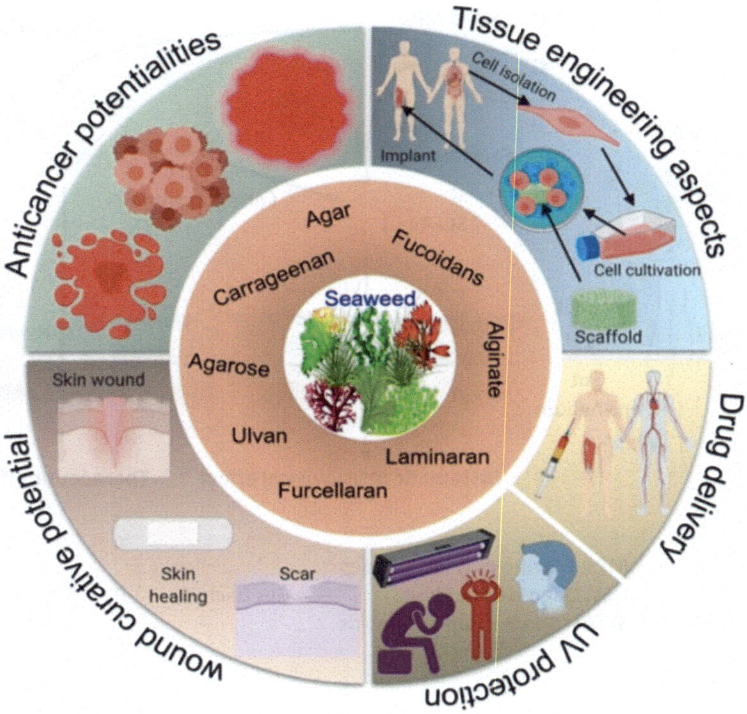

Fig. 10.5 Seaweed polysaccharides based engineered cues for the modern biomedical sector (Bilal and Iqbal 2020)

inhibit the growth of a variety of viruses (Trinchero et al. 2009). A well investigated proof-of-concept through a clinical study demonstrated lack of toxicity after 3 months of daily consumption of 5–6 g of brown seaweed (from whole dried *Undaria pinnatifida* or *Arthrospira platensis*) in HIV-positive individuals in South Carolina, United States. A clinically significant improvement in CD4 cell counts and a reduction in viral load were observed in one individual who remained in the study for 13 months in the absence of antiretroviral therapy (Teas and Irhimeh 2012; Cooper et al. 2002). A small dietary intervention study was performed in Tasmania to investigate the antiviral properties of a dried whole seaweed extract preparation from Tasmanian *Undaria pinnatifida* (GFS, a proprietary preparation) as founded by Wang et al. (2008).

4 Dietary Fiber in Seaweed Helps to Ameliorate Digestive Health

Seaweed supplementation can enhance water binding to the food pellet in the gut and facilitate stool bulking, and decrease transit time in the colon, that act as positive factors to prevent colon cancer (Brownlee et al. 2005). A study was carried out to determine apoptosis-inducing activity of fucoidan in cultured HT-29 and HCT116 human colon cancer cells and revealed that fucoidan can reduce the viability of tested cells in a dose-dependent manner through the inhibition of both tumor necrosis factor and caspase-induced cell signaling (Kim et al. 2010). The colonic microflora is a complex and co-existing microbial ecosystem of potentially pathogenic and beneficial bacteria associated with gut lymphoid tissue. Fermentation of fiber from brown seaweed with human fecal bacteria has indicated that probiotics follow their original fermentation pathways as exhibited with prebiotics from some other non-seaweed food sources (Mabeau and Fleurence 1993). This fermentable fiber stimulates the growth of bifidobacteria and lactobacilli, which are the most important probiotic genera in humans, and maintains a more favorable balance among the colonic microflorae. Laminarin, a less viscous phycolloid amply found in Laminaria and Saccharina, has shown its capability to promote higher production of butyric acid through bacterial fermentation (Deville et al. 2004). Butyrates are important energy-yielding metabolites for the colonial epithelial cells and account for about 70% of the energy requirement of the colon (Reilly et al. 2008). Prebiotic effects of laminarin studied in animal models reported that 1% dietary supplementation resulted in an in Bifidobacterium counts in the cecum of rats compared to a control diet, but there was no significant difference in Lactobacillus counts (Kuda et al. 2005). Studies carried out on seaweed extracts found that fucoidan also functions as a good prebiotic. Several other studies have also confirmed the positive dietary effects of alginates encouraging the growth of beneficial microbial fauna in fecal matter (Wang et al. 2006). Laminarin and fucoidan may offer a dietary means to modulate the gut environment and immunity, and thereby reducing the risk of pathogenic microorganisms in the gut. Inclusion of brown seaweed, Ascophyllum nodosum, to the diet of weanling pigs resulted in lower numbers of *Escherichia coli* in the small intestine (Dierick et al. 2009) (Fig. 10.6).

5 Unparalleled Inventory of the Human Gut Ecosystem

An international team of scientists has collated all known bacterial genomes from the human gut microbiome into a single large database, allowing researchers to explore the links between bacterial genes and proteins, and their effects on human health.

5.1 More Microbes Than Human Cells

Bacteria coat the human body, inside and out. They produce proteins that affect our digestion, our health, and our susceptibility to diseases. They are so prevalent that the body is estimated to contain more cells in its microbiome—the bacteria, fungi, and other microbes—than it has human cells.

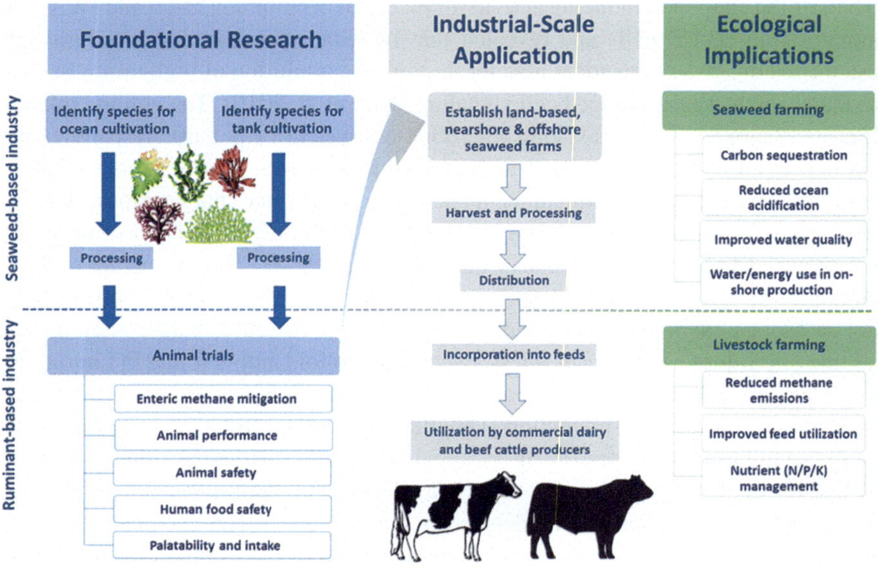

Fig. 10.6 The use of seaweeed to reduce enteric methane emissions from the cattle (Vijn et al. 2020)

To understand the role that bacterial species, play in human biology, scientists usually isolate and culture them in the lab before they sequence their DNA. However, many bacteria thrive in conditions that are not yet reproducible in a laboratory setting.

To obtain information on such species, researchers take another approach: they collect a single sample from the environment—in this case, the human gut—and sequence the DNA from the whole sample. They then use computational methods to reconstruct the individual genomes of thousands of species from that single sample. This method, called metagenomics, offers a powerful alternative to isolating and sequencing the DNA of individual species (Akbarzadeh et al. 2018).

5.2 *Biodiversity in the Human Gut*

The scientists have now compiled 200,000 genomes and 170 million protein sequences from more than 4600 bacterial species in the human gut. Their new databases, the Unified Human Gastrointestinal Genome collection, and the Unified Gastrointestinal Protein catalog, reveal the tremendous diversity in our guts and pave the way for further microbiome research (Fig. 10.7). This immense catalogue is a landmark in microbiome research, and will be an invaluable resource for scientists to start studying and hopefully understanding the role of each bacterial species in the human gut ecosystem, explains Nicola Segata, Principal Investigator at the

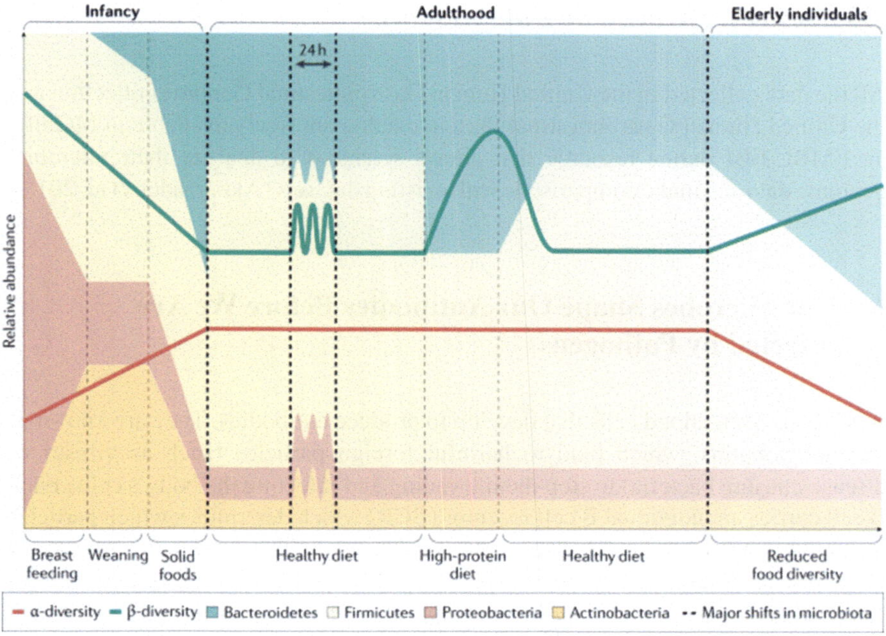

Fig. 10.7 Exploring human gut microbiome variations across life: from eubiosis to dysbiosis in Western populations (Barone 2020)

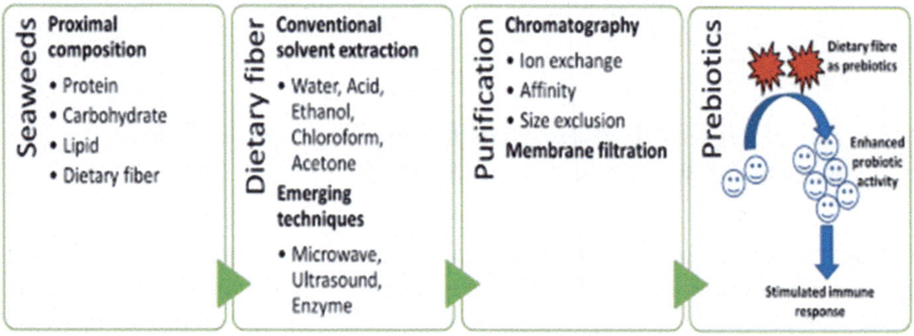

Fig. 10.8 Extraction and purification techniques seaweed dietary fibers for immuno modulation of gut microbiota (Praveen et al. 2019)

University of Trento (Fig. 10.8). The project revealed that more than 70% of the detected bacterial species had never been cultured in the lab—their activity in the body remains unknown. The largest group of bacteria that falls into that category is the Comantemales, an order of gut bacteria first described in 2019 in a study led by the Bork Group at EMBL Heidelberg (Akbarzadeh et al. 2018).

5.3 A Freely Accessible Data Resource

All the data collected in the Unified Human Gastrointestinal Genome collection and the Unified Human Gastrointestinal Protein catalog are freely available in MGnify, an EMBL-EBI online resource that allows scientists to analyze their microbial genomic data and make comparisons with existing datasets (Akbarzadeh et al. 2018).

6 Gut Microbes Shape Our Antibodies Before We Are Infected by Pathogens

B cells are white blood cells that develop to produce antibodies. They are known as immunoglobulins, which bind to harmful foreign particles (such as viruses or disease-causing bacteria) to stop them invading and infecting the body's cells. Each B cell carries an individual B cell receptor (BCR) which determines which particles it can bind, rather like each lock accepts a different key. There are many millions of B cells with different receptors in the body. This immense diversity comes from rearranging the genes that code these receptors, so the receptor is slightly different in every B cell resulting in billions of possibilities of different harmful molecules that could be recognized. Intestinal microbes trigger expansion of these B cell populations and antibody production (Li et al. 2020). Li et al. (2020) studied the billions of genes that code the antibodies in a system that allows the responses to individual benign intestinal microbes to be understood (Li et al. 2020).

6.1 The Range of Available Antibodies Depends on Where Beneficial Microbes Are in the Body

The number of benign microbes living in our intestines is about the same as the number of cells in our body. Mostly these bacteria stay within the intestinal tube rather than penetrate the body tissues. Unfortunately, some penetration is unavoidable, because the intestine only has a single layer of cells that separate the inside of the tube from blood vessels that we need to absorb our food. Limenitakis used specially designed computer programs to process millions of genetic sequences that compare the antibody repertoire from B cells, depending on whether the microbes stay in the intestine, or whether they reach the bloodstream (Li et al. 2020). In both cases the antibody repertoire is altered, but in rather different ways depending on how the exposure occurs. Interestingly, this is rather predictable depending on the microbe concerned and where it is in the body, indicating that the intestinal microbes direct the development of our antibodies before we get a serious infection and this process is certainly not random (Ganal-Vonarburg et al. 2017).

There are different sorts of antibodies in the lining of the intestine (IgA) compared with the bloodstream (IgM and IgG). Using the powerful genetic analysis, the researchers showed that the range of different antibodies produced in the intestine was far less that those produced in central body tissues. This means that once microbes get into the body, the immune system has many more possibilities to neutralize and eliminate them, whereas antibodies in the intestine mainly just bind the bacterial molecules that they can see at any one time.

6.2 *How the Antibodies Change When the Body Is Exposed to Different Microbes*

Over their life-span mammals face a huge variety of different microbial challenges. It was therefore important to know how once the antibody repertoire could change once had been shaped by a microbe when something else came along. The research team answered this question by testing what happened with the same microbe at different sites or with two different microbes on after another.

Although intestinal microbes do not directly produce an especially wide range of different antibodies, they sensitize the central immune tissues to produce antibodies if the microbe gets into the bloodstream. When a second microbes comes along, the rather limited intestinal antibody response changes to accommodate this microbe (rather like changing the lock in one's door). This is different from what happens when microbes get into the blood stream to reach the central body tissues when a second set of antibodies is made without compromising the first response to the original microbes (like installing another lock, so the door can be opened with different keys). This shows that central body tissues have the capacity to remember a range of different microbial species and to avoid the dangers of sepsis. It also shows that different B cell immune strategies in different body compartments are important for maintenance of our peaceful existence with our microbial passengers (Li et al. 2020). Dr. Li comments that Our data show for the first time that not only the composition of our intestinal microbiota, but also the timing and sequence of exposure to certain members of the commensal microbiota, happening predominantly during the first waves of colonization during early life, have an outcome on the resulting B cell receptor repertoire and subsequent immunity to pathogens (Li et al. 2020) .

7 Conclusions

It has recently been found that if we are deficient in the 8 essential sugars, known as glyconutrients, the cells will eventually lack the communication system necessary to maintain good health. The so-called fresh fruits and vegetables we buy today have few glyconutrients (or nutritional value at all) because they are often grown in nutrient-deficient soil, picked before they ripen naturally, gassed, irradiated, artificially ripened, stored for days, weeks, or months, cooked, frozen, canned, processed, refined, pasteurized, genetically engineered, etc. Cooking and processing deplete glyconutrients further. Glycobiology has also found that beneficial bacteria in the colon breakdown polysaccharides to monosaccharides (glyconutrients). But the bacterial content of modern people is different from our ancestors and so this process is less efficient. While our food has been reducing in nutrients, autoimmune diseases, cardiovascular disease, cancer, diabetes, and chronic degenerative disease have been increasing alarmingly and have been occurring in younger age groups.

A growing mountain of evidence shows that all these diseases are caused by a single dietary deficiency: glyconutrients that are missing from our diet. Glyconutrients are not vitamins, minerals, amino acids, or enzymes, but are in a class of their own as nutritional supplements derived from terrestrial as well as marine plants. Glyconutritional supplements are formulated based on new understanding in the biochemistry of how the human body maintains health at the cellular level. Healthy cells lead to healthy tissue—healthy tissue leads to healthy organs—and healthy organs lead to healthy bodies. Every cell in our body—all 600 trillion of them—needs glyconutrients (Fig. 10.9).

8 Future Perspectives

Despite the relatively recent discovery of glyconutrients and their functions, medical doctors and the public are becoming increasingly aware of their importance in treating underlying causes of disease and in maintaining good health. As good as allopathic medicine is it simply has NO answer to the increasing incidence of autoimmune diseases, cancers, and degenerative diseases in Western societies. Glyconutrients will soon become a part of standard care by medical practitioners. A glyconutritional approach gets at the root cause rather than treating only the symptoms. Our body has a remarkable ability to heal itself, but especially as we grow older the effects of daily stress and lack of proper nutrition reduce our body's ability to maintain good health. We do not have to get sick or grow old faster than we need to. We all live in a hostile environment were staying healthy is a major challenge for everyone especially those of us who have had more time to expose our bodies to toxins and inadequate nutritional intakes. New discoveries in biochemistry and specifically in glycobiology, provide us with knowledge on how to slow down the aging process and how to maintain optimum health into our 70s, 80s and beyond. No matter what our age, the addition of glyconutrients into our health regime will support our body's incredible ability to heal, repair, regenerate, regulate and protect itself. Science has proven that our bodies use glyconutrients to prevent infections and diseases and slow the aging process. Many chronic diseases that develop late in life have been found to be influenced by earlier poor eating habits or poor nutritional intake. The earlier a balanced nutrition supplementation program is undertaken the

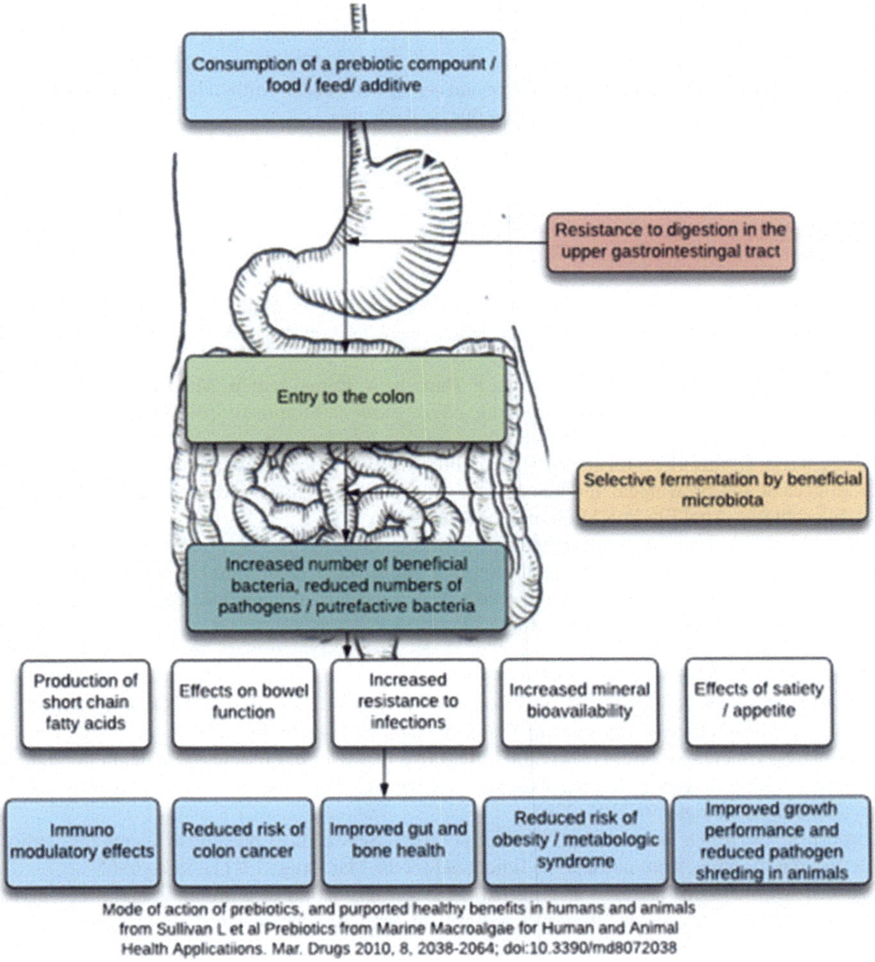

Fig. 10.9 Mode of action of prebiotics, and purported healthy in humans and animals (O'Sullivan et al. 2010)

greater the opportunity for prevention of the debilitating multi-diseases of aging. But even in later life when we are suffering the effects of earlier nutrition deficit and the debilitating effects of degenerative disease, the addition of nutritional supplements especially glyconutrients, can help to lessen the effects of diseases and improve the quality of life for people who are experiencing disease. This allows older people to maintain their independence for longer and shortening the recovery time from illnesses. Adopting a proactive approach to our wellness as we age, we find that we enter a beneficial recursive cycle. By taking nutritional supplements such as glyconutrients, we find we feel better—we have more energy and a greater sense of wellbeing. This leads us to want to be more physically active, which in turn enhances our positive attitude. Combined, these actions and attitudes lead to greatly improved wellness, which makes us feel better, and on, and on.

Acknowledgement Author express sincere thanks to Editors: *Gokare A. Ravishankar and Ambati Ranga Rao,* which upon the success of their volumes on Handbook of algal technologies and phytochemicals: Two Volume Set published in the year 2019 by CRC Press have ventured up with a specific volume on **Sustainable global resources of seaweeds: Industrial perspectives**. I am grateful for their invitation for contributing this book chapter.

References

Adan RA, van der Beek EM, Buitelaar JK, Cryan JF, Hebebrand J, Higgs S et al (2019) Nutritional psychiatry: towards improving mental health by what you eat. Eur Neuropsychopharmacol 29(12):1321–1332

Akbarzadeh S, Gholampour H, Farzadinia P, Daneshi A, Ramavandi B, Moazzeni A, Bargahi A (2018) Anti-diabetic effects of *Sargassum oligocystum* on Streptozotocin-induced diabetic rat. Iran J Basic Med Sci 21(3):342–346

An C, Kuda T, Yazaki T, Takahashi H, Kimura B (2013) FLX pyrosequencing analysis of the effects of the brownalgal fermentable polysaccharides alginate and laminarin on rat cecal Microbiotas. Appl Environ Microbiol 79(3):860–866

Barone M (2020) Exploring human gut microbiome variations across life: from eubiosis to dysbiosis in Western populations. Dissertation thesis, Alma Mater Studiorum Universita di Bologna

Bilal M, Iqbal HMN (2020) Marine seaweed polysaccharides-based engineered cues for the modern biomedical sector. Mar Drugs 18(1):7

Blaut M, Clavel T (2007) Metabolic diversity of the intestinal microbiota: implications for health and disease. J Nutr 137(Suppl):S751–S755

Brownlee IA, Allen A, Pearson JP, Dettmar PW, Havler ME, Atherton MR, Onsoyen E (2005) Alginate as a source of dietary fiber. Crit Rev Food Sci Nutr 45:497–510

Cani PD, Neyrinck AM, Fava F et al (2007) Selective increases of bifidobacteria in gut microflora improve high-fat-diet-induced diabetes in mice through a mechanism associated with endotoxaemia. Diabetologia 50:2374–2383

Cardoso S, Pereira O, Seca A, Pinto D, Silva A (2015) Seaweeds as preventive agents for cardiovascular diseases: from nutrients to functional foods. Mar Drugs 13(11):6838–6865

Chavarri M, Maranon I, Ares R et al (2010) Microencapsulation of a probiotic and prebiotic in alginate-chitosan capsules improves survival in simulated gastro-intestinal conditions. Int J Food Microbiol 142:185–189

Cooper R, Dragar C, Elliot K et al (2002) GFS, a preparation of Tasmanian *Undaria pinnatifida* is associated with healing and inhibition of reactivation of herpes. BMC Complement Altern Med 2:11

Deville C, Damas J, Forget P et al (2004) Laminarin in the dietary fibre concept. J Sci Food Agric 84:1030–1038

Deville C, Gharbi M, Dandrifosse G et al (2007) Study on the effects of laminarin, a polysaccharide from seaweed, on gut characteristics. J Sci Food Agric 87:1717–1725

Dierick N, Ovyn A, De Smet S (2009) Effect of feeding intact brown seaweed Ascophyllum nodosum on some digestive parameters and on iodine content in edible tissues in pigs. J Sci Food Agric 89:584–594

Ding WK, Shah NP (2009) Effect of various encapsulating materials on the stability of probiotic bacteria. J Food Sci 74:M100–M107

Ganal-Vonarburg SC, Fuhrer T, Gomez de Agüero M (2017) Maternal microbiota and antibodies as advocates of neonatal health. Gut Microbes 8(5):479–485

Hall AC, Fairclough AC, Mahadevan K et al (2012) *Ascophyllum nodosum* enriched bread reduces subsequent energy intake with no effect on post-prandial glucose and cholesterol in healthy, overweight males. A pilot studies. Appetite 58:379–386

Islam MA, Yun CH, Choi YJ et al (2010) Microencapsulation of live probiotic bacteria. J Microbiol Biotechnol 20:1367–1377

Kim EJ, Park SY, Lee JY, Park JH (2010) Fucoidan present in brown algae induces apoptosis of human colon cancer cells. BMC Gastroenterol 10:96

Kristensen M, Jensen MG (2011) Dietary fibres in the regulation of appetite and food intake. Importance of viscosity. Appetite 56:65–70

Kuda T, Yano T, Matsuda N, Nishizawa M (2005) Inhibitory effects of laminaran and low molecular alginate against the putrefactive compounds produced by intestinal microflora in vitro and in rats. Food Chem 91:745–749

Kulshreshtha G, Hincke MT, Prithiviraj B, Critchley A (2020) A review of the varied uses of macroalgae as dietary supplements in selected poultry with special reference to laying hen and broiler chickens. J Mar Sci Eng 8(7):536

Li H, Limenitakis JP, Greiff V et al (2020) Mucosal or systemic microbiota exposures shape the B cell repertoire. Nature 584(7820):274–278

Liu L, Heinrich M, Myers S et al (2012) Towards a better understanding of medicinal uses of the brown seaweed *Sargassum* in traditional Chinese medicine: a phytochemical and pharmacological review. J Ethnopharmacol 142:591–619

Lopez-Santamarina A, Miranda JM, Mondragon ADC, Lamas A, Cardelle-Cobas A, Franco CM, Cepeda A (2020) Potential use of marine seaweeds as prebiotics: a review. Molecules 25(4):1004

Mabeau S, Fleurence J (1993) Seaweed in food products: biochemical and nutritional aspects. Trends Food Sci Technol 4:103–107

Ondarza Beneitez MA (2012) Ficocoloides de Algas Rojas. Caracterización Bioquímica: Galactanos extraídos de la pared del Alga Gracilaria Verrucosa (HUDS) Papenfuss, cultivada bajo condiciones controladas (Spanish Edition) (Español) Tapa blanda—10 Junio 2012

O'Sullivan L, Murphy B, McLoughlin P et al (2010) Prebiotics from marine macroalgae for human and animal health applications. Mar Drugs 8:2038–2064

Penalver R, Lorenzo JM, Ros G, Amarowicz R, Pateiro M, Nieto G (2020) Seaweeds as a functional ingredient for a healthy diet. Mar Drugs 18(6):301

Praveen MA, Parvathy KK, Balasubramanian P, Jayabalan R (2019) An overview of extraction and purification techniques of seaweed dietary fibers for immunomodulation on gut microbiota. Trends Food Sci Technol 92:46–64

Ramnani P, Chitarrari R, Tuohy K et al (2012) In vitro fermentation and prebiotic potential of novel low molecular weight polysaccharides derived from agar and alginate seaweeds. Anaerobe 18:1–6

Rastall RA, Gibson GR, Gill HS et al (2005) Modulation of the microbial ecology of the human colon by probiotics, prebiotics and symbiotics to enhance human health: an overview of enabling science and potential applications. FEMS Microbiol Ecol 52:145–152

Reilly P, O'Doherty JV, Pierce KM, Callan JJ, O'Sullivan JT, Sweeney T (2008) The effects of seaweed extract inclusion on gut morphology, selected intestinal microbiota, nutrient digestibility, volatile fatty acid concentrations and the immune status of the weaned pig. Animal 2:1465–1473

Rowland I (1999) Optimal nutrition: fibre and phytochemicals. Proc Nutr Soc 58:415–419

Shang Q, Hao J, Chao C, Jiejie H, Li G, Yu G (2018) Gut microbiota fermentation of marine polysaccharides and its effects on intestinal ecology: an overview. Carbohydr Polym 179:173–185. ISSN 0144-8617

Teas J, Irhimeh MR (2012) Dietary algae and HIV/AIDS: proof of concept clinical data. J Appl Phycol 24:575–582

Trinchero J, Ponce NM, Cordoba OL et al (2009) Antiretroviral activity of fucoidans extracted from the brown seaweed *Adenocystis utricularis*. Phytother Res 23:707–712

Vaugelade P, Hoebler C, Bernard F et al (2000) Non-starch polysaccharides extracted from seaweed can modulate intestinal absorption of glucose and insulin response in the pig. Reprod Nutr Dev 40:33–47

Vijn S, Compart DP, Dutta N, Foukis A, Hess M, Hristov AN, Kalscheur KF, Kebreab E, Nuzhdin SV, Price NN, Sun Y, Tricarico JM, Turzillo A, Weisbjerg MR, Yarish C, Kurt TD (2020) Key considerations for the use of seaweed to reduce enteric methane emissions from cattle. Front Vet Sci 7:597430

Wang Y, Han F, Hu B, Li JB, Yu WG (2006) In vivo prebiotic properties of alginate oligosaccharides prepared through enzymatic hydrolysis of alginate. Nutr Res 26:597–603

Wang H, Ooi EV, Ang PO Jr (2008) Antiviral activities of extracts from Hong Kong seaweeds. J Zhejiang Univ Sci B 9:969–976

Wells ML, Potin P, Craigie JS, Raven JA, Merchant SS, Helliwell KE et al (2017) Algae as nutritional and functional food sources: revisiting our understanding. J Appl Phycol 29(2):949–982

Winberg P (2015) The contradictions of macroalgal applications; age, scale, and sophistication. Int Soc Appl Phycol Newslett 1:9–14

Chapter 11
Emerging Trends on the Integrated Extraction of Seaweed Proteins: Challenges and Opportunities

Tejal K. Gajaria and Vaibhav A. Mantri

Abbreviations

BCAA	Branched-chain amino acids
DW	Dry weight
EAA	Essential amino acids
FW	Fresh weight
IEA	International Energy Agency
LC-MS/MS	Liquid chromatography mass spectrometry
MALDI-TOF	Matrix assisted laser desorption ionization – Time of flight
NaOH	Sodium hydroxide
R-PC	R-phycocyanin
R-PE	R-phycoerythrin
TMG propionate	1,1,3,3-Tetramethylguanidine propionate
UAE	Ultrasound-assisted extraction

1 Introduction

A sustainable bioprocessing methodology is a key to an effective biomass utilisation followed by value addition to the nascent products in the developing markets. Downstream processing for proteins requires many cautions and manageable extraction schemes explicitly to support the finalized protein-based product yield and efficacy. The over-all process demands industrially viable and an environmentally supportive idea to thrive in the competitive markets. These efforts need alignment

T. K. Gajaria · V. A. Mantri (✉)
Academy of Scientific and Innovative Research, Ghaziabad, Uttar Pradesh, India

CSIR-Central Salt and Marine Chemicals Research Institute, Bhavnagar, Gujarat, India
e-mail: vaibhav@csmcri.res.in

A. Ranga Rao, G. A. Ravishankar (eds.), *Sustainable Global Resources of Seaweeds Volume 2*, https://doi.org/10.1007/978-3-030-92174-3_11

with the societal and community goals towards managing food scarcity and food security issues.

Extensive population rise and rapidly transforming climatic conditions call for a sustainable solution to food scarcity, malnutrition, and an affordable protein source globally. Protein sources like soya, egg and whey or poultry have become indispensable to our routine diet, fulfilling our daily protein demand. However, their feedstock supply chain is severely affected in the last few years due to many factors like the decreased freshwater resources for irrigation, floods, fertile soil erosion, and consequently, reduced yield and production levels. These challenges compelled exploration on potential alternatives like seaweeds without affecting health and fulfilling experience. Seaweeds considered as macroscopic aquatic photosynthetic organisms play a wide array of ecological roles including carbon sequestration. They are found mainly in three primary algal forms based on their pigment contents; Chlorophyta, Rhodophyta and Phaeophyta.

Seaweeds though popularly utilised raw in the form of sheets or salad sprinklers, their limited global usage could be due to fear of safety and chronic effects on human metabolic processes. However, their protein contents and amino acid profiles are significantly comparable to the traditional sources utilised so far, adding to their rapid growth rates (specifically *Ulva* spp.) that surpasses the over-all production yield of proteins compared to soya (Mata et al. 2016). The faster growth and higher productivity rates are additional advantages towards environment friendly aquaculture practices wherein there is no need for freshwater or land for the cultivation. This makes seaweeds a preferred choice over several alternative protein sources.

The definition of biorefinery involves integrating various biochemical processes to produce energy and other value-added products channelized through a single extraction process (Bikker et al. 2016). The concept was considered sustainable due to its efficient extraction of multi-products, leading to significantly reduced waste disposal and maximal biomass utilisation, reducing significant environmental burden and ecological disturbance. The International Energy Agency (IEA) announced the International Energy Agency (IEA) Bioenergy task 42 (IEA 2007) for accelerating biorefinery-based studies that are highly efficient with zero-waste leading towards a synergistic production of energy and commodity products. As against plant-based biorefineries (1G, 2G biofuels), the seaweed-based biorefinery approaches find greater appreciation due to their lack of food versus fuel conflict (Balina et al. 2017). To realise these benefits, bioprocessing of raw materials to provide the final products need to be efficient.

2 Marine Macroalgal Protein Biochemistry and Amino Acid Profile

Proteins are the fundamental nutritional requirement to all living beings to sustain their basic physiological functions, and their deficiency leads to many chronic ailments in humans. Hence, an alternative to proteins in routine dietary plan demands a thorough analysis before its utility in terms of the content of essential amino acids, total bioavailability and bioaccessibility.

2.1 Protein Biochemistry

Proteins made of a series of chemically diverse amino acids that contribute to their ultimate structure and properties. The charge distribution provides a specific iso-electric point directly related to proteins' solvation into the required media. Seaweed proteins show distinct solubility in buffers and media as reported by Galland-Irmouli et al. (1999). The proteins from red alga *Palmaria palmata* were obtained by aqueous extraction combined with sonication (Galland-Irmouli et al. 1999). An additional alkaline extraction significantly enhanced yield from 17.21 ± 0.03 to 35.87 ± 0.36 mg/g DW (Harnedy et al. 2014). Furthermore, alkaline extraction resulted in a 57% recovery from *Ascophyllum nodosum*; a yield reportedly the best protein/algae liquefaction ratio of 1.28 (Kadam et al. 2017). However, *Pyropia umbilicalis* and *Saccharina latissima* yielded 22.6 ± 7.3% and 25.1 ± 0.9% respectively along with enrichment of fatty acids (Harrysson et al. 2018). Furthermore, the extraction was found significantly effective to protein yield (6.28 ± 0.07% for aqueous extraction compared to 15.29 ± 0.64% DW for alkaline extraction) of *Ulva ohnoi* in addition to the enrichment of essential amino acids (39.07 ± 0.32% in aqueous extraction compared to 50.85 ± 0.87% DW in alkaline extraction) (Angell et al. 2017). The efficacy for extracting proteins at alkaline pH was also documented for *Kappaphycus alvarezii* and *Sargassum* spp. Along with functional properties i.e. digestibility (Wong and Cheung 2001a), emulsifying activity, water holding capacity, and foaming capacity (Suresh Kumar et al. 2014). However, a recent study by O'Connor et al. (2020) showed varying effects of extraction methodology on protein yields from *Fucus vesiculosus, Alaria esculenta, Palmaria palmata and Chondrus crispus.* Wherein the classical method yielded highest protein content in *F. vesiculosus* (35.1%), *A. esculenta* (18.2%) and *C. crispus* (35.2%) and *P. palmata* (21.5%) using autoclave processing method followed by significant variations in the respective amino acid profiles. The results indicated pre-feasibility analysis of extraction strategies based on the biochemical compositions of algae for extension to pilot study to obtain maximized extraction efficiencies (O'Connor et al. 2020).

2.2 Branched-Chain Amino Acids (BCAA)

Edible applications often require an enriched amount of essential amino acids; the importance of branched-chain amino acids is significantly higher for the applications intended for sports utility. Table 11.1 shows a comprehensive account of

Table 11.1 List of essential amino acid (EAA) and branched chain amino acid (BCAA) contents (mg/g DW) from the popular edible seaweeds

SN	Biomass	Product form	Extraction method	Estimation method	Yield	Co-products	Reference
1.	*Laminaria digitata*	Protein rich fraction	Enzymatic hydrolysis and fermentation	Total nitrogen	2.4 fold enrichment in residual biomass	Ethanol	Hou et al. (2015)
2.	*Ulva fasciata*	Protein concentrate	Thermo-alkaline treatment	CHNS elemental analysis	11% DW	Sap, lipids, ulvan and cellulose	Gajaria et al. (2017)
3.	*Ulva lactuca*	Protein rich fraction	High shear homogenisation	Spectrophotometric detection (DC protein assay)	39.0 ± 6.2 % DW	Carbohydrates	Postma et al. (2018)
4.	*Porphyra umbilicalis*	Protein rich fraction	Aqueous-alkaline extraction	Total nitrogen	7.5 ± 2.4% DW	Carrageenan, pectin, cellulose	Wahlström et al. (2018)
5.	*Ulva ohnoi*	Protein enriched biomass	Microwave assisted extraction	Total nitrogen	23.9 ± 0.9 (relative to original biomass)	Salts, ulvan	Magnusson et al. (2019)
6.	*Ulva ohnoi*	Protein rich extract	High-voltage pulsed electric field followed by alkaline extraction	Spectrophotometric detection (Lowry's assay)	1.26 ± 0.29% (15% yield relative to original biomass)	Starch, salts	Prabhu et al. (2019)
7.	*Sargassum vulgare*	Protein concentrate	Alkaline treatment	CHNS elemental analysis	2.53 ± 0.2% DW biomass	Sap, alginic acid, salts	Baghel et al. (2020)
8.	*Eucheuma denticulatum*	Protein precipitate	Enzyme assisted extraction and alkaline extraction	Protein analyser	59.4 ± 1.41* extraction efficiency	Carrageenan	Naseri et al. (2020)
9.	*Ulva* sp.	Protein isolate	Supercritical water extraction	Spectrophotometric detection (Lowry's assay)	5.8% DW	Ethanol and hydrochar	Polikovsky et al. (2020)

* Denotes the resultant yield obtained is due to the combined treatment of enzymes and alkaline extraction

BCAA from various seaweeds. The amino acid contents of seaweeds are now well established. Added with the importance of essential amino acids and branched-chain amino acids, they present an enormous potential of nitrogen source to health and wellness.

2.3 Protein Digestibility

The bioavailability and bioaccessibility of proteins are directly proportional to their digestibility as it renders proteins available for intestinal absorption hence their functional validation. The traditional inclusion of seaweeds in the diet did not show any adverse effect on the metabolism. Earliest reports on *ex-vivo* digestion suggest >80% digestibility of proteins from *Porphyra* sp. and *Enteromorpha* sp., while in the case of Sargassum *fuvellum, Undaria pinnatifida* and *Hizikia fusiforme* it varied between 70 and 80% (Ryu et al. 1982). The study also showed an additive influence of heat over protein digestibility. Protein digestibility was reported higher for red seaweeds (>84%; highest for *Sarcodiotheca gaudichaudii* 86.7 ± 0.21%) compared to brown seaweeds (78–82%; highest for *Saccharina latissima* 82.0 ± 0.78%) under the varying treatments (Tibbetts et al. 2016). Similar trends were reported for selective red and brown seaweeds in terms of digestible and unavailable proteins. The results showed significantly lower digestibility and higher unavailable proteins for *U. pinnatifida, L. digitata* and *F. vesiculosus* (4.23%, 1.55%, 1.21% digestible proteins and 2.51%, 2.09%, 2.21% unavailable protein respectively) compared to *C. crispus* and *P. tenera* (7.27%, 17.27% digestible proteins and 2.21%, 0.53% unavailable protein respectively) (Goñi et al. 2002).

3 Integrated Extraction and Its Implications on Protein Chemistry

The concept of biorefinery and integrated extraction of multiple products offers several advantages over the single product extraction from algal biomass, specifically from the carbon footprint scale. The waste flow and leftovers generate enormous amounts of effluents representing a potential challenge towards developing a sustainable economy and production chain. Unique biochemistry of proteins and other value-added macromolecules present various challenges while developing a multi-step extraction process; specifically, profound structure-function relationships are involved. The major products of the Seaweed-based industry are hydrocolloids, biostimulants and edible products, and due care needs to be taken to preserve their chemical stability and functional activities.

Proteins show strong structure-function relationships that are significantly influenced by the extraction process. Cell-wall polysaccharides and storage polysaccharides represent the first-line of interference during seaweed-based protein extraction

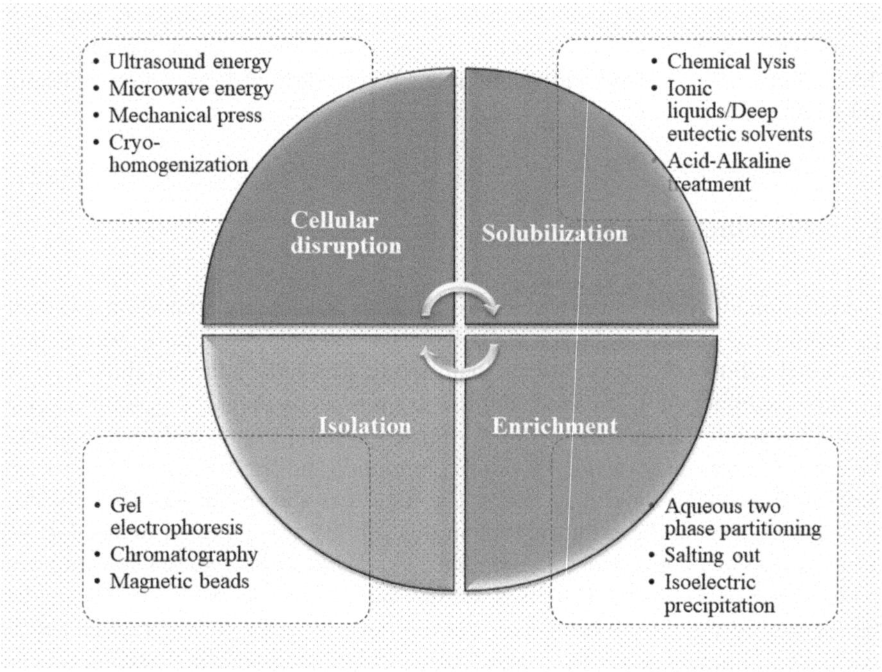

Fig. 11.1 Overview of seaweed protein extraction methodologies

whereas the protein-binding secondary metabolites remains a major challenge while precipitating total proteins aimed for gel-electrophoresis or the advanced molecular biology based studies like LC-MS/MS and MALDI-TOF analysis. Several combinations of chemical and physical techniques have been employed and analyzed to determine optimum extraction conditions concerning their yield and structural characteristics (Fig. 11.1). However, commercial product-oriented bioprocessing still faces challenges due to the complex chemistry of proteins. Such extracts are generally referred to as protein concentrates, hydrolysates, protein-enriched fractions or nitrogen-rich fractions that are based on either total nitrogen content or amino acid enrichment that renders the over-all presence of proteinaceous compounds (Table 11.2).

3.1 Role of Physical Methods During the Integrated Extraction of Proteins

Seaweed biomasses are highly resilient due to the presence of complex cell wall polysaccharides, and hence the complete physical disintegration remains a preliminary step of every process. Mechanical grinding, particle size, drying and dewatering are all known to influence the end product characteristics. There is a prior need to determine the pre-processing treatments depending on the type of seaweed and the chemical nature of the target compound being extracted.

Table 11.2 List of integrated protein extraction using biorefinery from various macroalgae (mg/gm DW biomass)

	EAA					BCAA			
	Phenylalanine	Methionine	Histidine	Lysine	Threonine	Isoleucine	Leucine	Valine	Reference
*Ulva intestinalis**	8.41 ± 0.55	2.90 ± 0.08	1.22 ± 0.02	3.90 ± 0.23	5.61 ± 0.14	3.66 ± 0.12	8.27 ± 0.20	7.25 ± 0.10	Jannat-Alipour et al. (2019)
Caulerpa lentillifera	19.9 ± 1.4	1.5 ± 0.08	1.4 ± 0.13	1.2 ± 0.05	5.8 ± 0.22	5.0 ± 0.12	7.79 ± 0.1	6.1 ± 0.02	Matanjun et al. (2009)
Eucheuma cottonii	19.0 ± 2.48	0.83 ± 0.1	0.25 ± 0.1	1.45 ± 0.4	2.0 ± 0.01	2.4 ± 0.04	3.3 ± 0.06	2.6 ± 0.07	Matanjun et al. (2009)
Laminaria digitata	2.8 ± 0.07	1.4 ± 0.11	2.3 ± 0.14	4.7 ± 0.23	3.4 ± 0.46	2.6 ± 0.35	4.4 ± 0.09	6.0 ± 0.76	Kolb et al. (2004)
Porphyra umbilicalis	8.56 ± 0.07	2.40 ± 0.04	1.66 ± 0.05	11.58 ± 0.19	12.13 ± 0.09	8.31 ± 0.07	16.12 ± 0.15	12.79 ± 0.09	Machado et al. (2020)
Undaria pinnatifida	7.8 ± 0.17	3.5 ± 0.42	5.2 ± 0.37	11.1 ± 1.07	7.3 ± 0.57	7.9 ± 0.18	13.7 ± 1.36	16.8 ± 0.32	Kolb et al. (2004)
Himanthalia elongate	2.28 ± 0.08	1.96 ± 0.03	2.01 ± 0.11	3.23 ± 0.05	3.25 ± 0.04	2.28 ± 0.07	2.05 ± 0.07	4.28 ± 0.18	Garcia-Vaquero et al. (2017)
Fucus vesiculosus	5.4 ± 0.2	2.1 ± 0.2	1.9 ± 0.8	8.0 ± 0.7	6.1 ± 0.3	5.0 ± 0.3	8.6 ± 0.5	5.8 ± 0.3	Lorenzo et al. (2017)
Alaria esculenta	5.2 ± 0.3	2.6 ± 0.4	2.8 ± 0.4	9.2 ± 1.1	5.3 ± 0.7	4.2 ± 0.6	8.1 ± 1.2	5.9 ± 0.4	Maehre et al. (2016a)
Saccharina latissima[a]	11.2 ± 0.7	2.9 ± 0.5	1.0 ± 0.09	5.3 ± 0.09	2.6 ± 0.2	12.0 ± 0.1	10.2 ± 0.1	2.5 ± 0.04	Marinho et al. (2015)
Hypnea charoides[a]	42.2	16.2	7.67	39.2	48.3	39.2	69.8	52.1	Wong and Cheung (2001b)
Palmaria palmata	8.7	4.0	2.6	11.0	8.9	8.2	14.3	12.4	Maehre et al. (2016b)
Egg albumin*	5.8	3.2	2.4	5.3	5.0	6.6	8.8	7.2	Becker (2007)
Soyabean*	5.0	1.3	2.6	6.4	4.0	5.3	7.7	5.3	
WHO recommendation*	6.0	3.5	–	5.5	–	4.0	7.0	5.0	

*Denotes g/100 gm protein
[a]Denotes amino acid content at mg/g DW biomass at ~10% and ~18% protein content of alga for Saccharina latissima and Hypnea charoides, respectively

3.1.1 Ultrasound-Assisted Extraction (UAE)

Ultrasound has been widely used for the extraction of various biomolecules from seaweeds. It also represents a promising green technology for the extraction of polyphenols, flavonoids and other essential compounds including proteins. This technique is based on the working principle of acoustic cavitation phenomenon that converts the energy of a sound wave to mechanical energy, thereby leading to the disintegration of biomass. The method has proven to be promising for the efficient extraction of phycobiliproteins from *G. pusillum*. Maximal extraction was achieved when combined with maceration of biomass before the treatment, i.e. 77 and 93% of R-PE and R-PC, respectively (Mittal et al. 2017). Similar enhancements were reported wherein 31% higher yield was obtained for R-PE from *F. lumbricali* (Saluri et al. 2019).

In contrast, UAE alone did not give higher yields for R-PE than maceration in *G. gracilis, i.e.* 3.58–4.15 mg/g for maceration v/s 1.60–1.88 mg/g for UAE (Pereira et al. 2020). Moreover, time-dependent increment in the yield was observed for R-PE extraction from *K. alvarezii* with the highest yield of 1.1 mg/mL and purity index of 0.2 (Uju et al. 2020). However, phycobiliproteins are the major proteins in red algae. UAE has proven to be equally efficient on a pilot-scale for protein extraction from *Ulva* sp. and *Gracilaria* sp. The protein concentrates obtained showed significant reducing activity with the recovery efficiency of 70% and 86% respectively.

Moreover, the process becomes more efficient when combined with enrichment processes such as ammonium sulfate mediated precipitation reported for *F. vesiculosus* and *C. Crispus* with 35.1% and 35.5% (w/w) protein yield, respectively (O'Connor et al. 2020). The efficient tissue disruption, reduced processing times and temperature control are the key advantages of ultrasound-assisted extraction processes, minimising protein denaturation and degradation during the multiple product extraction steps. It also reduces chemical inputs to the process, which is essential to make a process economically viable. However, challenges remain for integrated extraction of proteins from biomass at a larger scale. This is so because the technique leads to global tissue disruption which can account for co-extraction of sugar monomers and polysaccharides (Youssouf et al. 2017), pigments (Zhu et al. 2017) and other bioactives (Rodrigues et al. 2015; Ummat et al. 2020), demanding a systemic approach for the industrial-scale applications to extrapolate the scope of the method further.

3.1.2 Pulse Electric-Field Assisted Extraction

This technique is one of a unique innovation of the twenty-first century in the area of tissue disruption and extraction methods. The method has been tested on diverse biological samples such as microalgae, plants, microorganisms, mushrooms, vegetables, abalone, and the hard materials such as cocoa bean shell (Toepfl and Heinz 2011; Vorobiev and Lebovka 2011; Zeng and Zhang 2019). The preliminary study using green seaweed *Ulva* sp. showed a significant increase in the yield of proteins compared to control groups; 59.13 ± 3.82 μg mL^{-1} v/s 23.80 ± 1.33 μg mL^{-1} respectively with an over-all system temperature of 35.50 ± 2.02 °C (Polikovsky et al. 2016). Moreover, combined protein and carbohydrate extraction was performed from *Ulva lactuca*, producing 12% and 15% DW. These are significantly lesser than enzyme assisted extraction and high shear homogenization 39–25% DW, 51–28% DW, respectively (Postma et al. 2018). The core advantage lies in the greenness of method; no chemical inputs are required hence considerably reducing carbon footprint and energy consumption of the overall process. However, being a non-targeted method, i.e. the electric pulses disintegrate biomass leading to the liberation of co-products; specific extraction would require bioinformatic inputs such as response surface methodology or artificial neural networking to gain maximum separation. In addition to the protein concentrates (with yield <5% DW biomass) with seven to eightfold higher antioxidant activity from *Ulva* sp. (Robin et al. 2018), an advancement was reported with simultaneous extraction of proteins, starch and ash from *Ulva ohnoi* using a combination of pulse-field electroporation and mechanical press (Prabhu et al. 2019). The separation was achieved by collecting salts and proteins in soluble fraction after treatment, while starch was obtained from the residual biomass using further down-stream processing. The total protein extraction achieved 14.9% compared to the original biomass, i.e. 1.26 ± 0.29% DW biomass out of the 8.41 ± 0.11% DW of the original biomass. Authors also reported that this method achieved extraction of specific molecular weight proteins with the presence of polyphenols and sulfated polysaccharides that might serve as an advantage in nutritional perspectives (Prabhu et al. 2019).

3.1.3 Microwave-Assisted Extraction

Microwave-assisted extraction has been proven efficient on various tissue types and environmental samples. Seaweed protein extraction studies are yet to take the lead. However, limited studies are reported till date for either dedicated or integrated protein extraction from seaweeds using microwave energy (Magnusson et al. 2019). Some studies that have emerged so far include the extraction of seaweed-based bioactives such as bio-oil (Kostas et al. 2019), biostimulant (Michalak et al. 2015), polyphenols (Magnusson et al. 2017) and polysaccharides (Peñuela et al. 2018) and the recent ones on biofuel (Yuan and Macquarrie 2015; Remón et al. 2020). The technique is rapid, and the scale-up operations could be energy efficient as they require less solvent and a relatively decreased biomass/volume ratio. Furthermore, the major challenge for adopting this method is high temperatures which may completely disrupt protein structure, leading to the partial or complete loss of the protein activity.

3.2 Role of Chemical and Biochemical Methods During the Integrated Extraction of Proteins

3.2.1 Alkaline Treatment

Presence of sulfated heteropolymers in the cell-wall is among the critical challenges during protein extraction from seaweed biomass. The alkaline treatment has been a preliminary chemical disruption for algal biomass due to its robustness in depolymerising complex matrix for efficient protein extraction (Fleurence et al. 1995a). A lab-scale protein extraction using an integrated approach was reported for *U. fasciata* with 68.75 ± 4.01% efficiency (yield of 11 ± 2.12% DW as a protein pellet). The approach also reported 85.86 ± 5.92% *in-vitro* digestibility and minimum content of heavy metals, an emerging threat to algal-based edible products due to the inherent bioaccumulation characteristics (Gajaria et al. 2017). Moreover, a combination of alkali and enzymatic approach provided the highest protein extraction from E. denticulatum with 35.5 ± 2.12% yield and integrated recovery of carrageenan (Naseri et al. 2020). In an integrated process using *S. muticum*, alginate, mineral-rich sap, proteins, cellulose and salts were recovered as products. The procedure demonstrated complete utilization of biomass into commodity products and conversion of waste discharge into salts by solar evaporation (Baghel et al. 2020). As discussed in Sect. 2.1, alkaline treatment is a rapid tissue disruption and extraction method for complex macromolecules such as carbohydrates and proteins. However, a prior optimisation in terms of duration of treatment and concentration of alkali is needed to avoid excessive incubation of samples with alkali as, it may cause denaturation, non-specific linkages between hydrolysed peptides, and formation of modified amino acids that may pose a serious health risk.

3.2.2 Ionic Liquid/Deep Eutectic Solvents Mediated Extraction

These class of chemicals have brought a revolutionary change in the bioprocessing of materials. The "green" nature of these solvents makes them potential candidates for an integrated extraction of bioproducts from a variety of terrestrial as well as marine biomasses (Sequeira et al. 2020). The major challenge of integrated protein extraction lies with the stability of protein structure. Ionic liquids possess high solvation potential together with electrical conductivity and electrochemical stability. Hence, they are considered the preferred choice of protein solubilization from complex matrices such as seaweeds. Several efforts are reported for molecular dissolution from terrestrial biomass. The technology remains still in its nascent stage for its application in marine biomass. However, a recent study has shown effective solubilisation of carbohydrates (67 wt%, with 92 wt% recovery) from *Ulva rigida* biomass using 1,1,3,3-tetramethylguanidine (TMG) propionate (Pezoa-Conte et al. 2015). The study reported successful disruption of algal biomass into ash, proteins and carbohydrates/sugars and, 99.7 wt% removal of chlorophyll with the mass balance analysis. However, proteins least extracted during [$TMGH^+$][$EtCO2^-$] treatment rendered carbohydrates-specific molecular interaction. This presents with an opportunity for an integrated approach towards selective interaction between macromolecules to avoid cross-contamination between products to improve quality.

3.2.3 Enzyme Assisted Extraction

Enzymes are potential candidates in green technologies for integrated biorefinery approaches (Zollmann et al. 2019). They have been found efficient in extracting proteins and sugars from seaweed biomass for diverse applications. The significant advantages of enzyme-assisted extraction seem to lie in their recyclability and specificity that are the prime challenges while applying an integrated multi-product biorefinery protocol. Moreover, the enzymatic treatments target complex polysaccharides to bring about the release of proteins into suspension such as cellulase targeting cellulose (Vásquez et al. 2019), agarase targeting agar (Fleurence et al., 1995b), carrageenase targeting carrageenan (Denis et al. 2009) etc. Also, the protein-targeted enzymes are used to obtain peptic digest to assess such fractions' bioactivities for a medicinal purpose (Admassu et al. 2018). Integrated extraction could be achieved by digesting carbohydrate polymers to obtain monosaccharides for fermentation into biofuels.

In contrast, the proteins could be extracted under mild conditions with suitable buffers due to weakened cell-wall and depolymerised matrix carbohydrates. Sequential extractions have been performed to obtain phycoerythrin and agar from *Gracilaria verrucosa* and *G. lemaneiformis* without compromising agar yield and antioxidant potency (Mensi et al. 2012; Mensi 2019). Enzymatic treatments are also reported in post-agar extraction from *G. fisheri* to obtain protein hydrolysate as a source of amino-acids with umami compounds for food recipes (Laohakunjit et al. 2014). The treatment was found highly effective as a pre-treatment for protein extraction when combined with alkali treatment in *P. palmata* that increased extraction yield from 39.1% (using 0.1 M NaOH alone) to 75.6% (by using 0.1 M NaOH in addition to cellulase and xylanase) (Maehre et al. 2016b). In case of phaeophyceae; enzymatic extraction of proteins has been highly effective for *Macrocystis pyrifera* with 74.6 ± 21.3% yield (Vásquez et al. 2019). However, till date no integrated studies are reported for combined recovery of carbohydrates and proteins using enzymatic hydrolysis in this commercially valuable alga.

Furthermore, In *Ulva ohnoi*, salts and ulvan was sequentially extracted and the protein-enriched biomass residue was obtained with 23.6 ± 0.8% yield (Magnusson et al. 2019). Similarly, for *Eucheuma denticulatum,* improved protein yield and carrageenan extraction was achieved using a multi-extraction process. The methodology combined alkaline extraction with enzymatic treatment to achieve desirable gel characteristics of carrageenan and protein pellet (35.5 ± 2.12% yield) enriched with the essential amino acids comparable with meat and whey proteins (Naseri et al. 2020).

4 Concluding Remarks and Future Prospects

Seaweeds have been identified as an emerging source of botanical proteins with various functionalities. This has inspired their systemic exploration ranging from bioavailability to their nutritional potentials. Seaweeds have found identification as promising candidates for animal feed/food supplements, pharmaceuticals and cosmetic applications due to their unique content of sulfated polysaccharides and bioactive peptides. However, the structure of such bioactive molecules awaits full elucidation, and in this context, technological advances in high-resolution instruments and bioinformatics would help decipher with more pronounced reliability.

Marine macroalgal biorefineries' unique potential resides in their various bioproducts like hydrocolloids and proteins with a combined potential of CO_2 sequestration via open sea or offshore cultivation. However, it also involves critical steps to maintain the structure-function relationship among the extracted multi-products. Both cultivation and bioprocessing are still facing technological sustainability challenges including seasonal and geospatial variations in seaweed composition, dewatering of biomass, and modulating large amounts of ash. Hence, significant interventions to reach sustainable economic growth and market development are necessary for biochemical conversion and product formulation with consistency in feedstock's chemical composition. Subject to the realization of novel and advanced extraction methodologies, seaweed stands as one of the best promising feedstock for a sustainable blue economy.

Acknowledgements Authors duly acknowledge the support and encouragement from the Director, CSIR-CSMCRI. Authors thankfully acknowledge Dr. C.R.K. Reddy for offering comments on the first draft of the manuscript. TKG duly acknowledge Council of Scientific and Industrial Research (CSIR-India) for the Senior Research Fellowship grant (Grant no.: 31/28(0242)2018-EMR-I). This manuscript has CSIR-CSMCRI PRIS registration number 51/2021.

References

Admassu H, Gasmalla MAA, Yang R, Zhao W (2018) Bioactive peptides derived from seaweed protein and their health benefits: antihypertensive, antioxidant, and antidiabetic properties. J Food Sci 83:6–16

Angell AR, Paul NA, de Nys R (2017) A comparison of protocols for isolating and concentrating protein from the green seaweed Ulva ohnoi. J Appl Phycol 29:1011–1026. https://doi.org/10.1007/s10811-016-0972-7

Baghel RS, Suthar P, Gajaria TK et al (2020) Seaweed biorefinery: a sustainable process for valorising the biomass of brown seaweed. J Clean Prod 263:121359. https://doi.org/10.1016/j.jclepro.2020.121359

Balina K, Romagnoli F, Blumberga D (2017) Seaweed biorefinery concept for sustainable use of marine resources. In: Energy procedia. Elsevier, Amsterdam, pp 504–511

Becker EW (2007) Micro-algae as a source of protein. Biotechnol Adv 25:207–210

Bikker P, van Krimpen MM, van Wikselaar P et al (2016) Biorefinery of the green seaweed Ulva lactuca to produce animal feed, chemicals and biofuels. J Appl Phycol 28(6):3511–3525. https://doi.org/10.1007/s10811-016-0842-3

Denis C, Morancais M, Gaudin P, Fleurence J (2009) Effect of enzymatic digestion on thallus degradation and extraction of hydrosoluble compounds from Grateloupia turuturu. Bot Mar 52:262–267. https://doi.org/10.1515/BOT.2009.035

Fleurence J, Le Coeur C, Mabeau S et al (1995a) Comparison of different extractive procedures for proteins from the edible seaweeds Ulva rigida and Ulva rotundata. J Appl Phycol 7:577–582. https://doi.org/10.1007/BF00003945

Fleurence J, Massiani L, Guyader O, Mabeau S (1995b) Use of enzymatic cell wall degradation for improvement of protein extraction from Chondrus crispus, Gracilaria verrucosa and Palmaria palmata. J Appl Phycol 7:393–397. https://doi.org/10.1007/BF00003796

Gajaria TK, Suthar P, Baghel RS et al (2017) Integration of protein extraction with a stream of byproducts from marine macroalgae: a model forms the basis for marine bioeconomy. Bioresour Technol 243:867–873. https://doi.org/10.1016/j.biortech.2017.06.149

Galland-Irmouli AV, Fleurence J, Lamghari R et al (1999) Nutritional value of proteins from edible seaweed Palmaria palmata (Dulse). J Nutr Biochem 10:353–359. https://doi.org/10.1016/S0955-2863(99)00014-5

Garcia-Vaquero M, Lopez-Alonso M, Hayes M (2017) Assessment of the functional properties of protein extracted from the brown seaweed Himanthalia elongata (Linnaeus) S. F. Gray. Food Res Int 99:971–978. https://doi.org/10.1016/j.foodres.2016.06.023

Goñi I, Gudiel-Urbano M, Saura-Calixto F (2002) In vitro determination of digestible and unavailable protein in edible seaweeds. J Sci Food Agric 82:1850–1854. https://doi.org/10.1002/jsfa.1270

Harnedy PA, Soler-Vila A, Edwards MD, FitzGerald RJ (2014) The effect of time and origin of harvest on the in vitro biological activity of Palmaria palmata protein hydrolysates. Food Res Int 62:746–752. https://doi.org/10.1016/j.foodres.2014.04.035

Harrysson H, Hayes M, Eimer F et al (2018) Production of protein extracts from Swedish red, green, and brown seaweeds, Porphyra umbilicalis Kützing, Ulva lactuca Linnaeus, and Saccharina latissima (Linnaeus) J. V. Lamouroux using three different methods. J Appl Phycol 30:3565–3580. https://doi.org/10.1007/s10811-018-1481-7

Hou X, Hansen JH, Bjerre AB (2015) Integrated bioethanol and protein production from brown seaweed Laminaria digitata. Bioresour Technol 197:310–317. https://doi.org/10.1016/j.biortech.2015.08.091

IEA (2007) IEA bioenergy Task 42 on biorefineries: co-production of fuels, chemicals, power and materials from biomass. In: Minutes Third Task Meet

Jannat-Alipour H, Rezaei M, Shabanpour B, Tabarsa M (2019) Edible green seaweed, Ulva intestinalis as an ingredient in surimi-based product: chemical composition and physicochemical properties. J Appl Phycol 31:2529–2539. https://doi.org/10.1007/s10811-019-1744-y

Kadam SU, Álvarez C, Tiwari BK, O'Donnell CP (2017) Extraction and characterisation of protein from Irish brown seaweed Ascophyllum nodosum. Food Res Int 99:1021–1027. https://doi.org/10.1016/j.foodres.2016.07.018

Kolb N, Vallorani L, Milanović N, Stocchi V (2004) Evaluation of marine algae Wakame (Undaria pinnatifida) and Kombu (Laminaria digitata japonica) as food supplements. Food Technol Biotechnol 42:57–61

Kostas ET, Williams OSA, Duran-Jimenez G et al (2019) Microwave pyrolysis of Laminaria digitata to produce unique seaweed-derived bio-oils. Biomass Bioenergy 125:41–49. https://doi.org/10.1016/j.biombioe.2019.04.006

Laohakunjit N, Selamassakul O, Kerdchoechuen O (2014) Seafood-like flavour obtained from the enzymatic hydrolysis of the protein by-products of seaweed (Gracilaria sp.). Food Chem 158:162–170. https://doi.org/10.1016/j.foodchem.2014.02.101

Lorenzo JM, Agregán R, Munekata PES et al (2017) Proximate composition and nutritional value of three macroalgae: Ascophyllum nodosum, Fucus vesiculosus and Bifurcaria bifurcata. Mar Drugs 15(11):360. https://doi.org/10.3390/md15110360

Machado M, Machado S, Pimentel FB et al (2020) Amino acid profile and protein quality assessment of macroalgae produced in an integrated multi-trophic aquaculture system. Foods 9:1382. https://doi.org/10.3390/foods9101382

Maehre HK, Edvinsen GK, Eilertsen KE, Elvevoll EO (2016a) Heat treatment increases the protein bioaccessibility in the red seaweed dulse (Palmaria palmata), but not in the brown seaweed winged kelp (Alaria esculenta). J Appl Phycol 28:581–590. https://doi.org/10.1007/s10811-015-0587-4

Mæhre HK, Jensen IJ, Eilertsen KE (2016b) Enzymatic pre-treatment increases the protein bioaccessibility and extractability in dulse (palmaria palmata). Mar Drugs 14. https://doi.org/10.3390/md14110196

Magnusson M, Yuen AKL, Zhang R et al (2017) A comparative assessment of microwave assisted (MAE) and conventional solid-liquid (SLE) techniques for the extraction of phloroglucinol from brown seaweed. Algal Res 23:28–36. https://doi.org/10.1016/j.algal.2017.01.002

Magnusson M, Glasson CRK, Vucko MJ et al (2019) Enrichment processes for the production of high-protein feed from the green seaweed Ulva ohnoi. Algal Res 41:101555. https://doi.org/10.1016/j.algal.2019.101555

Marinho GS, Holdt SL, Angelidaki I (2015) Seasonal variations in the amino acid profile and protein nutritional value of Saccharina latissima cultivated in a commercial IMTA system. J Appl Phycol 27:1991–2000. https://doi.org/10.1007/s10811-015-0546-0

Mata L, Magnusson M, Paul NA, de Nys R (2016) The intensive land-based production of the green seaweeds Derbesia tenuissima and Ulva ohnoi: biomass and bioproducts. J Appl Phycol 28:365–375. https://doi.org/10.1007/s10811-015-0561-1

Matanjun P, Mohamed S, Mustapha NM, Muhammad K (2009) Nutrient content of tropical edible seaweeds, Eucheuma cottonii, Caulerpa lentillifera and Sargassum polycystum. J Appl Phycol 21:75–80. https://doi.org/10.1007/s10811-008-9326-4

Mensi F (2019) Agar yield from R-phycoerythrin extraction by-product of the red alga Gracilaria verrucosa. J Appl Phycol 31:741–751. https://doi.org/10.1007/s10811-018-1533-z

Mensi F, Ksouri J, Seale E et al (2012) A statistical approach for optimisation of R-phycoerythrin extraction from the red algae Gracilaria verrucosa by enzymatic hydrolysis using central composite design and desirability function. J Appl Phycol 24:915–926. https://doi.org/10.1007/s10811-011-9712-1

Michalak I, Tuhy Ł, Chojnacka K (2015) Seaweed extract by microwave assisted extraction as plant growth biostimulant. Open Chem 13:1183–1195. https://doi.org/10.1515/chem-2015-0132

Mittal R, Tavanandi HA, Mantri VA, Raghavarao KSMS (2017) Ultrasound assisted methods for enhanced extraction of phycobiliproteins from marine macro-algae, Gelidium pusillum (Rhodophyta). Ultrason Sonochem 38:92–103. https://doi.org/10.1016/j.ultsonch.2017.02.030

Naseri A, Jacobsen C, Sejberg JJP et al (2020) Multi-extraction and quality of protein and carrageenan from commercial spinosum (eucheuma denticulatum). Foods 9(8):1072. https://doi.org/10.3390/foods9081072

O'Connor J, Meaney S, Williams GA, Hayes M (2020) Extraction of protein from four different seaweeds using three different physical pre-treatment strategies. Molecules 25(8):2005. https://doi.org/10.3390/molecules25082005

Peñuela A, Robledo D, Bourgougnon N et al (2018) Environmentally friendly valorisation of solieria filiformis (Gigartinales, rhodophyta) from IMTA using a biorefinery concept. Mar Drugs 16(12):487. https://doi.org/10.3390/md16120487

Pereira T, Barroso S, Mendes S et al (2020) Optimisation of phycobiliprotein pigments extraction from red algae Gracilaria gracilis for substitution of synthetic food colorants. Food Chem 321:126688. https://doi.org/10.1016/j.foodchem.2020.126688

Pezoa-Conte R, Leyton A, Anugwom I et al (2015) Deconstruction of the green alga Ulva rigida in ionic liquids: closing the mass balance. Algae Res 12:262–273. https://doi.org/10.1016/j.algal.2015.09.011

Polikovsky M, Fernand F, Sack M et al (2016) Towards marine biorefineries: selective proteins extractions from marine macroalgae Ulva with pulsed electric fields. Innov Food Sci Emerg Technol 37:194–200. https://doi.org/10.1016/j.ifset.2016.03.013

Polikovsky M, Gillis A, Steinbruch E et al (2020) Biorefinery for the co-production of protein, hydrochar and additional co-products from a green seaweed Ulva sp. with subcritical water hydrolysis. Energy Convers Manag 225:113380. https://doi.org/10.1016/j.enconman.2020.113380

Postma PR, Cerezo-Chinarro O, Akkerman RJ et al (2018) Biorefinery of the macroalgae Ulva lactuca: extraction of proteins and carbohydrates by mild disintegration. J Appl Phycol 30:1281–1293. https://doi.org/10.1007/s10811-017-1319-8

Prabhu MS, Levkov K, Livney YD et al (2019) High-voltage pulsed electric field preprocessing enhances extraction of starch, proteins, and ash from marine macroalgae Ulva ohnoi. ACS Sustain Chem Eng 7:17453–17463. https://doi.org/10.1021/acssuschemeng.9b04669

Remón J, Danby SH, Clark JH, Matharu AS (2020) A new step forward nonseasonal 5G biorefineries: microwave-assisted, synergistic, co-depolymerization of wheat straw (2G biomass) and Laminaria saccharina (3G biomass). ACS Sustain Chem Eng 8:12493–12510. https://doi.org/10.1021/acssuschemeng.0c03390

Robin A, Kazir M, Sack M et al (2018) Functional protein concentrates extracted from the green marine macroalga Ulva sp., by high voltage pulsed electric fields and mechanical press. ACS Sustain Chem Eng 6:11. https://doi.org/10.1021/acssuschemeng.8b01089

Rodrigues D, Sousa S, Silva A et al (2015) Impact of enzyme- and ultrasound-assisted extraction methods on biological properties of red, brown, and green seaweeds from the central west coast of Portugal. J Agric Food Chem 63:3177–3188. https://doi.org/10.1021/jf504220e

Ryu H-S, Satterlee LD, Lee K-H (1982) Nitrogen conversion factors and invitro protein digestibility of some seaweeds. Bull Korean Fish Soc 15:263–270

Saluri M, Kaldmäe M, Tuvikene R (2019) Extraction and quantification of phycobiliproteins from the red alga Furcellaria lumbricalis. Algal Res 37:115–123. https://doi.org/10.1016/j.algal.2018.11.013

Sequeira RA, Bhatt J, Prasad K (2020) Recent trends in processing of proteins and DNA in alternative solvents: a sustainable approach. Sustain Chem 1:116–137. https://doi.org/10.3390/suschem1020010

Suresh Kumar K, Ganesan K, Selvaraj K, Subba Rao PV (2014) Studies on the functional properties of protein concentrate of Kappaphycus alvarezii (Doty) Doty—an edible seaweed. Food Chem 153:353–360. https://doi.org/10.1016/j.foodchem.2013.12.058

Tibbetts SM, Milley JE, Lall SP (2016) Nutritional quality of some wild and cultivated seaweeds: nutrient composition, total phenolic content and in vitro digestibility. J Appl Phycol 28:3575–3585. https://doi.org/10.1007/s10811-016-0863-y

Toepfl S, Heinz V (2011) Pulsed electric field assisted extraction-a case study. In: Nonthermal processing technologies for food. Wiley, Hoboken, pp 190–200

Uju, Dewi NPSUK, Santoso J et al (2020) Extraction of phycoerythrin from Kappaphycus alvarezii seaweed using ultrasonication. In: IOP conference series: earth and environmental science. Institute of Physics Publishing, Bristol, p 012028

Ummat V, Tiwari BK, Jaiswal AK et al (2020) Optimisation of ultrasound frequency, extraction time and solvent for the recovery of polyphenols, phlorotannins and associated antioxidant activity from brown seaweeds. Mar Drugs 18:250. https://doi.org/10.3390/md18050250

Vásquez V, Martínez R, Bernal C (2019) Enzyme-assisted extraction of proteins from the seaweeds Macrocystis pyrifera and Chondracanthus chamissoi: characterisation of the extracts and their bioactive potential. J Appl Phycol 31:1999–2010. https://doi.org/10.1007/s10811-018-1712-y

Vorobiev E, Lebovka N (2011) Pulse electric field-assisted extraction. In: Enhancing extraction processes in the food industry. CRC Press, Boca Raton, pp 25–84

Wahlström N, Harrysson H, Undeland I, Edlund U (2018) A strategy for the sequential recovery of biomacromolecules from red macroalgae Porphyra umbilicalis Kützing. Ind Eng Chem Res 57:42–53. https://doi.org/10.1021/acs.iecr.7b03768

Wong K, Cheung P (2001a) Influence of drying treatment on three Sargassum species 2. Protein extractability, in vitro protein digestibility and amino acid profile of protein concentrates. J Appl Phycol 13:51–58. https://doi.org/10.1023/A:1008188830177
Wong KH, Cheung PCK (2001b) Nutritional evaluation of some subtropical red and green seaweeds Part II. In vitro protein digestibility and amino acid profiles of protein concentrates. Food Chem 72:11–17. https://doi.org/10.1016/S0308-8146(00)00176-X
Youssouf L, Lallemand L, Giraud P et al (2017) Ultrasound-assisted extraction and structural characterisation by NMR of alginates and carrageenans from seaweeds. Carbohydr Polym 166:55–63. https://doi.org/10.1016/j.carbpol.2017.01.041
Yuan Y, Macquarrie DJ (2015) Microwave assisted acid hydrolysis of brown seaweed Ascophyllum nodosum for bioethanol production and characterization of alga residue. ACS Sustain Chem Eng 3:1359–1365. https://doi.org/10.1021/acssuschemeng.5b00094
Zeng X, Zhang Z (2019) Pulsed electric field assisted extraction of bioactive compounds. In: Advances in food processing technology. Springer, Singapore, pp 125–135
Zhu Z, Wu Q, Di X et al (2017) Multistage recovery process of seaweed pigments: investigation of ultrasound assisted extraction and ultra-filtration performances. Food Bioprod Process 104:40–47. https://doi.org/10.1016/j.fbp.2017.04.008
Zollmann M, Robin A, Prabhu M et al (2019) Green technology in green macroalgal biorefineries. Phycologia 58:516–534. https://doi.org/10.1080/00318884.2019.1640516

Chapter 12
Sustainable and Biodegradable Active Films Based on Seaweed Compounds to Improve Shelf Life of Food Products

Marlene A. Trindade, Cláudia Nunes, Manuel A. Coimbra, Fernando J. M. Gonçalves, João C. Marques, and Ana M. M. Gonçalves

Abbreviations

FAO Food and Agriculture Organization

M. A. Trindade
Department of Biology, University of Aveiro, Aveiro, Portugal
e-mail: trindade.mar@ua.pt

C. Nunes
CICECO, Department of Chemistry, University of Aveiro, Aveiro, Portugal

LAQV-REQUIMTE, Department of Chemistry, University of Aveiro, Aveiro, Portugal
e-mail: claudianunes@ua.pt

M. A. Coimbra
LAQV-REQUIMTE, Department of Chemistry, University of Aveiro, Aveiro, Portugal
e-mail: mac@ua.pt

F. J. M. Gonçalves
Department of Biology and CESAM, University of Aveiro, Aveiro, Portugal
e-mail: fjmg@ua.pt

J. C. Marques
Department of Life Sciences, MARE – Marine and Environmental Sciences Centre, University of Coimbra, Coimbra, Portugal
e-mail: jcmimar@ci.uc.pt

A. M. M. Gonçalves (✉)
Department of Life Sciences, MARE – Marine and Environmental Sciences Centre, University of Coimbra, Coimbra, Portugal

Department of Biology and CESAM, University of Aveiro, Aveiro, Portugal
e-mail: amgoncalves@uc.pt; anamartagoncalves@ua.pt

A. Ranga Rao, G. A. Ravishankar (eds.), *Sustainable Global Resources of Seaweeds Volume 2*, https://doi.org/10.1007/978-3-030-92174-3_12

1 Introduction

Macroalgae or seaweeds are a diverse and ubiquitous group of predominantly marine photosynthetic organisms, chemically, physiologically, and morphologically different from land plants (Shama et al. 2019). Compounds like carrageenan, alginate, agar, fucose, fucoidan, laminarin, and mannitol distinguish them from microalgae and land plants (Wang et al. 2020). Macroalgae together with microalgae are responsible for the production of over 50% of the oxygen globally (Henriques et al. 2017; Gonçalves 2020) and consume most of the carbon dioxide in the atmosphere (Marine Biotechnology ERA-NET 2019). Different natural pigments like chlorophylls, carotenes and xantophylls give the thallus different colors leading to the division of seaweed in three distinct groups: the phyla Ochrophyta (class Phaeophyceae; brown algae), Chlorophyta (green algae) and Rhodophyta (red algae) (Marques 2017; Gonçalves 2020; Pacheco et al. 2020).

Seaweeds are pivotal in the cuisine of several Asian countries, especially in Japan, Korea, and China, but they are also present in some specialties in the island nations of the Pacific, and in Celtic nations (Brittany, Scotland and Ireland) (Pereira 2018). Its consumption is increasing in several Western countries, due to the growing interest in adopting healthier diets (Delaney et al. 2016; Marine Biotechnology ERA-NET 2019). Seaweeds are richer in minerals, vitamins, proteins, trace elements, iodine, omega-3 and in soluble and insoluble dietary fibers (FAO 2020; Pereira 2020), comparing with land plants, therefore, seaweeds can be a healthier alternative to conventional food products (Shama et al. 2019). These organisms can help in the effort to mitigate hunger in underdeveloped and emerging countries, as they are an abundant and highly nutritious food source that grow everywhere (Pereira 2020). An additional advantage is the seaweeds can be consumed without being in its fresh form. Although the several applications of seaweeds in the food industry there is the need for more studies on their toxicological profile to ensure consumers' safety (Rengasamy et al. 2020).

The decrease of arable land and the increase of the world population pushes the search for food alternatives that are more nutritious and eco-friendly. Macroalgae cultivation does not compete with agriculture as it uses resources that are not suitable for that purpose; they show a greater photosynthetic efficiency, faster growth rates, and higher production yields per unit area comparing with land crops. They are also easier to process because they do not have lignin (Baweja et al. 2016; Marques 2017; Raikova et al. 2019; Sedayu et al. 2019). Besides its use as food, macroalgae are collected and are cultivated for the production of bio-fertilizers, animal feed, hydrocolloids, wastewater treatment and biofuels (García-Poza et al. 2020; Wang et al. 2020).

Macroalgae are a sustainable and rich resource of bioactive compounds with broad-spectrum antimicrobial activity, antioxidant and anti-inflammatory properties, among other properties. Therefore its use in food, medical, pharmaceutical and cosmetic industries brings many advantages and opportunities in the development of new active products and supplements. Its exploitation results in the increase of research and development of new sustainable and efficient techniques of exploitation and production. As an industrial crop, seaweed cultivation is still a relatively recent activity, but has the potential to address the long-term issue of environmental sustainability (Abdul Khalil et al. 2017).

The industry around macroalgae is worth several millions of dollars (Pereira 2018; Marine Biotechnology ERA-NET 2019) as they are more productive than any land crop. In 2018 the trade in seaweed and other aquatic plants accounted for 63% of the USD 2 billion in exports according to the Food and Agriculture Organization (FAO 2020). Wang et al. (2020) reported that "China, Japan, Korea and the Philippines account for about 72% of the global annual production of macroalgae". Industries around seaweed create a large variety of jobs since the growers/farmers, all the way up to, researchers, engineers, and also create marketing and financial services (Marine Biotechnology ERA-NET 2019). Hence, macroalgae are an eco-friendly source of food and raw materials for different industries, and can be included in the concepts of blue and circular economies as its cultivation, production, and harvesting increases local employment at several levels and its byproducts can be used as raw materials for different industries.

This chapter aims to explore and summarize the uses of macroalgae and its most explored bioactive compounds in different industries, particularly in the food packaging industry; how seaweed-based packaging is produced and their effects on the shelf life of food products.

2 Macroalgae Bioactive Compounds and Its Application in Different Industries

Seaweed occupy habitats exposed to constantly changing conditions, consequently they are pressured towards metabolic evolution leading to the production of several secondary metabolites with important biological, ecological and economical value. Their essential amino acids, polysaccharides, essential fatty acids, sterols, terpenes, carotenoids, phenolic compounds and minerals provide them with the capability of reducing or stopping cell growth, eliminating parasitic worms, preventing the growth or spread of fungi and bacteria, and are effective against viruses (de Almeida et al. 2011). Seaweeds also provide antioxidant and anti-inflammatory properties as mentioned before, and are explored in the treatment of several diseases. Thickening, gelling, stabilizing, encapsulating, and dispersant agents (Abdul Khalil et al. 2017) used widely in the food industry are also found in seaweeds. The composition and quantity of these biomolecules change according with the species, the season, and the region where they grow due to differences in ecological parameters (Marine Biotechnology ERA-NET 2019; Pacheco et al. 2020). A more detailed and extensive description of the properties of macroalgae bioactive compounds and on research in the pharmaceutical and cosmetics industries can be read in the publication of Rengasamy et al. (2020). In the biomedical field seaweed-based polysaccharides are being explored due to their capacity to form hydrogels, their biocompatibility and for being hypoallergenic; their hydrophilic groups, like carboxyl, sulfate, and hydroxyl, can easily interact with biological tissues. It was reported that seaweeds are used as carriers of therapeutic macromolecules and other molecules, controlled delivery and release systems, bone tissue engineering, regeneration and implantation, hydrogel wound dressings, removal of heavy metals from the body (Abdul Khalil et al. 2017).

In the cosmetic industry, algae are advantageous because they can be resourced fast and at low prices, and also because of all the benefits for the skin and the variety of forms that they can be used as oils, powders, flakes (Marine Biotechnology ERA-NET 2019). In this industry, seaweeds are used as bioactive extracts, coloring agents, texturing stabilizers or emulsifiers; and as photosynthetic organisms that produce compounds that absorb UV rays, being used in the formulation of sunscreens (García-Poza et al. 2020).

Taking advantage of the bioaccumulation process, seaweeds can be used in the remediation of contaminated waters with the advantage of being more efficient and cost-effective, because they thrive in nutrient-rich waters and can remove toxic compounds, like metals, whilst capturing carbon dioxide emissions, reducing the use of chemicals for the same purpose and with lower energy costs (Henriques et al. 2017). Seaweed biosorption activity can also diminish the toxic effect of different pollutants and provide a promising approach for integrated energy production (Ungureanu et al. 2017; Wang et al. 2020). This approach provides several useful byproducts that can be utilized in the production of fertilizers, feed additives, and biofuels (bioethanol and biobutanol) (Marine Biotechnology ERA-NET 2019; Michalak 2020; Wang et al. 2020; Mohammad et al. 2019; Hessami et al. 2019). Algal biomass is a promising alternative to fossil resources not only to produce energy but also chemicals, enhancing energy security and contributing to the mitigation of climate change (Baweja et al. 2016; Marine Biotechnology ERA-NET 2019; García-Poza et al. 2020).

Polysaccharides are of the most researched seaweed bioactive compounds. Alginate, agaran, carrageenan, and fucoidan are some of the most significant polysaccharides found in seaweed, and except for fucoidan, their viscosity and poor solubility makes them inefficient for pharmaceutical applications, but it makes them perfect to use in cosmetics and food products.

Fucoidan is a brown seaweed sulfated polysaccharide applied industrially in cosmetics, diet supplements and animal feeding (Pacheco et al. 2020). This polysaccharide has strong antioxidant and antiaging properties, and also anticancer, anticoagulant, antimicrobial, anti-inflammatory and antidiabetic properties. It can also be used as an emulsifying agent. By itself it cannot form gels, but the mixture with other polymers may provide some advantages (Gomaa et al. 2018), however it is not common in the development of film matrices. Fucoidan is a promising natural compound targeted by the biomedical and pharmaceutical industries due to its promising therapeutic properties. Its low toxicity and wide range of biologically active ingredients have been confirmed in numerous *in vitro* and *in vivo* studies (Pacheco et al. 2020).

3 Macroalgae Polysaccharides in the Food Industry

Carrageenan, agar and alginate are the three major phycocolloids used in the food industry and due to its composition, structure, and consequent properties they are extensively used in the food industry. Table 12.1 summarizes their main properties and consequent applications in the food industry. Further information can be read in Alba and Kontogiorgos (2018), Ozilgen and Bucak (2018) and Skurtys et al. (2010).

Carrageenan is a linear sulfated polysaccharide composed of alternating and repeating disaccharides units of β-D-galactopyranose and α-D-galactopyranose or 3,6-anhydro-α-D-galactopyranose, linked with β-(1 → 4) and α-(1 → 3) glycosidic linkages, obtained from different red algae. Depending if it has one, two, or three sulfate groups per disaccharide unit, carrageenan can be differentiated as κ-, ι-, and λ-carrageenan, respectively, and these are the most relevant types industrially (Alba and Kontogiorgos 2018; Ozilgen and Bucak 2018). Although the different types of carrageenan are all soluble in water, factors like temperature, pH, ionic strength of the medium, and the presence of cations influence its aqueous solubility (Alba and Kontogiorgos 2018). Regardless of being technically considered a dietary fiber, carrageenan has gelling, thickening, and stabilizing properties, therefore, they are mainly used in dairy and meat products. λ-carrageenan does not form gels, but polyelectrolyte solutions used as thickening agents in dairy products.

Agar is also a linear polysaccharide composed of alternating and repeating disaccharide units of β-D-galactopyranose and 3,6-anhydro-α-L-galactopyranose, linked with β-(1 → 4) and α-(1 → 3) glycosidic linkages. These units are agarobiose and neoagarobiose. The hydrogen bonding between the 3,6-anhydro-α-L-galactopyranose residues stabilizes the gel. Agar has a fraction with high gelling capacity, the neutral, low sulphate/methoxyl substituted one, and a fraction with low gelling capacity, the charged, heterogeneous, highly substituted one (Alba and Kontogiorgos 2018). Usually the non-gelling fraction is removed during its extraction to obtain a powder with a higher gelling strength (Mostafavi and Zaeim 2020).

Alginate refers to alginic acid and all its derivatives and salts; it is the biopolymer with more commercial interest (Paixão et al. 2019). It is a linear polysaccharide consisting of β-D-mannuronic acid and α-L-guluronic acid linked with 1 → 4 glycosidic linkages. The higher the content in guluronic acid the stronger the gelling properties (Alba and Kontogiorgos 2018). The presence of ions in the solution, the pH and ionic strength of the solvent will determine the solubility of alginate, since it is a charged polysaccharide and have a wide range of molecular weight distribution. The selective binding of divalent and multivalent cations determines its ability to form gels, but they also gel following the cation-independent gelation mechanism.

Table 12.1 Source, properties and applications of the three major phycocolloids in the food industry—carrageenan, agar and alginate

Phycocolloid	Source	Properties	Applications
Carrageenan (κ-, ι-, λ-)	Rhodophyta	• Gelling, thickening and stabilizing properties. • Shear thinning flow. • Thermoreversible gels on cooling. • Kappa type: firm, brittle gel with potassium. • Iota type: soft shear thinning, elastic gel with calcium. • Kappa type gives lower viscosity compared to iota type.	• Confectionary: vegan alternative to gelatin. • Meat industry: texture improving agent; binding agent; stabilizer. • Desserts (water-based or dairy-based): thickening and stabilizing agent; binding agent. • Dairy products: stabilizer and texture improving agent. • Salad dressings: stabilizers. • Drink industry: stabilizers; thickening and flavor adding agent; flocculating agent. • Processed cheese: binding agent; texture improving agent • Bread making: dough proofing improving agent; formulation of gluten-free bread. • Foods with encapsulation technology: added components encapsulation agent; controlled release of encapsulated components.
Agar	Rhodophyta	• Major gelling agent, even at low concentrations. • Thermoreversible gels on cooling. • Heat resistant. • Brittle gels, but addition of sugars improves its elasticity. • Toleration to acids (pH 2.5–10).	• Confectionary: vegan gelatin substitute; jelly production; stabilizing agent. • Baking industry: preservative; moisture binding agent; dough proofing improving agent. • Meat industry: structuring and fat reducing agent of gelled canned meat products. • Frozen desserts: whey-binding agent. • Beverage industry: emulsion and foam stabilizing agent: flocculating agent.
Alginate	Ochrophyta - Phaeophyceae	• Major gelling agent. • Thermoreversible gel in the presence of calcium ions. • Clear and transparent gels resistant to heat. • Shear thinning. • Thickening properties in the absence of calcium. • In.neutral solutions: low viscosity. • In lower pH (<5.5) solutions: high viscosity.	• Confectionary: jelly and bakery cream production. • Beverage industry: emulsion and foam stabilizing agent; flocculating agent. • Frozen desserts: whey-binding agent. • Food with encapsulation technology: added components encapsulation agent; controlled release of encapsulated components. • Edible coatings/films in meat products, fruits and vegetables.

3.1 Food Packaging Industry

The packaging industry is the largest market for plastic (Geyer et al. 2017), with food packaging having the bigger share in the total packaging sector (85%) (Silva et al. 2020). Petroleum-based synthetic polymers which are non-biodegradable polymers have been widely used in food packaging due to their low cost, durability and water resistance properties (Geyer et al. 2017; Karan et al. 2019; Farhan and Hani 2020; Silva et al. 2020). Because non-biodegradable plastic is cheap, resistant, and light, it is persistent in our lives, so it is challenging to reduce its use globally. Replacing traditional plastics with biodegradable materials could save the environment from a considerable amount of plastic waste, therefore, alternatives for single-use plastic packaging are being researched.

Bio-based plastics do not only have the advantages of traditional plastic but are also environmentally friendly, however, as they currently exist from crop plants, they compete for resources with crops used for food. This is where seaweed-based packaging enters in the effort of plastic pollution mitigation and for the innovation of new types of active packaging. As referred by Ozilgen and Bucak (2018) "Biopolymers are a group of molecules that are of natural resources and exhibit characteristic polymer properties". Food hydrocolloids are high molecular weight long-chain biopolymers commonly used as functional food additives. They are thickening, stabilizing, encapsulating, gelling and film forming agents, used in the food industry to produce films for packaging, maintain or improve the sensory properties of food and beverages, improve shelf life, make the production process easier and more effective, and produce functional foods (Skurtys et al. 2010; Ozilgen and Bucak 2018). Macroalgae can also be used indirectly as an alternative for largescale production of biopolymers, because its biomass can be the substrate to bacteria and *Archaea* capable of producing biopolymers like polyhydroxyalkanoate (Marques 2017; Ghosh et al. 2019).

Protecting food products during handling, storing, and transportation from physical, chemical and biological damage are the main roles of food packaging. By delaying food deterioration and keeping its integrity, the packaging maintains the quality and safety of food and increases its shelf life (Parreidt et al. 2018). Biofilms can exercise these functions, and the production of new biodegradable and edible biofilms consists in a matrix of hydrocolloids, lipids or a combination of both (a composite) (Skurtys et al. 2010; Paixão et al. 2019). Composites are a hybrid material with different properties from the ones of their individual components. Its mechanical and barrier properties depend on the properties of the individual polymers and their compatibility. Composite films made from polysaccharides and lipids, for example, have good mechanical and water barrier properties.

Consumers demand for environmentally friendly packaging, longer shelf life products, and better quality of fresh food has been growing, consequently the research and development of edible films for foodstuff are also growing. Edible films can be a solution for the environmental problems created by plastic packaging, and because they prevent moisture loss, oxidative rancidity and microbial spoilage, they extend the products' shelf life. Moreover, these films can be incorporated with antioxidant and other nutraceuticals supplementing the nutritional value of the food (Tavassoli-Kafrani et al. 2016; Gomaa et al. 2018).

3.2 Macroalgae Polysaccharide-Based Films for Food Packaging

Seaweed polysaccharides are an obvious choice material for edible films. The characteristics of the polysaccharides selected to produce a film will influence the disintegration rate, taste, and mechanical properties of the obtained film (Ozilgen and Bucak 2018).

Edible films need to be a selective barrier to gases and an effective barrier to water, water vapor, moisture, and temperature (Tavassoli-Kafrani et al. 2016). Films developed from algae polysaccharides have good oxygen vapor barrier properties and are impermeable to fats and oils, however, the major drawbacks of using polysaccharides as packing material are their poor mechanical and barrier properties. The most frequently used method to overcome these limitations is to produce composite films by mixing a polymer with another one and/or a hydrophobic component and/or nanoparticles (Abdul Khalil et al. 2017). The different characteristics of every component is used to improve the mechanical and barrier properties.

Carrageenan, agar and alginate are widely used in several food industries namely in the production of films for packaging, because they are nontoxic, biodegradable, and are derived from a renewable natural source (Kanmani and Rhim 2014; Tavassoli-Kafrani et al. 2016; Abdul Khalil et al. 2017; Gomaa et al. 2018; Parreidt et al. 2018).

Carrageenan solutions at certain temperatures and cation concentrations show shear thinning flow behavior and gel. Coil-to-helix transition upon cooling and cation-induced aggregation of helices are the two steps involved in the formation of gels from κ- and ι-carrageenan, however, these will have different properties (Farhan and Hani 2020). The addition of other hydrocolloids can improve the texture of carrageenan gels and its applications (Alba and Kontogiorgos 2018). However, the hydrophilic nature of carrageenan is a disadvantage for the manufacturing of packaging, because they would be susceptible to moisture transfer and have low water resistance, but blending carrageenan with hydrophobic compounds can solve this (Sedayu et al. 2019).

Agar films are biologically inert and can easily interact with different bioactive substances (Mostafavi and Zaeim 2020). Gels produced from agar are usually transparent and stiff, but the addition of sugars can increase its strength through the promotion of helices formation; the addition of other polysaccharides can expand the variety of textures of agar gels (Alba and Kontogiorgos 2018). Photodegradation and fluctuations in temperature and humidity in the surrounding environment promote formation of microfractures and embrittlement, therefore, agar has mainly been used in the development of edible films when mixed with film-forming emulsions, for example, to improve barrier performance (Skurtys et al. 2010). In the agar gel a network of agarose double helices is formed, as they aggregate by their external hydroxyl groups, and are stabilized by water molecules. During the casting, the temperature of the agar solution and casting surface needs to be higher than the agar gelling temperature to prevent the premature gelation of the solution. While drying the agarose structure allows its molecules to interact through hydrogen bonds, creating a continuous film network. Type and origin of agar, the film production method and the components incorporated in the matrix influence the needed concentration to produce an agar-based film.

Edible films prepared from alginate show poor water resistance due to its hydrophilic nature, but they are appropriate for loading additives and antibacterial compounds. Its mechanical properties can be improved mixing other polysaccharides in the alginate matrix (Skurtys et al. 2010). The β-D-mannuronic acid/α-L-guluronic acid ratio and the length of the blocks in the alginate structure impacts the mechanical properties of the gels; high ratios result in more elastic gels with good freeze-thaw stability, while low ratios result in strong and brittle gels with good heat stability but that shows syneresis after freeze-thaw processing (Alba and Kontogiorgos 2018). Alginate films have good oxygen barrier properties, prevent lipid oxidation, and improve food flavor and texture (Oussalah et al. 2007).

Comparing with carrageenan and alginate, agar has a less hygroscopic nature, which is an advantage in the production of packaging since one of its main functions is to reduce moisture transfer between the product and the surrounding environment (Sousa et al. 2010). The relatively high production cost of agar and its brittleness are drawbacks to its commercialization as food packaging, but the addition of plasticizers or the use of unpurified agar can be a way to overcome this. The plasticizers help to form softer and elastic gels, as well as the presence of proteins, polyphenols and other polysaccharides, considered as impurities in the agar extract, that additionally can give antioxidant properties to the gels (Martínez-Sanz et al. 2019). Alginate and carrageenan are highly hydrophilic, hence their weak moisture barrier properties. However, they are good barriers to fats and oils. These polysaccharides can prevent microbial proliferation in meat products and fruits. Additionally, films formulated with alginate or carrageenan prevent shrinkage and surface discoloration of meat products, and delay ripening of fruits, increasing its shelf life.

4 Processing of Macroalgae Polysaccharide-Based Films and Shelf Life Durability of Food Products

In the development of films for food packaging purposes the most common method is solvent casting; the biopolymer is dissolved in a constantly stirred solution (Abdul Khalil et al. 2017) and then casted. Figure 12.1 shows the general process of solvent casting. Extrusion is another method used, but usually in the production of synthetic polymer films. Even though this method is more applied to other non-seaweed hydrocolloids, some authors reported its use for the manufacturing of edible films (Skurtys et al. 2010).

Solvent casting is one of the most used methods for manufacturing films for food packaging. The chosen polymer is dissolved in a suitable solvent, centrifuged, or for a better result, filtered and degassed through vaccum, to remove any trapped air bubbles, poured into a mold, dried, and conditioned at a specific relative humidity environment, and then a film is obtained for analysis (Abdul Khalil et al. 2017; Ozilgen and Bucak 2018). When choosing a solvent for the manufacturing of an edible film, the preference is given to water-based solvents or to organic solvents listed safe, avoiding hazardous solvents not only to the environment but also to human health (Ozilgen and Bucak 2018). For analysis the American Society of Testing and Materials standard method is commonly utilized. The solvent casting method, or a similar one with the same principles, is the most recurrent in several studies on the development of edible seaweed-based packages.

The drying conditions influence the film structure as well as the wet casting thickness and the composition of the film solution. During the drying step the solubility of the polymer decreases and the film is formed when the polymer chains align themselves, therefore, the drying process cannot be fast. In the dried film a moisture content of 5–10% w/v is desirable (Skurtys et al. 2010).

The intrinsic properties of the food product (that determine their perishability), extrinsic factors (such as the storage conditions), and required shelf life are factors to be considered for the development of packaging for any type of food (Silva et al. 2020). In a general way, when these films are produced certain characteristics need to be evaluated to determine if and how they influence its efficiency. The characteristics are generally the following: physical properties (thickness, color, opacity), morphology, moisture barrier and water resistance properties, mechanical properties (tensile strength, Young's modulus, elongation at break), thermal properties, structural properties (X-ray diffraction patterns, Fourier-transform infrared spectra), diffusion of compounds (e.g., antimicrobial agents). In food packaging water resistance is important to prevent its disintegration when in contact with high moisture content food. Adequate mechanical properties also prevent the cracking or tears during the films' manufacture, when it is applied to the food product, handling, and storage. Uniformity, not only in its general aspect but also in the structure, is also important for its functionality, and the appearance important for the consumer. Thermal properties determine how the films react to heat fluctuations.

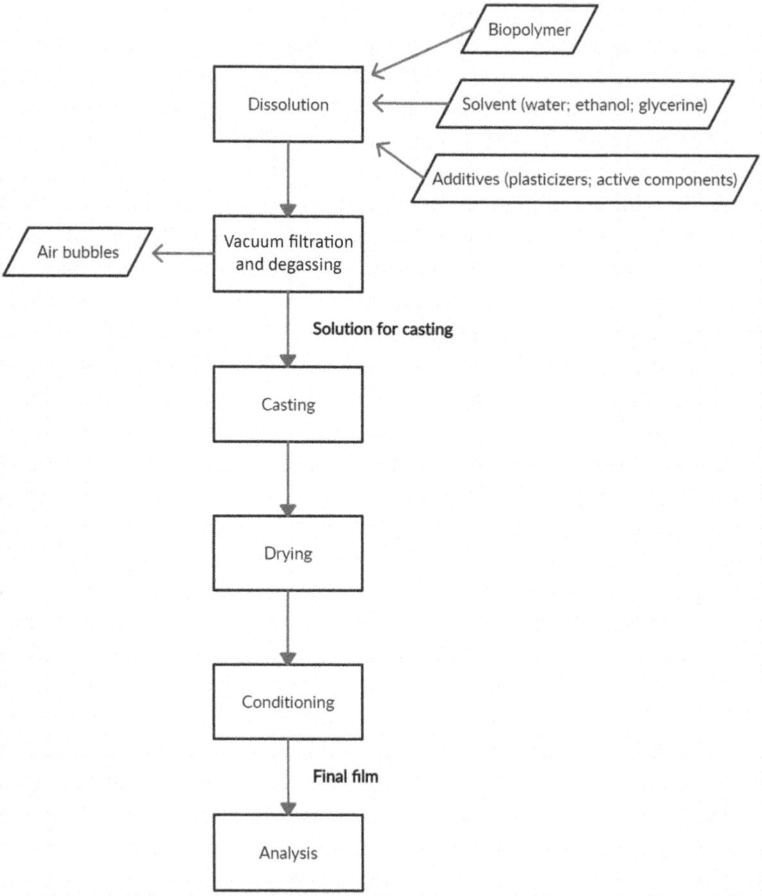

Fig. 12.1 Flowchart describing the general steps of the solvent casting method using a biopolymer as the main component of the matrix

The biological properties and the shelf life of food products are affected by the thickness of the film, that in turn is influenced by the concentrations of the polymer and additives, the spreading of the film solution, and its density, viscosity, and surface tension. A critical thickness must not be exceeded to avoid the decrease of oxygen and increase of carbon dioxide concentrations resulting from microbial activity. The food surface is dynamic and it modifies the film composition throughout storage time. The chemical composition and structure of the films, the characteristics of the products and the storage conditions depend on the barrier properties to gas, water vapor, aroma and oil which will influence the efficiency of the films (Skurtys et al. 2010). More detailed and extensive information on the properties of hydrocolloid edible films has been provided by Skurtys et al. (2010).

Films made of algae polysaccharides are usually brittle, they lack flexibility and toughness, therefore, they do not meet the criteria for commercial use edible films.

To overcome this problem plasticizers are added to improve its mechanical properties which in turn are affected by the type and concentration of plasticizer used. The most utilized plasticizers are polyols, like glycerol and sorbitol, sugars and water. What they do is increase the free volume in the matrix, intercalating between the biopolymer chains spreading them apart and disrupting hydrogen bonds, increasing flexibility, and water vapor and gas permeabilities (Skurtys et al. 2010). Although its advantages, the addition of plasticizers may alter the barrier properties of the films, decrease the ability of the film to attract water or decrease tensile strength (Oussalah et al. 2007; Skurtys et al. 2010). The elongation at break values is increased with the addition of plasticizers, however its increase in concentration decreases the tensile strength of the films. Therefore, the concentration of plasticizer added needs to be in accordance with the characteristics wanted/defined for the final package.

Active packaging has been one of the main areas of research in food packaging development to improve shelf life, add nutritional value, and improve sensory quality (Skurtys et al. 2010). Comparing to traditional packaging, it has several advantages because they can incorporate antimicrobial agents, antioxidants, enzymes, probiotics, minerals and vitamins (Skurtys et al. 2010). The addition of antioxidant and antimicrobial agents in the coatings or films makes their effect more efficient than when applied directly on the product, because its diffusion is slow.

Natural bio-preservatives are favored in the formulation of edible films, the most frequently used are essential oils, nisin and lysozyme. The incorporation of the enzyme lactoperoxidase in edible films has also been studied and showed bactericidal effects on Gram-negative bacteria, bacteriostatic effect on Gram-positive bacteria, and antifungal and antiviral activities. Oxidative rancidity, degradation, and discoloration can be avoided by the incorporation of antioxidant agents in the matrix, increasing the product shelf life as well. Most antimicrobial compounds have antioxidant properties, like the case of essential oils, phenolic compounds, vitamins E and C, that are some of the natural compounds used in the formulation of edible films (Oussalah et al. 2007; Raybaudi-Massilia et al. 2008; Alboofetileh et al. 2014; Azarakhsh et al. 2014; Kazemi and Rezaei 2015; Jalali et al. 2016).

The incorporation of essential oils of different spices and herbs provides a high antimicrobial activity to the composite material, expanding its application of food packaging. Rather than the type, the concentration of the essential oil is the factor that determines its antimicrobial efficiency. Several studies show the extension of shelf life and reduction on microbial counts in food products when these compounds are incorporated (Alboofetileh et al. 2014). Savory essential oil can act as a plasticizer as well as an antimicrobial agent, especially against Gram-positive ones. Because of their hydrophobicity and richness in fatty acids they improve water barrier properties and elongation at break values of the composites (Reboleira et al. 2020). Despite the advantages of integrating essential oils on food packaging films, they are usually extracted with toxic organic solvents implying the need for additional purification steps, thus the need for less toxic extraction techniques that can also reduce the production cost and increase the safety and sustainability of the product (Alboofetileh et al. 2014).

The formation of nanocomposites is currently one of the most effective ways to improve the properties of biopolymer films (Abdul Khalil et al. 2017). The most common types of inorganic nanomaterials used to develop seaweed-based nanocomposites are nanoclays and silver nanoparticles; and the most common organic nanomaterials are organically modified nanoclays, cellulose, carbon nanotubes and chitin nanofibers.

Several authors reported an improvement in mechanical strength and antimicrobial activity when films were reinforced with nanoclays, which are one of the most used additives in the development of seaweed-based nanocomposites. Mechanical strength of the films is improved as the stress is transferred to the nanoclay sheets, and these create a tortuous path in the matrix improving the barrier properties as well. The incorporation of nanoclays also decreases the rapid loss of other added components like essential oils. Chitin nanofibers can hypothetically be an allergen, so it might not be the best option to be incorporated in food packaging. On the other hand, silver in different chemical forms is toxic to several microorganisms. Nanocellulose can be added to food packaging to improve mechanical and barrier properties, and to serve as support for active compounds in active packaging. Its features include high crystallinity, high capacity of polymerization, high mechanical strength, low density, biocompatibility, non-toxicity and biodegradability. However, its use in food packaging, like the use of any other nanomaterial, raises potential safety concerns regarding their size, allergenicity and behavior, thus the need for more studies to demonstrate its safety.

Table 12.2 summarizes the main applications of several composites formulated with seaweed polysaccharides specifically in the food packaging industry. More detailed information, like concentrations and several examples of seaweed-based composites, and specifically for agar, alginate and carrageenan-based edible films and coatings for food packaging applications has been dealt with by Tavassoli-Kafrani et al. (2016) and Mostafavi and Zaeim (2020).

Several authors, like Jiang et al. (2013), Jalali et al. (2016), and Farhan and Hani (2020), to name a few, have reported for the use of different edible active films formulations where the characteristics of the food products were maintained for several days and at the end of the testing period the product was still good for consumption. The packaging reduced microbial spoilage, prevented moisture loss and maintained a good enough flavor, smell and color of the product. Oussalah et al. (2007) reported a controlled growth of foodborne pathogens (*Listeria monocytogenes* and *Salmonella Typhimurium*) in meat products packed in alginate-based films incorporated with essential oils. Other authors also report the decrease of microbial spoilage and maintenance of the organoleptic characteristics for several fruits (Rojas-Graü et al. 2007; Sipahi et al. 2013; Parreidt et al. 2018), mushrooms (Jiang et al. 2013), meat products (Oussalah et al. 2007; Parreidt et al. 2018; Surendhiran et al. 2019; Farhan and Hani 2020; Reboleira et al. 2020), and fish/seafood products (Kazemi and Rezaei 2015; Jalali et al. 2016; Albertos et al. 2019).

Table 12.2 Main seaweed-based polysaccharides composites and its applications in food packaging industry

Composite	Application
Alginate/nanocellulose	Films for packaging
Alginate/polysaccharides	Packaging of pre-cooked food
Alginate/polysaccharides/essential oils	Edible active films
Alginate/essential oils	Preservative coatings
	Shelf-life extension and quality retention of fresh cut fruits
Carrageenan/nanoclays	Shelf-life extension
Carrageenan/essential oils	Active packaging
Carrageenan/nanoclays/essential oils	Antimicrobial packaging
Agar/nanoclays	Biodegradable packaging
Agar/nanocellulose/essential oils	Active packaging for improving the safety and shelf-life of foodstuff

5 Conclusions and Future Perspectives

Macroalgae have been used for centuries, especially in Asian countries, however, its popularity and use have been increasing in several Western countries, mostly due to its reputation as a superfood. Macroalgae farming, aquaculture, and all the industries related to it are worth several millions of dollars and create jobs, especially on a local level. Seaweeds are a sustainable resource of several products and byproducts for the food, pharmaceutical, biomedical, cosmetic, and other industries; they can be used for wastewater treatment and production of biofuels. Therefore, it is a marine resource that can be included in sustainable economic models.

Seaweed-based polysaccharides can have several purposes on the industry previously mentioned due to its physical and chemical properties. One of its main uses is for the development of sustainable and biodegradable packaging for food, since traditional plastic packaging is the biggest contributor for plastic waste. Agar, carrageenan and alginate are the main phycocolloids used for the development of food packaging, since they have good gelling capacity and relatively good barrier properties. However, on their own these films are not commercially viable and do not have the properties that a good food packaging should have, like moisture and selective gases barrier, and resistance during handling. Therefore, they must be mixed with other components to improve these characteristics, and also with active components, like antioxidant, antimicrobial and nutraceuticals, to add some advantages comparing to traditional packaging. Nevertheless, the formulation of these composites need to be made suitable to the type of product, and the shelf life needed, since adding too much of one component to fix a problem can affect or create another one. Although the integration on the seaweed-based polysaccharide matrix of different components like plasticizers, biopolymers, lipids, nanoparticles, natural extracts, probiotic bacteria and bacteriocins, can improve its mechanical and chemical properties it also has some drawbacks affecting transparency, permeability, thermal stability or even mechanical strength. Therefore, further research to optimize the manufacturing of edible, biodegradable, nontoxic, seaweed-based packaging is needed. The probable toxicity and allergic effects of some additives like nanomaterials or natural extracts needs to be addressed as well. So, it is also important to study the synergy between components and the characteristics of each individual main component.

Although all the benefits of using seaweed polysaccharides to produce edible films, its commercial applications are still very limited. As there is still the need to optimize the production methods. Moreover, there is a need for guidelines and regulations in the packaging industry for these new types of packaging.

Acknowledgements This work was financed by national funds through FCT—Foundation for Science and Technology, I.P., within the scope of the projects UIDB/04292/2020 granted to MARE—Marine and Environmental Sciences Centre and UIDP/50017/2020 + UIDB/50017/2020 (by FCT/MTCES) granted to CESAM—Centre for Environmental and Marine Studies. CICECO-Aveiro Institute of Materials (FCT UIDB/50011/2020 and UIDP/50011/2020), and LAQV-REQUIMTE (UIDB/50006/2020) financed by national funds through the FCT/MEC and when appropriate co-financed by FEDER under the PT2020 Partnership Agreement. CN is funded by national funds (OE), through FCT—Fundação para a Ciência e a Tecnologia, I.P., in the scope of the framework contract foreseen in the numbers 4, 5 and 6 of the article 23, of the Decree-Law 57/2016, of August 29, changed by Law 57/2017, of July 19. Ana M. M. Gonçalves acknowledges University of Coimbra for the contract IT057-18-7253. Authors thank to the project MENU—Marine Macroalgae: Alternative recipes for a daily nutritional diet (FA_05_2017_011) which financed this research, funded by the Blue Fund under Public Notice No. 5—Blue Biotechnology.

References

Abdul Khalil HPS, Saurabh CK, Tye YY, Lai TK, Easa AM, Rosamah E, Fazita MRN et al (2017) Seaweed based sustainable films and composites for food and pharmaceutical applications: a review. Renew Sustain Energy Rev 77:353–362. https://doi.org/10.1016/j.rser.2017.04.025

Alba K, Kontogiorgos V (2018) Seaweed polysaccharides (agar, alginate carrageenan). Encyclopedia of food chemistry. Elsevier, Amsterdam, pp 240–250. https://doi.org/10.1016/B978-0-08-100596-5.21587-4

Albertos I, Martin-Diana AB, Burón M, Rico D (2019) Development of functional bio-based seaweed (*Himanthalia elongata* and *Palmaria palmata*) edible films for extending the shelflife of fresh fish burgers. Food Packag Shelf Life 22:100382. https://doi.org/10.1016/j.fpsl.2019.100382

Alboofetileh M, Rezaei M, Hosseini H, Abdollahi M (2014) Antimicrobial activity of alginate/clay nanocomposite films enriched with essential oils against three common foodborne pathogens. Food Control 36:1–7. https://doi.org/10.1016/j.foodcont.2013.07.037

de Almeida C, Layse F, Falcão HS, Lima GRM, Montenegro CA, Lira NS, Athayde-Filho PF, Rodrigues LC, Souza MFV, Barbosa-Filho JM, Batista LM (2011) Bioactivities from marine algae of the genus *Gracilaria*. Int J Mol Sci 12:4550–4573. https://doi.org/10.3390/ijms12074550

Azarakhsh N, Osman A, Ghazali HM, Tan CP, Adzahan NM (2014) Lemongrass essential oil incorporated into alginate-based edible coating for shelf-life extension and quality retention of fresh-cut pineapple. Postharv Biol Technol 88:1–7. https://doi.org/10.1016/j.postharvbio.2013.09.004

Baweja P, Kumar S, Sahoo D, Levine I (2016) Biology of seaweeds. In: Seaweed in health and disease prevention. Elsevier, Amsterdam, pp 41–106. https://doi.org/10.1016/B978-0-12-802772-1.00003-8

Delaney A, Frangoudes K, Ii S-A (2016) Society and seaweed. In: Seaweed in health and disease prevention, vol 2. Elsevier, Amsterdam, pp 7–40. https://doi.org/10.1016/B978-0-12-802772-1.00002-6

FAO (2020) The state of world fisheries and aquaculture 2020. FAO, Rome. https://doi.org/10.4060/ca9229en

Farhan A, Hani NM (2020) Active edible films based on semi-refined κ-carrageenan: antioxidant and color properties and application in chicken breast packaging. Food Packag Shelf Life 24:100476. https://doi.org/10.1016/j.fpsl.2020.100476

García-Poza S, Leandro A, Cotas C, Cotas J, Marques JC, Pereira L, Gonçalves AMM (2020) The evolution road of seaweed aquaculture: cultivation technologies and the industry 4.0. Int J Environ Res Public Health 17:6528. https://doi.org/10.3390/ijerph17186528

Geyer R, Jambeck JR, Law KL (2017) Production, use, and fate of all plastics ever made. Sci Adv 3:25–29. https://doi.org/10.1126/sciadv.1700782

Ghosh S, Gnaim R, Greiserman S, Fadeev L, Gozin M, Golberg A (2019) Macroalgal biomass subcritical hydrolysates for the production of polyhydroxyalkanoate (PHA) by *Haloferax mediterranei*. Bioresour Technol 271:166–173. https://doi.org/10.1016/j.biortech.2018.09.108

Gomaa M, Hifney AF, Fawzy MA, Abdel-Gawad KM (2018) Use of seaweed and filamentous fungus derived polysaccharides in the development of alginate-chitosan edible films containing fucoidan: study of moisture sorption, polyphenol release and antioxidant properties. Food Hydrocolloids 82:239–247. https://doi.org/10.1016/j.foodhyd.2018.03.056

Gonçalves AMM (2020) Sustainable premium ready meals for a daily nutritional diet: human population growing demand. In: Filho WL, Azul AM, Brandli L, Salvia AL, Wall T (eds) Decent work and economic growth. Encyclopedia of the UN sustainable development goals. Springer, Cham. https://doi.org/10.1007/978-3-319-71058-7_112-1

Henriques B, Rocha LS, Lopes CB, Paula Figueira AC, Duarte CV, Pardal MA, Pereira E (2017) A macroalgae-based biotechnology for water remediation: simultaneous removal of Cd, Pb and Hg by living *Ulva lactuca*. J Environ Manag 191:275–289. https://doi.org/10.1016/j.jenvman.2017.01.035

Hessami MJ, Cheng SF, Ranga Rao A, Yin YH, Phang SM (2019) Bioethanol production from agarophyte red seaweed, *Gelidium elegans* using a novel sample preparation method for analysing bioethanol content by gas chromatography. 3 Biotech J 9(1):25

Jalali N, Ariiai P, Fattahi E (2016) Effect of alginate/carboxyl methyl cellulose composite coating incorporated with clove essential oil on the quality of silver carp fillet and *Escherichia coli* O157:H7 inhibition during refrigerated storage. J Food Sci Technol 53:757–765. https://doi.org/10.1007/s13197-015-2060-4

Jiang T, Feng L, Wang Y (2013) Effect of alginate/nano-Ag coating on microbial and physicochemical characteristics of shiitake mushroom (*Lentinus edodes*) during cold storage. Food Chem 141:954–960. https://doi.org/10.1016/j.foodchem.2013.03.093

Kanmani P, Rhim J-W (2014) Development and characterization of carrageenan/grapefruit seed extract composite films for active packaging. Int J Biol Macromol 68:258–266. https://doi.org/10.1016/j.ijbiomac.2014.05.011

Karan H, Funk C, Grabert M, Oey M, Hankamer B (2019) Green bioplastics as part of a circular bioeconomy. Trends Plant Sci 24:237–249. https://doi.org/10.1016/j.tplants.2018.11.010

Kazemi SM, Rezaei M (2015) Antimicrobial effectiveness of gelatin-alginate film containing oregano essential oil for fish preservation. J Food Saf 35:482–490. https://doi.org/10.1111/jfs.12198

Marine Biotechnology ERA-NET (2019) Diverse applications of macroalgae

Marques, MM (2017) From macroalgae to bioplastic seaweed hydrolysates for polyhydroxyalkanoate production by marine bacteria biological engineering

Martínez-Sanz M, Martínez-Abad A, López-Rubio A (2019) Cost-efficient bio-based food packaging films from unpurified agar-based extracts. Food Packag Shelf Life 21:100367. https://doi.org/10.1016/j.fpsl.2019.100367

Michalak I (2020) The application of seaweeds in environmental biotechnology. Advances in botanical research, vol 95. Elsevier, Amsterdam. https://doi.org/10.1016/bs.abr.2019.11.006

Mohammad JH, Ranga Rao A, Ravishankar GA (2019) Opportunities and challenges in seaweeds as feed stock for biofuel production. In: Ravishnkar GA, Ranga Rao A (eds) Handbook of algal technologies and phytochemicals: volume II Phycoremediation, biofuels and global biomass production. CRC Press, Boca Raton, pp 39–50

Mostafavi FS, Zaeim D (2020) Agar-based edible films for food packaging applications—a review. Int J Biol Macromol 159:1165–1176. https://doi.org/10.1016/j.ijbiomac.2020.05.123

Oussalah M, Caillet S, Salmiéri S, Saucier L, Lacroix M (2007) Antimicrobial effects of alginate-based films containing essential oils on *Listeria monocytogenes* and *Salmonella typhimurium* present in bologna and ham. J Food Prot 70:901–908. https://doi.org/10.4315/0362-028X-70.4.901

Ozilgen S, Bucak S (2018) Functional biopolymers in food manufacturing. In: Biopolymers for food design. Elsevier, Amsterdam, pp 157–189. https://doi.org/10.1016/B978-0-12-811449-0.00006-2

Pacheco D, García-Poza S, Costas J, Gonçalves AMM, Pereira L (2020) Fucoidan—a valuable source from the ocean to pharmaceutical. Front Drug Chem Clin Res 3:1–4. https://doi.org/10.15761/FDCCR.1000141

Paixão LC, Lopes IA, Filho AKDB, Santana AA (2019) Alginate biofilms plasticized with hydrophilic and hydrophobic plasticizers for application in food packaging. J Appl Polym Sci 136:1–11. https://doi.org/10.1002/app.48263

Parreidt TS, Müller K, Schmid M (2018) Alginate-based edible films and coatings for food packaging applications. Foods 7:1–38. https://doi.org/10.3390/foods7100170

Pereira L (2018) Nutritional composition of the main edible algae. In: Therapeutic and nutritional uses of algae. CRC Press, Boca Raton, pp 65–127. https://doi.org/10.1201/9781315152844-2

Pereira L (2020) Characterization of bioactive components in edible algae. Mar Drugs 18:65. https://doi.org/10.3390/md18010065

Raikova S, Allen MJ, Chuck CJ (2019) Hydrothermal liquefaction of macroalgae for the production of renewable biofuels. Biofuels Bioprod Biorefin 13:1483–1504. https://doi.org/10.1002/bbb.2047

Raybaudi-Massilia RM, Rojas-Graü MA, Mosqueda-Melgar J, Martín-Belloso O (2008) Comparative study on essential oils incorporated into an alginate-based edible coating to assure the safety and quality of fresh-cut Fuji apples. J Food Prot 71:1150–1161. https://doi.org/10.4315/0362-028X-71.6.1150

Reboleira J, Adão P, Guerreiro SFC, Dias JR, Ganhão R, Mendes S, Andrade M et al (2020) Poultry shelf-life enhancing potential of nanofibers and nanoparticles containing *Porphyra dioica* extracts. Coatings 10:1–14. https://doi.org/10.3390/coatings10040315

Rengasamy KR, Mahomoodally MF, Aumeeruddy MZ, Zengin G, Xiao J, Kim DH (2020) Bioactive compounds in seaweeds: an overview of their biological properties and safety. Food Chem Toxicol 135:111013. https://doi.org/10.1016/j.fct.2019.111013

Rojas-Graü MA, Tapia MS, Rodríguez FJ, Carmona AJ, Martin-Belloso O (2007) Alginate and gellan-based edible coatings as carriers of antibrowning agents applied on fresh-cut Fuji apples. Food Hydrocoll 21:118–127. https://doi.org/10.1016/j.foodhyd.2006.03.001

Sedayu BB, Cran MJ, Bigger SW (2019) A review of property enhancement techniques for Carrageenan-based films and coatings. Carbohydr Polym 216:287–302. https://doi.org/10.1016/j.carbpol.2019.04.021

Shama A, Joyce SG, Mari FD, Ranga Rao A, Ravishankar GA, Hudaa N (2019) Macroalgae and microalgae: novel sources of functional food and feed. In: Ravishankar GA, Ranga Rao A (eds) Handbook of algal technologies and phytochemicals: volume I: food, health and nutraceutical applications. CRC Press, Boca Raton, pp 207–219

Silva FAGS, Dourado F, Gama M, Poças F (2020) Nanocellulose bio-based composites for food packaging. Nanomaterials 10:1–29. https://doi.org/10.3390/nano10102041

Sipahi RE, Castell-Perez ME, Moreira RG, Gomes C, Castillo A (2013) Improved multilayered antimicrobial alginate-based edible coating extends the shelf life of fresh-cut watermelon (*Citrullus lanatus*). LWT Food Sci Technol 51:9–15. https://doi.org/10.1016/j.lwt.2012.11.013

Skurtys O, Acevedo C, Pedreschi F, Enronoe J, Osorio F, Aguilera JM (2010) Food hydrocolloid edible films and coatings. Nova Science, New York

Sousa AMM, Sereno AM, Hilliou L, Gonçalves MP (2010) Biodegradable agar extracted from *Gracilaria vermiculophylla*: film properties and application to edible coating. Mater Sci Forum 636–637:739–744. https://doi.org/10.4028/www.scientific.net/MSF.636-637.739

Surendhiran D, Cui H, Lin L (2019) Encapsulation of phlorotannin in alginate/PEO blended nanofibers to preserve chicken meat from *Salmonella* contaminations. Food Packag Shelf Life 21:100346. https://doi.org/10.1016/j.fpsl.2019.100346

Tavassoli-Kafrani E, Shekarchizadeh H, Masoudpour-Behabadi M (2016) Development of edible films and coatings from alginates and carrageenans. Carbohydr Polym 137:360–374. https://doi.org/10.1016/j.carbpol.2015.10.074

Ungureanu G, Santos SCR, Volf I, Boaventura RAR, Botelho CMS (2017) Biosorption of antimony oxyanions by brown seaweeds: batch and column studies. J Environ Chem Eng 5:3463–3471. https://doi.org/10.1016/j.jece.2017.07.005

Wang S, Zhao S, Uzoejinwa BB, Zheng A, Wang Q, Huang J, Abomohra AE-F (2020) A state-of-the-art review on dual purpose seaweeds utilization for wastewater treatment and crude bio-oil production. Energy Convers Manag 222:113253. https://doi.org/10.1016/j.enconman.2020.113253

Chapter 13
Red Seaweeds: Their Use in Formulation of Nutraceutical Food Products

Diana Pacheco, Glacio Araújo, José W. A. Silva, João Cotas, Ana Marta Mendes Gonçalves, and Leonel Pereira

Abbreviations

FAO	Food and Agriculture Organization
PC	Protein concentrate
PUFAs	Polyunsaturated fatty acids

1 Introduction

Seaweeds are currently seen as promising species for providing new biologically active compounds to produce novel foods due to the large variety of compounds they contain, thus contributing to the development of nutraceutical food products (Kim 2011; Villanueva et al. 2014; Shama et al. 2019). Previous studies have shown that several seaweed-based compounds can improve human health by reducing symptoms of several diseases like cancer, asthma, diabetes, autoimmune, ocular, or cardiovascular (Lopes et al. 2013; Alves et al. 2018; Tanna and Mishra 2018). Seaweeds have been seen as a feedstock for bioactive molecules that can be incorporated in the daily diet as a supplement to promote human health, thus being considered nutraceutical food products (Alwaleed 2019; Shama et al. 2019).

D. Pacheco · J. Cotas · L. Pereira (✉)
Department of Life Sciences, MARE – Marine and Environmental Sciences Centre, University of Coimbra, Coimbra, Portugal
e-mail: leonel.pereira@uc.pt

G. Araújo · J. W. A. Silva
Federal Institute of Education, Science and Technology of Ceará – IFCE, Aracati, Ceará, Brazil

A. M. M. Gonçalves
Department of Life Sciences, MARE – Marine and Environmental Sciences Centre, University of Coimbra, Coimbra, Portugal

Department of Biology and CESAM, University of Aveiro, Aveiro, Portugal

A. Ranga Rao, G. A. Ravishankar (eds.), *Sustainable Global Resources of Seaweeds Volume 2*, https://doi.org/10.1007/978-3-030-92174-3_13

Among the algal phyla or classes, Chlorophyta (green), Phaeophyceae (brown) and Rhodophyta (red), the red seaweeds group contains great biodiversity (Torres et al. 2019). Red seaweeds contain pigments (i.e., phycoerythrin) that can be employed in food industry, as a natural substitute of synthetic food colorants, with antioxidant bioactivity (Ganesan and Shanmugam 2020). Besides that, these seaweeds present a rich nutritional profile, providing several important molecules, such as fatty acids, vitamins, essential amino acids, and minerals, through their direct incorporation in the daily diet. For instance, red seaweeds can contain up to 26 mg kg^{-1} of Vitamin E (Irene et al. 2018); and among the three phyla, Rhodophyta algal species exhibits the highest protein content (Černá 2011). Moreover, red seaweeds present several essential amino acids, exhibiting almost the same amount of plant-based protein sources (Barbier et al. 2019). For example, the species *Pyropia/Porphyra* has a score of 0.91 (on a scale between 0 and 1, whereas egg protein has a score of 1) (Murata and Nakazoe 2001). Furthermore, seaweeds lipidic profile, composed essentially by polyunsaturated fatty acids (ω-6 and ω-3), makes them an important source of these compounds, which humans acquire only through the daily diet (Belattmania et al. 2018).

However, their nutritional profile varies according to the harvest season, geolocation, and other biotic (i.e., temperature, salinity, light intensity, nutrients availability, and pollutants concentration) and abiotic parameters (i.e., predators). So, it is a current challenge to guarantee the safety and quality of seaweeds traded (Tanna and Mishra 2019; Cho and Rhee 2020; Rosa et al. 2020). For this reason, aquaculture techniques and technologies have evolved to tackle this need (Campbell et al. 2020). Currently, the most industrially exploited red seaweeds species are *Gracilaria* spp., *Kappaphycus alvarezii, Porphyra/Pyropia* spp. and *Eucheuma denticulatum*, being cultivated in several coastal countries.

Red seaweeds are pivotal for phycocolloids industry, for agar and carrageenan extraction, being widely employed in the food industry as thickener or emulsifier agent (Kraan 2012; Garcia-Vaquero et al. 2017). Furthermore, their physical capacity to emulsify and preserve water improves their technological functionality as food ingredients (Ruocco et al. 2016). The direct addition of seaweeds or their polysaccharides to meat and plant-based food products will improve nutritional, structural, antimicrobial, organoleptic, and shelf-life characteristics of these food products (Shannon and Abu-Ghannam 2019). Moreover, these phycocolloids can help to reduce glucose levels (BNF 2016). Besides the mentioned sulfated polysaccharides, porphyran can also be founded in some red seaweeds species, such *as Porphyra/Pyropia* sp., promoting health benefits, such as the reduction of cholesterol levels (Tsuge et al. 2004).

These polysaccharides, such as agar and carrageenan present several biological activities as it was forementioned, however their bioactivity varies according to their molecular weight, sulphation level and the quantity of sulphate esters groups present in the molecule (Tsuge et al. 2004; Cunha and Grenha 2016). Moreover, different seaweed species produce chemically distinct polysaccharides, and this can also be affected by the extraction method selected. Commonly, carrageenophytes contains a higher concentration of sulphate groups, in comparison to the agarophytes (Zhong et al. 2020). Still, carrageenan chemical structure is very heterogeneous, being the most economically important the carrageenan iota, lambda and kappa (Necas and Bartosikova 2013). So, what distinguish these different types of carrageenan is mainly their sulphate esters content and their position in the molecule (Sangha et al. 2015).

The goal of this book chapter is to highlight the role of red seaweeds for food industry as a key element for the development of innovative products with nutraceutical properties for the food industry as human health promoters.

2 Red Seaweed Nutritional Characteristics

The red seaweeds nutritional values are very dispersed due to a high number of species, and due to the variation of abiotic factors (salinity, nutrients, pH, luminosity, water temperature, profundity). However, these seaweeds can have an internal ratio of the nutrient composition (Table 13.1). Several eastern countries traditionally consume seaweeds as food (mainly Japan, China and Korea) (Dawes 1995), due to its high nutritional value as a source of proteins, carbohydrates, vitamins and minerals (Pereira 2011; Leandro et al. 2020a). Moreover, the seaweeds have interesting secondary metabolites with nutraceutical properties (Table 13.2).

These bioactivities are observed in the most cultivated species of red seaweeds, according to Food and Agriculture Organization (FAO) data (FAO 2020): *Gracilaria* spp., *Kappaphycus alvarezii*, *Porphyra* spp./*Porphyra tenera* and *Eucheuma denticulatum*. Moreover, in 2018 32.4 million tons of seaweeds were produced. In 2016, the volume of aquatic plants collected or produced was 31.2 million tons, where practically all this production was related to seaweeds, with growth from 13.5 million tons in 1995 to over 30 million tons in 2016 (FAO 2018). Also, according to FAO (Ferdouse et al. 2018), some red seaweed species, such as *Porphyra*/*Pyropia* spp. (Rhodophyta) were produced in the East and Southeast Asia, are produced almost exclusively for direct human food consumption. Indonesia increased seaweed production from less than four million tons in 2010 to more than 11 million tons in 2015 and 2016, mainly in red seaweed species (FAO 2018). This exponential growth in the production of tropical red seaweeds species, such as *Kappaphycus alvarezii* and *Eucheuma* spp. was due to their commercial

Table 13.1 Red seaweed nutritional values (Pereira 2011)

Nutrient	Concentration range (% of dry weight)
Protein	6.9–47
Lipid	0.3–3.3
Carbohydrates	43–68
Dietary fiber	10–46
Ash	7.8–37

Table 13.2 Secondary metabolites with nutraceutical properties, based in (Sonani 2016; Cotas et al. 2020a, b; Leandro et al. 2020a, b)

Compound	Bioactivity
Vitamins A, B_{12}, C, E, K	Immunological and human development
Carotenoids	Antioxidative, anti-inflammatory, antitumor
Allophycocyanin	Antioxidative, anti-inflammatory, antitumor, anti-enterovirus, hepatoprotective
Phycocyanin	Antitumor, anti-inflammatory, anti-oxidative, anti-irradiative
Phycoerythrin	Antioxidant, antitumor, neuroprotective, anti-inflammatory, hepatoprotective, hypocholesterolemic
Mycosporine-like amino acid	Anti-inflammatory, immunomodulatory
Phenolic terpenoids	Antioxidant, anti-inflammatory
Flavonoids	Antioxidant
Bromophenols	Antioxidant, antitumor, anti-angiogenesis, anti-diabetic, anti-obesity, antimicrobial, anti-fungal, anti-viral, neuroprotective

exploitation for carrageenan extraction and direct food consumption, which has been the main contributor to the growth in seaweed cultivation in recent years. The high economic interest is also justified by the growing demand for phycocolloids for different uses in the pharmaceutical, food, and cosmetics industries (Smit 2004; Leandro et al. 2020b; Morais et al. 2021), which has led several countries to cultivate seaweed (García-Poza et al. 2020).

2.1 **Gracilaria** ***sp. Nutritional Profile***

Gracilaria sp. is a red seaweed that produces agar as main structural polysaccharide, and their content have interesting values for direct food consumption, due to high content in protein (10.86%) and carbohydrates (63.13%) (Rasyid et al. 2019).

The study of Kazir et al. (2019) aimed to study *Gracilaria* sp. protein content for the development of food products, obtaining interesting protein levels (25% dry weight (DW)). The authors also obtained, in the highest concentrations, the amino acids: glutamic acid (13.01%), aspartic acid (12.81%), arginine (10.32%) and alanine (10.03%). In order to develop a seaweed protein-based product, the authors employed the ion exchange technique, enabling a high protein yield (70%) and a reduced co-extraction of carbohydrates (1%) in the seaweed protein extract. Showing a potential of *Gracilaria* sp. as protein food alternative source as a raw or processed ingredient.

This potential is also demonstrated in lipidic profile (Table 13.3), where the content of lipids and polyunsaturated fatty acids (PUFAs) in several *Gracilaria* species exhibited a high content of PUFAs, mainly *G. corticata* and *G. dura* (Kumari et al. 2013).

Table 13.3 Lipidic profile in *Gracilaria* sp. (Kumari et al. 2013)

Species	Lipids (%)	PUFAs (%)
Gracilaria dura	6.3 ± 0.3	62.8 ± 1.2
Gracilaria salicornia	7.6 ± 0.5	11.5 ± 0.8
Gracilaria textorii	7.3 ± 1.2	28.6 ± 2.9
Gracilaria corticata	8 ± 2.0	65.6 ± 2.5
Gracilaria corticata var. *cylindrica*	5.2 ± 1.1	29.8 ± 2.8
Gracilaria debilis	2.9 ± 0.2	48.4 ± 3.4

Khan et al. (2019) studied the polysaccharides of *Gracilaria chouae* and found the presence of a heteropolysaccharide (agar). Agar contained a sulphate content of 7.9%, in addition to 52.63% total sugar (mainly galactose) and 9.62% galacturonic acid. Galactose and 3,6-anhydrogalactose were found in a molar ratio of 1.0: 0.6. On further analysis, this polysaccharide exhibited jellification and melting points at 41.3 and 71.7 °C, respectively, which makes it a suitable candidate for industrial processing where further heating is required and/or where the end product needs to have an extended shelf life in hot climate.

2.2 K. alvarezii *Nutritional Profile*

K. alvarezii is a red seaweed that produces kappa-carrageenan as a main polysaccharide, and its nutritional profile has interesting values of minerals (58%) and carbohydrates (38%) and lower content in lipids and proteins (Wanyonyi et al. 2017), which is similar to other *Kappaphycus* species (Adharini et al. 2019). Moreover, these seaweeds can have an interesting prebiotic effect as a food supplement (Wanyonyi et al. 2017).

Kumar et al. (2014) analyzed the protein content of *K. alvarezii*, grown on the west coast of India, and found that this species contained 62.3 ± 1.62%. To further exploit this high yield in the food industry, this protein content can be transformed into a concentrate (PC), by the increase of the pH up to 12 and adding a solution of NaCl (0.5 M in the final solution). The emulsifier and foaming properties of this PC varied with time and pH. Thus, the results obtained in this study suggest the possibility of this seaweed and their based PC as an inexpensive source of protein; thus, this PC could be incorporated into several value-added food products.

2.3 Euchema *sp. Nutritional Profile*

Eucheuma sp. is a red seaweed that produces iota-carrageenan as main polysaccharide, which is mainly exploited from these species (Naseri et al. 2020). From an industrial perspective, the seaweed powder can be added into the food products to preserve and add nutritional value (Huang and Yang 2019). However, there is a lack of nutritional studies in this genus.

The red macroalgae *Eucheuma denticulatum*, also known by the common name "Spinosum", develops naturally in coral reefs with moderately strong currents in tropical and subtropical regions. This species has a high commercial value, as it contains iota-carrageenan, a compound widely used in the nutraceutical and manufacturing industries. Due to the high demand, the cultivation of *Eucheuma denticulatum* has significantly expanded (Othman et al. 2019).

Balasubramaniam et al. (2020) studied *E. denticulatum* carotenoids (mg/100 g of DW extract) and detected in various samples maximum content of each of the pigments viz. lutein 87.7, zeaxanthin 21.3, coxanthin 4.0, β-cryptoxanthin 3.6, canthaxanthin <0.001, astaxanthin 3.0, and β-carotene 4.7. These results indicate that *E. denticulatum* has an excellent carotenoid profile (as vitamin A) composition and hence rich in antioxidant potential.

De Corato et al. (2017) evaluated *E. denticulatum* composition in terms of concentration of fatty acids, polysaccharides and phenolic compounds. *E. denticulatum* presented in its composition 20.5 ± 0.5% of lipids, water-soluble polysaccharides 16 ± 0.6% and phenolic compounds 0.2 ± 0.001%, demonstrating that the species presents a good percentage of fatty acids and polysaccharides, with potential in several applications as direct food source and as ingredient for food industry.

2.4 Porphyra/Pyropia/Neopyropia/Neoporphyra *Nutritional Profile*

The genus *Porphyra* is evolving and modified/divided into four different genera due to the genetic analysis (Yang et al. 2020; Kavale et al. 2021).

Porphyra/Pyropia/Neopyropia/Neoporphyra sp. is a red seaweed that produces porphyran, as main polysaccharide, and actually is the seaweed most consumed in the world, due to their presence in the Japanese cuisine as "nori" (Levine and Sahoo 2010; Bito et al. 2017) Nori can ameliorate the deficiency of iron and vitamin B_{12} in vegan diet. Thus, *Porphyra/Pyropia/Neopyropia/Neoporphyra* sp. are one of the most economically important species, as it has functional bioactivities such as porphyrans, dietary fibers, PUFAs, minerals, phycoerythrin, mycosporine-like Amino Acids and vitamins (Bito et al. 2017). Porphyran and oligo-porphyran have a range of biological functions, such as antioxidant, anticancer, anti-aging, anti-allergic, immunomodulatory, hypoglycemic and hypolipemic effects. Consequently, these species' demonstrate several potential applications in the food, medicinal and cosmetic fields (Qiu et al. 2021). The most consumed and cultivated seaweeds are *Porphyra tenera*, *Neopyropia tenera*, *Neopyropia yezoensis*, *Neoporphyra dentata* and *Neoporphyra haitanensis* (Levine and Sahoo 2010; Niu et al. 2010). In fact, for these genera cultivation was of the value US Dollar 0.9 billion (Kim et al. 2017) mainly for direct food consumption.

Table 13.4 Macro and trace elements in *Porphyra* commercial samples from different European and Asian countries based in Larrea-Marín et al. (2010)

Element (mg g^{-1} DM)	France	Spain	Korea	Japan
Ca	7.06 ± 0.30a	6.04 ± 0.47[b]	7.26 ± 0.11[a]	2.90 ± 0.22[c]
Mg	7.94 ± 0.11b	7.10 ± 0.16[b]	3.73 ± 0.08[d]	4.24 ± 0.20[c]
P	1.49 ± 0.07c	5.60 ± 0.43[b]	8.59 ± 0.15[a]	8.47 ± 0.38[a]
Na	43.7 ± 0.57a	41.4 ± 5.41[a]	6.54 ± 0.16[b]	2.34 ± 0.38[c]
K	23.6 ± 0.60b	23.1 ± 0.67[b]	29.8 ± 0.10[a]	29.8 ± 1.91[a]
Sr	0.12 ± 0.02b	0.13 ± 0.02[b]	0.22 ± 0.02[a]	0.06 ± 0.01[c]
Al	21.5 ± 0.37c	15.0 ± 2.55[c]	220.8 ± 7.95[a]	94.0 ± 5.43[b]
Ba	0.53 ± 0.05c	0.85 ± 0.05[c]	3.97 ± 0.16[a]	2.5 ± 0.29[b]
Cu	9.98 ± 0.59c	20.2 ± 2.40[b]	19.7 ± 0.91[b]	37.0 ± 5.00[a]
Fe	149.2 ± 9.83d	201.2 ± 6.30[b]	285.9 ± 12.20[a]	165.8 ± 3.90[c]
Mn	23.0 ± 0.50b	32.5 ± 2.60[a]	34.3 ± 2.41[a]	32.0 ± 3.54[a]
Zn	82.4 ± 2.88b	52.5 ± 1.80[c]	85.4 ± 3.65[b]	94.2 ± 5.31[a]

[*] Values in the same row bearing different superscript letters are significantly different ($p < 0.05$)

Kim et al. (2018), studied *Neopyropia tenera*, determining the composition of carbohydrates, lipids, and proteins of *N. tenera* procured from a local market. The composition ratio (% DW) of carbohydrates, lipids, and proteins in *N. tenera* was 41.4%, 1.7% and 39.6%, respectively were determined. The study showed similar levels of carbohydrates and proteins, but with a low lipid content.

Holdt and Kraan (2011) and Rioux and Turgeon (2015), evaluated several seaweed species of the genus *Porphyra/Pyropia/Neopyropia/Neoporphyra*, and studied their bioactive compounds with economic importance. Among the determined compounds, the total polysaccharides had higher concentrations, with 40 and 76%, while the lipid content was 0.12 and 2.8% and protein of 7 and 50%. The results demonstrate that the *Porphyra/Pyropia/Neopyropia/Neoporphyra* genus, due to its higher concentration of polysaccharides, namely the hybrid porphyran/carrageenan/agar, are essential compounds in the food industry, as they are characterized by their solubility, gelation, viscosity, stability, reactivity with proteins and thixotropy properties (Hongfeng et al. 1993; Sasuga et al. 2017; Wahlström et al. 2018).

Larrea-Marín et al. (2010), evaluated the macro and trace elements in commercially grown *Porphyra* from four different countries (Table 13.4). The 12 elements determined were Al, Ba, Ca, Cu, Fe, K, Mg, Mn, Na, P, Sr and Zn, in the seaweed used as human food. However, differential compositions were due to the origin of the seaweeds, and impact of seawater mineral composition.

Although, these four genera are being cultivated and consumed, there is a lack of nutritional values of these seaweeds, which are understandable, due to the variation and influence of extrinsic factor that make seaweed composition fluctuate (García-Poza et al. 2020; Leandro et al. 2020b). However, the principal traits are normally identical as demonstrated above. Still, *Porphyra/Pyropia/Neopyropia/Neoporphyra* are the most exploited seaweeds for food and not for polysaccharide extraction, mainly due to the presence of a hybrid polysaccharide.

3 Red Seaweeds Consumption and Commercial Based Products

Seaweeds and their components already hold a market positioning worldwide (Ścieszka and Klewicka 2019; Rahikainen and Yang 2020). Still, several seaweeds remain as undiscovered resources, showing a very promising potential for the food industry, with nutraceutical properties (Pereira et al. 2020; John et al. 2020).

For instance, the red seaweeds dulse (*Palmaria palmata*) and nori (*Porphyra/Pyropia/Neopyropia/Neoporphyra*) are the major commercial products in the nutraceutical industry, as a food ingredient. These can be marketed fresh or dried, in a form of sheet, powder or capsule (Griffiths et al. 2016). The incorporation of milled seaweeds, such as *Porphyra umbilicalis* in meat products can indeed enrich these products in phenolic compounds and other nutrients likewise manganese, calcium and magnesium, which are pivotal for the homeostasis of the human organism (López-López et al. 2009). Besides, the bioactive and nutritional components of this seaweed species are associated with several health benefits, such as anticancer, cardiovascular disease prevention, antioxidant and anti-inflammatory (Cho and Rhee 2020). Moreover, the enrichment of cereal-based products (i.e., bread or pasta) with dried and milled seaweeds, such as *Kappaphycus alvarezii* was found to improve the nutritional profile of commercial noodles (Kumoro et al. 2016). However, heavy metals, toxic isotopes, dioxins, or pesticides are all risks associated with the whole seaweed intake. To address these drawbacks, thorough testing of seaweed for food application is needed prior to its use (Garcia-Vaquero and Hayes 2016).

In counterpart, industrial phycocolloids are under strict regulation, and several analyses are required for their incorporation as food additives in commercially available products (Mortensen et al. 2016; Younes et al. 2018). Thus, red seaweeds phycocolloids are already exploited and employed in several commercial products in food and nutraceutical industries. For instance, WavePure is a product based on *Gracilaria* sp., whereas the phycocolloid (agar) is extracted in order to be commercialized for food proposes as a gelling and thickening agent for desserts confection (Cargill 2021a). While carrageenan is a key element of the products Satiagel™, Satiagum™, Aubygel™ and Seabrid™, which can be used in a wide range of food products, such as dairy, fruit meat, ice-creams, powder products, pharmaceuticals and nutraceuticals (Cargill 2021b). Moreover, the vegetable jelly sold by Condi (based on carrageenan), shown to be also a nutraceutical promoter, due to its anticholesterolemic properties (Valado et al. 2020).

4 Conclusions and Future Perspectives

As novel foods and nutraceutical products demand for red seaweeds has risen. From the seaweeds belonging to the phylum Rhodophyta, only a few of them are industrially exploited. Thus, red seaweeds constitute a pool of unexplored biodiversity in several areas; for food and nutraceutical implying that a wide range of innovative products can be developed with these seaweeds.

Acknowledgments This work is financed by national funds through FCT—Foundation for Science and Technology, I.P., within the scope of the projects UIDB/04292/2020 granted to MARE—Marine and Environmental Sciences Centre and UIDP/50017/2020 + UIDB/50017/2020 (by FCT/MTCES) granted to CESAM—Centre for Environmental and Marine Studies. João Cotas thanks to the European Regional Development Fund through the Interreg Atlantic Area Program, under the project NASPA (EAPA_451/2016). Diana Pacheco thanks to PTDC/BIA-CBI/31144/2017-POCI-01 project-0145-FEDER-031144-MARINE INVADERS, co-financed by the ERDF through POCI (Operational Program Competitiveness and Internationalization) and by the Foundation for Science and Technology (FCT, IP). Ana M.M. Gonçalves acknowledges University of Coimbra for the contract IT057-18-7253, and also thanks to the project MENU—Marine Macroalgae: Alternative recipes for a daily nutritional diet (FA_05_2017_011), funded by the Blue Fund under Public Notice No. 5—Blue Biotechnology, which partially financed this research.

References

Adharini RI, Suyono EA, Suadi, Jayanti AD, Setyawan AR (2019) A comparison of nutritional values of *Kappaphycus alvarezii*, *Kappaphycus striatum*, and *Kappaphycus spinosum* from the farming sites in Gorontalo Province, Sulawesi, Indonesia. J Appl Phycol 31:725–730. https://doi.org/10.1007/s10811-018-1540-0

Alves C, Silva J, Pinteus S, Gaspar H, Alpoim MC, Botana LM, Pedrosa R (2018) From marine origin to therapeutics: the antitumor potential of marine algae-derived compounds. Front Pharmacol 9:777. https://doi.org/10.3389/fphar.2018.00777

Alwaleed EA (2019) Biochemical composition and nutraceutical perspectives Red Sea seaweeds. Am J Appl Sci 16:346–354. https://doi.org/10.3844/ajassp.2019.346.354

Balasubramaniam V, June Chelyn L, Vimala S, Mohd Fairulnizal MN, Brownlee IA, Amin I (2020) Carotenoid composition and antioxidant potential of *Eucheuma denticulatum, Sargassum polycystum* and *Caulerpa lentillifera*. Heliyon 6:e04654. https://doi.org/10.1016/j.heliyon.2020.e04654

Barbier M, Charrier B, Araujo R, Holdt SL, Jacquemin B, Rebours C (2019) PEGASUS—PHYCOMORPH European guidelines for a sustainable aquaculture of seaweeds, Roscoff, France. https://doi.org/10.21411/2c3w-yc73

Belattmania Z, Engelen AH, Pereira H, Serrão EA, Custódio L, Varela JC, Zrid R, Reani A, Sabour B (2018) Fatty acid composition and nutraceutical perspectives of brown seaweeds from the Atlantic coast of Morocco. Int Food Res J 25:1520–1527

Bito T, Teng F, Watanabe F (2017) Bioactive compounds of edible purple laver *Porphyra* sp. (nori). J Agric Food Chem 65:10685–10692. https://doi.org/10.1021/acs.jafc.7b04688

BNF (2016) Nutrition requirements: reference nutrient intakes for minerals

Campbell I, Kambey CSB, Mateo JP, Rusekwa SB, Hurtado AQ, Msuya FE, Stentiford GD, Cottier-Cook EJ (2020) Biosecurity policy and legislation for the global seaweed aquaculture industry. J Appl Phycol 32:2133–2146. https://doi.org/10.1007/s10811-019-02010-5

Cargill (2021a) WavePure® ADG for gelled dairy desserts

Cargill (2021b) Carrageenan for innovative textures

Černá M (2011) Chapter 24 - Seaweed proteins and amino acids as nutraceuticals. In: Kim S-K (ed) Advances in food and nutrition research. Academic, San Diego, pp 297–312

Cho TJ, Rhee MS (2020) Health functionality and quality control of laver (*Porphyra, Pyropia*): current issues and future perspectives as an edible seaweed. Mar Drugs 18:1–31. https://doi.org/10.3390/md18010014

Cotas J, Leandro A, Monteiro P, Pacheco D, Figueirinha A, Gonçalves AMM, da Silva GJ, Pereira L (2020a) Seaweed phenolics: from extraction to applications. Mar Drugs 18:384. https://doi.org/10.3390/md18080384

Cotas J, Leandro A, Pacheco D, Gonçalves AMMM, Pereira L (2020b) A comprehensive review of the nutraceutical and therapeutic applications of red seaweeds (Rhodophyta). Life 10:19. https://doi.org/10.3390/life10030019

Cunha L, Grenha A (2016) Sulfated seaweed polysaccharides as multifunctional materials in drug delivery applications. Mar Drugs 14:42. https://doi.org/10.3390/md14030042

Dawes CJ (1995) Marine Botany. John Wiley & Sons, New York

De Corato U, Salimbeni R, De Pretis A, Avella N, Patruno G (2017) Antifungal activity of crude extracts from brown and red seaweeds by a supercritical carbon dioxide technique against fruit postharvest fungal diseases. Postharvest Biol Technol 131:16–30. https://doi.org/10.1016/j.postharvbio.2017.04.011

FAO (2018) The state of the world fisheries and aquaculture—meeting the sustainable development goals, vol 3. FAO, Rome

FAO (2020) The state of world fisheries and aquaculture. Sustainability in action. Rome

Ferdouse F, Holdt SL, Smith R, Murúa P, Yang Z, et al (2018) The global status of seaweed production, trade and utilization. FAO Globefish Research Programme 124, p 120

Ganesan AR, Shanmugam M (2020) Isolation of phycoerythrin from *Kappaphycus alvarezii*: a potential natural colourant in ice cream. J Appl Phycol 32:4221–4233. https://doi.org/10.1007/s10811-020-02214-0

García-Poza S, Leandro A, Cotas C, Cotas J, Marques JC, Pereira L, Gonçalves AMMM (2020) The evolution road of seaweed aquaculture: cultivation technologies and the industry 4.0. Int J Environ Res Public Health 17(8):6528. https://doi.org/10.3390/ijerph17186528

Garcia-Vaquero M, Hayes M (2016) Red and green macroalgae for fish and animal feed and human functional food development. Food Rev Int 32:15–45. https://doi.org/10.1080/87559129.2015.1041184

Garcia-Vaquero M, Rajauria G, O'Doherty JV, Sweeney T (2017) Polysaccharides from macroalgae: recent advances, innovative technologies and challenges in extraction and purification. Food Res Int 99:1011–1020. https://doi.org/10.1016/j.foodres.2016.11.016

Griffiths M, Harrison STL, Smit M, Maharajh D (2016) Major commercial products from micro- and macroalgae. In: Bux F, Chisti Y (eds) Algae biotechnology. Green energy and technology. Springer, Cham, pp 269–300. https://doi.org/10.1007/978-3-319-12334-9_14

Holdt SL, Kraan S (2011) Bioactive compounds in seaweed: functional food applications and legislation. J Appl Phycol 23:543–597. https://doi.org/10.1007/s10811-010-9632-5

Hongfeng G, Minghou J, Wenda C (1993) Comparative studies on structural feature of agar polysaccharides from *Porphyra haitanensis* grown in south and North China. Chin J Oceanol Limnol 11:25–33. https://doi.org/10.1007/BF02850725

Huang M, Yang H (2019) *Eucheuma* powder as a partial flour replacement and its effect on the properties of sponge cake. LWT 110:262–268. https://doi.org/10.1016/j.lwt.2019.04.087

Irene B, Ikram B, Christian BG, Nina LS, Rune W, Heidi A, Svenja H, Erik-Jan L (2018) Chemical characterization of 21 species of marine macroalgae common in Norwegian waters: benefits of and limitations to their potential use in food and feed. J Sci Food Agric 98:2035–2042. https://doi.org/10.1002/jsfa.8798

John OD, du Preez R, Panchal SK, Brown L (2020) Tropical foods as functional foods for metabolic syndrome. Food Funct 11:6946–6960. https://doi.org/10.1039/D0FO01133A

Kavale MG, Kazi MA, Brodie J (2021) *Phycocalidia* species (Bangiales, Rhodophyta), from the warm west coast of India. Eur J Phycol 56(3):337–347. https://doi.org/10.1080/09670262.2020.1829714

Kazir M, Abuhassira Y, Robin A, Nahor O, Luo J, Israel A, Golberg A, Livney YD (2019) Extraction of proteins from two marine macroalgae, *Ulva* sp. and *Gracilaria* sp., for food application, and evaluating digestibility, amino acid composition and antioxidant properties of the protein concentrates. Food Hydrocoll 87:194–203. https://doi.org/10.1016/j.foodhyd.2018.07.047

Khan BM, Qiu H-M, Wang X-F, Liu Z-Y, Zhang J-Y, Guo Y-J, Chen W-Z, Liu Y, Cheong K-L (2019) Physicochemical characterization of *Gracilaria chouae* sulfated polysaccharides and their antioxidant potential. Int J Biol Macromol 134:255–261. https://doi.org/10.1016/j.ijbiomac.2019.05.055

Kim S (2011) In: Kim S-K (ed) Handbook of marine macroalgae. John Wiley & Sons, Chichester. https://doi.org/10.1002/9781119977087

Kim JK, Yarish C, Hwang EK, Park M, Kim Y (2017) Seaweed aquaculture: cultivation technologies, challenges and its ecosystem services. Algae 32:1–13. https://doi.org/10.4490/algae.2017.32.3.3

Kim Y-M, Han TU, Lee B, Watanabe A, Teramae N, Kim J-H, Park Y-K, Park H, Kim S (2018) Analytical pyrolysis reaction characteristics of *Porphyra tenera*. Algal Res 32:60–69. https://doi.org/10.1016/j.algal.2018.03.003

Kraan S (2012) Algal polysaccharides, novel applications and outlook. In: Carbohydrates—comprehensive studies on glycobiology and glycotechnology. InTech, London. https://doi.org/10.5772/51572

Kumari P, Bijo AJ, Mantri VA, Reddy CRK, Jha B (2013) Fatty acid profiling of tropical marine macroalgae: an analysis from chemotaxonomic and nutritional perspectives. Phytochemistry 86:44–56. https://doi.org/10.1016/j.phytochem.2012.10.015

Kumoro AC, Johnny D, Alfilovita D (2016) Incorporation of microalgae and seaweed in instant fried wheat noodles manufacturing: nutrition and culinary properties study. Int Food Res J 23:715–722

Larrea-Marín MT, Pomares-Alfonso MS, Gómez-Juaristi M, Sánchez-Muniz FJ, Ródenas de la Rocha S (2010) Validation of an ICP-OES method for macro and trace element determination in *Laminaria* and *Porphyra* seaweeds from four different countries. J Food Compos Anal 23:814–820. https://doi.org/10.1016/j.jfca.2010.03.015

Leandro A, Pacheco D, Cotas J, Marques JC, Pereira L, Gonçalves AMM (2020a) Seaweed's bioactive candidate compounds to food industry and global food security. Life 10:140. https://doi.org/10.3390/life10080140

Leandro A, Pereira L, Gonçalves AMMM (2020b) Diverse applications of marine macroalgae. Mar Drugs 18:17. https://doi.org/10.3390/md18010017

Levine IA, Sahoo D (2010) Porphyra: harvesting gold from the sea. Journal of Chemical Information and Modeling. I. K. International Publishing House Pvt. Ltd., New Delhi

Lopes G, Sousa C, Valentão P, Andrade PB (2013) Sterols in algae and health. In: Bioactive compounds from marine foods: plant and animal sources. Wiley, Oxford, pp 173–191. https://doi.org/10.1002/9781118412893.ch9

López-López I, Bastida S, Ruiz-Capillas C, Bravo L, Larrea MTT, Sánchez-Muniz F, Cofrades S, Jiménez-Colmenero F (2009) Composition and antioxidant capacity of low-salt meat emulsion model systems containing edible seaweeds. Meat Sci 83:492–498. https://doi.org/10.1016/j.meatsci.2009.06.031

Morais T, Cotas J, Pacheco D, Pereira L (2021) Seaweeds compounds: an ecosustainable source of cosmetic ingredients? Cosmetics 8:8. https://doi.org/10.3390/cosmetics8010008

Mortensen A, Aguilar F, Crebelli R, Di Domenico A, Frutos MJ, Galtier P, Gott D et al (2016) Re-evaluation of agar (E 406) as a food additive. EFSA J 14:e04645. https://doi.org/10.2903/j.efsa.2016.4645

Murata M, Nakazoe JI (2001) Production and use of marine algae in Japan. Jpn Agric Res Q 35(4):281–290. https://doi.org/10.6090/jarq.35.281

Naseri A, Jacobsen C, Sejberg JJP, Pedersen TE, Larsen J, Hansen KM, Holdt SL (2020) Multi-extraction and quality of protein and carrageenan from commercial spinosum (*Eucheuma denticulatum*). Foods 9:1072. https://doi.org/10.3390/foods9081072

Necas J, Bartosikova L (2013) Carrageenan: a review. Vet Med 58:187–205. https://doi.org/10.17221/6758-VETMED

Niu J-F, Chen Z-F, Wang G-C, Zhou B-C (2010) Purification of phycoerythrin from *Porphyra yezoensis* Ueda (Bangiales, Rhodophyta) using expanded bed absorption. J Appl Phycol 22:25–31. https://doi.org/10.1007/s10811-009-9420-2

Othman R, Rasib AAA, Ilias MA, Murthy S, Ismail N, Hanafi NM (2019) Transcriptome data of the carrageenophyte *Eucheuma denticulatum*. Data Brief 24:103824. https://doi.org/10.1016/j.dib.2019.103824

Pereira L (2011) A review of the nutrient composition of selected edible seaweeds. In: Seaweed: ecology, nutrient composition and medicinal uses. Nova Science, Hauppauge, pp 15–47

Pereira SA, Kimpara JM, Valenti WC (2020) A bioeconomic analysis of the potential of seaweed *Hypnea pseudomusciformis* farming to different targeted markets. Aquacult Econ Manag 24:507–525. https://doi.org/10.1080/13657305.2020.1803445

Qiu Y, Jiang H, Fu L, Ci F, Mao X (2021) Porphyran and oligo-porphyran originating from red algae Porphyra: preparation, biological activities, and potential applications. Food Chem 349:129209. https://doi.org/10.1016/j.foodchem.2021.129209

Rahikainen M, Yang B (2020) Macroalgae as food and feed ingredients in the Baltic Sea region—regulation by the European Union. Finland

Rasyid A, Ardiansyah A, Pangestuti R (2019) Nutrient composition of dried seaweed Gracilaria gracilis. ILMU KELAUTAN Indones J Mar Sci 24(1):1–16. https://doi.org/10.14710/ik.ijms.24.1.1-6

Rioux L, Turgeon SL (2015) Chapter 7 - Seaweed carbohydrates. In: Seaweed sustainability: food and non-food applications. Elsevier, Amsterdam. https://doi.org/10.1016/B978-0-12-418697-2/00007-6

Rosa J, Lemos MFL, Crespo D, Nunes M, Freitas A, Ramos F, Pardal MÂ, Leston S (2020) Integrated multitrophic aquaculture systems—potential risks for food safety. Trends Food Sci Technol 96:79–90. https://doi.org/10.1016/j.tifs.2019.12.008

Ruocco N, Costantini S, Guariniello S, Costantini M (2016) Polysaccharides from the marine environment with pharmacological, cosmeceutical and nutraceutical potential. Molecules 21(5):551. https://doi.org/10.3390/molecules21050551

Sangha JS, Kandasamy S, Khan W, Bahia NS, Singh RP, Critchley AT, Prithiviraj B (2015) λ-Carrageenan suppresses tomato chlorotic dwarf viroid (TCDVd) replication and symptom expression in tomatoes. Mar Drugs 13:2875–2889. https://doi.org/10.3390/md13052875

Sasuga K, Yamanashi T, Nakayama S, Ono S, Mikami K (2017) Optimization of yield and quality of agar polysaccharide isolated from the marine red macroalga *Pyropia yezoensis*. Algal Res 26:123–130. https://doi.org/10.1016/j.algal.2017.07.010

Ścieszka S, Klewicka E (2019) Algae in food: a general review. Crit Rev Food Sci Nutr 59:3538–3547. https://doi.org/10.1080/10408398.2018.1496319

Shama A, Joyce SG, Mari FD, Ranga Rao A, Ravishankar GA, Hudaa N (2019) Macroalgae and microalgae: novel sources of functional food and feed. In: Ravishankar GA, Rao AR (eds) Handbook of algal technologies and phytochemicals: Volume I: food, health and nutraceutical applications. CRC Press, Boca Raton, pp 207–219

Shannon E, Abu-Ghannam N (2019) Seaweeds as nutraceuticals for health and nutrition. Phycologia 58:563–577. https://doi.org/10.1080/00318884.2019.1640533

Smit AJ (2004) Medicinal and pharmaceutical uses of seaweed natural products: a review. J Appl Phycol 16:245–262. https://doi.org/10.1023/B:JAPH.0000047783.36600.ef

Sonani RR (2016) Recent advances in production, purification and applications of phycobiliproteins. World J Biol Chem 7:100. https://doi.org/10.4331/wjbc.v7.i1.100

Tanna B, Mishra A (2018) Metabolites unravel nutraceutical potential of edible seaweeds: an emerging source of functional food. In: Comprehensive reviews in food science and food safety. Blackwell, Malden. https://doi.org/10.1111/1541-4337.12396

Tanna B, Mishra A (2019) Nutraceutical potential of seaweed polysaccharides: structure, bioactivity, safety, and toxicity. Compr Rev Food Sci Food Saf 18:817–831. https://doi.org/10.1111/1541-4337.12441

Torres P, Nagai A, Teixeira DIA, Marinho-Soriano E, Chow F, dos Santos DYACAC (2019) Brazilian native species of *Gracilaria* (Gracilariales, Rhodophyta) as a source of valuable compounds and as nutritional supplements. J Appl Phycol 31:3163–3173. https://doi.org/10.1007/s10811-019-01804-x

Tsuge K, Okabe M, Yoshimura T, Sumi T, Tachibana H, Yamada K (2004) Dietary effects of Porphyran from *Porphyra yezoensis* on growth and lipid metabolism of Sprague-Dawley rats. Food Sci Technol Res 10:147–151. https://doi.org/10.3136/fstr.10.147

Valado A, Pereira M, Caseiro A, Figueiredo JP, Loureiro H, Almeida C, Cotas J, Pereira L (2020) Effect of carrageenans on vegetable jelly in humans with hypercholesterolemia. Mar Drugs 18:19. https://doi.org/10.3390/md18010019

Villanueva MJ, Morcillo M, Tenorio MD, Mateos-Aparicio I, Andrés V, Redondo-Cuenca A (2014) Health-promoting effects in the gut and influence on lipid metabolism of *Himanthalia elongata* and *Gigartina pistillata* in hypercholesterolaemic Wistar rats. Eur Food Res Technol 238:409–416. https://doi.org/10.1007/s00217-013-2116-5

Wahlström N, Harrysson H, Undeland I, Edlund U (2018) A strategy for the sequential recovery of biomacromolecules from red macroalgae Porphyra umbilicalis Kützing. Ind Eng Chem Res 57:42–53. https://doi.org/10.1021/acs.iecr.7b03768

Wanyonyi S, du Preez R, Brown L, Paul N, Panchal S (2017) *Kappaphycus alvarezii* as a food supplement prevents diet-induced metabolic syndrome in rats. Nutrients 9:1261. https://doi.org/10.3390/nu9111261

Yang L-E, Deng Y-Y, Xu G-P, Russell S, Lu Q-Q, Brodie J (2020) Redefining *Pyropia* (Bangiales, Rhodophyta): four new genera, resurrection of *Porphyrella* and description of *Calidia pseudolobata* sp. nov. from China. Edited by K. Müller. J Phycol 56:862–879. https://doi.org/10.1111/jpy.12992

Younes M, Aggett P, Aguilar F, Crebelli R, Filipič M, Frutos MJ, Galtier P et al (2018) Re-evaluation of carrageenan (E 407) and processed Eucheuma seaweed (E 407a) as food additives. EFSA J 16(4):e05238. https://doi.org/10.2903/j.efsa.2018.5238

Zhong H, Gao X, Cheng C, Liu C, Wang Q, Han X (2020) The structural characteristics of seaweed polysaccharides and their application in gel drug delivery systems. Mar Drugs 18:658. https://doi.org/10.3390/md18120658

Chapter 14
Seaweed-Based Recipes for Food, Health-Food Applications, and Innovative Products Including Meat and Meat Analogs

Daina Yesuraj, Charu Deepika, Gokare A. Ravishankar, and Ambati Ranga Rao

Abbreviations

ACE	Angiotensin converting enzyme
CO_2	Carbon-dioxide
FDA	Food and drug administration
PKA	Protein kinase A
PUFA	Polyunsaturated fatty acids
Tsp	Teaspoon
UN	United Nations

1 Introduction

Earth constitutes about 71% water and 29% land. Around 96% of water is marine while 3.5% is freshwater. According to (Mora et al. 2011) the oceans inhabit roughly 2.2 million species, whose superficial layer is occupied by algae. This indicates

D. Yesuraj
General Undergraduate Curriculum Requirements, University of Dubai, Dubai, UAE
e-mail: dyesuraj@ud.ac.ae

C. Deepika
Institute of Molecular Bioscience, The University of Queensland, Brisbane, QLD, Australia

G. A. Ravishankar
Dr. C. D. Sagar Centre for Life Sciences, Dayananda Sagar College of Engineering, Dayananda Sagar Institutions, Bengaluru, Karnataka, India

A. Ranga Rao (✉)
Department of Biotechnology, Vignan's Foundation of Science, Technology and Research Deemed to be University, Vadlamudi, Guntur, Andhra Pradesh, India

A. Ranga Rao, G. A. Ravishankar (eds.), *Sustainable Global Resources of Seaweeds Volume 2*, https://doi.org/10.1007/978-3-030-92174-3_14

Fig. 14.1 Benefits—seaweed in health food applications (Forster and Radulovich 2015)

there is ample surface to avail and cultivate these photoautotrophic species. These species are cultivated in seawater and have a positive environmental impact by re-establishing marine life, once planted after harvesting. More than 30,000 species have been reported in the literature by Gomez-Zavaglia et al. (2019). On land, seaweed extracts can substantially boost farm productivity without any negative consequences to the environment, making them a sustainable resource in agricultural production as well (Layek et al. 2015). Additionally, seaweeds capture and store CO_2 in their tissues making seaweed farming a sustainable adaptation (Chung et al. 2013) (Fig. 14.1). Impacts of seaweeds on climate change have encouraged relevant research studies on both the Caribbean and Pacific coasts of Costa Rica (Radulovich et al., 2015). Furthermore, it was reported that seaweed cultivation could be a potential lifesaver as they require zero land and surface-level water to thrive. Moreover, seaweed captures carbon, phosphorus, and nitrogen, hence seaweed cultivation effectively reduces carbon emission and thus proves itself a sustainable investment.

Seaweed cultivation can be targeted towards setting them up as nutritive food sources for human consumption (Shama et al. 2019). Edible seaweeds have long been harvested and consumed in Asian cuisine, particularly in Japan and Korea. The Western gastronome differs greatly that of Asia, in terms of seaweed usage where the primary approach has been its utility in gelling agents, food emulsions, and medicinal drugs (https://guide.michelin.com/en/article/dining-in/6-edible-delicious-varieties-of-seaweed). However, Canadaian and American cuisines seem to limit them to sushi and other foreign Asian dishes (Penalver et al. 2020). Lately, France has tried to firmly experiment with European cuisine by introducing seaweeds, but it was regarded too exotic to be included as revealed by Van den Burg et al. (2021). Another inclusion is 'Algenbrot', an overpriced bread in Austria and Germany which constitutes 3% seaweed (McHugh, 2003). Penalver et al. (2020) studied seaweeds as a useful ingredient for a healthy diet in many South-East Asian countries like Japan, China, Korea, Malaysia, Thailand, Indonesia and the Philippines due to its high protein content. These valuable seaweeds are more popular for their nutritional richness in fiber and minerals (Turan and Cırık 2018).

But lack of knowledge of the flavor and texture of seaweeds have limited them from entering the regular kitchen cuisine. From a sensory standpoint, seaweeds can be improvised and seasoned to enhance the flavor, and hence broadly support culinary applications. Due to increased appeal towards sustainable resources and the rise of digital media and online recipes hosted by various great chefs, seaweeds became noticeable. A seaweed recipes are more indicated by their common names in the food sector, for example., Nori or purple laver (*Porphyra* spp.), *aonori* or green laver (*Enteromorpha* spp.), kombu (*Laminaria japonica*), winged kelp (*Alaria esculenta*) wakame (*Undaria pinnatifida*), irish moss (*Chondrus crispus*), mozuku (*Cladosiphon okamuranus*), sea grapes (*Caulerpa lentillifera*), hiziki (*Hizikia fusiforme*), Dulse (*Palmaria palmata*), and ogo (*Gracilaria* spp.). Most common seaweeds used in the preparation of soup, salad and curry across countries include *Ulva* sp., *Sargassum* sp., *Enteromorpha* sp., *Monostroma* sp., *Hydroclathrus* sp., *Laminaria* sp., *Macrocystis* sp., *Caulerpa* sp., *Codium* sp., *Porphyra* sp., *Eucheuma* sp., *Undaria* sp., and *Acanthophora* sp. (Gomez-Zavaglia et al. 2019).

2 Ancient Seaweed Recipes

Seaweeds are renowned for strong biological properties such as anti-inflammatory (Lee et al. 2020) and anti-microbial properties (Cabral et al. 2021). The Romans made good use of this property in treating tissue injuries and skin ailments and evidence show that the Egyptians have used them as a treatment for breast cancer (Pati et al. 2016). Archaeological evidence states that seaweeds were cooked and partially consumed at a 14,000-year-old site in southern Chile (Tom et al. 2008). Sushi is a popular food now and it is about 1500 years old wherein raw fish and sticky rice were mixed with a seaweed called Nori. Even though South Asian nationals like Indonesians, Japanese, and Koreans have understood the nutritional properties, Indians are yet to avail themselves and exploit the benefits. According to Sumayaa and Kavitha (2015) this difference and unpopularity is broadly because of the varied range of spiciness and extreme flavor in Indian cuisines. Southern coastal states of India such i.e. Kerala and Tamil Nadu cultivate seaweeds abundantly and consume them in the form of porridge made from *Gracilaria* species and *Acanthophora* species, but indirect consumption is widespread in the form of phycocolloids added in chocolate, ice cream, jellies, and as stabilizers in food products (Dhargalkar and Verlecar 2009). Onset and incidence of diseases like cardiovascular, obesity, cancer, osteoarthritis and diabetes mellitus are global concerns have been reported to be prevented according to a research conducted in western and other Asian countries (Brown et al. 2014). South India's, Mandapam coast on the Gulf of Mannar, is a dominant harbor for lavish growth of seaweeds (Krishnan et al. 2015).

3 Green Chemistry of Seaweeds

The chemical composition of seaweed is altered upon the type of species, place of cultivation, climatic conditions, and harvesting techniques and periods (Garcia-Poza et al. 2020). According to Penalver et al. (2020) polysaccharides are one of the most important nutritionally relevant components of seaweeds, most of which are indigestible by humans due to lack of cellulase. They can be regarded as soluble dietary fiber (33–75% of the total composition) as reported (Gomez-Ordonez et al. 2010; Ramnani et al. 2012).

Seaweeds have a high concentration of essential vitamins, trace elements, proteins, lipids, polysaccharides, enzymes, and minerals as compared to terrestrial foodstuffs (Lozano Muñoz and Díaz 2020; Penalver et al. 2020). Edible seaweeds act as a good source of minerals like sodium, magnesium, calcium, potassium, chlorine, sulphur, phosphorus, and micronutrients such as iron, iodine, zinc, copper, selenium, molybdenum, fluoride, manganese, boron, nickel and cobalt (Dharmananda 2002). An experiment conducted on 34 edible seaweed products of the *Laminaria* sp., *Undaria pinnatifida*, *Hizikia fusiforme,* and *Porphyra* species revealed that all the nine essential amino acids were detected in most of the seaweed species (Dawczynski et al. 2007). Most rhodophyta species were found to have higher levels of taurine compared to phaeophyceae. The protein content of algae varies greatly between large groups of brown, red, and green algae. In brown algae, the protein content is generally low (5–24% of dry weight), while red and green algae have a higher protein content (10–47% of dry weight) according to reported by Mohamed et al. (2012).

Algae are the prime producers of carotenoid pigments and comprise a family of over 600 natural fat-soluble pigments (Ambati et al. 2019). Carotenoids essentially enhance anti-oxidant effects, but individual carotenoids may also act to give pro-vitamin A function and constitution of macular pigment in the eye (Fiedor and Burda 2014). According to (Eggersdorfer and Wyss 2018) humans cannot produce carotenoids hence need to ingest them in dietary form. The daily iodine requirement of 150 μg/day can be provided by seaweeds which are a chief source of iodine (Ryu et al. 2013). Owing to their high mineral content (Fig. 14.2), they can also be used as a dietary supplement to help achieve the recommended daily amounts of other essential macrominerals and trace elements as stated in an article by Penalver et al. (2020). Algae are a rich source of B-group vitamins in

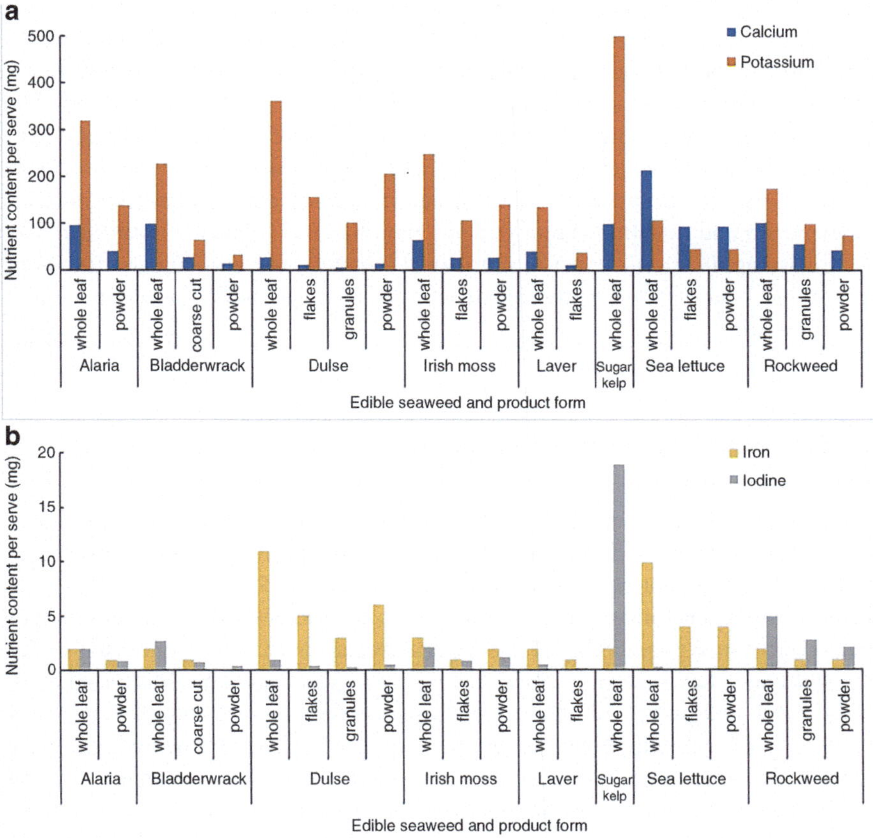

Fig. 14.2 Nutrient content of seaweed based on the product form of edible seaweeds (serve size as defined by FDA). (**a**) Essential minerals—calcium and potassium vs. seaweed. (**b**) Iron and Iodine verses seaweed (Smith et al. 2010; Flores et al. 2015)

particular vitamins B_1 and B_{12}, as well as the lipophilic vitamin A which is derived from β-carotene along with vitamin E also known as tocopherols according to (Skrovankova 2011; Jaime Ortiz 2009). Finally, seaweed nutrition offers one of the few vegetarian alternatives for cyanocobalamin also commonly known as Vitamin B_{12} in the diet (Croft et al. 2005). The red, brown, and green seaweeds have been shown to have medicinal properties for health and disease (treatment and prevention), such as anti-cancer, anti-obesity, anti-diabetic, anti-hypertensive, anti-hyperlipidemic, anti-oxidant, anti-coagulant, anti-inflammatory, immuno-modulatory, anti-estrogenic, thyroid stimulating, neuro-protective, anti-viral, anti-fungal, anti-bacterial and tissue healing properties in *in vivo* studies reported by Mohamed et al. (2012).

Seaweeds provide an ideal combination of low-calorie consumption with high nutrient density which make them a crucial inclusion to everyone's diet (https://

stevesmithhealthfoods.medium.com/discover-healthy-and-enchanting-seaweed-superfoods-7d4aac284752). It is regarded to be an edible substance with exceptional nutritious qualities, counting it in the list of the famous "superfood". Nowadays, recipe books are inclusive of sea vegetables in kitchen cuisine. The use of these valuable seaweeds is increasingly becoming popular as an ingredient in salads and wraps, which is a great indicator of the steady evolution of ordinary cuisine. This also emphasizes a rise in health literacy and information on consumer health and nutrition (Sumayaa and Kavitha 2015). Thus, a traditional mix of culinary applications, boost in nutritional benefits, and growing conditions of zero arable land or fresh water, make seaweeds a sustainable option and future food reservoir.

4 Seaweeds Flavor and Texture

Extensive utilization of seaweeds will not just need their advancement as nutritious and healthy foods but will also require their inclusion in savory culinary dishes and food product formulations that attracts the consumers (Rioux et al. 2017). The umami components (umami refers to the fifth basic taste describing the sensation of deliciousness in Western cuisine) in seaweed species provide a unique burst of flavors and play a key role in sensorial and physiological functions (Mouritsen et al. 2012). Fresh and processed macroalgae have a certain texture and nuance. Novel applications of seaweeds in the culinary field include fresh and fermented seaweeds, redesigned recipes of traditional products, powdered ingredients savored as healthy additives, salty and flavorful seasonings, processed and packed crackers, and gastronomic creations of chefs in cutting-edge restaurants (Figueroa et al. 2021). Nordic seaweeds, especially the large brown seaweed, sugar kelp also known as *Saccharina latissimi* and the red seaweed dulse known as *Palmaria palmata*, dispense synergy in the umami taste when fused with bacon, chicken meat, or dried mushrooms (Mouritsen et al. 2012). These findings also explored the potential of other local Nordic seaweeds like sugar kelp and dulse, which are high in free glutamate and therefore are a great candidate for umami season. The Japanese seaweed kombu has high levels of glutamate, which is the main stimulatory agent in umami and it is exclusively used to prepare the soup broth dashi (Milinovic et al. 2020). Some of the most common seaweeds used in different health foods across the world are listed in Table 14.1.

5 Preparation of Seaweed Recipes

Seaweeds can be eaten raw, cooked, baked, toasted, pureed, dried, powdered, and/or fried. They can be eaten on their own or combined in countless ways with other hot or cold ingredients. So, it can be considered a multi-faceted ingredient.

Table 14.1 Seaweed recipes in various countries and their possible applications

Seaweed species	Country	Seaweed recipes	Applications	References
Kappaphycus alvarezii	Indonesia	Gluten-free pasta	Iron bioavailability and antioxidant properties	Sholichah et al. (2021), Mohamed Fayaz et al. (2005)
Saccharina japonica	Japan	Scallops with seaweed and miso kombu broth	Prebiotic effect and use as functional food	Zhang et al. (2020)
Laminaria japonica	Vietnam	Ice dessert drink with seaweeds	Inhibitory effect against noroviruses	Kim et al. (2020)
Pyropia columbina	New Zealand	Kale, karengo and cucumber salad	Source of biomedical compounds with high concentrations of porphyran, vitamin B_{12} and taurine	Cho and Rhee (2020)
Caulerpa lentillifera	Philippines	Sea grapes salad (green caviar)	High contents of vitamins and minerals	Du Preez et al. (2020)
Laminaria japonica	Korea	Seaweed rice roll (wrap)	Inhibitory effect against noroviruses	Kim et al. (2020)
Laminaria japonica	India	Saag tofu with kombu	Urinary biomarkers for dietary intake of arsenic from seaweeds	Cherry et al. (2019)
Caulerpa lentillifera	Philippines	Lato salad	Improves cardiovascular health	Du Preez et al. (2020)
Laminaria japonica	Ethiopia	Vegan scampi wot	Inhibitory effect against noroviruses	Kim et al. (2020)
Palmaria palmata	Ireland	Dulse/Dillisk bread	High protein content	Cherry et al. (2019)
Palmaria palmata	New Zealand	Dulse fish pie	Antioxidant activity	Lopes et al. (2019)
Pyropia tenera	Japan, Italy	Seaweed seasoning seaweed risotto	Immuno-modulatory, antihypertensive, anticoagulant and anticancer properties of bioactive compounds	Venkatraman and Mehta (2019)
Pyropia tenera	Canada	Nori chips	Immuno-modulatory, anti-hypertensive, anti-coagulant and anticancer properties of bioactive compounds	Venkatraman and Mehta (2019)
Undaria pinnatifida	USA	Seaweed smoothie	Reduced glycemic drive in pre-diabetes	Yoshinaga and Mitamura (2019)

(continued)

Table 14.1 (continued)

Seaweed species	Country	Seaweed recipes	Applications	References
Pyropia tenera	Turkey	Turkish bagel with seaweeds	Immuno-modulatory, antihypertensive, anticoagulant and anticancer properties of bioactive compounds	Venkatraman and Mehta (2019)
Pyropia tenera	Spain	Nori wraps with cauliflower pate	Immuno-modulatory, anti-hypertensive, anti-coagulant and anticancer properties of bioactive compounds	Venkatraman and Mehta (2019)
Pyropia tenera	Netherlands, Mexico	Meat seaweed soup, crunchy seaweed tacos	Immuno-modulatory, antihypertensive, anticoagulant and anticancer properties of bioactive compounds	Venkatraman and Mehta (2019)
Pyropia tenera	Italy, Germany	Gnocchi with seaweed and crab sauce Seaweed tartare	Immuno-modulatory, antihypertensive, anticoagulant and anticancer properties of bioactive compounds	Venkatraman and Mehta (2019)
Pyropia tenera	Japan, China	Chungmu kimbap, the seaweed rice rolls, seaweed popcorn	Immuno-modulatory, antihypertensive, anticoagulant and anticancer properties of bioactive compounds	Venkatraman and Mehta (2019)
Undaria pinnatifida	Korea	Korean seaweed soup (Miyeok Guk)	Antioxidant, anticancer, and anti-coagulant properties	Zhao et al. (2018)
Undaria pinnatifida	New Zealand, Brazil	Seaweed salsa, Vegan stew recipe with wakame	Antioxidant, anticancer, and anticoagulant properties	Zhao et al. (2018)
Undaria pinnatifida	California	Miso coated eggplant brochette	Antioxidant, anticancer, and anticoagulant properties	Zhao et al. (2018)
Undaria pinnatifida	Scotland	Oatmeal with soy nuts and wakame seaweed	Anti-oxidant, anticancer, and anti-coagulant properties	Zhao et al. (2018)
Undaria pinnatifida	USA	Wakame brown rice, seaweed tempura	Antioxidant, anticancer, and anticoagulant properties	Zhao et al. (2018)
Laminaria hyperborea	China	Chinese seaweed salad	Increased quantities of PUFA that enhances human cardiovascular health	Foseid et al. (2017)

(continued)

Table 14.1 (continued)

Seaweed species	Country	Seaweed recipes	Applications	References
Ecklonia radiata	Australia	Pickled seaweed	Gut health benefits	Charoensiddhi et al. (2017)
Sargassum fusiforme	Japan	Healthy avocado and hijiki seaweed sandwich	Anti-tumor effects of saccharides from *Sargassum fusiforme*	Fan et al. (2017)
Himanthalia elongata	Portugal, Spain and Norway	Seaweed pasta	Medicinal and nutritional ingredient	Rajauria et al. (2017)
Chondrus crispus	Cambodia	Irish moss pudding	Neuro-protective effects in elderly people affected with Parkinson's disease	Liu et al. (2015)
Chondrus crispus	Caribbean	Sea moss drink	Neuro-protective effects in elderly people affected with Parkinson's disease	Liu et al. (2015)
Osmundea pinnatifida	Cornwall, England	Pea and seaweed dip	Diverse in amino acids and anti-oxidant property	Paiva et al. (2014)
Pyropia tenera	South Africa	Sesame seed fried seaweed	Source of cobalamin	Yamada et al. (2013)
Pyropia tenera	Japan	Seaweed cookies	Source of cobalamin	Yamada et al. (2013)
Pyropia tenera	China	Chinese dried seaweed used in multiple recipes	Source of cobalamin	Yamada et al. (2013)
Pyropia tenera	Australia	Toasted seaweed chips	Source of cobalamin	Yamada et al. (2013)
Corallina vancouveriensis	Sweden	Coral seaweed jelly recipe	PKA-inhibitory activity	Zivanovic and Skropeta (2012)
Fucus vesiculosus	Korea, Japan, China	Used as a pickled side-dish	Antioxidant capacity	Wang et al. (2012)
Saccharina japonica	London	Kombu seaweed martini	Anticancer effects on certain human prostate cancerous cells	Jo et al. (2012)
Eisenia bicyclis	Croatia	Brown arame rice	Potent antioxidant compounds	Skrovankova (2011).
Eucheuma	India	Used as food ingredient in the preparation of spice	Functional food applications	Senthil et al. (2011)

(continued)

Table 14.1 (continued)

Seaweed species	Country	Seaweed recipes	Applications	References
Alaria marginata	California	Grilled oysters wrapped in Alaria seaweed	Best protein content among kelp seaweeds, rich in niacin and trace elements.	Pereira (2011)
Laminaria japonica	Maine	Maine seaweed salad	Anti-proliferative effects on HT-29 colon cancerous cells.	Go et al. (2010)
Ecklonia kurome	China	Steamed seaweed meat rolls	Antioxidant and antibacterial properties	Kuda et al. (2007)
Enteromorpha	India	Pakoda traditional snack food	Food applications	Mamatha et al. (2007)
Ecklonia kawa	Japan	Okinawan seaweed salad	Antioxidant and anti-inflammatory effects	Shin et al. (2006)
Palmaria palmata	Egypt	Seaweed dukkah	Effective antioxidants	Yuan et al. (2005)
Palmaria palmata	Japan	Seaweed snacks	Effective antioxidants	Yuan et al. (2005)
Undaria pinnatifida	Japan	Seaweed salad using wakame	Anti-hypertensive effects	Sato et al. (2002)
Porphyra umbilicalis	China	Chinese dried seaweed soup	Excellent source of vitamin B_{12}	Watanabe et al. (1999)
Crambe maritima	China	Crispy seaweed	Reduces risk of development of chronic diseases	Quinsac et al. (1994)

Additionally, the heath-effect of seaweeds is diverse because seaweeds have distinct nutrient profiles. Some of the popular recipes of seaweeds are shown in Fig. 14.3. According to Mamatha et al. (2007), *Enteromorpha* compressa, a green seaweed, is utilised in the manufacture of *Pakoda*, a popular Indian snack dish. Senthil et al. (2011) investigated the functional qualities of *Eucheuma* (*Kappaphycus* alvarezzi), a red seaweed used in the manufacture of spices, including water holding capacity, oil holding capacity, and solubility. Here we describes few specific recipes in the following sub-section.

5.1 *Seaweed Dumplings*

The ingredients used are 50 dumpling skins, 0.25 cup enoki mushrooms, dried wakame seaweed strands about ½ cup when, soaked, a bunch of scallions, a bunch of fresh cilantro, half a cup of cabbage, a tbsp. of soy sauce, 2 tsp. any cooking oil,

Fig. 14.3 Seaweed recipes for food and health food applications; seaweed dumpling (**a**), seaweed butter (**b**), seaweed salt (**c**), seaweed soufflé (**d**), seaweed ice cream (**e**), seaweed bread (**f**), seaweed chocolate (**g**), seaweed biscuits (**h**), and seaweed cheese (**i**)

few garlic cloves, 2 tsp. cornstarch, salt and pepper to taste. All vegetables are finely chopped. The preparation involves soaking the seaweed in warm water until soft and drying it. Mix all ingredients except the dumpling skins. Place a dumpling skin on the hand and fill 1 tsp. of the mixture in the center. The skin must be folded in a way that the edges are sealed. Heat oil in a pan and line the dumplings in a way that they don't stick to each other. Cook on high flame for around 2 min. Then add water to the pan to cover ¼ of the height of dumplings. Reduce heat to medium flame and cover the pan. Let it cook for about 4 min, or until the skins are translucent. Then remove the lid cover and cook until all the water has evaporated. Carefully remove the dumplings with a spatula and serve with soy sauce and/or any sauce of your choice.

Dumplings are steamed edibles that use very little oil. This when combined with nutritive and gluten-free seaweeds is a great healthy alternative. Not only do you get access to easy carbohydrates, but you also support gut health. Seaweeds are high in several vitamins and amino acids and ensures help with diabetes as well. This is a considerable recommendation if you want to lose weight, satisfy taste-buds and keep healthy.

5.2 Seaweed Butter

The ingredients used are 100 g unsalted and softened butter and 2.5 g of any seaweed. The preparation method involves toasting the seaweed in a hot dry pan until it turns crunchy and emits a savory smell. Powder it and stir it along with the softened butter. The butter can be smoothly laid out onto a kitchen film and sealed to prevent contamination and freeze it. Before use, thaw it for about 10 min. Seaweed butter is entirely natural and free of add-ons. The taste of this novel butter is fresh and savory and has been positively reported among people. In addition to the flavor, seaweeds are also rich in iron, vitamin C, antioxidants, soluble and insoluble fiber (Mohamed Fayaz et al. 2005). Thus, a perfectly healthy ingredient to have in the kitchen.

5.3 Seaweed Salt

The contents are 1 L of water and 3–5 strands of kelp seaweed. Sediment harvested seawater and filter the top-layer clear water to remove any debris. Boil this saltwater on high flame for 2–3 h. Cut the kelp strands into small pieces and add them to the water. After ½ hour, scrape the bottom of the vessel to remove the salt that gets sedimented. Scrape the sides and bottom of the pan often and turn the heat down. Repeat this every ½ h until almost all the water is evaporated and the seaweed salt is found to settle at the bottom layer of the vessel.

Seaweeds have a significant amount of iodine, which is a key mineral for healthy functioning of the thyroid gland and vitamin B_{12} a key precursor for healthy nerves. Consumption of seaweed salt is a plant-based option to take care of iodine deficiencies and improve brain function.

5.4 Seaweed Souffle

The ingredients are a cup of Nori seaweed strands, ¼ cup butter, 1/3 cup plain flour (seaweed-based flour can also be used), 1/3 cup fresh cream, ½ cup chopped scallions, 4 egg yolks, and egg whites separate, seaweed salt and black pepper to taste, ½ cup cheese and coriander leaves. The preparation involves pre-heating the oven to 180 °C. Stir-fry Nori and scallions in butter on low-flame and add flour gradually and stir until smooth. Cook for a minute and add cream and cook over medium-flame until mixture thickens. Set the mixture aside to cool. Whisk the egg yolks one by one into the smooth sauce. Beat the egg whites until they're stiff and quickly and gently fold them into the yolk mixture so as not to collapse the volume. Transfer the mixture to buttered souffle cup and bake for around 20 min garnish and serve.

Souffle is a dish that has sufficient fats and proteins if consumed in right amounts. It not only has a great flavor but also a smooth texture by getting in just the right amount of air. But because it is a high-calorie edible and people tend to overdo it, seaweed-based soufflé is the healthier option to choose. Seaweed soufflé is a low-calorie version as it is full of polyunsaturated fatty acids. The digestibility of seaweed proteins relaxes the gut and its soluble fibrous content not only keeps you feel full for a longer time but also prevents chronic constipation and heart conditions. Nori seaweed is just sautéed for a while which retains all polysaccharides that improve overall digestive and cardiovascular health.

5.5 *Seaweed Soup*

The contents are a bunch of wakame seaweed, a liter of water, ½ cup chopped vegetables, ½ tbsp miso, ¼ cup minced green onions, ½ cup sliced white onions, and a bunch of freshly chopped cilantro. For the preparation, rinse and soak wakame in cold water for a few minutes and chop. Boil water and add white onions and wakame. Cook on medium heat for 5 min. Add the vegetables and cook for 5 more minutes. Reduce the flame to low heat, add miso and cook for 2 min. Garnish with green onions and cilantro.

Seaweed soup is a great combination because the goodness of this healthy dish brings forth the rich nutritive contents of seaweeds. The macronutrient and micronutrients are in significant quantities. Not only is it easy to digest and absorb, but it also prevents deficiencies arising due to nutrient imbalances. Certain seaweed polysaccharides which are difficult to digest serve as soluble fibers and aid in the smooth transit of food in the alimentary canal. The antioxidants protect the body from free radicals and decrease cellular inflammation. The only care that should be taken is to prevent excess iodine intake as seaweeds are mostly naturally salty.

5.6 *Seaweed Ice-Cream*

The ingredients are 2½ cup milk, a cup of fresh cream, a cup of sugar, ½ cup of unsweetened cocoa, ½ tsp. salt, 5 sheets of Nori seaweed, 4 egg yolks, and 4 oz. bitter-sweet chocolate. The preparation involves whisking all ingredients in a pan together except Nori sheets and egg yolks. Whisk continuously on medium heat until sugar and cocoa dissolve. Cook for another 5 min and add Nori. Cook until the seaweed sheets become soft and remove from heat. Allow it to rest for 30 min. Strain half of the mixture into another bowl and reserve the rest. In a large bowl, place bittersweet chocolate. In another small bowl, whisk the egg yolks and slowly add the reserved milk mixture into it. On medium heat, continuously stir this whole mixture. In around 15 min when the mixture starts to thicken, remove from heat and pour over bitter sweet chocolate. Combine them well and cool them to room temperature. Refrigerate it and allow sufficient aeration. After about 3 h pour the mixture into a freezer bowl of an ice-cream maker and proceed as per labeled instructions. Once it's done, freeze until ready to serve.

Seaweeds and seaweed extracts like carrageenan and alginate are a source of natural thickeners and gelling and stabilizing agents that can be particularly useful in making desserts. Stabilizers in seaweeds reduce the amount of free water in ice cream binding it to the gel structure, resulting in a softer texture of the ice cream, unlike conventional stabilizers which add to the overall fat content of traditional ice creams (Irawan and Fitriyana 2021). Seaweed ice cream is a low-calorie ice cream which is highly nutritious due to diverse minerals, vitamins and proteins in it.

5.7 Seaweed Bread

The ingredients are 2 tbsp. Seaweed flakes, 500 g flour (seaweed-based flour can also be used), 20 g fresh yeast, 1½ tsp. dried active yeast, 1 tsp. salt, 100 mL cooking oil, and 400 mL warm water. Begin by mixing the flour, seaweed flakes, and salt along with fresh yeast in a bowl. Knead well using oil and water and make a soft and smooth dough. Water can be added based on the dryness of the dough cover the dough with a moist cloth. Leave it to rise. After about 2 h, knead the dough again and divide it into two balls. Roll them in smaller circles and place them in bread tins. Cover it with a moist cloth and leave it to rise for another hour. Poke the dough with a fork or finger, oil it, and cover it with a moist cloth to prevent dryness. After ½ h, add the seaweed flakes and salt and bake at 200 °C for ½ h.

Seaweed bread is another example of sustainable food innovation. Kombu flakes used here add to the nutritional profile offering a healthier diet as compared to conventional bread. It is reported by Brownlee et al. (2011) that incorporation of seaweed decreases the use of added salt which increases the shelf-life of the bread and prevents fungal growth. It is nutrient-rich preservative-free bread that can be preserved longer without the actual use of preservatives. This vouches for health benefits and consumers have reported positive results in seaweed bread taste.

5.8 Seaweed Chocolate

The ingredients are 3 tsp. seaweed flakes, 300 g dark chocolate and 2 handfuls of dry fruits. The preparation involves melting the chocolate in a dry bowl over hot water. Take care to not burn the chocolate. Once melted, add the chocolate into molds and tap the molds to remove any air bubbles and spread the chocolate uniformly. Dust the seaweed flakes onto the chocolate and refrigerate instantly. The significance here is that seaweeds are rich in minerals than any other class of food which makes seaweed chocolate better than plain chocolates. Thahira Banu and Uma Mageswari (2015) reported that consumption of seaweed chocolate significantly increased the hemoglobin levels in anemic adolescent girls. This emphasizes that not only can it be used as a gluten-free version of chocolate but it can also remove nutritional deficiencies. As Dulse seaweed is a promising vegetable resource of iron, it can be used as an effective nutrient supplement for all types of people. In countries like India, where iron deficiency is on the high, seaweed chocolates can deliver great results in preventing and curing iron deficiencies.

5.9 Seaweed Pickle

The ingredients are 1 kg of seaweed flakes, 1 tbsp. Fenugreek powder, ½ tbsp. Mustard seeds, 3–4 tbsps. Gingelly oil, 1 tsp. asafoetida, 1 tbsp. Tamarind extract, salt and chilli powder as per taste. The preparation involves heating the oil and spluttering mustard seeds. Then add seaweed flakes, fenugreek powder, chilli powder, asafoetida and tamarind extract and mix well. Add the contents in a jar and cover with a lid. Allow the contents to rest for a day and then wrap the jar with a muslin cloth and keep in sunlight for few hours. Add oil in the jar to cover the pickle level by at least ½ an inch. Cover airtight and let the contents rest for a week before use.

Seaweed pickles can relieve one of the hypertensive effects due to reduced iodine intake while preparation because seaweeds are naturally salty. Seaweed consumption by humans and reduced incidence of goiter and thyroid imbalances have been reported since long. Seaweed pickles can be consumed as an effective iron deficiency remover as revealed by Wells et al. (2017). The only concern is avoiding over-consumption of seaweed pickle to avoid excess iodine intake.

5.10 Seaweed Flour for Biscuits

The ingredients are seaweed flour, flour, butter, egg yolk, sugar, milk powder, baking powder and corn starch. The preparation steps involve washing and soaking seaweeds for 3 days so that the raw odor and color of seaweed is lost. Slice them for faster sun-drying. Powder the dried slices and strain to get a fine and uniform seaweed flour. For preparing the biscuit, mix well the butter and egg yolk for 10 min. Next add the remaining ingredients and stir again for 10 min. Then mix the flour with freshly processed seaweed flour and knead until the dough is smooth and soft. The dough is enhanced with taste enhancers on a butter-spread pan. Then bake the dough in an oven with a temperature of 150 °C for 15 min. Due to its high protein content, seaweed usage enhances the nutritive formulation of biscuit flour. This is especially beneficial to school-going children to prevent and cure protein disorders. The diverse amino acid profile highlights its potential in health benefits and prevention of malnutrition. A study conducted by Jenifer and Kanjana (2018) on malnourished children aged 5–6 years proved that consumption of seaweed flour-based biscuits increased serum protein levels and mineral absorption in the body. Significant increases in height and weight were observed in the earlier study which leaves us with the conclusion that seaweed-based flours can be increasingly used in biscuit and cookie manufacturing as well as in the baking industry. It can also be used as an effective supplementation for increasing carbohydrate and protein levels and minimal lipid levels.

5.11 Fresh Cheese with Dulse Seaweed

The ingredients are 1.25 kg dulse seaweed, 1.5 L of milk, 50 g of cream, 25 g of buttermilk and 5 g of rennet. The preparation involves mixing seaweed and milk in a plastic bag. Create vacuum and seal it. Leave it in refrigerator overnight and allow cold infusion. Next day, strain the milk to get a uniform mixture, heat it to 33 °C and add the remaining ingredients. Pour the mixture into plastic containers, cover tightly and cook in an oven at 36 °C for 45 min. The incorporation of seaweeds in cheese is a creative way of preparing functional food. Controlled consumption of cheese has health benefits and is clinically proven to show improvements in cardiovascular health. The addition of seaweeds into cheese formulations and dairy products is interesting because it not only elevates the nutritional profile of cheese but also improves its appeal to the general public. This is because, seaweeds in addition to having high peptides, dietary fiber, vitamins and mineral content, also has bioactive properties like anti-microbial, anti-inflammatory, ACE-inhibitory and cardio-protection as reported by Hell et al. (2018). This develops the bio-functional properties of cheese.

6 Seaweed-Based Meat Analogs

The consumption of animal-derived products is declined due to dietary restrictions as vegan, vegetarian or flexitarian, health, or environmental issues (Chen et al. 2019). As a result, we may offer a dietary choice that substitutes proteins from natural sources such as algae or seaweed, allowing consumers to have the same feeling of eating meat without really consuming it. Meat analogs are designed for food products, and the majority of non-vegetarians consume them as meat (Smetana et al. 2015, 2019). It is also known as a meat substitute. Meat analog is the most sustainable alternative, as it allows us to replace proteins with plant-based materials like soya, algae, and seaweed (Bhuva et al. 2021; Rubio et al. 2020). It is required to develop an analogue that has similar taste, texture and flavor as meat. According to Egbert and Borders (2006), meat analogue contains water (50–80%), plant proteins (10–25%), other proteins (4–29%), 3–10% flavoring agents, 0–15% oil, 1–5% binding agent and less than 0.55 % coloring agent. Water is an important component of the constituents because it provides desired juiciness, shape and acts as emulsifier for the ingredients. Protein is important for nutrition and it is responsible for the mouth feel, texture and appearance. Aside from the aforementioned benefits, protein also has good solubility, hydration, interfacial and gelation properties and flavor binding. The plant based proteins are chosen on the texture desired (Rubio et al. 2020). The plant based proteins such as soybean, cottonseed, and wheat gluten to make vegan meat analogue (Riaz 2011). The utilization of algal protein in meat substitutes is one of the most difficult propositions in the food industry. However, other processing procedures like as emulsification, extrusion, and beaching can be used to alleviate these difficulties (Caporgno et al. 2020). However, we need to do a lot of research to find a viable algal protein for this purpose. Another key consideration is to choose a heat-stable protein that will remain stable during the cooking of meat analogues (Becker 2007). However, considering the numerous health benefits linked to seaweed consumption, there has been an increasing interest in introducing seaweeds into western diets in recent years. The increased demand for seaweed products in the culinary and medicinal sectors, according to the seafood research, will enhance the industry, which is predicted to reach USD 22.1 billion by 2024 (Afonso et al. 2019) In Europe, the usage of seaweed as a food ingredient has increased, and consumers are increasingly looking for seaweed-based foods such as processed meats, fish, baked goods, and dairy products (Roohinejad et al. 2017).

6.1 Use of Seaweeds as Salt Substitutes in Processed Meat

Developing innovative low-salt meat products while maintaining physicochemical, microbiological, and sensory qualities is important for various reasons. The salt reduction had an unintended effect on texture, frankfurters, and salami, resulting in a diminished salty taste and less intense colour (Pietrasik and Gaudette 2014; McGough et al. 2012; Zanardi et al. 2010). The meat business faces a dual challenge: developing healthier processed foods with appropriate technological and organoleptic properties while maintaining the final product's economic viability. The substitution of calcium, potassium, and magnesium for sodium chloride in the manufacturing of such products is considered in several ways (Cittadini et al. 2020).

Manufacturers of meat products have also looked at using salt substitutes, salt boosters, or natural salty tasting ingredients like yeast extract, hydrolyzed vegetable protein, and seaweeds (Inguglia et al. 2017). Seaweeds are being proposed for use in the development of healthier foods, particularly in the formulation of novel sodium-reduced meat products (Lorenzo et al. 2015). Polysaccharides, fatty acids, carotenoids, phenolics, vitamins, proteins, and minerals are bioactive components produced by seaweeds (Lomartire et al. 2021; Deepika et al. 2016; Deepika 2017a, b). These compounds have been found to have anti-hyperlipidemic, anti-oxidant, anti-cancer, and anti-hypertensive properties (Saeed et al. 2021; Chen et al. 2021). Despite the numerous nutritional benefits of seaweeds, their use in the meat industry to create functional meat products with potential health benefits has yet to be extensively explored (Cofrades et al. 2017; Parniakov et al. 2018; Zugcic et al. 2018; Marti-Quijal et al. 2019). With customer desires for new functional meals, the market for seaweed-based meat products is likely to grow in Europe. Several research projects have been conducted in order to promote the usage of seaweeds as new food additives. Seaweeds are a good source of minerals such as Ca, P, Na, Mg, P K, Mn, I, Zn, and Fe (Jayakody et al. 2021), and their mineral presence allows them to be used as salt substitutes in processed meat, lowering sodium consumption while increasing intake of other minerals not found in sodium salted meat products (Circuncisao et al. 2018). Seaweeds have a low Na/K ratio, making them ideal for lowering blood pressure and reducing the risk of cardiovascular disease. Given the importance of seaweeds, including them into consumables could be a great way for the meat industry to create low-salt meat products with added health advantages. The suitability of seaweeds to replace salt in meat products, as well as the effects of salt replacement on sensorial attributes and physicochemical properties of reformulated meat products, has sparked increased interest due to their functional properties and the growing demand for healthier processed meat products. Seaweeds are eaten for nutrition in Asia, but their application in Western countries has been limited to the extraction of hydrocolloids like alginate, agar, and carrageenan (Catarino et al. 2018). Seaweed constituents are processed into products like AlgySalt®, PureSea and Seagreens® which are used as salt substitutes in processed meals. These innovative components enable low-salt goods to retain their organoleptic and technical qualities. Seaweeds were included in low-salt reformulated meat products, as shown in Table 14.2.

Triki et al. (2017) investigated the sensory, microbiologic and technological effects of replacing sodium chloride in fresh and cooked sausages with a combination of salts or by AlgySalt®. The results indicated that the sample formulated with AlgySalt® showed increased solidity and less cooking losses than the sausages prepared with a mixture of salts. AlgaySalt® provides good binding properties among the meat constituents and water molecules. Microbiological quality was examined in both formulations, and a similar preservative effect was seen when compared to a control sample containing sodium chloride. Fresh meat products made with AlgySalt®, a seaweed-based component, had improved texture, juiciness, and colour, whilst cooked sausages made with the salt mixture had greater sensory acceptability. Choi et al. (2015) reported on frankfurters in which the sodium chloride content was replaced by the inclusion of edible seaweeds. They examined how addition of seaweeds to low-salt frankfurters affected cooking loss, pH, colour, textural profile, and sensory qualities. In comparison to the normal salt sample, frankfurters prepared with sea tangle and sea mustard produced better results in terms of cooking loss, emulsion stability and sensory qualities. The addition of sea mustard to low-NaCl meat emulsion systems was also studied by Kim et al. (2015). The presence of seaweed, resulted in samples with emulsion stability and cooking loss

Table 14.2 Seaweeds effects on reformulated low-salt meat products

Seaweed	Meat products	Properties
Sea mustard, hijiki and glasswort	Frankfurters	Decrease in moisture content, salinity, cooking loss, light ness, redness, gumminess, and chewiness
Porphyra umbilicalis, Palmaria palmate, Himanthalia elongate and *Undaria pinnatifida*		Reduced hardness and chewiness
		Decreased sensory acceptance
		Decreased color values
Eucheuma spinosum	Chicken breast batters	Increased water holding capacity
Sea mustard	Meat emulsion	Increased apparent viscosity
		Reduction in salt content
Wakame	Black puddings	Decreased cooking loss, appearance and color similar to the control; higher species
Palmaria palmate	Cooked ham	Unchanged in yield, texture and colour
		Reduction of 30% salt content
AlgySalt®	Sausages	Decreased cooking loss
		Increased hardness

Source from Gullon et al. (2020)

comparable to commercial formulations. The textural parameters like hardness, springiness, and cohesiveness remained constant when sea mustard was applied in the emulsion-based meat product. Further, it also stipulated that the reformulated meat emulsion exhibited a good viscosity apparently and improves its elasticity. This is an important parameter of quality in emulsion meat products. Barbieri et al. (2016) investigated the efficacy of using a water soluble extract from *Palmaria palmata* as a salt replacement in the preparation of cooked ham. These authors discovered that the addition of seaweed extract had no effect on the yield, texture and color of the reformulated product. Fellendorf et al. (2016) revealed that using Wakame in black puddings with decreased salt content resulted in lesser cooking loss and products with a similar appearance and colour to the control. Another study conducted by Vilar et al. (2020), the impact of the incorporation of *Porphyra umbilicalis, Palmaria palmate*, *Himanthalia elongate* and *Undaria pinnatifida* in the preparation of frankfurters with 50% less salt than their conventional recipe was evaluated. The physicochemical, microbiological and sensory attributes of the reformulated products were compared with the control sample during the chilled storage. The results revealed that the presence of seaweeds resulted in decreasing of the lightness, redness and yellowness values in comparison to the control sample. The authors attributed these changes in the colour to the formation of dark compounds from the Maillard reaction that is occurring between seaweeds and pork meat or to the colouration of the used algae. In terms of textural properties, partial salt replacement with seaweeds reduced hardness and chewiness in all formulations, while increasing springiness and adhesiveness in frankfurters with *P. umbilicalis* or *P. Plamate*. He et al. (2019) investigated the use of *Eucheuma spinosum* in the preparation of chicken breast batters when combined with high pressure. In comparison to the samples without *Eucheuma spinosum* and untreated, the scientists claim that the combined application of seaweeds and high pressure treatment resulted in an improvement in water-holding capacity. Furthermore, image analysis of the samples revealed that the pressure may promote the interaction between meat protein and seaweed carrageenan, which leads to a three-dimensional network structure with small cavities, which favors the trapping of water, ultimately resulting in an increase of water holding capacity of the meat matrix. The authors observed that the addition of algae in combination with high pressure is a suitable strategy for the elaboration of low-salt seaweed chicken breast.

7 Conclusion

The nutritional standards of edible seaweeds owing to its low calorie count, high concentration of proteins, minerals, vitamins, dietary fiber, fatty acids, polysaccharides, and bioactive compounds with broad medicinal potential could further contribute to the improvement of the quality of human life. Daily intake of seaweeds could exert beneficial effects such as anti-cancer, anti-viral, anti-coagulant, hypocholesterolemic, anti-tyrosinase, anti-obesity and anti-oxidant properties (Deepika 2017c). In addition, macroalgae have a quality feature, which allows their inclusion in dairy, fish, and meat, preserving or enhancing its sensitive, nutritive, and quality well-being. This is the case of the *Rhodophyta* species; *Gracilaria domingensis* used for agar production. Reports have supported its use as a texture modifier in fermented milk, as well as its possible use as an alternative to gelatin (Tavares Estevam et al. 2016). The gastronomic future of brown and red seaweeds has emerged from local waters and extended towards provide umami flavoring in the New Nordic Cuisine. An explosion of functional characteristics may conduct seaweeds to a variety of culinary dishes and recipe re-formulations. Seaweeds may be used along with or as a replacement of certain daily-use vegetables to further boost healthy food. Partnering with culinary experts along with scientific technology innovation can increase the perspective and adoption of seaweeds as an edible resource by presenting foods where seaweeds are presented. The addition of seaweeds to meat products as a substitute appears to be a practical way to improve the high quality functional characteristics of food ingredients. However, based on the research, we must overcome a few obstacles, including the selection of seaweeds for desired features in meat or food products, as well as its dosage. Texture, sensory, nutritional, physical and chemical characteristics, stability, microbiological and shelf-life features of seaweed incorporation in reformulated products must all be evaluated. As a result, significant research is required to fill gaps that will allow the expansion of new functional seaweed-based meat products that are in line with customer desires.

Acknowledgments DY would also like to thank Mr. Hatim Khalil from the University of Dubai for his constant belief and encouragement. CD thank The University of Queensland, Australia for International Graduate Scholarship. ARR acknowledge fund for Improvement of Science & Technology Infrastructure in Higher Educational Institutions (FIST Project no: LSI-576/2013), Department of Science and Technology, Government of India and Centre of Excellence, Department of Biotechnology, Vignan's Foundation for Science, Technology and Research for the support and resources.

References

Afonso NC, Catarino MD, Silva A, Cardoso SM (2019) Brown macroalgae as valuable food ingredients. Antioxidants 8(9):365

Ambati RR, Gogisetty D, Aswathanarayana RG, Ravi S, Bikkina PN, Bo L, Yuepeng S (2019) Industrial potential of carotenoid pigments from microalgae: current trends and future prospects. Crit Rev Food Sci Nutr 59(12):1880–1902

Barbieri G, Barbieri G, Bergamaschi M, Francheschini M, Berizi E (2016) Reduction of NaCl in cooked ham by modification of the cooking process and addition of seaweed extract (*Palmaria palmata*). LWT Food Sci Technol 73:700–706

Becker EW (2007) Micro-algae as a source of protein. Biotechnol Adv 25(2):207–210

Bhuva V, Morya S, Borah A (2021) A review on meat analogue: is this time to see the algal proteins as a sustainable substitute for the meat proteins? Pharma Innov J 10(5):1160–1168

Brown ES, Allsopp PJ, Magee PJ, Gill CI, Nitecki S, Strain CR, McSorley EM (2014) Seaweed and human health. Nutr Rev 72(3):205–216

Brownlee IA, Fairclough AC, Hall AC, Paxman JR (2011) The potential health benefits of seaweed and seaweed extracts. In: Pomin Vitor H (ed) Seaweed: ecology, nutrient composition and medicinal uses. Marine biology: earth sciences in the 21st century. Nova Science, Hauppauge, pp 119–136

Cabral EM, Oliveira M, Mondala JRM, Curtin J, Tiwari BK, Garcia-Vaquero M (2021) Antimicrobials from seaweeds for food applications. Mar Drugs 19(4):211

Caporgno MP, Bocker L, Mussner C, Stirnemann E, Haberkorn I, Adelmann H, Handschin S, Windhab EJ, Mathys A (2020) Extruded meat analogues based on yellow, heterotrophically cultivated Auxenochlorella protothecoides microalgae. Innov Food Sci Emerg Technol 59:102275

Catarino MD, Silva A, Cardoso SM (2018) Phycochemical constituents and biological activities of *Fucus* spp. Mar Drugs 16(8):249

Charoensiddhi S, Conlon M, Methacanon P, Franco C, Su P, Zhang W (2017) Gut health benefits of brown seaweed *Ecklonia radiata* and its polysaccharides demonstrated in vivo in a rat model. J Funct Foods 37:676–684

Chen C, Chaudhary A, Mathys A (2019) Dietary change scenarios and implications for environmental, nutrition, human health and economic dimensions of food sustainability. Nutrients 11(4):856

Chen L, Liu R, He X, Pei S, Li D (2021) Effects of brown seaweed polyphenols, a class of phlorotannins, on metabolic disorders via regulation of fat function. Food Funct 12:2378–2388

Cherry P, O'Hara C, Magee PJ, McSorley EM, Allsopp PJ (2019) Risks and benefits of consuming edible seaweeds. Nutr Rev 77(5):307–329

Cho CJ, Rhee MS (2020) Health functionality and quality control of laver (*Porphyra, Pyropia*): current issues and future perspectives as an edible seaweed. Mar Drugs 18(1):14

Choi YS, Kum JS, Jeon KH, Park JD, Choi HW, Hwang KE, Jeong TJ, Kim YB, Kim CJ (2015) Effects of edible seaweed on physicochemical and sensory characteristics of reduced-salt frankfurters. Korean J Food Sci Anim Resour 35:748–756

Chung IK, Oak JH, Lee JA, Shin JA, Kim JG, Park KS (2013) Installing kelp forests/seaweed beds for mitigation and adaptation against global warming: Korean project overview. ICES J Mar Sci 70(5):1038–1044

Circuncisao AR, Catarino MD, Cardoso SM, Silva A (2018) Minerals from macroalgae origin: health benefits and risks for consumers. Mar Drugs 16(11):400

Cittadini A, Domínguez R, Gómez B, Pateiro M, Pérez-Santaescolástica C, López- Fernández O, Sarriés MV, Lorenzo JM (2020) Effect of NaCl replacement by other chloride salts on physicochemical parameters, proteolysis and lipolysis of dry-cured foal "cecina". J Food Sci Technol 57:1628–1635

Cofrades S, Benedí J, Garcimartin A, Sánchez-Muniz FJ, Jimenez-Colmenero F (2017) A comprehensive approach to formulation of seaweed-enriched meat products: from technological development to assessment of healthy properties. Food Res Int 99:1084–1094

Croft MT, Lawrence AD, Raux-Deery E, Warren MJ, Smith AG (2005) Algae acquire vitamin B12 through a symbiotic relationship with bacteria. Nature 438(7064):90–93

Dawczynski C, Schubert R, Jahreis G (2007) Amino acids, fatty acids, and dietary fibre in edible seaweed products. Food Chem 103(3):891–899

Deepika C (2017a) Fucose-containing sulfated polysaccharides from *sargassum wightii* extraction technology and anticancer activity assessment. Int J Pharm Chem Biol Sci 7(3):272–277

Deepika C (2017b) Phytochemical characterization of *sargassum wightii*, *kappaphycus alvarezii* and *gracilaria corticata* after phycocolloid extraction. Int J Pharm Chem Biol Sci 7(3):223–232

Deepika C (2017c) Antidiabetic, anticoagulant, anticholinesterase inhibition, anti-tyrosinase and antiinflammatory activity of three selected seaweeds after eradication of phycocolloids from vedalai. Int J Pharma Biosci 9:114–121

Deepika C, Bhaskar TCJ, Madhusudhanan J (2016) Antioxidant activities of a few common seaweeds from Gulf of Mannar and the effect of drying as the method of preservation. Indian J Eng Sci Technol 10(1):37

Dhargalkar VK, Verlecar XN (2009) Southern ocean seaweeds: a resource for exploration in food and drugs. Aquaculture 287(3–4):229–242

Dharmananda S (2002) ITM online. Retrieved from Institute for Traditional Medicine Organization. http://www.itmonline.org/arts/seaweed.htm

Du Preez R, Majzoub ME, Thomas T, Panchal SK, Brown L (2020) *Caulerpa lentillifera* (sea grapes) improves cardiovascular and metabolic health of rats with diet-induced metabolic syndrome. Metabolites 10(12):500

Egbert R, Borders C (2006) Achieving success with meat analogs. Food Technol Chicago 60:28–34

Eggersdorfer M, Wyss A (2018) Carotenoids in human nutrition and health. Arch Biochem Biophys 652:18–26

Fan S, Zhang J, Nie W, Zhou W, Jin L, Chen X, Lu J (2017) Antitumor effects of polysaccharide from *Sargassum fusiforme* against human hepatocellular carcinoma HepG2 cells. Food Chem Toxicol 102:53–62

Fellendorf S, O'Sullivan MG, Kerry JP (2016) Impact of ingredient replacers on the physicochemical properties and sensory quality of reduced salt and fat black puddings. Meat Sci 113:17–25

Fiedor J, Burda K (2014) Potential role of carotenoids as antioxidants in human health and disease. Nutrients 6(2):466–488

Figueroa V, Farfan M, Aguilera JM (2021) Seaweeds as novel foods and source of culinary flavors. Food Rev Int. https://doi.org/10.1080/87559129.2021.1892749

Flores SR, Dobbs J, Dunn MA (2015) Mineral nutrient content and iron bioavailability in common and Hawaiian seaweeds assessed by an in vitro digestion/Caco-2 cell model. J Food Compos Anal 43:185–193

Forster J, Radulovich R (2015) Seaweed and food security. Seaweed sustainability. Elsevier, Amsterdam, pp 289–313

Foseid L, Devle H, Stenstrøm Y, Naess-Andresen CF, Ekeberg D (2017) Fatty acid profiles of stipe and blade from the Norwegian brown macroalgae *Laminaria hyperborea* with special reference to acyl glycerides, polar lipids, and free fatty acids. J Lipids 2017:11029702

Garcia-Poza S, Leandro A, Cotas C, Cotas J, Marques JC, Pereira L, Goncalves AMM (2020) The evolution road of seaweed aquaculture: cultivation technologies and the industry 4.0. Int J Environ Res Public Health 17(18):6528

Go H, Hwang HJ, Nam TJ (2010) A glycoprotein from *Laminaria japonica* induces apoptosis in HT-29 colon cancer cells. Toxicol In Vitro 24(6):1546–1553

Gomez-Ordonez E, Jimenez-Escrig A, Rupérez P (2010) Dietary fibre and physicochemical properties of several edible seaweeds from the northwestern Spanish coast. Food Res Int 43(9):2289–2294

Gomez-Zavaglia A, Prieto Lage MA, Jimenez-Lopez C, Mejuto JC, Simal-Gandara J (2019) The potential of seaweeds as a source of functional ingredients of prebiotic and antioxidant value. Antioxidants 8(9):406

Gullon B, Gagaoua M, Barba FJ, Gullón P, Zhang W, Lorenzo JM (2020) Seaweeds as promising resource of bioactive compounds: overview of novel extraction strategies and design of tailored meat products. Trends Food Sci Technol 100:1–18

He D, Wang X, Ai M, Kong Y, Fu L, Zheng B, Song H, Huang Q (2019) Molecular mechanism of high-pressure processing for improving the quality of low-salt *Eucheuma spinosum* chicken breast batters. Poult Sci 98:2670–2678

Hell A, Labrie S, Beaulieu L (2018) Effect of seaweed flakes addition on the development of bioactivities in functional Camembert-type cheese. Int J Food Sci Technol 53:1054–1064

Inguglia ES, Zhang Z, Tiwari BK, Kerry JP, Burgess CM (2017) Salt reduction strategies in processed meat products—a review. Trends Food Sci Technol 59:70–78

Irawan I, Fitriyana (2021) Ice cream properties affected by carrageenan form seaweed deference type drying methods. Earth Environ Sci 679:012022

Jaime Ortiz EU (2009) Functional and nutritional value of the Chilean seaweeds Codium fragile, Gracilaria chilensis and Macrocystis pyrifera. Eur J Lipid Sci Technol 111:320–327

Jayakody MM, Vanniarachchy MPG, Wijesekara WLI (2021) Mineral content of selected seaweed varieties in Southern and North Western Sea of Sri Lanka. Vidyodaya J Sci 24:31–37

Jenifer A, Kanjana K (2018) Effect of seaweed based biscuit supplementation on anthropometric profile of malnourished children residing at Tuticorin. J Sci Technol 4(2):30–39. ISSN: 2349-5456

Jo MJ, Kim HR, Kim GD (2012) The anticancer effects of *Saccharina japonica* on 267B1/K-ras human prostate cancer cells. Int J Oncol 41:1789–1797

Kim CJ, Hwang KE, Song DH, Jeong TJ, Kim HW, Kim YB, Jeon KH, Choi YS (2015) Optimization for reduced-fat/low-NaCl meat emulsion systems with sea mustard (*Undaria pinnatifida*) and phosphate. Korean J Food Sci Anim Resour 35:515–523

Kim H, Lim CY, Lee DB, Seok JH, Kim KH, Chung MS (2020) Inhibitory effects of *Laminaria japonica* Fucoidans against noroviruses. Viruses 12:997

Krishnan P, Abhilash KR, Sreeraj CR, Deepak Samuel V, Purvaja R, Anand A, Mahapatra M, Sankar R, Raghuraman R, Ramesh R (2015) Balancing livelihood enhancement and ecosystem conservation in seaweed farmed areas: a case study from Gulf of Mannar biosphere reserve. India Ocean Coast Manag 207:105590

Kuda T, Kunii T, Goto H, Suzuki T, Yano T (2007) Varieties of antioxidant and antibacterial properties of *Ecklonia stolonifera* and *Ecklonia kurome* products harvested and processed in the Noto peninsula, Japan. Food Chem 103(3):900–905

Layek J, Das A, Ramkrushna GI, Trivedi K, Yesuraj D, Chandramohan M, Kubavat D, Agarwa PK, Ghosh A (2015) Seaweed Sap: a sustainable way to improve productivity of maize in North-East India. Int J Environ Stud 72(2):305–315

Lee HG, Lu YA, Li X, Hyun JM, Kim HS, Lee JJ, Kim TH, Kim HM, Kang MC, Jeon YJ (2020) Anti-obesity effects of *Grateloupia elliptica*, a red seaweed, in mice with high-fat diet-induced obesity via suppression of adipogenic factors in white adipose tissue and increased thermogenic factors in brown adipose tissue. Nutrients 12(2):308

Liu J, Banskota AH, Critchley AT, Hafting J, Prithiviraj B (2015) Neuroprotective effects of the cultivated *Chondrus crispus* in a *C. elegans* model of Parkinson's disease. Mar Drugs 13(4):2250–2266

Lomartire S, Cotas J, Pacheco D, Marques JC, Pereira L, Gonçalves AM (2021) Environmental impact on seaweed phenolic production and activity: an important step for compound exploitation. Mar Drugs 19(5):245

Lopes D, Melo T, Meneses J, Abreu MH, Pereira R, Domingues P, Lillebø AI, Calado R, Domingues MR (2019) A new look for the red macroalga *Palmaria palmata*: a seafood with polar lipids rich in EPA and with antioxidant properties. Mar Drugs 17(9):533

Lorenzo JM, Bermúdez R, Domínguez R, Guiotto A, Franco D, Purriños L (2015) Physicochemical and microbial changes during the manufacturing process of drycured lacón salted with potassium, calcium and magnesium chloride as a partial replacement for sodium chloride. Food Control 50:763–769

Lozano Muñoz I, Díaz NF (2020) Minerals in edible seaweed: health benefits and food safety issues. Crit Rev Food Sci Nutr 60:1–16

Mamatha BS, Namitha KK, Senthil A, Smitha J, Ravishankar GA (2007) Studies on use of *Enteromorpha* in snack food. Food Chem 101(4):1707–1713

Marti-Quijal FJ, Zamuz S, Tomašević I, Gomez B, Rocchetti G, Lucini L, Remize F, Barba FJ, Lorenzo JM (2019) Influence of different sources of vegetable, whey and microalgae proteins on the physicochemical properties and amino acid profile of fresh pork sausages. LWT 110:316–323

McGough M, Sato T, Rankin S, Sindelar J (2012) Reducing sodiumlevels in frankfurters using naturally brewed soy sauce. Meat Sci 91:69–78

McHugh D (2003) A guide to the seaweed industry. FAO fisheries, technical paper no. 441. FAO, Rome

Milinovic J, Mata P, Diniz M, Noronha JP (2020) Umami taste in edible seaweeds: the current comprehension and perception. Int J Gastronomy Food Sci 23:100301

Mohamed Fayaz K, Namitha K, Chidambara Murthy KN, Mahadeva Swamy M, Sarada R, Khanam S, Subbarao PV, Ravishankar GA (2005) Chemical composition, iron bioavailability, and antioxidant activity of *Kappaphycus alvarezzi* (Doty). J Agric Food Chem 53(3):792–797

Mohamed S, Hashim SN, Rahman HA (2012) Seaweeds: a sustainable functional food for complementary and alternative therapy. Trends Food Sci Technol 23(2):83–96

Mora C, Tittensor DP, Adl S, Simpson AGB, Worm B (2011) How many species are there on earth and in the ocean? PLoS Biol 9(8):e1001127

Mouritsen OG, Williams L, Bjerregaard R, Duelund L (2012) Seaweeds for umami flavour in the new Nordic cuisine. Flavour 1:4

Paiva L, Lima E, Patarra RF, Neto AI, Baptista J (2014) Edible Azorean macroalgae as source of rich nutrients with impact on human health. Food Chem 164:128–135

Parniakov O, Toepfl S, Barba FJ, Granato D, Zamuz S, Galvez F, Lorenzo J (2018) Impact of the soy protein replacement by legumes and algae based proteins on the quality of chicken rotti. J Food Sci Technol 55:2552–2559

Pati MP, Sharma SD, Nayak L, Panda CR (2016) Uses of seaweed and its application to human welfare: a review. Int J Pharm Pharm Sci 8(10):12–20

Penalver R, Lorenzo JM, Ros G, Amarowicz R, Pateiro M, Nieto G (2020) Seaweeds as a functional ingredient for a healthy diet. Mar Drugs 18:301

Pereira L (2011) A review of the nutrient composition of selected edible seaweeds. In: Pomin VH (ed) Seaweed: ecology, nutrient composition and medicinal uses. Nova Science, Coimbra, pp 15–47

Pietrasik Z, Gaudette NJ (2014) The impact of salt replacers and flavor enhancer on the processing characteristics and consumer acceptance of restructured cooked hams. Meat Sci 96:1165–1170

Quinsac A, Charrier A, Ribaillier D, Peron JY (1994) Glucosinolates in etiolated sprouts of sea-kale (*Crambe maritima L)*. J Sci Food Agric 65(2):201–207

Radulovich R, Umanzor S, Cabrera R, Rebeca M (2015) Tropical seaweeds for human food, their cultivation and its effect on biodiversity enrichment. Aquaculture 436:40–46

Rajauria G, Foley B, Abu-Ghannam N (2017) Characterization of dietary fucoxanthin from Himanthalia elongata brown seaweed. Food Res Int 99:995–1001

Ramnani P, Chitarrari R, Tuohy K, Grant J, Hotchkiss S, Philp K, Campbell R, Gill C, Rowland I (2012) In vitro fermentation and prebiotic potential of novel low molecular weight polysaccharides derived from agar and alginate seaweeds. Anaerobe 18(1):1–6

Riaz MN (2011) Texturized vegetable proteins. In: Philips GO, Williams PA (eds) Handbook of food proteins. Woodhead Publishing, Sawston, pp 395–418

Rioux LE, Beaulieu L, Turgeon SL (2017) Seaweeds: a traditional ingredients for new gastronomic sensation. Food Hydrocoll 68:255–265

Roohinejad S, Koubaa M, Barba FJ, Saljoughian S, Amid M, Greiner R (2017) Application of seaweeds to develop new food products with enhanced shelf-life, quality and health related beneficial properties. Food Res Int 99:1066–1083

Rubio NR, Xiang N, Kaplan DL (2020) Plant-based and cell-based approaches to meat production. Nat Commun 11:6276

Ryu MJ, Kim AD, Kang KA, Chung HS, Kim HS, Suh IS, Chang WY, Hyun JW (2013) The green algae Ulva fasciata Delile extract induces apoptotic cell death in human colon cancer cells. In Vitro Cellular Dev Biol Anim 49(1):74–81

Saeed M, Arain MA, Fazlani A et al (2021) A comprehensive review on the health benefits and nutritional significance of fucoidan polysaccharide derived from brown seaweeds in human, animals and aquatic organisms. Aquac Nutr 27:633–654

Sato M, Oba T, Yamaguchi T, Nakano T, Kahara T, Funayama K, Kobayashi A, Nakano T (2002) Antihypertensive effects of hydrolysates of wakame (*Undaria pinnatifida*) and their angiotensin-I-converting enzyme inhibitory activity. Ann Nutr Metab 46(6):259–267

Senthil A, Mamatha BS, Vishwanath P, Bhat KK, Ravishankar GA (2011) Studies on development and storage stability of instant spice adjunct mix from seaweed (*Eucheuma*). J Food Sci Technol 48(6):712–717

Shama A, Joyce SG, Mari FD, Ranga Rao A, Ravishankar GA, Hudaa N (2019) Macroalgae and microalgae: novel sources of functional food and feed. In: Ravishankar GA, Rao AR (eds) Handbook of algal technologies and phytochemicals: volume I: food, health and nutraceutical applications. CRC Press, Boca Raton, pp 207–219

Shin HC, Hwang HJ, Kang KJ, Lee BH (2006) An antioxidative and antiinflammatory agent for potential treatment of osteoarthritis from *Ecklonia cava*. Arch Pharm Res 29(2):165–171

Sholichah E, Kumalasari R, Indrianti N, Ratnawati L, Restuti A, Munandar A (2021) Physicochemical, sensory, and cooking qualities of gluten-free pasta enriched with Indonesian edible red seaweed (*Kappaphycus alvarezii*). J Food Nutr Res 9(4):187–192

Skrovankova S (2011) Seaweed vitamins as nutraceuticals. Adv Food Nutr Res 64:357–369

Smetana S, Mathys A, Knoch A, Heinz V (2015) Meat alternatives: life cycle assessment of most known meat substitutes. Int J Life Cycle Assess 20:1254–1267

Smetana S, Schmitt E, Mathys A (2019) Sustainable use of Hermetia illucens insect biomass for feed and food: attributional and consequential life cycle assessment. Resour Conserv Recycl 144:285–296

Smith J, Summers G, Wong R (2010) Nutrient and heavy metal content of edible seaweeds in New Zealand. N Z J Crop Hortic Sci 38(1):19–28

Sumayaa S, Kavitha K (2015) Preparation of novel seaweed recipes and standardization for the human consumption. Int J Adv Res 3(10):159–167

Tavares Estevam AC, Alonso Buriti FC, de Oliveira TA, Pereira EV, Florentino ER, Porto AL (2016) Effect of aqueous extract of the seaweed *Gracilaria domingensis* on the physicochemical, microbiological, and textural features of fermented milks. J Food Sci 81(4):C874–C880

Thahira Banu A, Uma Mageswari S (2015) Nutritional status and effect of seaweed chocolate on anemic adolescent girls. Food Sci Human Wellness 4(1):28–34

Tom DD, Ramirez C, Pino M, Collins MB, Rossen J, Pino-Navarro JD (2008) Monte Verde: seaweed, food, medicine, and the peopling of south America. Science 320(5877):784–786

Triki M, Khemakhem I, Trigui I, Ben Salah R, Jaballi S, Ruiz-Capillas C, Ayadi MA, Attia H, Besbes S (2017) Free-sodium salts mixture and AlgySalt® use as NaCl substitutes in fresh and cooked meat products intended for the hypertensive population. Meat Sci 133:194–203

Turan G, Cırık S (2018) Sea vegetables. In: Asaduzzaman M, Asao T (eds) Vegetables—importance of quality vegetables to human health. IntechOpen, London, pp 85–102

Van den Burg SWK, Dagevos H, Helmes RJK (2021) Towards sustainable European seaweed value chains: a triple P perspective. ICES J Mar Sci 78(1):443–450

Venkatraman KL, Mehta A (2019) Health benefits and pharmacological effects of *Porphyra* species. Plant Foods Hum Nutr 74(1):10–17

Vilar EG, Ouyang H, O'Sullivan MG, Kerry JP, Hamill RM, O'Grady MN, Mohammed HO, Kilcawley KN (2020) Effect of salt reduction and inclusion of 1% edible seaweeds on the chemical, sensory and volatile component profile of reformulated frankfurters. Meat Sci 161:108001

Wang T, Jónsdóttir R, Liu H, Gu L, Kristinsson HG, Raghavan S, Olafsdóttir G (2012) Antioxidant capacities of phlorotannins extracted from the brown algae Fucus vesiculosus. J Agric Food Chem 60(mam23):5874–5883

Watanabe F, Takenaka S, Katsura H, Masumder SA, Abe K, Tamura Y, Nakano Y (1999) Dried green and purple lavers (nori) contain substantial amounts of biologically active vitamin B_{12} but less of dietary iodine relative to other edible seaweeds. J Agric Food Chem 47(6):2341–2343

Wells ML, Potin P, Craigie JS, Raven JA, Merchant SS, Helliwell KE, Smith AG, Camire ME, Brawley SH (2017) Algae as nutritional and functional food sources: revisiting our understanding. J Appl Phycol 29(2):949–982

Yamada K, Yamada Y, Fukuda M, Yamada S (2013) Bioavailability of dried asakusanori (*Porphyra tenera*) as a source of cobalamin (vitamin B_{12}). Int J Vit Nutr Res 69(6):412–418

Yoshinaga K, Mitamura R (2019) Effects of *Undaria pinnatifida* (Wakame) on postprandial glycemia and insulin levels in humans: a randomized crossover trial. Plant Foods Hum Nutr 74(4):461–467

Yuan YV, Carrington MF, Walsh NA (2005) Extracts from dulse (Palmaria palmata) are effective antioxidants and inhibitors of cell proliferation in vitro. Food Chem Toxicol 43(7):1073–1081

Zanardi E, Ghidini S, Conter M, Ianieri A (2010) Mineral composition of Italian salami and effect of NaCl partial replacement on compositional, physico-chemical and sensory parameters. Meat Sci 86:742–747

Zhang X, Liu Y, Chen XQ, Aweya JJ, Cheong KL (2020) Catabolism of Saccharina japonica polysaccharides and oligosaccharides by human fecal microbiota. LWT Food Sci Technol 130:10965

Zhao Y, Zheng Y, Wang J, Ma S, Yu Y, White WL, Yang S, Yang F, Lu J (2018) Fucoidan extracted from *Undaria pinnatifida*: source for nutraceuticals/functional foods. Mar Drugs 16(9):321

Zivanovic A, Skropeta D (2012) c-AMP dependent protein kinase A inhibitory activity of six algal extracts from southeastern Australia and their fatty acid composition. Nat Prod Commun 7(7):923–926

Zugcic T, Abdelkebir R, Barba FJ, Rezek-Jambrak A, Gálvez F, Zamuz S, Granato D, Lorenzo JM (2018) Effects of pulses and microalgal proteins on quality traits of beef patties. J Food Sci Technol 55:4544–4553

Chapter 15
Issues Regarding Toxicity and Safety of Foods from Seaweeds

Lydia Ferrara

1 Introduction

Seaweed-based foods are gaining importance in the human diet due to the presence of proteins, minerals, vitamins, soluble dietary fibre, omega 3 fatty acids, flavonoids, which are considered preventive agents against diseases related to the modern lifestyle. The health benefits attributed to algae-based diet related to antiviral, anticancer, and anticoagulant properties, as well as the ability to modulate intestinal health and risk factors for obesity and diabetes as evidenced by many research reports (Brown et al. 2014; Rajapakse and Kim 2011; Jiménez-Escrig and Goñi 1999; Peñalver et al. 2020).

Algae are commonly consumed in East Asian countries, especially Japan. Recent Japanese studies have reported the association between algae consumption and a reduced risk of cardiovascular disease (Willcox et al. 2009; Fukuda et al. 2007). Green algae are widely used in the food (Shama et al. 2019; Ranga Rao et al. 2019) and pharmaceutical sector. Brown algae and red algae are important raw materials for the industrial sector. They contain phyco-colloids, polysaccharides present in the cell wall of algae, the composition of which varies according to species. Alginic acid is present in brown algae, while red algae are rich in agar and carrageenan. These polysaccharides are used as thickeners and as food gelling agents both in the composition of beverages and as additives to improve the consistency of solid foods, also prolonging shelf life.

L. Ferrara (✉)
Department of Pharmacy, University of Naples Federico II, Naples, Italy
e-mail: lyferrar@unina.it

A. Ranga Rao, G. A. Ravishankar (eds.), *Sustainable Global Resources of Seaweeds Volume 2*, https://doi.org/10.1007/978-3-030-92174-3_15

2 Food Algae

2.1 Microalgae

Blue or blue-green algae are single-celled organisms also known as cyanobacteria: they are not all edible, indeed some of them can be highly toxic to both fish and humans due to the presence of substances such as anatoxin, cylinderspermosin, domoic acid, microcystin, nodularine, saxitoxin (Beasley et al. 1989; Cook et al. 1989; Briand et al. 2003). Edible blue green and green algae containing many biological substances of nutritional interest such as polyunsaturated fatty acids, essential amino acids, pigments, vitamins, and minerals have been studied for their antioxidant, inflammatory, anticancer activity. The consumption of these could reduce the risk of aging-related disorders such as cataracts and macular degeneration, coronary heart disease, joint diseases, Parkinson's disease (Ku et al. 2013; Abu-Taweel et al. 2019; Pabon et al. 2012).

Chlorella pyrenoidosa is a green microalga in the class Chlorophyceae, which is a popular food all over the world for its excellent content in amino acids, proteins, fibers, antioxidant and bioactive substances. Chlorella powder can be consumed directly dissolved in water or fruit juices; it can be added to fresh fruit smoothies or form the basis for sauces to accompany toast, or be added in soups, soy ice cream, vegetable milk for breakfast.

Chlorella is formulated in the form of chewable tablets that contain all active ingredients and is widely used as a nutritional supplement. The properties of *Chlorella* are manifold so much that it turns out to be an interesting food both from a nutritional and therapeutic point of view. Due to its high content of mineral salts and vitamins—A-C-D-B_{12} is used as a supplement and as an additive in fortified foods (Grossman 2016). It has effective anti-inflammatory properties and also able to eliminate heavy metals, by complexation reaction, thus preventing intoxication phenomena for the organism (Merchant and Andre 2001; Nakashima et al. 2009). Moreover, *Chlorella* is known to possess antioxidant properties to scavenge free radicals, hence useful in preventing aging and enhancing repair functions of tissues (Bito et al. 2020).

Spirulina, *Arthrospira platensis,* is particularly rich in proteins, essential amino acids, and lipids of the mono and polyunsaturated type; minerals including iron and natural pigments such as phycocyanin, β-carotene and chlorophyll; Vitamins including E, group B, and vitamin B_{12}. Its cell wall is of mucopolysaccharides, due to the absence of cellulose, and gives this alga a good digestibility. Spirulina is sold in the form of tablets, powders, flakes, or as an ingredient in processed foods. It also seems to have effects on appetite control, accelerating the sense of satiety, when taken before meals.

Spirulina-based supplements are particularly taken by sportsmen for the high content of vitamins and minerals that act as enzyme cofactors to support energy production during a particularly intense muscular effort. It also protects the liver, decreases blood cholesterol levels, increases the production of immune system cells involved in the fight against infections, prevents allergies and some tumors (Gutiérrez-Salmeán et al. 2015; El-Baky et al. 2003; Grosshagauer et al. 2020).

Studies have been carried out on the use of spirulina in animal feed and aquaculture. It was noted that, supplementing with this alga up to 10% of the feed of chickens, for 2 weeks the weight of animals and egg production did not differ from those bred with non-integrated feed. The yolks, however, contained 24% less cholesterol, a decrease to be attributed to the content of polyunsaturated fatty acids and ω-3 with antioxidant activity. They assumed a more intense color due to the high content of β-carotene, zeaxanthin, xanthophylls, and other carotenoid pigments present in Spirulina. These pigments also accumulate in the muscle tissue enhancing the nutritional value besides the aesthetic appeal of the food (Muhling et al. 2005; Sujatha and Narahari 2011; Toyomizu et al. 2001; Anderson et al. 1991). It also has been shown to be a effective pig feed (Fevrier and Seve 1975) and cattle feed for the health of the animals and the output of healthy foods from them (Kulpys et al. 2009; Christaki et al. 2012).

For the richness of iron and protein, in 1974, the World Health Organization (WHO) referred to Spirulina as a superfood, very suitable for feeding children. Spirulina and Chlorella have also been used in human nutrition as supplements in very common products such as pasta and biscuits, of which they have enhanced the nutritional and sensory properties, being very welcome to tasters (Fradique et al. 2010; Wandurraga et al. 2020; Niccolai et al. 2019).

Haematococcus pluvialis and *Dunaliella salina* are the richest source of natural carotenoids with constant and intense antioxidant activity, protecting the cells from molecular damage caused by solar radiation. They find applications in the nutraceutical, cosmetic, food and aquaculture industries.

Astaxanthin found in *H. pluvialis* is available in the form of dietary supplements, soluble in oils or as dried aplanospores. In recent years, a lot of research has been interested in adding astaxanthin in food, achieving good results for whole cakes and feed (Hossain et al. 2017; Mercke Odeberg et al. 2003). Astaxanthin is well tolerated by the human organism and numerous studies have revealed no toxic effects, indeed they have shown considerable potential for the protection of the organism from a wide range of diseases (Palozza et al. 2009).

Excessive consumption of astaxanthin leads to the accumulation of pigment in tissues and skin, a desirable effect especially in aquaculture because this pigment improves the organoleptic properties of fish. *H. pluvialis* and its pigment, astaxanthin, can also be used for feeding livestock: chickens and pigs have not only shown greater muscle growth and an improvement in the organoleptic and nutritional properties of meat, but also a better state of health and greater resistance to bacterial infections.

Dunaliella salina is very rich in β-carotene which, in addition to the antioxidant action, plays an important role in the physiological balance of our organism through the increase in visual capacity and in the enhancement of immune defenses due to T lymphocytes (Ye et al. 2008; Mendiola et al. 2008).

2.2 Brown Algae

Arame alga, *Eisenia bicyclis* (Fig. 15.1), also known as "sea oak", is a brown alga that is abundant in the coasts of Japan and Korea; also found in Peru and the north Pacific coast. It lives attached to rocks in shallow water and is collected manually since it is quite strong. It has rather wide leaves, up to 30 cm in length, which is cut into thin strips, boiled, and dried in the sun. It is rich in minerals, such as potassium, calcium, iodine, copper, iron and zinc, and vitamins and for this reason, they are often used as supplements for those who follow a vegetarian or vegan diet, but also for those who practice a lot of sports and need to replenish the mineral salts lost with effort, reducing the risk of muscle cramps. Among algae from the east, Arame is the most welcome on the Western palate, has a more delicate taste, due to mannitol, and can be boiled and consumed as a side dish; it is ideal accompanied by tofu, tempeh, or seitan and with vegetables.

Arame alga is particularly rich in β-carotene, a powerful antioxidant that helps cell regeneration and prevents aging. Also, much attention has been paid to fucoxanthin, a carotenoid belonging to the class of xanthophylls, for its anti obesity effect obtained by improving insulin resistance and reducing blood glucose levels, through the regulation of cytokine secretions in abdominal adipose tissue, resulting in oxidation of fatty acids and heat production (Airanthi et al. 2011; Gammone and D'Orazio 2015).

A methanolic extract of brown algae Arame, showed a significant inhibitory effect against pancreatic lipase attributed to the presence, isolation and identification of six florotannines. Two florotannines identified as fucofuroeckol A and dioxynodehydroeckol showed strong inhibitory activity against α-glucosidase and α-amilase and may be indicated for the formulation of a dietary supplement or anti-diabetic pharmaceutical product.

In vitro, antibacterial activity of florotannines towards *Listeria monocytogenes*, a streptomycin-resistant pathogen, responsible for listeriosis and cases of encephalitis and meningitis, has been demonstrated. The good results obtained indicate a potential use of natural florotannines present in brown algae to control the spread of

Fig. 15.1 Arame—*Eisenia bicyclis*

pathogenic infections from *Listeria monocytogenes* in food (Eom et al. 2012, 2013; Kim et al. 2018). It is also rich in alginic acid with a profound chelating effect on heavy metals and radioactive substances (Wang et al. 2020; Waldron-Edward et al. 1964; Paul et al. 1966) Low caloric intake and high fiber content increase the sense of satiety, promote metabolism, reducing fat accumulation. Fibers also regularize intestinal transit and fight constipation. The ability to reduce hypertension and prevent anemia are other properties of such alga. Its iodine content regularizes thyroid and metabolic activity; has antiviral properties and has been used in the past to treat disorders of female and male reproductive organs.

Owing to the presence of two florotannines identified as fucofuroeckol A and dioxynodehydroeckol exhibiting strong inhibitory activity against α-glucosidase and α-amylase it holds promise as a dietary supplement and as a ingredient of anti-diabetic pharmaceutical product.

It content of alginates also helps in bowel movement to fight constipation. The ability to reduce hypertension and prevent anemia are other properties of this alga. Its iodine content regularizes thyroid and metabolic activity; has antiviral properties and has been used in the past to treat disorders of female and male reproductive organs.

Hiziki, *Sargassum fusiforme* (Fig. 15.2), is a brown algae, almost black in color, has its habitat in the deep sea, very rich in minerals and in particular calcium (1.4 g/100 g). It is considered the "alga of regeneration", for its healing power: it facilitates the lowering of the cholesterol rate, promotes the growth of nails and hair, prevents caries, is also recommended in cases of anemia for the high iron content.

This alga is consumed a few times during the week due to the presence of inorganic arsenic. After harvesting it is put in large tanks full of water and cooked for several hours in boiling water to make it more tender. Further to this treatment the thallus expand considerably, increasing their volume five times, then dried, selected and packed for sale.

Hiziki *Sargassum fusiform* contains various health-beneficial bioactive components, including polysaccharides with antioxidant properties, such as alginate and fucoidan, a sulfate polysaccharide that has also shown anticoagulant activity of human plasma (Wu et al. 2013; Dobashi et al. 1989). Polysaccharide extracts have

Fig. 15.2 Hiziki—*Sargassum fusiforme*

a protective effect on gastrointestinal damage induced by ethanol and immunomodulating activity, activating B cells, macrophages and the production of nitric oxide and pro-inflammatory cytokines (Hwang et al. 2011; Okai et al. 1998; Jeong et al. 2015; Choi et al. 2009). The fucosterol isolated from the alga suppressed the proliferation of cells derived from osteosarcoma (Huh et al. 2012).

The anti-inflammatory activity of *H. fusiformis* suggests applications for different disorders including allergic, autoimmune and synonasal diseases. Allergy is a pathology caused by hyperactivity to environmental allergens that cause irritation of the airways and skin. It is generally treated with antihistamine drugs and intranasal/oral steroids that you take for a long time are the cause of undesirable adverse effects. The percutaneous and oral administration of an ethereal fraction of *H. fusiformis* has highlighted anti-inflammatory effects on rat ear swelling, meaning that such alga could be a viable alternative to the drugs currently in use (Zhang et al. 2019).

Kombu, *Laminaria Japonica* (Fig. 15.3) is a brown alga of leathery consistency that grows along the coasts of the Pacific and Atlantic oceans, also present in the Mediterranean. It is very rich in vitamins, mineral salts, iodine and an amino acid, the glutamic acid, from which monosodium glutamate is obtained, a powerful flavor enhancer that interacts with specific taste receptors producing umami. It is the richest seaweed containing alginate. Among the sugars there is mannitol in concentrations of 65%, a natural low-calorie and acariogenic sweetener with a mild laxative effect, with a metabolism independent of insulin. It is used, in addition to antidepressant action, as an osmotic diureic in cases of acute renal failure; for control of kidney function; in the presence of the intracranial hypertension, of the spinal hypertension and brain masses, to decrease the pressure inside the eye, to facilitate the elimination of toxic substances (Li and Chen 2015; Fink 2012).This alga is very rich in iodine and is highly appreciated as a regulator of metabolism and normalizer of disorders related to the functioning of the thyroid gland, preventing the obesity, the congestion of the lymphatic system, the excessive weight loss. It is widely used for the preparation of broths and soups, to flavor seitan, to make miso soup tastier, to prepare excellent pâtés (Kapp and Sumner 2019; Vīna et al. 2014).

Fig. 15.3 Kombu–*Laminaria Japonica*

2.3 Red Algae

Eucheuma algae (*Eucheuma cottonii, Eucheuma spinosum*) (Fig. 15.4), *Kappaphycus alvarezii,* are widely used not only in nutrition but also in the industrial sector for the presence of k-carrageenan a substance that finds very wide use in the food, in medicine and industrial field. It swells in cold water and completely dissolves in hot water, around 50 °C, giving for cooling a gelatinous mass, transparent, consistent and devoid of particular flavors or smells (Suganya et al. 2016; Younes et al. 2018; Bercea and Wolf 2019).

These gelatinous substances present in algae are known as carrageenans being made up of different compounds, but with similar properties, which differ according to the species of algae from which they are extracted and the processing process. In fact, three types of carrageenans indicated by Greek letters kappa, iota, gamma, chemically are, polymers of D-galactose are identified. K-carrageenan forms very strong gels in the presence of potassium ions; *Chondrus crispus* (Fig. 15.5) and *Gigartina mamitiosa,* two red algae, known as carrageenans in Ireland and Great Britain, after boiling provide the λ-carrageenan that is cold soluble and does not form gel. The combination of stearoyl–2–sodium lacthylate makes it possible to obtain stable emulsions both hot and cold. Iota-carrageenan consists of calcium salts, it provides a gel characterized by softness and elasticity that makes it very suitable in the cosmetic sector. Commercially, mixtures of carrageenan (kappa, iota, lamda) standardized with sucrose, destrose or dehydrated glucose syrup widely used in the agri-food industry are often found. (De Ruiter and Rudolph 1997). The carrageenan is highly valued in the dairy industry, canned meat products, jams, ice cream, and pet food. As a supplement, it is taken as an adjuvant of restrictive diets increasing the sense of satiety and for the symptomatic treatment of constipation. Carrageenan is widely used as an inactive excipient also in pharmaceutical technique, for the preparation of pastes, gels, and emulsions, and in the cosmetic one, where it is part of the composition of kinds of toothpaste, hair fixers and shampoos.

In addition to carrageenan, red algae also contain agar, a polysaccharide consisting of agarose and agaropectin used, mainly in food technology as a thickener for

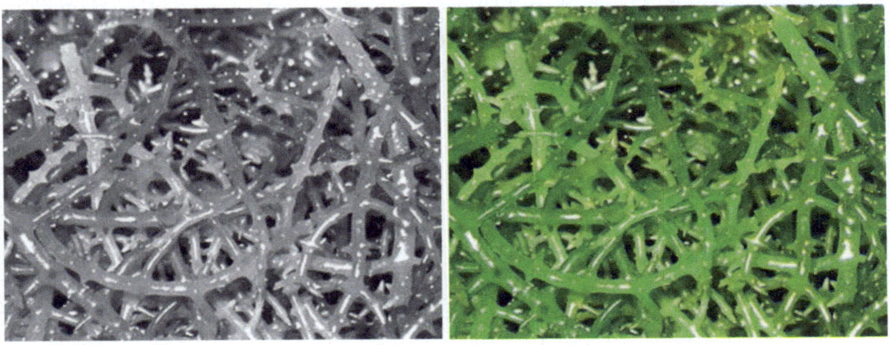

Fig. 15.4 *Eucheuma spinosum*

Fig. 15.5 *Chondrus crispus*

Fig. 15.6 *Hypnea charoides*

soups, vegetable jellies, ice creams, and some desserts and as a clarification for beer. This natural vegetable jelly is present in the cell wall of algae of the order Gelidiales and Gracilariales, from which it is extracted by boiling, is used as a food additive, and is referred to as E406.

Two edible red algae *Hypnea charoides* (Fig. 15.6) and *Hypnea japonica* are very rich in plant dietary protein and fiber and could be added to foods as functional ingredients to supplement the diet of patients debilitated by long illnesses. K-carrageenan isolated from the red *Hypnea musciformis* has revealed an antibacterial and antifungal action against Staphylococcus aureus and Candida albicans, as well as anticancer and neutral activities on the cell lines of breast cancer and human neuroblastoma (Lahaye and Kaeffer 1997; Souza et al. 2018; Brito et al. 2016; Benevides 2018).

3 Toxicity and Safety of Algae

The increased presence of commercial algae-based products such as food/supplements, cosmetics, and consumer preference for natural products, because they are considered healthy and suitable for intake for long periods, has given rise to the problem of the benefits and safety of such products. Algae contain a high concentration of essential mineral elements such as sodium, calcium, magnesium, iron, potassium, chloride, sulfate, phosphorus, the variability of which is related to the type of alga, geographical location, climatic conditions, time of stay in the water, the type of processing to which they are subjected. Numerous research has also shown that algae can accumulate pollutants, iodine, and heavy metals in high concentrations, which are very dangerous to human health, the presence of which is not declared on labels, there is no strict legislation on this matter (Desideri et al. 2016; Wang et al. 2016).

Selenium is a micronutrient present in an organic form that protects cell membranes from oxidative stress and plays an antagonistic role against heavy metals, such as mercury, cadmium, and silver (Schiavon et al. 2017; Shreenath et al. 2020).

Arsenic is present at low concentration in inorganic, very toxic form, such as arsenites and arseniates; in organic form as free arsenic, as arsenobetaine, nontoxic, and as arsenoglucides, toxic (Francesconi 2010; Garcia-Salgado et al. 2012).

The concentration of iodine, element necessary for the synthesis and functionality of thyroid hormones, in algae is very variable, until it reaches in the Laminaria typed over 600 mg/100 g dry weight. EFSA recommends a daily intake of 150 μg by setting the maximum tolerable limit at 600 μg (Zimmermann 2008) and reduced use over time of both these algae and supplements (Teas et al. 2004).

The presence and accumulation of heavy metals such as Mercury Lead, Cadmium Arsenic as a result of contamination of the marine environment are the cause of toxicity of food algae. Although molluscs and fish accumulate heavy metals in their meat, higher values of these elements have been found in algae as a result of higher longevity of these and longer stay in the water. In order to protect the health of consumers, some states have set limits for trade in algae, considered "sea vegetables": for cadmium 0.5 mg/kg (dry weight), for mercury 0.1 mg/kg (dry weight) and for lead 5 mg/kg (dry weight) (Hwang et al. 2010; Dawczynski et al. 2007).

The objective of ensuring food safety has been an important priority for the European Union, which, following the release of the "White Paper on Food Safety" in 2000, issued Regulation (EC) No 466/2001, which defines the maximum levels of certain contaminants contained in food, including lead, cadmium and mercury. It also stipulated that foodstuffs intended for infants and children under the age of 18 must contain the minimum possible level of contaminants in order not to cause damage to their health. Since its entry into force on 5 April 2001, this Regulation has been the subject of numerous amendments, because the list of products examined has been considered to be not exhaustive.

The European Union has considered the possibility of collaboration between all the states belonging to it, to guarantee healthy and safe food throughout the production chain. Regulation (EC) No 178/2002 of the European Parliament and the Council issued on 28 January 2002 establishes the general principles and requirements of food legislation, establishes the European Food Safety Authority (EFSA) and establishes procedures in the field of food safety. Average amount of a chemical contaminant a Tolerable Daily Intake (TDI), and a Provisional Tolerable Weekly Intake (PTWI) for contaminants, in particular in fish species, expressed in mg/kg body weight has been published.

In the Ministerial Decree of 16 May 2002 reported in Official Journal No. 165 of 16 July 2002, the maximum levels and methods of analysis of algal biotoxins in live bivalve mollusks, echinoderms, tunicates, and marine gastropods are established.

In 2003, the UK Independent Committee on the Toxicity of Chemicals in Food, Consumer Products and the Environment (COT) concluded that it would have been more appropriate, as arsenic is a genotoxic carcinogen instead of PTWI, to define a value as the lowest reasonably practicable (ALARP) establishing as a safety level 0.05–0.1 μg/kg body weight/day.

In November 2004, Food Standards Australia New Zealand (FSANZ) issued a press release advising Australian consumers to avoid hijiki algae due to the high arsenic content much higher than the maximum limit (ML) established for algae of 1 mg/kg body weight.

Regulation (EC) No 396/2005 of the European Parliament and of the Council for algae and prokaryotic organisms sets the maximum residue level (MRR) for mercury of 0.01 mg/kg.

Regulation (EC) No 1881/2006 defines the maximum levels of arsenic, cadmium and lead in various foodstuffs, not in seaweed and halophyte, whereas these maximum levels have been defined for supplements composed exclusively or mainly of seaweed or seaweed products. Also, the Scientific Committee for Food Safety (EFSA) in 2006 set a maximum iodine intake limit of 600 μg/day for adults and 200 μg/day for children aged 1–3 years.

In 2002, the French Agency for Food Safety (AFSSA) expressed its opinion on the acceptable iodine content for edible algae up to a maximum of 6 g/kg dry weight, then proposing, for arsenic, the limit of 3 mg/kg dry weight, previously recommended by the High Council French for Public Hygiene (CSHPF) in 1997.

The Food Safety Agency French (ANSES) arose from the merger of AFSSA (French Agency for Food Safety) and AFSSET (Agency French for The Safety of the Environment, Health, and Labour) due to the presence of cadmium in algae in quantities higher than the maximum recommended by the Superior Council of Public Hygiene of France of 0.5 mg/ kg, it decided to reduce the safety threshold to 0.35 mg/kg, also taking into account the average cadmium intake in the Diet of the French.

In 2010, a survey was conducted on iodine levels in algae and algae products by Food Standards Australia New Zealand (FSANZ) and the adult limit of 1100 μg/day was set; for children (1–8 years old) 200 μg to 300 μg/day and for children (9–18 years) 600-μg to 900-μg/day. Iodine levels varied between red and brown algae, but were generally higher in brown algae; iodine concentrations in the most consuming algae such as wakame and nori and products containing such algae were generally low; other types of dried algae had very high levels of iodine, considered unsafe for humans. Safety for the latter was established at the level of $\leq$1000 mg of iodine/dried kg.

In Europe, algae have traditionally been used to add minerals and vitamins to food and their presence in the diet has been very low, so that European legislation is deficient on the safety of food and feed containing algae.

Species of macro-algae placed on the EU market are regulated by new food legislation restricting the entry of new species (Regulation (EC) 2283/2015/ and Regulation (EC) No 258/97). "New species" are considered to be those whose consumption was very limited, before the 1997 regulation awarded algae as a sustainable raw material for food and feed came into force.

To place on the market, Recommendation (EU) 2018/464 on the monitoring of metals and iodine in seaweed, halophytes, and seaweed products state: the method of sampling (Regulation (EC) No 333/2007); analytical methods (Annex III to Regulation (EC) No 882/2004); a list of algae that need to be monitored. It also stipulates that the determination of individual metals and that of their compounds shall be carried out on mercury and arsenic.

4 Future Prospects and Conclusions

Microalgae have entered the food market such as supplements, tablets, capsules, powders, suspensions or solutions; have been included in the category of "superfoods", meeting the approval of consumers aware of the benefits deriving from a healthy diet. Macroalgae are an important source of food for eastern countries where their industrial cultivation is also developed. Food use is not only reserved for fresh or dried algae, but they enter into the preparation of various dishes, they are used as fortifying agents and food preservatives, they are used in pastry and in the beverage industry.

Looking to the future, algae is our new food source capable of making up for the current food shortage. Research, however, must continue to establish whether the prolonged consumption of algae and their bioactive components can be harmful to health; the mechanisms of action must also be clarified in order to consider whether their use is effective and safe for industrial applications in the therapeutic field. In this regard, it is important to keep a constant check on the environment in which algae grow, in order to avoid the negative impact of algae in the human diet, due to the excessive presence of heavy metals, especially iodine and arsenic.Conflict of InterestThe author declares she has no conflicts of interest.

References

Abu-Taweel GM, Mohsen GAM, Antonisamy P, Arokiyaraj S, Kim HJ, Kim SJ, Park KH, Kim YO (2019) Spirulina consumption effectively reduces anti-inflammatory and pain related infectious diseases. J Infect Public Health 12(6):777–782. https://doi.org/10.1016/j.jiph.2019.04.014

AFSSA (2002) Agence Française de Sécurité Sanitaire des Aliments. Avis de l'Agence française de sécurité alimentaire des aliments relatif à une demande d'évaluation sur la teneur maximale en iode acceptable pour les algues alimentaires faisant suite à un message d'alerte émanant des autorités allemandes concernant le retrait 15 du marché d'algues séchées d'origine chinoise et contenant 4988 et 5655 mg d'iode par kg de poids sec. AFSSA Saisine No. 2002-SA0144

AFSSA (2009) Saisine No 2002-SA-0144. Agence Française de Sécurité Sanitaire des Aliments. Opinion of the French Food Safety Agency on the recommended maximum inorganic arsenic content of laminaria and consumption of these seaweeds in light of their iodine content. AFSSA. Request no 2007-SA-0007

Airanthi MK, Hosokawa M, Miyashita K (2011) Comparative antioxidant activity of edible Japanese brown seaweeds. J Food Sci 76(1):C104–C111. https://doi.org/10.1111/j.1750-3841.2010.01915.x

Ambati RR, Gogisetty D, Aswathanarayana RG, Ravi S, Bikkina PN, Bo L, Yuepeng S (2019) Industrial potential of carotenoid pigments from microalgae: current trends and future prospects. Crit Rev Food Sci Nutr 59(12):1880–1902

Anderson DW, Tang CS, Ross E (1991) The xanthophylls of *Spirulina* and their effect on egg yolk pigmentation. J Poult Sci 70:115–119

ANSES (n.d.) Opinion of the French Agency for Food, Environmental and Occupational Health & Safety on "maximum cadmium levels for seaweed intended for human consumption". Request No 2017-SA-0070

Beasley VR, Dahlem AM, Cook WO, Valentine WM, Lovell RA, Hooser SB, Harada K, Suzuki M, Carmichael WW, Diagnvell RA (1989) Diagnostic and clinically important aspects of cyanobacterial (blue-green algae) toxicoses. J Vet Diagn Investig 1(4):359–365. https://doi.org/10.1177/104063878900100417

Benevides NMB (2018) In vitro activities of kappa-carrageenan isolated from red marine alga *Hypnea musciformis*: antimicrobial, anticancer and neuroprotective potential. Int J Biol Macromol 112:1248–1256. https://doi.org/10.1016/j.ijbiomac.2018.02.029

Bercea M, Wolf BA (2019) Associative behavior of κ-carrageenan in aqueous solutions and its modification by different monovalent salts as reflected by viscometric parameters. Int J Biol Macromol 140:661–667. https://doi.org/10.1016/j.ijbiomac.2019.08.144

Bito T, Okumura E, Fujishima M, Watanabe F (2020) Potential of chlorella as a dietary supplement to promote human health. Nutrients 12(9):2524. https://doi.org/10.3390/nu12092524

Briand JF, Jacquet S, Bernard C, Humbert JF (2003) Health hazards for terrestrial vertebrates from toxic cyanobacteria in surface water ecosystems. Vet Res 34(4):361–377. https://doi.org/10.1051/vetres:2003019

Brito VT, Barros FCN, Silva RO, Dias Júnior GJ, Júnior GJD, Franco ÁX, Soares PMG, Chaves LS, Abreu CMWS, de Paula RCM, Souza MHLP, Freitas ALP, Barbosa ALR (2016) Sulfated polysaccharide from the marine algae *Hypnea musciformis* inhibits TNBS-induced intestinal damage in rats. Carbohydr Polym 151:957–964. https://doi.org/10.1016/j.carbpol.2016.06.047

Brown ES, Allsopp PJ, Magee PJ, Gill CI, Nitecki S, Strain CR, McSorley EM (2014) Seaweed and human health. Nutr Rev 72(3):205–216. https://doi.org/10.1111/nure.12091

Choi EY, Hwang HJ, Kim IH, Nam TJ (2009) Protective effect of polysaccharide from *Hizikia fusiformis* against ethanol-induced cytoxicity. Food Chem Toxicol 47(1):134–139

Christaki E, Karatzia M, Bonos E, Florou-Paneri P, Karatzias C (2012) Effect of dietary *Spirulina platensis* on milk fatty acid profile of dairy cows. Asian J Anim Vet Adv 7:597–604

Cook WO, Beasley VR, Lovell RA, Dahlem AM, Hooser SB, Na M, Carmichael WW (1989) Consistent inhibition of peripheral cholinesterases by neurotoxins from the freshwater cyanobacterium *Anabaena flos-aquae*: studies of ducks, swine, mice and a steer. Environ Toxicol Chem 8(10):915–922. https://doi.org/10.1002/etc.5620081010

Dawczynski C, Schäfer U, Leiterer M, Jahreis G (2007) Nutritional and toxicological importance of macro, trace, and ultra-trace elements in algae food products. J Agric Food Chem 55(25):10470–10475. https://doi.org/10.1021/jf0721500.-

De Ruiter GA, Rudolph B (1997) Carrageenan biotechnology. Trends Food Sci Technol 8:389–395

Decree Ministry of Health of 16 May 2002 (2002) Maximum and methodical levels of analysis of algal biotoxins in live bivalve molluscs, echinoderms, tunicates and marine gastropods. Official Journal No 165 of 16 July 2002

Desideri D, Cantaluppi C, Ceccotto F, Meli MA, Roselli C, Feduzi L (2016) Essential and toxic elements in seaweeds for human consumption. J Toxicol Environ Health A 79:112–122

Dobashi K, Nishino T, Fujihara M, Nagumo T (1989) Isolation and preliminary characterization of fucose-containing sulfated polysaccharides with blood-anticoagulant activity from the brown seaweed *Hizikia fusiforme*. Carbohydr Res 194:315–320. https://doi.org/10.1016/0008-6215(89)85032

El-Baky HHA, El-Baz FK, El-Baroty GS (2003) Spirulina species as a source of carotenoids and α-tocopherol and its anticarcinoma factors. Biotechnology 2:222–240

Eom SH, Lee SH, Yoon NY, Jung WK, Jeon YJ, Kim SK, Lee MS, Kim YM (2012) α-Glucosidase- and α-amylase-inhibitory activities of phlorotannins from Eisenia bicyclis. J Sci Food Agric 92(10):2084–2090. https://doi.org/10.1002/jsfa.5585

Eom SH, Lee MS, Lee EW, Kim YM, Kim TH (2013) Pancreatic lipase inhibitory activity of phlorotannins isolated from Eisenia bicyclis. Phytother Res 27(1):148–151. https://doi.org/10.1002/ptr.4694

Fevrier C, Seve B (1975) Essais d'incorporation de spiruline (Spirulina maxima) dans les aliments des porcins [Incorporation of a spiruline (Spirulina maxima) in swine food]. Ann Nutr Aliment 29(6):625–650

Fink ME (2012) Osmotherapy for intracranial hypertension: mannitol versus hypertonic saline. Continuum (Minneap Minn) 18:640–654

Fradique M, Batista AP, Nunes MC, Gouveia L, Bandarra NM, Raymundo A (2010) Incorporation of *Chlorella vulgaris* and *Spirulina maxima* biomass in pasta products. Part 1: preparation and evaluation. J Sci Food Agric 90(10):1656–1664. https://doi.org/10.1002/jsfa.3999

Francesconi KA (2010) Arsenic species in seafood: origin and human health implications. Pure Appl Chem 82:373–381

FSANZ (2008) Proposal P1003—consideration of mandatory iodine fortification for Australia and New Zealand. Dietary intake assessment report

FSANZ (2011) The survey of iodine in seaweed and seaweed containing products in Australia. Food Standards Australia New Zealand, Canberra. https://www.foodstandards.gov.au/science/surveillance/documents/iodine%20in%20Seaweed.pdf

FSANZ (n.d.) Survey of inorganic arsenic in seaweed and seaweed-containing products available in Australia. http://www.foodstandards.gov.au/consumerinformation/arsenic.cfm

Fukuda S, Saito H, Nakaji S, Yamada M, Ebine N, Tsushima E, Oka E, Kumeta K, Tsukamoto T, Tokunaga S (2007) Pattern of dietary fiber intake among the Japanese general population. Eur J Clin Nutr 61(1):99–103. https://doi.org/10.1038/sj.ejcn.1602505

Gammone MA, D'Orazio N (2015) Anti-obesity activity of the marine carotenoid fucoxanthin. Mar Drugs 13(4):2196–2214. https://doi.org/10.3390/md13042196

Garcia-Salgado S, Raber G, Raml R, Magnes C, Francesconi KA (2012) Arsenosugar phospholipids and arsenic hydrocarbons in two species of brown macroalgae. Environ Chem 9:63–66

Grosshagauer S, Kraemer K, Somoza V (2020) The true value of spirulina. J Agric Food Chem 68(14):4109–4115. https://doi.org/10.1021/acs.jafc.9b08251

Grossman A (2016) Nutrient acquisition: the generation of bioactive vitamin B_{12} by microalgae. Curr Biol 26(8):R319–R321. https://doi.org/10.1016/j.cub.2016.02.047

Gutiérrez-Salmeán G, Fabila-Castillo L, Chamorro-Cevallos G (2015) Nutritional and toxicological aspect of spirulina (Arthrospira). Nutr Hosp 32(1):34–40. https://doi.org/10.3305/nh.2015.32.1.9001

Hossain AKMM, Brennan MA, Mason SL, Guo X, Zeng XA, Brennan CS (2017) The effect of astaxanthin rich microalgae "Haematococcus pluvialis" and wholemeal flours incorporation in improving the physical and functional properties of cookies. Foods 6(8):57. https://doi.org/10.3390/foods6080057

Huh GW, Lee DY, In SJ, Lee DG, Park SY, Yi TH (2012) Fucosterols from Hizikia fusiformis and their proliferation activities on osteosarcoma-derived cell MG63. J Korean Soc Appl Biol Chem 55(4):551–555

Hwang YO, Park SG, Park GY, Choi SM, Kim MY (2010) Total arsenic, mercury, lead, and cadmium contents in edible dried seaweed in Korea. Food Addit Contam Part B Surveill 3(1):7–13. https://doi.org/10.1080/19440040903532079

Hwang HJ, Kim IH, Nam TJ (2011) Protective effect of polysaccharide from *Hizikia fusiformis* against ethanol-induced toxicity. Adv Food Nutr Res 64:143–161. https://doi.org/10.1016/B978-0-12-387669-0.00011-9

Jeong SC, Jeong YT, Lee SM, Kim JH (2015) Immune-modulating activities of polysaccharides extracted from brown algae *Hizikia fusiforme*. Biosci Biotechnol Biochem 79(8):1–4. https://doi.org/10.1080/09168451.2015.1018121

Jiménez-Escrig A, Goñi Cambrodón I (1999) Evaluación nutricional y efectos fisiológicos de macroalgas marinas comestibles [Nutritional evaluation and physiological effects of edible seaweeds]. Arch Latinoam Nutr 49(2):114–120

Kapp JM, Sumner W (2019) Kombucha: a systematic review of the empirical evidence of human health benefit. Ann Epidemiol 30:66–70. https://doi.org/10.1016/j.annepidem.2018.11.001

Kim HJ, Dasagrandhi C, Kim SH, Kim BG, Eom SH, Kim YM (2018) In vitro antibacterial activity of phlorotannins from edible brown algae, *Eisenia bicyclis* against streptomycin-resistant *Listeria monocytogenes*. Indian J Microbiol 58(1):105–108. https://doi.org/10.1007/s12088-017-0693-x

Ku CS, Yang Y, Park Y, Lee J (2013) Health benefits of blue-green algae: prevention of cardiovascular disease and nonalcoholic fatty liver disease. J Med Food 16(2):103–111. https://doi.org/10.1089/jmf.2012.246

Kulpys J, Paulauskas E, Pilipavicius V, Stankevicius R (2009) Influence of cyanobacteria Arthrospira (*Spirulina*) platensis biomass additive towards the body condition of lactation cows and biochemical milk indexes. Agric Res 7:823–835

Lahaye M, Kaeffer B (1997) Seaweed dietary fibers: structure, physico-chemical and biological properties relevant to intestinal physiology. Sci Alim 17:563–594

Li M, Chen T. (2015) Comparison of equimolar doses of mannitol and hypertonic saline for the treatment of elevated intracranial pressure after traumatic brain injury: a systematic review and meta-analysis. Medicine 94(17):e668. p 8. https://doi.org/10.1097/MD.0000000000000668

Mendiola JA, Santoyo S, Cifuentes A, Reglero G, Ibáñez E, Señoráns FJ (2008) Antimicrobial activity of sub- and supercritical CO2 extracts of the green alga *Dunaliella salina*. J Food Prot 71(10):2138–2143. https://doi.org/10.4315/0362-028x-71.10.2138

Merchant RE, Andre CA (2001) A review of recent clinical trials of the nutritional supplement *Chlorella pyrenoidosa* in the treatment of fibromyalgia, hypertension, and ulcerative colitis. Altern Ther Health Med 7(3):79–91

Mercke Odeberg J, Lignell A, Pettersson A, Höglund P (2003) Oral bioavailability of the antioxidant astaxanthin in humans is enhanced by incorporation of lipid based formulations. Eur J Pharm Sci 19(4):299–304. https://doi.org/10.1016/s0928-0987(03)00135-0

Muhling M, Belay A, Whitton BA (2005) Variation in fatty acid composition of Arthrospira (Spirulina) strains. J Appl Phycol 17:137–146

Nakashima Y, Ohsawa I, Konishi F, Hasegawa T, Kumamoto S, Suzuki Y, Ohta S (2009) Preventive effects of Chlorella on cognitive decline in age-dependent dementia model mice. Neurosci Lett 464(3):193–198. https://doi.org/10.1016/j.neulet.2009.08.04

Niccolai A, Venturi M, Galli V, Pini N, Rodolfi L, Biondi N, D'Ottavio M, Batista AP, Raymundo A, Granchi L, Tredici MR (2019) Development of new microalgae-based sourdough "crostini": functional effects of *Arthrospira platensis* (spirulina) addition. Sci Rep 9(1):19433. p 12. https://doi.org/10.1038/s41598-019-55840-1

Okai Y, Higashi-Okai K, Ishizaka S, Ontani K, Matsui-Yuada I, Yamashita U (1998) Possible immunodulating activities in an extract of edible brown alga, *Hijikia fusiforme* (Hijiki). J Sci Food Agric 76:56–62

Pabon MM, Jernberg JN, Morganti J, Contreras J, Hudson CE, Klein RL, Bickford PC (2012) A spirulina-enhanced diet provides neuroprotection in an α-synuclein model of Parkinson's disease. PLoS One 7(9):e45256. https://doi.org/10.1371/journal.pone.0045256

Palozza P, Torelli C, Boninsegna A, Simone R, Catalano A, Mele MC, Picci N (2009) Growth-inhibitory effects of the astaxanthin-rich alga Haematococcus pluvialis in human colon cancer cells. Cancer Lett 283(1):108–117. https://doi.org/10.1016/j.canlet.2009.03.031

Paul TM, Skoryna SC, Waldron-Edward D (1966) Studies on the inhibition of intestinal absorption of radioactive strontium. V. The effect of administration of calcium alginate. Can Med Assoc J 95(19):957–960

Peñalver R, Lorenzo JM, Ros G, Amarowicz R, Pateiro M, Nieto G (2020) Seaweeds as a functional ingredient for a healthy diet. Mar Drugs 18(6):301–327. https://doi.org/10.3390/md18060301

Rajapakse N, Kim SK (2011) Nutritional and digestive health benefits of seaweed. Adv Food Nutr Res 64:17–28. https://doi.org/10.1016/B978-0-12-387669-0.00002-8

Recommendation (EU) 2018/464 of 19 March 2018 on the monitoring of metals and iodine in seaweed, halophytes and products based on seaweed. https://eur-lex.europa.eu/legal-content/EN/TXT/?uri=uriserv%3AOJ.L_.2018.078.01.0016.01.ENG

Regulation (EC) No 258/97 of the European Parliament and of the Council of 27 January 1997 concerning novel foods and novel food ingredients

Regulation (EC) No 466/2001 of 8 March 2001 setting maximum levels for certain contaminants in foodstuffs

Regulation (EC) No 178/2002 of the European Parliament and of the Council of 28 January 2002 laying down the general principles and requirements of food law, establishing the European Food Safety Authority and laying down procedures in matters of food safety

Regulation (EC) n. 882/2004 of the European Parliament and of the Council, of 29 April 2004, on official controls aimed at verifying compliance with feed and food law and with the rules on animal health and welfare (OJ L 165 of 30.4.2004, page 1)

Regulation (EC) No 396/2005 of the European Parliament and of the Council of 23 February 2005 on maximum residue levels of pesticides in or on food and feed of plant and animal origin and amending Council Directive 91/414/EEC. https://eur-lex.europa.eu/legal-content/EN/TXT/?uri=CELEX:02005R0396-20190501

Regulation (EC) No 1881/2006 of 19 December 2006 setting maximum levels for certain contaminants in foodstuff. https://eur-lex.europa.eu/legal-content/EN/TXT/?uri=CELEX:02006R1881-20180319

Regulation (EC) n. 333/2007 of the Commission of 28 March 2007 on the methods of sampling and analysis for the control of the levels of trace elements and process contaminants in food products (OJ L 88 of 29.3.2007, p 29)

Regulation (EU) No 2283/2015 of the European Parliament and of the Council of 25 November 2015 on novel foods, amending Regulation (EU) No 1169/2011 of the European Parliament and of the Council and repealing Regulation (EC) No 258/97 of the European Parliament and of the Council and Commission Regulation (EC) No 1852/2001. https://eur-lex.europa.eu/legal-content/EN/TXT/HTML/?uri=CELEX:32015R2283

Schiavon M, Ertani A, Parrasia S, Vecchia FD (2017) Selenium accumulation and metabolism in algae. Aquat Toxicol 189:1–8. https://doi.org/10.1016/j.aquatox.2017.05.011

Shama A, Joyce SG, Mari FD, Ranga Rao A, Ravishankar GA, Hudaa N (2019) Macroalgae and microalgae: novel sources of functional food and feed. In: Ravishankar GA, Ranga Rao A (eds) Handbook of algal technologies and phytochemicals: volume I: food, health and nutraceutical applications. CRC Press, Boca Raton, pp 207–219

Shreenath AP, Ameer MA, Dooley J (2020) Selenium deficiency. In: StatPearls [Internet]. StatPearls Publishing, Treasure Island

Souza RB, Frota AF, Silva J et al (2018) In vitro activities of kappa-carrageenan isolated from red marine alga Hypnea musciformis: antimicrobial, anticancer and neuroprotective potential. Int J Biol Macromol 112:1248–1256. https://doi.org/10.1016/j.ijbiomac.2018.02.029

Suganya AM, Sanjivkumar M, Chandran MN, Palavesam A, Immanuel G (2016) Pharmacological importance of sulphated polysaccharide carrageenan from red seaweed *Kappaphycus alvarezii* in comparison with commercial carrageenan. Biomed Pharmacother 84:1300–1312. https://doi.org/10.1016/j.biopha.2016.10.067

Sujatha T, Narahari D (2011) Effect of designer diets on egg yolk composition of 'white Leghorn' hens. J Food Sci Technol 48:494–497

Teas J, Pino S, Critchley A, Braverman LE (2004) Variability of iodine content in common commercially available edible seaweeds. Thyroid 14(10):836–841. https://doi.org/10.1089/thy.2004.14.836

Toyomizu M, Sato K, Taroda H, Kato T, Akiba Y (2001) Effects of dietary Spirulina on meat colour in muscle of broiler chickens. Br Poult Sci 42:197–202

Vīna I, Semjonovs P, Linde R, Denina I (2014) Current evidence on physiological activity and expected health effects of kombucha fermented beverage. J Med Food 17(2):179–188. https://doi.org/10.1089/jmf.2013.0031

Waldron-Edward D, Skorina SC, Paul TM (1964) Studies on the inhibition of intestinal absorption of radioactive strontium. III. The effect of administration of sodium alginate in food and in drinking water. Can Med Assoc J 91(19):1006–1010

Wandurraga ZNU, Iguail M, Segovia PG, Martínez-Monzó J (2020) *In vitro* accessibility of mineral from microalgae-enriched cookies. Food Funct 11(3):2186–2194. https://doi.org/10.1039/c9fo02603g

Wang C, Yatsuya H, Li Y, Ota A, Tamakoshi K, Fujino Y, Mikami H, Iso H, Tamakoshi A (2016) JACC Study Group. Prospective study of seaweed consumption and thyroid cancer incidence in women: the Japan collaborative cohort study. Eur J Cancer Prev 25(3):239–245. https://doi.org/10.1097/CEJ.0000000000000168

Wang Q, Zhang L, Liu Y, Zhang G, Zhu P (2020) Characterization and functional assessment of alginate fibers prepared by metal-calcium ion complex coagulation bath. Carbohydr Polym 232:115693–115702. https://doi.org/10.1016/j.carbpol.2019.115693

Willcox DC, Willcox BJ, Todoriki H, Suzuki M (2009) The Okinawan diet: health implications of a low-calorie, nutrient-dense, antioxidant-rich dietary pattern low in glycemic load. J Am Coll Nutr 28(Suppl):500S–516S. https://doi.org/10.1080/07315724.2009.10718117

Wu M, Wu Y, Qu M, Li W, Yan X (2013) Evaluation of antioxidant activities of water-soluble polysaccharides from brown alga Hizikia fusiformis. Int J Biol Macromol 56:28–33. https://doi.org/10.1016/j.ijbiomac.2013.01.017

Ye ZW, Jiang JG, Wu GH (2008) Biosynthesis and regulation of carotenoids in Dunaliella: progresses and prospects. Biotechnol Adv 26(4):352–360. https://doi.org/10.1016/j.biotechadv.2008.03.004

Younes M, Aggett P, Aguilar F, Crebelli R, Filipič M et al (2018) Re-evaluation of carrageenan (E 407) and processed Eucheuma seaweed (E 407a) as food additives. EFSA J 16(4):e05238. https://doi.org/10.2903/j.efsa.2018.5238

Zhang YL, Shin HJ, Lee JH, Lee J (2019) Antiallergic effect of *Hizikia fusiformis* in an ovalbumin-induced allergic rhinitis mouse model. Clin Exp Otorhinolaryngol 12(2):196–205. https://doi.org/10.21053/ceo.2019.00094

Zimmermann MB (2008) Iodine requirements and the risks and benefits of correcting iodine deficiency in populations. J Trace Elem Med Biol 22:81–92. https://doi.org/10.1016/j.jtemb.2008.03.001

Chapter 16
Seaweed as Food: How to Guarantee Their Quality?

Pedro Monteiro, João Cotas, Diana Pacheco, Artur Figueirinha, Gabriela Jorge da Silva, Leonel Pereira, and Ana Marta Mendes Gonçalves

1 Introduction

Algae are a promising next-generation bio-based/food feedstock, due to their lipid composition, high protein content, and other bioactive molecules. Compounds such as soluble proteins, phenols, polyunsaturated fatty acids, pigments or carbohydrates are some of the molecules with potential for implementation in several markets, such as food/feed, pharmaceutical, and cosmetic industries (Eppink et al. 2019; Shama et al. 2019; Mohammad et al. 2019).

The trend towards increasing nutritional demand for algal products stems from a focus on health and sustainability (Wells et al. 2017) and has led to an exponential

P. Monteiro · G. J. da Silva
Faculty of Pharmacy and Center for Neurosciences and Cell Biology, Health Sciences Campus, University of Coimbra, Coimbra, Portugal
e-mail: pmonteiro@student.uc.pt; gjsilva@ci.uc.pt

J. Cotas · D. Pacheco · L. Pereira
Department of Life Sciences, MARE – Marine and Environmental Sciences Centre, University of Coimbra, Coimbra, Portugal
e-mail: jcotas@uc.pt; diana.pacheco@uc.pt; leonel.pereira@uc.pt

A. Figueirinha
Faculty of Pharmacy of the University of Coimbra, University of Coimbra, LAQV, REQUIMTE, Coimbra, Portugal

Faculty of Pharmacy of University of Coimbra, University of Coimbra, Coimbra, Portugal
e-mail: amfigueirinha@ff.uc.pt

A. M. M. Gonçalves (✉)
Department of Life Sciences, MARE – Marine and Environmental Sciences Centre, University of Coimbra, Coimbra, Portugal

Department of Biology and CESAM, University of Aveiro, Aveiro, Portugal
e-mail: amgoncalves@uc.pt; anamartagoncalves@ua.pt

A. Ranga Rao, G. A. Ravishankar (eds.), *Sustainable Global Resources of Seaweeds Volume 2*, https://doi.org/10.1007/978-3-030-92174-3_16

increase and interest in the development of seaweed farms in the western countries (Blikra et al. 2019).

Some seaweed species are known to contain protein levels similar to some of those from traditional sources, such as egg, meat, soybean, and milk, in addition to natural secondary compounds with health benefits for the consumer (Bleakley and Hayes 2017). Although seaweeds have substantial evidence regarding their health-promoting benefits, they also possess potential health hazards, such as accumulation of heavy metals, toxins, pesticide residues, and pathogens. The presence of contaminants, antinutritional factors (ANFs), allergens, and accumulation/modification of substances in the protein matrices of these new products can have severe impacts on human health. Iron, copper, and magnesium are found in high concentrations in red and green seaweeds. Brown seaweeds have shown to accumulate higher arsenic levels than red and green seaweeds. Contaminants such as dioxins and pesticides can also be found in seaweeds, such as polychlorinated dibenzo-*p*-dioxins (PCDDs) and polychlorinated dibenzofurans (PCDFs) of industrial origin. Pesticides and other organic micropollutants, such as polychlorinated biphenyls (PCBs), chlorinated pesticides and polycyclic aromatic hydrocarbons (PAHs) can also accumulate in seaweeds (van der Spiegel et al. 2013).

Thus, the safe implementation of seaweeds for human consumption is accompanied by a safe and precise biochemical analysis of the components present in seaweed products. In addition to algae characteristics, the effects of harvest, storage, and processing can dramatically influence the potential benefits of seaweed-derived foods, as well as, modify potential and promote hazard compounds present in these organisms (Wells et al. 2017).

Furthermore, for the secure and successful implementation of these novel food sources in the market, clear regulation regarding consumption thresholds must be defined. So far, the regulation regarding the production and consumption of these new foods are unclear. Legislative barriers result in cautious attitudes from investors regarding these new products, limiting their development (Probst et al. 2015).

In this chapter, we will review the current methods for the biochemical analysis of beneficial/hazardous components in seaweeds, identify crucial legislations that encompass the development and exploration of seaweed products and address the current bottlenecks facing market implementation of these novel foods.

2 Current Chemical Techniques to Certify Seaweeds as Food: Road for Innovation?

For natural nutrient composition as direct food, the techniques used are based on methods of analysis of the international Association of Official Analytical Chemists (AOAC) as for other food sources (AOAC 2021). Thus, the basis of seaweed nutritional value is identical to other foods, such as plants. However, seaweeds do not have only essential nutrients but also compounds such phenolics compounds, pigments, fatty acids, minerals and possibly heavy metals. This last ones have biological properties and can give also toxicity to the seaweed biomass (Cherry et al. 2019).

2.1 Chromatography

Chromatography is a technique to isolate compounds or fractions from a complex matrix by dispersal amongst two phases (stationary and mobile) and comprehends a wide range of methodologies that can be used in a sequential way to isolate high yields of material. These techniques are often costly, but very efficient to characterize seaweed quality when coupled with appropriate methods of detection (Misra et al. 2015). It includes liquid chromatography (LC), gas chromatography (GC), and the thin layer chromatography (TLC). LC can be coupled to an ultraviolet and visible (LC–UV/Vis) detector (quantitative analysis), a photodiode array (LC–PDA) and/or mass spectrometer (LC–MS) (identification of compounds in mixtures). GC can be coupled with a mass spectrometer (GC–MS) and/or Flame ionized diode (GC-FID) (quantitative analysis of volatile compounds) (Misra et al. 2015). Compounds separated by TLC are usually detected using visualization under visible or UV light with or without application of appropriate visualization reagents.

Liquid chromatography is mainly done by high-performance liquid chromatography (HPLC) and ultrahigh-performance liquid chromatography (UPHLC), which are the most common and consistent systems to study several seaweeds compounds, for example pigments, phenolics and amino acids (Tierney et al. 2014). Moreover, these techniques can detect oxidation in pigments and phenolic compounds (Norziah and Ching 2000). The LC diverse techniques require specific columns and reagents as mobile phases which depends on the compound to be analyzed, as the seaweed samples need to be carefully prepared and safe checked to be analysed without errors.

GC-FID is already used for quantifying and characterizing fatty acids, thus being the principal norm in Europe to do lipidic profile and quantification (Comite Europeen de Normalisation 2021). Also, this technique can be applied to quantify the monosaccharide concentration in seaweeds (Araujo et al. 2020). GC-MS can also be used to determine the seaweed fatty acid profile after respective derivatization (Araujo et al. 2020). Thus, GC is already used to analyze foods fatty acids, and can be further exploited to other seaweed compounds (Arulkumar et al. 2018). GC is a flexible and effective tool for separating similar chemical groups. However, it should be noted that one of the most important requirements for GC is that the molecules have enough volatility to pass through the GC column when heated. The non-volatile compounds can be derivatized to make it them volatile compounds (Misra et al. 2015).

Other chromatography methods, such as the high-performance anion exchange chromatography (HPAEC) - MS, can be applied to quantify and characterize the seaweed polysaccharide fraction (Nishino et al. 1989), which are the most difficult to analyze by liquid extracts due to their viscosity properties (Leandro et al. 2020). This technique exploits the acidic nature of the seaweed polysaccharides to separate fractions by pH and a strong anion exchange (Lim et al. 2014).

Thin layer chromatography (TLC) is the easiest technique used for an initial assessment of the presence of a certain type/class of compounds, for example, to detect, identify and isolate the seaweed pigments/phenolic compounds form various types of seaweed samples (Cotas et al. 2019). This technique advantage is only need small amounts of sample, easy to implement and fast, without the need for expensive equipment. It allows to make qualitative (using appropriate reagents and visualization methods) and quantitative analysis using devices that measure the density of stains.

2.2 UV/Vis Spectrophotometry

UV/Vis spectrophotometry is also employed in seaweed (aqueous fractions) quality check, as a complement to ensure seaweed quality mainly by evaluating the oxidation of the seaweed pigment, phenolics and proteins, using spectrum analysis with seaweed extracts to diminish the complexity and noise in the analysis (Cotas et al. 2019). This technique is already used in the measurement of phosphorus (José et al. 2007), and can be applied to analyze the oxidation of biomass and pigment concentration with or without TLC, before the analysis to separate the pigments (with the analysis of pigment standards). Protein content can be analyzed by colorimetric techniques (CT) with the addition of colorimetric reagents, such as the Bradford or Kjeldahl methods (Mæhre et al. 2018). The phenolic content can be done by CT technique, the Folin-Ciocalteau assay (Ford et al. 2019), however, this method can generate irregular values due to interference of other seaweed compounds (Sumampouw et al. 2021).

Mineral detection, even the harmful, can be analyzed by atomic absorption spectrophotometry with dry, digested or seaweed ashes, which is the standard method to analyze and quantify the total and specific mineral content and heavy metals (Assubaie 2015).

2.3 IR Spectroscopy

IR spectroscopical techniques uses infrared light frequencies to analyze the light absorbance by the sample, and vibration by chemical bonds which are the main target (Unkown 2021).

The FTIR (Fourier-transform infrared spectroscopy) is a low-cost technique that analyze the chemical bonds of dried samples in liquid a extract solution (Pereira et al. 2003). This technique uses infrared light to make the chemical bonds vibrate (Coates 2004), and can be used to analyze polysaccharides, pigments and phenolic fractions, compound oxidation and microplastics in the seaweed before commercialization (Cotas et al. 2019; Bogolitsyn et al. 2020; Li et al. 2020). The cost of FTIR technique is lower and it is easier to operate, when compared to chromatography, but the biochemical quantification and quality analysis is less sensible. Nevertheless, this technique accoupled to chromatographic can enhance seaweed quality characterization (Sim et al. 2014).

Nevertheless, for the certification of seaweed quality, these analytical techniques are still evolving, due to the complexity of seaweeds' composition. Seaweeds characterization is crucial to guarantee quality check and consumer safety. Safety frameworks and policies are still evolving in order to regulate seaweed food quality control, using specific analysis for seaweeds, considering their chemical variability and its compounds complexity (Leandro et al. 2020).

3 Seaweed Food Legislation Around the Globe

In general, the seaweed production, commercialization and consumption in Europe has increased in the past decades (Grebe et al. 2019; Banach et al. 2020b). Consequently, there is an overall lack of information regarding the seaweed consumption threshold, as well as regulation of seaweed commercialization (Rahikainen and Yang 2020; Lähteenmäki-Uutela et al. 2021). Nevertheless, the increasing demand led to a rising supply in the seaweed market (Bostock et al. 2016). However, seaweed are well-known phytobioremediators, which can accumulate and bioabsorb several noxious compounds, and concerning micronutrients (Gianello et al. 2019; Rosa et al. 2020), which lead to competent authorities showing an increasing concern regarding seaweed consumption and concentration of these noxious compounds (SaMonteiro et al. 2019). In this context, the European Union requested risk assessment assays to establish maximum levels for seaweed consumption.

There is a set of European regulations that seaweed-based food products and supplements must fulfill. For instance, the regulation (EC) No 178/2002 (Parliament European 2002), (EC) No 1881/2006 (European Comission 2006), (EC) 2073/2005 (European Comission 2005), (EC) 396/2005 (European Commission 2005) establishes the measures for food safety, concerning the biological and chemical parameters (European Parliament and Council 2002). Moreover, seaweed producers or importers must ensure that the marketed species are approved under the Novel Food Law ((EC) 2015/2283) (European Parliament 2015) to be sold as food in the EU market.

As the seaweed market is still evolving, is expected that the regulatory framework is adjusted accordingly. Since 2017, there has been an effort by the European Union to surpass some barriers of legislation applicable to seaweed production and commercialization. In this context, the Board of European Committee for Standardization elected a Technical Committee (CEN/TC 454) for algae and algae-based products specification, algae identification, processing, and methods standardization (Rahikainen and Yang 2018).

The regulation (EC) 710/2009 (European Comission 2009) already controls the organic certification of seaweed. While this legislation has provided the basis for the labeling of seaweed in some European countries such as France, there are no certified producers of seaweed in other countries (Banach et al. 2020a).

The United States Department of Agriculture (USDA) and Food and Drug Administration (FDA) are the competent authorities empowered for regulate food, feed, and fertilizer applications. In the United States of America, seaweed aquaculture and commercialization is increasing, leading to a higher demand and offer (Augyte et al. 2017). However, there are no federal laws or guidelines regarding the commercialization of seaweed as a direct food item (Janesie 2019). States will act, to fill the regulatory void, in these cases. Federal laws and other acts related to other uses of seaweed exist, but are only applicable for some seaweed species and their compounds (Food and Drug Administration 2020).

The organic classification of farmed kelp and other algae is governed by the USDA, but only when it is used as an ingredient for food/feed or fertilizer. The USDA approves the authorization of four non-organic substances derived from farmed aquatic plants and algae in products labeled 'organic' where the algae product is otherwise not commercially available in organic form (Janesie 2019; Concepcion et al. 2020). Due to this gap on the legislation, the Connecticut Sea Grant in cooperation with the Connecticut Department of Agriculture Bureau of Aquaculture, developed a set of guidelines for food hygiene and safety, regarding seaweed production, storage, handling, processing, and transportation (Concepcion et al. 2020).

Japanese food products production and commercialization are under several regulations that enables food security (Food and Agriculture Standards 2018). For instance, the Food Safety Basic Act lays down the guidelines for the establishment of a food safety regime and sets out the position of the Food Safety Commission (FSC), a competent authority responsible for food risk assessment. Under the authority of the Ministry of Health, Labor and Welfare (MHLW), a food risk management agency, the Food Sanitation Act aims to protect the health of people by ensuring food and beverage protection and sanitation, through the establishment of standardized measures for quality and safety food and beverages production, regulating parameters such as packaging, additives, contaminants and other pollutants concentration (Holdt and Kraan 2011).

The Japanese Ministry of Health, Labor and Welfare established a list of Food for Specified Health Uses (FOSHU), whereas seaweed is included (i.e., seaweed pickles, processed, canned or bottled) within a certain threshold (Murata and Nakazoe 2001). Nevertheless, seaweeds are considered functional ingredients, meaning that affect the structure and/or function of the body and are used to maintain or control specific health conditions (Murata and Nakazoe 2001).

China is one of the most ancient population that incorporate seaweeds on a daily diet (SaMonteiro et al. 2019). For this reason, since 1950, the Chinese government encouraged the seaweed industry through regulations and funding to develop aquaculture. Nevertheless, due to environmental issues that aquatic resources are under, the government started to appeal to more sustainable and quality seaweed aquaculture practices. Since then, investment in water treatment plants is mandatory and strictly regulated and inspected by all environmental protection agencies at all levels (Zhang 2018). The Asian Integrated Food Security (AIFS) framework has released a strategic action plan for fisheries cooperation, including aquaculture products, however, it is not clear if and which seaweed species will be included (Campbell et al. 2020) .

4 Seaweed Food Quality Check: Bottlenecks and Future Perspectives

There are technical and legislative bottlenecks hampering the implementation of macroalgae in daily diets. It is paramount to guarantee consumer safety in exposure to these products and in that scenario, seaweeds can be a source of exposure to harmful contaminants (Desideri et al. 2016). In fact, due to their specific cell wall structure and characteristics, seaweeds are accumulators for minerals and elements present in the surrounding waters and consequently, elements present in algae are often several orders of magnitude higher than the values found in the surrounding environment (SaMonteiro et al. 2019). Toxic elements, like lead (Pb), cadmium (Cd) and arsenic (As) have noxious health impacts, even at low concentrations, if ingested over long periods of time (Desideri et al. 2016). Even though their intake is usually below dangerous dosages, the content of Pb, Cd and inorganic Arsenic (iAs) still poses a consumer risk, being advisable the surveillance of concentrations within consumed products. It has been shown that arsenic generally occurs in seaweeds as organic forms of arsenosugars and methylated species, that are rapidly excreted and have low potential for toxicity (Ho and Redan 2020). However, to assess the real danger of As-mediated toxicity, chemical characterization and biochemical properties of individual As species are crucial (Desideri et al. 2016). Another special concern regarding macroalgae consumption if the overexposure to elements such as iodine (González et al. 2020) . Algae stand out for their iodine content that, whilst being an essential element when ingested at recommended levels, it can cause harmful effects when ingested in high quantities (González et al. 2020). The excessive intake of this element is associated with increased production of thyroid hormones, which lead to diseases such as hyperthyroidism, manifested by symptoms such as increased metabolic rate (SaMonteiro et al. 2019).

Technically, post harvesting techniques have yet to fully understand the behavior of the biochemical fraction after treatment. The literature has mostly considered the effects of thermal processing on the concentration of the bioactive components, without regards to the bioavailability of these compounds after thermal processing. It is critical to characterize how post-harvesting processing, inclusive of storage, thermal processing and cooking affects these nutrients, phytochemicals and contaminants (Ho and Redan 2020).

It is still unclear how gut microbiota modulates metabolism of dietary components and the extent to which this modulation varies between individuals. Gut microbes transform many classes of dietary compounds, such as complex polysaccharides, lipids, proteins and phytochemicals, and these metabolic reactions are linked to a myriad of health benefits as well as disease susceptibilities. Additionally, gut microbiota also transform industrial chemicals and pollutants, altering toxicities and lifetimes in the body (Koppel et al. 2017). There is still limited literature regarding the specific interactions and metabolic modulation of the gut microbiome and macroalgae components. The direct and indirect metabolic influence of the gastrointestinal microbiota, in regards to the chemical modifications of a wide range of compounds, can potentially have consequent implications for the host health (Clarke et al. 2019).

The actual and official characterization techniques in seaweeds are based in plant methodologies and consequently have more variability and errors, as compounds in seaweeds do not react the same way as equivalent compound in plants, such as carbohydrates/dietary fibers. Thus, for the correct detection of bioactive compounds there is a need for equipment with high precision, which is costly and hard to obtain for the food industry (Leandro et al. 2020).

In the future, the bottlenecks associated with these techniques can be ameliorated through their optimization and standardization, allowing accurate assessment of seaweeds nutritional advantages and risks (Holdt and Kraan 2011). None of the issues is the legislation about assays and the food safety sheet (Holdt and Kraan 2011); thus, seaweed producers and industry do the required analysis, which can give deviated results due to the methodology-applied sensibility, also because most laboratory tests are standard for plants and not for seaweed.

In this case, there is a need to create a proper legislation to certify seaweeds with the best method in a cost effective way (Salehi et al. 2019). The above cited chemical methods can be further exploited to have a full seaweed food safety sheet. Moreover, there are several problems to be addressed when analysing the seaweeds, namely "when and how can be seaweed be analyzed?". For the majority of the techniques, the seaweeds need to be in liquid form, and if the pre-commercial seaweed is dried, it will be necessary to evaluate how the seaweed preservation techniques affects the nutrients and compounds along the time of shelf life (Badmus et al. 2019).

Overall, seaweed potential for food development is recognized and cultivation of algal biomass for several markets is transitioning to commercial-scale systems. During this period, it is crucial that institutional frameworks promote the development and commercialization, stimulating the evolution of algal biomass industry to an important and source of high-value food (Trentacoste et al. 2015). Seaweed business is dependent on elements of the regulatory environment, such as aquaculture/agriculture subsidies, maritime spatial planning or land use spatial planning (for offshore or inshore cultivation, respectively), licensing and other legislative agreements (Lähteenmäki-Uutela et al. 2021).

5 Conclusion

As a novel food product, seaweed industry for human consumption still faces several challenges related to standard methods for quality and safety procedures, as well as lack of regulatory policies. There are several analytical techniques that can be used currently for seaweeds nutritional evaluation, Nevertheless, improvements to standardize the techniques and results are urgently needed. Throughout the Asian, European, and American continents, a will to work with the scientific community to develop the guidelines and the legislation applicable to the seaweed producers for food industry is emerging.

Acknowledgments This work is financed by national funds through FCT—Foundation for Science and Technology, I.P., within the scope of the projects UIDB/04292/2020—MARE—Marine and Environmental Sciences Centre and UIDP/50017/2020 + UIDB/50017/2020 (by FCT/MTCES) granted to CESAM—Centre for Environmental and Marine Studies. João Cotas thanks to the European Regional Development Fund through the Interreg Atlantic Area Program, under the project NASPA (EAPA_451/2016). Diana Pacheco thanks to PTDC/BIA-CBI/31144/2017-POCI-01 project-0145-FEDER-031144-MARINE INVADERS, co-financed by the ERDF through POCI (Operational Program Competitiveness and Internationalization) and by the Foundation for Science and Technology (FCT, IP). Ana M. M. Gonçalves acknowledges University of Coimbra for the contract IT057-18-7253 and also thanks to the project MENU—Marine Macroalgae: Alternative recipes for a daily nutritional diet (FA_05_2017_011), funded by the Blue Fund under Public Notice No. 5—Blue Biotechnology. Artur Figueirinha thanks FCT/MCTES, Fundação para a Ciência e Tecnologia and Ministério da Ciência, Tecnologia e Ensino Superior through grant UIDB/50006/2020.

References

AOAC (2021) Scientific Standards & Methods—AOAC International. https://www.aoac.org/scientific-solutions/. Accessed 3 March

Araujo GS, Cotas J, Morais T, Leandro A, García-Poza S, Gonçalves AMM, Pereira L (2020) *Calliblepharis jubata* cultivation potential—a comparative study between controlled and semi-controlled aquaculture. Appl Sci 10:7553. https://doi.org/10.3390/app10217553

Arulkumar A, Rosemary T, Paramasivam S, Rajendran RB (2018) Phytochemical composition, in vitro antioxidant, antibacterial potential and GC-MS analysis of red seaweeds (*Gracilaria corticata* and *Gracilaria edulis*) from Palk Bay, India. Biocatal Agric Biotechnol 15:63–71. https://doi.org/10.1016/j.bcab.2018.05.008

Assubaie FN (2015) Assessment of the levels of some heavy metals in water in Alahsa Oasis farms, Saudi Arabia, with analysis by atomic absorption spectrophotometry. Arab J Chem 8:240–245. https://doi.org/10.1016/j.arabjc.2011.08.018

Augyte S, Yarish C, Redmond S, Kim JK (2017) Cultivation of a morphologically distinct strain of the sugar kelp, *Saccharina latissima* f. *angustissima*, from coastal Maine, USA, with implications for ecosystem services. J Appl Phycol 29:1967–1976. https://doi.org/10.1007/s10811-017-1102-x

Badmus UO, Taggart MA, Boyd KG (2019) The effect of different drying methods on certain nutritionally important chemical constituents in edible brown seaweeds. J Appl Phycol 31:3883–3897. https://doi.org/10.1007/s10811-019-01846-1

Banach JL, van den Burg SWK, van der Fels-Klerx HJ (2020a) Food safety during seaweed cultivation at offshore wind farms: an exploratory study in the North Sea. Mar Policy 120:104082. https://doi.org/10.1016/j.marpol.2020.104082

Banach JL, Hoek-van den Hil EF, Fels-Klerx HJ (2020b) Food safety hazards in the European seaweed chain. Compr Rev Food Sci Food Saf 19:332–364. https://doi.org/10.1111/1541-4337.12523

Bleakley S, Hayes M (2017) Algal proteins: extraction, application, and challenges concerning production. Foods 6:33. https://doi.org/10.3390/foods6050033

Blikra MJ, Løvdal T, Vaka MR, Roiha IS, Lunestad BT, Lindseth C, Skipnes D (2019) Assessment of food quality and microbial safety of brown macroalgae (*Alaria esculenta* and *Saccharina latissima*). J Sci Food Agric 99:1198–1206. https://doi.org/10.1002/jsfa.9289

Bogolitsyn K, Parshina A, Druzhinina A, Ovchinnikov D, Krasikov V, Khviyuzov S (2020) Physicochemical characteristics of the active fractions of polyphenols from Arctic macrophytes. J Appl Phycol 32:4277–4287. https://doi.org/10.1007/s10811-020-02226-w

Bostock J, Lane A, Hough C, Yamamoto K (2016) An assessment of the economic contribution of EU aquaculture production and the influence of policies for its sustainable development. Aquacult Int 24:699–733. https://doi.org/10.1007/s10499-016-9992-1

Campbell I, Kambey CSB, Mateo JP, Rusekwa SB, Hurtado AQ, Msuya FE, Stentiford GD, Cottier-Cook EJ (2020) Biosecurity policy and legislation for the global seaweed aquaculture industry. J Appl Phycol 32:2133–2146. https://doi.org/10.1007/s10811-019-02010-5

Cherry P, O'Hara C, Magee PJ, McSorley EM, Allsopp PJ (2019) Risks and benefits of consuming edible seaweeds. Nutr Rev 77:307–329. https://doi.org/10.1093/nutrit/nuy066

Clarke G, Sandhu KV, Griffin BT, Dinan TG, Cryan JF, Hyland NP (2019) Gut reactions: breaking down xenobiotic–microbiome interactions. Pharmacol Rev 71:198–224. https://doi.org/10.1124/pr.118.015768

Coates J (2004) Interpretation of infrared spectra, a practical approach. In: Encyclopedia of analytical chemistry. Wiley, Hoboken, pp 1–23

Comite Europeen de Normalisation (2021) ISO—ISO 12966-2:2017—animal and vegetable fats and oils—gas chromatography of fatty acid methyl esters—part 2: preparation of methyl esters of fatty acids. https://www.iso.org/standard/72142.html. Accessed 3 March

Concepcion A, DeRosia-Banick K, Balcom N (2020) Seaweed production and processing in Connecticut: a guide to understanding and controlling potential food safety hazards. Connecticut

Cotas J, Figueirinha A, Pereira L, Batista T (2019) The effect of salinity on *Fucus ceranoides* (Ochrophyta, Phaeophyceae) in the Mondego River (Portugal). J Oceanol Limnol 37:881–891. https://doi.org/10.1007/s00343-019-8111-3

Desideri D, Cantaluppi C, Ceccotto F, Meli MA, Roselli C, Feduzi L (2016) Essential and toxic elements in seaweeds for human consumption. J Toxicol Environ Health Part A Curr Issues 79:112–122. https://doi.org/10.1080/15287394.2015.1113598

Eppink MHM, Olivieri G, Reith H, van den Berg C, Barbosa MJ, Wijffels RH (2019) From current algae products to future biorefinery practices: a review. Adv Biochem Eng Biotechnol 166:99–123. https://doi.org/10.1007/10_2016_64

European Comission (2005) Commission Regulation (EC) No 2073/2005 of 15 November 2005 on microbiological criteria for foodstuffs. Official journal of the European Union OJ L 338, 22.12.2005, pp 1–26. http://data.europa.eu/eli/reg/2005/2073/oj

European Comission (2006) Regulamento (CE) n. o 1881/2006 da Comissão, de 19 de Dezembro de 2006, que fixa os teores máximos de certos contaminantes presentes nos géneros alimentícios. Jornal Oficial da União Europeia. OJ L 364, 20.12.2006, pp 5–24. http://data.europa.eu/eli/reg/2006/1881/oj

European Comission (2009) Commission regulation (EC) No 710/2009 of 5 August 2009 amending Regulation (EC) No 889/2008 laying down detailed rules for the implementation of Council Regulation (EC) No 834/2007, as regards laying down detailed rules on organic aquaculture animal and seaweed production. Official Journal of the European Union L, pp 204–215. http://data.europa.eu/eli/reg/2009/710/oj

European Commission (2005) Regulation (EC) No 396/2005 of the European Parliament and of the Council of 23 February 2005 on maximum residue levels of pesticides in or on food and feed of plant and animal origin and amending Council Directive 91/414/EEC. Official Journal of the European Union OJ L 70, 16.3.2005, pp 1–16. http://data.europa.eu/eli/reg/2005/396/oj

European Parliament (2015) Regulation (EU) 2015/2283 of the European Parliament and of the Council of 25 November 2015 on novel foods, amending Regulation (EU) No 1169/2011 of the European Parliament and of the Council and repealing Regulation (EC) No 258/97 of the European Parliament and of the Council and Commission Regulation (EC) No 1852/2001. Official Journal of the European Union OJ L 327, 11.12.2015, pp 1–22. http://data.europa.eu/eli/reg/2015/2283/oj

European Parliament and Council (2002) Regulation (EC) No 178/2002 of the European Parliament and of the Council of 28 January 2002 laying down the general principles and requirements of food law, establishing the European Food Safety Authority and laying down procedures in matters of food safety. Official Journal L 031, 01/02/2002, pp 0001–0024. http://data.europa.eu/eli/reg/2002/178/oj

Food and Agriculture Standards (2018) Food and agricultural import regulations and standards report. Tokyo, Japan

Food and Drug Administration (2020) Food and Drug Adminstration—Department of Health and Human Services 21CFR184.1120

Ford L, Theodoridou K, Sheldrake GN, Walsh PJ (2019) A critical review of analytical methods used for the chemical characterisation and quantification of phlorotannin compounds in brown seaweeds. Phytochem Anal 30:587–599. https://doi.org/10.1002/pca.2851

Gianello D, Ávila-Hernández E, Aguer I, Crettaz-Minaglia MC (2019) Water quality assessment of a temperate urban lagoon using physico-chemical and biological indicators. SN Appl Sci 1:470. https://doi.org/10.1007/s42452-019-0469-5

González A, Paz S, Rubio C, Gutiérrez ÁJ, Hardisson A (2020) Human exposure to iodine from the consumption of edible seaweeds. Biol Trace Elem Res 197:361–366. https://doi.org/10.1007/s12011-019-01996-w

Grebe GS, Byron CJ, Gelais AS, Kotowicz DM, Olson TK (2019) An ecosystem approach to kelp aquaculture in the Americas and Europe. Aquacult Rep 15:100215. https://doi.org/10.1016/j.aqrep.2019.100215

Ho KKHY, Redan BW (2020) Impact of thermal processing on the nutrients, phytochemicals, and metal contaminants in edible algae. Crit Rev Food Sci Nutr:1–19. https://doi.org/10.1080/10408398.2020.1821598

Holdt SL, Kraan S (2011) Bioactive compounds in seaweed: functional food applications and legislation. J Appl Phycol 23:543–597. https://doi.org/10.1007/s10811-010-9632-5

Janesie C (2019) Regulation of seaweed as a food source. Connecticut

José J, Milagres M, Alvarez H, Cantarutti RB, César J, Neves L (2007) Determinação de Fe, Zn, Cu E Mn Extraídos do Solo por Diferentes Extratores e Dosados por Espectofotometria de Emissão Ótica em Plasma Induzido e Espectrofotometria de Absorção Atômica. Rev Bras Ciênc Solo [online] 31(2):237–245. https://doi.org/10.1590/S0100-06832007000200006

Koppel N, Rekdal VM, Balskus EP (2017) Chemical transformation of xenobiotics by the human gut microbiota. Science 356:1246–1257. https://doi.org/10.1126/science.aag2770

Lähteenmäki-Uutela A, Rahikainen M, Camarena-Gómez MT, Piiparinen J, Spilling K, Yang B (2021) European Union legislation on macroalgae products. Aquacult Int 29:487–509. https://doi.org/10.1007/s10499-020-00633-x

Leandro A, Pacheco D, Cotas J, Marques JC, Pereira L, Gonçalves AMMM (2020) Seaweed's bioactive candidate compounds to food industry and global food security. Life 10:140. https://doi.org/10.3390/life10080140

Li Q, Feng Z, Zhang T, Ma C, Shi H (2020) Microplastics in the commercial seaweed nori. J Hazard Mater 388:122060. https://doi.org/10.1016/j.jhazmat.2020.122060

Lim SJ, Aida WMW, Maskat MY, Mamot S, Ropien J, Mohd DM (2014) Isolation and antioxidant capacity of fucoidan from selected Malaysian seaweeds. Food Hydrocoll 42:280–288. https://doi.org/10.1016/j.foodhyd.2014.03.007

Mæhre HK, Dalheim L, Edvinsen GK, Elvevoll EO, Jensen I-J (2018) Protein determination—method matters. Foods 7:5. https://doi.org/10.3390/foods7010005

Misra NN, Rai DK, Hossain M (2015) Analytical techniques for bioactives from seaweed. In: Seaweed sustainability. Elsevier, Amsterdam, pp 271–287. https://doi.org/10.1016/B978-0-12-418697-2.00010-6

Mohammad JH, Ranga Rao A, Ravishankar GA (2019) Opportunities and challenges in seaweeds as feed stock for biofuel production. In: Ravishnkar GA, Rao AR (eds) Handbook of algal technologies and phytochemicals: volume II Phycoremediation, biofuels and global biomass production. CRC Press, Boca Raton, pp 39–50

Murata M, Nakazoe JI (2001) Production and use of marine algae in Japan. Jpn Agric Res Quart 35:281–290. https://doi.org/10.6090/jarq.35.281

Nishino T, Yokoyama G, Dobashi K, Fujihara M, Nagumo T (1989) Isolation, purification, and characterization of fucose-containing sulfated polysaccharides from the brown seaweed *Ecklonia kurome* and their blood-anticoagulant activities. Carbohydr Res 186:119–129. https://doi.org/10.1016/0008-6215(89)84010-8

Norziah MH, Ching CY (2000) Nutritional composition of edible seaweed *Gracilaria changgi*. Food Chem 68:69–76. https://doi.org/10.1016/S0308-8146(99)00161-2

Parliament European (2002) Regulation (EC) No 178/2002 of the European Parliament and of the council of 28 January 2002 laying down the general principles and requirements of food law, establishing the European Food Safety Authority and laying down procedures in matters of food safety

Pereira L, Sousa A, Coelho H, Amado AM, Ribeiro-Claro PJA (2003) Use of FTIR, FT-Raman and ^{13}C-NMR spectroscopy for identification of some seaweed phycocolloids. Biomol Eng 20:223–228. https://doi.org/10.1016/S1389-0344(03)00058-3

Probst L, Frideres L, Pedersen B, Amato F (2015) Business innovation observatory sustainable, safe and nutritious food food processing technologies. Report from European Union, City of Luxembourg, Luxembourg. https://ec.europa.eu/docsroom/documents/13425/attachments/1/translations/en/renditions/native

Rahikainen M, Yang B (2018) Macroalgae as food and feed ingredients in the Baltic Sea region—regulation by the European Union, pp 1–20

Rahikainen M, Yang B (2020) Macroalgae as food and feed ingredients in the Baltic Sea region—regulation by the European Union. Finland

Rosa J, Lemos MFL, Crespo D, Nunes M, Freitas A, Ramos F, Pardal MÂ, Leston S (2020) Integrated multitrophic aquaculture systems—potential risks for food safety. Trends Food Sci Technol 96:79–90. https://doi.org/10.1016/j.tifs.2019.12.008

Salehi B, Sharifi-Rad J, Seca AML, Pinto DCGA, Michalak I, Trincone A, Mishra AP, Nigam M, Zam W, Martins N (2019) Current trends on seaweeds: looking at chemical composition, phytopharmacology, and cosmetic applications. Molecules 24:4182. https://doi.org/10.3390/molecules24224182

SaMonteiro M, Sloth J, Holdt S, Hansen M (2019) Analysis and risk assessment of seaweed. EFSA J 17:e170915. https://doi.org/10.2903/j.efsa.2019.e170915

Shama A, Joyce SG, Mari FD, Ranga Rao A, Ravishankar GA, Hudaa N (2019) Macroalgae and microalgae: novel sources of functional food and feed. In: Ravishankar GA, Rao AR (eds) Handbook of algal technologies and phytochemicals: volume I: food, health and nutraceutical applications. CRC Press, Boca Raton, pp 207–219

Sim SF, Lee TZE, Nurul Aida L, Mohd Irwan L, Samling B (2014) Synchronized analysis of FTIR spectra and GCMS chromatograms for evaluation of the thermally degraded vegetable oils. J Anal Methods Chem 2014:1–9. https://doi.org/10.1155/2014/271970

van der Spiegel M, Noordam MY, van der Fels-Klerx HJ (2013) Safety of novel protein sources (insects, microalgae, seaweed, duckweed, and rapeseed) and legislative aspects for their

application in food and feed production. Compr Rev Food Sci Food Saf 12:662–678. https://doi.org/10.1111/1541-4337.12032
Sumampouw GA, Jacobsen C, Getachew AT (2021) Optimization of phenolic antioxidants extraction from *Fucus vesiculosus* by pressurized liquid extraction. J Appl Phycol 33(2):1195–1207. https://doi.org/10.1007/s10811-020-02362-3
Tierney MS, Soler-Vila A, Rai DK, Croft AK, Brunton NP, Smyth TJ (2014) UPLC-MS profiling of low molecular weight phlorotannin polymers in *Ascophyllum nodosum, Pelvetia canaliculata* and *Fucus spiralis*. Metabolomics 10:524–535. https://doi.org/10.1007/s11306-013-0584-z
Trentacoste EM, Martinez AM, Zenk T (2015) The place of algae in agriculture: policies for algal biomass production. Photosynth Res 123:305–315. https://doi.org/10.1007/s11120-014-9985-8
Unkown (2021) Spectroscopy | COSMOS. https://astronomy.swin.edu.au/cosmos/s/Spectroscopy. Accessed 3 Mar 2021
Wells ML, Potin P, Craigie JS, Raven JA, Merchant SS, Helliwell KE, Smith AG, Camire ME, Brawley SH (2017) Algae as nutritional and functional food sources: revisiting our understanding. J Appl Phycol 29:949–982. https://doi.org/10.1007/s10811-016-0974-5
Zhang J (2018) Seaweed industry in China. Submariner-NetworkEu: 1–31

Part II
Pharmaceutical Applications of Seaweeds and Health Benefits

Chapter 17
Global Trade of Seaweed Foods

Sara García-Poza, João Cotas, Tiago Morais, Diana Pacheco, Leonel Pereira, João C. Marques, and Ana M. M. Gonçalves

Abbreviations

BNF	British Nutrition Foundation
RNI	Adult reference nutrient intake
WHO	World Health Organization

1 Introduction

Marine seaweeds comprise a wide biodiversity in the ocean, being categorized in phyla Ochrophyta—Phaeophyceae, Rhodophyta, and Chlorophyta (Rindi et al. 2012). Traditionally, seaweeds have been used as food, natural colorants, fertilizers, and as folk remedies. The Era of industrialization in the early 1900, lead to the increase of seaweed capacity of production. Since the major seaweed component

S. García-Poza · J. Cotas · D. Pacheco · L. Pereira · J. C. Marques
Department of Life Sciences, MARE – Marine and Environmental Sciences Centre, University of Coimbra, Coimbra, Portugal
e-mail: jcotas@uc.pt; diana.pacheco@uc.pt; leonel.pereira@uc.pt; jcmimar@ci.uc.pt

T. Morais
Lusalgae, Lda., Incubadora de Empresas da Figueira da Foz, Figueira da Foz, Portugal
e-mail: tsmorais@lusalgae.pt

A. M. M. Gonçalves (✉)
Department of Life Sciences, MARE – Marine and Environmental Sciences Centre, University of Coimbra, Coimbra, Portugal

Department of Biology and CESAM, University of Aveiro, Aveiro, Portugal
e-mail: amgoncalves@uc.pt; anamartagoncalves@ua.pt

A. Ranga Rao, G. A. Ravishankar (eds.), *Sustainable Global Resources of Seaweeds Volume 2*, https://doi.org/10.1007/978-3-030-92174-3_17

are their polysaccharides, the exploitation of the phycocolloids, such as alginate, agar and carrageenan, significantly increased (Tanna and Mishra 2019). Due to the rheological and biological properties of these algal hydrocolloids, their application reached several industries, from the foods to pharmaceuticals (Panzella and Napolitano 2017; Zollmann et al. 2019).

Marine organisms, particularly, seaweeds unveil several compounds with biological activity, exhibiting nutraceutical and therapeutical potential (Shannon and Abu-Ghannam 2019). For instance, seaweeds contain fatty acids, phenolic compounds, vitamins, minerals, and carbohydrates, which are essential for the good functioning of the human body (Zhao et al. 2018). The mentioned algal compounds have already been isolated and characterized, with confirmed bioactivities, such as antitumor, antioxidant, anti-inflammatory, anti-obesity, or antidiabetic (Ganesan et al. 2019). These characteristics made seaweeds to be seen as a rich source of nutraceuticals (Nunes et al. 2018; Tanna and Mishra 2018).

Concomitantly, the societal awareness regarding natural food products, led to the growth of seaweed markets. Despite their traditional consumption in Asian countries, the Western countries opened up consumption only in the past few decades started to look to seaweeds as a healthy and appetizing food product/ingredient (Barbier et al. 2019). Changing our eating habits and adding seaweed-derived ingredients to our diet will help us to prevent several health problems. This will also promote dietary diversification needed for meeting the wide range of nutrients for better health upkeep.

One of the limitation being availability of the fresh seaweed feedstock in the local market. For this reason, the availability of seaweeds needs to be scaled up, adopting cost-effective cultivation technologies to produce biomass of high quality and to develop downstream processing methods to efficiently extract valuable metabolites for food applications and other uses (Tanna and Mishra 2018).

In contrast to terrestrial plants, seaweeds's application in food industry is in a limited proportion only as compared to the total number of edible species. Owing to a lack of regulatory mechanisms and information relating to the technical issues of cultivation, such as aquaculture and upscaling technologies, many algal species with nutraceutical potential are not well explored (Barbier et al. 2019).

Seaweeds's health benefits, particularly as ingredients for novel flavoring agents or as a salt substitute, appear to be appealing to European consumers. According to Mintel's Global New Products Database, 37% of seaweed-flavored food and drink products launched in Europe, between 2011 and 2015, were in the snack category (Barbier et al. 2019). Between 2011 and 2015, the amount of food and drink items flavored with seaweed ingredients (namely, Kombu, Nori, and Wakame) (Fig. 17.1h) increased by 147% in the European market (Mintel 2016). Because of this development, Europe is now the second most creative region in the world when it comes to the introduction of seaweed-seasoned foods and beverages (Barbier et al. 2019).

Seaweeds and their compounds offer a wide range of products, worldwide, intended for human use (Craigie 2011). Thus, seaweeds commercialization for human consumption already constitutes 83% of total global seaweed production (FAO 2016).

Fig. 17.1 *Ulva lactuca* (**a**) (Chlorophyta); *Gracilaria* sp. (**b**); *Kappaphycus alvarezii* (**c**); *Palmaria palmata* (**d**) (Rhodophyta); *Ascopyllum nodosum* (**e**); *Fucus vesiculosus* (**f**); *Saccharina latissima* (**g**); *Undaria pinnatifida* (**h**) (Ochrophyta, Phaeophyceae); Wakame (*U. pinnatifida*) rice with grapes and pine nuts (**i**)

Studies to evaluate the effect of various functional foods, individually or in combination with seaweeds and their extracts, are always required. Despite the abundance of published evidence, only a few studies have come to a reasonable conclusion regarding seaweeds as actual functional food items. There is also a need for systematic studies with extensive supporting data to determine the full potential of seaweeds and to establish cost-effective functional food products for the market. Thus, seaweeds can only become a modern functional food with a long shelf life and health-related benefits if further research is done (Tanna and Mishra 2018).

Thus, the main goal of this chapter is to elucidate the application of seaweed in the global food market, as well as the nutraceutical role they hold. Moreover, it will be illustrated strategies to enhance seaweed global production and trade.

2 Seaweed as a Food Product/Ingredient

Coastal communities around the world have sought and consumed algae since the beginning of human civilization (Dillehay et al. 2008) and since ancient times, seaweed has been used as food, home remedies, dyes, and mineral-rich fertilizers. Traditionally seaweeds are consumed in many Asian countries such as China, Indonesia, the Philippines, South Korea, North Korea, Japan, and Malaysia for centuries. In recent decades, seaweeds have gained more popularity in Western countries as well. Nowadays they are widely used as food in the United States, South America, and European countries due to their functional properties and the introduction of Asian cuisine (Bocanegra et al. 2009). Currently, algae can be incorporated directly or indirectly in the preparation of food and beverages. They are a versatile form of food ingredient as their physical ability to emulsify and retain water enhances the functional properties.

Seaweed is now an important industrial research and development concept as a nutraceutical or functional food with dietary benefits beyond its fundamental macronutrient content (Shannon and Abu-Ghannam 2019). The onset of diet and lifestyle related diseases, especially type 2 diabetes, obesity, cancer and metabolic syndrome, has become a widespread health problem being a health epidemic in developed regions such as Europe, USA, and Australia (WHO 2019). It has been shown through worldwide global dietary studies that countries, where seaweed is consumed regularly, have significantly fewer cases of obesity and diet-related diseases (Iso 2011; Nanri et al. 2017).

Certain seaweeds contain 10–100 times more minerals and vitamins per unit dry mass than terrestrial plants or foods derived from animals (Rupérez 2002). They also have vitamins A, D, E, K, C, B_1, B_2, B_9, B_{12} soluble in water and fat and essential minerals, such as calcium, iron, iodine, magnesium, phosphorus, potassium, zinc, copper, manganese, selenium and fluoride (Misurcova 2011; Qin 2018), however the content varies between species and geographical area. From a portion perspective, taking 8 g of dried weight (DW) as a typical serving size, many seaweeds

Table 17.1 Seaweed nutritional composition (%)

Biomolecule	Proportion	Reference
Total protein content	5–47% of seaweed dry mass	Černá (2011)
Total amino acids content	Approximately 42–48% are essential amino acids	Wong and Cheung (2001)
Total polysaccharide or sugar content	4–76% (DW)	Paniagua-Michel et al. (2014)
Total lipid content	0.60–4.14%	Rodrigues et al. (2015)

perform better than plant and animal foods in terms of adult reference nutrient intake (RNI) (Astorga-España et al. 2015).

As an example, *Palmaria palmata* (Fig. 17.1d), the red seaweed contains on average 6.4 mg of iron per 8 g serving, compared to only 1.2–3.1 mg in a 100 g portion of lean beef (Branscheid and Judas 2011). In the same way, 8 g of the green seaweed *Ulva lactuca* (Fig. 17.1a) contains on average 260 mg of calcium or 37% of the RNI, while 8 g of cheddar cheese provides on average 5% of the RNI (MacArtain et al. 2007).

Protein constitutes a high quantity of the seaweed dried biomass (Table 17.1). Red seaweeds have the greatest protein content, while brown seaweeds the least (Černá 2011). For example, *Undaria pinnatifida* (Phaeophyceae) has an amino acid score of 1.0, similarly to an egg and soy, *Pyropia/Porphyra* 0.91 (Rhodophyta), and *Saccharina latissima* (Fig. 17.1g) (previously known as *Laminaria saccharina*) 0.82 (Phaeophyceae) (Murata and Nakazoe 2001). However, the high polyphenolic content of algae can reduce their protein digestibility (Wong and Cheung 2001). Despite this, algae are still a viable alternative to animal protein.

The cell wall structure of many algae is constituted by cellulose, which is a non-digestible and non-nutritive polysaccharide and constitutes 2–10% of the total polysaccharides. Digestible polysaccharides vary between phyla. Brown algae have alginates, fucoidans and laminarin as main polysaccharides (Rodrigues et al. 2015); red algae have carrageenans and agarans; and green algae have ulvans (Jiao et al. 2011). Most algal polysaccharides are non-starchy fibre, which helps balance normal blood glucose levels and may contribute to the RNI of 30 g fiber day^{-1} (BNF 2016).

Algae have a majority of polyunsaturated lipids, made up of fatty acids (ω-3 or omega-3), such as docosahexaenoic (DHA) and eicosapentaenoic (EPA) acids. The most common monounsaturated fatty acids in algae are linoleic and arachidonic acids (ω-6 or omega-6) (Belattmania et al. 2018). While palmitic and myristic acids are predominant as saturated fatty acids. Both ω-6 and ω-3 fatty acids are essential from a dietary perspective. However, the consumption of these compounds in unbalanced proportions can lead to chronic inflammatory diseases such as obesity, rheumatoid arthritis, non-alcoholic fatty liver and cardiovascular diseases (Patterson et al. 2012). Seaweed maintains this low ω-6: ω-3 ratio of fatty acids (Irene et al. 2018), making them excellent sources of dietary lipids (Dawczynski et al. 2007).

A nutraceutical is defined as any nutritional food product with additional health benefits being a combination of a nutrient and a pharmaceutical product (Tanna and Mishra 2018). Urbanization, economic development and changes in lifestyle seriously affect eating habits and there are several reports that have confirmed a direct relationship between diet and health, and people believe that diet has more impact on health, than exercise and inheritance (Espín et al. 2007).

Nutraceuticals are expected to increase life expectancy steadily, outperform expensive health treatments, and improve the quality of life for older adults and are therefore in high demand (Bigliardi and Galati 2013). Due to increased consumer awareness, the potential and treatment for chronic diseases of nutraceuticals is being investigated on a large scale (Choudhary and Grover 2012).

Besides the benefits of regular consumption of algae in the diet, the medicinal properties of algae bioactivities have historically been recognized. For example, seaweeds are used for the treatment and prevention of goiter, which is caused by a lack of iodine in the diet (Rosenfeld 2000). The various curative effects of algal species against noncommunicable diseases, such as inflammation, obesity, diabetes, hypertension, and viral infections have been demonstrated in several studies (Rajauria et al. 2016). A clinical study indicated that regular consumption of the brown seaweed *U. pinnatifida* can effectively minimize the risk of breast cancer in women (Teas et al. 2013), while an oral administration of seaweed extracts (*Fucus vesiculosus* (Fig. 17.1f), *Macrocystis pyrifera* and *Saccharina/ Laminaria japonica*) (Phaeophyceae) with zinc, manganese and vitamin B_6, potentially reduces the symptoms of osteoarthritis in a mixed population (Myers et al. 2016). In addition to their extensive medicinal properties, algae are also recognized for their antioxidant capabilities and bioactive polyphenolic compounds (Stephens et al. 2017).

Studies have also shown potential functions in protection against HIV, mainly related to compounds present in algae such as phlorotannins, sulfated polysaccharides, certain diterpenes and lectins (Nagarajan and Mathaiyan 2015). In addition, insoluble and fermentable dietary fibers are main components found in macroalgae and help improve digestive health, including colorectal cancer, gastrointestinal inflammation, and probiotics and other adverse health conditions (Lange et al. 2015). Cancer prevention and metabolic syndrome (METS) associated with obesity, cardiovascular disease, diabetes, and chronic inflammation are also principal attributes of algae in relation to human health and well-being (Lowenthal and Fitton 2015). Although some evidence suggests that the effect of bioactive compounds in the human body is moderate and relatively short, if they are consumed routinely as part of the daily diet, they could contribute significantly (Wang et al. 2019) (Fig. 17.1i).

The world population is projected to increase to nine billion by 2050 (Zhou et al. 2018). Traditional farming and herding practices increase competition for land use. In addition, livestock farming is also one of the largest generators of greenhouse gas emissions of carbon dioxide and methane. While algae reduce the carbon dioxide emissions generated by the production of animal proteins. As a functional food, seaweed offers a sustainable, alternative, and low-cost source of protein without the saturated fats associated with meat (Shannon and Abu-Ghannam 2019). Several studies have been conducted to evaluate the food, pharmaceutical and nutraceutical properties, although more research is needed on seaweed for its commercial expansion and to understand the safety, toxicity and environmental impact of cultivation, processing and its bioactive extraction (Ganesan et al. 2019). In addition to this, functional and nutraceutical food products based on seaweed ingredients should be analyzed for the presence of contaminants, allergens, heavy metals, or hazardous substances generated during the cultivation or processing of algae (Ganesan et al. 2019). To boost their business development and production, these products must meet and adhere to strict safety legislation.

3 Seaweed Harvesting and Cultivation

Seaweed for human consumption can be obtained by either by wild harvesting or by seaweed cultivation. The latter one is being considered due to sustainability and safety issues as opposed to the wild harvest.

The wild harvest is practiced in case of fast growing seaweeds such as *Sargassum* sp., *Ascophyllum nodosum* (Phaeophyceae) (Fig. 17.1e) or *Gracilaria* sp. (Rhodophyta) (Barry et al. 2015). However, some of these seaweeds can be also cultivated also for polysaccharide extraction. Thus, most of the seaweed for food industry is cultivated, while a minor portion is harvested from naturally growing seaweeds (Lähteenmäki-Uutela et al. 2021). The seaweeds cultivation is popular in countries where seaweeds are traditionally consumed (Fei et al. 1999; Hwang and Park 2020). However, in other parts of the world, the cultivation methods are being developed and optimized by adopting new techniques, technologies, and new species (García-Poza et al. 2020).

The most cultivated species are the red seaweeds: *Eucheuma* sp., *Gracilaria* spp. (Fig. 17.1b), *Kappaphycus* spp. (Fig. 17.1c), and *Porphyra/Pyropia/Neopyropia* spp. (Japanese Nori); brown algae: *Saccharina japonica* (Japanese Kelp or Kombu) and *Undaria pinnatifida* (Japanese Wakame). This species nearly represents 90% of the total of the seaweed cultivated in the world (FAO 2018). The wild biomass harvest practice are adopted in Norway and Chile and not in the Asiatic countries (Mac Monagail et al. 2017; Ferdouse et al. 2018).

The seaweeds cultivation do not pose adverse impacts in the coastal ecosystems when compared to the wild harvest activity. Thus, seaweed cultivation systems are being recommended to meet the Sustainable Development Goals by UN (Gouvello et al. 2017; Stead 2018).

3.1 Are the Seaweeds Beneficial or Problematic: A Solution to the Seaweed Exploitation?

The benefits of seaweed cultivation mainly rests in the prevention of overexploitation of natural ecosystem of seaweeds which support ecological balance of the dependent biotic components of seas besides balancing climate change. They nurture aquatic animals, microorganisms and all living forms in their environment preserving the health of the sea (Reisewitz et al. 2006; Bertocci et al. 2015). Whereas, seaweed cultivation guarantees food safety more easily as it is grown cleaner environment, free from ecological disturbances (García-Poza et al. 2020; Leandro et al. 2020). In contrast, wild seaweed can assimilate or absorb dangerous contaminants, heavy metals, and toxins, which have a very negative effect if consumed by the humans (Henriques et al. 2017; Cherry et al. 2019). Thus, the consumers prefer the cultivated seaweeds which are safe and is of reliable quality (Holdt and Kraan 2011; Leandro et al. 2020).

Consequently, the seaweeds' aquaculture is the only way to produce enough quantity of quality- biomass to meet the demands from various industries, besides the food industry's requirements (Ashkenazi et al. 2019). However, the food industry is the maximum demand of seaweed rawmaterial when compared to other applications (FAO 2018; Campbell et al. 2019).

Therefore, there is an increase in the efforts to develop these aquaculture techniques, to enhance the biomass production and to promote seaweed consumption in the western countries coupled to the expansion of product range as per the consumer demands (Hafting et al. 2015; García-Poza et al. 2020). In this regard, the water quality and the measurement of the abiotic parameters governing seaweed production assumes utmost importance (Uribe et al. 2018; Pliego-Cortés et al. 2019; García-Poza et al. 2020).

4 Seaweed Food Global Market: Economics and Tendencies

In 2016, aquaculture was responsible for the production of 30.1 million tons of biomass, worldwide (FAO 2018). The biomass production has grown progressively from 13.5 million tons (in 1995) to over 30 million tons in 2016 (FAO 2018). The farming of tropical species, such as *Kappaphycus alvarezii* and *Eucheuma* spp. (Rhodophyta), for carrageenan extraction is usually pointed out as the main reason for this fast growth. Evidently in Indonesia seaweed production enhanced from less than 4 million tons to 11 million tons over a period of 6 years, turning it into the major contributor for the global seaweed aquaculture production (FAO 2018). In later years, such increase tendency has regressed, registering a 0.7% fall in 2018. However, in the same year, 32.4 million tons of seaweeds were produced, where around 97% of which were farmed. The main reason for this happening is linked with a slow growth in the output of tropical species and reduced production in Southeast Asia, while the farming of temperate and cold-water species was still rising (FAO 2020).

Thus, seaweeds' compound add higher market margin into seaweed as food source, due to the exploitation of seaweed polysaccharides as emulsifiers and thickening agents in the food industry. Consequently, the seaweeds are industrially handled to extract polysaccharide such as alginate, agar, and carrageenan (Leandro et al. 2020).

The overall market of seaweed has 5.9 USD billion in 2019, and it is expected to grow steadily up to 9.1% until 2027. This increase in market would be felt due to increase in consumption mostly in Europe and North America regions. Whereas, the Asian countries (mainly, China, Japan, and Korea) are leading this market segment due to a high number of traditional consumers (Unknown 2021). The base of the industry is now shifting the wild harvest into industry. It is expected to 99% of the seaweed demand be produced in aquaculture system in a short time, due to the concerns of seawater pollution and to the shift in the mindset to reduce the impact in coastal ecosystems (Unknown 2021).

Interestingly, the market share of various seaweed groups is as follows, 50% constituting red seaweed, 45% the brown seaweeds, and only 5% are green seaweeds. This segmentation are predominantly ruled by the polysaccharide extraction industry, and the traditional food sectors (Ferdouse et al. 2018). However, this estimated growth can be augmented to the wide range of seaweed-flavored foods, ice creams and drinks, who are being developed and launched, mainly in the Asia and the Pacific markets which would eventually impact the markets in Europe and North America (FAO 2018). This enhancement of the seaweed food applications needs to be supported by the seaweed cultivation and production enhancement (García-Poza et al. 2020).

5 Conclusions and Future Perspectives

Seaweeds are evaluated as enriched nutritional foods, many of them with a high potential for application as food supplements, due to their rich nutritional profile; thus, promote human health. However, there are compounds that are already exploited into the food industry, which are being used as ingredients in processed foods or as a food additive.

Seaweeds ecological relevance, coupled with biomass safety issues, lead to the evolution of aquaculture technologies, which are necessary to maintain the seaweed feedstock safety and quality without harmful compounds (i.e., metal, pesticides, fertilizers) that could cause health hazards in humans.

The seaweed market value is increasing due to the high demand of seaweed consumption around the world. Moreover, food industry development is evolving in order to develop new seaweed-based foods and beverages.

Acknowledgments This work is financed by national funds through FCT—Foundation for Science and Technology, I.P., within the scope of the projects UIDB/04292/2020 granted to MARE—Marine and Environmental Sciences Centre and UIDP/50017/2020 + UIDB/50017/2020 (by FCT/MTCES) granted to CESAM—Centre for Environmental and Marine Studies. Sara García-Poza thanks to the project MENU—Marine Macroalgae: Alternative recipes for a daily nutritional diet (FA_05_2017_011) which co-financed this research, funded by the Blue Fund under Public Notice No. 5—Blue Biotechnology. João Cotas thanks to the European Regional Development Fund through the Interreg Atlantic Area Program, under the project NASPA (EAPA_451/2016). Diana Pacheco thanks to PTDC/BIA-CBI/31144/2017-POCI-01 project-0145-

FEDER-031144-MARINE INVADERS, co-financed by the ERDF through POCI (Operational Program Competitiveness and Internationalization) and by the Foundation for Science and Technology (FCT, IP). Ana M. M. Gonçalves acknowledges University of Coimbra for the contract IT057-18-7253.

References

Ashkenazi DY, Israel A, Abelson A (2019) A novel two-stage seaweed integrated multi-trophic aquaculture. Rev Aquac 11:246–262. https://doi.org/10.1111/raq.12238

Astorga-España MS, Rodríguez Galdón B, Rodríguez Rodríguez EM, Díaz Romero C (2015) Mineral and trace element concentrations in seaweeds from the sub-Antarctic ecoregion of Magallanes (Chile). J Food Compos Anal 39:69–76. https://doi.org/10.1016/j.jfca.2014.11.010

Barbier M, Charrier B, Araujo R, Holdt SL, Jacquemin B, Rebours C (2019) PEGASUS: PHYCOMORPH European guidelines for a sustainable aquaculture of seaweeds. Roscoff, France. https://doi.org/10.21411/2c3w-yc73

Barry AN, Starkenburg SR, Sayre RT (2015) Strategies for optimizing algal biology for enhanced biomass production. Front Energy Res 3:1–5. https://doi.org/10.3389/fenrg.2015.00001

Belattmania Z, Engelen AH, Pereira H, Serrão EA, Custódio L, Varela JC, Zrid R, Reani A, Sabour B (2018) Fatty acid composition and nutraceutical perspectives of brown seaweeds from the Atlantic coast of Morocco. Int Food Res J 25:1520–1527

Bertocci I, Araújo R, Oliveira P, Sousa-Pinto I (2015) Potential effects of kelp species on local fisheries. J Appl Ecol 52:1216–1226. https://doi.org/10.1111/1365-2664.12483

Bigliardi B, Galati F (2013) Innovation trends in the food industry: the case of functional foods. Trends Food Sci Technol 31:118–129. https://doi.org/10.1016/j.tifs.2013.03.006

Bocanegra A, Bastida S, Benedí J, Ródenas S, Sánchez-Muniz FJ (2009) Characteristics and nutritional and cardiovascular-health properties of seaweeds. J Med Food 12:236–258. https://doi.org/10.1089/jmf.2008.0151

Branscheid W, Judas M (2011) Detection of bone in meat. In: Nollet L, Toldra F (eds) Handbook of analysis of edible animal by-products. CRC Press, Boca Raton, p 278

British Nutrition Foundation (2016) Nutrition requirements: reference nutrient intakes for minerals

Campbell I, Macleod A, Sahlmann C, Neves L, Funderud J, Øverland M, Hughes AD, Stanley M (2019) The environmental risks associated with the development of seaweed farming in Europe-prioritizing key knowledge gaps. Front Mar Sci 6:107. https://doi.org/10.3389/fmars.2019.00107

Černá M (2011) Chapter 24 - Seaweed proteins and amino acids as nutraceuticals. In: Kim S-K (ed) Advances in food and nutrition research. Academic, San Diego, pp 297–312

Cherry P, O'Hara C, Magee PJ, McSorley EM, Allsopp PJ (2019) Risks and benefits of consuming edible seaweeds. Nutr Rev 77:307–329. https://doi.org/10.1093/nutrit/nuy066

Choudhary M, Grover K (2012) Development of functional food products in relation to obesity. Funct Foods Health Dis 2:188. https://doi.org/10.31989/ffhd.v2i6.90

Craigie JS (2011) Seaweed extract stimuli in plant science and agriculture. J Appl Phycol 23:371–393. https://doi.org/10.1007/s10811-010-9560-4

Dawczynski C, Schubert R, Jahreis G (2007) Amino acids, fatty acids, and dietary fibre in edible seaweed products. Food Chem 103:891–899. https://doi.org/10.1016/j.foodchem.2006.09.041

Dillehay TD, Ramirez C, Pino M, Collins MB, Rossen J, Pino-Navarro JD (2008) Monte Verde: seaweed, food, medicine, and the peopling of South America. Science 320:784–786. https://doi.org/10.1126/science.1156533

Espín JC, García-Conesa MT, Tomás-Barberán FA (2007) Nutraceuticals: facts and fiction. Phytochemistry 68:2986–3008. https://doi.org/10.1016/j.phytochem.2007.09.014

FAO (2016) The state of world fisheries and aquaculture—contributing to food security and nutrition for all. FAO, Rome

FAO (2018) The state of the world fisheries and aquaculture—meeting the sustainable development goals, vol 3. FAO, Rome

FAO (2020) The state of world fisheries and aquaculture. Sustainability in action. FAO, Rome

Fei X, Bao Y, Lu S (1999) Seaweed cultivation: traditional way and its reformation. Chin J Oceanol Limnol 17:193–199

Ferdouse F, Holdt SL, Smith R, Murúa P, Yang Z (2018) The global status of seaweed production, trade and utilization. In: FAO globefish research programme, vol 124, p 120

Ganesan AR, Tiwari U, Rajauria G (2019) Seaweed nutraceuticals and their therapeutic role in disease prevention. Food Sci Hum Wellness 8:252–263. https://doi.org/10.1016/j.fshw.2019.08.001

García-Poza S, Leandro A, Cotas C, Cotas J, Marques JC, Pereira L, Gonçalves AMM (2020) The evolution road of seaweed aquaculture: cultivation technologies and the industry 4.0. Int J Environ Res Public Health 17:6528. https://doi.org/10.3390/ijerph17186528

Gouvello R, Le C, Andrade D, Fezzardi L, Hochart E, Laffoley D, Callier M, Angel D, Haroun R, Harris A (2017) Aquaculture and marine protected areas: potential opportunities and synergies. Aquat Conserv Mar Freshw Ecosyst 27:138–150. https://doi.org/10.1002/aqc.2821

Hafting JT, Craigie JS, Stengel DB, Loureiro RR, Buschmann AH, Yarish C, Edwards MD, Critchley AT (2015) Prospects and challenges for industrial production of seaweed bioactives. Edited by M. Graham. J Phycol 51:821–837. https://doi.org/10.1111/jpy.12326

Henriques B, Lopes CB, Figueira P, Rocha LS, Duarte AC, Vale C, Pardal MA, Pereira E (2017) Bioaccumulation of Hg, Cd and Pb by Fucus vesiculosus in single and multi-metal contamination scenarios and its effect on growth rate. Chemosphere 171:208–222. https://doi.org/10.1016/j.chemosphere.2016.12.086

Holdt SL, Kraan S (2011) Bioactive compounds in seaweed: functional food applications and legislation. J Appl Phycol 23:543–597. https://doi.org/10.1007/s10811-010-9632-5

Hwang EK, Park CS (2020) Seaweed cultivation and utilization of Korea. Algae 35:107–121. https://doi.org/10.4490/algae.2020.35.5.15

Irene B, Ikram B, Christian BG, Nina LS, Rune W, Heidi A, Svenja H, Erik-Jan L (2018) Chemical characterization of 21 species of marine macroalgae common in Norwegian waters: benefits of and limitations to their potential use in food and feed. J Sci Food Agric 98:2035–2042. https://doi.org/10.1002/jsfa.8798

Iso H (2011) Lifestyle and cardiovascular disease in Japan. J Atheroscler Thromb 18:83–88. https://doi.org/10.5551/jat.6866

Jiao G, Yu G, Zhang J, Ewart H (2011) Chemical structures and bioactivities of sulfated polysaccharides from marine algae. Mar Drugs 9:196–223. https://doi.org/10.3390/md9020196

Lähteenmäki-Uutela A, Rahikainen M, Camarena-Gómez MT, Piiparinen J, Spilling K, Yang B (2021) European Union legislation on macroalgae products. Aquac Int 29:487–509. https://doi.org/10.1007/s10499-020-00633-x

Lange KW, Hauser J, Nakamura Y, Kanaya S (2015) Dietary seaweeds and obesity. Food Sci Hum Wellness 4:87–96. https://doi.org/10.1016/j.fshw.2015.08.001

Leandro A, Pacheco D, Cotas J, Marques JC, Pereira L, Gonçalves AMM (2020) Seaweed's bioactive candidate compounds to food industry and global food security. Life 10:140. https://doi.org/10.3390/life10080140

Lowenthal RM, Fitton JH (2015) Are seaweed-derived fucoidans possible future anti-cancer agents? J Appl Phycol 27:2075–2077

Mac Monagail M, Cornish L, Morrison L, Araújo R, Critchley AT (2017) Sustainable harvesting of wild seaweed resources. Eur J Phycol 52:371–390. https://doi.org/10.1080/09670262.2017.1365273

MacArtain P, Gill CIR, Brooks M, Campbell R, Rowland IR (2007) Nutritional value of edible seaweeds. Nutr Rev 1:535–543. https://doi.org/10.1111/j.1753-4887.2007.tb00278.x

Mintel (2016) Seaweed-flavoured food and drink launches increased by 147% in Europe between 2011 and 2015. https://www.mintel.com/press-centre/food-and-drink/seaweed-flavoured-food-and-drink-launches-increased-by-147-in-europe-between-2011-and-2015. Searched 2020 FEB 26

Misurcova L (2011) Chemical composition of seaweeds. In: Kim S-K (ed) Handbook of marine macroalgae: biotechnology and applied phycology. John Wiley & Sons, Hoboken, pp 171–192

Murata M, Nakazoe JI (2001) Production and use of marine algae in Japan. Jpn Agric Res Quart 35(4):281–290. https://doi.org/10.6090/jarq.35.281

Myers SP, Mulder AM, Baker DG, Robinson SR, Rolfe MI, Brooks L, Fitton JH (2016) Effects of fucoidan from Fucus vesiculosus in reducing symptoms of osteoarthritis: a randomized placebo-controlled trial. Biol Targets Ther 10:81–88. https://doi.org/10.2147/BTT.S95165

Nagarajan S, Mathaiyan M (2015) Emerging novel anti HIV biomolecules from marine algae: an overview. J Appl Pharm Sci 5:153–158. https://doi.org/10.7324/JAPS.2015.50928

Nanri A, Mizoue T, Shimazu T, Ishihara J, Takachi R, Noda M, Iso H, Sasazuki S, Sawada N, Tsugane S (2017) Dietary patterns and all-cause, cancer, and cardiovascular disease mortality in Japanese men and women: the Japan public health center-based prospective study. PLoS One 12:1–15. https://doi.org/10.1371/journal.pone.0174848

Nunes N, Valente S, Ferraz S, Barreto MC, Pinheiro de Carvalho MAA (2018) Nutraceutical potential of Asparagopsis taxiformis (Delile) Trevisan extracts and assessment of a downstream purification strategy. Heliyon 4:e00957. https://doi.org/10.1016/j.heliyon.2018.e00957

Paniagua-Michel J, Olmos-Soto J, Morales-Guerrero ER (2014) Algal and microbial exopolysaccharides: new insights as biosurfactants and bioemulsifiers. Adv Food Nutr Res 73:221–257. https://doi.org/10.1016/B978-0-12-800268-1.00011-1

Panzella L, Napolitano A (2017) Natural phenol polymers: recent advances in food and health applications. Antioxidants 6:1–24. https://doi.org/10.3390/antiox6020030

Patterson E, Wall R, Fitzgerald GF, Ross RP, Stanton C (2012) Health implications of high dietary omega-6 polyunsaturated fatty acids. J Nutr Metab 2012. https://doi.org/10.1155/2012/539426

Pliego-Cortés H, Bedoux G, Boulho R, Taupin L, Freile-Pelegrín Y, Bourgougnon N, Robledo D (2019) Stress tolerance and photoadaptation to solar radiation in *Rhodymenia pseudopalmata* (Rhodophyta) through mycosporine-like amino acids, phenolic compounds, and pigments in an Integrated Multi-Trophic Aquaculture system. Algal Res 41:101542. https://doi.org/10.1016/j.algal.2019.101542

Qin Y (2018) Applications of bioactive seaweed substances in functional food products. In: Qin Y (ed) Bioactive seaweeds for food applications: natural ingredients for healthy diets. Academic, San Diego, pp 111–134. https://doi.org/10.1016/B978-0-12-813312-5.00006-6

Rajauria G, Foley B, Abu-Ghannam N (2016) Identification and characterization of phenolic antioxidant compounds from brown Irish seaweed *Himanthalia elongata* using LC-DAD–ESI-MS/MS. Innov Food Sci Emerg Technol 37:261–268. https://doi.org/10.1016/j.ifset.2016.02.005

Reisewitz SE, Estes JA, Simenstad CA (2006) Indirect food web interactions: sea otters and kelp forest fishes in the Aleutian archipelago. Oecologia 146:623–631. https://doi.org/10.1007/s00442-005-0230-1

Rindi F, Soler-Vila A, Guiry MD (2012) Taxonomy of marine macroalgae used as sources of bioactive compounds. In: Hayes M (ed) Marine bioactive compounds. Springer, New York, pp 1–53

Rodrigues D, Freitas AC, Pereira L, Rocha-Santos TAP, Vasconcelos MW, Roriz M, Rodríguez-Alcalá LM, Gomes AMP, Duarte AC (2015) Chemical composition of red, brown and green macroalgae from Buarcos bay in central west coast of Portugal. Food Chem 183:197–207. https://doi.org/10.1016/J.FOODCHEM.2015.03.057

Rosenfeld L (2000) Discovery and early uses of iodine. J Chem Educ 77:984–987. https://doi.org/10.1021/ed077p984

Rupérez P (2002) Mineral content of edible marine seaweeds. Food Chem 79:23–26. https://doi.org/10.1016/S0308-8146(02)00171-1

Shannon E, Abu-Ghannam N (2019) Seaweeds as nutraceuticals for health and nutrition. Phycologia 58:563–577. https://doi.org/10.1080/00318884.2019.1640533

Stead SM (2018) Rethinking marine resource governance for the United Nations sustainable development goals. Curr Opin Environ Sustain 34:54–61. https://doi.org/10.1016/j.cosust.2018.12.001
Stephens PRS, Cirne-Santos CC, de Souza Barros C, Teixeira VL, Carneiro LAD, Amorim LSC, Ocampo JSP, Castello-Branco LRR, de Palmer Paixão ICN (2017) Diterpene from marine brown alga *Dictyota friabilis* as a potential microbicide against HIV-1 in tissue explants. J Appl Phycol 29:775–780
Tanna B, Mishra A (2018) Metabolites unravel nutraceutical potential of edible seaweeds: an emerging source of functional food. Compr Rev Food Sci Food Saf 17(6):1613–1624. https://doi.org/10.1111/1541-4337.12396
Tanna B, Mishra A (2019) Nutraceutical potential of seaweed polysaccharides: structure, bioactivity, safety, and toxicity. Compr Rev Food Sci Food Saf 18:817–831. https://doi.org/10.1111/1541-4337.12441
Teas J, Vena S, Cone DL, Irhimeh M (2013) The consumption of seaweed as a protective factor in the etiology of breast cancer: proof of principle. J Appl Phycol 25:771–779. https://doi.org/10.1007/s10811-012-9931-0
Unknown (2021) lobal Commercial Seaweeds Market Size Report, 2020-2027. https://www.grandviewresearch.com/industry-analysis/commercial-seaweed-market. Searched 2020 February 28
Uribe E, Vega-Gálvez A, Heredia V, Pastén A, Di Scala K (2018) An edible red seaweed (*Pyropia orbicularis*): influence of vacuum drying on physicochemical composition, bioactive compounds, antioxidant capacity, and pigments. J Appl Phycol 30:673–683. https://doi.org/10.1007/s10811-017-1240-1
Wang Y, Chen G, Peng Y, Rui Y, Zeng X, Ye H (2019) Simulated digestion and fermentation *in vitro* with human gut microbiota of polysaccharides from *Coralline pilulifera*. LWT 100:167–174. https://doi.org/10.1016/j.lwt.2018.10.028
WHO (2019) Double burden of malnutrition. Searched on 13 Apr 2019
Wong KH, Cheung PCK (2001) Nutritional evaluation of some subtropical red and green seaweeds part II. In vitro protein digestibility and amino acid profiles of protein concentrates. Food Chem 72:11–17. https://doi.org/10.1016/S0308-8146(00)00176-X
Zhao Y, Yizhou Z, Wang J, Ma S, Yu Y, Lindsey White W, Yang S, Yang F, Jun L (2018) Fucoidan extracted from *Undaria pinnatifida*: source for nutraceuticals/functional foods. Mar Drugs 16:321. https://doi.org/10.3390/md16090321
Zhou C, Elshkaki A, Graedel TE (2018) Global human appropriation of net primary production and associated resource decoupling: 2010-2050. Environ Sci Technol 52:1208–1215. https://doi.org/10.1021/acs.est.7b04665
Zollmann M, Robin A, Prabhu M, Polikovsky M, Gillis A, Greiserman S, Golberg A (2019) Green technology in green macroalgal biorefineries. Phycologia 58:516–534. https://doi.org/10.1080/00318884.2019.1640516

Chapter 18
Seaweeds as a Source of Vitamin B_{12}

Tomohiro Bito and Fumio Watanabe

Abbreviations

B_{12}	Vitamin B_{12} or cobalamin
HPLC	High-performance liquid chromatography
LC/MS-MS	Liquid chromatography/electrospray ionization–tandem mass spectrometry

1 Introduction

Vitamin B_{12} (B_{12}), also known as cobalamin, has the largest molecular mass (1355.4) and the most complex structure of all the vitamins (Watanabe and Bito 2016). It is commonly known as the red-colored vitamin, but the scientific use of the term "B_{12}" is restricted to cyanocobalamin. In this chapter, B_{12} refers to all potentially biologically active B_{12} compounds (Fig. 18.1).

With 5,6-dimethylbenzimidazole as the base, cyanocobalamin has a cobalt-coordinated nucleotide as the lower axial ligand. Cyanocobalamin or hydroxocobalamin is readily converted into methylcobalamin, a coenzyme of methionine synthase (EC 2.1.1.13) involved in methionine biosynthesis, and 5′-deoxyadenosylcobalamin, a coenzyme of methylmalonyl-CoA mutase (EC 5.4.99.2) involved in amino acid and odd-chain fatty acid metabolism in mammalian cells (Chen et al. 1994; Fenton et al. 1982).

T. Bito · F. Watanabe (✉)
Department of Agricultural, Life and Environmental Sciences, Faculty of Agriculture, Tottori University, Tottori, Japan
e-mail: bito@tottori-u.ac.jp; watanabe@tottori-u.ac.jp

A. Ranga Rao, G. A. Ravishankar (eds.), *Sustainable Global Resources of Seaweeds Volume 2*, https://doi.org/10.1007/978-3-030-92174-3_18

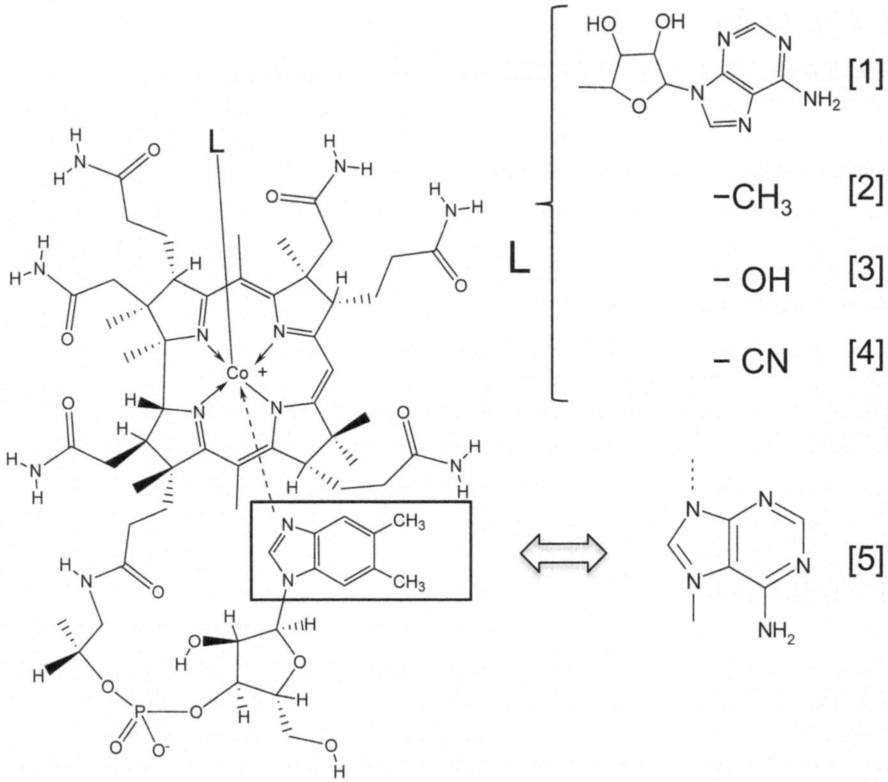

Fig. 18.1 The structural formula of vitamin B_{12} and partial structures of vitamin B_{12}-related compounds. (1) 5′-Deoxyadenosylcobalamin. (2) Methylcobalamin. (3) Hydroxocobalamin. (4) Cyanocobalamin (vitamin B_{12}). (5) Pseudovitamin B_{12}

During B_{12} deficiency, methylmalonic acid and homocysteine are significantly accumulated. B_{12} deficiency inhibits the B_{12}-dependent methionine biosynthesis, leading to the accumulation of homocysteine (Bito et al. 2013), which has pro-oxidant activity (Andrzej and Kilmer 1993). As a result, B_{12} deficiency disrupts cellular redox homeostasis and induces oxidative stress, which is implicated in various human diseases, including atherosclerosis and neurodegenerative diseases (Jessica et al. 2012). Also, the significant accumulation of methylmalonic acid results from B_{12} deficiency (Toyoshima et al. 1996). Methylmalonic acid is a potent inhibitor of succinate dehydrogenase (EC 1.3.99.1), which participates in both the TCA cycle and the respiratory chain (Toyoshima et al. 1995). Thus, methylmalonic acid increases due to B_{12} deficiency blocked mitochondrial respiration and consequently disrupts various metabolic pathways (Toyoshima et al. 1995). Therefore, methylmalonic acid and homocysteine are usually used as the indices of vitamin deficiency (Bito and Watanabe 2016) (Fig. 18.2).

B_{12} is synthesized only by certain bacteria and archaea but not by plants. B_{12} accumulates in animal tissues through microbial interaction in the natural food

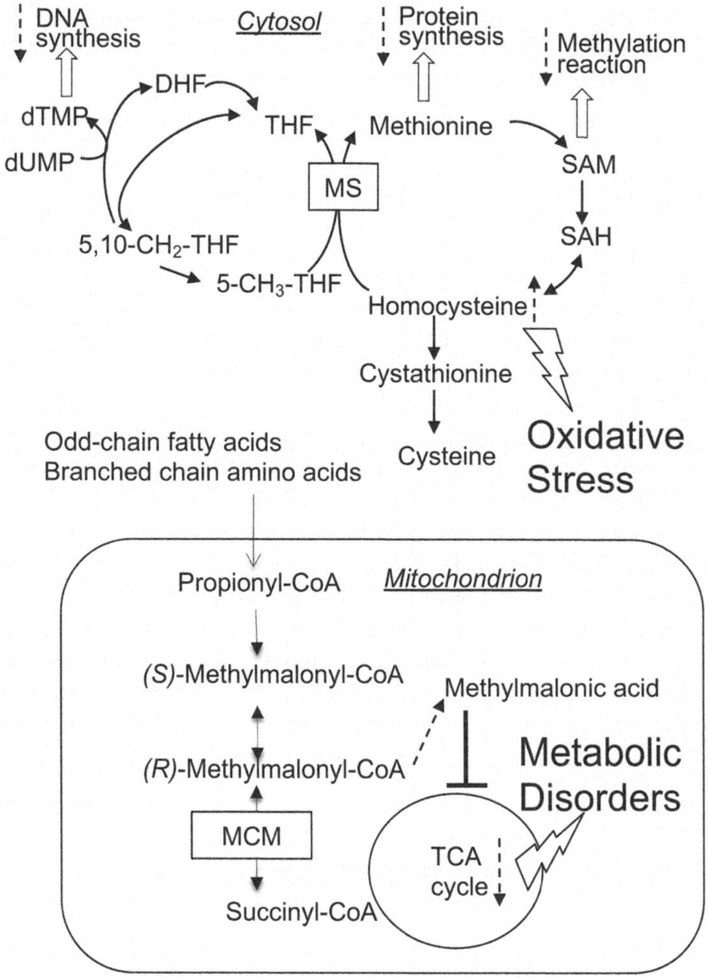

Fig. 18.2 The physiological function of vitamin B_{12} in mammals. *DHF* dihydrofolate, *THF* tetrahydrofolate, *5-CH$_3$-THF* 5-methyltetrahydrofolate, *5,10-CH$_2$-THF* 5,10-methylenetetrahydrofolate, *SAM* S-adenosylmethionine, *SAH* S-adenosylhomocysteine, *MS* cobalamin-dependent methionine synthase, *MCM* methylmalonyl-CoA mutase

chain (Watanabe 2007). For example, ruminants, such as cattle and sheep, acquire B_{12} through a symbiotic relationship with the bacteria in their stomach (Watanabe and Bito 2018). Thus, animal-based foods, such as meat, milk, and fish, but not plant foods, are the primary dietary sources of B_{12} (Watanabe and Bito 2018; Watanabe 2007). Therefore, vegans, or strict vegetarians, who cannot ingest any animal-based foods, are reportedly at risk of developing B_{12} deficiency (Pawlak et al. 2013).

We have identified edible seaweeds that naturally contain large quantities of B_{12} (Watanabe et al. 2002, 2014). This chapter summarizes the up-to-date information on the bioavailability of B_{12}-compounds from edible seaweeds.

2 Vitamin B_{12} Content in Edible Seaweeds

Raw and dried edible seaweeds contain various amounts of B_{12} (Table 18.1). The edible brown algae, such as *Laminaria angustata* (Kombu), *Undaria pinnatifida* (Wakame), *Eisenia bicyclis* (Arame), and *Sargassum fusiforme* (Hijiki), contain none or a trace amount of B_{12}. On the other hand, a substantial amount of B_{12}, at more than 30 μg/100 g dry weight, was found in some green algae, such as *Enteromorpha* spp. and *Ulva* spp., and red algae, such as *Porphyra* spp.

The genus *Porphyra* is widely consumed in Japan, Korea, and China (Levine and Sahoo 2010). Wild *Porphyra* spp. is harvested in these counties. On the other hand, *Porphyra tenera* and *Porphyra yesoensis* are cultivated in Korea and China, respectively (Nui et al. 2010). Various species of *Porphyra* are most usually consumed as packages of dried nori sheets that are available in Japan and Korea.

Studies have been conducted to determine whether *Porphyra* spp. contains the true B_{12} or inactive corrinoids. The B_{12} compound purified from various purple lavers, or Porphyra spp., was identified as B_{12} and not as the inactive pseudovitamin B_{12} (PseudoB_{12}) found in humans (Miyamoto et al. 2009; Watanabe et al. 1999b, 2000).

3 The Origin of Vitamin B_{12} in Edible Seaweeds

Purple lavers (*Porphyra* spp.) contain a substantial amount of B_{12} (Watanabe et al. 1999a, 2002). They can also accumulate exogenous B_{12} (Yamada et al. 1996). The B_{12} found in the photic zone, where seaweeds live, come from the major B_{12} producers in the deeper zone, the B_{12}-synthesizing bacteria and Thaumarchaeota (Doxey et al. 2015; Heal et al. 2017) (Fig. 18.3). B_{12}-synthesizing bacteria colonize macrophytic algae (Iguchi et al. 2015). Since almost half the algae species require B_{12}, such microbial interactions play a pivotal role in algal growth (Croft et al. 2005).

4 Vitamin B_{12} Content of *Porphyra* Products

The nutritional value of purple laver products, such as toasted or seasoned products, is reportedly similar. However, different drying methods can affect the various nutrients and bioactive components (Cho and Rhee 2020). While lyophilization, or

Table 18.1 Vitamin B_{12} contents of edible seaweeds

	Scientific name	English name	Products	B_{12} content (mg/100 g dry weight)	B_{12} content (mg/100 g wet weight)	Reference	Remarks
Chlorophyta	*Ulva* spp.	Sea lettuce	Raw		6.3	MacArtain et al. (2007)	
	〃	〃	Dried	1.3		[a]STFC (2015)	
	Ulva lactuca	〃	Dried	60		Yagame et al. (2017)	
	Ulva prolifera	〃	Dried	31 ± 2.5		Hwang et al. (2008)	From Korea
	〃	〃	Dried	99 ± 4.7		Hwang et al. (2008)	From Japan
	Enteromorpha spp.	Green laver	Dried	31.8		STFC (2015)	
	〃	〃	Dried	63.58 ± 2.90		Watanabe et al. (1999a)	
	Caulerpa lentillifera	Green caviar	Raw	0		STFC (2015)	
	Prasiola japonica	Kawa-nori	Dried	5.7		STFC (2015)	
	Monostroma nitidum	Hitoegusa	Dried	0.3		STFC (2015)	
	Capsosiphon fulvescens	Maesaengi	Dried	61 ± 6.3		Hwang et al. (2008)	

(continued)

Table 18.1 (continued)

	Scientific name	English name	Products	B_{12} content (mg/100 g dry weight)	(mg/100 g wet weight)	Reference	Remarks
Chromophyta	*Laminaria digitata*	Oar weed, tangle	Raw		0.495	MacArtain et al. (2007)	
	〃	〃	Dried	5		Yagame et al. (2017)	
	〃	Kelp	Dried	0.1		STFC (2015)	
	Laminaria diabolica	Enaga-oni-kombu	Dried	0.1		STFC (2015)	
	Kjellmaniella crassifolia	Gagome-kombu	Dried	0		STFC (2015)	
	Laminaria longissima	Naga-kombu	Dried	0.1		STFC (2015)	
	Laminaria religiosa	Hosome-kombu	Dried	0		STFC (2015)	
	Laminaria japonica	Ma-kombu	Dried	0		STFC (2015)	
	Laminaria angustata	Mitsuishi-kombu	Dried	0		STFC (2015)	
	Laminaria ochotensis	Rishiri-kombu	Dried	0		STFC (2015)	
	Undaria pinnatifida	Wakame	Raw	0.3	0.345	MacArtain et al. (2007), STFC	
	〃	〃	Dried	0.2		STFC (2015)	
	〃	〃	Dried	36		Yagame et al. (2017)	
	〃	〃	Dried	0.6		STFC (2015)	
	Sargassum fusiformis	Hiziki	Dried	0 or trace		STFC (2015)	
	Analipus japonicus	Matsumo	Dried	0		STFC (2015)	
	Ascophyllum nodosum	Egg wrack, knotted wrack	Raw		0.131	MacArtain et al. (2007)	
	Colpomenia sinuosa	Oyster thief, sinuous ballweed	Dried	5.7		Manam and Subbaiah (2020)	

	Scientific name	English name	Products	B_{12} content (mg/100 g dry weight)	(mg/100 g wet weight)	Reference	Remarks
Rhodophyta	*Porphyra* spp.	Purple laver	Dried	77.6		STFC (2015)	
	〃	〃	Dried	32.36 ± 1.16		Watanabe et al. (1999a)	
	〃	〃	Dried	133.8		Miyamoto et al. (2009)	Korean purple laver
	〃	Iwa-nori	Dried	39.9		STFC (2015)	
	〃	Iwa-nori	Dried	86.5, 120.7		Miyamoto et al. (2009)	
	Porphyra tenera	Purple laver	Dried	—		Yagame et al. (2017)	
	Porphyra tumbilcalis	〃	Dried	290		Yagame et al. (2017)	
	Porphyra yezoensis	〃	Dried	52		Yagame et al. (2017)	
	Porphyra umbilicalis	〃	Raw		0.769	MacArtain et al. (2007)	
	Campylaephora hypnaeoides	Ego-nori	Dried	5.1		STFC (2015)	
	Gelidium elegans	Tengusa	Dried	0.5		STFC (2015)	
	Meristotheca senegalense		Dried	2000		Yagame et al. (2017)	
	Gloiopeltis spp.	Fu-nori	Dried	0		STFC (2015)	
	Chondrus crispus	Irish moss	Dried	6–40		Yagame et al. (2017)	
	Palmaria palmata	Dulse	Raw		1.84	MacArtain et al. (2007)	
	〃	〃	Dried	90		Yagame et al. (2017)	
	Halymenia porphyroides		Dried	33		Manam and Subbaiah (2020)	

[a] Standard Table of Food Composition in Japan 2015 cited in Reference is abbreviated as STFC 2015

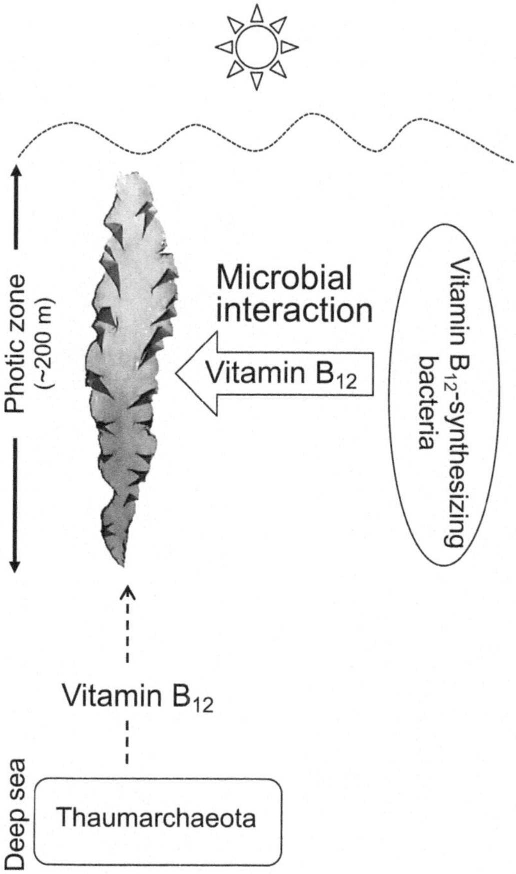

Fig. 18.3 The origin of vitamin B_{12} in the purple laver, *Porphyra* spp.

freeze-drying, prevents the loss of B_{12}, air-drying seems to convert B_{12} to inactive B_{12} analogs (Yamada et al. 1999). There is no information on the chemical structures of the inactive B_{12} analogs that form from air-drying.

The commercially available purple laver products, such as toasted, seasoned and toasted, and fermented nori-based products, contained a substantial amount of B_{12} (Table 18.2).It was unclear whether the toasting process destroyed B_{12} so that the seasoned and toasted lavers had less B_{12} than the dried lavers. In one study, the dried laver was toasted until its color changed from purple to green; it was found toasting did not affect the B_{12} content in the dried purple lavers (Miyamoto et al. 2009). Also, the decreased B_{12} content in the seasoned and toasted lavers was not due to the destruction of B_{12} during toasting but a decreased amount of laver per 100 g of the product by adding various seasonings, such as salt, sesame oil, and others (Fig. 18.4).

Table 18.2 Vitamin B_{12} contents of purple laver products

Scientific name	Products	B_{12} content (μg/100 g dry weight)	(μg/100 g wet weight)	Reference	Remarks
Porphyra spp.	Toasted	57.6		[a]STFC (2015)	
″	Seasoned and toasted	58.1		STFC (2015)	
″	Seasoned and toasted	51.7		Miyamoto et al. (2009)	Korean purple laver
Porphyra yezoensis	Fermended nori source		14	Uchida et al. (2017)	
″	Fermended nori source		4.18	Uchida et al. (2018)	Low-quality nori
″	Fermended nori source		15.4	Uchida et al. (2018)	High-quality nori

[a]Standard Table of Food Composition in Japan 2015 cited in Reference is abbreviated as STFC 2015

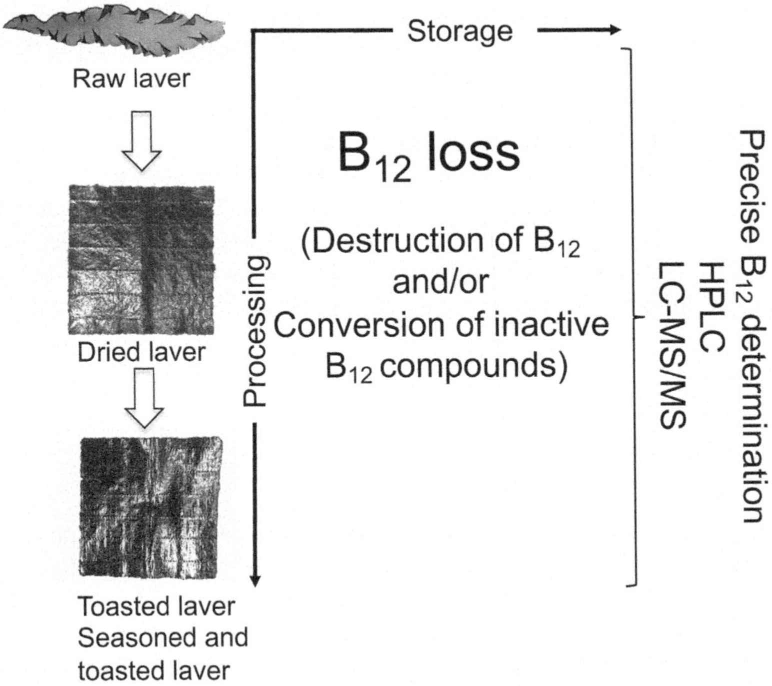

Fig. 18.4 The loss of vitamin B_{12} in purple laver products during processing and storage

5 Bioavailability of Vitamin B_{12} from Edible Seaweeds

It is also important to study how much of the ingested B_{12} was absorbed by studying the bioavailability of B_{12}. The bioavailability of the B_{12} found in purple lavers was determined using an *in vitro* gastrointestinal digestion model. Phycoerythrobilin of the purple pigment protein was significantly released from the laver by the *in vitro* digestion at pH 2.0, indicating that the dried purple lavers could be well-digested under the normal gstric conditions of pH 2.0. During the *in vitro* digestion, approxi-altey half of the B_{12} in the dried purple laver was recovered in the free B_{12} fraction (Miyamoto et al. 2009). These results suggest that the digestion rate of B_{12} in the purple laver was approximately 50% in persons with normal gastric function.

The bioavailability of B_{12} in nori, a dried purple laver product, was examined by studying the effects of feeding nori to B_{12}-deficient rats (Takenaka et al. 2001). The B_{12}-deficient rats excrete large amounts of methylmalonic acid in the urine. They were fed a diet supplemented with nori at 10 g/kg diet for 20 days. As a result, the amount of methylmalonic acid in the urine became undetectable. On the other hand, hepatic B_{12} levels in the B_{12}-deficient rats increased, indicating that B_{12} from dried purple laver was bioavailable to rats (van den Berg et al. 1991).

In addition, a nutritional analysis of vegans, who had been on a vegan diet that included brown rice and dried nori for 4–10 years, suggested that the consumption of dried nori prevented B_{12} deficiency in this group (Suzuki 1995). Also, the nori-consuming vegans had a higher serum or plasma B_{12} concentration than those not consuming nori (Dagnelie et al. 1991; Rauma et al. 1995). The vegans who did not consume nori also had lower mean corpuscular volumes (Dagnelie et al. 1991). However, when they started consuming dried nori for 8 months, their serum total B_{12} level and mean corpuscular volume became normal. Their serum level of holo-transcobalamin, a B_{12}-transport protein, and homocysteine was within tolerable levels, but the methylmalonic acid values were elevated (Schwarz et al. 2014). Thus, the bioavailability of dried purple laver B_{12} in humans remains to be investigated in detail.

6 Conclusion

This chapter indicates that some edible seaweeds, especially *Porphyra* spp., contain a substantial amount of biologically available B_{12}, which is absent from plant-derived foods. However, B_{12} may be converted into inactive B_{12} compounds or destroyed during the drying and storage of purple laver products. Thus, the B_{12} compounds found in purple laver products must be identified and studied more precisely using HPLC or LC-MS/MS or both.Author ContributionsAll authors contributed equally to the preparation of this manuscript and have approved the final version.

Notes The authors declare that there are no competing financial interests.

References

Andrzej JO, Kilmer SM (1993) Homocysteine metabolism and the oxidative modification of proteins and lipids. Free Radic Biol Med 14:683–693

van den Berg H, Brandsen L, Sinkeldam BJ (1991) Vitamin B_{12} content and bioavailability of spirulina and nori in rats. J Nutr Biochem 2:314–318

Bito T, Watanabe F (2016) Biochemistry, function, and deficiency of vitamin B_{12} in *Caenorhabditis elegans*. Exp Biol Med 241:1663–1668

Bito T, Matsunaga Y, Yabuta Y, Kawano T, Watanabe F (2013) Vitamin B_{12} deficiency in *Caenorhabditis elegans* results in loss of fertility, extended life cycle, and reduced lifespan. FEBS Open Bio 3:112–117

Chen Z, Crippen K, Gulati S, Banerjee R (1994) Purification and kinetic mechanism of a mammalian methionine synthase from pig liver. J Biol Chem 269:27193–27197

Cho TJ, Rhee MS (2020) Health functionality and quality control of laver (*Porphyra*, *Pyropia*): current issues and future perspectives as an edible seaweed. Mar Drugs 18:14

Croft MT, Lawrence AD, Raux-Deery E, Warren MJ, Smith AG (2005) Algae acquire vitamin B_{12} through a symbiotic relationship with bacteria. Nature 438:90–93

Dagnelie PC, van Staveren WA, van den Berg H (1991) Vitamin B_{12} from algae appears not to be bioavailable. Am J Clin Nutr 53:695–697

Doxey AC, Kurtz DA, Lynch MDJ, Sauder LA, Neufeld JD (2015) Aquatic metagenomes implicate Thaumarchaeota in global cobalamin production. ISME J 9:461–471

Fenton WA, Hack AM, Willard HF, Gertler A, Rosenberg LE (1982) Purification and properties of methylmalonyl coenzyme A mutase from human liver. Arch Biochem Biophys 214:815–823

Heal KR, Qin W, Ribalet F, Bertagnolli AD, Coyote-Maestas W, Hmelo LR, Moffett JW, Devol AH, Armbrust EV, Stahl DA, Ingalls AE (2017) Two distinct pools of B_{12} analogs reveal community interdependencies in the ocean. Proc Natl Acad Sci U S A 114:364–369

Hwang EK, Amano H, Park CS (2008) Assessment of the nutritional value of *Capsosiphon fulvescens* (Chlorophyta): developing a new species of marine macroalgae for cultivation in Korea. J Appl Phycol 20:147–151

Iguchi H, Yurimoto H, Sakai Y (2015) Interactions of methylotrophs with plants and other heterotrophic bacteria. Microorganims 3:137–151

Jessica AS, Nynke EH, Henk JB, Sinead ML, Coen DAS, Jan AR, Paul JK, Victor WMH, Hans WMN (2012) S-adenosylhomocysteine induces apoptosis and phosphatidylserine exposure in endothelial cells independent of homocysteine. Atherosclerosis 221:48–54

Levine IA, Sahoo D (2010) Porphyra: harvesting gold from the sea. I.K. International Publishing House Pvt. Ltd, Delhi

MacArtain P, Gill CIR, Brooks M, Cambell R, Rowland IR (2007) Nutritional value of edible seaweeds. Nutr Rev 65:535–543

Manam VK, Subbaiah M (2020) Phytochemical, amino acid, fatty acid and vitamin investigation of marine seaweeds colpomenia sinuosa and halymenia Porphyroides collected along southeast coast of Tamilnadu, India. World J Pharm Res 9:1088–1102

Miyamoto E, Yabuta Y, Kwak CS, Enomoto T, Watanabe F (2009) Characterization of vitamin B_{12} compounds from Korean purple laver (*Porphyra* sp.) products. J Agric Food Chem 57:2793–2796

Nui JF, Chen ZF, Wang GC, Zhou BC (2010) Purification of phycoerythrin from *Porphyra yezoensis* Ueda (Bangiales, Rhodophyta) using expanded bed absorption. J Appl Phycol 22:25–31

Pawlak R, Parrott SJ, Raj S, Cullum-Dugan D, Lucus D (2013) How prevalent is vitamin B_{12} deficiency among vegetarians? Nutr Rev 71:110–117

Rauma A-L, Törrönen R, Hänninen O, Mykkänen H (1995) Vitamin B_{12} status of long-term adherents of a strict uncooked vegan diet ("living food diet") is compromised. J Nutr 125:2511–2515

Schwarz J, Dschietzig T, Schwarz J, Dura A, Nelle E, Watanabe F, Wintgens KF, Reich M, Armbruster FP (2014) The influence of a whole food vegan diet with nori algae and wild mushrooms on selected blood parameters. Clin Lab 60:2039–2050

Standard Tables of Food Composition in Japan 2015 (Seventh Revised Version). Ministry of Education, Culture, Sports, Science and Technology, Tokyo, pp 114–119

Suzuki H (1995) Serum vitamin B_{12} levels in young vegans who eat brown rice. J Nutr Sci Vitaminol 41:587–594

Takenaka S, Sugiyama S, Ebara S, Miyamoto E, Abe K, Tamura Y, Watanabe F, Tsuyama S, Nakano Y (2001) Feeding dried purple laver (nori) to vitamin B_{12}-deficient rats significantly improves vitamin B_{12} status. Br J Nutr 85:699–703

Toyoshima S, Watanabe F, Saido H, Miyatake K, Nakano Y (1995) Methylmalonic acid inhibits respiration in rat liver mitochondria. J Nutr 125:2846–2850

Toyoshima S, Watanabe F, Saido H, Pezacka EH, Jacobsen DW, Miyatake K, Nakano Y (1996) Accumulation of methylmalonic acid caused by vitamin B_{12} deficiency disrupts normal cellular metabolism in rat liver. Br J Nutr 75:929–938

Uchida M, Kurushima H, Ishihara K, Murata Y, Touhata K, Ishida N, Niwa K, Araki T (2017) Characterization of fermented seaweed sauce prepared from nori (*Pyropia yezoensis*). J Biosci Bioeng 123:327–332

Uchida M, Kurushima H, Hideshima N, Araki T, Murata Y, Touhata K, Ishida N (2018) Preparation and characterization of fermented seaweed sauce manufactured from low-quality nori (dried and fresh fronds of *Phyropia yezoensis*). Fish Sci 84:589–596

Watanabe F (2007) Vitamin B_{12} sources and bioavailability. Exp Biol Med 232:1266–1274

Watanabe F, Bito T (2016) Corrinoids in food and biological samples. Front Nat Prod Chem 2:229–244

Watanabe F, Bito T (2018) Vitamin B_{12} sources and microbial interaction. Exp Biol Med 243:148–158

Watanabe F, Takenaka S, Katsura H, Masumder ZH, Masumder SA, Abe K, Tamura Y, Nakano Y (1999a) Dried green and purple lavers (nori) contain substantial amounts of biologically active vitamin B_{12} but less of dietary iodine relative to other edible seaweeds. J Agric Food Chem 47:2341–2343

Watanabe F, Katsura H, Takenaka S, Fujita T, Abe K, Tamura Y, Nakatsuka T, Nakano Y (1999b) Pseudovitamin B_{12} is the predominate cobamide of an algal health food, spirulina tablets. J Agric Food Chem 47:4736–4741

Watanabe F, Takenaka S, Katsura H, Miyamoto E, Abe K, Tamura Y, Nakatsuka T, Nakano Y (2000) Characterization of a vitamin B_{12} compound in the edible purple laver, Porphyra yezoensis. Biosci Biotechnol Biochem 64:2712–2715

Watanabe F, Takenaka S, Kittaka-Katsura H, Ebara S, Miyamoto E (2002) Characterization and bioavailability of vitamin B_{12}-compounds from edible algae. J Nutr Sci Vitaminol 48:325–331

Watanabe F, Yabuta Y, Bito T, Teng F (2014) Vitamin B_{12}-containing plant food sources for vegetarians. Nutrients 6:1861–1873

Yagame BM, Mensah ANC, Mady C, Cheikh N, Noba K (2017) Nutritional composition of *Meristotheca senegalense* (Rhodophyta): a new nutrient source. Afr J Food Sci 11:12–17

Yamada S, Sasa M, Yamada K, Fukuda M (1996) Release and uptake of vitamin B_{12} by Asakusanori (*Porphyra tenera*) seaweed. J Nutr Sci Vitaminol 42:507–515

Yamada K, Yamada Y, Fukuda M, Yamada S (1999) Bioavailability of dried asakusanori (*Porphyra tenera*) as a source of cobalamin (vitamin B_{12}). Int J Vitam Nutr Res 69:412–418

Chapter 19
Health Benefits of Seaweeds

Conrad O. Perera and Mona Al-Zahrani

Abbreviations

ABTS	2,2′-Azino-bis(3-ethylbenzothiazoline-6-sulfonic) acid (a method to determine antioxidant activity)
ACE	Angiotensin converting enzyme
AMPK	AMP activated protein kinase
CVD	Cardio-vascular disorders
FOSHU	Foods for specified health uses
FRAP	Ferric ion reducing antioxidant power
IL	Interleukin
LHCs	Light–harvesting complexes
LPS	Liposaccharides
NDC	Non-communicable diseases
PDT	Photodynamic therapy
RNS	Reactive nitrogen species
ROS	Reactive oxygen species
SPS	Sulphated polysaccharides
TNF-α	Tumour necrosis factor-α

C. O. Perera (✉) · M. Al-Zahrani
School of Chemical Sciences, The University of Auckland, Auckland, New Zealand

A. Ranga Rao, G. A. Ravishankar (eds.), *Sustainable Global Resources of Seaweeds Volume 2*, https://doi.org/10.1007/978-3-030-92174-3_19

1 Introduction

Marine algae can be broadly classified into two classes based on their physical size. They are the microalgae and macroalgae. Marine macroalgae are generally referred to as seaweeds (Shama et al. 2019). The presence of phytopigments other than chlorophyll is a characteristic of a particular algal classification (Qin 2018). Based on the pigmentation, the most commonly found macroalgal divisions include red (Rhodophyta), brown (Phaeophyta), and green (Chlorophyta) seaweeds (Fig. 19.1) (Shannon and Abu-Ghannam 2019). The nature of the reserve polymer synthesized as a result of photosynthesis is also a key variable used in algal classification. Important differences are seen in the storage products they utilize as well as in their cell wall chemistry (Ciancia et al. 2020).

Red seaweeds (Rhodophyceae) and green seaweed (Chlorophyceae) are generally small, ranging from a few centimeters to about a meter in length. Brown seaweeds (Phaeophyceae) on the other hand are large, ranging from a few centimeters to several meters in length (Hamid et al. 2019). Kelp belongs to the brown seaweed family. Some of the main products derived from seaweeds are food hydrocolloids. Carrageenan, Agar (red seaweed) and alginate (brown seaweed) are the most common hydrocolloids used in foods, pharmaceutical and biotechnological applications for their bioactivity and gelling properties (Zollmann et al. 2019). The recent growing interest in bioactive compounds from seaweed rose from their vast potential applications in nutraceutical and pharmaceutical industries, especially in alleviating metabolic risk factors like hyperglycemica, hypercholesterolemia, and hyperlipidemia (Collins et al. 2016; Cherry et al. 2019).

Alginate is one of the major compounds extracted from brown seaweed. They are use extensively in the food industry as a viscosity and texture modifier, and as a gelling agent (Alba and Kontogiorgos 2019). Carrageenan (Irish Moss) and agar are

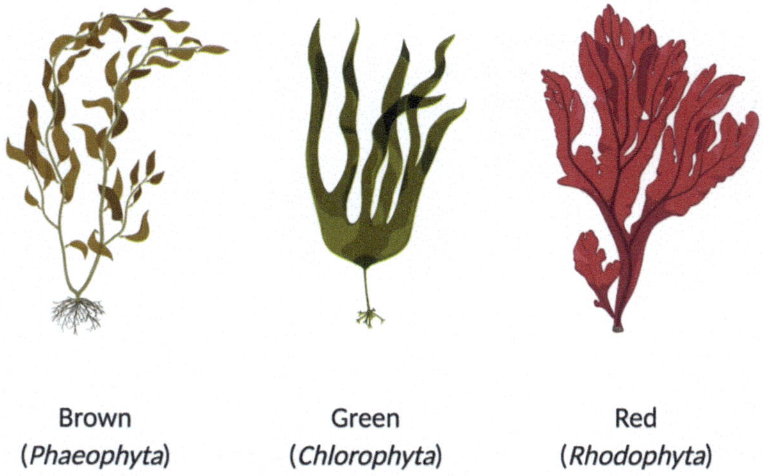

Fig. 19.1 Broad classification of seaweeds based on color

derived from red seaweeds. Today most of the red seaweeds used are derived from two species (*K. alvarezii and E. denticulatum*) originally from the Philippines (Qin 2018).

All three main types of seaweeds (red, brown, and green) contain Sulphated polysaccharides (SPS) which are recognized to possess anticoagulant, antiviral, and anti-inflammatory activities (Pradhan et al. 2020; Phull and Kim 2017). Fucoidans, laminarans, and ulvans are useful in nutraceuticals and functional food products.

Seaweeds are a sustainable and rich source of macro- and micronutrients in human and animal diets. In Japan and other coastal communities, seaweed contribute significantly to regular meals (Rebours et al. 2014; Cherry et al. 2019).

2 Seaweed Composition

The nutritional composition of seaweed varies between species, seasons, and ecology of the harvesting location (Cherry et al. 2019). In general, the moisture and ash content of seaweed are reported to be higher when compared to terrestrial plant species. Seaweed contains all essential amino acids, vitamins A, B-complex, C, and E, and several minerals such as magnesium (Mg), phosphorous (P), potassium (K), iodine (I), iron (F) and zinc (Zn) (Circuncisão et al. 2018). *Palmaria palmata* (Rhodophyta) contains on average 0.8 mg iron/g compared to 0.012–0.031 mg iron/g of lean beef (Shannon and Abu-Ghannam 2019).

Seaweed contains several bioactive proteinaceous compounds, including proteins, linear and cyclic peptides, depsipeptides, peptide derivatives, amino acids, and amino acid-like components (Harnedy and FitzGerald 2011). The protein content of seaweed varies among different phyla. Generally, the protein content of Rhodophyta is reported to be higher than that of Chlorophyta and Phaeophyta, and accounts for 10–50% of the dry weight. The highest protein content was found in *Porphyra* sp., followed by *Palmaria* sp. which both belong to Rhodophyta, followed by *Ulva* sp. in Chlorophyta, and *Undaria* sp. in Phaeophyta (Harnedy and FitzGerald 2011; Vieira et al. 2018).

Seaweed contains a significant amount of carbohydrates, which are involved in the structural integrity, and storage, the total content may range from 20 to 76% of dry weight depending on the phylum and species (Cesario et al. 2018), most of which are in the form of dietary fibers. The soluble fiber makes up 55–70% of the total polysaccharides in seaweeds, and they are mainly agar, carrageenan and alginate at varying amounts (Cesario et al. 2018). In addition, other bioactive sulphated polysaccharides (SPS) such as fucoidans, laminarin, porphyrin and alvan are also found in seaweeds.

The lipid content of seaweed is quite low in comparison to terrestrial plants, and ranges from 1 to 5% of dry weight. Glycolipids and neutral lipids are the predominant types of lipids found in seaweed, and the essential fatty acid content is higher in seaweed than that in land plant (Kendel et al. 2015).

3 Bioactive Compounds Found in Seaweed

In recent years, chronic, non-communicable diseases (NCD), such as cancer, cardiovascular diseases, and diabetes mellitus have replaced infectious diseases as the number one cause of mortality among humans. While treatment of NCDs depends mainly on synthesized and natural compounds derived from terrestrial regions, seas and oceans remain untapped reservoir for natural functional compounds useful in food and medicine (Collins et al. 2016). Some seaweeds are known for their functionality and have been used to treat a wide variety of conditions, while many more are still under investigation. Table 19.1 encompasses a selection of seaweeds and their bioactivity from the literature in the last decade.

4 Bioactive Properties

There are numerous reported studies on many bioactive compounds found in seaweed that are used to control communicable and non-communicable diseases that are plaguing the humankind currently. However, for brevity, only five types of the major diseases are discussed below.

4.1 Antiviral Properties

The anti-virus properties of algal polysaccharides have been reported by many researchers (Jiao et al. 2012; Pereira 2018a; Rosa et al. 2020; Gentile et al. 2020). Carrageen has been used in Ireland to make traditional medicinal teas and cough medicines to combat colds and coughs. It is said to be particularly useful for dislodging mucus and has antiviral properties (Pereira 2018a). In a recent review, Pereira and Critchley (2020) reported that the low initial levels of COVID-19 infection in Hokkaido, Japan were probably associated with the widespread consumption of seaweed and the regular supply of iodine in their diet. Carrageenan nasal spray used to treat the common cold in children and adults is known to reduce the duration of the disease, increase viral clearance and reduce relapse of symptoms (Pereira and Critchley 2020). Polysaccharides extracted from the red seaweed *Gelidium robustum* (Rhodophyta) are known to have protective effect of embryonic eggs against influenza B or mumps virus. Many species of marine algae contain significant quantities of complex structural sulphated polysaccharides that have been shown to inhibit the replication of enveloped viruses including members of the Nidovirales (Zeng et al. 2014).

Other compounds extracted from red algae (griffithsin), green algae (ulvans), and brown algae (fucoidans) could be potential antiviral therapeutic agents against SARS-CoV-2 (Shi et al. 2017; Pereira 2018b; Lee 2019; Gentile et al. 2020).

Table 19.1 Bioactive compounds from seaweeds from the literature 2010–2020

Phylum	Species	Compound	Bioactivity	Reference
Brown algae (*Phaeophyta*)	*Cladosiphon okamuranus*	Fucoidan (polysaccharide)	Cardio-protective	Thomes et al. (2010)
	Fucus vesiculosus	Fucoidan	Anti-inflammatory	Park et al. (2011)
	*Sargassum horneri, Ecklonia cava, Costaria costata (*C. Agardh*)*	Fucoidan	Anti-cancer	Ermakova et al. (2011)
	Canistrocarpus cervicornis	Heterofucans	Anti-coagulant, antioxidant	Gomes Camara et al. (2011)
	Sargassum wightii	Ethanolic extracts	Antibacterial, antioxidant	Devi et al. (2012)
	Himanthalia elongata	Fucoxanthin	Antibacterial, antioxidant	Rajauria and Abu-Ghannam (2013)
	Dictyota dichotoma thalli	Fucoidan: Galactofucan fraction	Anti-viral	Rabanal et al. (2014)
	Sargassum polycystum	Powdered seasweed	Anti-obesity	Awang et al. (2014)
	Turbinaria tricostata	Fucoidan	Heptaprotective, antioxidant	Chale-Dzul et al. (2015)
	Padina tetrastromatica	Dried seaweed	Antioxidant	Ismail et al. (2016)
	Sargassum fusiforme (Hijiki)	Fucoidan	Anti-cancer	Chen et al. (2016)
	Sargassum thunbergii	Polysaccharides	Anti-tumor	Jin et al. (2017)
	Sargassum longifolium	Polysaccharides	Anti-cancer	Shofia et al. (2018)
	Egregia menziesii (Feather boa kelp)	Hexane extracts	Anti-proliferative	Olivares-Bañuelos et al. (2019)
	Padina, Sargassum	Aqueous extracts	Anti-diabetic	Chin et al. (2020)

(continued)

Table 19.1 (continued)

Phylum	Species	Compound	Bioactivity	Reference
Green algae (*Chlorophyta*)	*Cladophora glomerata, Ulva lactuca, Ulva reticulata*	Methanolic extracts and aqueous extracts	Antifungal	Aruna et al. (2010)
	Enteromorpha compressa, Enteromorpha linza *and* Enteromorpha tubulosa	Alcoholic extracts	Antioxidant	Ganesan et al. (2011)
	Ulva rigida C. Agardh	Aqueous and alcoholic extracts	Antioxidant	Yildiz et al. (2012)
	Ulva fasciata, Ulva lactuca	Methanolic extracts	Antibacterial, immunostimulation	Thirunavukkarasu et al. (2013)
	Caulerpa racemosa	Methyl 3-bromo-1-adamantaneacetate, Chola-5, 22-Dien-3-Ol, 3 Beta	Antibacterial, antilarval	Nagaraj and Osborne (2014)
	Ulva intestinalis	Alcoholic and aqueous extracts	Antimicrobial	Srikong et al. (2015)
	Ulva armoricana	Enzymatic extracts	Anti-viral, antioxidant	Hardouin et al. (2016)
	Cladophora pellucida	Dried seaweed	Antioxidant	Ismail et al. (2016)
	Cladophora rupestris, Codium fragile	Crude hydroalcoholic extracts	Antioxidant, mineralogenic, anti-proliferative	Surget et al. (2017)
	Caulerpa spp.	Alcoholic extracts	Antioxidant, anti-proliferative	Tanna et al. (2018)
	Enteromorpha prolifera	Ethanolic extracts	Anti-diabetic	Yan et al. (2019)
	Halimeda	Aqueous extracts	Anti-diabetic	Chin et al. (2020)

(continued)

Table 19.1 (continued)

Phylum	Species	Compound	Bioactivity	Reference
Red algae (*Rhodophyta*)	*Gracilaria corticata* (J. Agardh), *Kappaphycus alvarezii*	Methanolic extracts and aqueous extracts	Antifungal	Aruna et al. (2010)
	Dichotomaria obtusata	Aqueous extracts	Anti-inflammatory, analgesic	Vázquez et al. (2011)
	Gracilaria corticata	Acetone extracts	Antimicrobial	Govindasamy et al. (2012)
	Gracilaria gracilis	Freeze-dried seaweed	Antioxidant	Francavilla et al. (2013)
	Laurencia snackeyi	5 ß-hydroxy palisadin B	Anti-inflammatory	Wijesinghe et al. (2014)
	Gracilaria changii	Ethyl acetate extracts	Antioxidant	Chan et al. (2015)
	Laurencia papillosa	Dried seaweed	Antioxidant	Ismail et al. (2016)
	Jania rubens, *Corallina mediterranea*, *Pterocladia capillacea*	Methanolic extracts	Antibacterial	El-Din and El-Ahwany (2016)
	Mastocarpus stellatus	Freeze-dried seaweed	Antioxidant	Nguyen et al. (2017)
	Pyropia orbicularis	Dried seaweed	Antioxidant	Uribe et al. (2018)
	Halymenia durvilae	Water and alcoholic extracts	Anti-diabetic	Sanger et al. (2019)
	Kappaphycus alvarezii	Hot water and ethanolic extracts	Antioxidant, antibacterial	Bhuyar et al. (2020)
	Kappaphycus spp.	Aqueous extracts	Anti-diabetic	Chin et al. (2020)

Most brown seaweeds contain carotenoid pigment fucoxanthin, and other polysaccharides such as alginates, laminarin, fucans and cellulose. In addition, some of them also contain a range of unique secondary metabolites such as phlorotannins, phloroglucinol, terpenes and tocopherol which are known to have antiviral properties (Remya and Rajasree 2016).

4.2 *Anti-tumor Properties*

Cancer is a leading cause of death world-wide accounting for an estimated 9.6 million deaths in 2018 (WHO 2018). Chlorophyll breakdown products are known for their antioxidative and anti-inflammatory activities. Pheophorbide *a* (PPBa), is a chlorophyll derivative that has photosensitizing activity that can induce significant anti-proliferative effects in several human cancer cell lines. Four types of chlorophylls are found in marine algae. The light–harvesting complexes (LHCs) in chloroplasts of plants and algal cells usually include chlorophyll *a*, *b*, *c*, and *d* (Larkum and Kuhl 2005).

In 1996, Schuitmaker et al., introduced a promising new model for treating cancer called photodynamic therapy (PDT). They discovered that two relatively innocuous agents, light and photosensitizing agent when used in combination caused selective tumour destruction. As stated by Saide et al. (2020), the three major mechanisms by which PDT mediates tumour destruction are:

(1) The photosensitizer can transfer energy from light to molecular oxygen to produce Reactive Oxygen Species (ROS) and ROS generated can directly kill the cancer cells, (2) Destruction of the tumour vasculature, (3) PDT can induce an immune response against the tumour.

Photosensitizer PPBa has been found to possess antiproliferative activity against a number of cancer cell-lines, with or without PDT (Saide et al. 2020). Ahn et al. (2017) demonstrated that PPBa/PDT inhibited cell proliferation of human oral squamous cell line in a dose dependent manner up to 2 μM.

Salhi et al. (2020) found that extracts of the Moroccan Mediterranean Sea red alga *Sphaerococcus coronopifolius* have an antitumoral effect on human cervix (HeLa), breast (SKBR-3), and pancreatic (MIA PaCa-2) cancer cell lines in a dose and time dependent manners. The secretion of Interleukin-8 (IL-8) in LPS- and Tumour Necrosis Factor α (TNF-α)-stimulated HUVEC-tert endothelial cells was found to be significantly inhibited by *S. coronopifolius* extracts.

Similarly, Zbakah et al. (2020) found antitumor activities of extracts from the marine green alga—*Codium decorticatum*. They found dramatic inhibition of the expression of the pro-inflammatory cytokine IL-8 in LPS- and TNF-α-stimulated endothelial cells. Various solvent extracts of green, brown, and red algae showed strong antitumor effects against various tumour cell lines, K562 (chronic myelocytic leukemia), HEp-2 (laryngeal epidermoid carcinoma) and NCI-H292 (human lung mucoepidermoid carcinoma) (Guedes et al. 2013; Saadaoui et al. 2020).

Various secondary metabolites of seaweeds include polysaccharides, lipids, and proteins. Fucoidans are sulfated polysaccharides (SPS) generally produced by brown algae. They have high anticancer activity against several cancer types, including lung cancer, by targeting the key apoptotic molecules (Saadaoui et al. 2020). They have the ability to act synergistically with anticancer drugs currently in use (Sakthivel et al. 2016). Fucoidans also have the ability to mitigate toxic effects associated with conventional cancer therapies (Saadaoui et al. 2020).

Phycobiliproteins derived from seaweed are proteins covalently linked to chromophore phycobilins have been reported to show anticancer properties (Deniz et al. 2016). Senthilkumar and Jayanthi (2016) purified glycoproteins from *Codium decorticatum*, which showed anticancer effect against lung cancer cell A549.

4.3 Anti-hypertensive Properties

Hypertension is a significant factor that contributes to the onset and exacerbation of a cascade of mechanisms, including activation of the sympathetic and renin-angiotensin systems, oxidative stress, and release of inflammatory mediators, that promote systemic dysfunction leading to clinical manifestations of cardiovascular diseases (Seca and Pinto 2018).

The beneficial effects of various polyphenols of plant origin on different cardiovascular disorders (CVD), such as hypertension, and other metabolic syndromes are well known. Seaweeds are a rich source of polyphenols. Marine derived drugs have emerged recently as potential treatments for CVD. The active compounds responsible for these activities are identified as seaweed polyphenols (Gómez-Guzmán et al. 2018). The dominant polyphenols found only in marine brown algae are a unique group of complex polymers of phloroglucinol (1,3,5-trihydroxybenzene) called phlorotannins (Gómez-Guzmán et al. 2018).

Of four seaweeds, nori (*Phorphyra* sp.), kombu (*Laminaria* sp.), wakame (*Undaria* sp.) and sea spaghetti (*Himanthalia elongata*), studied by Fernández-Segovia et al. (2018), sea spaghetti had the highest phenolic content and the highest corresponding antioxidant capacity as measured by FRAP and ABTS methods. The antioxidant properties of polyphenols and their ability to scavenge free radicals by activating the Nrf2/ARE pathway is shown in Fig. 19.2. The antihypertensive and antioxidant properties are strongly associated with the chemical structure of the peptides or carbohydrates (Lafarga et al. 2020).

Bioactive peptides are sequences of 2–30 amino acids in length that display hormone-like beneficial properties when released from their parent protein. Peptides showing antihypertensive, angiotensin converting enzyme (ACE) inhibition, and antioxidative activities have been successfully isolated from seaweed (Admassu et al. 2018). Lafarga et al. (2020) reported that the Japanese Ministry of Health and Welfare have established a policy for approving some selected functional food products as Foods for Specified Health Uses (FOSHU) whose health claims are legally permitted. Seaweed-derived peptide-containing products with FOSHU approved antihypertensive claims include Wakame peptide jelly (Riken Vitamin Co., Ltd., Tokyo, Japan) and Nori peptide S (Shirako Co., Ltd., Tokyo, Japan). In-vivo trials of the tridecapeptide IRLIIVLMPILMA derived from marine seaweeds exhibited ACE inhibition properties similar to the drug captopril and could be a potential drug for the control of high blood pressure (Rosa et al. 2020). These studies prove beyond doubt the possible use of seaweed derived polyphenols, peptides, carbohydrates, and other extracts as potential future drugs for the control of cardiovascular-related diseases.

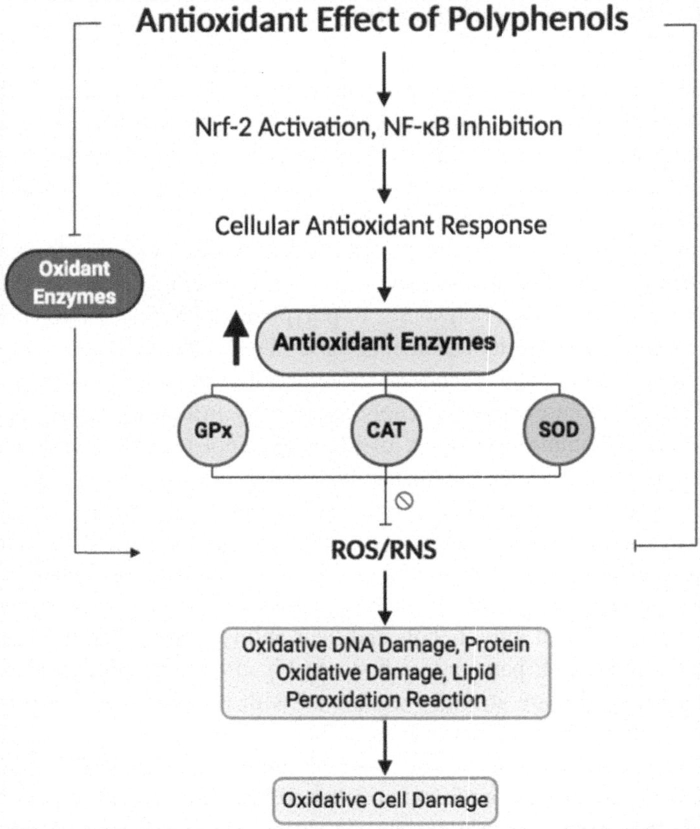

Fig. 19.2 Antioxidant properties of polyphenols (Adapted from Gómez-Guzmán, et al. 2018). Glutathione peroxidase (GPx), catalase (CAT), and superoxide dismutase (SOD), reactive oxygen species (ROS), reactive nitrogen species (RNS)

4.4 Antioxidant Properties

The antioxidant activity of seaweeds is due to several bioactive compounds present in them. Macroalgae are rich in polyphenols, carotenoids, several bioactive polysaccharides, phycobiliproteins and peptides resulting in high antioxidant activity. Antioxidant substances of very different nature are found in algae, among which vitamin E (α-tocopherol) and carotenoids are present within the fat-soluble fraction, whereas the most powerful water-soluble antioxidants found are polyphenols, phycobiliproteins and vitamins (vitamin C).

The phenolic compounds from terrestrial plants are mainly derived from gallic and ellagic acid, however, those in algae are mainly derived from polymerised phloroglucinol units (1,3,5-trihydroxybenzene) (Mateos et al. 2020). In addition to the major phenolic compounds found in terrestrial plants, seaweeds also contain more complex phlorotannin polymeric structures. The brown *Ecklonia* species was found to contain considerable quantities of phlorotannins (phloroglucinol, eckol, 7-phloroeckol, 6,6-bieckol, phlorofucofuroeckol A, fucodiphloroethol) which are powerful antioxidants (Santos et al. 2019).

Phloroglucinol, the basic unit of phlorotannins found in seaweeds was able to reduce H_2O_2-induced toxicity in zebrafish. The augmented survival rate was attributed to the antioxidant activity of this phenolic compound which reduced the H_2O_2-induced cell death, lipid peroxidation, and ROS formation (Cha et al. 2017). Compared to epigallocatechin gallate (EGCG), the extracted phlorotannins displayed higher antioxidant activity (Dong et al. 2019).

As shown in Fig. 19.2, the antioxidant effect of polyphenols is related to their capacity to enhance the enzymatic activity of glutathione peroxidase (GPx), catalase (CAT), and superoxide dismutase (SOD), their potent free radical scavenging properties, and their ability to interact with other molecular targets, as they are capable of activating the Nrf2 pathways (Gómez-Guzmán et al. 2018).

Eckol, a three membered phenolic ring compound found in seaweed is demonstrated to have hepatoprotective effects on mice by modulating antioxidant mechanism and suppressing the expression of pro-inflammatory cytokines, like tumor necrosis factor (TNF), interleukin (IL)-1, and IL-6, and by upregulating the expression of IL-10, an anti-inflammatory interleukin (Rosa et al. 2020).

Dieckol, a six membered phenolic ring compound, in addition to having hepatoprotective effects, also promoted the increase of the activity of antioxidant enzymes, including superoxide dismutase (SOD), catalase (CAT), and glutathione peroxidase (GPx) in liver tissues, and it increased levels of the phosphorylation of AMPK and Akt in muscle tissues (Rosa et al. 2020).

Phycoerythin, a red protein pigmented complex abundant in Rhodophyta has strong antioxidant properties (Rosa et al. 2020). One of the most studied seaweed metabolites is Fucoxanthin, a xanthophyll-like carotenoid, because of its powerful antioxidant properties (Rajauria and Abu-Ghannam 2013).

Marine polysaccharides such as carrageenan and agaro-oligosaccharides from red algae possess powerful antioxidative properties by scavenging hydroxyl free radicals and superoxide anion radicals, and inhibiting lipid peroxidation (Plaza et al. 2006).

Marine macroalgae have high concentrations of sulphated polysaccharides (SPS). Studies have demonstrated their antioxidant and anti-inflammatory effects. Brown seaweeds are rich source of functional polysaccharides, fucoidan, laminaran, and alginate that exhibit antioxidant activity (Fauziee et al. 2021).

4.5 Anti-aging Properties

Senescence of cells is a hallmark of ageing. Under normal conditions, senescent cells are recognized and removed by the body's immune system. However, accumulative senescent cells, which cannot be eliminated by the immune system in time, arise from multiple mechanisms (Childs et al. 2015). Senescence-associated beta-galactosidase (SA-β-Gal) activity is the most widely used biomarker for identifying senescent cells. Cao et al. (2020) recently reported that fucoidans reduced senescence in long-term cultured endothelial colony-forming cells, by alleviation of SA-β-Gal activity. They also reported that the treatment with fucoidan also reduces the SA-β-Gal activity induced by p-cresol, a major uremic toxin, in mesenchymal stem cells. They found that porphyrin isolated from seaweed showed anti-SA-β-Gal activity in stress-induced senescent fibroblasts.

Phycoerythrin-derived peptide from the red alga *Pyropia yezoensis* was shown to downregulate the activity of SA-β-Gal in aged primary hippocampal neuron cells and attenuated age-dependent degeneration of neurites (Oh et al. 2018). A growing body of evidence suggests that several pathways, such as Sirtuin, AMP-activated protein kinase, insulin-like growth factor, autophagy, and nuclear factor erythroid 2-related factor 2 (Nrf2) play critical roles in regulating aging (Cao et al. 2020). Fucoidans isolated from seaweed were found to rescue cells from both replicative and stress-induced senescence.

5 Conclusions

Over 70% of the earth's surface is covered by water and over 95% of that is sea water in which seaweeds grow. The presence of some unique bioactive components not found in terrestrial biomass makes seaweed an excellent sustainable resource. Because seaweeds are virtually untapped, they are receiving increasing global attention as potentially sustainable resources of ingredients for food, feed, cosmetics, pharmaceutical applications, or as raw materials for chemical, and biomaterials. Current knowledge is limited to only a few species and a few biochemicals from seaweeds. There is tremendous potential to harness the plants growing wild in the seas for the benefit of humankind. The unique bioactive compounds that have been isolated from seaweeds show tremendous potential for their future applications not only as functional foods but as medicinal cures for many diseases both communicable and non-communicable in origin.

Further research: Further research is needed to understand the chemical, biological nature of the diverse compounds that can be extracted from seaweeds. In this regard, efficient extraction, purification, and advanced analytical methods to validate the claims as functional foods and pharmaceutical applications are needed. Further research is also needed to determine their safety to validate the health benefits.

References

Admassu H, Gasmalla MAA, Yang R, Zhao W (2018) Bioactive peptides derived from seaweed protein and their health benefits: antihypertensive, antioxidant, and antidiabetic properties. J Food Sci 83(1):6–16. https://doi.org/10.1111/1750-3841.14011

Ahn MY, Yoon HE, Moon SY, Kim YC, Yoon JH (2017) Intratumoral photodynamic therapy with newly synthesized pheophorbide a in murine oral cancer. Oncol Res 25:295–304

Alba K, Kontogiorgos V (2019) Seaweed polysaccharides (Agar, Alginate and Carrageenan). In: Encyclopaedia of food chemistry. Elsevier, Amsterdam, pp 240–250. https://doi.org/10.1016/B978-0-08-100596-5.21587-4

Aruna P, Mansuyal P, Sridhar S, Kumar JS, Babu S (2010) Pharmacognostical and antifungal activity of selected seaweeds from the gulf of Mannar region. Recent Res Sci Technol 2(1):115–119

Awang AN, Ng JL, Matanjun P, Sulaiman MR, Tan TS, Yasmin BHO (2014) Anti-obesity property of the brown seaweed, *Sargassum polycystum* using an in vivo animal model. J Appl Phycol 26(2):1043–1048

Bhuyar P, Rahim M, Sundararaju S, Maniam G, Govindan N (2020) Antioxidant and antibacterial activity of red seaweed Kappaphycus alvarezii against pathogenic bacteria. Global J Environ Sci Manage 6(1):47–58

Cao L, Lee SG, Lim KT, Kim H-R (2020) Potential anti-aging substances derived from seaweeds. Mar Drugs 18(11):564. https://doi.org/10.3390/md18110564

Cesario MT, da Fonseca MMR, Marques MM, de Almeida MCMD (2018) Marine algal carbohydrates as carbon sources for the production of biochemicals and biomaterials. Biotechnol Adv 36:798–817

Cha S-H, Lee J-H, Kim E-A, Shin CH, Jun H-S, Jeon Y-J (2017) Phloroglucinol accelerates the regeneration of liver damaged by H_2O_2 or MNZ treatment in zebrafish. RSC Adv 7:46164–46170

Chale-Dzul J, Moo-Puc R, Robledo D, Freile-Pelegrín Y (2015) Hepatoprotective effect of the fucoidan from the brown seaweed Turbinaria tricostata. J Appl Phycol 27(5):2123–2135

Chan PT, Matanjun P, Yasir SM, Tan TS (2015) Antioxidant activities and polyphenolics of various solvent extracts of red seaweed, Gracilaria changii. J Appl Phycol 27(6):2377–2386

Chen H, Cong Q, Du Z, Liao W, Zhang L, Yao Y, Ding K (2016) Sulfated fucoidan FP08S2 inhibits lung cancer cell growth in vivo by disrupting angiogenesis via targeting VEGFR2/VEGF and blocking VEGFR2/Erk/VEGF signaling. Cancer Lett 382(1):44–52

Cherry P, O'Hara C, Magee PJ, McSorley EM, Allsopp PJ (2019) Risks and benefits of consuming edible seaweeds. Nutr Rev 77(5):307–329

Childs BG, Durik M, Darren J, Baker DJ, van Deursen JM (2015) Cellular senescence in aging and age-related disease: from mechanisms to therapy. Nat Med 21(12):1424–1435. https://doi.org/10.1038/nm.4000

Chin YX, Chen X, Cao WX, Sharifuddin Y, Green BD, Lim PE et al (2020) Characterization of seaweed hypoglycemic property with integration of virtual screening for identification of bioactive compounds. J Funct Foods 64:103656. https://doi.org/10.1016/j.jff.2019.103656

Ciancia M, Fernández PV, Leliaert F (2020) Diversity of sulfated polysaccharides from cell walls of coenocytic green algae and their structural relationships in view of green algal evolution. Front Plant Sci 11:554585. https://doi.org/10.3389/fpls.2020.554585

Circuncisão AR, Catarino MD, Cardoso SM, Silva A (2018) Minerals from macroalgae origin: health benefits and risks for consumers. Mar Drugs 16(11):400

Collins KG, Fitzgerald GF, Stanton C, Ross RP (2016) Looking beyond the terrestrial: the potential of seaweed derived bioactives to treat non-communicable diseases. Mar Drugs 14(3):60

Deniz I, Ozen MO, Yesil-Celiktas O (2016) Supercritical fluid extraction of phycocyanin and investigation of cytotoxicity on human lung cancer cells. J Supercrit Fluids 108:13–18

Devi KN, Kumar TA, Dhaneesh K, Marudhupandi T, Balasubramanian T (2012) Evaluation of antibacterial and antioxidant properties from brown seaweed, Sargassum wightii (greville, 1848) against human bacterial pathogens. Int J Pharm Pharm Sci 4(3):143–149

Dong X, Bai Y, Xu Z, Shi Y, Sun Y, Janaswamy S, Yu C, Qi H (2019) Phlorotannins from Undaria pinnatifids sporophyll: extraction, antioxidant, and anti-inflammatory activities. Mar Drugs 2019(17):434. https://doi.org/10.3390/md17080434

El-Din SMM, El-Ahwany AM (2016) Bioactivity and phytochemical constituents of marine red seaweeds (Jania rubens, Corallina mediterranea and Pterocladia capillacea). J Taibah Univ Sci 10(4):471–484

Ermakova S, Sokolova R, Kim S, Um B, Isakov V, Zvyagintseva T (2011) Fucoidans from brown seaweeds Sargassum hornery, Eclonia cava, Costaria costata: structural characteristics and anticancer activity. Appl Biochem Biotechnol 164(6):841–850

Fauziee NAM, Chang LS, Mustapha WAW, Nor AR Md, Lim SJ (2021) Functional polysaccharides of fucoidan, laminaran and alginate from Malaysian brown seaweeds (*Sargassum polycystum, Turbinaria ornata* and *Padina boryana*). Int J Biol Macromol 167:1135–1145. https://doi.org/10.1016/j.ijbiomac.2020.11.067

Fernández-Segovia I, Lerma-García MJ, Fuentes A, Barat JM (2018) Characterization of Spanish powdered seaweeds: Composition, antioxidant capacity and technological properties. Food Res Int 111:212–219

Francavilla M, Franchi M, Monteleone M, Caroppo C (2013) The red seaweed Gracilaria gracilis as a multi products source. Mar Drugs 11(10):3754–3776

Ganesan K, Kumar KS, Rao PS (2011) Comparative assessment of antioxidant activity in three edible species of green seaweed, Enteromorpha from Okha, northwest coast of India. Innov Food Sci Emerg Technol 12(1):73–78

Gentile D, Patamia V, Scala A, Sciortino MY, Piperno A, Rescifina A (2020) Putative inhibitors of SARS-CoV-2 main protease from a library of marine natural products: a virtual screening and molecular modeling study. Mar Drugs 18:225. https://doi.org/10.3390/md18040225

Gomes Camara BR, Silva Costa L, Pereira Fidelis G, Barreto Nobre LTD, Dantas-Santos N, Cordeiro SL, Pereira Costa MSS, Alves LG, Oliveira Rocha HA (2011) Heterofucans from the brown seaweed *Canistrocarpus cervicornis* with anticoagulant and antioxidant activities. Mar Drugs 9(1):124–138. https://doi.org/10.3390/md9010124

Gómez-Guzmán M, Rodríguez-Nogales A, Algieri F, Gálvez J (2018) Potential role of seaweed polyphenols in cardiovascular-associated disorders. Mar Drugs 16:250. https://doi.org/10.3390/md16080250

Govindasamy C, Arulpriya M, Ruban P (2012) Nuclear magnetic resonance analysis for antimicrobial compounds from the red seaweed Gracilaria corticata. Asian Pac J Trop Biomed 2(1):S329–S333

Guedes EAC, da Silva TG, Aguiar JS, de Barros LD, Pinotti LM, Sant'Ana, A.E.G. (2013) Cytotoxic activity of marine algae against cancerous cells. Braz J Pharmacogn 23(4):668–673

Hamid SS, Wakayama M, Ichihara K, Sakurai K, Ashino Y, Kadowaki R, Soga T, Tomita M (2019) Metabolome profiling of various seaweed species discriminates between brown, red, and green algae. Planta 249:1921–1947. https://doi.org/10.1007/s00425-019-03134-1

Hardouin K, Bedoux G, Burlot A, Donnay-Moreno C, Bergé J, Nyvall-Collén P, Bourgougnon N (2016) Enzyme-assisted extraction (EAE) for the production of antiviral and antioxidant extracts from the green seaweed Ulva armoricana (Ulvales, Ulvophyceae). Algal Res 16:233–239

Harnedy PA, FitzGerald RJ (2011) Bioactive proteins, peptides, and amino acids from macroalgae 1. J Phycol 47(2):218–232

Ismail MM, Gheda SF, Pereira L (2016) Variation in bioactive compounds in some seaweeds from Abo Qir Bay, Alexandria, Egypt. Rendiconti Lincei 27(2):269–279

Jiao G, Yu G, Wang W, Zhao X, Zhang J et al (2012) Properties of polysaccharides in several seaweeds from Atlantic Canada and their potential anti-influenza viral activities. J Ocean Univ China 11:205–212. https://doi.org/10.1007/s11802-012-1906-x

Jin W, Zhang W, Liu G, Yao J, Shan T, Sun C, Zhang Q (2017) The structure-activity relationship between polysaccharides from Sargassum thunbergii and anti-tumor activity. Int J Biol Macromol 105:686–692

Kendel M, Wielgosz-Collin G, Bertrand S, Roussakis C, Bourgougnon N, Bedoux G (2015) Lipid composition, fatty acids and sterols in the seaweeds Ulva armoricana, and Solieria chordalis from Brittany (France): an analysis from nutritional, chemotaxonomic, and antiproliferative activity perspectives. Mar Drugs 13(9):5606–5628

Lafarga T, Acién-Fernández FG, Garcia-Vaquero M (2020) Bioactive peptides and carbohydrates from seaweed for food applications: natural occurrence, isolation, purification, and identification. Algal Res 48:101909. https://doi.org/10.1016/j.algal.2020.101909

Larkum AW, Kuhl M (2005) Chlorophyll d: the puzzle resolved. Trends Plant Sci 10:355–357

Lee C (2019) Griffithsin, a highly potent broad-spectrum antiviral lectin from red algae: from discovery to clinical application. Mar Drugs 17:567

Mateos R, Pérez-Correa JR, Domínguez H (2020) Bioactive properties of marine phenolics. Mar Drugs 2020(18):501. https://doi.org/10.3390/md18100501

Nagaraj SR, Osborne JW (2014) Bioactive compounds from Caulerpa racemosa as a potent larvicidal and antibacterial agent. Front Biol 9(4):300–305

Nguyen HPT, Morançais M, Fleurence J, Dumay J (2017) *Mastocarpus stellatus* as a source of R-phycoerythrin: optimization of enzyme assisted extraction using response surface methodology. J Appl Phycol 29(3):1563–1570

Oh JH, Kim E-Y, Nam T-J (2018) Phycoerythrin-derived tryptic peptide of a red alga *Pyropia yezoensis* attenuates glutamate-induced ER stress and neuronal senescence in primary rat hippocampal neurons. Mol Nutr Food Res 62(8):1700469. https://doi.org/10.1002/mnfr.201700469

Olivares-Bañuelos T, Gutiérrez-Rodríguez AG, Méndez-Bellido R, Tovar-Miranda R, Arroyo-Helguera O, Juárez-Portilla C, Meza-Menchaca T, Aguilar-Rosas LE, Hernández-Kelly LCR, Ortega A, Zepeda RC (2019) Brown seaweed *Egregia menziesii's* cytotoxic activity against brain cancer cell lines. Molecules 24:260. https://doi.org/10.3390/molecules24020260

Park HY, Han MH, Park C, Jin C, Kim G, Choi I et al (2011) Anti-inflammatory effects of fucoidan through inhibition of NF-κB, MAPK and akt activation in lipopolysaccharide-induced BV2 microglia cells. Food Chem Toxicol 49(8):1745–1752

Pereira L (2018a) Biological and therapeutic properties of the seaweed polysaccharides. Int Biol Rev 2:1–50

Pereira L (2018b) Antiviral activity of seaweeds and their extracts. In: Pereira L (ed) Therapeutic and nutritional uses of algae. Science Publishers, Boca Raton, pp 175–211

Pereira L, Critchley AT (2020) The COVID 19 novel coronavirus pandemic 2020: seaweeds to the rescue? Why does substantial, supporting research about the antiviral properties of seaweed polysaccharides seem to go unrecognized by the pharmaceutical community in these desperate times? J Appl Phycol 32:1875–1877

Phull AR, Kim SJ (2017) Fucoidan as bio-functional molecule: insights into the anti-inflammatory potential and associated molecular mechanisms. J Funct Foods 38:415–426

Plaza M, Cifuentes A, Ibáñez E (2006). In search of new functional food ingredients from algae. https://core.ac.uk/download/pdf/36023759.pdf. Accessed 2 Feb 2021

Pradhan B, Patra S, Nayak R, Behera C, Dash SR, Nayak S, Sahu BB, Bhutia SK, Jena M (2020) Multifunctional role of fucoidan, sulfated polysaccharides in human health and disease: a journey under the sea in pursuit of potent therapeutic agents. Int J Biol Macromol 164:4263–4278

Qin Y (2018) Seaweed bioresources. In: Qin Y (ed) Bioactive seaweeds for food applications: natural ingredients for healthy diets. Academic, New York, pp 3–24

Rabanal M, Ponce NM, Navarro DA, Gómez RM, Stortz CA (2014) The system of fucoidans from the brown seaweed Dictyota dichotoma: chemical analysis and antiviral activity. Carbohydr Polym 101:804–811

Rajauria G, Abu-Ghannam N (2013) Isolation and partial characterization of bioactive fucoxanthin from Himanthalia elongata brown seaweed: a TLC-based approach. Int J Anal Chem 2013:802573

Rebours C, Marinho-Soriano E, Zertuche-González JA, Hayashi L, Vásquez JA, Kradolfer P et al (2014) Seaweeds: an opportunity for wealth and sustainable livelihood for coastal communities. J Appl Phycol 26(5):1939–1951

Remya RR, Rajasree SRR (2016) A study on bioactive compounds derived from brown seaweeds and their therapeutic applications towards various diseases. Res J Pharm Technol 9(4):369–372

Rosa GP, Tavares WR, Sousa PMC, Pages AK, Seca AML, Pinto D (2020) Seaweed secondary metabolites with beneficial health effects: an overview of successes in in vivo studies and clinical trials. Mar Drugs 18:8. https://doi.org/10.3390/md18010008

Saadaoui I, Rasheed R, Abdulrahman N, Bounnit T, Cherif M, Al Jabri H, Mraiche F (2020) Algae-derived bioactive compounds with anti-lung cancer potential. Mar Drugs 18:197. https://doi.org/10.3390/md18040197

Saide A, Lauritano C, Ianora A (2020) Pheophorbide a: state of the art. Mar Drugs 18:257. https://doi.org/10.3390/md18050257

Sakthivel R, Muniasamy S, Archunan G, Devi KP (2016) *Gracilaria edulis* exhibit antiproliferative activity against human lung adenocarcinoma cell line A549 without causing adverse toxic effects *in vitro* and *in vivo*. Food Funct 7:1155–1165

Salhi G, Zbakh H, Moussa H, Hassoun M, Bochkov V, Ciudad CJ, Véronique Noé V, Riadi H (2020) Antitumoral and anti-inflammatory activities of the red alga *Sphaerococcus coronopifolius*. Eur J Integr Med 18:66–74

Sanger G, Rarung L, Damongilala L, Kaseger B, Montolalu L (2019) Phytochemical constituents and antidiabetic activity of edible marine red seaweed (Halymenia durvilae). In: Paper presented at the IOP conference series: earth and environmental science, vol 278(1), p 012069. https://doi.org/10.1088/1755-1315/278/1/012069

Santos S, Félix R, Pais A, Rocha SM, Silvestre A (2019) The quest for phenolic compounds from macroalgae: A review of extraction and identification methodologies. Biomolecules 9(12):847. https://doi.org/10.3390/biom9120847

Seca AML, Pinto DCGA (2018) Overview on the antihypertensive and anti-obesity effects of secondary metabolites from seaweed. Mar Drugs 16:237. https://doi.org/10.3390/md16070237

Senthilkumar D, Jayanthi S (2016) Partial characterization and anticancer activities of purified glycoprotein extracted from green seaweed Codium decorticatum. J Funct Foods 25:323–332

Shama A, Joyce SG, Mari FD, Ranga Rao A, Ravishankar GA, Hudaa N (2019) Macroalgae and microalgae: novel sources of functional food and feed. In: Ravishankar GA, Ranga Rao A (eds) Handbook of algal technologies and phytochemicals: volume I: food, health and nutraceutical applications. CRC Press, Boca Raton, pp 207–219

Shannon E, Abu-Ghannam N (2019) Seaweeds as nutraceuticals for health and nutrition. Phycologia 58(5):563–577

Shi Q, Wang A, Lu Z, Qin C, Hu J, Yin J (2017) Overview on the antiviral activities and mechanisms of marine polysaccharides from seaweeds. Carbohydr Res 453–454:1–9

Shofia SI, Jayakumar K, Mukherjee A, Chandrasekaran N (2018) Efficiency of brown seaweed (Sargassum longifolium) polysaccharides encapsulated in nanoemulsion and nanostructured lipid carrier against colon cancer cell lines HCT 116. RSC Adv 8(29):15973–15984

Srikong W, Mittraparp-arthorn P, Rattanaporn O, Bovornreungroj N, Bovornreungroj P (2015) Antimicrobial activity of seaweed extracts from Pattani, southeast coast of Thailand. Food Appl Biosci J 3(1):39–49

Surget G, Roberto VP, Le Lann K, Mira S, Guérard F, Laizé V et al (2017) Marine green macroalgae: a source of natural compounds with mineralogenic and antioxidant activities. J Appl Phycol 29(1):575–584

Tanna B, Choudhary B, Mishra A (2018) Metabolite profiling, antioxidant, scavenging and anti-proliferative activities of selected tropical green seaweeds reveal the nutraceutical potential of Caulerpa spp. Algal Res 36:96–105

Thirunavukkarasu R, Pandiyan P, Balaraman D, Subaramaniyan K, Edward Gnana Jothi G, Manikkam S, Sadaiyappan B (2013) Isolation of bioactive compound from marine seaweeds against fish pathogenic bacteria Vibrio alginolyticus (VA09) and characterisation by FTIR. J Coastal Life Med 1(1):26–33

Thomes P, Rajendran M, Pasanban B, Rengasamy R (2010) Cardioprotective activity of Cladosiphon okamuranus fucoidan against isoproterenol induced myocardial infarction in rats. Phytomedicine 18(1):52–57

Uribe E, Vega-Gálvez A, Heredia V, Pastén A, Di Scala K (2018) An edible red seaweed (Pyropia orbicularis): influence of vacuum drying on physicochemical composition, bioactive compounds, antioxidant capacity, and pigments. J Appl Phycol 30(1):673–683

Vázquez AIF, Sánchez CMD, Delgado NG, Alfonso AMS, Ortega YS, Sánchez HC (2011) Anti-inflammatory and analgesic activities of red seaweed Dichotomaria obtusata. Braz J Pharm Sci 47(1):111–118

Vieira EF, Soares C, Machado S, Correia M, Ramalhosa MJ, Oliva-Teles MT et al (2018) Seaweeds from the Portuguese coast as a source of proteinaceous material: total and free amino acid composition profile. Food Chem 269:264–275

WHO (2018) Cancer fact sheet WHO 28 September 2018. https://www.who.int/news-room/fact-sheets/detail/cancer#:~:text=Key%20facts,%2D%20and%20middle%2Dincome%20countries. Accessed 3 Mar 2021

Wijesinghe W, Kim E, Kang M, Lee W, Lee H, Vairappan CS, Jeon Y (2014) Assessment of anti-inflammatory effect of 5β-hydroxypalisadin B isolated from red seaweed Laurencia snackeyi in zebrafish embryo in vivo model. Environ Toxicol Pharmacol 37(1):110–117

Yan X, Yang C, Lin G, Chen Y, Miao S, Liu B, Zhao C (2019) Antidiabetic potential of green seaweed Enteromorpha prolifera flavonoids regulating insulin signaling pathway and gut microbiota in type 2 diabetic mice. J Food Sci 84(1):165–173

Yildiz G, Celikler S, Vatan O, Dere S (2012) Determination of the anti-oxidative capacity and bioactive compounds in green seaweed Ulva rigida C. agardh. Int J Food Prop 15(6):1182–1189

Zbakah H, Salhi G, Bochkov V, Ciudad CJ, Noe V, Hassoun M, Riadi H (2020) Insights on the anti-inflammatory and antitumor activities of extracts from the marine green alga *Codium decorticatum*. Eur J Inter Med 37:101170. https://doi.org/10.1016/j.eujim.2020.101170

Zeng Q-H, Zhang X-W, Xu K-P, Jiang J-G (2014) Application of fluorescently labelled tracer technique for detection of natural active macromolecules in Chinese medicine. Drug Metab Rev 46(1):57–71. https://doi.org/10.3109/03602532.2013.839699

Zollmann M, Robin A, Prabhu M, Polikovsky M, Gillis A, Greiserman S, Golberg A (2019) Green technology in green macroalgal biorefineries. Phycologia 58(5):516–534

Chapter 20
Seaweeds as Prospective Marine Resources for the Development of Bioactive Pharmacophores and Nutraceuticals

Kajal Chakraborty

1 Introduction

Marine-derived bioactive components and functional food ingredients with potential health benefits are an emerging area of research. The rich diversity of mollusks, seaweeds, marine algae, heterotrophic microbiota, echinoderms, and sponges in the coastal and marine waters represent an untapped reservoir of bioactive compounds with valuable pharmaceutical and biomedical use. The natural products from marine organisms, which are adapted to the adverse living conditions in the saline ecosystem, were reported to biosynthesize bioactive secondary metabolites as an adaptive mechanism and were recognized as valuable pharmacophores (Blunt et al. 2007; Carroll et al. 2020). The bioactive properties of marine organisms, mainly anti-inflammatory, antioxidant, antitumor properties etc. were reported in the previous literature (Driggers et al. 2008; Faulkner 2002; Blunt et al. 2015; Winter et al. 2013). One of the most interesting marine phyla concerning pharmacologically active marine compounds includes the abundantly available seaweed species, which are potential sources of bioactive substances (Blunt et al. 2015). Seaweeds constitute a large assemblage of species that predominate the coastal shelf areas and are often termed as the *wonder herbs of the ocean* owing to their potential pharmaceutical properties. Recent research has provided evidence for the immense pharmaceutical potential of seaweeds for their structurally diverse bioactive compounds with anti-hypertensive (Chakraborty and Dhara 2022), anti-osteoporotic (Chakraborty et al. 2021a, b; Chakraborty and Dhara 2021), immune-boosting, anti-hyperglycemic (Antony et al. 2021), anti-inflammatory (Chakraborty et al. 2019), and anticarcinogenic activities (Carroll et al. 2020). India's annual production of seaweed is

K. Chakraborty (✉)
Bioprospecting Section of Marine Biotechnology Division, Central Marine Fisheries Research Institute, Cochin, Kerala, India
e-mail: kajal.chakraborty@icar.gov.in

A. Ranga Rao, G. A. Ravishankar (eds.), *Sustainable Global Resources of Seaweeds Volume 2*, https://doi.org/10.1007/978-3-030-92174-3_20

approximately 22,000 tonnes (wet weight) including aquaculture and wild-sourced. The development of value-added products from underutilized seaweed species also has the potential to expand the opportunities for their downstream value chain augmentation thereby boosting the livelihoods of resource-poor fisher folk and budding entrepreneurs across the coastal belt.

Seaweed has long been part of the traditional diet of coastal communities. It is widely consumed in East Asia, particularly in Japan, China, and Korea. A total of 900 species of green seaweed, 1500 species of brown seaweed, and 4000 species of red seaweed are present worldwide (Dawes 1998; Khan and Satam 2003).

2 Seaweeds or Marine Macroalgae: A Valuable Source of High-Value Compounds

Seaweeds are a rich source of nutrients, bioactive compounds, and secondary metabolites. The nutrients in seaweed has the potential to supply sustainable nutrients to meet the increasing nutritional requirements, whereas the bioactive compounds form the basis to develop high-value nutraceuticals, functional foods, cosmeceuticals, and pharmacophore agents (Shama et al. 2019). The secondary metabolites of seaweeds have interesting functional properties and offer enormous possibilities to develop novel drugs and future pharmaceuticals. The discovery of metabolites with biological activity from seaweeds increased substantially in the last three decades (Blunt et al. 2015). These substances exhibit an appreciable number of distinct biological activities, such as anti-tumoral, anti-viral, antifungal, insecticidal, cytotoxic, phytotoxic, and anti-proliferative actions. Most of the bioactive substances isolated from seaweeds are chemically classified as brominated, aromatics, nitrogen-heterocyclic, nitrosulphuric-heterocyclic, sterols, dibutanoids, proteins, peptides, sulphated polysaccharides, terpenes, acetogenins, alkaloids and polyphenolics. Seaweeds are the only sources for industrially important phycocolloids like agar, carrageenan and alginate. They have a lot of applications as stabilizer, viscosifier, gelling and emulsifying agents.

Studies on natural product chemistry and chemical defense systems of seaweeds recognized the presence of bioactive leads with prominent pharmacological activities. Phlorotannins, sulfated polysaccharides, and polyphenols from the seaweeds have revealed their activities against the proliferation of cancer cells and produce some effects on the anti-inflammatory and anti-diabetic response (Chakraborty et al. 2010a, b, c, 2012, 2015a, b, c, 2016, 2017a, b, 2018). Bioactive components in seaweeds were recognized to modulate glucose-induced oxidative stress and their ability to control the presence of starch-digestive enzymes. Among various abundantly available seaweed species, brown and red seaweeds (classes Phaeophyceae and Rhodophyceae, respectively) were found to be the potential sources of bioactive substances (Chakraborty et al. 2019; Chakraborty and Antony 2020; Antony and Chakraborty 2020a, b, c). These bioactive compounds are mainly produced under various stress conditions. The major bioactive compounds isolated from the seaweeds include polysaccharides, phenolics, proteins, peptides, terpenes, terpenoids, carotenoids, sterols, dibutanoids, acetogenins alkaloids, etc. (Balboa et al. 2013; Anusree and Chakraborty 2017a, b; Chakraborty and Anusree 2020).

3 Nutraceuticals from Seaweeds

The term "nutraceutical" was coined by Dr. Stephen DeFelice, founder, and chairman of the Foundation for Innovation in Medicine. He defined nutraceutical as "... any substance that is a food or a part of a food and provides medical or health benefits, including the prevention and treatment of disease". The rich diversity of seaweeds represents an untapped reservoir of bioactive compounds with valuable pharmaceutical and biomedical use. The pioneering research work at ICAR-Central Marine Fisheries Research Institute envisages a systematic approach involving chemical profiling of major species of seaweeds for lead pharmacophores coupled with the evaluation of target biological activities against different disease models, for example, 3-hydroxy-3-methylglutaryl coenzyme A reductase, type-2 diabetes modulators (dipeptidyl peptidase-4, protein tyrosine phosphatase 1B), angiotensin-converting enzyme, inflammatory cyclooxygenase-2, and 5-lipoxygenase. Optimized physical/chromatographic procedures have been developed to isolate and purify the molecules with target bioactivities. Further applications of these compounds have led to the development of an array of nutraceutical products/formulations against arthritis, type-2 diabetes, dyslipidemia, hypothyroidism, osteoporosis, hypertension, and immunity-boosting.

4 Seaweeds as Potential Sources of Health Food

Seaweeds were consumed by the coastal communities since pre-historic times, particularly in Japan and China (McHugh 2003), and also used up traditionally in Indonesia, the Philippines, South Korea, North Korea, and Malaysia (Ganesan et al. 2019). Seaweeds are used in the traditional Japanese cuisine "shojin ryori" (Tsuji and Ishige 1983; Tsuji 1983), whereas Kombu, wakame and nori are integral part of the Japanese seaweed diet (Griffin 2015). Lately, consumption of seaweeds has gained wide attention in the Americas and Europe because of their functional properties and introduction of Asian cuisine (Bocanegra et al. 2009). In India, *Ulva, Gracilaria* and *Acanthophora* are used in preparing food items with the coastal states of Kerala and Tamil Nadu (Dhargakar 2014). Seaweeds are deliberated as food supplements attributable to availability of valuable macro-nutrients and micronutrients and bioactive compounds.

Seaweeds are considered as a good source of food fiber, protein and minerals for human consumption (Table 20.1). Mineral content of several edible brown (*Fucus*

Table 20.1 Nutritional composition of seaweeds

Seaweeds	Protein (%)	Carbohydrate (%)	Lipid (%)	Ash (%)	Area/habitat	Reference
Galaxaura rugosa	5.34	16.91	0.53	72.97	Ujung Genteng coastal waters Indonesia	Rasyid and Handayani (2019)
Gelidiella acerosa	8.66	68.67	0.54	13.42	Eritrean Red Sea coast of Gurgussum and Hirgigo Bay Marina Park (Sesostris Bay)	Kasimala et al. (2020) Rasyid and Handayani (2019) Banu and Mishra (2018)
Gracilaria canaliculata	11.18	19.15		37.52		
Gracilaria gracilis	10.86	63.13	0.19	6.78		
Tricleocarpa fragilis	4.07	28.76	0.84	42.29		
Gracilaria folifera	6.98	22.32	3.23		Mandapam coastal regions	Pati et al. (2016) Manivannan (2008)
Gracilaria verrucosa	9.47		1.29	21.90	Mandapam coastal regions	Pati et al. (2016)
Hypnea valentiae	8.34	23.60			Mandapam coastal regions	Pati et al. (2016) Manivannan (2008)
Kappaphycus alvarezii	18.78	5.24	1.09	27.49	Rameshwaram	Pati et al. (2016) Rajasulochana et al. (2012)
Acanthophora spicifera	20.2	23.54	0.55	28.38	Mandapam coastal regions	Manivannan (2008) Ganesan et al. (2020)
Gracilaria edulis	18.04	24.80	4.71	7.36	Mandapam coastal regions	Ganesan et al. (2020)
Gracilaria corticata	22.84	8.30	7.07	8.10	Thondi coast of Palk Bay, southeast India	Rosemary et al. (2019)
Sargassum wightii	19.2	26.5	13.6		Palk Bay region of India	Bharathi et al. (2021)
Sargassum muticum	22.1	26.5	14.7			
Turbinaria ornata	19.8	26.2	15.1			
Sargassum tenerrimum	7.55	48.9	2.2	11.5	Coastal regions of Cox's Bazar and St. Martin's Island of Bangladesh	Hossain et al. (2021)
Spatoglossom asperum	10.4	46.8	3.8	7.4		

(continued)

Table 20.1 (continued)

Seaweeds	Protein (%)	Carbohydrate (%)	Lipid (%)	Ash (%)	Area/habitat	Reference
Ascophyllum nodosum	5.9	31.7		20.2	Swedish west coast	Olsson et al. (2020)
Chorda filum	6.3	29.2		39.0		
Desmarestia aculeata	11.5	30.1		25.4		
Laminaria digitata	6.6	51.9		16.8		
Saccharina latissima	6.9	55.7		11.8		

vesiculosus, Laminaria digitata, Undaria pinnatifida) and red (*Chondrus crispus, Porphyra tenera*) algae was determined. Edible brown and red seaweeds could be used as a food supplement to fulfill the daily recommended intake of some essential minerals and trace elements. The protein content of seaweed products varied widely with 26.6 g/100 g in red algae and 12.9 g/100 g in brown seaweeds. Total lipid content of edible seaweeds ranged from 0.70 to 1.80 g/100 g dry weight. Unsaturated fatty acids were mostly found in all the brown seaweeds and the four most abundant fatty acids were 16:0, C18:1*n*-9, and C20:5*n*-3. Ash content ranged from 19.07 to 34.0 g/100 g dry weight. Seaweeds comprise with large amounts of polysaccharides, notably cell wall structural polysaccharides that are utilized by the hydrocolloid industry (alginate from brown algae and agar from red algae). Other polysaccharides, fucoidans from brown seaweeds, xylans from certain red and green seaweeds and ulvans in green seaweeds are also found in the cell wall. Other than this, seaweeds also contain storage polysaccharides like laminarin (in brown seaweeds), floridean starch (in red seaweeds). Most of these polysaccharides (agars, carrageenans, ulvans and fucoidans) are not digested directly by humans, and therefore, can be recognized as dietary fibers. Water-soluble and insoluble fibers have been associated with different physiological effects, for example, some soluble fibers have been correlated with hypocholesterolemic and hypoglycemic effects. Fucoidans were particularly studied for its interesting biological activities (anti-thrombotic, anti-coagulant, anticancer, anti-proliferative, anti-viral and anti-complementary agent, anti-inflammatory). Seaweeds are known as an excellent source of vitamins and minerals, especially sodium and iodine, due to their high polysaccharide content. Muthuraman and Ranganathan (2004) had selected six algae species (*Caulerpa scalpelliformis*, *Cladophora vagabunda*, *Enteromorpha compressa*, *Halimeda macroloba*, *Ulva fasciata* and *Chaetomorpha antennina*) to investigate protein, amino acids, total sugars and lipid contents. Fatty acid composition of *Ulva lactuca*, *Caulerpa chemnitzia*, *Padina tetrastromatica*, *Sargassum longifolium*, *Acanthophora spicifera* and *Gelidium micropterum* collected from the Mandapam coast in Tamil Nadu, India, was also reported.

An average of 10–30% of protein was reported in red seaweeds. *Palmaria palmata* and *Porphyra tenera* has been reported to have 35 and 47% of protein content, respectively. It was also observed that red algae could contain non-protein nitrogen, such as free nitrates, which might result in the overestimation of their protein content. Hence, nitrogen-to-protein conversion factors proposed for green, brown and red seaweeds were 5.13, 5.38 and 4.92, respectively. Seaweeds contain minerals 10–20 folds as the minerals in land plants (Moreda-Piñeiro et al. 2012). Hence, they are a significant source of valuable minerals like sodium, calcium, magnesium, potassium, chlorine, sulphur, phosphorus and micronutrients such as iodine, iron, zinc, copper, selenium, molybdenum, fluoride, manganese, boron, nickel and cobalt for nutrition. Red seaweed *P. palmata* is rich in Na, K and Cl. Due to high phycocolloid content in the cell wall of red seaweeds, it was noted with maximum value of carbohydrate content among other seaweeds. However, the lipid composition of red seaweed is too small, majority of which are polyunsaturated *n*-3 and *n*-6 fatty acids particularly with 20 carbon atoms, such as eicosapentanoïc acid (EPA, *n*-3 20:5) and arachidonic acid (AA, *n*-6 20:4), which play a major role in controlling the levels of low density lipoprotein (LDL) cholesterol.

Various studies have evaluated the effect of integration of seaweeds on the nutritional quality of food products. Effectiveness of seaweed supplementation in improving nutritional quality of food products, such as spice adjunct (*Kappaphycus alvarezii*), fish jerky (*Sargassum wightii*), chicken sausages (semi-refined carrageenan), biscuits (*Caulerpa racemosa*) etc. were assessed. Different studies have explored the potential of seaweed extracts to improve food preservation and storage, thereby ensuring the quality of foods. Effects of antioxidative substances from seaweed on quality of refined liver oil of leafscale gulper shark, *Centrophorus squamosus* during an accelerated stability study was previously reported (Chakraborty and Joseph 2017). The capability of seaweeds to improve the storage stability of C_{20-22}*n*-3 fatty acid methyl ester was evaluated by analyzing the combined effect of ethyl acetate fractions of *Kappaphycus alvarezii*, *Hypnea musciformis*, and *Jania rubens* (Chakraborty and Joseph 2017). Effect of antioxidant compounds from seaweeds on storage stability of C_{20-22} polyunsaturated fatty acid concentrate prepared from dogfish liver oil was reported (Chakraborty and Joseph 2018).

5 Therapeutic Applications of Seaweeds

Other than the primary metabolites, the seaweeds also possess structurally diverse secondary metabolites, with promising biological activities. Though in earlier times, therapeutic uses of seaweeds were found only in traditional medicines, lately at the end of 1990s, with the discovery of bioactive compounds from seaweeds further widened their utilization in pharmaceutical industries. Noticeably, the long life expectancy and lower rate of cardiovascular diseases of Japanese people might be associated with their regular intake of seaweeds. As a result, most of the Western countries have also started to include algae in their diet. Above all, these have been identified as a new source of ingredients for developing nutraceuticals (Pati et al. 2016; Bilal and Iqbal 2020).

Seaweeds were used in the traditional folk medicine for many years in Japan (13,000–300 BC), Egypt (1550 BC), China (2700 BC), and India (300 BC) (NAAS 2003; Tease 2005). The ancient Romans used seaweed to treat burns and wounds (Pati et al. 2016). *Sargassum* has been used in traditional Chinese medicine for more than 1000 years to treat different diseases (Liu et al. 2012), whereas *Capsosiphon fulvescens* was used to treat stomach disorders (Go et al. 2011). *Codium fragile* was used in traditional medicine to treat dropsy, dysuria, and enterobiasis (Sanjeewa et al. 2018). *Ulva pertusa* has been used as the traditional medicine for urinary ailments and dyslipidemia (Qi et al. 2006). *Gloiopeltis tenax* is traditionally used against diarrhoea (Zheng et al. 2012). *Caulerpa lentillifera* was used against hypertension, rheumatism, microbial infection, and diabetes (Sharma et al. 2015). Among the harvested seaweeds, approximately 13% have been used for the production of hydrocolloids, such as agar, alginate and carrageenan while 75% are used for food. Oral administration of some seaweed extracts (*Fucus vesiculosus, Macrocystis pyrifera* and *Laminaria japonica*) with zinc, manganese and vitamin B6 could lead to decrease osteoarthritis symptoms in a mixed population. A diet rich in seaweeds in many Asian countries with a low incidence of cancers has been reported on account of rich seaweed diet, and other possible health benefits including cardioprotective, neuroprotective and anti-inflammatory effects. These studies supported the utility of seaweeds in functional food development and several food products. Except its general medicinal properties, seaweeds are also recognized for antioxidant capacities, bioactive polyphenolic compounds and potential roles for the treatment of HIV (Blunt et al. 2015).

5.1 Anti-oxidant Activity

The organic extracts of seaweeds along with sulfated polysaccharides have exhibited anti-oxidant capacity. Most of them displayed similar results with that of reference standards (ascorbic acid and butylated hydroxyl toluene—BHT) (Seedevi et al. 2017). Several *in vivo* studies recognized potential anti-oxidant activity of seaweeds (Murakami et al. 2005). Antioxidant activities and phenolic contents of seaweeds harvested from the Gulf of Mannar of Peninsular India were documented (Chakraborty et al. 2015a, b, c, 2017a, b). An *in vitro* evaluation of solvent extracts of seaweed *Porteiria hornemanii* against reactive oxygen species revealed antioxidant activities, which might be due to the higher content of polyphenols and flavonoids. These compounds could donate electrons, and also reduce the lipid peroxidation process through oxidation (Cojandaraj et al. 2020). Antioxidant potential and phenolic compounds of brown seaweeds *Turbinaria conoides* and *Turbinaria ornata* (class: Phaeophyceae) were reported previously (Chakraborty and Joseph 2016). Crude extracts from more than 30 seaweed species were evaluated for antioxidant activity. Some of the seaweed species studied include *Halimeda tuna, Turbinaria conoides, Gracilaria foliifera, Enteromorpha compressa, Caulerpa veravelensis, Hypnea musciformis, Jania rubens, Chaetomorpha linum, Gelidiella acerosa, Kappaphycus alvarezii* etc. Sulfated polysaccharide purified from seaweeds like *Sargassum swartzii, S.tenerrimum, Turbinaria conoides* etc. were evaluated for antioxidant potential. Kang et al. (2012) studied *Saccharina japonica* (as *Laminaria*), which is used as a folk remedy in Korea for centuries, for antioxidant potential. A human clinical trial examined whether *S. japonica* could enhance the antioxidant defence system of 48 Korean men (aged 25–60). The seaweed was fermented with a lactic-acid-producing bacterium, *Lactobacillus brevis*, for 5 days to potentially augment seaweed bioactivity and digestibility. The fermented seaweed was dried and 250 mg of the powder was encapsulated. For 1 month, six capsules per day (1.5 g of alga) were administered to healthy subjects.

5.2 Anti-cancer Activity

The US Food and Drug Administration, European Pharmacopoeia, and European Food Safety Authority consider seaweeds as health food and nutraceuticals (USFDA 2017). In many Asian countries seaweeds are considered as medicinal, and are packaged with details of their effects and directions for use. Examples are included in the Japanese and Korean pharmacopoeias and the Chinese Marine Materia Medica (CMMM). In traditional Chinese medicine, seaweeds and other marine organisms are documented separately from terrestrial Materia Medica. In the CMMM, 171 species of medicinal seaweeds are listed. Since seaweeds have been used as a regular part of the diet and accepted as medicine for millennia in Asia, this might account for the majority of epidemiological evidence originating there. A direct relationship has been identified between high levels of seaweed consumption and lower instances of dietary-related disease, such as cancer. At the National Cancer Centre of South Korea, Park et al. (2016) assessed the dietary patterns of 923 men and women with an average age of 56 who had previously undergone surgery for colorectal cancer (plus 1846 control participants). Three dietary types were identified, such as prudent, traditional Korean, and Westernized. A highly significant reduction of risk factors for colorectal cancer was identified in subjects in the prudent group who consumed the most marine algae and vegetables, followed by the traditional diet (slightly less algae), with the highest risk found in the Westernized diet group who consumed little or no seaweeds but high levels of red meat and processed foods. Nelson et al. (2017) found the same association in a study of 627 people (age 35–74) across 42 hospitals in China. Risk factors for the development of biliary tract cancer were measured using 39 food groups. Only four food groups had either a significantly positive or negative association with risk factors for biliary tract cancer. These groups were seaweeds, allium (onions and garlic), salted meats, and preserved vegetables (pickled with salt-brine). Reports described that 29% of the anti-cancer studies conducted in red seaweed *Gracilaria* species were found promising (Torres et al. 2019). Among them, a study conducted by da Costa et al. (2017) revealed the activity of lipid extracts of an unknown species of *Gracilaria*. They obtained an IC_{50} of 12.2 and 12.9 μg mL^{-1} against a human breast cancer cell line (T-47D) and human bladder carcinoma cell line (5637), respectively. Organic crude extracts were the most studied and active extracts compared to others. From the studies of Sakthivel et al. (2016) and Sheeja et al. (2016), phytol could be considered as a potential candidate responsible for the anti-cancer activity of ethyl acetate fraction obtained from *G. edulis*. *Sargassum plagiophyllum* was found to be active against liver cancer (HepG2 cell line), whereas *Turbinaria conoides*, *Acanthophora spicifera* were assessed against lung cancer (A549 cell line), and *Sargassum polycystum, S. wightii* against breast cancer (MCF-7; MDA-MB-231). *Sargassum cinereum* and *S. longifolium* were potentially active against colon cancer (HCT-15; HCT 116; Caco-2 cell line). Two cytotoxic squalenoid-derived triterpenoids, laurenmariannol and (21a)-21-hydroxythyrsiferol from the marine red alga *Laurencia mariannensis* have exhibited significant cytotoxic activity against P-388 tumor cells with IC_{50} values of 0.6 and 6.6 mg/mL, respectively (Ji et al. 2008). Simultaneously, the red seaweed *Laurencia viridis* was found to be the rich source of many squalene derived secondary metabolites possessing anti-cancer activity (Norte et al. 1997). Japanese red alga *Laurencia obtusa* was found to contain 5 cytotoxic triterpenoids 28-anhy-drothyrsiferyl diacetate, 15-anhy-drothyrsiferyl diacetate diacetate, magireol-A, magireol B and magireol C (Suzuki et al. 1987; Gamal 2010). Among the various secondary metabolites isolated, terpenes, polysaccharides and polyphenols were the most, which displayed high potential against cancer. Suppression of cancer by seaweeds might be due to their high anti-oxidant properties since most of the processes in carcinogenesis lead via oxidative phases (Liu et al. 2012).

5.3 Anti-inflammatory Activity

There are numerous studies that demonstrated the anti-inflammatory and analgesic activities of seaweeds. Several anti-inflammatory studies conducted on seaweed *Gracilaria* were on the aqueous extracts or sulfated polysaccharides, which showed potential inhibition of pro-inflammatory mediators (Chaves et al. 2013). However, the results with organic extracts were quiet not promising (Shu et al. 2013). Although, aqueous extracts and sulfated polysaccharides could inhibit inflammation by reducing edema, migration of leukocytes and suppressing important compounds responsible for inflammation, their mode of mechanism was different, such as inhibition of NF-κB (factor nuclear kappa B) and MAPK (mitogen-activated protein kinase) pathways (Tseng et al. 2014), and act on mast cells, thus preventing the release of their content (Coura et al. 2015). A sulfated galactan from *Gracilaria opuntia* exhibited greater anti-inflammatory activity as determined by *in vitro* cyclo-oxygenase (COX-1, COX-2) and lipoxygenase (5-LOX) inhibition assays (Makkar and Chakraborty 2017a, b). Besides these, the heme oxygenase-1 pathway also found to be important in the inhibition activity of extracts (Vanderlei et al. 2011). Antioxidative *2H*-chromenyl derivatives, *abeo*-labdane type diterpenoid, and polyether triterpenoids from the intertidal red seaweed *Gracilaria salicornia* as potential anti-inflammatory agents were reported (Antony and Chakraborty 2020a, b, c). 2*H*-pyranoids from brown seaweed *Turbinaria conoides* with antioxidant and anti-inflammatory activities were reported (Chakraborty and Dhara 2020). Anti-inflammatory concentrate enriched with substituted oligofucans derived from brown seaweed *Turbinaria conoides* was developed (Chakraborty et al. 2016). da Matta et al. (2011) demonstrated the pronounced anti-inflammatory and antinociceptive activity of seaweed *Caulerpa mexicana* and *Caulerpa sertularioides*, although the mechanism responsible for the action is still under research. Neorogioltriol a tricyclic brominated diterpenoid derived from *Laurencia glandulifera* displayed anti-inflammatory activity both *in vitro* and *in vivo* explained by the inhibition of LPS-induced NF-κB activation and TNFα production. The anti-inflammatory activity arrayed by other seaweed species were reviewed in a previous report of literature (Lee et al. 2013). Unprecedented antioxidative and anti-inflammatory aryl polyketides, and sulfated polygalactopyranosyl-fucopyranan from the brown seaweed *Sargassum wightii* were characterized (Anusree and Chakraborty 2017a, 2018a). Anti-inflammatory activities of brown seaweed *Sargassum wightii* using different *in vitro* models were reported (Anusree et al. 2016).

Complete inhibition of phospholipase A2 activity was reported by *Asparagopsis armata*, *Chondrus crispus* and *Gelidium sesquipedale* extracts on a study of 23 seaweed species, while more than 95% inhibition of elastase was achieved by extracts of *Corallina elongata*, *Chondrus crispus*, *Gelidium sesquipedale* and *Laurencia pinnatifida* (Oumaskour et al. 2013). Crude extracts of seaweed species, such as *Gracilaria opuntia*, *Turbinaria ornata*, *Padina* spp., *Gracilaria salicornia*, *Kappaphycus alvarezii, Sargassum wightii* etc. were evaluated for its anti-inflammatory potential (Antony and Chakraborty 2019a, b). The studies indicated that the anti-inflammatory properties of the seaweeds were due to the presence of bioactive compounds such as azocinyl morpholinone, fucoidan, oxocine carboxylate cyclic ether, 2H-chromen derivative, aryl polyketide lactones, furanyl compounds etc. (Makkar and Chakraborty 2018a, b). Unprecedented antioxidative cyclic ether, halogen derivatives, and oxygenated meroterpenoids from the red seaweed *Kappaphycus alvarezii* with anti-cyclooxygenase and lipoxidase activities was reported (Makkar and Chakraborty 2017a, b, 2018a, b, c, d, e, f). Highly oxygenated antioxidative 2*H*-chromen derivative, furanyl derivatives and azocinyl morpholinone alkaloid from the red seaweed *Gracilaria opuntia* with pro-inflammatory cyclooxygenase and lipoxygenase inhibitory properties were reported (Makkar and Chakraborty 2018a, b, c, d, e, f, 2019).

5.4 Anti-microbial Activities

Anti-microbial agents' help in killing the disease causing causative organisms are frequently followed by side effects. Meantime, most of the microorganisms might develop its resistance to drugs, which in turn failing to cure diseases. Hence, requirement of new anti-microbial agents compatible for fighting against infectious diseases is necessary. Seaweeds have largely evaluated for their anti-microbial activity along with fungi and protozoa pathogens by using *in vitro* disk or well diffusion assay against gram positive and gram negative bacteria. A study on 21 species of seaweeds revealed that their lipophilic extract had both the highest values and broadest spectrum of bioactivities among other species (Chingizova et al. 2017). Antibacterial guaiane sesquiterpenes and labdane diterpenoids from seaweed *Ulva fasciata* were reported (Chakraborty et al. 2010a, b). A study on the anti-microbial activity of ethyl acetate extracts of *Portieria hornemanii* revealed a good zone inhibition against *Klebsiella pneumonia* and *Staphylococcus aureus* (Cojandaraj et al. 2020). Kulshreshtha et al. firstly reported the utilization of seaweed in antibiotic potentiation on existing industry standards, which in turn, demonstrated the increased life-time of patented antibiotics, and a way to reduce costly, therapeutic and prophylactic use of antibiotics. Seaweed-associated heterotrophic bacteria were reported as promising antibacterial agents, such as polyketide-derived macrobrevins, macrocyclic lactones, difficidin class of polyketide antibiotics, and aryl-crowned polyketide compounds for use against pathogens causing nosocomial infections (Chakraborty et al. 2017, 2020, 2021a, b, 2022; Kizhakkekalam and Chakraborty 2018). Feed supplementation with red seaweeds, *Chondrus crispus* and *Sarcodiotheca gaudichaudii*, reduce *Salmonella Enteritidis* in laying hens (Kulshreshtha et al. 2017).

5.5 Effect of Seaweeds on COVID-19 Prevention

Seaweeds have produced enough evidence to fight against COVID-19 infections. The infection of COVID-19 in Hokkaido was very low, and it was attributed to seaweed consumption, which is prevalent traditionally. The anti-oxidant property of seaweed-derived compounds could support numerous immune systems, act against oxidative damage and accelerate cell signalling (Kavitha 2020). The aqueous extracts of seaweed containing sulfated polysaccharides also proved to battle against COVID-19 since it could inhibit replication of enveloped viruses belonging to the order *Nidovirales*, comprising the genera Coronavirus (Pereira and Critchley 2020). A recent study by Bansal et al. (2020) suggested that iota-carrageenan (a thickening agent), which has been utilized in foods, could be used as a nasal spray for prevention of COVID-19. This was also supported by a study led by Morokutti-Kurz et al. (2021), testing the ability of several sulfated polysaccharides to inhibit viral attachment and entry in which iota-carrageenan presented the same rate of inhibition as that of other respiratory viruses. A sulfated polysaccharide obtained from *Saccharina japonica* was found to be significantly more potent than Remdesivir, the antiviral drug currently approved for emergency aid for severe COVID-19 infections (Kwon et al. 2020). The bioactivities of another polysaccharide obtained *Fucus vesiculosus* was screened and found that it could be considered as a potential candidate while treating COVID-19 patients (Pozharitskaya et al. 2020). Carbohydrate-binding proteins, lectins have also been emerged as tools against COVID-19 as they have shown anti-viral activities (O'Keefe et al. 2010; Barton et al. 2014; Barre et al. 2019; Lee 2019; Cheepsattayakorn and Cheepsattayakorn 2020). ICAR-Central Marine Fisheries Research Institute developed Cadalmin® Immuno-boost extract (IBe) as a novel immunity-boosting nutraceutical from seaweeds; helps to improve the non-specific innate immune system. The bioactive ingredients in Cadalmin® IBe increase innate immune response in animal models by regulating nuclear factor kappa-B along with oxidative stress markers. Cadalmin™ IBe has great market potential, particularly when strong immunity is the key in the fight against Covid-19 (Chakraborty et al. 2020; Indian Patent Application Number 202011054632).

5.6 Anti-diabetic Activity

Metabolic syndrome involving hyperglycemia, hypocholesterolemia, hypertension, hypertriglyceridemia, and obesity increases the risk of heart diseases as well as diabetes mellitus (Cornier et al. 2008). Anti-hyperglycemic or anti-diabetic studies were conducted on seaweeds by *in vitro* α-glucosidase, α-amylase and dipeptidyl peptidase (DPP-4) inhibition assays. Crude extracts of seaweeds were tested for their anti-diabetic potential using enzyme inhibitory assays α-amylase, α-glucosidase, and dipeptidyl peptidase-4 (Anusree et al. 2016). Various extracts (petroleum ether, ethyl acetate, methanol, acetone, etc.) of seaweeds *Sargassum polycystum* and *Sargassum wightii* were investigated for their anti-diabetic potential using *in vitro* enzyme inhibitory assays, and the study revealed the presence of the anti-diabetic compound fucosterol in the seaweeds. Tyrosine phosphatase-1B inhibitory activity of *frido* oleanene triterpenoids and labdane diterpenoids isolated from *Sargassum wightii* was studied (Anusree and Chakraborty 2017a, b). Antioxidative *abeo*-oleanenes from red seaweed *Gracilaria salicornia* as dual inhibitors of starch digestive enzymes were reported (Chakraborty and Antony 2019). In human clinical trials, daily supplementation with *Undaria pinnatifida* and *Sacchariza polyschides* (as *Gigantea bulbosa*) balanced the blood glucose levels, increases high-density lipoprotein cholesterol, and decreased serum triglyceride concentrations (Kim et al. 2008). The effect of a commercial seaweed concentrate (InSea2) on postprandial plasma glucose and insulin concentrations of patients (aged between 19 and 59 years with mean BMI 24.9 kg m^2) was studied (Paradis et al. 2011). The InSea2 extract of *Fucus vesiculosus* and *Ascophyllum nodosum* (contained natural α-amylase and α-glucosidase inhibitors), had previously demonstrated *in vitro* anti-diabetic properties by inhibiting these digestive enzymes that convert polysaccharides into simple sugars in the intestine and increase blood glucose levels (Roy et al. 2011). However, a significant insulin reduction of 12.1% was seen in plasma levels of the seaweed treated group, in addition, the peripheral insulin sensitivity and muscular glucose uptake (cederholm index) increased by 7.9%. The improvements in glucose and insulin profiles produced by the seaweed extracts suggest that they could be used to maintain insulin homeostasis in subjects with type 2 diabetes. The InSea2 extract significantly enhanced postprandial cognitive performance in 33 women and 27 men (age 18–65 years) who commonly experienced acute postprandial drowsiness (Haskell-Ramsay et al. 2018). Tanemura et al. (2014) examined the postprandial blood glucose profiles of 12 healthy adults (8 men, 4 women, average age 25) after a meal with and without the inclusion of fresh, cooked, whole *U. pinnatifida* (wakame), or sporophylls of *U. pinnatifida* (mekabu). The authors surmised that it was the fucoxanthin fraction and the polysaccharide-rich content of the mekabu sporophylls that exerted the antiglycaemic effect. The small mekabu shoots that grew from the parent wakame seaweed thallus were more viscous, when cooked. This was due to their higher content of soluble viscous fibre, which could affect carbohydrate metabolism and delays gastric emptying (Tanemura et al. 2014). It was concluded that the addition of fresh, brown seaweed to meals could be useful in controlling blood glucose levels for people with type 2 diabetes. The mechanisms of anti-diabetic action by seaweeds were attributed to the compounds including phlorotannins, fucoxanthin, polyphenolics, and polysaccharides (Kellogg et al. 2014; Murray et al. 2018), which inhibit hepatic gluconeogenesis and reduce the activity of digestive enzymes such as α-amylase, α-glucosidase, lipase, and aldose reductase (Sharifuddin et al. 2015). Seaweeds and their extracts might improve the health epidemic of type 2 diabetes leading to nephropathy, blindness, and peripheral neuropathy (Yamazaki et al. 2018).

6 Miscellaneous Bioactivities of Seaweeds

Hepatoprotective effect of fucoidan extracted from *Turbinaria decurrens* was evaluated (Meenakshi et al. 2014). The effects of crude extract of *Ulva lactuca* on D-galactosamine (D-Gal)-induced DNA damage, hepatic oxidative stress, and necrosis in rats were investigated. The neuroprotective effect of fucoidan, extracted from *Turbinaria decurrens*, was studied (Meenakshi et al. 2016). Neuroprotective potential of the seaweed *Gelidiella acerosa* against Aβ-25–35 peptide mediated toxicity was explored under *in vivo* conditions (Nisha and Devi 2017). Anti-Alzheimer potential of *Sargassum wightii* was evaluated, and found that the presence of high amount of terpenoids could be the possible reason for potential cholinesterase inhibitory activity (Syad et al. 2013). Organic extract of *Sargassum wightii* was examined for its angiotensin converting enzyme (ACE) inhibition potential for application in anti-hypertensive therapeutics, and found that the biological activity of the extracts was due to the presence of phlorotannin compounds and *O*-heterocyclic analogues (Anusree and Chakraborty 2018a, b).

Sulfated polysaccharide isolated from *Schizymenia pacifica* could selectively inhibit *in vitro* HIV replication and reverse transcriptase (Nakashima et al. 1987a, b). Oral delivery of seaweed derived oligomannate (developed by Shanghai Green Valley Pharmaceuticals) has received approval from the NMPA (National Medical Products Administration, China) during November 2019 to treat mild to moderate Alzheimer's disease, and to improve cognitive function (Syed 2020). Kishida et al. (2020) found an opposite association between seaweed intake and cardiovascular morbidity among Japanese. Sulfated polysaccharides extracted from the seaweed showed potential against diabetic retinopathy, age-related macular degeneration, chronic stomach inflammation, and gastric cancers (Chua et al. 2015; Klettner 2016). Myers et al. (2010) found that oral intake of seaweeds for over 12 weeks decreased the symptoms of osteogenesis. Seaweeds have been found to improve insulin resistance, decrease blood glucose levels, and improve antioxidant enzyme activities, thus reducing risk factors for cardiovascular ailments (Sørensen et al. 2019; Kim et al. 2008). Iota-carrageenan from the red seaweed was used to develop an antiviral drug Carragelose® (Marinomed Biotech AG, Austria), which is marketed as an OTC drug since 2008, as a treatment for viral infections of the upper respiratory tract, and could trap the viruses entering the nasal and pharyngeal cavity.

Seaweeds are considered as a good reservoir of molecules with varied biological properties (Chakraborty et al. 2018). Seaweed-derived alginates and laminarin were found to protect stomach and intestine from potential carcinogens (Brownlee et al. 2005; Szekalska et al. 2016; Déléris et al. 2016). Seaweed protein hydrolysates were used to develop nutraceuticals for controlling hypertension and/or oxidative stress (Paiva et al. 2017). Seaweed-derived mannitol was used as a natural sweetener for people with diabetes (Qin 2018a, b). Exploring the ACE-inhibitors from seaweeds displayed a growing interest in pharmaceuticals, nutraceuticals, and functional foods industries (Chakraborty et al. ICAR-CMFRI, Indian Patent Application No 202011011489; Anusree and Chakraborty 2018a, b; Makkar and Chakraborty 2018a, b, c, d, e, f). The effectiveness of crude extract of seaweed *Sargassum wightii*, in combination with *Bacillus thuringiensis* var. israelensis, in controlling malaria vector *Anopheles sundaicus* Liston was also determined. The anti-plasmodial activity of the extracts of seaweeds such as *Caulerpa toxifolia, C. peltata, Chaetomorpha antennina, Gracilaria verrucosa, Hypnea espera* etc. were studied. Anti-plasmodial effect of fucosterol was investigated. K-carrageenan was used as an emulsifier in reformulated meat products (Peñalver et al. 2020), and seaweed incorporated meat products displayed significant improvement in lipoprotein metabolism (Olivero-David et al. 2011). Incorporation of *Caulerpa racemosa* could improve the antioxidant and nutritive values of biscuits (Kumar et al. 2008). Supplementation of seaweeds in meat products considerably lowered the saturated fat contents, and increased the fibre and polyunsaturated fatty acids (Cofrades et al. 2017). Incorporation of *Sargassum wightii* in the ready-to-eat fish and cereal-based food enriched the product with fiber, minerals, besides improving antioxidant attributes (Hanjabam et al. 2017; Shannon and Ghannam 2019). The high mineral in seaweeds offer a prospect of using them to reduce the salt in processed meat products (Cofrades et al. 2009). Antibacterial labdane diterpenoids of *Ulva fasciata* Delile from the southwestern coast of Indian Peninsula were reported in a previous literature (Chakraborty et al. 2010c). The nutaceutical products from various companies across the world indicates the potential global demand (Table 20.2).

Table 20.2 Nutraceutical products and their manufacturing companies/developing organizations

Products	Description	Company/ Trademark/ Organization	Country
AstaFirst	Astaxanthin a red carotenoid to be used as an nutraceutical	Wefirst Biotechnology Co., Ltd	China
AstaPure	Astaxanthin a red carotenoid to be used as an nutraceutical	Alga Technologies	Israel
Red alage capsules	Extract of red algae *Gracilaria gracilis* to be used for blood circulation	Ahana Nutrition	USA
Red marine algae	Immune support, digestive health	Bio Nutrition Inc.	USA
Red algae calcium powder	Multi-mineral complex from marine red algae	Now Foods	USA
Red seaweed	Nutrition source	Omega One	USA
Cadalmin®ATe	Thyroid	Cadalmin® (ICAR-CMFRI)	India
Cadalmin®AOe	Osteoporosis	Cadalmin® (ICAR-CMFRI)	India
Cadalmin®GAe	Arthritis	Cadalmin® (ICAR-CMFRI)	India
Cadalmin®ACe	For lowering cholesterol levels	Cadalmin® (ICAR-CMFRI)	India
Cadalmin®ATe	Thyroid support	Cadalmin® (ICAR-CMFRI)	India
Cadalmin®AOe	For curing osteoporosis	Cadalmin® (ICAR-CMFRI)	India
Cadalmin®AHe	Regulation of hypertension	Cadalmin® (ICAR-CMFRI)	India
Cadalmin®IBe	Immunoboost	Cadalmin® (ICAR-CMFRI)	India
Cadalmin®ADe	Regulation of blood glucose level	Cadalmin® (ICAR-CMFRI)	India
Brown seaweed extract	Supports Healthy Weight & Healthy Immune System	NusaPure	USA
Brown seaweed plus	Immune stimulating properties	Only Natural	USA
Kelp iodine supplement	Thyroid support, strengthen immune system, regulates metabolism, boost cognitive ability	Natural Nutra	Russia
Marine-D3	Anti-aging	Marine Essentials	USA
Brown seaweed extract	Boost the immune system	Modifilan	USA
Brown seaweed extract	Weight loss, lowing cholesterol and hypertension, relief of constipation, and boosting the immune system.	SeaHerb	Korea

(continued)

Table 20.2 (continued)

Products	Description	Company/ Trademark/ Organization	Country
Xanthadrene	Promotes and supports healthy metabolic rate and breakdown of fat	Newton-Everett Nutraceuticals	USA
Organic Irish Carragheen Moss	Traditionally used to set jellies and puddings, can be used as thickening agent also for soups and sauces	AlgAran Seaweed Products	Ireland
Organic Irish kombu/kelp	High in calcium and magnesium	AlgAran Seaweed Products	Ireland
Savory seaweed Thai rice chips	Made with nutrient-packed seaweed flakes for an umami-rich flavor in this crunchy, savory snack.	Dang Foods Company	USA
Seaweed snacks Full sized seaweed sheets	Gluten free, vegetarian	gimMe Health Foods Inc.	USA
Irish moss & Irish spirulina capsules	The blend of red and green seaweeds provide a complete protein with all of the essential amino acids, many vitamins, minerals and other trace elements	Irish Seaweeds	Ireland
Triple blend seaweed capsules	Source of over 70 vitamins and minerals (nothing added) also contains proteins, selenium, antioxidants, essential fatty acids (including omega 3) phenol's, enzymes, trace elements, amino acids, and all the rare nutrients like B12, zinc	Irish Seaweeds	Ireland
Kelp seaweed capsules	Natural source of iodine which supports healthy thyroid function and is good for general health and skin condition	Irish Seaweeds	Ireland

7 ICAR-Central Marine Fisheries Research Institute: A Pioneering Institute in India to Develop Nutraceutical Products from Seaweeds

ICAR-Central Marine Fisheries Research Institute devoted research program for the development of promising bioactive molecules for human health and medication from seaweeds towards their utilization based on the National Policy to harness the potential of this natural wealth of Indian coastal waters. ICAR-Central Marine Fisheries Research Institute is the pioneering marine research institute in India to work in the frontier area of marine bioprospecting/bioactive molecule discovery from seaweeds and development of high-value nutraceutical products as dietary supplements and health management. This prestigious marine fisheries research institute of the Indian Council of Agricultural Research (ICAR) has developed and commercialized the nutraceutical products Cadalmin™ Green Algal extract (Cadalmin™ GAe) and Antidiabetic extract (Cadalmin™ ADe) as green alternatives to synthetic drugs to combat rheumatic arthritic pains and type-2 diabetes, respectively. Cadalmin™ Antihypercholesterolemic extract (Cadalmin™ ACe) and Cadalmin™ Antihypothyroidism extract (Cadalmin™ ATe) developed from seaweeds to combat dyslipidemia and hypothyroid disorder, respectively were out-licensed to a wellness company. Semisynthetic C-4/C-6 methylene-polycarboxylate cross-linked hybrid drug delivery system and a topical antibacterial formulation have been developed from seaweeds, and were found to be comparable with commercially available products. The lead molecules with action against angiotensin-converting enzyme-I, from seaweeds were isolated and added to a nutraceutical product Cadalmin™ Antihypertensive extract (Cadalmin™ AHe) that was out-licensed to a biopharmaceutical company. The latest efforts in this line of research have yielded the anti-osteoporotic and immune-boost nutraceuticals, which were out-licensed to the pharmaceutical company. Seaweed-derived natural template-inspired synthetic derivatives as potential pharmacophores with potential antibacterial activities against methicillin-resistant *Staphylococcus aureus* and anti-angiotensin-I inhibitory activities were designed and developed. Several cosmeceutical products from seaweeds are in pipeline, and are being commercialized.

8 Seaweeds: Nutraceutical Products from India

Combined with high throughput screening through a large number of drug targets, bioactivity research against hypertension, type-2 diabetes, hypercholesterolemia, dyslipidemia, type-2 diabetes, hypothyroid disorder, and inflammation will be effective in revealing the potentially useful biological properties of seaweed-derived marine natural products. Furthermore, the discovery of new bioactive compounds from seaweeds would form the basis for new drug leads.

The following nutraceutical products were developed from seaweeds:

Cadalmin® Green Algal extract (Cadalmin®GAe) to combat rheumatic arthritic pains (Indian Patent Grant number 294451)

Cadalmin®Antidiabetic extract (Cadalmin®ADe) for use against Type II diabetes (Indian Patent Grant number 346531)

Cadalmin®Antihypercholesterolemic extract (Cadalmin®ACe) for dyslipidemia (Indian Patent Application number 201711018741)

Cadalmin®Antihypothyroidism extract (Cadalmin®ATe) to combat hypothyroid disorders (Indian Patent Application number 201911036205)

Cadalmin®Antihypertensive extract (Cadalmin®AHe) for use against hypertension (Indian Patent Application number 201911038055)

Cadalmin®Antiosteoporotic extract (Cadalmin®AOe) to treat osteoporosis (Indian Patent Application number 201911053105)

Cadalmin®Immunoboost extract (Cadalmin®IBe) to boost innate immunity (Indian Patent Application number 202011054632)

9 Patents in Nutraceuticals from Seaweeds

Investigation into the filed patents on seaweeds presented a considerable increase throughout the years. A comparative study on red algal patents from 1990 to 2020 was conducted, mainly focusing on pharmacological activities and nutraceuticals being developed. The data were obtained from the patent search site 'Derwent Innovations' as of 15/10/2020. The number of patents was found to be between 0 and 10 during 1990–2000, however, it witnessed a rapid surge after 2000, especially patents related to red algal polysaccharides, nutraceuticals and food supplements. Graphical representation of the patents granted in the subject area of "seaweed nutraceuticals" during 1995–2019 was depicted. An increasing interest in seaweed-based bioactive compounds and functional food ingredients as evidenced by the scientific publications and patents in the last decade has appropriately demonstrated the possibilities of bioactive compounds from seaweeds to maintain and improve human health and well-being. Considerable numbers of granted patents during 1995–2019 (retrieved from Google patents, accessed on 12/01/202021) recognized the increasing importance of bioactive compounds from seaweeds (Table 20.3). The increasing trend in the number of patents in the field of marine natural product research, particularly in the pharmacological effects of seaweeds attracted the attention of marine natural product chemists and medical researchers to focus on this diverse phylum.

10 Future Prospects and Conclusions

The bioactive compounds, derived from seaweeds, with potential health benefits, are an emerging area of research. The rich diversity of seaweeds in the Indian marine biosphere represents an untapped reservoir of bioactive compounds with valuable pharmaceutical and biomedical use. Over the last few years, the use of seaweeds for the development of new products as well as a source for obtaining high-value compounds have attracted much interest from both the food and pharmaceutical industries. The research work developed a hitherto unraveled database of seaweeds with small molecular weight bioactive molecules responsible to combat various life-threatening diseases. This subsequently paved the way for the development of several nutraceutical products for use against arthritis, type-2 diabetes, dyslipidemia, hypothyroidism, osteoporosis, immunoboost agent, and hypertension. The marine macroflora are gaining immense attention in nutraceutical industries due to their protective function against various chronic diseases. Indian nutraceuticals market has been growing at a compound annual growth rate of 20% for the past 3 years, particularly in the segments of functional food products, antioxidants, and immunity boosters (https://www.nuffoodsspectrum.in; https://www.investindia.gov.in). These developments point towards the fast proliferation of the specific segments of nutraceuticals in India and their acceptance by Indian consumers and healthcare

Table 20.3 List of some important patents granted worldwide in the area of seaweed nutraceuticals

Patent number	Publication year	Seaweed species	Description	Assignee/ applicant
CN111629733A	2020	Not specified	Anti-cancer, anti-inflammatory composition	Korea University Research and Business Foundation
TW201927320A	2019	*Gracilaria blodgettii, G. coforvoides, G. gigas, G. chorda, G. lichenoides, G. compressa*	Treatment of nervous diseases including depression, bipolar disorder, anxiety, autism or dementia.	Industrial Technology Research Institute, Taiwan
US10772933B2	2020	Not specified	Medicament for treating pulmonary tuberculosis	Shandong Zhonghai Pharmaceutical Co. Ltd., China
CN111568801A	2020	Not specified	Inhibition of 5-α-reductase activity, inhibition of oil and metabolites from sebaceous glands	Guangzhou Danke Network Technology Co. Ltd, China
WO2020124167A1	2020	*Asparagopsis* species	Boosting innate immunity and preventing diseases like mastitis, pasteurellosis, clostridial disease, pleuropneumonia, and exudative dermatitis	University of the Sunshine Coast, Australia
CN110812364A	2020	Not specified	Treatment of anti-diabetic nephropathy, anti-cardiovascular diseases	Ocean University of China
CN111150701A	2020	Not specified	Composition used in cosmetics for masking wrinkles	BAO Ji-xian, China
CN111588667A	2020	Not specified	Anti-wrinkle essence	JIANG Yan-wei, Panjin, Liaoning, China
ES2753218T3	2020	*Undaria pinnatifida*	Nasal composition for treating respiratory tract conditions	Gerolymatos International SA, Greece
CN107904223B	2020	Not specified	Alginate lyase useful in alginate oligosaccharide, food flavoring, health supplement, cosmetic or seaweed fertilizer	Tianjin Institute of Industrial Biotechnology Chinese Academy of Sciences, China

(continued)

Table 20.3 (continued)

Patent number	Publication year	Seaweed species	Description	Assignee/ applicant
CN107050013B	2020	Not specified	Application of phloroglucinol derivative as drug for treating Alzheimer's disease	Ningbo University, China
CN110013447A	2019	Not specified	Cream used for removing striae gravidarum of pregnancy	Guangzhou Best Clean Cosmetic Manufacturing Co. Ltd., China

providers. With increasing health awareness and the shift towards preventative health care India's future in this segment is promising.

Acknowledgments The author thanks the Director, Central Marine Fisheries Research Institute (CMFRI) and Head, Marine Biotechnology Division of CMFRI for facilitating the research activities. This work was funded by the Indian Council of Agricultural Research, New Delhi, India under Central Marine Fisheries Research Institute supported project "Development of Bioactive Pharmacophores from Marine Organisms" (grant number MBT/HLT/SUB23).

References

Antony T, Chakraborty K (2019a) Pharmacological properties of seaweeds against progressive lifestyle diseases. J Aquat Food Prod Technol 28(10):1092–1104

Antony T, Chakraborty K (2019b) Xenicanes attenuate pro-inflammatory 5-lipoxygenase: prospective natural anti-inflammatory leads from intertidal brown seaweed *Padina tetrastromatica*. Med Chem Res 28(4):591–607

Antony T, Chakraborty K (2020a) First report of antioxidative *2H*-chromenyl derivatives from the intertidal red seaweed *Gracilaria salicornia* as potential anti-inflammatory agents. Nat Prod Res 34(24):3470–3482

Antony T, Chakraborty K (2020b) Anti-inflammatory polyether triterpenoids from the marine macroalga *Gracilaria salicornia*: newly described natural leads attenuate pro-inflammatory 5-lipoxygenase and cyclooxygenase-2. Algal Res 47:101791

Antony T, Chakraborty K (2020c) First report of antioxidant abeo-labdane type diterpenoid from intertidal red seaweed *Gracilaria salicornia* with 5-lipoxygenase inhibitory potential. Nat Prod Res 34(10):1409–1416

Antony T, Chakraborty K, Joy M (2021) Antioxidative dolabellanes and dolastanes from brown seaweed *Padina tetrastromatica* as dual inhibitors of starch digestive enzymes. Nat Prod Res 35:614–626

Anusree M, Chakraborty K (2017a) Previously undescribed fridooleanenes and oxygenated labdanes from the brown seaweed *Sargassum wightii* and their protein tyrosine phosphatase-1B inhibitory activity. Phytochemistry 144:19–32

Anusree M, Chakraborty K (2017b) Unprecedented antioxidative and anti-inflammatory aryl polyketides from the brown seaweed *Sargassum wightii*. Food Res Int 100:640–649

Anusree M, Chakraborty K (2018a) Previously undescribed antioxidative O-heterocyclic angiotensin converting enzyme inhibitors from the intertidal seaweed *Sargassum wightii* as potential antihypertensives. Food Res Int 113:474–486

Anusree M, Chakraborty K (2018b) Pharmacological potential of sulfated polygalactopyranosylfucopyranan from the brown seaweed *Sargassum wightii*. J Appl Phycol 30:1971–1988

Anusree M, Chakraborty K, Makkar F (2016) Pharmacological activities of brown seaweed *Sargassum wightii* (family Sargassaceae) using different *in vitro* models. Int J Food Prop 20(4):931–945

Balboa EM, Conde E, Moure A, Falqué E, Domínguez H (2013) In vitro antioxidant properties of crude extracts and compounds from brown algae. Food Chem 138:1764–1785

Bansal S, Jonsson CB, Taylor SL, Figueroa JM, Dugour AV, Palacios C, Vega JC (2020) Iota-carrageenan and xylitol inhibit SARS-CoV-2 in cell culture. bioRxiv 2020.08.19.225854. https://doi.org/10.1101/2020.08.19.225854

Banu SV, Mishra JK (2018) Fatty acid, micronutrient, proximate composition and phytochemical analysis of red seaweed *Tricleocarpa fragilis* (L.) Huisman & RA towns from Andaman Sea, India. J Pharmacogn Phytochem 7(4):2143–2148

Barre A, Simplicien M, Benoist H, Van Damme EJM, Rouge P (2019) Mannose-specific lectins from marine algae: diverse structural scaffolds associated to common virucidal and anti-cancer properties. Mar Drugs 17:440

Barton C, Kouokam JC, Lasnik AB, Foreman O, Cambon A, Brock G, Montefiori DC, Vojdani F, McCormick AA, O'Keefe BR, Palmer KE (2014) Activity of and effect of subcutaneous treatment with the broad-spectrum antiviral lectin griffithsin in two laboratory rodent models. Antimicrob Agents Chemother 58(1):120–127

Bharathi S, Kumar SD, Sekar S, Santhanam P, Divya M, Krishnaveni N, Pragnya M, Dhanalakshmi B (2021) Experimental evaluation of seaweeds liquid extracts as an alternative culture medium on the growth and proximate composition of Picochlorum maculatum. Proc Natl Acad Sci India Sect B Biol Sci 91(1):205–215

Bilal M, Iqbal HMN (2020) Marine seaweed polysaccharides-based engineered cues for the modern biomedical sector. Mar Drugs 18:7

Blunt JW, Copp BR, Hu WP, Munro MH, Northcote PT, Prinsep MR (2007) Marine natural products. Nat Prod Rep 24(1):31–86

Blunt JW, Copp BR, Keyzers RA, Munro MH, Prinsep MR (2015) Marine natural products. Nat Prod Rep 32(2):116–211

Bocanegra A, Bastida S, Benedi J, Rodenas S, Sanchez-Muniz FJ (2009) Characteristics and nutritional and cardiovascular-health properties of seaweeds. J Med Food 12(2):236–258

Brownlee IA, Allen A, Pearson JP, Dettmar PW, Havler ME, Atherton MR, Onsøyen E (2005) Alginate as a source of dietary fiber. Crit Rev Food Sci Nutr 45:497–510

Carroll AR, Copp BR, Davis RA, Keyzers RA, Prinsep MR (2020) Marine natural products. Nat Prod Rep 37(2):175–223

Chakraborty K, Antony T (2019) First report of antioxidative abeo-oleanenes from red seaweed *Gracilaria salicornia* as dual inhibitors of starch digestive enzymes. Med Chem Res 28(5):696–710

Chakraborty K, Antony T (2020) Salicornolides A-C from *Gracilaria salicornia* attenuate pro-inflammatory 5-lipoxygense: prospective natural anti-inflammatory leads. Phytochemistry 172:112259

Chakraborty K, Anusree M (2020) Marine-derived polygalactofucan and its β-2-deoxy-amino-substituted glucopyranan composite attenuate 3-hydroxy-3-methylglutaryl-CoA reductase: prospective natural anti-dyslipidemic leads. Med Chem Res 29(2):281–300

Chakraborty K, Dhara S (2020) First report of substituted 2*H*-pyranoids from brown seaweed *Turbinaria conoides* with antioxidant and anti-inflammatory activities. Nat Prod Res 34(24):3451–3461

Chakraborty K, Dhara S (2021) Polygalacto-fucopyranose biopolymer structured nanoparticle conjugate attenuates glucocorticoid-induced osteoporosis: an in vivo study. Int J Biol Macromol 190:739–753

Chakraborty K, Dhara S (2022) Spirornatas A-C from brown alga Turbinaria ornata: anti-hypertensive spiroketals attenuate angiotensin-I converting enzyme. Phytochemistry 195:113024
Chakraborty K, Joseph D (2016) Antioxidant potential and phenolic compounds of brown seaweeds *Turbinaria conoides* and *Turbinaria ornata* (class: Phaeophyceae). J Aquat Food Prod Technol 25(8):1249–1265
Chakraborty K, Joseph D (2017) Effects of antioxidative substances from seaweed on quality of refined liver oil of leafscale gulper shark, Centrophorus squamosus during an accelerated stability study. Food Res Int 103:450–461
Chakraborty K, Joseph D (2018) Effect of antioxidant compounds from seaweeds on storage stability of C20-22 polyunsaturated fatty acid concentrate prepared from dogfish liver oil. Food Chem 260:135–144
Chakraborty K et al (2010a) A process to prepare antioxidant and anti-inflammatory concentrates from brown and red seaweeds and a product thereof. Indian Patent Grant number 294451
Chakraborty K, Lipton AP, Paul RR, Chakraborty RD (2010b) Guaiane sesquiterpenes from seaweed *Ulva fasciata* Delile and their antibacterial properties. Eur J Med Chem 45(6):2237–2244
Chakraborty K, Lipton AP, Paul RR, Vijayan KK (2010c) Antibacterial labdane diterpenoids of *Ulva fasciata* Delile from the southwestern coast of Indian Peninsula. Food Chem 119(4):1399–1408
Chakraborty K. et al (2012) A product containing anti-inflammatory principles from brown seaweeds and a process thereof. Indian Patent Grant number 5199/CHE/2012
Chakraborty K et al (2015a) Anti-inflammatory principles in a preparation of brown seaweeds. Indian Patent Grant number 333392 (4254/DEL/2015-Indian Patent)
Chakraborty K et al (2015b) A process to prepare antidiabetic concentrates from seaweeds and a product thereof. Indian Patent Appl no. 3366/DEL/2015
Chakraborty K, Joseph D, Praveen NK (2015c) Antioxidant activities and phenolic contents of three red seaweeds (division: Rhodophyta) harvested from the Gulf of Mannar of Peninsular India. J Food Sci Technol 52(4):1189–2002
Chakraborty K, Joseph D, Raola VK (2016) Anti-inflammatory concentrate enriched with substituted oligofucans derived from brown seaweed *Turbinaria conoides* (J. Agardh) Kützing and its safety assessment on wistar rats. J Aquat Food Prod Technol 25(8):1323–1338
Chakraborty K et al (2017a) A process to prepare anti-dyslipidemic concentrate from seaweed and a product thereof. Indian Patent Appl. no. 201711013741
Chakraborty K, Anusree M, Makkar F (2017b) Antioxidant activity of brown seaweeds. J Aquat Food Prod Technol 26(4):406–419
Chakraborty K, Vijayagopal P, Gopalakrishnan A (2018) Nutraceutical products from seaweeds, wonder herbs of the oceans. Mar Fish Inf Serv Tech Ext Ser 237:7–12
Chakraborty K, Antony T, Joy M (2019) Prospective natural anti-inflammatory drimanes attenuating pro-inflammatory 5-lipoxygenase from marine macroalga *Gracilaria salicornia*. Algal Res 40:101472
Chakraborty K et al (2020) A process to prepare a stabilized composition from seaweed to improve innate immune system and a product thereof. Application no 202011054632 (Indian Patent application)
Chakraborty K, Antony T, Dhara S (2021a) Marine macroalgal polygalactan-built nanoparticle construct for osteogenesis. Biomacromolecules 22:2197–2210
Chakraborty K, Kizhakkekalam VK, Joy M, Dhara S (2021b) Difficidin class of polyketide antibiotics from marine macroalga-associated *Bacillus* as promising antibacterial agents. Appl Microbiol Biotechnol 105:6395–6408
Chakraborty K, Kizhakkekalam VK, Joy M (2022) Polyketide-derived macrobrevins from marine macroalga-associated *Bacillus amyloliquefaciens* as promising antibacterial agents against pathogens causing nosocomial infections. Phytochemistry 193:112983
Chakraborty K, Kizhakkekalam VK, Joy M, Chakraborty RD (2020) Moving away from traditional antibiotic treatment: can macrocyclic lactones from marine macroalga-associated heterotrophy be the alternatives? Appl Microbiol Biotechnol 104:7117–7130

Chakraborty K, Thilakan B, Kizhakkekalam VK (2017) Antibacterial aryl-crowned polyketide from *Bacillus subtilis* associated with seaweed *Anthophycus longifolius*. J Appl Microbiol 124:108–125

Chaves LDSL, Nicolau LAD, Silva RO, Barros FCN, Freitas ALP, Aragão KS, Ribeiro RDA, Souza MHLP, Barbosa ALDR, Medeiros J-VR (2013) Antiinflammatory and antinociceptive effects in mice of a sulfated polysaccharide fraction extracted from the marine red algae *Gracilaria caudata*. Immunopharmacol Immunotoxicol 35:93–100

Cheepsattayakorn A, Cheepsattayakorn R (2020) Promising drug candidates for 2019-novel coronavirus (COVID-19) pneumonia and related acute respiratory syndrome treatment. Acta Sci Microbiol 3(4):01

Chingizova EA, Skriptsova AV, Anisimov MM, Aminin DL (2017) Antimicrobial activity of marine algal extracts. Int J Phytomed 9(1):113–122

Chua EG, Verbrugghe P, Perkins TT, Tay CY (2015) Fucoidans disrupt adherence of *Helicobacter pylori* to AGS cells in vitro. Evid Based Complement Alternat Med 2015:120981

Cofrades S, López-López I, Ruiz-Capillas C, Jiménez-Colmenero F (2009) Nutritional properties of potential functional frankfurter with healthier lipid profile, seaweed and low salt content. In: Proceedings of 55th International Congresses of Meat Science and Technology (ICoMST), 16–21 August 2009, Copenhagen, Denmark

Cofrades S, Benedì J, Garcimartin A, Sànchez-Muniz F, andJiménez-Colmenero F. (2017) A comprehensive approach to formulation of seaweed-enriched meat products: from technological development to assessment of healthy properties. Food Res Int 99:1084–1094

Cojandaraj L, Surya PU, Shyamala ME (2020) Antioxidant activity of marine red algae–*Portieria hornemannii*. Plant Arch 20(2):1075–1081

Cornier M-A, Dabelea D, Hernandez TL, Lindstrom RC, Steig AJ, Stob NR, Van Pelt RE, Wang H, Eckel RH (2008) The metabolic syndrome. Endocr Rev 29:777–782

da Costa E, Melo T, Moreira ASP, Bernardo C, Helguero L, Ferreira I, Cruz MT, Rego AM, Domingues P, Calado R, Abreu MH, Domingues MR (2017) Valorization of lipids from *Gracilaria sp.* through lipidomics and decoding of antiproliferative and anti-inflammatory activity. Mar Drugs 15:62

Coura CO, Souza RB, Rodrigues JAG, Vanderlei EDSO, de Araújo IWF, Ribeiro NA, Frota AF, Ribeiro KA, Chaves HV, Pereira KMA, da Cunha RMS, Bezerra MM, Benevides NMB (2015) Mechanisms involved in the anti-inflammatory action of a polysulfated fraction from *Gracilaria cornea* in rats. PLoS One 10:e0119319

Dawes CJ (1998) Marine botany. John Wiley & Sons, New York. 496 P

Déléris P, Nazih H, Bard JM (2016) Seaweeds in human health. In: Fleurence J (ed) Seaweed in health and disease prevention. Elsevier, New York, pp 319–367

Dhargakar VK (2014) Uses of seaweeds in the Indian diet for sustenance and well-being. Sci Cult 80(7–8):192–202

Driggers EM, Hale SP, Lee J, Terrett NK (2008) The exploration of macrocycles for drug discovery—an underexploited structural class. Nat Rev Drug Discov 7(7):608–624

Faulkner DJ (2002) Marine natural products. Nat Prod Rep 19:1–48

Gamal A (2010) Biological importance of marine algae. Saudi Pharm J 18:1–25

Ganesan AR, Tiwari U, Rajauria G (2019) Seaweed nutraceuticals and their therapeutic role in disease prevention. Food Sci Human Wellness 8(3):252–263

Ganesan A, Subramani K, Shanmugam M, Seedevi P, Park S, Alfarhan A, Rajagopal R, Balasubramanian B (2020) A comparison of nutritional value of underexploited edible seaweeds with recommended dietary allowances. J King Saud Univ Sci 32:1206–1211

Go H, Hwang HJ, Nam TJ (2011) Polysaccharides from *Capsosiphon fulvescens* stimulate the growth of IEC-6 cells by activating the MAPK signaling pathway. Mar Biotechnol 13:433–440

Griffin J (2015) An investigative study into the beneficial use of seaweed in bread and the broader food industry. Dissertation presented to Dublin Institute of Technology, School of Culinary Arts and Food Technology in partial fulfillment of the requirements for the bachelor's degree BSc (Hons) Baking and Pastry Arts Management, 132 p

Hanjabam MD, Zynudheen AA, Ninan G, Panda S (2017) Seaweed as an ingredient for nutritional improvement of fish jerky. J Food Process Preserv 41:1–8

Haskell-Ramsay CF, Jackson PA, Dodd FL, Forster JS, Bérubé J, Levinton C, Kennedy DO (2018) Acute post-prandial cognitive effects of brown seaweed extract in humans. Nutrients 10:85

Hossain MT, Sohag AAM, Haque MN, Tahjib-Ul-Arif M, Dash R, Chowdhury MTH, Hossain MA, Moon IS, Hannan MA (2021) Nutritional value, phytochemical profile, antioxidant property and agar yielding potential of macroalgae from Coasts of Cox's Bazar and St. Martin's Island of Bangladesh. J Aquat Food Prod Technol 30(2):217–227

Ji NY, Li XM, Xie H, Ding J, Li K, Ding LP, Wang BG (2008) Highly oxygenated triterpenoids from the marine red alga *Laurencia mariannensis* (Rhodomelaceae). Helv Chim Acta 91(10):1940–1946

Kang YM, Lee B-J, Kim JI, Nam B-H, Cha J-Y, Kim Y-M, Ahn C-B, Choi J-S, Choi IS, Je J-Y (2012) Antioxidant effects of fermented sea tangle (*Laminaria japonica*) by *Lactobacillus brevis* BJ20 in individuals with high level of γ-GT: a randomized, double-blind, and placebo-controlled clinical study. Food Chem Toxicol 50:1166–1169

Kasimala M, Mogos GG, Negasi KT, Bereket GA, Abdu MM, Melake HS (2020) Biochemical composition of selected seaweeds from intertidal shallow waters of Southern Red Sea, Eritrea

Kavitha K (2020) Boost immunity with seaweed against Covid-19. GIS Bus 15(5):420–426

Kellogg J, Grace M, Lila M (2014) Phlorotannins from Alaskan seaweed inhibit carbolytic enzyme activity. Mar Drugs 12:5277

Khan SI, Satam S (2003) Seaweed mariculture: scope and potential in India. Aquat Asia 8:26–29

Kim MS, Kim JY, Choi WH, Lee SS (2008) Effects of seaweed supplementation on blood glucose concentration, lipid profile, and antioxidant enzyme activities in patients with type 2 diabetes mellitus. Nutr Res Pract 2(2):62–67

Kishida R, Yamagishi K, Muraki I, Sata M, Tamakoshi A, Iso H (2020) Frequency of seaweed intake and its association with cardiovascular disease mortality: the JACC study. J Atheroscler Thromb 27:1340–1347. https://doi.org/10.5551/jat.53447

Kizhakkekalam VK, Chakraborty K (2018) Pharmacological properties of marine macroalgae-associated heterotrophic bacteria. Arch Microbiol 201:505–518

Klettner A (2016) Fucoidan as a potential therapeutic for major blinding diseases—a hypothesis. Mar Drugs 1:31

Kulshreshtha G, Rathgeber B, MacIsaac J, Boulianne M, Brigitte L, Stratton G, Thomas NA, Critchley AT, Hafting J, Prithiviraj B (2017) Feed supplementation with red seaweeds, *Chondrus crispus* and *Sarcodiotheca gaudichaudii*, reduce *Salmonella Enteritidis* in laying hens. Front Microbiol 8:1–12

Kumar CS, Ganesan P, Suresh PV, Bhaskar N (2008) Seaweeds as a source of nutritionally beneficial compounds—a review. J Food Sci Technol 45:1–13

Kwon PS, Oh H, Kwon SJ, Jin W, Zhang F, Fraser K, Hong JJ, Linhardt RJ, Dordick JS (2020) Sulfated polysaccharides effectively inhibit SARS-CoV-2 in vitro. Cell Discov 6:50

Lee C (2019) Griffithsin, a highly potent broad-spectrum antiviral lectin from red algae: from discovery to clinical application. Mar Drugs 17:567

Lee JC, Hou MF, Huang HW, Chang FR, Yeh CC, Tang JY, Chang HW (2013) Marine algal natural products with anti-oxidative, anti-inflammatory, and anti-cancer properties. Cancer Cell Int 13(1):55

Liu L, Heinrich M, Myers S, Dworjanyn SA (2012) Towards a better understanding of medicinal uses of the brown seaweed *Sargassum* in Traditional Chinese Medicine: a phytochemical and pharmacological review. J Ethnopharmacol 142:591–619

Makkar F, Chakraborty K (2017a) Antidiabetic and anti-inflammatory potential of sulphated polygalactans from red seaweeds *Kappaphycus alverizii* and *Gracilaria opuntia*. Int J Food Prop 20(6):1326–1337

Makkar F, Chakraborty K (2017b) Unprecedented antioxidative cyclic ether from the red seaweed *Kappaphycus alvarezii* with anti-cyclooxygenase and lipoxidase activities. Nat Prod Res 31(10):1131–1141

Makkar F, Chakraborty K (2018a) Antioxidant and anti-inflammatory oxygenated meroterpenoids from the thalli of red seaweed *Kappaphycus alvarezii*. Med Chem Res 27(8):2016–2026

Makkar F, Chakraborty K (2018b) Antioxidative sulphated polygalactans from marine macroalgae as angiotensin-I converting enzyme inhibitors. Nat Prod Res 32(17):2100–2106

Makkar F, Chakraborty K (2018c) First report of dual cyclooxygenase-2 and 5-lipoxygenase inhibitory halogen derivatives from the thallus of intertidal seaweed *Kappaphycus alvarezii*. Med Chem Res 27:2331–2340

Makkar F, Chakraborty K (2018d) Highly oxygenated antioxidative 2*H*-chromen derivative from the red seaweed *Gracilaria opuntia* with pro-inflammatory cyclooxygenase and lipoxygenase inhibitory properties. Nat Prod Res 32(23):2756–2765

Makkar F, Chakraborty K (2018e) Novel furanyl derivatives from the red seaweed *Gracilaria opuntia* with pharmacological activities using different *in vitro* models. Med Chem Res 27:1245–1259

Makkar F, Chakraborty K (2018f) Previously undescribed antioxidative azocinyl morpholinone alkaloid from red seaweed *Gracilaria opuntia* with anti-cyclooxygenase and lipoxygenase properties. Nat Prod Res 32(10):1150–1160

Makkar F, Chakraborty K (2019) Highly oxygenated antioxidative 2*H*-chromen derivative from the red seaweed *Gracilaria opuntia* with pro-inflammatory cyclooxygenase and lipoxygenase inhibitory properties. Nat Prod Res 32(23):2756–2765

Manivannan K, Thirumaran G, Karthikai DG, Hemalatha A, Anantharaman P (2008) Biochemical composition of seaweeds from mandapam coastal regions along the southeast coast of India. Am Eurasian J Bot 1:32–37

da Matta CBB, De Souza ÉT, De Queiroz AC, De Lira DP, De Araújo MV, Cavalcante-Silva LHA, De Miranda GEC, de Araújo-Júnior JX, Barbosa-Filho JM, de Oliveira SBV, Alexandre-Moreira MS (2011) Antinociceptive and anti-inflammatory activity from algae of the genus *Caulerpa*. Mar Drugs 9(3):307

McHugh DJ (2003) A guide to the seaweed industry. FAO fisheries technical paper 441. Food and Agriculture Organization of the United Nations, Rome

Meenakshi S, Umayaparvathi S, Saravanan R, Manivasagam T, Balasubramanian T (2014) Hepatoprotective effect of fucoidan isolated from the seaweed *Turbinaria decurrens* in ethanol intoxicated rats. Int J Biol Macromol 67:367–372

Meenakshi S, Umayaparvathi S, Saravanan R, Manivasagam T, Balasubramanian T (2016) Neuroprotective effect of fucoidan from *Turbinaria decurrens* in MPTP intoxicated Parkinsonic mice. Int J Biol Macromol 86:425–433

Moreda-Piñeiro A, Peña-Vásquez E, Bermejo-Barrera P (2012) Significance of the presence of trace and ultratrace elements in seaweeds. In: Kim S-K (ed) Handbook of marine macroalgae: biotechnology and applied phycology. Wiley, New York, p 567

Morokutti-Kurz M, Fröba M, Graf P, Große M, Grassauer A, Auth J, Schubert U, Prieschl-Grassauer E (2021) Iota-carrageenan neutralizes SARS-CoV-2 and inhibits viral replication in vitro. PLoS One 16(2):e0237480

Murakami A, Ishida H, Kubo K, Furukawa I, Ikeda Y, Yonaha M, Aniya Y, Ohigashi H (2005) Suppressive effects of Okinawan food items on free radical generation from stimulated leukocytes and identification of some active constituents: implications for the prevention of inflammation-associated carcinogenesis. Asian Pac J Cancer Prev 6:437–448

Murray M, Dordevic A, Ryan L, Bonham M (2018) The impact of a single dose of a polyphenol-rich seaweed extract on postprandial glycaemic control in healthy adults: a randomised cross-over trial. Nutrients 10:270

Muthuraman B, Ranganathan R (2004) Biochemical studies on some green algae of Kanyakumari coast. Seaweed Res Utilisation 26(1&2):69–71

Myers SP, O'Connor J, Fitton JH, Brooks L, Rolfe M, Connellan P, Wohlmuth H, Cheras PA, Morris C (2010) A combined phase I and II open label study on the effects of a seaweed extract nutrient complex on osteoarthritis. Biologics 24(4):33–44. https://doi.org/10.2147/btt.s8354

NAAS (2003) Seaweed cultivation and utilization. Policy paper 22. National Academy of Agricultural Sciences, New Delhi

Nakashima H, Kido Y, Kobayashi N, Motoki Y, Neushul M, Yamamoto N (1987a) Purification and characterization of an avian myeloblastosis and human immunodeficiency virus reverse transcriptase inhibitor, sulfated polysaccharides extracted from sea algae. Antimicrob Agents Chemother 31:1524–1528

Nakashima H, Kido Y, Kobayashi N, Motoki Y, Neushul M, Yamamoto N (1987b) Antiretroviral activity in a marine algal: reverse transcriptase inhibition by an aqueous extract of *Schizymenia pacifica*. J Cancer Res Clin Oncol 113:413–416

Nelson SM, Gao YT, Nogueira LM, Shen MC, Wang B, Rashid A, Hsing AW, Koshiol J (2017) Diet and biliary tract cancer risk in Shanghai, China. PLoS One 12(3):e0173935

Nisha SA, Devi KP (2017) *Gelidiella acerosa* protects against Aβ 25-35-induced toxicity and memory impairment in Swiss Albino mice: an *in vivo* report. Pharm Biol 55(1):1423–1435

Norte M, Fernández J, Souto ML, Gavín J, García-Grávalos MD (1997) Thyrsenols A and B, two unusual polyether squalene derivatives. Tetrahedron 53(9):3173–3178

O'Keefe BR, Giomarelli B, Barnard DL, Shenoy SR, Chan PKS, McMahon JB, Palmer KE, Barnett BW, Meyerholz DK, Wohlford-Lenane CL, McCray PB Jr (2010) Broad-spectrum in vitro activity and in vivo efficacy of the antiviral protein griffithsin against emerging viruses of the family Coronaviridae. J Virol 84:2511–2521

Olivero-David R, Schultz-Moreira A, Vázquez-Velasco M, GonzálezTorres L, Bastida S, Benedí J, Isabel S-RM, José G-MM, Sánchez-Muniz FJ (2011) Effects of noriand wakame-enriched meats with or without supplementary cholesterol on arylesterase activity, lipaemia and lipoproteinaemia in growing Wistar rats. Br J Nutr 106:1476–1486

Olsson J, Toth GB, Albers E (2020) Biochemical composition of red, green and brown seaweeds on the Swedish west coast. J Appl Phycol 32:3305–3317

Oumaskour K, Boujaber N, Etahiri S, Assobhel O (2013) Anti-inflammatory and antimicrobial activities of twenty-three marine red algae from the coast of Sidi Bouzid (El Jadida-Morocco). Int J Pharm Pharm Sci 5:145–149

Paiva L, Lima E, Neto AI, Baptista J (2017) Angiotensin I-converting enzyme (ACE) inhibitory activity, antioxidant properties, phenolic content and amino acid profiles of *Fucus spiralis* L. protein hydrolysate fractions. Mar Drugs 15:311

Paradis M-E, Couture P, Lamarche B (2011) A randomised crossover placebo-controlled trial investigating the effect of brown seaweed (*Ascophyllum nodosum* and *Fucus vesiculosus*) on postchallenge plasma glucose and insulin levels in men and women. Appl Physiol Nutr Metab 36:913–919

Park Y, Lee J, Oh JH, Shin A, Kim J (2016) Dietary patterns and colorectal cancer risk in a Korean population: a case-control study. Medicine 95:e3759

Pati MP, Sharma SD, Lakshman N, Panda CR (2016) Uses of seaweed and its application to human welfare: a review. Int J Pharm Pharm Sci 8:12–20

Peñalver R, Lorenzo JM, Ros G, Amarowicz R, Pateiro M, Nieto G (2020) Seaweeds as a functional ingredient for a healthy diet. Mar Drugs 18(6):301

Pereira L, Critchley AT (2020) The COVID 19 novel coronavirus pandemic 2020: seaweeds to the rescue? Why does substantial, supporting research about the antiviral properties of seaweed polysaccharides seem to go unrecognized by the pharmaceutical community in these desperate times? J Appl Phycol 32:1875–1877

Pozharitskaya ON, Obluchinskaya ED, Shikov AN (2020) Mechanisms of bioactivities of fucoidan from the brown seaweed *Fucus vesiculosus* L. of the Barents Sea. Mar Drugs 18(5):275

Qi H, Zhang Q, Zhao T, Hu R, Zhang K, Li Z (2006) *In vitro* antioxidant activity of acetylated and benzoylated derivatives of polysaccharide extracted from *Ulva pertusa* (Chlorophyta). Bioorg Med Chem Lett 16:2441–2445

Qin Y (2018a) Applications of bioactive seaweed substances in functional food products. In: Qin Y (ed) Bioactive seaweeds for food applications: natural ingredients for healthy diets. Elsevier, Amsterdam, pp 111–134

Qin Y (2018b) Production of seaweed-derived food hydrocolloids. In: Qin Y (ed) Bioactive seaweeds for food applications. Academic, New York, pp 53–69

Rajasulochana P, Krishnamoorthy P, Dhamotharan R (2012) Biochemical investigation on red algae family of *Kappahycus* sp. J Chem Pharm Res 4:4637–4631

Rasyid A, Handayani T (2019) Evaluation of the biochemical composition of tropical red seaweeds *Galaxaura rugosa* and *Gelidiella acerosa* from Ujung Genteng waters, Indonesia. Aquacult Aquarium Conserv Legislation 12(2):601–609

Rosemary T, Arulkumar A, Paramasivam S, Mondragon-Portocarrero A, Miranda JM (2019) Biochemical, micronutrient and physicochemical properties of the dried red seaweeds *Gracilaria edulis* and *Gracilaria corticata*. Molecules 24(12):2225

Roy M-C, Anguenot R, Fillion C, Beaulieu M, Bérubé J, Richard D (2011) Effect of a commercially-available algal phlorotannins extract on digestive enzymes and carbohydrate absorption *in vivo*. Food Res Int 44:3026–3029

Sakthivel R, Muniasamy S, Archunan G, Devi KP (2016) *Gracilaria edulis* exhibit antiproliferative activity against human lung adenocarcinoma cell line A549 without causing adverse toxic effect *in vitro* and *in vivo*. Food Funct 7:1155–1165

Sanjeewa KKA, Lee W, Jeon Y (2018) Nutrients and bioactive potentials of edible green and red seaweed in Korea. Fish Aquat Sci 21:19

Seedevi P, Moovendhan M, Viramani S, Shanmugam A (2017) Bioactive potential and structural characterization of sulfated polysaccharide from seaweed (*Gracilaria corticata*). Carbohydr Polym 155:516–524

Shama A, Joyce SG, Mari FD, Ranga Rao A, Ravishankar GA, Hudaa N (2019) Macroalgae and microalgae: novel sources of functional food and feed. In: Ravishankar GA, Ranga Rao A (eds) Handbook of algal technologies and phytochemicals: volume I: food, health and nutraceutical applications. CRC Press, Boca Raton, pp 207–219

Shannon E, Ghannam NA (2019) Seaweeds as nutraceuticals for health and nutrition. Phycologia 58(5):563–577

Sharifuddin Y, Chin Y-X, Lim P-E, Phang S-M (2015) Potential bioactive compounds from seaweed for diabetes management. Mar Drugs 13:5447–5491

Sharma BR, Kim HJ, Rhyu DY (2015) *Caulerpa lentillifera* extract ameliorates insulin resistance and regulates glucose metabolism in C57BL/KsJ-db/db mice via PI3K/AKT signaling pathway in myocytes. J Transl Med 13:62

Sheeja L, Lakshmi D, Bharadwaj S, Parveen KS (2016) Anticancer activity of phytol purified from *Gracilaria edulis* against human breast cancer cell line (MCF-7). Int J Curr Sci 19:36–46

Shu M-H, Appleton D, Zandi K, Abubakar S (2013) Anti-inflammatory, gastroprotective and anti-ulcerogenic effects of red algae *Gracilaria changii* (Gracilariales, Rhodophyta) extract. BMC Complement Altern Med 13:61

Sørensen LE, Jeppesen PB, Christiansen CB, Hermansen K, Gregersen S (2019) Nordic seaweed and diabetes prevention: exploratory studies in KK-Ay mice. Nutrients 11:1435

Suzuki T, Takeda S, Suzuki M, Kurosawa E, Kato A, Imanaka Y (1987) Constituents of marine plants. Part 67. Cytotoxic squalene-derived polyethers from the marine red alga *Laurencia obtusa* (Hudson) Lamouroux. Chem Lett:361–364

Syad AN, Shunmugiah KP, Kasi PD (2013) Antioxidant and anti-cholinesterase activity of *Sargassum wightii*. Pharm Biol 51(11):1401–1410

Syed YY (2020) Sodium Oligomannate: first approval. Drugs 80:441–444

Szekalska M, Puciłowska A, Szymańska E, Ciosek P, Winnicka K (2016) Alginate: current use and future perspectives in pharmaceutical and biomedical applications. Int J Polym Sci Article ID 7697031, 17 pages

Tanemura Y, Yamanaka-Okumura H, Sakuma M, Nii Y, Taketani Y, Takeda E (2014) Effects of the intake of *Undaria pinnatifida* (wakame) and its sporophylls (mekabu) on postprandial glucose and insulin metabolism. J Med Investig 61:291–297

Tease J (2005) Dietary brown seaweeds and human health effects. In: Critchley AT, Masao O, Danilo M (eds) Seaweed resources. Expert Centre for Taxonomic Identification, Amsterdam, pp 124–137

Torres MD, Flórez-Fernández N, Domínguez H (2019) Integral utilization of red seaweed for bioactive production. Mar Drugs 17(6):314

Tseng C-K, Lin C-K, Chang H-W, Wu Y-H, Yen F-L, Chang F-R, Chen W-C, Yeh C-C, Lee J-C (2014) Aqueous extract of *Gracilaria tenuistipitata* suppresses LPSinduced NF-κB and MAPK activation in RAW 264.7 and rat peritoneal macrophages and exerts hepatoprotective effects on carbon tetrachloride-treated rat. PLoS One 9:e86557

Tsuji S (1983) Cooking, Japanese. In: Kodansha encyclopedia of Japan, vol 2. Kodansha International, Tokyo, pp 20–25

Tsuji S, Ishige N (1983) Food and eating. In: Kodansha encyclopedia of Japan, vol 2. Kodansha International, Tokyo, pp 304–307

USFDA (2017) Code of federal regulations title 21. Subchapter B—food for human consumption. PART 184 direct food substances affirmed as generally recognized as safe. Subpart B listing of specific substances affirmed as GRAS. Sec 184.1120 Brown algae

Vanderlei ESO, de Araújo IWF, Quinderé ALG, Fontes BP, Eloy YRG, Rodrigues JAG, Silva AARE, Chaves HV, Jorge RJB, de Menezes DB, Evangelista JSAM, Bezerra MM, Benevides NMB (2011) The involvement of the HO-1 pathway in the anti-inflammatory action of a sulfated polysaccharide isolated from the red seaweed *Gracilaria birdiae*. Inflamm Res 60:1121–1130

Winter JM, Chiou G, Bothwell IR, Xu W, Garg NK, Luo M, Tang Y (2013) Expanding the structural diversity of polyketides by exploring the cofactor tolerance of an inline methyltransferase domain. Org Lett 15(14):3774–3777

Yamazaki D, Hitomi H, Nishiyama A (2018) Hypertension with diabetes mellitus complications. Hypertens Res 41:147–156

Zheng J, Chen Y, Yao F, Chen W, Shi G (2012) Chemical composition and antioxidant/antimicrobial activities in supercritical carbon dioxide fluid extract of *Gloiopeltis tenax*. Mar Drugs 10:2634–2647

Chapter 21
Applications of Seaweed Derived Polymeric Fibrous Materials

Yimin Qin

Abbreviations

EGF	Epidermal growth factor
FGF	Fibroblast growth factor
G	Guluronic acid
GM-CSF	Granulocyte-macrophage colony-stimulating factor
IGF	Insulin-like growth factor
M	Mannuronic acid
PDGF	Platelet derived growth factor
TGF	Transforming growth factor
TNFα	tumor necrosis factor-α

1 Introduction

Seaweeds are an important source of biopolymers such as alginate, carrageen and agar (Bixler and Porse 2011; Das 2015; FAO 2016). Whilst these natural polymers are traditionally used as food hydrocolloids, their hydrophilic and bioactive characteristics are increasingly explored in other health related fields such as medical and hygiene industries where they are made into hydrogels, foams, films, fibers, fabrics and many other forms of novel materials capable of delivering heath enhancing benefits. Fibers are a particularly important field of applications for alginate and other seaweed derived polymers, which have been used as raw materials for the manufacture of modern wound management and cosmetic products. In these

Y. Qin (✉)
State Key Laboratory of Bioactive Seaweed Substances, Qingdao, China

Jiaxing College, Jiaxing, China

A. Ranga Rao, G. A. Ravishankar (eds.), *Sustainable Global Resources of Seaweeds Volume 2*, https://doi.org/10.1007/978-3-030-92174-3_21

applications, the biocompatibility, hydrophilicity, absorbency and other important material properties are advantageous for seaweed derived polymers compared to other natural and synthetic polymers (Hu et al. 2012; Ioannou and Roussis 2009; Qin 2016).

In recent years, seaweed derived fibrous materials have been widely used in wound healing, drug delivery, and tissue engineering applications, since fibrous materials and hydrogels possess structural similarity to extracellular matrices in tissues and can be manipulated to play several critical roles in wound repair and tissue engineering (Qin 2005, 2006, 2008b). Because of the unique ion exchange between calcium alginate fibers and sodium ions in the body fluid, nonwoven fabrics made from alginate fibers can form a soft hydrogel when in contact with body fluid, a property highly valuable in wound dressings, face masks, absorbent pads and other medical and hygiene textile materials (Qin 2005, 2006, 2008a). In addition, alginic acid can combine with many types of metal ions to form fibrous materials containing a high concentration of metal ions, enabling the fibers to have flame retardant and magnetic wave shielding properties (Wang et al. 2009). Carrageenan and agar based fibers are also showing potential as high absorbency materials (Yin et al. 2010; Shi et al. 2013).

This chapter summarizes the past and present development of alginate and other seaweed derived fibers and related products, and presents a critical analysis of the future research in seaweed derived fibrous materials.

2 Historical Development of Alginate Fibers

Alginate was discovered by Scottish chemist E C C Stanford in 1881 (Stanford 1881, 1883) later commercial production started in 1929 by Kelco in the USA. The first detailed study on calcium alginate fibers was published by Speakman and Chamberlain (1944) long before the chemical structure of alginate was fully understood, in particular before the copolymer nature of guluronic acid and mannuronic acid was revealed by a number of scientists in the 1960s and 1970s (Haug and Larsen 1962; Haug et al. 1966, 1967, 1974; Penman and Sanderson 1972; Grasdalen et al. 1979).

Although alginate fibers can be easily made from sodium alginate by extruding its aqueous solution into an aqueous bath containing calcium ions, they have limited use as a textile material since these fibers are relatively expensive, and tend to dissolve in the alkali conditions of many textile processes such as bleaching, dyeing and finishing. In the early stages of development, alginate fibers were used principally in the production of water soluble yarns that would dissolve in a scouring process. These yarns were used as a support during the manufacture of fine lace, or as draw threads in the production of hosiery. Fabrics from alginate fibers were once produced commercially for their fire-resistant property, because of the high content of metal ions in the fiber (Chamberlain et al. 1945, 1949). Alginate fibers were also used for the manufacture of bags used for the transportation of soiled hospital linen that were designed to dissolve in the wash. However, by the 1970s, they were replaced for these applications by cheaper synthetic fibers.

The first person in modern times to recognize the potential value of alginate fibers in surgery and wound management was George Blaine, a major in the Royal Army Medical Corps. He showed them to be absorbable in tissue, sterilizable by heat, and compatible with penicillin (Blaine 1946). He also described how he had used alginate films clotted *in situ* for the treatment of wounds and burns in troop ship hospitals in the Far East and described the use of alginate, sometimes in combination with plasma as an alginate-plasma film, as ‘puncture patches’ over **septal defects.**

During a subsequent assessment of the use of alginate as haemostats and wound dressings, Blaine reported their apparent lack of toxicity following a series of animal studies in which fibers were implanted into animal tissues, and gels made from alginate were used to treat experimentally produced burns (Blaine 1947). Successful use of alginate derived materials in aural surgery and neurosurgery was reported by Passe and Blaine (Passe and Blaine 1948) and Oliver and Blaine (1950) respectively. Bray et al. (1948) described the results of a three-month trial into the use of alginate in the casualty department of Croydon Hospital where alginates in the form of films, wool, gauzes, and clots formed *in situ* by mixing sterile solutions of calcium chloride and sodium alginate, were applied to a wide range of wounds, including burns, lacerations, ulcers and amputations. In all cases, healing was rapid and uneventful. By the late 1940s and early 1950s, alginates were being used in some 70 hospitals in the UK over a range of surgical specialties (Blaine 1951).

3 Development of Modern Alginate Wound Dressings

The revival of alginate fibers began with a new theory in wound management. Although it has a long history, the standard practice in wound management remained fairly static until Winter published his work in 1962 on acute superficial wounds in the domestic pig (Winter 1962), which showed that wounds covered with a film dressing maintained a moist environment and healed faster than those left to dry out, heralding the advent of modern wound dressings based on the principle of moist healing (Harding et al. 2002; Russell 1999; Qin 2016). In the early 1980s, as wound dressings based on the moist healing principle were expanding, alginate fibers and nonwoven fabrics were found to have unique gel forming characteristics whereby on contact with wound exudates, sodium ions in the exudates can exchange with calcium ions in the fibers, and as the ion-exchange process proceeds, the fibers absorb more and more water to form a fibrous gel. For the alginate wound dressings, as water enters the fiber structure, the entire textile structure is transformed into a sheet of moist gel, thus providing an ideal moist healing environment for the underlining wound. Many clinical trials have shown that alginate wound dressings not only have the high absorption capacities, but also possess the ability to promote wound healing (Fraser and Gilchrist 1983; Groves and Lawrence 1986; Attwood

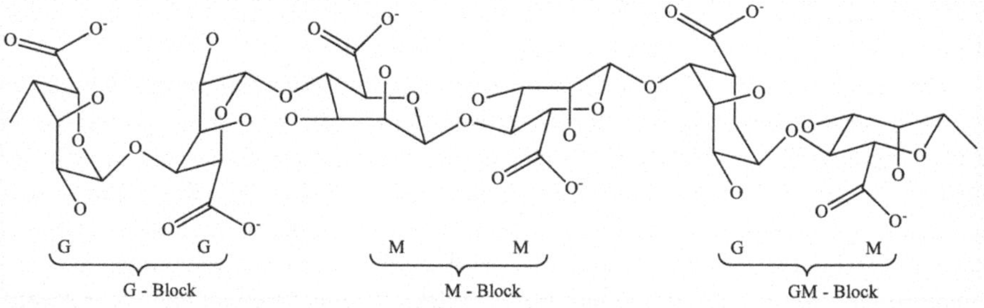

Fig. 21.1 Chemical structures of β-D-mannuronic acid (M) and α-L-guluronic acid (G) and their distribution in the polymeric chain

1989; Jeter and Tintle 1990; McMullen 1991; Young 1993; Bettinger et al. 1995; Sayag et al. 1996; Doyle et al. 1996; Williams 1999).

As can be seen in Fig. 21.1, alginate is a linear polymeric acid composed of 1, 4-linked β-D-mannuronic acid (M) and α-L-guluronic acid (G) residues. During the fiber making process, sodium alginate is dissolved in water to form a viscous solution, which is then extruded through spinneret holes into an aqueous calcium chloride bath, where upon the ion exchange between calcium ions in the bath and sodium ions in the as-spun filament, a swollen calcium alginate filament is formed. Calcium alginate fibers can be produced upon further stretching, washing and drying (Qin 2008a).

Figure 21.2 shows the structural changes during the production and application of calcium alginate fibers as a wound dressing material. During the production process, sodium alginate is converted into calcium alginate when the sodium alginate filament emerging from a spinneret is coagulated by calcium ions in the calcium chloride solution. On contact with wound fluid, calcium ions in the fiber exchange with sodium ions, slowly converting the fiber into sodium alginate. Since sodium alginate is water soluble, water is drawn into the fiber, turning it into a fibrous gel (Qin 2004, 2008a,b, 2006).

Figure 21.3 shows the photomicrographs of a piece of calcium alginate fiber when wet in normal saline solution. The fiber absorbed a large amount of water during the wetting process, effectively turning itself into a piece of hydrogel. In clinical applications, when alginate fiber based wound dressings are placed in contact with wound exudate, the exchange of sodium and calcium ions gradually transforms calcium alginate fiber into sodium alginate fiber, resulting in the formation of a fibrous gel in situ on the wound surface, which helps maintain a moist environment highly beneficial to wound healing.

Figure 21.1 shows the use of nonwoven alginate dressing on a leg ulcer wound. During applications, wound exudate is absorbed into the fibers and upon the subsequent swelling, the capillary structure in the nonwoven fabric is closed, which inhibits the lateral spreading of wound exudate. This unique property is known as "gel blocking" in that when highly exuding wounds such as leg ulcers, pressure sores, etc., are covered by alginate dressings, the exudates would go upwards through the dressings, whilst maintaining a relatively dry condition for the healthy skin surrounding the wound site, as clearly illustrated in Fig. 21.1.

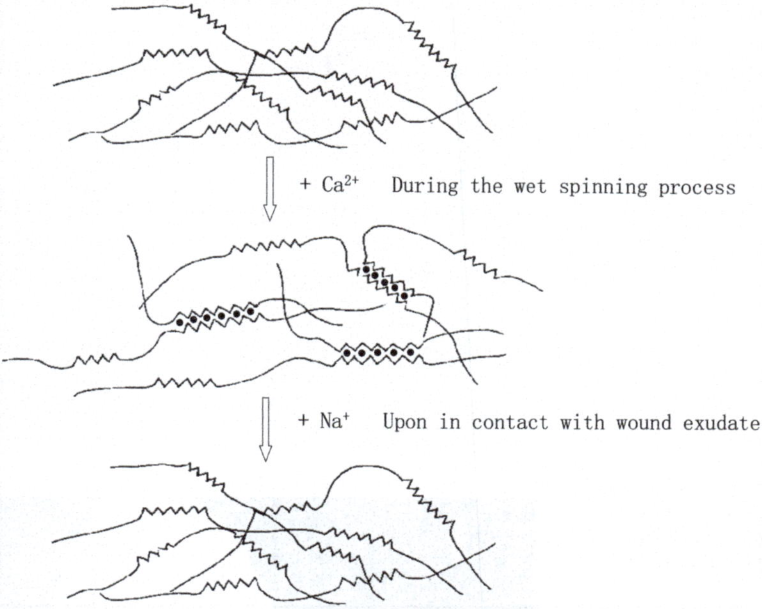

Fig. 21.2 Structural changes during the production and application of calcium alginate fibers as a wound dressing material

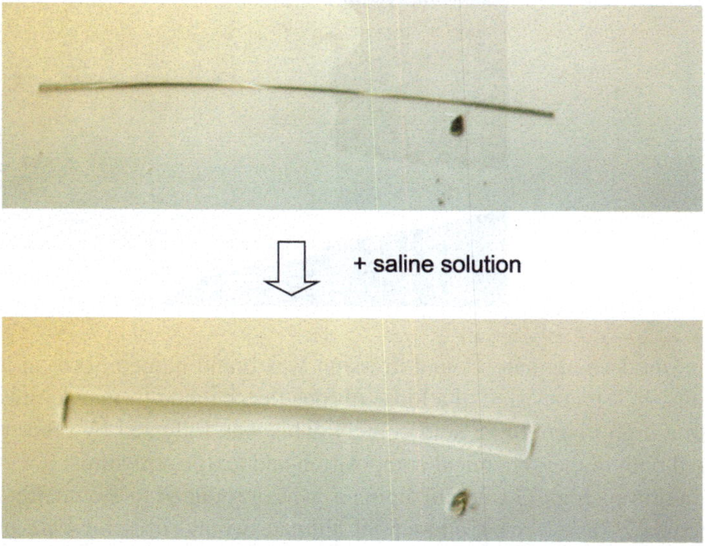

Fig. 21.3 Photomicrographs of calcium alginate fiber wet in saline solution, ×200

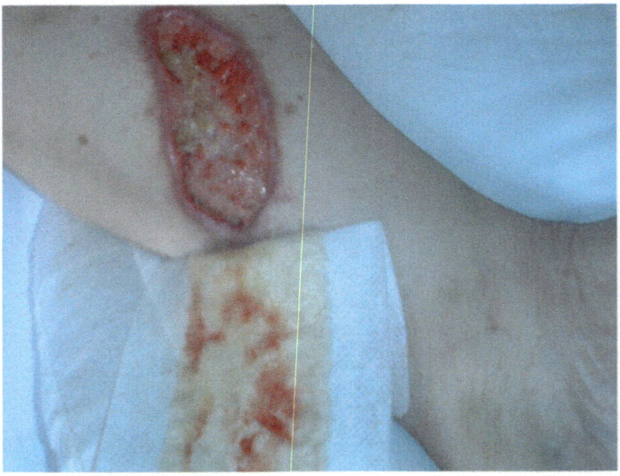

Fig. 21.4 A leg ulcer wound covered by alginate wound dressing

Fig. 21.5 Face mask made from hydroentangled alginate nonwoven fabric

The first modern alginate wound dressing was brand named Sorbsan, and was launched in 1983. It consisted of a loose fibrous fleece made from calcium alginate fibers with a high mannuronic acid content. This was followed by other products that differed both in their chemical composition and textile structures. For example, Kaltostat, a fibrous high-G calcium alginate, was introduced to the market place in 1986 (Thomas 2000a, b, c). Commercial alginate wound dressings are produced either as flat sheets, which are used to cover superficial wounds, or as cavity fillers, usually in the form of ribbon or rope. The flat dressings are normally made in a nonwoven fabric process in which the fibers are carded to form a web that is then cross-lapped to form a felt. In some products, the felt is then needled or entangled by means of high-pressure water jets to give the dressing a coherent structure (Qin and Gilding 1996).

Table 21.1 summarizes the main types of alginate wound dressings and outlines their key structural features (Qin 2006).

In addition to their high absorbency and gel forming properties, alginate wound dressings can also stimulate macrophage activities with a favorable effect on wound healing when the interaction between alginate molecule and macrophage cells play a key role in many physiological and pathophysiological processes by synthesizing various biologically active molecules called cytokines, for example tumor necrosis factor-α (TNFα), which is an important cytokine with cytotoxic properties against both tumor cells and normal cells infected with intracellular pathogens. Otterlei et al. (1991) compared the ability of different alginates to stimulate macrophage cells to produce TNF-α, interleukin-1 and interleukin-6. They reported that high M alginates were approximately 10 times more potent in inducing cytokine production than high G alginates and therefore proposed that mannuronic acid residues are the active cytokine-inducers in alginates. Other authors have also produced evidence to suggest that it is MM blocks rather than GG that is responsible for cytokine stimulation and anti-tumor activity. It is shown that a treatment of high M alginate with C-5 epimerase, which converts β-D-mannuronic acid into α-L-guluronic acid, results in a loss of TNF-inducing ability (Skjak-Braek and Espevik 1996). In another study, it was found that a low concentration of an extract of alginate dressing stimulated human fibroblasts on extended contact (Bettinger et al. 1995), indicating that alginate wound dressings can promote wound healing. Many clinical studies also described the ability of alginate dressings to promote the healing process for various types of wounds (Fraser and Gilchrist 1983; Gilchrist and Martin 1983; Thomas 1985; Odugbesan and Barnett 1987). Table 21.2 summarizes some of the clinical findings for alginate wound dressings.

Table 21.1 Main types of alginate wound dressings

Product	Product description	Supplier
Algisite™	Nonwoven high M calcium alginate	Smith & Nephew
AlgitoCare™	Nonwoven high M calcium alginate	Qingdao Brightmoon
Algosteril™	Nonwoven high G calcium alginate	Beiersdorf
Comfeel SeaSorb™	Freeze dried calcium alginate	Coloplast
Curasorb™	Nonwoven high G calcium alginate	Kendall
Kaltogel™	Nonwoven high M calcium sodium alginate	ConvaTec
Kaltostat™	Nonwoven high G calcium sodium alginate	ConvaTec
Melgisorb™	Nonwoven high M calcium alginate	Molnlycke
Sorbsan™	Nonwoven high M calcium alginate	Maesk medical
Tegagel™	Nonwoven mid M calcium alginate	3M health care
Tegagen HG™	Nonwoven high M calcium	3M health care
Urgosorb™	Nonwoven high G calcium alginate containing CMC	Urgo

Table 21.2 Clinical findings for alginate wound dressings

Wounds	Key findings	Authors
Leg ulcers	High absorbency and long duration on wound, higher healing rate than traditional paraffin gauze	Thomas and Tucker (1989), Smith and Lewis (1990), Moffatt et al. (1992), Scurr et al. (1994), Armstrong and Ruckley (1997)
Pressure ulcers	Faster healing compared to traditional dextranomer paste, good filling performance	Chapuis and Dollfus (1990), Sayag et al. (1996)
Diabetic foot ulcers	High absorbency, low adherence	Fraser and Gilchrist (1983), Bradshaw (1989), Smith (1992)
Burns and donor sites	Good haemostatic and absorbent properties compared to standard gauze pad, much lower blood loss, faster healing than paraffin gauze, less pain upon dressing change	Groves and Lawrence (1986), Attwood (1989), Lawrence and Blake (1991), Basse et al. (1992), Rives et al. (1997), Ye and Chen (2001), Higgins et al. (2012)
Surgical bleeding wounds	Less bleeding and reduced pain than traditional gauze dressing	Gupta et al. (1991), Dawson et al. (1992), Han et al. (2002), Xie et al. (2003), Fang and Lan (2004)

4 Future Development of Novel Alginate Based Fibrous Materials

As a natural polymer of marine origin, alginate is biocompatible, non-toxic and hydrophilic, making it an ideal material for many medical and hygiene products. The unique polymeric acid structure enables a novel ion-exchange and gelling characteristics that have already been widely exploited in functional wound management products. Looking into the future, alginate fibers and fibrous materials have a number of high valued applications for further development. Some of these applications are outlined below.

4.1 Absorption of Heavy Metal Ions

Alginic acid can bind metal ions to form salt, with the ion-exchange coefficient in the order of $Pb^{2+}>Cu^{2+}>Cd^{2+}>Ba^{2+}>Sr^{2+}>Ca^{2+}>Co^{2+}=Ni^{2+}=Zn^{2+}>Mn^{2+}$ (Haug and Smidsrod 1965). When calcium alginate fibers are placed in contact with aqueous solutions containing Pb^{2+}, Cu^{2+}, Cd^{2+}, Ba^{2+} and other heavy metal ions, these metal ions are quickly absorbed onto the fibers, which can then be removed with a treatment in acidic solution, making the fibers as re-usable ion-exchange material. This unique property can be used in water treatment and in the purification of bio-extracts with heavy metal contamination, for example in the after-treatment of traditional herb medicine.

4.2 Composite Materials

In addition to nonwoven wound dressings, alginate fibers can be used to make a number of novel composite materials for wound management and other biomedical applications. For example, Qin and Gilding (2001, Gilding and Qin 2003) made fiber reinforced alginate gel by dispersing chopped calcium alginate fibers in a solution of sodium alginate. During the drying process, calcium ions released from the alginate fibers cross-link the sodium alginate to form a gel network, resulting in a fiber reinforced sheet of pure alginate. When in contact with water, the sodium alginate component in the gel absorbs water into the sheet and a hydrogel is formed. If the mixture of calcium alginate fiber and sodium alginate solution is freeze dried, a porous fiber reinforced foam can be made, which is capable of absorbing more than 20 g of normal saline for a gram of the foam material (Qin and Gilding 2001; Gilding and Qin 2003; Renn et al. 2003). Calcium alginate fibers can also be mixed with sodium alginate solution to form amorphous hydrogels that can be used to donate water to dry wound (Thompson 1996; Gilding and Qin 2001).

4.3 Carrier for Controlled Release of Active Agents

In recent years, advances in micro-encapsulation technologies and novel fiber making processes have made textile fibers an attractive alternative for the encapsulation and delivery of therapeutic drugs. For example, microfiber-based medical textiles such as sutures and wound dressings can be made to contain a drug or an antimicrobial agent. Nanofibrous drug delivery systems have also been developed which have several advantages due to their large surface area to volume ratio, high porosity and flexibility, and the electrospinning technique have been modified to include multiple needles, needleless and coaxial forms of electrospinning, which are suitable for the production of drug loaded fibrous materials.

As a water soluble polymer, alginate can be mixed with pharmaceutical components in the aqueous phase and spun into fibers through electrospinning. In addition, drugs can also be attached to the fiber surface by utilizing the opposite charges present on the fiber surfaces and the drug molecules to produce ion complexes. In such complexes, drug elution is controlled by the ability of the fiber surface to preferentially exchange with counter-ions, where drug molecules are released in exchange for ions in the physiological system. A variety of bioactive molecules, such as dexamethasone, bovine serum albumin, growth factor and avidin can be mixed with alginate solution to produce drug loaded fibers.

A variety of growth factors such as epidermal growth factor(EGF), platelet derived growth factor(PDGF), fibroblast growth factor(FGF), transforming growth factor(TGF-b1), insulin-like growth factor(IGF-1), human growth hormone and granulocyte-macrophage colony-stimulating factor(GM-CSF) can be loaded into alginate fibers and wound dressing to offer enhanced wound healing functions (Boateng et al. 2008; Joshi et al. 2014).

4.4 Protective Materials against Radiation

The human body is exposed to radiation both at work and in daily life. For example, ultraviolet rays in the solar spectrum can influence human being by causing a range of effects from simple tanning to highly malignant skin cancers. Sunscreen lotions, clothing and shade structures have been developed to provide protection from the deleterious effects of ultraviolet radiations. Since alginate fibers are rich in metal ions, they can be made into textile fabrics that can protect the body against harmful radiation. In particular, barium alginate fibers can be made into nonwoven fabrics and converted into appropriate coveralls, aprons, smocks, etc. These products can be used to offer protection against certain levels of radioactivity such as for operators working with X-ray equipment and in the shielding of electronic equipment.,

4.5 Face Mask Materials

Cosmetotextiles are textile material with cosmetic properties, although these types of textiles can also harbor other functions and ingredients, such as medical properties, mosquito repellents, odor reducers, antimicrobials or UV-protection agents. Cosmetic textile is an industry that has grown along with consumer interest in wellness and well-being. It involves the use of fiber and textile materials to deliver a wide range of microencapsulated ingredients such as aloe vera, vitamin E, retinol, and caffeine that can offer moisturizing, firming, or slimming benefits. The next generation of cosmetic textile products could potentially go beyond beauty, by utilizing innovative new methods to deliver medical, anti-aging, and stress-relieving benefits through apparel textiles and other products. In this respect, alginate fibers and nonwoven fabrics are highly hydrophilic and biocompatible, which are ideal materials for the production of face masks. In addition, the fibers and fabrics can be used to carry various bioactive ingredients to achieve sustained release to the skin (Wijesinghe and Jeon 2011). Figure 21.1 shows face mask made from hydroentangled alginate nonwoven fabric.

4.6 Hygiene Products

Over the past 50 years, absorbent hygiene products such as baby diapers, adult incontinence products, feminine protection pads and personal care wipes have all become essential features of modern day life. Their increased use has been accompanied by dramatic improvements in skin health and hygiene, particularly in the incidence of diaper dermatitis. Alginate fibers can be made into hydroentangled nonwoven fabric and used as the contact layer. Their gel forming characteristics and antimicrobial properties can help maintain a fresh and dry interface between the body and the absorbent materials. In addition, they can help reduce leakage and prevent contamination and the transmission of infectious diseases.

4.7 Carrageenan and Agar Fibers

Both carrageenan and agar are marine biopolymers extracted from red seaweeds. Similar to alginate, they have excellent gel forming properties and can be made into fibers through wet spinning (Yin et al. 2010). In order to increase the mechanical strength of carrageenan fibers, it is necessary to cross-link the carrageenan macromolecules with cross-linking agents, for example with aqueous $BaCl_2$ solution. Alternatively, the hydroxyl groups in the carrageenan macromolecules can react with epichlorohydrin, succinic anhydride, maleic anhydride and other conventional cross-linking agents to form stable fiber structure. It was found that in order to obtain high quality carrageenan fibers, the optimal cross-linking reaction condition is at 90 °C with epichlorohydrin concentration at 6.25% and pH at 10.

In the case of agar fiber production, since agar is soluble in boiling water and gels when the solution temperature drops below 50 °C, it is not convenient to make fibers using water as the solvent. It was found that agar can easily dissolve in DMSO at 50–60 °C to form a stable solution, which can be extruded into a coagulation bath composed of 50% ethanol (v/v) (Shi et al. 2013).

5 Conclusions and Future Perspectives

Seaweed derived alginate, carrageen and agar are natural biopolymers that can be made into fibers and fibrous materials through wet spinning and related textile processing. These fibers are highly biocompatible and hydrophilic and are useful in the production of medical and hygiene products such as wound dressings, personal care products, female hygiene materials and other medical textile products. In particular, alginate fibers are now widely used because of their novel performance characteristics such as gelling, high absorbency, gel blocking, wound healing promotion and other novel properties.

Acknowledgments This chapter was sponsored by National Key R&D Program of China, Project No. 2018YFC1105600.

References

Armstrong SH, Ruckley CV (1997) Use of a fibrous dressing in exuding leg ulcers. J Wound Care 6(7):322–324

Attwood AI (1989) Calcium alginate dressing accelerate split graft donor site healing. Br J Plast Surg 42:373–379

Basse P, Siim E, Lohmann M (1992) Treatment of donor sites: calcium alginate versus paraffin gauze. Acta Chir Plast 34(2):92–98

Bettinger D, Gore D, Humphries Y (1995) Evaluation of calcium alginate for skin graft donor sites. J Burn Care Rehabil 16(1):59–61

Bixler HJ, Porse H (2011) A decade of change in the seaweed hydrocolloids industry. J Appl Phycol 23:321–335
Blaine G (1946) The use of plastics in surgery. Lancet 1946:525–528
Blaine G (1947) Experimental observations on absorbable alginate products in surgery. Ann Surg 125(1):102–114
Blaine G (1951) A comparative evaluation of absorbable haemostatics. Postgrad Med J 27:613–620
Boateng JS, Matthews KH, Stevens HNE (2008) Wound healing dressings and drug delivery systems: a review. J Pharm Sci 97:2892–2923
Bradshaw T (1989) The use of Kaltostat in the treatment of ulceration in the diabetic foot. Chiropodist 9:204–207
Bray C, Blaine G, Hudson P (1948) New treatment for burns, wounds and haemorrhage. Nurs Mirror 86:239–242
Chamberlain NH, Johnson A, Speakman JB (1945) Some properties of alginate rayons. J Soc Dye Colour 61(1):13–20
Chamberlain NH, Lucas F, Speakman JB (1949) The action of light on calcium alginate rayon. J Soc Dye Colour 65(12):682–692
Chapuis A, Dollfus P (1990) The use of calcium alginate dressings in the management of decubitus ulcers in patients with spinal cord lesions. Paraplegia 28(4):269–271
Das D (2015) Algal biorefinery: an integrated approach. Springer, New York, pp1–34
Dawson C, Armstrong MW, Fulford SC (1992) Use of calcium alginate to pack abscess cavities. J Roy Coll Surg Edinb 37(3):177–179
Doyle JW, Roth TP, Smith RM (1996) Effects of calcium alginate on cellular wound healing processes modeled in vitro. J Biomed Mater Res 32(4):561–568
Fang M, Lan L (2004) Observation of the filling of nose by calcium alginate dressing. Nurs Res 18(3):517–518
FAO (2016) Fishery and aquaculture statistics 2014. FAO, Rome
Fraser R, Gilchrist T (1983) Sorbsan calcium alginate fibre dressings in footcare. Biomaterials 4:222–224
Gilchrist T, Martin AM (1983) Wound treatment with Sorbsan–an alginate fibre dressing. Biomaterials 4:317–320
Gilding DK, Qin Y (2001) Wound treatment composition. USA patent 2001/6,258,995
Gilding DK, Qin Y (2003) Sheet hydrogel. USA Patent 2003/6,534,083
Grasdalen H, Larsen B, Smidsrod O (1979) A PMR study of the composition and sequence of uronate residues in alginate. Carbohydr Res 68:23–31
Groves AR, Lawrence JC (1986) Alginate dressing as a donor site haemostat. Ann R Coll Surg Engl 68:27–28
Gupta R, Foster ME, Miller E (1991) Calcium alginate in the management of acute surgical wounds and abscesses. J Tiss Viab 1(4):115–116
Han F, Liang F, Wang Q (2002) Application of calcium alginate rope in nose operation. Shandong J Biomed Eng 21:47–48
Harding KG, Morris HL, Patel GK (2002) Science, medicine and the future: healing chronic wounds. Br Med J 324:160–163
Haug A, Larsen B (1962) Quantitative determination of the uronic acid composition of alginates. Acta Chem Scand 16:1908–1918
Haug A, Smidsrod O (1965) Effect of divalent ions on solution property. Acta Chem Scand 19:341–351
Haug A, Larsen B, Smidsrod O (1966) A study of the constitution of alginic acid by partial acid hydrolysis. Acta Chem Scand 20:183–190
Haug A, Larsen B, Smidsrod O (1967) Studies on the sequence of uronic acid residues in alginic acid. Acta Chem Scand 21:691–704
Haug A, Larsen B, Smidsrod O (1974) Uronic acid sequence in alginate from different sources. Carbohydr Res 32:217–225

Higgins L, Wasiak J, Spinks A, Cleland H (2012) Split-thickness skin graft donor site management: a randomized controlled trial comparing polyurethane with calcium alginate dressings. Int Wound J 9(2):126–131

Hu Y, Chen J, Hu G (2012) Statistical research on the bioactivity of new marine natural products discovered during the 28 years from 1985 to 2012. Mar Drugs 13(1):202–221

Ioannou E, Roussis V (2009) Natural products from seaweeds. In: Osbourn AE, Lanzotti V (eds.) Plant-derived natural products. Springer Science+Business Media, Cham, pp51–81

Jeter KF, Tintle TE (1990) Early experience with a calcium alginate dressing. Ostomy and Wound Management 28:75–81

Joshi M, Butola BS, Saha K (2014) Advances in topical drug delivery system: micro to nanofibrous structures. J Nanosci Nanotechnol 14(1):853–867

Lawrence JE, Blake GB (1991) A comparison of calcium alginate and scarlet red dressings in the healing of split thickness skin donor sites. Br J Plast Surg 44(4):247–249

McMullen D (1991) Clinical experience with a calcium alginate dressing. Dermatol Nurs 3(4):216–219

Moffatt CJ, Oldroyd MI, Franks PJ (1992) Assessing a calcium alginate dressing for venous ulcer of the leg. J Wound Care 1(4):22–24

Odugbesan O, Barnett AH (1987) Use of a seaweed-based dressing in management of leg ulcers in diabetics: a case report. Pract Diabet 4:46–47

Oliver LC, Blaine G (1950) Haemostasis with absorbable alginates in neurosurgical practice. Br J Surg 37:307–310

Otterlei M, Ostgaard K, Skjak-Braek G (1991) Induction of cytokine production from human monocytes stimulated with alginate. J Immunother 10:286–291

Passe ERG, Blaine G (1948) Alginates in endaural wound dressing. Lancet 1948:651

Penman A, Sanderson GR (1972) A method for the determination of uronic acid sequence in alginates. Carbohydr Res 25:273–282

Qin Y (2004) Gel swelling properties of alginate fibers. J Appl Polym Sci 91(3):1641–1645

Qin Y (2005) The ion exchange properties of alginate fibers. Text Res J 75(2):165–168

Qin Y (2006) The characterization of alginate wound dressings with different fiber and textile structures. J Appl Polym Sci 100(3):2516–2520

Qin Y (2008a) Alginate fibers: an overview of the production processes and applications in wound management. Polym Int 57(2):171–180

Qin Y (2008b) The gel swelling properties of alginate fibers and their application in wound management. Polym Adv Technol 19(1):6–14

Qin Y (2016) Medical Textile Materials. Elsevier, New York, pp89–107

Qin Y, Gilding DK (1996) Alginate fibers and wound dressings. Med Device Technol 7(9):32–41

Qin Y, Gilding DK (2001) Dehydrated hydrogels. USA Patent 2001/6,203,845

Renn DW, Qin Y, Rossetto C (2003) Foam materials. USA patent 2003/6,656,974

Rives JM, Pannier M, Castede JC (1997) Calcium alginate versus paraffin gauze in the treatment of scalp graft donor sites. Wounds 9(6):199–205

Russell L (1999) Understanding physiology of wound healing and how dressings help. Br J Nurs 9:10–21

Sayag J, Meaume S, Bohbot S (1996) Healing properties of calcium alginate dressings. J Wound Care 5(8):357–362

Scurr JH, Wilson LA, Coleridge-Smith PD (1994) A comparison of calcium alginate and hydrocolloid dressings in the management of chronic venous ulcers. Wounds 6(1):1–8

Shi G, Xue Z, Yan T (2013) Research on preparation and characterization of agar fiber. Sci Technol Eng 13(8):2182–2185

Skjak-Braek G, Espevik T (1996) Application of alginate gels in biotechnology and biomedicine. Carbohydr Eur 14:19–25

Smith J (1992) Comparing sorbsan and polynoxylin melolin dressing after toenail removal. J Wound Care 1(3):17–19

Smith J, Lewis JD (1990) Sorbsan and leg ulcer. Pharm J 244:468

Speakman JB, Chamberlain NH (1944) The production of rayon from alginic acid. J Soc Dye Colour 60:264–272
Stanford ECC (1881) Improvements in the manufacture of useful products from seaweeds British Patent 1881/142
Stanford ECC (1883) New substance obtained from some of the commoner species of marine algae, Algin. Chemical News 47:254–257
Thomas S (1985) Use of a calcium alginate dressing. Pharm J 235:188–190
Thomas S (2000a) Alginate dressings in surgery and wound management--Part 1. J Wound Care 9(2):56–60
Thomas S (2000b) Alginate dressings in surgery and wound management--Part 2. J Wound Care 9(3):115–119
Thomas S (2000c) Alginate dressings in surgery and wound management--Part 3. J Wound Care 9(4):163–166
Thomas S, Tucker CA (1989) Sorbsan in the management of leg ulcers. Pharm J 243:706–709
Thompson J (1996) Fibrous gel. USA Patent 1996/5,482,932
Wang B, Kong Q, Ji Q (2009) Preparation and characterization of barium alginate fibers. Funct Mater 40(2):345–347
Wijesinghe WAJP, Jeon YJ (2011) Biological activities and potential cosmeceutical applications of bioactive components from brown seaweeds: a review. Phytochem Rev 10:431–443
Williams C (1999) Alginate wound dressings. Br J Nurs 8(5):313–317
Winter GD (1962) Formation of the scab and the rate of epithelialization of superficial wounds in the skin of the young domestic pig. Nature 193:293–294
Xie M, Xu G, Li Y (2003) Comparison of the clinical efficacy of four filling materials for nose. China J Endosc 9(12):19–22
Ye Q, Chen T (2001) The application of alginate dressings in skin donor site for burn wounds. Zhejiang Med 23(4):248–249
Yin L, Xue Z, Xia Y (2010) Preparation and characterization of new seaweed fiber-carrageenan fiber. Synthetic Fiber of China 39(3):27–30
Young MJ (1993) Alginate wound dressings. Dermatol Nurs 5(5):359–363

Chapter 22
Challenges and Recent Progress in Seaweed Polysaccharides for Industrial Purposes

Guilherme Augusto Colusse, Jaqueline Carneiro, Maria Eugênia Rabello Duarte, Ambati Ranga Rao, Gokare Aswathanarayana Ravishankar, Julio Cesar de Carvalho, and Miguel Daniel Noseda

Abbreviations

3,6-AnGal	3,6-anhydrogalactose
Ara	Arabinose
Fuc	Fucose
Gal	Galactose
Glc	Glucose
GlcA	Glucuronic acid
IdoA	Iduronic acid
MW	Molecular weight

G. A. Colusse
Biochemistry and Molecular Biology Department, Federal University of Paraná, PO, Curitiba, Brazil

Bioprocess Engineering and Biotechnology Department, Federal University of Paraná, PO, Curitiba, Brazil

J. Carneiro · M. E. R. Duarte · M. D. Noseda (✉)
Biochemistry and Molecular Biology Department, Federal University of Paraná, PO, Curitiba, Brazil
e-mail: mdn@ufpr.br

A. Ranga Rao
Department of Biotechnology, Vignan's Foundation of Science, Technology and Research Deemed to be University, Vadlamudi, Guntur, Andhra Pradesh, India

G. A. Ravishankar
Dr. C. D. Sagar Centre for Life Sciences, Dayananda Sagar College of Engineering, Dayananda Sagar Institutions, Bengaluru, Karnataka, India

J. C. de Carvalho
Bioprocess Engineering and Biotechnology Department, Federal University of Paraná, PO, Curitiba, Brazil

A. Ranga Rao, G. A. Ravishankar (eds.), *Sustainable Global Resources of Seaweeds Volume 2*, https://doi.org/10.1007/978-3-030-92174-3_22

Rha	Rhamnose
Sor	Sorbose
VCA	Value chain analysis
Xyl	Xylose

1 Introduction

Natural molecule producers associated with biotechnological processes have increased scientific interest due to the global population growth. Macroalgae or seaweeds are natural, reliable, and sustainable organisms that have been used for direct human consumption or processed into functional foods, nutraceuticals, cosmetics, and medicines worldwide (Abu-Ghannam and Cox 2013; Hasan 2017; Shama et al. 2019). In 2019, the global seaweed market reached a production value of US$ 8.8 billion, and the market is expected to increase by 7.5% during 2020–2025 (IMARC Group 2020). Seaweeds are marine organisms that have been explored as promising sources of novel molecules such as significant amounts of proteins, minerals, and mainly polysaccharides (Lafarga et al. 2020; Salehi et al. 2019). Seaweeds harbor several polysaccharides that are extremely important in medical, pharmaceutical, cosmetic, and nutraceutical areas. These polysaccharides have a wide range of chemical structures and biological applications (Cosenza et al. 2017).

In biotechnological fields, seaweed production and seaweed-derived polysaccharides have presented immense potential. Moreover, the use of non-digestible carbohydrates as a source of dietary fiber (Garcia-Vaquero et al. 2017), biological activity such as anticoagulant effect from sulfated fucoidan isolated from *Sargassum vulgare* (Dore et al. 2013), and anti-tumor activity of galactofucan obtained from *Alaria angusta* (Menshova et al. 2015) are few examples of seaweed-derived polysaccharides of importance. Seaweeds were also investigated as natural manipulation of rumen fermentation, mitigating ruminal methane production (Maia et al. 2016).

Current information and economic assessment are essential to reinforce the quality of the decision-making process in the industry. In this chapter, the industrial seaweed-derived polysaccharide potential was evaluated and assessed through its chemical structure and biological activity in several key areas. Besides, seaweed cultivation features and their economic impacts were also evaluated to present a current vision of the global scenario. Also, advanced technologies for massive seaweed production are discussed and analyzed.

2 Potential of Seaweed Polysaccharide Production

Since marine macroalgae present a wide range of natural chemicals with several potential applications, the development of large-scale seaweed aquaculture becomes essential for biomass production. Targeted research and monitoring are required to help the decision-making, improvement on cultivation, and market development during a large project in this area (Campbell et al. 2019). Moreover, in industrial seaweed-derived polysaccharide production, several biological and environmental variables can cause influences on macroalgae tissue, biochemical content and bioactivity. Factors such as reproductive stage, blade age, salinity, sampling season, light climate, nutrient availability, sampling location, biomass density, water motion, temperature, and grazing pressure can directly influence the algal metabolism (Hafting et al. 2015).

According to Santelices (1999), seaweeds can be cultivated in one or multiple-step farming systems. Several seaweed industrial propagation techniques have been used, such as fragmentation followed by propagation directly for growth in the cultivation system for Gracilaria and Kappaphycus, or the use of a hatchery for the propagation of unitary seaweeds such as kelp (Buschmann et al. 2017). Seaweed propagation is done at different production scales: intensive on-land tanks, ponds, or open-sea culture systems. Also, the farming method and agronomic requirements are strictly associated with the target seaweed species. A few examples are nets at surface level (Porphyra and Ulva), bottom planting (Gracilaria and Sarcodiotheca), middle water (Kappaphycus), and rafts in deeper water (Macrocystis, Laminaria, and Undaria) (Santelices 1999). It is worth highlighting that open-sea culture systems are considered the most commercially successful system due to lower operational and capital costs (Sahoo and Yarish 2005).

Polysaccharides can represent up to 76% of most seaweeds, on a dry weight basis (Holdt and Kraan 2011). According to these authors, the highest contents are found in Ascophyllum, Porphyra, and Palmaria. Furthermore, brown algae (Laminaria, Fucus, Ascophyllum, and Sargassum) present a polysaccharide content variation from 35 to 70% on a dry weight basis. Also, green algae (Ulva) show a 15–65% variation, whereas red algae (Chondrus, Porphyra, Gracilaria, and Palmaria) represent a variation on polysaccharide content from 40 to 74% (Holdt and Kraan 2011; Venugopal 2019). Holdt and Kraan (2011) reviewed and described some seaweed-derived polysaccharides such as alginic acid, fucoidan, laminaran, porphyran, and floridean starch. According to the authors, these polysaccharides present variation in the cellular content based on climate change. Taking the large polysaccharide content into account, several methodologies are used to extract, purify, and characterize these molecules. Figure 22.1 describes an industrial process to obtain high-value molecules from macroalgae biomass. Here we focused on a general schematic diagram of polysaccharide production.

To obtain a good quality polysaccharide its extraction, purification, and quality control are necessary. The extraction of polysaccharides requires fine-tuning several operating parameters such as extraction time, the ratio of solvent to sample, temperature, microwave power, and type of solvent followed by a rigid purification step. DEAE-Sephadex A-50 and A103S ion-exchange chromatography are a few examples of existing techniques of polysaccharide purification (He et al. 2016; Shi et al. 2018).

In downstream processes, reliable methodologies are necessary to characterize the polysaccharide molecule. Usov (2011) reviewed and explained several

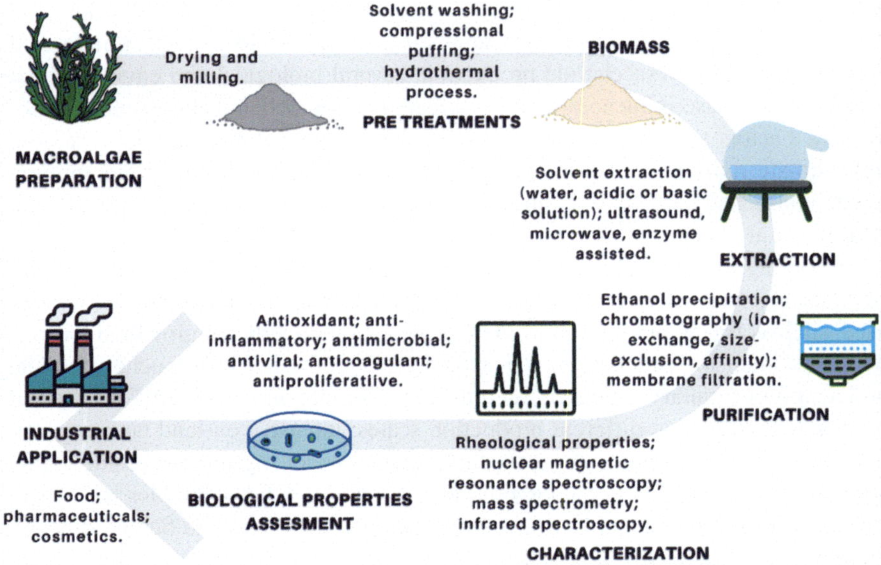

Fig. 22.1 Steps in seaweed-derived polysaccharide extraction and purification

important techniques used to characterize polysaccharides obtained from seaweeds, mainly red algae. Infrared spectroscopy, gas chromatography, and high-performance gel permeation chromatography (He et al. 2016) are useful to identify functional groups, evaluate the monosaccharide composition, and molecular weight (MW), respectively. Total sugars and uronic acids hint at the polysaccharide chemical structure (Rioux et al. 2007). Also, nuclear magnetic resonance (NMR) spectroscopy (1D and 2D analyses) is an essential analytical technique in resolving the polysaccharide structure, as mentioned by Ferreira et al. (2019) and explored by Gonçalves et al. (2002). To determine the pattern of glycosidic linkages present in the polysaccharide, it is common to use the methylation approach that includes, permethylation of the polysaccharide followed by total hydrolysis and GC-MS analysis of the partially methylated derivatives (Shi et al. 2018). In addition to polysaccharide chemical structure, several studies showed the potential of these macromolecules in pharmaceutical, food, and medical areas (de Jesus Raposo et al. 2015; Holdt and Kraan 2011; Venugopal 2016).

3 Chemical Structure and Biological Activity

Generally, a polysaccharide biological activity is associated with several chemical characteristics such as monosaccharide composition, functional groups, molecular weight (MW), anomeric configuration (α or β), and branching degree (Duarte et al. 2004; Guo et al. 2017; Lahaye 2001; Shi et al. 2018; Wu et al. 2012). A

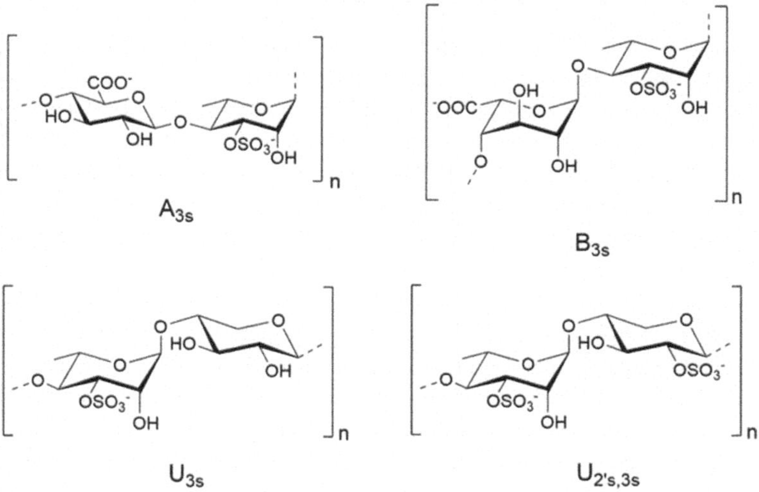

Fig. 22.2 Diads present in ulvan, sulfated heteropolysaccharide isolated from green seaweeds of Ulva genus. A_{3s}: [→4)-β-D-GlcA*p*-(1→4)-α-L-Rha*p* 3-sulfate-(1→]; B_{3s}: [→4)-α-L-IdoA*p*-(1→4)-α-L-Rha*p* 3-sulfate-(1→]; U_{3s}: [→4)-α-L-Rha*p* 3-sulfate-(1→4)-β-D-Xyl*p*-(1→]; $U_{2's,3s}$: [→4)-α-L-Rha*p* 3-sulfate-(1→4)-β-D-Xyl*p* 2-sulfate-(1→]. These structures were elucidated by Lahaye's group (Lahaye and Robic 2007)

polysaccharide (synonym: glycan) consists of several units of monosaccharides linked by glycosidic bonds. Heteropolysaccharides have two or more different types of monosaccharides in their composition, while homopolysaccharides have just one type of monosaccharide unit. Polysaccharides may be neutral (e.g. laminaran) or acidic, with the presence of sulfate groups (e.g., carrageenans, fucoidans, and ulvan) or carboxylic acid groups (as in alginate and ulvan) (Ngo and Kim 2013), or present other substituent groups, such as methyl or pyruvyl (Venugopal 2019).

The polysaccharide chemical structure, as well as its above-mentioned features, are strictly associated with a particular biological activity. Ulvan represents the main polysaccharide found in green seaweeds from Ulva genus (Fig. 22.2). This sulfated heteropolysaccharide is constituted by rhamnose, xylose, and glucuronic and iduronic acids (Jiao et al. 2011; Lahaye and Ray 1996). According to Costa et al. (2010), biological effects as antiproliferative efficacy of polysaccharide correlate with its sulfate content. A variety of other sulfated seaweed-derived polysaccharides are synthesized by green macroalgae (Ghosh et al. 2004; Matsubara et al. 2001), such as a branched heteropolysaccharide from *Caulerpa racemosa,* that presents in vitro anti-herpetic activity. This polysaccharide contains 1,3- and 1,3,6-linked galactose, 1,3,4-linked arabinose, 1,4-linked glucose, and terminal- and 1,4-linked xylose residues (Ghosh et al. 2004). Figure 22.2 shows the disaccharide repeating units (diads) of the green macroalgae polysaccharide ulvan.

Additionally, chemical modification of the native polysaccharides such as sulfation, partial depolymerization, over-sulfation, and cyclization reaction can enhance the polysaccharide's biological activity or rheological properties. The latter one (formation of 3,6-anhydro-α-galactose or 3,6-anhydro-α-galactose 2-sulfate units from α-galactose 6-sulfate or α-galactose 2,6-disulfate, respectively) is an important reaction used in carrageenans and agarans to increase their rheological properties and it was applied to several polysaccharides from red seaweeds (Ciancia et al. 1993, 1997; Noseda et al. 2000; Noseda and Cerezo 1995; Viana et al. 2004; Zibetti et al. 2005, 2009). Carlucci et al. (1997) showed the relationship between the native carrageenans obtained from *Gigartina skottsbergii* and their cyclized derivatives with antiviral activity. The authors reported that the sulfate position in carrageenans molecules can influence the antiviral behavior. Furthermore, according to Patel (2012), chemical modification on carrageenan oligosaccharides increases its anti-tumor effect and anti-tumor immunity, whereas depolymerization and over-sulfation of fucoidan enhance its anti-angiogenic and anti-tumor activities (Koyanagi et al. 2003; Silchenko et al. 2018). Furthermore, polysaccharide features and biological activity present an intrinsic relationship that is essential to understand the biochemical pathway in biological assays. Table 22.1 presents a list of polysaccharide traits, structural features, and their biological activity.

Anticoagulant activity is the most studied biological property of seaweed-derived sulfated polysaccharides (Costa et al. 2010; Ciancia et al. 2010). Moreover, sulfated polysaccharides isolated from *Padina tetrastromatica* and *Gracilaria cervicornis* exhibits anti-angiogenic properties (Jose and Kurup 2017) and antidiarrheal activity (Bezerra et al. 2018), respectively, whereas polysaccharide obtained from *Laminaria japonica* presented renoprotective effect in rats (Li et al. 2017). Furthermore, therapeutic agents for atherosclerosis are also reviewed and reported associated with the use of fucoidan, sulfated laminaran, and ulvan (Patil et al. 2018).

Most polysaccharides are considered non-toxic, these molecules have been studied in several biological assays in pharmaceutical and medical areas (Bilal and Iqbal 2019). Undoubtedly, seaweed-derived polysaccharides enhance the possibilities of new drug development and therefore contribute to the environment by carbon storage and the reduction of greenhouse emissions during seaweed cultivation (Buschmann et al. 2017; Chung et al. 2011). Additionally, industrial steps – mainly downstream processes – are critical to support a high-quality polysaccharide extraction in a large-scale seaweed facility.

4 Downstream Process: A Critical Step on Industrial Seaweed-Derived Polysaccharide Production

Downstream processes are essential for any biotechnological bioprocess due to their influence on total production cost, energy input, and product quality. For seaweed polysaccharides, these activities comprise of biomass pre-treatment, extraction procedures, purification, enzymatic modification processes, marketing, trading, and transportation (Nor et al. 2019; Zayed and Ulber 2020). According to Zayed and Ulber (2020), in a seaweed-derived fucoidan facility, the downstream process

Table 22.1 Summary of existing seaweed-derived polysaccharides, main structure features, and biological application

Seaweed specie	Polysaccharide main structural features	Biological activity	References
Sarcodia ceylonensis (Plocamiales, Rhodophyta)	MW 466 kDa. Main monosaccharides: Man, Glc, sor and Ara (molar ratio: 14.367:5.339:2.829:1.213)	Antioxidant activity	He et al. (2016)
Ulva lactuca (Ulvales, Chlorophyta)	MW 404 kDa. Main monosaccharides: Man, Glc, Ara, sor, gal, and Fuc (molar ratio: 6.659:1.931:0.519:0.461:0.277:0.222:0.194)		
Durvillaea Antarctica (Fucales)	MW 482 kDa. Main monosaccharides: Glc, man, sor, Fuc, and Xyl (molar ratio: 26.238:2.936:2.704:1.060:0.892)		
Gracilaria lemaneiformis (Gracilariales, Rhodophyta)	MW 591 kDa. Main monosaccharides: Gal, Fuc, Glc, and Xyl (molar ratio: 18.76:5.968:4.48:1.811)		
	GLP1 (MW 5.5 kDa), GLP2 (MW 85 kDa, 10.8% sulfate) and GLP3 (MW 82 kDa, 23.2% sulfate). All fractions were mainly composed of gal and had a (1→3)-Gal*p* and (1→6)-Gal*p* backbone.	Anti-tumor activity	Shi et al. (2018)
	MW > 152 kDa and contains 3,6-anhydrogalactose (8.23%), sulfate (11.26%) and protein (0.98%).	Anti-allergy	Liu et al. (2016)
Laurencia dendroidea (Ceramiales, Rhodophyta)	DHS-4 (181.3 x10^3 g.Mol^{-1}, 21.3% of $NaSO_3$) presented unit A 2-sulfated (18.9 Mol%), nonsubstituted (15.3 Mol%) and 6-*O*-methylated (10.1 Mol%), and unit B composed mainly by galactose 6-sufate precursor units (19.2 Mol%) and 3,6-anhydrogalactose (13.8 Mol%).	Snake antivenom	Ferreira et al. (2019)
Ulva fasciata (Ulvales, Chlorophyta)	F2 ($NaSO_3^-$14.1%; COO^-1.23 mmol.g^{-1}; MW 8.1 kg.Mol^{-1}) presented a compact sphere conformation with a helical motif as secondary structure. C3 showed a $NaSO_3^-$21.0%; COO^-1.81 mmol.g^{-1}; MW 49 kg.Mol^{-1} whereas C3b presented $NaSO_3^-$14.1%; COO^-1.23 mmol.g^{-1}; MW 8.1 kg.Mol^{-1} and C3c showed $NaSO_3^-$21.0%; COO^-1.81 mmol.g^{-1}; MW 18 kg.Mol^{-1}	Anticoagulant	de Carvalho et al. (2020)
Ulva rigida (Ulvales, Chlorophyta)	MW 56.7 kDa, 41% neutral sugars, 34% uronic acids and 4% proteins. Main monosaccharides: Glc/gal (12.2 Mol%), Xyl (8 Mol%), Rha (42.6 Mol%), GlcN (6.9 Mol%) and GlcA/IdoA (30.3 Mol%).		Adrien et al. (2019)

(continued)

Table 22.1 (continued)

Seaweed specie	Polysaccharide main structural features	Biological activity	References
Sphaerococcus coronopifolius (Gigartinales, Rhodophyta)	MW 308,700 Da. Main monosaccharides: Gal (33.1%), Xyl (1.8%) and Glc (1.7%), contains 3,6-AnGal (11%), GlcA (6.7%), GalA (1%), sulfate (24%) and pyruvic acid (0.34%).	Antiviral	Bouhlal et al. (2011)
Boergeseniella thuyoides (Ceramiales, Rhodophyta)	MW 360,300 Da. Main monosaccharides: Gal (25.4%), Xyl (2.8%), Glc (3%) and Rha (0.3%), contains 3,6-AnGal (16%), GalA (3.2%), sulfate (7.6%) and proteins (6%).		

Note: GlcN = glucosamine; GalA = galacturonic acid; GlcA = glucuronic acid; IdoA = iduronic acid; 3,6-AnGal = 3,6-anhydrogalactose

composes of pre-treatment (washing, drying, and milling), followed by extraction procedures (hot- or cold-water incubation and solvent precipitation), separation techniques, and purification to obtain a high-quality final product. Figure 22.3 shows manufacturing process flow charts for carrageenan and alginate production.

Different extraction and purification methodologies can directly influence the quantitative yield and quality of seaweed-derived polysaccharides. The choice of these parameters is based on the physicochemical properties of the polysaccharide and its interactions with the cell wall, as well as the seaweed species, the period of collection, and the biomass conservation (Robic et al. 2009). Furthermore, variables such as extraction temperature, solvents, solvent to biomass ratio, biomass particle size, and extraction time are also extremely important to limit the co-extraction of macromolecular impurities and reduce the necessity of refined purification techniques (Kidgell et al. 2019).

A relevant example of an industrial downstream process concern is the correct distribution of seaweed intended for the processing versus that destined to be used as seedlings. Malaysia's seaweed industry is based on 90% dried *Kappaphycus* spp. to be used in industry on carrageenan production and 10% fresh seaweed as seedlings to maintain the cultivation cycle (Nor et al. 2019). In summary, the decision-making process in these industrial production steps requires a great understanding of seaweed processing, innovative technologies, and lab-scale experimental tests.

5 Economic Impact and Commercial Importance

The global market of industrial marine biotechnology applications is expected to grow to US$6.4 billion by 2025 (Hurst et al. 2016). This increase is mainly due to the potential use of polysaccharides in food, nutraceuticals, and pharmaceutical industries in Asian, Western, and European countries (Tanna and Mishra 2019). Seaweed-derived polysaccharide production by the large-scale cultivation at open sea or using a biorefinery approach is dependent on the sustainable development of the process economy. According to the European Commission, concepts such as

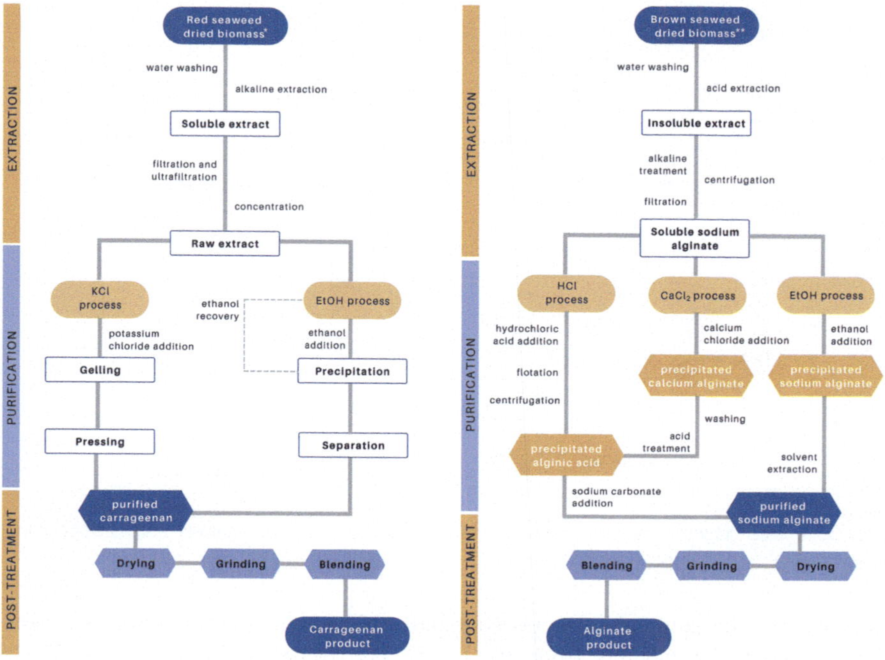

Fig. 22.3 General extraction and purification technology applied on carrageenan and alginate manufacturing (adapted from Blakemore and Harpell 2010; Peteiro 2018). **Note:** main genera industrially used: **Chondrus*, *Euchema*, *Gigartina*, *Iridaeae*, and *Kappaphycus* (see item 5.1). ***Ascophyllum*, *Fucus*, *Ecklonia*, *Cystoseira*, *Macrocystis*, and *Laminaria* (see item 5.3).

food security, guarantee sustainable use of resources, climate impact reduction, job creation, and competitiveness are key factors to strengthening the economy (Balina et al. 2017; Mathijs et al. 2015).

Like all natural product industries, large-scale seaweed market requires a level of standardization, efficacy, and traceability that can directly influence the scenario (Hafting et al. 2015). It is therefore not surprising that the increase of scientific studies based on the seaweed-derived polysaccharide centers around biological properties, improvement of quality, and low cost cultivation of biomass (Hafting et al. 2015; Hurst et al. 2016). Here we will briefly discuss important concepts such as species, characteristics, and biological properties of some established industrial seaweed-derived polysaccharides.

5.1 *Carrageenans*

Most commercial carrageenans are obtained by *Kappaphycus alvarezii*, *Eucheuma spinosum*, *Chondrus crispus*, *Gigartina stellata*, and other *Gigartina* spp. and *Chondrus* spp. (Chang et al. 2017; McHugh 2003). This family of polysaccharides

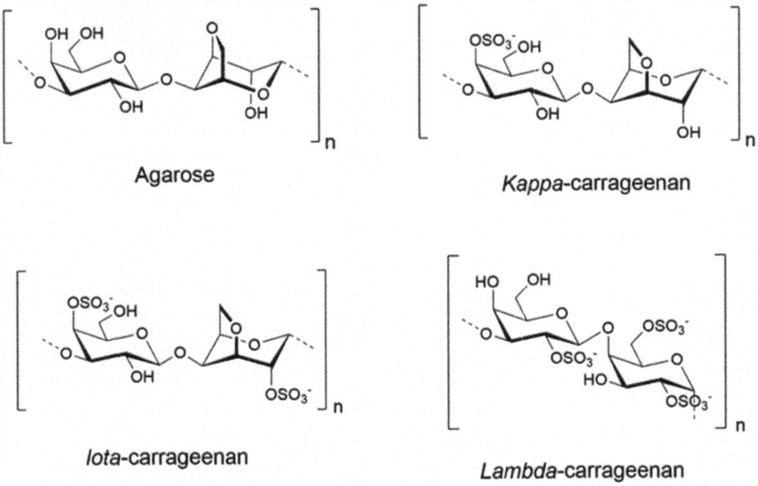

Fig. 22.4 Red seaweed polysaccharides. Main industrial types of agarans and carrageenans (Usov 2011)

is constituted by sulfated galactans containing 15 to 40% of ester-sulfate and presenting molecular weights from 10^5 to 10^6 Da (Necas and Bartosikova 2013; Venugopal 2019). Carrageenans are constituted by alternative units of β-D-galactose and α-D-galactose (or 3,6-anhydro-α-D-galactose) joined by 1,3 and 1,4-glycosidic linkages (Usov 2011). Moreover, it presents several biological activities as anti-tumor (Zhou et al. 2006), anti-viral (Carlucci et al. 1997, 2004), and immunomodulatory properties (Zhou 2004), whereas applications in the non-food industry such as textile, pharmaceutical, and cosmetic sectors have also been reported (Balina et al. 2017). The use of carrageenan is approved by the Food and Drug Administration (FDA) in the United States (Burges Watson 2008) and it is widely used in many sectors with different applications such as the addition in sausages (Ayadi et al. 2009) and application in drug delivery (Li et al. 2014). Figure 22.4 represents the disaccharide repeating units of the three main commercial carrageenans, kappa-, iota-, and lambda-carrageenans.

5.2 *Agarans*

Agarans are extracted and purified from seaweed species belonging to the Rhodophyceae (red algae), mainly *Gelidium sesquipedale*, *Hydropuntia cornea*, and *Gracilaria* species (Carmona et al. 1998; Martínez-Sanz et al. 2019; Pereira-Pacheco et al. 2007; Rodríguez et al. 2009). Agarans are constituted by a disaccharide-based repeating unit [→3)-*β*-D-galactopyranose-(1→4)-α-L-galactopyranose-(1→]. The α-L-Gal*p* units can be partially or totally substituted by 3,6-anhydro-α-L-galactopyranose. This backbone can be partially substituted in different positions by variable degrees of one or more of the following substituents, sulfate, methyl, pyruvyl, and glycosyl groups. Agar and agarose are the two main types of industrially relevant agarans and are those presenting predominantly neutral structures (Fig. 22.4).

Agar is considered an extremely important polysaccharide to the industrial sector due to its excellent thickening and gelling properties (Pereira-Pacheco et al. 2007). Alternative purposes, such as the development of biodegradable films and encapsulation structures, have been also reported (Alehosseini et al. 2018; Kanmani and Rhim 2014).

5.3 *Alginate*

Alginate is a linear polyuronide constituted by 1,4-linked *β*-D-mannuronic acid (M) and 1,4 *α*-L-guluronic acid (G) residues (Tanna and Mishra 2019). The residues are distributed in blocks, G-block, M-block, and MG-block according to Fig. 22.5.

The copolymer composition and block distribution are species-specific. The commercial application of alginate is mainly in the food and pharmaceutical industries. It can be used as an emulsifier, stabilizer, flavoring adjuvant, surfactant, viscosity increasing agent, tablet disintegrant, and diluent in capsule formulation (Colusse et al. 2021; Nause et al. 2009). Alginate is obtained from edible brown seaweed, particularly from *Cystoseira barbata* (Trica et al. 2019), *Macrocystis pyrifera*, *Laminaria hyperborea*, and *Ascophyllum nodosum* (Szekalska et al. 2016).

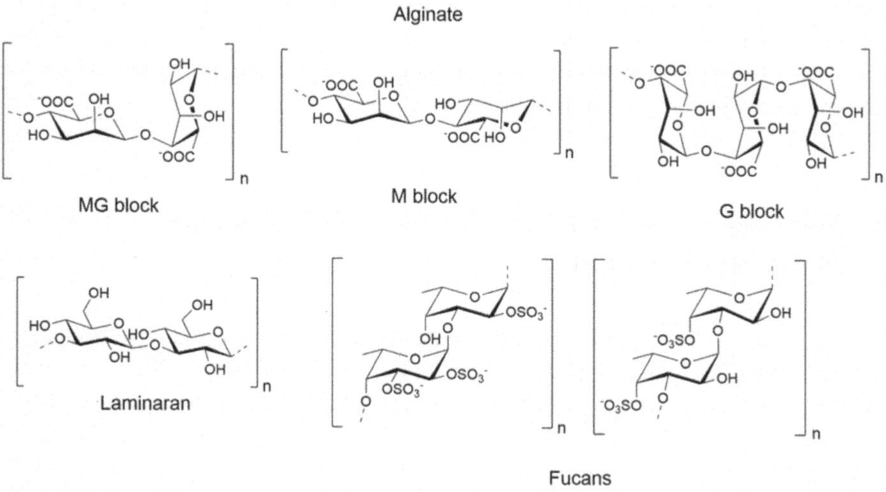

Fig. 22.5 Main polysaccharides produced by brown seaweeds. Alginate, blocks distribution is species specific (Duarte et al. 1991); Laminaran, storage polysaccharide (Rinaudo 2007); Fucans from: *Ascophyllum nodosum*/*Fucus vesiculosus* (left) and *Ecklonia kurome* (right) (Berteau and Mulloy 2003)

5.4 *Laminaran*

Laminaran is a β-glucan, with low molecular weight (approx. 20–30 glucosyl residues), constituted by β-1,3-linked-D-glucopyranose units (see Fig. 22.5). This backbone can also present some interspersed β-1,6-linked units and β-1,6-linked glucose side chains (Rinaudo 2007). Laminaran is the carbohydrate reserve of brown seaweeds and can be obtained from species such as *Laminaria hyperborea, L. digitata, Ascophyllum nodosum,* and *Eisenia bicyclis* in average levels of 35% on a dry weight basis (Jönsson et al. 2020; Kadam et al. 2015; Maeda and Nisizawa 1968; Read et al. 1996). Commercial applications of laminaran are limited, however, products with this polysaccharide claiming plant defense stimulant capacity, such as Vacciplant® are available in the agricultural market (Stadnik and de Freitas 2014; UPL 2021). Several authors have studied and reported the disease resistance elicitor property of laminaran (Aziz et al. 2003; Tziros et al. 2021; Xin et al. 2019). Moreover, laminaran also presents the potential for functional foods, nutraceuticals, and cosmeceuticals (Jönsson et al. 2020), thus showing biological properties such as antioxidant, anti-inflammatory, and anticoagulant activities (Kadam et al. 2015).

Potential seaweeds-based biorefinery has been described based on the use of agar, cellulose, laminaran, and starch to be converted into bioethanol through the saccharification process (Balina et al. 2017; Mohammad et al. 2019; Hessami et al. 2019). An important tool to be used to assess the commercial viability of seaweed aquaculture in a biorefinery approach is the value chain analysis (VCA) (Nor et al. 2019). This tool can identify the most valuable activities and improvements, help with financial accounts related to consumption, recycling, and disposal. Furthermore, the economic impact of those polysaccharides is strictly related to biological application and large-scale production. However, it has been suggested the use of smaller seaweed-derived polysaccharide facilities that may contribute to define molecules for pharmaceutical applications (Jönsson et al. 2020).

6 Seaweed Polysaccharides: Advanced Technologies in Massive Production

Several seaweed-derived polysaccharide industries around the world have been producing polysaccharides with advanced technology. Dobrinčić et al. (2020) reviewed and discussed some recently advanced extraction techniques that can be used for industrial purposes. The techniques are microwave-assisted extraction, ultrasound-assisted extraction, pressurized liquid extraction, and enzymatic assisted extraction. It is worth mentioning that laboratorial parameters such as temperature, extraction time, power, and sample to solvent ratio should also be optimized to reach improvements on the results. In order to obtain a better understanding of industry and technology used to the seaweed-derived polysaccharide, Table 22.2 presents some companies around the world that produce and market polysaccharides obtained from seaweeds.

Table 22.2 Global companies involved in production of seaweed-derived polysaccharides

Company	Polysaccharide	Seaweed species	Country	References
Gelymar	Carrageenan (kappa II and lambda carrageenan) and alginate	Red seaweeds and various species of brown seaweeds	Chile and Indonesia	gelymar.com
Iro alginate industry	Alginate	Brown seaweeds	China	iroalginate.com
MCPI corporation	Carrageenan	Red seaweeds	Philippines	mcpicarrageenan.com
Elicityl	Alginate, fucoidan, galactan, ulvan, and xylan	*Laminaria japonica*, *Fucus vesiculosus*, *Undaria pinnatifida*, *Codium fragile*, *Ulva* sp., *Enteromorpha* sp., *Palmaria palmata*	France	elicityl-oligotech.com
Agargel	Agar and carrageenan	Red seaweeds	Brazil	agargel.com.br/en
Cargill	Carrageenan	Red seaweeds (Rhodophyceaea) from the Gigartinales group	United States	cargill.com
Gather Great Ocean algae industry group	Alginate, carrageenan, and agar	Brown and red seaweeds	China	en.judayang.com
Algaia	Carrageenan and alginate	Red and brown seaweeds	France	algaia.com
Compañia Española de Algas marinas (CEAMSA)	Carrageenan and alginate	Red and brown seaweeds	Spain	ceamsa.com
TBK: Manufacturing corporation	Carrageenan	*Kappaphycus alvarezii*[a] and *Eucheuma denticulatum*[b]	Philippines	tbk.com.ph

[a] Formerly known as *Eucheuma cottonii*

[b] Formerly known as *Eucheuma spinosum*

Although there are several studies about seaweed-derived polysaccharides and their biological activities, existing bottlenecks need some attention due to their influence on the economics of the production process in the industrial sector. Issues related to this approach include:

- Novel extraction and purification methods are not optimized for a large-scale seaweed-derived polysaccharide facilities;
- Extraction yields are still low;
- Lack of information about ecological consequences of implementing large-scale offshore seaweed biorefineries;

- The potential of offshore seaweed farming is rare to estimate;
- Difficulty in controlling cultivation parameters;
- The complex structure of glycans;
- The necessity for new strain development by breeding tools;
- Vulnerability to novel diseases;
- The requirement of expensive techniques for polysaccharide characterization.

The use of biotechnological approaches can reinvent the production cycle and decrease the influence of bottlenecks of large-scale seaweed facility. Mathematical models can be applied in industrial processes to identify the potential productivity of the seaweed. Lehahn et al. (2016) integrated climatological oceanographic data associated with seaweed metabolism and growth rate to provide a global potential for offshore production of seaweed-derived biomass, proteins, platform chemicals, transportation fuels, and energy. This methodology could be used to calculate the production potential of an area and estimate the profit of the project.

Another feature with a dramatic impact on large-scale seaweed facilities is the possibility to create new strains that present light and thermal tolerance and resistant to disease (Kim et al. 2017). Seaweed breeding programs are standard in countries such as Korea, Japan, and China through a consecutive selection of individuals that present outstanding performance in the cultivation system (Hwang et al. 2019). According to the authors, breeding methodologies as the generation of mutant strains need a careful evaluation since they can disrupt the genetic structure of natural populations.

Moreover, advanced applications such as the production of nanoengineered injectable hydrogels in tissue regeneration therapy (Lokhande et al. 2018) or the use as the polymer matrix in oral extended-release tablets (Li et al. 2014) should increase the use of seaweed polysaccharides on a commercial scale. Although seaweed-derived molecules have been discussed in several studies, the current scenario invites a significant investment due to the necessity of new molecule sources. However, it might pay off with species with greater polysaccharide productivity, and molecules with high added value to be used in the pharmaceutical, medical, and food sectors.

7 Conclusions and Future Perspectives

Biological and environmental factors associated with the cultivation system influences seaweed productivity and polysaccharide biological activity. It is worth highlighting that, both in lab- and in pilot-scale seaweed farms, it is vital to evaluate growth parameters that affect production. Scientific efforts should be implanted in large-scale facilities to increase data reliability and reproducibility since environmental variables such as sunlight and temperature present variability during cultivation. Also, appropriate breeding strategy and optimization on cultivation systems is a valuable tool to increase production in the cultivation system.

It is important also to consider the role in new drug development to take into account the biological application diversity obtained from seaweed-derived polysaccharides, which include antiviral, antioxidant, anti-tumor, antidiarrheal, renoprotective effect, among other biological properties. It must be emphasized that studies related to seaweed-derived polysaccharides such as extraction procedures, purification methods, and chemical modifications are critical since they can address new methodologies that can be applied in large-scale production to build a promising future for polysaccharides obtained from seaweeds.

Acknowledgments The authors would like to thank the Coordenação de Aperfeiçoamento de Pessoal de Nível Superior (CAPES, Brazil) and Conselho Nacional de Desenvolvimento Científico e Tecnológico (CNPq, Brazil) for funding. GAC acknowledges a doctoral fellowship from CAPES (Brazil). JC acknowledges a postdoctoral fellowship from CAPES. MRD, JCC, and MDN are Research Members of CNPq (Brazil). ARR acknowledge fund for Improvement of Science & Technology Infrastructure in Higher Educational Institutions (FIST Project no: LSI-576/2013), Department of Science and Technology, Government of India and Centre of Excellence, Department of Biotechnology, Vignan's Foundation for Science, Technology and Research for the support and resources.

References

Abu-Ghannam N, Cox S (2013) Seaweed-based functional foods. In: Bioactive compounds from marine foods. Wiley, Chichester, UK, pp 313–327. https://doi.org/10.1002/9781118412893.ch14

Adrien A, Bonnet A, Dufour D, Baudouin S, Maugard T, Bridiau N (2019) Anticoagulant activity of sulfated ulvan isolated from the green macroalga *Ulva rigida*. Mar Drugs 17:291. https://doi.org/10.3390/md17050291

Agargel (2021). https://agargel.com.br/en/. Accessed 5 Apr 2021

Alehosseini A, Gomez del Pulgar E-M, Gómez-Mascaraque LG, Martínez-Sanz M, Fabra MJ, Sanz Y et al (2018) Unpurified Gelidium-extracted carbohydrate-rich fractions improve probiotic protection during storage. LWT 96:694–703. https://doi.org/10.1016/j.lwt.2018.06.043

Algaia (2021). https://www.algaia.com/. Accessed 15 Apr 2021

Ayadi MA, Kechaou A, Makni I, Attia H (2009) Influence of carrageenan addition on Turkey meat sausages properties. J Food Eng 93:278–283. https://doi.org/10.1016/j.jfoodeng.2009.01.033

Aziz A, Poinssot B, Daire X, Adrian M, Bézier A, Lambert B et al (2003) Laminarin elicits defense responses in grapevine and induces protection against *Botrytis cinerea* and *Plasmopara viticola*. Mol Plant-Microbe Interact 16:1118–1128. https://doi.org/10.1094/MPMI.2003.16.12.1118

Balina K, Romagnoli F, Blumberga D (2017) Seaweed biorefinery concept for sustainable use of marine resources. Energy Procedia 128:504–511. https://doi.org/10.1016/j.egypro.2017.09.067

Berteau O, Mulloy B (2003) Sulfated fucans, fresh perspectives: structures, functions, and biological properties of sulfated fucans and an overview of enzymes active toward this class of polysaccharide. Glycobiology 13:29–40. https://doi.org/10.1093/glycob/cwg058

Bezerra FF, Lima GC, de Sousa NA, de Sousa WM, Costa LEC, da Costa DS et al (2018) Antidiarrheal activity of a novel sulfated polysaccharide from the red seaweed *Gracilaria cervicornis*. J Ethnopharmacol 224:27–35. https://doi.org/10.1016/j.jep.2018.05.033

Bilal and Iqbal (2019) Marine seaweed polysaccharides-based engineered cues for the modern biomedical sector. Mar Drugs 18:7. https://doi.org/10.3390/md18010007

Blakemore WR, Harpell AR (2010) Carrageenan. In: Imeson A (ed) *Food Stabilizers, Thickeners and Gelling Agents*. Wiley-Blackwell, London, pp 73–94

Bouhlal R, Haslin C, Chermann J-C, Colliec-Jouault S, Sinquin C, Simon G et al (2011) Antiviral activities of sulfated polysaccharides isolated from *Sphaerococcus coronopifolius* (Rhodophytha, Gigartinales) and *Boergeseniella thuyoides* (Rhodophyta, Ceramiales). Mar Drugs 9:1187–1209. https://doi.org/10.3390/md9071187

Burges Watson D (2008) Public health and carrageenan regulation: a review and analysis. J Appl Phycol 20:505–513. https://doi.org/10.1007/s10811-007-9252-x

Buschmann AH, Camus C, Infante J, Neori A, Israel Á, Hernández-González MC et al (2017) Seaweed production: overview of the global state of exploitation, farming and emerging research activity. Eur J Phycol 52:391–406. https://doi.org/10.1080/09670262.2017.1365175

Campbell I, Macleod A, Sahlmann C, Neves L, Funderud J, Øverland M et al (2019) The environmental risks associated with the development of seaweed farming in Europe - prioritizing key knowledge gaps. Front Mar Sci 6. https://doi.org/10.3389/fmars.2019.00107

Cargill (2021). https://www.cargill.com/. Accessed 6 Apr 2021

Carlucci MJ, Pujol CA, Ciancia M, Noseda MD, Matulewicz MC, Damonte EB et al (1997) Antiherpetic and anticoagulant properties of carrageenans from the red seaweed *Gigartina skottsbergii* and their cyclized derivatives: correlation between structure and biological activity. Int J Biol Macromol 20:97–105. https://doi.org/10.1016/S0141-8130(96)01145-2

Carlucci M, Scolaro L, Noseda M, Cerezo A, Damonte E (2004) Protective effect of a natural carrageenan on genital herpes simplex virus infection in mice. Antivir Res 64:137–141. https://doi.org/10.1016/j.antiviral.2004.07.001

Carmona R, Vergara JJ, Lahaye M, Niell FX (1998) Light quality affects morphology and polysaccharide yield and composition of *Gelidium sesquipedale* (Rhodophyceae). J Appl Phycol 10:323–332. https://doi.org/10.1023/A:1008042904972

de Carvalho MM, Noseda MD, Dallagnol JCC, Ferreira LG, Ducatti DRB, Gonçalves AG et al (2020) Conformational analysis of ulvans from *Ulva fasciata* and their anticoagulant polycarboxylic derivatives. Int J Biol Macromol 162:599–608. https://doi.org/10.1016/j.ijbiomac.2020.06.146

Chang V-S, Okechukwu PN, Teo S-S (2017) The properties of red seaweed (*Kappaphycus alvarezii*) and its effect on mammary carcinogenesis. Biomed Pharmacother 87:296–301. https://doi.org/10.1016/j.biopha.2016.12.092

Chung IK, Beardall J, Mehta S, Sahoo D, Stojkovic S (2011) Using marine macroalgae for carbon sequestration: a critical appraisal. J Appl Phycol 23:877–886. https://doi.org/10.1007/s10811-010-9604-9

Ciancia M, Noseda MD, Matulewicz MC, Cerezo AS (1993) Alkali-modification of carrageenans: mechanism and kinetics in the kappa/iota-, mu/nu- and lambda-series. Carbohydr Polym 20:95–98. https://doi.org/10.1016/0144-8617(93)90083-G

Ciancia M, Matulewicz MC, Cerezo AS (1997) Alkaline modification of carrageenans. Part III. Use of mild alkaline media and high ionic strengths. Carbohydr Polym 32:293–295. https://doi.org/10.1016/S0144-8617(96)00130-0

Ciancia M, Quintana I, Cerezo A (2010) Overview of anticoagulant activity of sulfated polysaccharides from seaweeds in relation to their structures, focusing on those of green seaweeds. Curr Med Chem 17:2503–2529. https://doi.org/10.2174/092986710791556069

Colusse GA, Santos AO, Rodrigues JM, Barga MC, Duarte MER, Carvalho JC et al (2021) Rice vinasse treatment by immobilized *Synechococcus pevalekii* and its effect on *Dunaliella salina* cultivation. Bioprocess Biosyst Eng. https://doi.org/10.1007/s00449-021-02531-9

Compañia Española de Algas Marinas [CEAMSA] (2021). https://www.ceamsa.com/. Accessed 9 Apr 2021

Cosenza VA, Navarro DA, Ponce NMA, Stortz CA (2017) Seaweed polysaccharides: Structure and applications. In: Cosenza VA (ed) Industrial applications of renewable biomass products. Springer International Publishing, Cham, pp 75–116. https://doi.org/10.1007/978-3-319-61288-1_3

Costa LS, Fidelis GP, Cordeiro SL, Oliveira RM, Sabry DA, Câmara RBG et al (2010) Biological activities of sulfated polysaccharides from tropical seaweeds. Biomed Pharmacother 64:21–28. https://doi.org/10.1016/j.biopha.2009.03.005

Dobrinčić A, Balbino S, Zorić Z, Pedisić S, Bursać Kovačević D, Elez Garofulić I et al (2020) Advanced technologies for the extraction of marine brown algal polysaccharides. Mar Drugs 18:168. https://doi.org/10.3390/md18030168

Dore CMPG, das Faustino Alves CMG, Will LSEP, Costa TG, Sabry DA, de Souza Rêgo LAR, Accardo CM, Rocha HAO et al (2013) A sulfated polysaccharide, fucans, isolated from brown algae *Sargassum vulgare* with anticoagulant, antithrombotic, antioxidant and anti-inflammatory effects. Carbohydr Polym 91:467–475. https://doi.org/10.1016/j.carbpol.2012.07.075

Duarte MER, Gorin PAJ, Duarte JH (1991) Homogeneous guluronic and mannuronic acid blocks in the alginate of the brown seaweed *Laminaria brasiliensis*. Phytochemistry 30:1707–1708. https://doi.org/10.1016/0031-9422(91)84239-O

Duarte MER, Cauduro JP, Noseda DG, Noseda MD, Gonçalves AG, Pujol CA et al (2004) The structure of the agaran sulfate from *Acanthophora spicifera* (Rhodomelaceae, Ceramiales) and its antiviral activity. Relation between structure and antiviral activity in agarans. Carbohydr Res 339:335–347. https://doi.org/10.1016/j.carres.2003.09.028

Elicityl (2021). elicityl-oligotech.com. Accessed 6 Apr 2021

Ferreira LG, da Silva ACR, Noseda MD, Fuly AL, de Carvalho MM, Fujii MT et al (2019) Chemical structure and snake antivenom properties of sulfated agarans obtained from *Laurencia dendroidea* (Ceramiales, Rhodophyta). Carbohydr Polym 218:136–144. https://doi.org/10.1016/j.carbpol.2019.04.066

Garcia-Vaquero M, Rajauria G, O'Doherty JV, Sweeney T (2017) Polysaccharides from macroalgae: recent advances, innovative technologies and challenges in extraction and purification. Food Res Int 99:1011–1020. https://doi.org/10.1016/j.foodres.2016.11.016

Gather Great Ocean Algae Industry Group (2021). https://en.judayang.com/. Accessed 12 Apr 2021

Gelymar SA (2021). gelymar.com Accessed 5 Apr 2021

Ghosh P, Adhikari U, Ghosal PK, Pujol CA, Carlucci MJ, Damonte EB et al (2004) In vitro antiherpetic activity of sulfated polysaccharide fractions from *Caulerpa racemosa*. Phytochemistry 65:3151–3157. https://doi.org/10.1016/j.phytochem.2004.07.025

Gonçalves AG, Ducatti DRB, Duarte MER, Noseda MD (2002) Sulfated and pyruvylated disaccharide alditols obtained from a red seaweed galactan: ESIMS and NMR approaches. Carbohydr Res 337:2443–2453. https://doi.org/10.1016/S0008-6215(02)00318-X

Guo MQ, Hu X, Wang C, Ai L (2017) Polysaccharides: structure and solubility. In: Solubility of polysaccharides. InTech, London. https://doi.org/10.5772/intechopen.71570

Hafting JT, Craigie JS, Stengel DB, Loureiro RR, Buschmann AH, Yarish C et al (2015) Prospects and challenges for industrial production of seaweed bioactives. J Phycol 51:821–837. https://doi.org/10.1111/jpy.12326

Hasan MM (2017) Algae as nutrition, medicine and cosmetics: the forgotten history, present status and future trends. World J Pharm Pharm Sci 14:1934–1959. https://doi.org/10.20959/wjpps20176-9447

He J, Xu Y, Chen H, Sun P (2016) Extraction, structural characterization, and potential antioxidant activity of the polysaccharides from four seaweeds. Int J Mol Sci 17:1988. https://doi.org/10.3390/ijms17121988

Hessami MJ, Cheng SF, Ranga Rao A, Yin YH, Phang SM (2019) Bioethanol production from agarophyte red seaweed, *Gelidium elegans* using a novel sample preparation method for analysing bioethanol content by gas chromatography. Biotech J 9(1):25

Holdt SL, Kraan S (2011) Bioactive compounds in seaweed: functional food applications and legislation. J Appl Phycol 23:543–597. https://doi.org/10.1007/s10811-010-9632-5

Hurst D, Børresen T, Almesjo L, Raedemaecker F, Bergseth S (2016) Marine biotechnology strategic research and innovation roadmap: insights to the future direction of European marine biotechnology

Hwang EK, Yotsukura N, Pang SJ, Su L, Shan TF (2019) Seaweed breeding programs and progress in eastern Asian countries. Phycologia 58:484–495. https://doi.org/10.1080/00318884.2019.1639436

IMARC Group (2020) Seaweed market: global industry trends, share, size, growth, opportunity and forecast 2020-2025. 109. https://www.researchandmarkets.com/reports/5009064/seaweed-market-global-industry-trends-share?utm_source=dynamic&utm_medium=BW&utm_code=ggqjm7&utm_campaign=1375254+-+Seaweed+Market+Insights+Report%2C+2020-2025+Featuring+Profiles+of+Key+Players+Acadian+Se. Accessed 8 Oct 2020

IRO Alginate Industry Co. L (2021) IRO Alginate Industry Co., Ltd. Accessed 2 Apr 2021

de Jesus Raposo M, de Morais A, de Morais R (2015) Marine polysaccharides from algae with potential biomedical applications. Mar Drugs 13:2967–3028. https://doi.org/10.3390/md13052967

Jiao G, Yu G, Zhang J, Ewart H (2011) Chemical structures and bioactivities of sulfated polysaccharides from marine algae. Mar Drugs 9:196–223. https://doi.org/10.3390/md9020196

Jönsson M, Allahgholi L, Sardari RRR, Hreggviðsson GO, Nordberg Karlsson E (2020) Extraction and modification of macroalgal polysaccharides for current and next-generation applications. Molecules 25:930. https://doi.org/10.3390/molecules25040930

Jose GM, Kurup GM (2017) Sulfated polysaccharides from *Padina tetrastromatica* arrest cell cycle, prevent metastasis and downregulate angiogenic mediators in HeLa cells. Bioact Carbohydrates Diet Fibre 12:7–13. https://doi.org/10.1016/j.bcdf.2017.10.001

Kadam SU, Tiwari BK, O'Donnell CP (2015) Extraction, structure and biofunctional activities of laminarin from brown algae. Int J Food Sci Technol 50:24–31. https://doi.org/10.1111/ijfs.12692

Kanmani P, Rhim J-W (2014) Antimicrobial and physical-mechanical properties of agar-based films incorporated with grapefruit seed extract. Carbohydr Polym 102:708–716. https://doi.org/10.1016/j.carbpol.2013.10.099

Kidgell JT, Magnusson M, de Nys R, Glasson CRK (2019) Ulvan: a systematic review of extraction, composition and function. Algal Res 39:101422. https://doi.org/10.1016/j.algal.2019.101422

Kim JK, Yarish C, Hwang EK, Park M, Kim Y (2017) Seaweed aquaculture: cultivation technologies, challenges and its ecosystem services. Algae 32:1–13. https://doi.org/10.4490/algae.2017.32.3.3

Koyanagi S, Tanigawa N, Nakagawa H, Soeda S, Shimeno H (2003) Oversulfation of fucoidan enhances its anti-angiogenic and antitumor activities. Biochem Pharmacol 65:173–179. https://doi.org/10.1016/S0006-2952(02)01478-8

Lafarga T, Acién-Fernández FG, Garcia-Vaquero M (2020) Bioactive peptides and carbohydrates from seaweed for food applications: natural occurrence, isolation, purification, and identification. Algal Res 48:101909. https://doi.org/10.1016/j.algal.2020.101909

Lahaye M (2001) Chemistry and physico-chemistry of phycocolloids. Cah Biol Mar 42:137–157. https://doi.org/10.21411/cbm.a.a7aade12

Lahaye M, Ray B (1996) Cell-wall polysaccharides from the marine green alga *Ulva "rigida"* (Ulvales, Chlorophyta) — NMR analysis of ulvan oligosaccharides. Carbohydr Res 283:161–173. https://doi.org/10.1016/0008-6215(95)00407-6

Lahaye M, Robic A (2007) Structure and functional properties of ulvan, a polysaccharide from green seaweeds. Biomacromolecules 8:1765–1774. https://doi.org/10.1021/bm061185q

Lehahn Y, Ingle KN, Golberg A (2016) Global potential of offshore and shallow waters macroalgal biorefineries to provide for food, chemicals and energy: feasibility and sustainability. Algal Res 17:150–160. https://doi.org/10.1016/j.algal.2016.03.031

Li L, Ni R, Shao Y, Mao S (2014) Carrageenan and its applications in drug delivery. Carbohydr Polym 103:1–11. https://doi.org/10.1016/j.carbpol.2013.12.008

Li X, Wang J, Zhang H, Zhang Q (2017) Renoprotective effect of low-molecular-weight sulfated polysaccharide from the seaweed *Laminaria japonica* on glycerol-induced acute kidney injury in rats. Int J Biol Macromol 95:132–137. https://doi.org/10.1016/j.ijbiomac.2016.11.051

Liu Q-M, Yang Y, Maleki SJ, Alcocer M, Xu S-S, Shi C-L et al (2016) Anti-food allergic activity of sulfated polysaccharide from *Gracilaria lemaneiformis* is dependent on immunosuppression and inhibition of p38 MAPK. J Agric Food Chem 64:4536–4544. https://doi.org/10.1021/acs.jafc.6b01086

Lokhande G, Carrow JK, Thakur T, Xavier JR, Parani M, Bayless KJ et al (2018) Nanoengineered injectable hydrogels for wound healing application. Acta Biomater 70:35–47. https://doi.org/10.1016/j.actbio.2018.01.045

Maeda M, Nisizawa K (1968) Fine structure of laminaran of *Eisenia bicyclis*. J Biochem 63:199–206. https://doi.org/10.1093/oxfordjournals.jbchem.a128762

Maia MRG, Fonseca AJM, Oliveira HM, Mendonça C, Cabrita ARJ (2016) The potential role of seaweeds in the natural manipulation of rumen fermentation and methane production. Sci Rep 6:32321. https://doi.org/10.1038/srep32321

Martínez-Sanz M, Gómez-Mascaraque LG, Ballester AR, Martínez-Abad A, Brodkorb A, López-Rubio A (2019) Production of unpurified agar-based extracts from red seaweed *Gelidium sesquipedale* by means of simplified extraction protocols. Algal Res 38:101420. https://doi.org/10.1016/j.algal.2019.101420

Mathijs E, Brunori G, Carus M, Griffon M, Last L (2015) Sustainable agriculture, forestry and fisheries in the bioeconomy - a challenge for Europe. https://op.europa.eu/en/publication-detail/-/publication/7869030d-6d05-11e5-9317-01aa75ed71a1

Matsubara K, Matsuura Y, Bacic A, Liao M-L, Hori K, Miyazawa K (2001) Anticoagulant properties of a sulfated galactan preparation from a marine green alga, *Codium cylindricum*. Int J Biol Macromol 28:395–399. https://doi.org/10.1016/S0141-8130(01)00137-4

McHugh D (2003) A guide to the seaweed industry. FAO Fish Tech Pap No 441:105

MCPI Corporation (2021). https://www.mcpicarrageenan.com/. Accessed 5 Apr 2021

Menshova RV, Anastyuk SD, Ermakova SP, Shevchenko NM, Isakov VI, Zvyagintseva TN (2015) Structure and anticancer activity in vitro of sulfated galactofucan from brown alga *Alaria angusta*. Carbohydr Polym 132:118–125. https://doi.org/10.1016/j.carbpol.2015.06.020

Mohammad JH, Ranga Rao A, Ravishankar GA (2019) Opportunities and challenges in seaweeds as feed stock for biofuel production. In: Ravishnkar GA, Rao AR (eds) Handbook of algal technologies and phytochemicals: Volume II Phycoremediation, biofuels and global biomass production. CRC, New York, pp 39–50

Nause RG, Reddy RD, Soh JLP (2009) Propylene glycol alginate. In: Handbook of pharmaceutical excipients. Pharmaceutical Press, London, pp 594–595

Necas J, Bartosikova L (2013) Carrageenan: a review. Vet Med (Praha) 58:187–205. https://doi.org/10.17221/6758-VETMED

Ngo D-H, Kim S-K (2013) Sulfated polysaccharides as bioactive agents from marine algae. Int J Biol Macromol 62:70–75. https://doi.org/10.1016/j.ijbiomac.2013.08.036

Nor AM, Gray TS, Caldwell GS, Stead SM (2019) A value chain analysis of Malaysia's seaweed industry. J Appl Phycol. https://doi.org/10.1007/s10811-019-02004-3

Noseda MD, Cerezo A (1995) Alkali modification of carrageenans—II. The cyclization of model compounds containing nonsulfated β-D-galactose units. Carbohydr Polym 26:1–3. https://doi.org/10.1016/0144-8617(95)98826-3

Noseda MD, Viana AG, Duarte MER, Cerezo AS (2000) Alkali modification of carrageenans. Part IV. Porphyrans as model compounds. Carbohydr Polym 42:301–305. https://doi.org/10.1016/S0144-8617(99)00176-9

de Paniagua-Michel J, Olmos-Soto J, Morales-Guerrero ER (2014) Algal and microbial exopolysaccharides: new insights as biosurfactants and bioemulsifiers. 12:221–257. https://doi.org/10.1016/B978-0-12-800268-1.00011-1

Patel S (2012) Therapeutic importance of sulfated polysaccharides from seaweeds: updating the recent findings. 3 Biotech 2:171–185. https://doi.org/10.1007/s13205-012-0061-9

Patil NP, Le V, Sligar AD, Mei L, Chavarria D, Yang EY et al (2018) Algal polysaccharides as therapeutic agents for atherosclerosis. Front Cardiovasc Med 5. https://doi.org/10.3389/fcvm.2018.00153

Pereira-Pacheco F, Robledo D, Rodríguez-Carvajal L, Freile-Pelegrín Y (2007) Optimization of native agar extraction from *Hydropuntia cornea* from Yucatán, México. Bioresour Technol 98:1278–1284. https://doi.org/10.1016/j.biortech.2006.05.016

Peteiro C (2018) Alginate production from marine macroalgae, with emphasis on kelp farming. Algin Their Biomed Appl 12:27–66. https://doi.org/10.1007/978-981-10-6910-9_2
Read SM, Currie G, Bacic A (1996) Analysis of the structural heterogeneity of laminarin by electrospray-ionisation-mass spectrometry. Carbohydr Res 281:187–201. https://doi.org/10.1016/0008-6215(95)00350-9
Rinaudo M (2007) Seaweed polysaccharides. In: Comprehensive glycoscience. Elsevier. 691–735. https://doi.org/10.1016/B978-044451967-2/00140-9
Rioux L-E, Turgeon SL, Beaulieu M (2007) Characterization of polysaccharides extracted from brown seaweeds. Carbohydr Polym 69:530–537. https://doi.org/10.1016/j.carbpol.2007.01.009
Robic A, Rondeau-Mouro C, Sassi J-F, Lerat Y, Lahaye M (2009) Structure and interactions of ulvan in the cell wall of the marine green algae *Ulva rotundata* (Ulvales, Chlorophyceae). Carbohydr Polym 77:206–216. https://doi.org/10.1016/j.carbpol.2008.12.023
Rodríguez MC, Matulewicz MC, Noseda MD, Ducatti DRB, Leonardi PI (2009) Agar from *Gracilaria gracilis* (Gracilariales, Rhodophyta) of the Patagonic coast of Argentina – content, structure and physical properties. Bioresour Technol 100:1435–1441. https://doi.org/10.1016/j.biortech.2008.08.025
Sahoo D, Yarish C (2005) Mariculture of seaweeds. In: Andersen R (ed) Phycological methods: algal culturing techniques. Elsevier/Academic, New York, pp 219–237
Salehi S-R, Seca P, Michalak T et al (2019) Current trends on seaweeds: looking at chemical composition, phytopharmacology, and cosmetic applications. Molecules 24:4182. https://doi.org/10.3390/molecules24224182
Santelices B (1999) A conceptual framework for marine agronomy. Hydrobiologia 398–399:15–23. https://doi.org/10.1007/978-94-011-4449-0_3
Shama A, Joyce SG, Mari FD, Ranga Rao A, Ravishankar GA, Hudaa N (2019) Macroalgae and microalgae: novel sources of functional food and feed. In: Ravishankar GA, Rao AR (eds) Handbook of algal technologies and phytochemicals: Volume-I: food, health and nutraceutical applications. CRC, New York, pp 207–219
Shi F, Yan X, Cheong K-L, Liu Y (2018) Extraction, purification, and characterization of polysaccharides from marine algae *Gracilaria lemaneiformis* with anti-tumor activity. Process Biochem 73:197–203. https://doi.org/10.1016/j.procbio.2018.08.011
Silchenko AS, Rasin AB, Kusaykin MI, Malyarenko OS, Shevchenko NM, Zueva AO et al (2018) Modification of native fucoidan from *Fucus evanescens* by recombinant fucoidanase from marine bacteria *Formosa algae*. Carbohydr Polym 193:189–195. https://doi.org/10.1016/j.carbpol.2018.03.094
Stadnik MJ, de Freitas MB (2014) Algal polysaccharides as source of plant resistance inducers. Trop Plant Pathol 39:111–118. https://doi.org/10.1590/S1982-56762014000200001
Szekalska M, Puciłowska A, Szymańska E, Ciosek P, Winnicka K (2016) Alginate: current use and future perspectives in pharmaceutical and biomedical applications. Int J Polym Sci 2016:1–17. https://doi.org/10.1155/2016/7697031
Tanna B, Mishra A (2019) Nutraceutical potential of seaweed polysaccharides: structure, bioactivity, safety, and toxicity. Compr Rev Food Sci Food Saf 18:817–831. https://doi.org/10.1111/1541-4337.12441
TBK Manufacturing Corporation (2021). https://www.tbk.com.ph/. Accessed 7 Apr 2021
Trica D, Gros U, Dobre D et al (2019) Extraction and characterization of alginate from an edible brown seaweed (*Cystoseira barbata*) harvested in the Romanian black sea. Mar Drugs 17:405. https://doi.org/10.3390/md17070405
Tziros GT, Samaras A, Karaoglanidis GS (2021) Laminarin induces defense responses and efficiently controls olive leaf spot disease in olive. Molecules 26:1043. https://doi.org/10.3390/molecules26041043
UPL (2021). https://www.upl-ltd.com/ke. Accessed 31 Mar 2021
Usov AI (2011) Polysaccharides of the red algae. pp. 115–217. https://doi.org/10.1016/B978-0-12-385520-6.00004-2
Venugopal V (2016) Marine Polysaccharides. https://doi.org/10.1201/b10516

Venugopal V (2019) Sulfated and non-sulfated polysaccharides from seaweeds and their uses: an overview. EC Nutr 12:126–141

Viana A, Noseda M, Duarte M, Cerezo A (2004) Alkali modification of carrageenans. Part V. the iota-nu hybrid carrageenan from *Eucheuma denticulatum* and its cyclization to iota-carrageenan. Carbohydr Polym 58:455–460. https://doi.org/10.1016/j.carbpol.2004.08.006

Wu Y, Li W, Cui W, Eskin NAM, Goff HD (2012) A molecular modeling approach to understand conformation–functionality relationships of galactomannans with different mannose/galactose ratios. Food Hydrocoll 26:359–364. https://doi.org/10.1016/j.foodhyd.2011.02.029

Xin Z, Cai X, Chen S, Luo Z, Bian L, Li Z et al (2019) A disease resistance elicitor laminarin enhances tea defense against a piercing herbivore *Empoasca (Matsumurasca) onukii* Matsuda. Sci Rep 9:814. https://doi.org/10.1038/s41598-018-37424-7

Zayed A, Ulber R (2020) Fucoidans: downstream processes and recent applications. Mar Drugs 18:170. https://doi.org/10.3390/md18030170

Zhou G (2004) In vivo antitumor and immunomodulation activities of different molecular weight lambda-carrageenans from *Chondrus ocellatus*. Pharmacol Res 50:47–53. https://doi.org/10.1016/j.phrs.2003.12.002

Zhou G, Sheng W, Yao W, Wang C (2006) Effect of low molecular λ-carrageenan from *Chondrus ocellatus* on antitumor H-22 activity of 5-Fu. Pharmacol Res 53:129–134. https://doi.org/10.1016/j.phrs.2005.09.009

Zibetti RGM, Noseda MD, Cerezo AS, Duarte MER (2005) The system of galactans from *Cryptonemia crenulata* (Halymeniaceae, Halymeniales) and the structure of two major fractions. Kinetic studies on the alkaline cyclization of the unusual diad G2S→D(L)6S. Carbohydr Res 340:711–722. https://doi.org/10.1016/j.carres.2005.01.010

Zibetti RGM, Duarte MER, Noseda MD, Colodi FG, Ducatti DRB, Ferreira LG et al (2009) Galactans from Cryptonemia species. Part II: studies on the system of galactans of Cryptonemia seminervis (Halymeniales) and on the structure of major fractions. Carbohydr Res 344:2364–2374. https://doi.org/10.1016/j.carres.2009.09.003

Chapter 23
Industrial Potential of Seaweeds in Biomedical Applications: Current Trends and Future Prospects

Eko Susanto, Yanuariska Putra, and Ratih Pangestuti

Abbreviations

FA	Fatty acids
LMW	Low-molecular-weight
MAAs	Mycosporine like amino acids
SIRT1	Sirtuin 1
UV	Ultraviolet

1 Introduction

Ocean covers more than 70 percent of the Earth's surface, holds more than 92 percent of the Earth's water. It is also a habitat for almost three-quarters of all known species (around 500,000 species) (Gomez-Zavaglia et al. 2019). Among marine organisms, seaweeds represent a large and diverse group of marine flora, with sizes ranging from a few centimeters up to 100 m in length. Seaweeds can be classified based on their pigment profiles, into red (Rhodophyta), brown (Phaeophyta), and green seaweeds (Chlorophyta) (García-Poza et al. 2020; Pangestuti and Kim 2011).

Presently, majority seaweeds production were used to support phycocolloids industry, food, animal feeds and fertilizers (Pangestuti et al. 2019). However, recent data revealed an increase in global seaweed productions worldwide, indicating that the popularity and consumption of seaweed are increasing not only in East Asian countries but also in other countries (Chen and Roca 2019; Nova et al. 2020;

E. Susanto
Department of Fisheries Technology Product, Faculty of Fisheries and Marine Sciences, Diponegoro University, Semarang, Central Java, Republic of Indonesia

Y. Putra · R. Pangestuti (✉)
Research and Development Divisions for Marine Bio-industry, Indonesian Institute of Science (BBIL-LIPI), Lombok, West Nusa Tenggara, Republic of Indonesia
e-mail: ratih.pangestuti@lipi.go.id

A. Ranga Rao, G. A. Ravishankar (eds.), *Sustainable Global Resources of Seaweeds Volume 2*, https://doi.org/10.1007/978-3-030-92174-3_23

Pangestuti and Kim 2014). This phenomenon is due to the increase consumer knowledge and awareness of seaweed's nutritional and medicinal value. In addition, seaweeds are excellent sources of functional materials including polysaccharides, amino acids, essential fatty acids (FAs) and other bioactive materials (Chen and Roca 2019; Nova et al. 2020).

Large numbers of studies have demonstrated biological activities as well as health benefits effects of seaweeds-derived functional materials. For example, sulphated polysaccharides derived from seaweeds (i.e carrageenan, ulvan and fucoidan) presents several potentially valuable biological properties such as anticoagulant activities (heparin like), antioxidant activities, anti-thrombotic activities, immunomodulating activities, neuroprotective, strain-specific anti-influenza activities, anticoagulant activities antihyperlipidemic activities, and antitumoral activities. Therefore, seaweed-derived functional materials are potential to be developed in biomedical industries. This contribution presents an overview of the developments of current trends of seaweeds for biomedical applications, fundamental and potential of seaweeds products in biomedical industries.

2 Potential Applications of Seaweeds in Biomedical Industries

Many seaweeds species have been used as traditional remedies in many countries (Murakami et al. 2011; Kim et al. 2011). In addition, seaweeds-derived functional materials have been used from generations to generations as food and folk medicine. The relation of seaweeds and human health have attracted great interest (Tanna and Mishra 2019). Italian physicians of the School of Salerno were the first to report the medicinal value of dried seaweed to treat goitre. Further, the Swiss physician Coindet, in 1813, practised the traditional treatment of goitre with seaweeds (Zimmermann 2008). Up to now, numerous studies have demonstrated the biomedical applications of seaweed bioactive compounds from different seaweeds species.

2.1 Dietary Supplements

Dietary supplements comprise a wide array of products intended for ingestion to meet essential nutritional requirements (Rautiainen et al. 2016). Dietary supplements are well recognized to offer the potential to maintain health and disease prevention. Large numbers of high quality scientific literature supporting significant bioactivity of seaweeds functional materials and their potential to be applied as dietary supplements. Seaweed's functional materials have been continuously shown to possess wide spectrum of biological activities including antioxidant, anti-cancer, neuroprotective, anti-inflammatory, anti-microbial, anti-viral and many other health benefit effects.

Many studies have demonstrated the antioxidant potentials of seaweeds functional materials such as carotenoids, polysaccharides, phenolic compounds and peptides (Pangestuti and Kim 2011; Wijesekara et al. 2010; Pangestuti et al. 2013). In addition, seaweeds also showed another potential as neuroprotective agents. Recently, seaweeds-derived oligosaccharide (sodium oligomannate) has been showed to harnesses neuroinflammation and reverses the cognition impairment. These finding has enlightened the development of seaweeds-based dietary supplements and drug discovery in Alzheimer's diseases.

Fucoidan have been demonstrated to inhibit SARS-CoV-2 in vitro (Kwon et al. 2020). These sulphated polysaccharides have been reported to inhibit the binding and entry of viruses and bacteria into human cells. Fucoidan provide special saccharides and biological sugars, which are important for cell-to-cell communication through glycoproteins and glycolipids (Pangestuti et al. 2021). Currently, seaweeds functional materials based dietary supplements products are available in the markets and the numbers are growing. Seaweeds based dietary supplements are available in different forms such as capsules, tablets, soft gels, liquids, chewable preparations and powders.

2.2 Microbicides for Sexually Transmitted Diseases

Sulfated polysaccharides from seaweeds gained a large amount of attention due to their broad-spectrum antimicrobial activities. Carrageenans were expected to be the first clinically applicable microbicides among certain polysaccharides classes, owing to their superior antiviral potency and favorable safety profiles in subclinical studies (Lee 2020). It has been reported that carrageenan especially iota (ι)–carrageenan inhibitory activity on human papilloma virus (HPV) infection is higher than heparin (an effective model for HPV inhibitors) (Roberts et al. 2007). The HPV inhibitory activity of carrageenan was mediated by the prevention of the binding of HPV virions to cells and further blocks HPV infection. In addition, chemical structures of carrageenan resemble heparan sulphate. These sulphated polysaccharides has also demonstrated to inhibit genital transmission of HPV in female mouse model of cervicovaginal (Schiller and Davies 2004). Carrageenan was able to generate antigen-specific immune responses and antitumor effects in female mice vaccinated with HPV-16 E7 peptide (Zhang et al. 2010). Previous work indicates carrageenan-based microbicides are safe for vaginal use, as shown by the Carraguard trial in South Africa (Carraguard Phase IISAST 2010).

2.3 Photoprotective Agents against Cancers

Ultraviolet (UV) radiation possess beneficial effects for human skins; however, prolonged exposure to UVR could be an aggressive factor for skin photoaging and mutations which causes skin cancer and other disorders (Solano 2020). The UVR (direct and or indirect) induced the activation of complex signalling cascade in human skin (Xu and Fisher 2005). In addition, prolonged UVR exposure have been reported as leading cause of photoaging which can be characterized by wrinkles, loss of skin tone, pigmentation (hypo- or hyperpigmentation), rough skin, dryness, sallowness, etc. (Wlaschek et al. 2001; Pandel et al. 2013; Wang et al. 2019). Therefore, it is important to search for natural photoprotective agents to prevent photoaging and other skin disorders due to the deleterious effect of UVR.

Functional materials from seaweeds has been proven as great source of novel materials for incorporating into anti-photoaging formulations (Kim and Pangestuti 2011; Pangestuti et al. 2019). Seaweeds polysaccharides including fucoidan, laminarin, and carrageenan showed potential anti-photoaging properties which were mediated by intra-cellular ROS scavenging activity in vitro and in vivo (Wijesekara et al. 2010; Pangestuti and Kim 2014; Zargarzadeh et al. 2020; Ku et al. 2010). In addition, other seaweed-derived materials such as mycosporine like amino acids (MAAs) are known as the most potential natural UVA-absorbing molecules (Pangestuti et al. 2018). Seaweeds extracts are also continuously reported as potential anti-photoaging agents (Freitas et al. 2020).

Seaweeds photoprotective products have been developed and currently available in the market. For example, extract from brown seaweeds (*Undaria pinnatifida Fucus vesiculosus*) have been demonstrated to increase the expression of sirtuin 1 (SIRT1, a protein known for anti-ageing and longevity). Furthermore, clinical testing established the efficacy of the extracts in a range of tested applications, relative to placebo. The *U. pinnatifida* extract modulated skin immunity while *Fucus vesiculosus* extract reduced age spot and increased brightness (Fitton et al. 2015). Both brown seaweed extract provide skin soothing and protection and currently available in the market under Marinova's (Biotech Company from Autralia). The other anti-photoaging extract containing MAAs are also available in the market is presented in Table 23.1. Seaweed derived functional materials are potentials as active photoprotective ingredients in sunscreen, anti-photoaging cream, and other biomedical applications.

2.4 Drug Delivery

Drug-delivery has attracted great interest, since they can deliver molecules from low-molecular-weight (LMW) to macromolecules. In addition, extensive research articles have been published on seaweeds-derived functional materials (i.e alginates and carrageenan) for drug delivery. Alginate for example has been reported as a

Table 23.1 Anti-photoaging agents from seaweeds currently available in the market

Commercial name	Company	Active ingredients	Anti-photoaging activity
Helionori®	Gelyma, French	MAAs from *Poprphyra umbilicalis*	Photoprotective (UV-A) DNA protection Prevention of sunburn
Helioguard365	Mibelle Biochemistry, Switzerland	Porphyra-334 & Shinorine from *Poprphyra umbilicalis*	Photoprotective (UV-A)
Algae gorria; alga marris	Laboratoires de Biarritz, French	NA from *Poprphyra umbilicalis*	Photoprotective (UV-A)
Fucorich	Marinova, Australia	Fucoidan from *Undaria pinnatifida*	Anti-aging
Maritech reverse	Marinova, Australia	Fucoidan from *Fucus vesiculosus*	Anti-aging; antioxidant; anti-inflammation
Maritech synergy	Marinova, Australia	Fucoidan & polyphenol complex from *Fucus vesiculosus*	Anti-aging; antioxidant; anti-inflammation
Maritech synergy	Marinova, Australia	Fucoidan & polyphenol complex from *Fucus vesiculosus*	Anti-aging; brightening, antioxidant

MAAs Mycosporine like amino acids, UV ultraviolet

carrier to immobilize or encapsulate drugs, bioactive molecules, proteins and cells (Venkatesan et al. 2015). Sulfated polysaccharides from red seaweeds, carrageenans also potential to be applied in drug delivery due to their unique structiure such as glycosidic bonds, sulphate groups and the presence of hydroxyl groups (Pacheco-Quito et al. 2020). Currently, seaweeds functional materials have been applied in different forms such as hydrogels, colloidal particles, nanoparticle, microspheres, microstabilizers and polyelectrolyte complexes.

2.5 *Wound Healing*

Wound healing is a complex biological process that requires various forms of cell rejuvenation (Venkatesan et al. 2015). Explorations of novel functional materials with remarkable tissue repairing and scar formation-limiting abilities have attracted great attentions. Polysaccharides extracted from seaweeds such as fucoidan, alginates, laminarin, and carrageenan have gained a more and more attention in the wound healing owing to their diversity and other potencies (i.e non-toxicity, biodegradability, and bio-renewable characteristics) (Bilal and Iqbal 2020). To date, seaweeds polysaccharides have proven to be particularly appealing in wound healing applications. It can be applied to products that have wound-healing properties. For example, alginate has been used to make a variety of wound dressing products, including hydrogels, gels, and foams.

In 2010, Murakami and his colleagues demonstrated that hydrogel sheet fabricated by combination of alginate, chitin/chitosan, and fucoidan is favourable as wound healing agent by repairing impaired wounds through the stimulation of repair of mitocycin treated healing impaired wounds in rats model (Murakami et al. 2010). In addition, recently an innovative study examining polyelectrolyte multilayers strengthens the case for fucoidan as a promising candidate for the development of new materials for use in biomedical sectors (Benbow et al. 2020).

2.6 *Tissue Engineering*

Tissue engineering is an emerging field of studies aimed to prepare artificial materials for the treatment or replacement of infected organs. Alginate is seaweeds-derived biopolymer that has been commonly used in tissue engineering because of its favourable properties, such as non-toxic, biocompatibility and ease of gelation. Alginates have been used to create tissue engineering constructs with a variety of structures such as porous scaffolds, microspheres, films, and other applications. This seaweeds biopolymer is one of the best known biomaterials to form scaffold forming properties. In addition, due to their biocompatibility in a minimally invasive way and ability to fill irregularly shaped defects, alginate gels showed advantages for bone and cartilage regeneration.

3 Conclusions and Future Prospectives

Functional materials from seaweeds have been shown to possess medicinal properties; however, there are still large opportunities to explore seaweeds in pharmacy and biomedical industries. In addition, seaweeds-derived functional materials have several advantages to be applied in biomedical industries such as bio-renewable characteristics, non-toxic properties, low-cost productions, biodegradability and seaweeds diversity. Further studies towards bioavailability of seaweeds-derived functional materials, clinical trials and development of biomedical products are required. Observational studies, randomized trials are important to advancing our understanding of the role of seaweeds functional materials use in disease prevention. Future research towards promising seaweeds based dietary supplements will help to fill the existing research gaps and inform future public-health recommendations. In addition, to support biomedical industries, seaweed aquaculture and environmental friendly extraction techniques need to be explored. Collectively, the development of seaweeds functional materials in biomedical industries is feasible and offer great opportunities but also possess as a great challenge for many sectors (i.e scientist, engineers, pharmacies, medical doctors, and product developers).

Acknowledgments This study is supported by National Research Priority (*Program Prioritas Riset Nasional*, PRN)-Marine Micro and Macro-algae from Indonesia (MALSAI).

References

Benbow NL, Karpiniec S, Krasowska M, Beattie DA (2020) Incorporation of FGF-2 into pharmaceutical grade Fucoidan/chitosan polyelectrolyte multilayers. Mar Drugs 18(11):531

Bilal M, Iqbal H (2020) Marine seaweed polysaccharides-based engineered cues for the modern biomedical sector. Mar Drugs 18(1):7

Carraguard Phase IISAST (2010) Expanded safety and acceptability of the candidate vaginal microbicide Carraguard® in South Africa. Contraception 82(6):563–571. https://doi.org/10.1016/j.contraception.2010.04.019

Chen K, Roca M (2019) Cooking effects on bioaccessibility of chlorophyll pigments of the main edible seaweeds. Food Chem 295:101–109

Fitton JH, Dell'Acqua G, Gardiner V-A, Karpiniec SS, Stringer DN, Davis E (2015) Topical benefits of two fucoidan-rich extracts from marine macroalgae. Cosmetics 2(2):66–81

Freitas R, Martins A, Silva J, Alves C, Pinteus S, Alves J, Teodoro F, Ribeiro HM, Gonçalves L, Petrovski Ž (2020) Highlighting the biological potential of the Brown seaweed Fucus spiralis for skin applications. Antioxidants 9(7):611

García-Poza S, Leandro A, Cotas C, Cotas J, Marques JC, Pereira L, Gonçalves AM (2020) The evolution road of seaweed aquaculture: cultivation technologies and the industry 4.0. Int J Environ Res Public Health 17 (18):6528

Gomez-Zavaglia A, Prieto Lage MA, Jimenez-Lopez C, Mejuto JC, Simal-Gandara J (2019) The potential of seaweeds as a source of functional ingredients of prebiotic and antioxidant value. Antioxidants 8(9):406

Kim S-K, Pangestuti R (2011) 15 biological properties of cosmeceuticals derived from marine algae. Marine Cosmeceuticals: Trends and Prospects:191

Kim SK, Pangestuti R, Rahmadi P (2011) Sea lettuces: culinary uses and nutritional value. Adv Food Nutr Res 64:57–70

Ku M-J, Jung J-W, Lee M-S, Cho B-K, Lee S-R, Lee H-S, Vischuk OS, Zvyagintseva TN, Ermakova SP, Lee Y-H (2010) Effect of Fucus evanescens fucoidan on expression of matrix metalloproteinase-1 promoter, mRNA, protein and signal pathway. Journal of Life Science 20(11):1603–1610

Kwon PS, Oh H, Kwon S-J, Jin W, Zhang F, Fraser K, Hong JJ, Linhardt RJ, Dordick JS (2020) Sulfated polysaccharides effectively inhibit SARS-CoV-2 in vitro. Cell discovery 6(1):1–4

Lee C (2020) Carrageenans as broad-Spectrum microbicides: current status and challenges. Mar Drugs 18(9):435

Murakami K, Aoki H, Nakamura S, Nakamura S-i, Takikawa M, Hanzawa M, Kishimoto S, Hattori H, Tanaka Y, Kiyosawa T, Sato Y, Ishihara M (2010) Hydrogel blends of chitin/chitosan, fucoidan and alginate as healing-impaired wound dressings. Biomaterials 31(1):83–90. https://doi.org/10.1016/j.biomaterials.2009.09.031

Murakami K, Yamaguchi Y, Noda K, Fujii T, Shinohara N, Ushirokawa T, Sugawa-Katayama Y, Katayama M (2011) Seasonal variation in the chemical composition of a marine brown alga, Sargassum horneri (turner) C. Agardh. J Food Compos Anal 24(2):231-236

Nova P, Martins AP, Teixeira C, Abreu H, Silva JG, Silva AM, Freitas AC, Gomes AM (2020) Foods with microalgae and seaweeds fostering consumers health: a review on scientific and market innovations. J Appl Phycol 32:1789–1802

Pacheco-Quito E-M, Ruiz-Caro R, Veiga M-D (2020) Carrageenan: drug delivery systems and other biomedical applications. Mar Drugs 18(11):583

Pandel R, Poljšak B, Godic A, Dahmane R (2013) Skin Photoaging and the role of antioxidants in its prevention. ISRN Dermatology 2013:930164. https://doi.org/10.1155/2013/930164

Pangestuti R, Kim S-K (2011) Biological activities and health benefit effects of natural pigments derived from marine algae. J Funct Foods 3 (4):255–266. doi:http://dx.doi.org/https://doi.org/10.1016/j.jff.2011.07.001

Pangestuti R, Kim S-K (2014) Biological activities of carrageenan. Marine Carbohydr Fund Appl 113

Pangestuti R, Vo T-S, Ngo D-H, Kim S-K (2013) Fucoxanthin ameliorates inflammation and oxidative Reponses in microglia. J Agric Food Chem 61(16):3876–3883

Pangestuti R, Siahaan E, Kim S-K (2018) Photoprotective substances derived from marine algae. Mar Drugs 16(11):399

Pangestuti R, Getachew AT, Siahaan EA, Chun B-S (2019) Characterization of functional materials derived from tropical red seaweed Hypnea musciformis produced by subcritical water extraction systems. J Appl Phycol 31(4):2517–2528. https://doi.org/10.1007/s10811-019-1754-9

Pangestuti R, Shin K-H, Kim S-K (2021) Anti-Photoaging and potential skin health benefits of seaweeds. Mar Drugs 19(3):172

Rautiainen S, Manson JE, Lichtenstein AH, Sesso HD (2016) Dietary supplements and disease prevention—a global overview. Nat Rev Endocrinol 12(7):407–420

Roberts JN, Buck CB, Thompson CD, Kines R, Bernardo M, Choyke PL, Lowy DR, Schiller JT (2007) Genital transmission of HPV in a mouse model is potentiated by nonoxynol-9 and inhibited by carrageenan. Nat Med 13(7):857–861

Schiller JT, Davies P (2004) Delivering on the promise: HPV vaccines and cervical cancer. Nat Rev Microbiol 2(4):343–347

Solano F (2020) Photoprotection and skin pigmentation: melanin-related molecules and some other new agents obtained from natural sources. Molecules 25(7):1537

Tanna B, Mishra A (2019) Nutraceutical potential of seaweed polysaccharides: structure, bioactivity, safety, and toxicity. Compr Rev Food Sci Food Saf 18(3):817–831

Venkatesan J, Lowe B, Anil S, Manivasagan P, Al Kheraif AA, Kang KH, Kim SK (2015) Seaweed polysaccharides and their potential biomedical applications. Starch-Stärke 67(5–6):381–390

Wang M, Charareh P, Lei X, Zhong JL (2019) Autophagy: multiple mechanisms to protect skin from ultraviolet radiation-driven Photoaging. Oxidative Med Cell Longev 2019:8135985. https://doi.org/10.1155/2019/8135985

Wijesekara I, Pangestuti R, Kim SK (2010) Biological activities and potential health benefits of sulfated polysaccharides derived from marine algae. Carbohydr Polym 84(1):14–21

Wlaschek M, Tantcheva-Poór I, Naderi L, Ma W, Schneider LA, Razi-Wolf Z, Schüller J, Scharffetter-Kochanek K (2001) Solar UV irradiation and dermal photoaging. J Photochem Photobiol B Biol 63 (1):41–51. doi:https://doi.org/https://doi.org/10.1016/S1011-1344(01)00201-9

Xu Y, Fisher GJ (2005) Ultraviolet (UV) light irradiation induced signal transduction in skin photoaging. J Dermatol Sci Suppl 1(2):S1–S8

Zargarzadeh M, Amaral AJR, Custódio CA, Mano JF (2020) Biomedical applications of laminarin. Carbohydr Polym 232:115774. doi:https://doi.org/https://doi.org/10.1016/j.carbpol.2019.115774

Zhang YQ, Tsai YC, Monie A, Hung CF, Wu TC (2010) Carrageenan as an adjuvant to enhance peptide-based vaccine potency. Vaccine 28(32):5212–5219

Zimmermann MB (2008) Research on iodine deficiency and goiter in the 19th and early 20th centuries. J Nutr 138(11):2060–2063. https://doi.org/10.1093/jn/138.11.2060

Chapter 24
Antiviral Compounds from Seaweeds: An Overview

João Cotas, Diana Pacheco, Ana Marta Mendes Gonçalves, and Leonel Pereira

1 Introduction

Algae has fueled a lot of interest mostly, as a source of fertilizers, feeds, foods, and pharmaceutical precursors. However, there is a growing interest in using algal diversity for anti-viral purposes. A summary of 50 years of scientific and technical advancements in the field of algae antivirals was recently reviewed by Pagarete et al. 2021.

Marine algae, or seaweeds, have proven to be a prolific source of secondary metabolites with complex and special chemical structures and a wide range of possible biological activities among marine organisms. Antibacterial, antifungal, antiprotozoal, antituberculosis, anticoagulant, antithrombotic, and antiviral effects are among them (Vonthron-Sénécheau 2016).

Seaweeds were thought to have medicinal value in Asian cultures as early as 3000 B.C. For over 1000 years, the Romans and some British populations have used them to treat wounds and as effective vermifuges and anthelmintics (Smit 2004). Currently, the quest for cures for human diseases seems to be a never-ending process. Many algal organisms, including Rhodophyta, Phaeophyceae, and Chlorophyta, produce a variety of secondary metabolites that are essential for chemical defenses. In the last 50 years, more than 3000 natural compounds with pharmacological

J. Cotas · D. Pacheco · L. Pereira (✉)
Department of Life Sciences, University of Coimbra, MARE – Marine and Environmental Sciences Centre, Coimbra, Portugal
e-mail: jcotas@uc.pt; diana.pacheco@uc.pt; leonel.pereira@uc.pt

A. M. M. Gonçalves
Department of Life Sciences, University of Coimbra, MARE – Marine and Environmental Sciences Centre, Coimbra, Portugal

Department of Biology and CESAM, University of Aveiro, Aveiro, Portugal
e-mail: amgoncalves@uc.pt; anamartagoncalves@ua.pt

A. Ranga Rao, G. A. Ravishankar (eds.), *Sustainable Global Resources of Seaweeds Volume 2*, https://doi.org/10.1007/978-3-030-92174-3_24

properties have been discovered, many of them identified as novel compounds with specific chemical structures, according to excellent reports on bioactive metabolites from seaweeds recently released (Blunt et al. 2016; Freile-Pelegrín and Tasdemir 2019).

Antiviral properties in macroalgae against the transmission of the human immunodeficiency virus (HIV) and other sexually transmitted viruses including herpes simplex virus (HSV) and genital warts (Ismail et al. 2020). Algal molecules, such as sulfated polysaccharides (SPs) and phenolic compounds present antiviral activity by preventing viral adsorption (simultaneous-treatment assay) and replication (post-treatment assay) (Kwon et al. 2013). The mechanism of macroalgal polysaccharides against viral diseases focuses on the viral attachment process, where they bind to virions and/ or link to the appropriate protein receptors and the immunomodulators stimulate immune responses or trigger natural killer cells (NK) (Shi et al. 2017).

Historically, Irish moss or carrageen (a mixture of naturally occurring red algae, consisting of *Chondrus crispus* and *Mastocarpus stellatus*) has many medical applications, some of which date back to the 1830s. In fact, it is still used in Ireland to make traditional medicines, teas, and cough medicine to fight colds, bronchitis, and chronic cough. It is said to be particularly useful for dislodging mucus and has several antiviral properties (Pereira 2018; Pereira and Critchley 2020).

Carrageenans and oligo-polysaccharides agars from different red seaweed species showed antiviral properties (Torres et al. 2019). *In vitro*, crude polysaccharides from red algae (*Pterocladia capillacea* and *Laurencia obtusa*) from Abo Qir, Alexandria, were found to be antiviral agents against hepatitis C virus (HCV) (Gheda et al. 2016). Carrageenans extracted from *Gigartina skottsbergii* (currently known as *Sarcopeltis skottsbergii*) may be used as models for designing new anti-HSV 1 and 2 agents that block viral attachment. It can also improve the efficiency of NK cells and increase the number of lymphocytes generated (Diogo et al. 2015). Moreover, the polysaccharide extracted from the red seaweed *Solieria chordalis* (iota-carrageenan) had antiviral activity against HSV1 (Boulho et al. 2017).

2 Seaweed Compounds as Antiviral

2.1 Phenolic Compounds

The bromophenol (2,3,6-tribromo-4,5-dihydroxybenzyl methyl ether) extracted from the red seaweed *Symphyocladia latiuscula* has been shown to have antiviral properties against different forms of herpes simplex type 1. This assay was carried out in mice with antiviral activity at a dose of 20 mg/kg for 6–10 days, with no significant side effects reported. In the skin lesions and the concentration of the virus particles in the brain, the antiviral activity behaved similarly to acyclovir (Park et al. 2005).

Dieckol (isolated from the brown seaweed *Ecklonia cava*) has been shown to interfere with SARS-CoV viral replication, specifically in SARS-CoV 3CL protease trans/ cis-cleavage with a high association rate and a dose-dependent effect on 3CL protease hydrolysis (Park et al. 2013). Furthermore, this compound has shown to be effective against influenza. The most promising target is a virus neuraminidase, which plays an important role in the viral life cycle; however, phlorofucofuroeckol was also the best phlorotannin in the assay, equaling dieckol. Furthermore, these compounds work in tandem with oseltamivir to boost the inhibitory effects (Ryu et al. 2011).

The phenolic compounds (Eckol, dieckol, 8,8′-bieckol, 6,6′-bieckol, and phlorofucofuroeckol-A) extracted from the brown seaweed from *Eisenia bicyclis* also demonstrated antiviral activity against human papilloma virus (Kim and Kwak 2015).

Particularly, dieckol and phlorofucofuroeckol-A, both extracted from *E. bicyclis*, were demonstrated to have antiviral action against murine norovirus (Eom et al. 2015). Lee et al. (2016) demonstrated that dieckol (isolated from *E. cava*) suppresses liver fibrosis, indicating that this phlorotannin could be used to treat chronic liver inflammation caused by alcohol misuse, metabolic diseases, viral hepatitis, cholestatic liver diseases, and autoimmune diseases.

Diterpene derivatives isolated from the red species *Dictyota pfaffii* (currently known as *Dictyota friabilis*) and *Dictyota menstrualis* showed anti-HIV activity with reduced toxicity, making them promising drug candidates (Garrido et al. 2017).

2.2 *Polysaccharides*

Sulfated polysaccharides (SPs) are a type of polyanionic molecule found in algae that has been studied for the treatment of a variety of viral infections (Ahmadi et al. 2015), such as HIV, Herpes Simplex Virus (HSV), African swine fever virus (ASFV), and influenza A virus (Flu-A), in particular. Carrageenan's are the most studied algal polysaccharides and are considered healthy and safe for human consumption (Weiner 2016). Other algal polysaccharides, such as fucans and ulvans, have been studied and are thought to be promising candidates for antiviral drug production (Rosales-Mendoza et al. 2020).

2.2.1 Carrageenan

Carrageenans' antiviral mechanisms include viral attachment and uncoating inhibition, as well as transcription, replication, and immune function modulation. The molecular weight of carrageenans and the degree of sulfation have an effect on viral attachment blocking (Jiao et al. 2011). Low molecular weight (LMW) carrageenan derivatives, for example, had antiviral properties. LMW carrageenans can be obtained by enzymatic degradation, free radical depolymerization, mild acid hydrolysis, or free radical depolymerization (Liang et al. 2009; Zhang 2018). However, the antiviral activity may be affected by the depolymerization technique. For instance, it was recorded that the antiviral activity of LMW derivatives of kappa and kappa/ beta-carrageenans was more effective through mild acid hydrolysis; followed by free radical depolymerization and enzymatic degradation (Kalitnik et al. 2013). LMW can enter the host cell and prevent viral replication. Kapa-carrageenan oligosaccharides (KCO), for example, had this effect on the influenza A virus (Wang et al. 2011, 2012).

While, iota-carrageenans not only inhibited viral attachment, but also viral internalization; blocking the attachment of HSV and the Dengue virus (Talarico et al. 2007, 2011). Moreover, iota-carrageenan was able to inhibit the viral duplication of rhinovirus (HRV) (Grassauer et al. 2008), and substantially reduce the viral replication of the influenza virus H1N1 strain, while increasing the survival rate of the infected cells (Leibbrandt et al. 2010).

2.2.2 Fucoidan

The anti-HIV activity of fucoidans from several brown algae has been documented. For instance, fucan A and B, from the brown algae *Spatoglossum schroederi* and *Dictyota mertensii,* may stop HIV replication and transcription (Rocha et al. 2005; Queiroz et al. 2008). Other fucans, extracted from species, such as *Lobophora variegata* and *Fucus vesiculosus*, were found to have a significant inhibitory effect on HIV-1 reverse transcriptase. The HIV-1 viral entry point on the host cell was inhibited by fucoidans from three brown algae (*Sargassum mcclurei*, *Sargassum polycystum* and *Turbinaria ornata*) (Thuy et al. 2015). Fucoidans have been tested for anti-influenza A virus (IAV) activity *in vitro* and *in vivo*; for instance, fucoidan extracted from *Undaria pinnatifida* has exhibited anti-HSV activity (Hayashi et al. 2008) and anti-IAV activity *in vitro* and in mice with normal and compromised immune system (Hayashi et al. 2013). Jiao et al. (2012) tested antiviral activity against the influenza A/PR/8/34 (H1N1) virus and found that fucoidans from *Fucus vesiculosus* and *Ascophyllum nodosum* had the highest antiviral activity. Fucoidan was isolated from the brown alga *Kjellmaniella crassifolia* and found to increase the survival rate and lifespan of mice infected with influenza viruses while also reducing viral load, according to Wang et al. (2017). Furthermore, the H1N1 (Ca109) strain was the most susceptible, and the antiviral mechanism may be blocking the viral penetration by inhibiting epidermal growth factor receptor activation. Fucoidan should be administered via nasal and oral, and it should be used as soon as possible after the infection.

In vitro experiments with two fucoidans derived from edible brown algae recently revealed that these molecules are effective SARS-CoV-2 inhibitors (Song et al. 2020). Fucoidans clearly outperformed Remdesivir, which is currently approved for use in serious COVID-19 infections as an emergency treatment (Kwon et al. 2020).

2.2.3 Ulvan

Ulvans isolated from the green algae *Ulva armoricana, Ulva clathrata, Ulva compressa* (formerly *Enteromorpha compressa*), *Ulva intestinalis, Ulva pertusa*, and *Ulva lactuca* have shown to have antiviral activity. *In vitro, U. armoricana* extracts prepared with enzymes showed antiviral activity against HSV-1 (Hardouin et al. 2016). HSV-1 was inhibited by the ulvan SU1F1 extracted from *U. compressa*, presenting virucidal properties (Lopes et al. 2017).

In vivo, the SPs from *U. pertusa* significantly increased the production of avian influenza virus specific antibodies (Song et al. 2015). Chiu et al. (2012) found that SPs extract from *U. lactuca* showed antiviral activity against the Japanese encephalitis virus. The anti-Newcastle disease viral mechanism of ulvans from *U. clathrata* avoids the cleavage of the viral protein F0, and the activity was increased when fucoidan from the brown alga *Cladosiphon okamuranus* was added (Aguilar-Briseño et al. 2015). Ulvans and fucoidans function in the same way by preventing viral attachment. When two or more molecules are used together, a synergistic effect may occur (Rosales-Mendoza et al. 2020).

2.3 *Peptide*

Griffithsin was detected for the first time in an aqueous extract of *Griffithsia* sp., a red seaweed (Mori et al. 2005). According to a study, a protein with very little homology to other recognized proteins and a strong ability to inhibit HIV-1 *in vitro* was discovered (EC_{50} range 0.043–0.63 nM) (Mori et al. 2005).

In terms of HIV, the *Griffithsia* lectin remains one of the most active HIV entry inhibitors discovered to date. Griffithsin is more effective than broadly neutralizing antibodies (bNAbs), such as the high-mannose-binding 2G12. Griffithsin has been shown to be not only a highly effective HIV entry inhibitor *in vitro* and *in vivo* trials (Emau et al. 2007), but it also enhance antibody responses (Banerjee et al. 2012; Fischer et al. 2019) and prevents cell fusion and cell-to-cell transmission of HIV (Zeitlin et al. 2009).

Griffithsin's antiviral activity has been demonstrated *in vivo* against a variety of other enveloped viruses, including the Hepatitis C virus (Meuleman et al. 2012), Herpes simplex virus (Nixon et al. 2013), Japanese encephalitis virus (Ishag et al. 2013), Nipah virus (Lo et al. 2020), Hantaviruses (Shrivastava-Ranjan et al. 2020) and coronaviruses MERS-CoV (Millet et al. 2016) and SARS-CoV (Barry R. O'Keefe et al. 2010; Cai et al. 2020).

2.4 Diterpene

Diterpenes from Dictyotales seaweed species have been shown to be very successful against HIV in several *in vitro* studies (von Ranke et al. 2020). For instance, the brown seaweed *Dictyota friabilis* (previously known as *Dictyota pfaffii*) (Barbosa et al. 2004), *Dictyota menstrualis* (Abrantes et al. 2010) and *Canistrocarpus cervicornis* (Vallim et al. 2010) have all shown to strongly inhibit HsV-1 infection in Vero cells. Later, it was tested in mice infected with HsV-1 *in vivo*. Untreated animals had slightly more serious lesions than those given a diterpene extract ointment or acyclovir ($p < 0.05$) or acyclovir ($p < 0.01$). These observations, together with the lack of side effects, indicate that *C. cervicornis* extract is a promising anti-HsV agent for cutaneous use (de Souza Barros et al. 2017). The highly pathogenic Asian Avian Influenza A (H5N1) virus was also inhibited by diterpenes from *Dictyota plectens*, with inhibition rates of 50–62% at 30.0 μM (Cheng et al. 2014).

The Zika virus is another significant virus that *Dictyota* sp. diterpenes have shown to be effective against (Cirne-Santos et al. 2020). The virucidal potential and inhibition of viral adsorption represent two different mechanisms of action for these diterpenes. What's more intriguing is that they can fully prevent Zika virus replication when paired with the antiviral Ribavirin at inefficient doses (Cirne-Santos et al. 2019).

3 Clinical and Commercial Exploitation

As reviewed by Pagarete et al. (2021), there was a total of 16 clinical trials on the use of algae-derived antivirals. The goal of a significant percentage of these trials (63%) was to find a viable HIV treatment. Carrageenan was used in seven of the trials and, although all the applications were found to be well received by the body, the treatments were found to be unsuccessful. Recently, the attention turned to the protein Griffithsin, which has shown promise in a couple of clinical trials. Papillomaviruses are another sexual transmitted virus that has been the target of human trials (3 trials). The use of carrageenan-based formulations to avoid the spread of these viruses appears to be promising as well (Pagarete et al. 2021).

Carrageenans are the algae-derived molecules that have had the most clinical trials for antiviral activity. Studies on the sexual transmitted viruses (i.e., HIV, HsV, and HpV) as well as rhinoviruses, have been performed. The ability of iota-carrageenan to interact directly with HpV adsorption from human sperm cells has led to two trials that have so far indicated that a carrageenan-based gel is both effective and well tolerated in preventing HpV transmission (Magnan et al. 2019; Perino et al. 2019).

These remarkable results led to carrageenan being accepted as safe for medical use by the Joint Expert Committee on Food Additives (JECFA), as well as patent approval and commercialization (Coldamaris) by Marinomed Biotechnologie

GmbH. (Austria). To our knowledge, carrageenans are the only algae-derived compounds that have successfully completed all clinical trials and have been commercialized for their antiviral properties, especially in the treatment of common cold symptoms and viruses (Graf et al. 2018).

Iota-carrageenans were developed as nasal spray and clinically approved for common cold in Europe due to their low solubility and inhibition of viral attachment. The nasal spray significantly reduced the effects of the common cold, reduced viral load, and reduced inflammation in patients in clinical trials (Eccles et al. 2010; Fazekas et al. 2012; Ludwig et al. 2013). Carrageenan nasal spray was found to shorten the length of common cold symptoms in patients (Koenighofer et al. 2014). The addition of zanamivir (an antiviral drug) to the carrageenan nasal spray was effective against the Influenza virus in a synergistic manner. (Morokutti-Kurz et al. 2015).

Although trials in other systems continue, recombinant Griffithsin is now being developed in *Escherichia coli* and in larger amounts in the plant *Nicotiana benthamiana* (Giomarelli et al. 2006; O'Keefe et al. 2009; Petrova et al. 2018; Vafaee and Alizadeh 2018).

4 Conclusions and Future Perspectives

Seaweeds obviously have a lot of potential for being used as specific platforms or sources for detecting and developing new antivirals for a wide range of viruses in a variety of situations. Until now, such efforts have been dominated by a desire to cure diseases that affect humans.

However, as most of seaweeds compounds remains uncharacterized, these organisms unveil a wide range of biomolecules with several biotechnological applications, amongst them antiviral properties.

Acknowledgments This work is financed by national funds through FCT - Foundation for Science and Technology, I.P., within the scope of the projects UIDB/04292/2020 granted to MARE - Marine and Environmental Sciences Centre and UIDP/50017/2020 + UIDB/50017/2020 (by FCT/MTCES) granted to CESAM - Centre for Environmental and Marine Studies. João Cotas thanks to the European Regional Development Fund through the Interreg Atlantic Area Program, under the project NASPA (EAPA_451/2016). Diana Pacheco thanks to PTDC/BIA-CBI/31144/2017-POCI-01 project -0145-FEDER-031144-MARINE INVADERS, co-financed by the ERDF through POCI (Operational Program Competitiveness and Internationalization) and by the Foundation for Science and Technology (FCT, IP). Ana M. M. Gonçalves acknowledges University of Coimbra for the contract IT057-18-7253 and also thanks to the project MENU – Marine Macroalgae: Alternative recipes for a daily nutritional diet (FA_05_2017_011), funded by the Blue Fund under Public Notice No. 5 – Blue Biotechnology, which partially financed this research.

References

Abrantes J, Barbosa J, Cavalcanti D, Pereira R, Fontes CF, Teixeira V, Souza TM, Paixão I (2010) The effects of the Diterpenes isolated from the Brazilian Brown algae Dictyota pfaffii and Dictyota menstrualis against the herpes simplex Type-1 replicative cycle. Planta Med 76:339–344. https://doi.org/10.1055/s-0029-1186144

Aguilar-Briseño J, Cruz-Suarez L, Sassi J-F, Ricque-Marie D, Zapata-Benavides P, Mendoza-Gamboa E, Rodríguez-Padilla C, Trejo-Avila L (2015) Sulphated polysaccharides from Ulva clathrata and Cladosiphon okamuranus seaweeds both inhibit viral attachment/entry and cell-cell fusion, in NDV infection. Mar Drugs 13:697–712. https://doi.org/10.3390/md13020697

Ahmadi A, Moghadamtousi SZ, Abubakar S, Zandi K (2015) Anti-viral potential of algae polysaccharides isolated from marine sources: a review. Biomed Res Int 2015:1–10. https://doi.org/10.1155/2015/825203

Banerjee K, Michael E, Eggink D, van Montfort T, Lasnik AB, Palmer KE, Sanders RW, Moore JP, Klasse PJ (2012) Occluding the mannose moieties on human immunodeficiency virus type 1 gp120 with Griffithsin improves the antibody responses to both proteins in mice. AIDS Res Hum Retrovir 28:206–214. https://doi.org/10.1089/aid.2011.0101

Barbosa JP, Pereira RC, Abrantes JL, César C, dos Santos M, Rebello A, Christina I, de Palmer Paixão Frugulhetti, and Valéria Laneuville Teixeira. (2004) In vitro anti-viral Diterpenes from the Brazilian Brown alga Dictyota pfaffii. Planta Med 70:856–860. https://doi.org/10.1055/s-2004-827235

Blunt JW, Copp BR, Keyzers RA, Munro MHG, Prinsep MR (2016) Marine natural products. Nat Prod Rep 33:382–431. https://doi.org/10.1039/C5NP00156K

Boulho R, Marty C, Freile-Pelegrín Y, Robledo D, Bourgougnon N, Bedoux G (2017) Antiherpetic (HSV-1) activity of carrageenans from the red seaweed Solieria chordalis (Rhodophyta, Gigartinales) extracted by microwave-assisted extraction (MAE). J Appl Phycol 29:2219–2228. https://doi.org/10.1007/s10811-017-1192-5

Cai Y, Wei X, Chenjian G, Cai X, Di Q, Lu L, Xie Y, Jiang S (2020) Griffithsin with a broad-Spectrum anti-viral activity by binding Glycans in viral glycoprotein exhibits strong synergistic effect in combination with a pan-coronavirus fusion inhibitor targeting SARS-CoV-2 spike S2 subunit. Virol Sin 35:857–860. https://doi.org/10.1007/s12250-020-00305-3

Cheng S, Zhao M, Sun Z, Yuan W, Zhang S, Xiang Z, Cai Y, Dong J, Huang K, Yan P (2014) Diterpenes from a Chinese collection of the Brown alga Dictyota plectens. J Nat Prod 77:2685–2693. https://doi.org/10.1021/np5006955

Chiu YH, Chan YL, Li TL, Chang Jer W (2012) Inhibition of Japanese encephalitis virus infection by the sulfated polysaccharide extracts from Ulva lactuca. Mar Biotechnol 14:468–478. https://doi.org/10.1007/s10126-011-9428-x

Cirne-Santos CC, Caroline de S, Barros MWL, Gomes RG, Cavalcanti DN, Obando JMC, Ramos CJB, Villaça RC, Teixeira VL, de Paixão ICNP (2019) In vitro anti-viral activity against Zika virus from a natural product of the Brazilian Brown seaweed Dictyota menstrualis. Nat Prod Commun 14:1934578X1985912. https://doi.org/10.1177/1934578X19859128

Cirne-Santos CC, de Souza C, Barros MC, de Oliveira V, Rabelo W-H, Azevedo RC, Teixeira VL, Ferreira DF, Nunes IC, de Palmer Paixão. (2020) In vitro studies on the inhibition of replication of Zika and chikungunya viruses by Dolastane isolated from seaweed Canistrocarpus cervicornis. Sci Rep 10:8263. https://doi.org/10.1038/s41598-020-65357-7

Diogo JV, Novo SG, González MJ, Ciancia M, Bratanich AC (2015) Anti-viral activity of lambda-carrageenan prepared from red seaweed (Gigartina skottsbergii) against BoHV-1 and SuHV-1. Res Vet Sci 98:142–144. https://doi.org/10.1016/j.rvsc.2014.11.010

Eccles R, Meier C, Jawad M, Weinmüllner R, Grassauer A, Prieschl-Grassauer E (2010) Efficacy and safety of an anti-viral iota-carrageenan nasal spray: a randomized, double-blind, placebo-controlled exploratory study in volunteers with early symptoms of the common cold. Respir Res. https://doi.org/10.1186/1465-9921-11-108

Emau P, Tian B, O'keefe BR, Mori T, McMahon JB, Palmer KE, Jiang Y, Bekele G, Tsai CC (2007) Griffithsin, a potent HIV entry inhibitor, is an excellent candidate for anti-HIV microbicide. J Med Primatol 36:244–253. https://doi.org/10.1111/j.1600-0684.2007.00242.x

Eom S-H, Moon S-Y, Lee D-S, Kim H-J, Park K, Lee E-W, Kim TH, Chung Y-H, Lee M-S, Kim Y-M (2015) In vitro anti-viral activity of dieckol and phlorofucofuroeckol-a isolated from edible brown alga Eisenia bicyclis against murine norovirus. Algae 30:241–246. https://doi.org/10.4490/algae.2015.30.3.241

Fazekas T, Eickhoff P, Pruckner N, Vollnhofer G, Fischmeister G, Diakos C, Rauch M et al (2012) Lessons learned from a double-blind randomised placebo-controlled study with a iota-carrageenan nasal spray as medical device in children with acute symptoms of common cold. BMC Complement Altern Med 12:1262. https://doi.org/10.1186/1472-6882-12-147

Fischer K, Nguyen K, LiWang PJ (2019) Griffithsin retains anti-HIV-1 potency with changes in gp120 glycosylation and complements broadly neutralizing antibodies PGT121 and PGT126. Antimicrob Agents Chemother 64. https://doi.org/10.1128/AAC.01084-19

Freile-Pelegrín Y, Tasdemir D (2019) Seaweeds to the rescue of forgotten diseases: a review. Bot Mar 62:211–226. https://doi.org/10.1515/bot-2018-0071

Garrido V, Barros C, Melchiades VA, Fonseca RR, Pinheiro S, Ocampo P, Teixeira VL et al (2017) Subchronic toxicity and anti-HSV-1 activity in experimental animal of dolabelladienetriol from the seaweed, Dictyota pfaffii. Regul Toxicol Pharmacol 86:193–198. https://doi.org/10.1016/j.yrtph.2017.03.007

Gheda SF, El-Adawi HI, El-Deeb NM (2016) Anti-viral profile of Brown and red seaweed polysaccharides against hepatitis C virus. Iranian J Pharmac Res 15:483–491

Giomarelli B, Schumacher KM, Taylor TE, Sowder RC, Hartley JL, McMahon JB, Mori T (2006) Recombinant production of anti-HIV protein, griffithsin, by auto-induction in a fermentor culture. Protein Expr Purif 47:194–202. https://doi.org/10.1016/j.pep.2005.10.014

Graf C, Bernkop-Schnürch A, Egyed A, Koller C, Prieschl-Grassauer E, Morokutti-Kurz M (2018) Development of a nasal spray containing xylometazoline hydrochloride and iota-carrageenan for the symptomatic relief of nasal congestion caused by rhinitis and sinusitis. Int J General Med 11:275–283. https://doi.org/10.2147/IJGM.S167123

Grassauer A, Weinmuellner R, Meier C, Pretsch A, Prieschl-Grassauer E, Unger H (2008) Iota-carrageenan is a potent inhibitor of rhinovirus infection. Virol J 5:107. https://doi.org/10.1186/1743-422X-5-107

Hardouin K, Bedoux G, Burlot A-S, Donnay-Moreno C, Bergé J-P, Nyvall-Collén P, Bourgougnon N (2016) Enzyme-assisted extraction (EAE) for the production of anti-viral and antioxidant extracts from the green seaweed Ulva armoricana (Ulvales, Ulvophyceae). Algal Res 16:233–239. https://doi.org/10.1016/j.algal.2016.03.013

Hayashi K, Nakano T, Hashimoto M, Kanekiyo K, Hayashi T (2008) Defensive effects of a fucoidan from brown alga Undaria pinnatifida against herpes simplex virus infection. Int Immunopharmacol 8:109–116. https://doi.org/10.1016/j.intimp.2007.10.017

Hayashi K, Lee J-B, Nakano T, Hayashi T (2013) Anti-influenza a virus characteristics of a fucoidan from sporophyll of Undaria pinnatifida in mice with normal and compromised immunity. Microbes Infect 15:302–309. https://doi.org/10.1016/j.micinf.2012.12.004

Ishag HZA, Li C, Huang L, Sun M-x, Wang F, Ni B, Malik T, Chen P-y, Mao X (2013) Griffithsin inhibits Japanese encephalitis virus infection in vitro and in vivo. Arch Virol 158:349–358. https://doi.org/10.1007/s00705-012-1489-2

Ismail MM, Alotaibi BS, Mostafa M, EL-Sheekh. (2020) Therapeutic uses of red macroalgae. Molecules 25:4411. https://doi.org/10.3390/molecules25194411

Jiao G, Guangli Y, Zhang J, Ewart H (2011) Chemical structures and bioactivities of sulfated polysaccharides from marine algae. Marine drugs 9. Molecular diversity preservation. International 196–223:doi:10.3390/md9020196

Jiao G, Guangli Y, Wang W, Zhao X, Zhang J, Ewart SH (2012) Properties of polysaccharides in several seaweeds from Atlantic Canada and their potential anti-influenza viral activities. J Ocean Univ China 11:205–212. https://doi.org/10.1007/s11802-012-1906-x

Kalitnik AA, Byankina Barabanova AO, Nagorskaya VP, Reunov AV, Glazunov VP, Soloveva TF, Yermak IM (2013) Low molecular weight derivatives of different carrageenan types and their anti-viral activity. J Appl Phycol 25:65–72. https://doi.org/10.1007/s10811-012-9839-8

Kim EB, Kwak JH (2015) Anti-viral phlorotannin from Eisenia bicyclis against human papilloma virus in vitro. Planta Med 81. https://doi.org/10.1055/s-0035-1565646

Koenighofer M, Lion T, Bodenteich A, Prieschl-Grassauer E, Grassauer A, Unger H, Mueller CA, Fazekas T (2014) Carrageenan nasal spray in virus confirmed common cold: individual patient data analysis of two randomized controlled trials. Multidiscip Respir Med. https://doi.org/10.1186/2049-6958-9-57

Kwon H-J, Ryu YB, Kim Y-M, Song N, Kim CY, Rho M-C, Jeong J-H, Cho K-O, Lee WS, Park S-J (2013) In vitro anti-viral activity of phlorotannins isolated from Ecklonia cava against porcine epidemic diarrhea coronavirus infection and hemagglutination. Bioorg Med Chem 21:4706–4713. https://doi.org/10.1016/j.bmc.2013.04.085

Kwon PS, Hanseul O, Kwon S-J, Jin W, Zhang F, Fraser K, Hong JJ, Linhardt RJ, Dordick JS (2020) Sulfated polysaccharides effectively inhibit SARS-CoV-2 in vitro. Cell Discovery 6:50. https://doi.org/10.1038/s41421-020-00192-8

Lee SY, Lee J, Lee HJ, Kim B, Lew J, Baek N, Kim S-H (2016) MicroRNA134 mediated upregulation of JNK and downregulation of NF k B Signalings are critically involved in Dieckol induced Antihepatic fibrosis. J Agric Food Chem 64:5508–5514. https://doi.org/10.1021/acs.jafc.6b01945

Leibbrandt A, Meier C, König-Schuster M, Weinmüllner R, Kalthoff D, Pflugfelder B, Graf P et al (2010) Iota-carrageenan is a potent inhibitor of influenza a virus infection. PLoS One 5:e14320. https://doi.org/10.1371/journal.pone.0014320

Liang Y, Sarkany N, Cui Y (2009) Biomass and lipid productivities of Chlorella vulgaris under autotrophic, heterotrophic and mixotrophic growth conditions. Biotechnol Lett 31:1043–1049. https://doi.org/10.1007/s10529-009-9975-7

Lo MK, Spengler JR, Krumpe LRH, Welch SR, Chattopadhyay A, Harmon JR, Coleman-McCray JAD et al (2020) Griffithsin inhibits Nipah virus entry and fusion and can protect Syrian Golden hamsters from lethal Nipah virus challenge. J Infect Dis 221:S480–S492. https://doi.org/10.1093/infdis/jiz630

Lopes N, Ray S, Espada SF, Bomfim WA, Ray B, Faccin-Galhardi LC, Linhares REC, Nozawa C (2017) Green seaweed Enteromorpha compressa (Chlorophyta, Ulvaceae) derived sulphated polysaccharides inhibit herpes simplex virus. Int J Biol Macromol 102:605–612. https://doi.org/10.1016/j.ijbiomac.2017.04.043

Ludwig M, Enzenhofer E, Schneider S, Rauch M, Bodenteich A, Neumann K, Prieschl-Grassauer E, Grassauer A, Lion T, Mueller CA (2013) Efficacy of a carrageenan nasal spray in patients with common cold: a randomized controlled trial. Respir Res 14:124. https://doi.org/10.1186/1465-9921-14-124

Magnan S, Tota JE, El-Zein M, Burchell AN, Schiller JT, Ferenczy A, Tellier P-P et al (2019) Efficacy of a carrageenan gel against transmission of cervical HPV (CATCH): interim analysis of a randomized, double-blind, placebo-controlled, phase 2B trial. Clin Microbiol Infect 25:210–216. https://doi.org/10.1016/j.cmi.2018.04.012

Meuleman P, Albecka A, Belouzard S, Vercauteren K, Verhoye L, Wychowski C, Leroux-Roels G, Palmer KE, Dubuisson J (2012) 861 the lectin griffithsin has anti-viral activity against hepatitis C virus in vitro and in vivo. J Hepatol 56:S335–S336. https://doi.org/10.1016/S0168-8278(12)60873-3

Millet JK, Séron K, Labitt RN, Danneels A, Palmer KE, Whittaker GR, Dubuisson J, Belouzard S (2016) Middle East respiratory syndrome coronavirus infection is inhibited by griffithsin. Antivir Res 133:1–8. https://doi.org/10.1016/j.anti-viral.2016.07.011

Mori T, O'Keefe BR, Sowder RC, Bringans S, Gardella R, Berg S, Cochran P et al (2005) Isolation and characterization of Griffithsin, a novel HIV-inactivating protein, from the red alga Griffithsia sp. J Biol Chem 280:9345–9353. https://doi.org/10.1074/jbc.M411122200

Morokutti-Kurz M, König-Schuster M, Koller C, Graf C, Graf P, Kirchoff N, Reutterer B et al (2015) The intranasal application of Zanamivir and carrageenan is synergistically active against influenza a virus in the murine model. PLoS One 10:e0128794. https://doi.org/10.1371/journal.pone.0128794

Nixon B, Stefanidou M, Mesquita PMM, Fakioglu E, Segarra T, Rohan L, Halford W, Palmer KE, Herold BC (2013) Griffithsin protects mice from genital herpes by preventing cell-to-cell spread. J Virol 87:6257–6269. https://doi.org/10.1128/JVI.00012-13

O'Keefe BR, Vojdani F, Buffa V, Shattock RJ, Montefiori DC, Bakke J, Mirsalis J et al (2009) Scaleable manufacture of HIV-1 entry inhibitor griffithsin and validation of its safety and efficacy as a topical microbicide component. Proc Natl Acad Sci 106:6099–6104. https://doi.org/10.1073/pnas.0901506106

O'Keefe BR, Giomarelli B, Barnard DL, Shenoy SR, Chan PKS, McMahon JB, Palmer KE et al (2010) Broad-Spectrum in vitro activity and in vivo efficacy of the anti-viral protein Griffithsin against emerging viruses of the family Coronaviridae. J Virol 84:2511–2521. https://doi.org/10.1128/JVI.02322-09

Pagarete A, Ramos AS, Puntervoll P, Allen MJ, Verdelho V (2021) Anti-viral potential of algal metabolites—a comprehensive review. Mar Drugs 19:94. https://doi.org/10.3390/md19020094

Park H-J, Kurokawa M, Shiraki K, Nakamura N, Choi J-S, Hattori M (2005) Anti-viral activity of the marine alga Symphyocladia latiuscula against herpes simplex virus (HSV-1) in vitro and its therapeutic efficacy against HSV-1 infection in mice. Biol Pharm Bull 28:2258–2262. https://doi.org/10.1248/bpb.28.2258

Park J-Y, Kim JH, Kwon JM, Kwon H-J, Jeong HJ, Kim YM, Kim D, Lee WS, Ryu YB (2013) Dieckol, a SARS-CoV 3CLpro inhibitor, isolated from the edible brown algae Ecklonia cava. Bioorg Med Chem 21:3730–3737. https://doi.org/10.1016/j.bmc.2013.04.026

Pereira L (2018) Biological and therapeutic properties of the seaweed polysaccharides. Int Biol Rev 2:1–50. https://doi.org/10.18103/ibr.v2i2.1762

Pereira L, Critchley AT (2020) The COVID 19 novel coronavirus pandemic 2020: seaweeds to the rescue? Why does substantial, supporting research about the antiviral properties of seaweed polysaccharides seem to go unrecognized by the pharmaceutical community in these desperate times? J Appl Phycol 32:1875–1877. https://doi.org/10.1007/s10811-020-02143-y

Perino A, Consiglio P, Maranto M, De Franciscis P, Marci R, Restivo V, Manzone M, Capra G, Cucinella G, Calagna G (2019) Impact of a new carrageenan-based vaginal microbicide in a female population with genital HPV-infection: first experimental results. Eur Rev Med Pharmacol Sci 23:6744–6752. https://doi.org/10.26355/eurrev_201908_18567

Petrova MI, van den Broek MFL, Spacova I, Verhoeven TLA, Balzarini J, Vanderleyden J, Schols D, Lebeer S (2018) Engineering lactobacillus rhamnosus GG and GR-1 to express HIV-inhibiting griffithsin. Int J Antimicrob Agents 52:599–607. https://doi.org/10.1016/j.ijantimicag.2018.07.013

Queiroz KCS, Medeiros VP, Queiroz LS, Abreu LRD, Rocha HAO, Ferreira CV, Jucá MB, Aoyama H, Leite EL (2008) Inhibition of reverse transcriptase activity of HIV by polysaccharides of brown algae. Biomed Pharmacother 62:303–307. https://doi.org/10.1016/j.biopha.2008.03.006

von Ranke NL, Ribeiro MMJ, Miceli LA, de Souza NP, Abrahim-Vieira BA, Castro HC, Teixeira VL, Rodrigues CR, Souza AMT (2020) Structure-activity relationship, molecular docking, and molecular dynamic studies of diterpenes from marine natural products with anti-HIV activity. J Biomol Struct Dyn 14:1–11. https://doi.org/10.1080/07391102.2020.1845977

Rocha HAO, Moraes FA, Trindade ES, Franco CRC, Torquato RJS, Veiga SS, Valente AP et al (2005) Structural and hemostatic activities of a sulfated Galactofucan from the Brown alga Spatoglossum schroederi. J Biol Chem 280:41278–41288. https://doi.org/10.1074/jbc.M501124200

Rosales-Mendoza S, García-Silva I, González-Ortega O, Sandoval-Vargas JM, Malla A, Vimolmangkang S (2020) The potential of algal biotechnology to produce anti-viral compounds and biopharmaceuticals. Molecules 25:4049. https://doi.org/10.3390/molecules25184049

Ryu YB, Jeong HJ, Yoon SY, Park J-Y, Kim YM, Park S-J, Rho M-C, Kim S-J, Lee WS (2011) Influenza virus neuraminidase inhibitory activity of Phlorotannins from the edible Brown alga Ecklonia cava. J Agric Food Chem 59:6467–6473. https://doi.org/10.1021/jf2007248

Shi Q, Wang A, Lu Z, Qin C, Hu J, Yin J (2017) Overview on the anti-viral activities and mechanisms of marine polysaccharides from seaweeds. Carbohydr Res 14:453–454. https://doi.org/10.1016/j.carres.2017.10.020

Shrivastava-Ranjan P, Lo MK, Chatterjee P, Flint M, Nichol ST, Montgomery JM, O'Keefe BR, Spiropoulou CF (2020) Hantavirus infection is inhibited by Griffithsin in cell culture. Front Cell Infect Microbiol 10. https://doi.org/10.3389/fcimb.2020.561502

Smit AJ (2004) Medicinal and pharmaceutical uses of seaweed natural products: a review. J Appl Phycol 16:245–262. https://doi.org/10.1023/B:JAPH.0000047783.36600.ef

Song L, Chen X, Liu X, Zhang F, Linfeng H, Yue Y, Li K, Li P (2015) Characterization and comparison of the structural features, immune-modulatory and anti-avian influenza virus activities conferred by three algal sulfated polysaccharides. Mar Drugs 14:4. https://doi.org/10.3390/md14010004

Song S, Peng H, Wang Q, Liu Z, Dong X, Wen C, Ai C, Zhang Y, Wang Z, Zhu B (2020) Inhibitory activities of marine sulfated polysaccharides against SARS-CoV-2. Food Funct 11:7415–7420. https://doi.org/10.1039/D0FO02017F

de Souza Barros C, Caroline VG, Melchiades V, Gomes R, Gomes MWL, Teixeira VL, de Paixão ICNP (2017) Therapeutic efficacy in BALB/C mice of extract from marine alga Canistrocarpus cervicornis (Phaeophyceae) against herpes simplex virus type 1. J Appl Phycol 29:769–773. https://doi.org/10.1007/s10811-016-0865-9

Talarico L, Duarte M, Zibetti R, Noseda M, Damonte E (2007) An algal-derived DL-Galactan hybrid is an efficient preventing agent for in vitro dengue virus infection. Planta Med 73:1464–1468. https://doi.org/10.1055/s-2007-990241

Talarico LB, Noseda MD, Ducatti DRB, Duarte MER, Damonte EB (2011) Differential inhibition of dengue virus infection in mammalian and mosquito cells by iota-carrageenan. J Gen Virol 92:1332–1342. https://doi.org/10.1099/vir.0.028522-0

Thuy TT, Thu BM, Ly TT, Van T, Van Quang N, Ho Cam T, Zheng Y, Seguin-Devaux C, Mi B, Ai U (2015) Anti-HIV activity of fucoidans from three brown seaweed species. Carbohydr Polym 115:122–128. https://doi.org/10.1016/j.carbpol.2014.08.068

Torres MD, Flórez-Fernández N, Domínguez H (2019) Integral utilization of red seaweed for bioactive production. Mar Drugs 314:doi:10.3390/md17060314

Vafaee Y, Alizadeh H (2018) Heterologous production of recombinant anti-HIV microbicide griffithsin in transgenic lettuce and tobacco lines. Plant cell. Tiss Organ Cult 135:85–97. https://doi.org/10.1007/s11240-018-1445-2

Vallim MA, Barbosa JE, Cavalcanti DN, De-Paula JC, da Silva VAGG, Teixeira VL, de Palmer Paixão ICN (2010) In vitro anti-viral activity of diterpenes isolated from the Brazilian brown alga Canistrocarpus cervicornis. J Med Plants Res 4:2379–2382. https://doi.org/10.5897/JMPR10.564

Vonthron-Sénécheau C (2016) Medicinal Properties. In: Seaweed in health and disease prevention. Elsevier, London, pp 369–388. https://doi.org/10.1016/B978-0-12-802772-1.00011-7

Wang W, Zhang P, Hao C, Zhang X-E, Cui Z-Q, Guan H-S (2011) In vitro inhibitory effect of carrageenan oligosaccharide on influenza a H1N1 virus. Antivir Res 92:237–246. https://doi.org/10.1016/j.anti-viral.2011.08.010

Wang W, Zhang P, Guang-Li Y, Li C-X, Hao C, Qi X, Zhang L-J, Guan H-S (2012) Preparation and anti-influenza a virus activity of κ-carrageenan oligosaccharide and its sulphated derivatives. Food Chem 133:880–888. https://doi.org/10.1016/j.foodchem.2012.01.108

Wang W, Jiandong W, Zhang X, Hao C, Zhao X, Jiao G, Shan X, Tai W, Guangli Y (2017) Inhibition of influenza a virus infection by Fucoidan targeting viral neuraminidase and cellular EGFR pathway. Sci Rep 7:40760. https://doi.org/10.1038/srep40760

Weiner ML (2016) Parameters and pitfalls to consider in the conduct of food additive research, carrageenan as a case study. Food Chem Toxicol 87:31–44. https://doi.org/10.1016/j.fct.2015.11.014

Zeitlin L, Pauly M, Whaley KJ (2009) Second-generation HIV microbicides: continued development of griffithsin. Proc Natl Acad Sci 106:6029–6030. https://doi.org/10.1073/pnas.0902239106

Zhang X (2018) Anti-retroviral drugs: current state and development in the next decade. Acta Pharm Sin B 8:131–136. https://doi.org/10.1016/j.apsb.2018.01.012

Chapter 25
Antiviral Applications of Macroalgae

Shivdayal Singh and Maushmi S. Kumar

Abbreviations

ASFV	African swine fever virus
Banlec	Banana lectin
BHK-21	Baby hamster kidney (strain-21)
BVDV	Bovine viral diarrhea virus
Ca-SP	Calcium spirulan
CV-N	Cyanovirin
DENV	Dengue Virus
DENV-2	Dengue virus-2
EV 71	Enterovirus 71
gp120	Glycoprotein 120
GRFT	Griffithsin
H5N1	Avian influenza virus
HBV	Hepatitis B virus
HCMV	Human cytomegalovirus
HCV	Hepatitis C virus
HIV	Human immunodeficiency virus
HIV-1	Human immunodeficiency virus-1
HIV-2	Human immunodeficiency virus-2
HPV	Human papilloma virus
HRV	Human rhinovirus
HSV	Herpes simplex virus
HSV-1	Herpes simplex virus-1

S. Singh · M. S. Kumar (✉)
Shobhaben Pratapbhai Patel School of Pharmacy and Technology Management, SVKM'S NMIMS, Mumbai, Maharashtra, India
e-mail: maushmi.kumar@nmims.edu

A. Ranga Rao, G. A. Ravishankar (eds.), *Sustainable Global Resources of Seaweeds Volume 2*, https://doi.org/10.1007/978-3-030-92174-3_25

HSV-2	Herpes simplex virus-2
JEV	Japanese Encephalitis Virus
MERS	Middle east respiratory syndrome
NSF	Nostoflan
RTase	Reverse transcriptase
SARS	Severe acute respiratory syndrome
SPMG	Sulfated polymannuroguluronate

1 Introduction

Viral infections are responsible for causing severe acute respiratory syndrome (SARS), Ebola fever, influenza, hepatitis, and several other serious diseases in humans. They are one of the leading causes of death worldwide, among others like cardiovascular diseases and cancer. Enzootic and epizootic viral transmissions are the main sources of emerging human viral diseases. Parrish et al. (2008) emphasized an increase in epizootic transmissions, and which is also the cause of SARS CoV-2 pandemic that has taken its toll on the worldwide population. There is an immediate need for the development of new antiviral drugs as viral epidemics are estimated to increase in the future due to the increase in the interactions between humans and wildlife populations. Figure 25.1 represents a graph that indicates the number of global incidences of common viral infections in the last ten years. Various antiviral drugs have been developed and are being used but they are prone to drug resistance due to their extensive clinical use and side effects (Kim et al. 2011). Hence, exploring non-traditional resources for new compounds with a wide range of applications is being explored.

About 71% of the earth's surface is covered in water; oceans consist of about 97% of the water volume available on earth (Charette and Walter 2010). Macroalgae are multicellular plants found in marine ecosystem and can belong to either eukaryotic or prokaryotic group of organisms. There is no particular definition of macroalgae as it is of polyphyletic origin. They are part of Asian culture in the form of food or traditional remedies and are consumed in dry or wet forms since earlier times. In the last few decades, there is a surge in the studies focused on obtaining biologically active metabolites from them (Kandale et al. 2011). They are a source of various primary and secondary metabolites, which have diverse chemical structures and a wide range of uses (Anil et al. 2017). The secondary metabolites profile is subjected to various physical and biological factors (Gallimore 2017); which can be used either in their natural form as an antiviral drug or their chemical structure can be evaluated to develop their semi-synthetic derivatives.

Various bioactive metabolites such as polysaccharides, tannins, phenolic acids, flavonoids, carotenoids, and bromophenols are derived from macroalgae. The amount of these metabolites in the macroalgae varies with the species. Some of the chemical compounds from these classes of metabolites have shown antiviral properties (Anil et al. 2017).

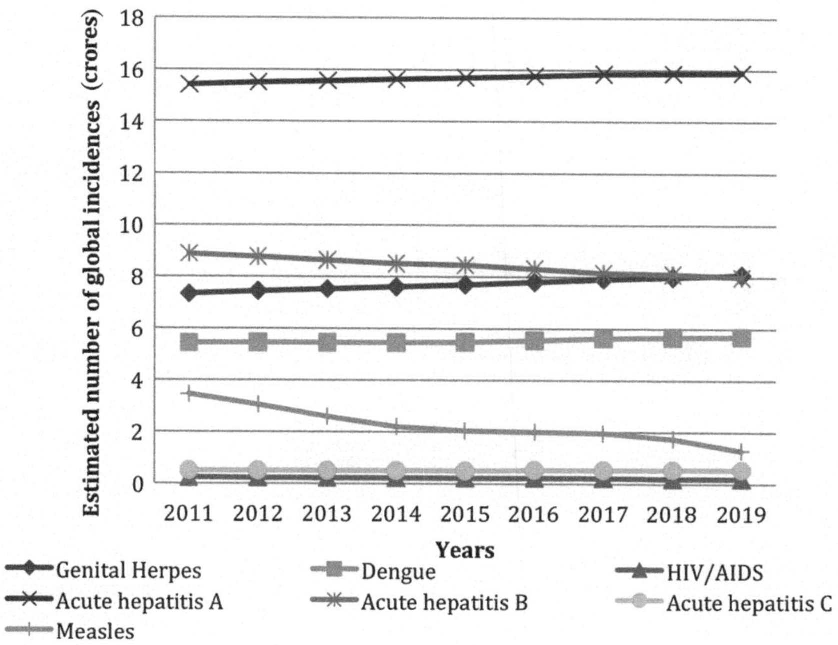

Fig. 25.1 A graph depicting the number of global incidences of genital herpes, dengue, HIV/AIDS, acute hepatitis A, acute hepatitis B, acute hepatitis C, and measles. (Institute for Health Metrics and Evaluation 2021)

2 Antiviral Compounds from Macroalgae

2.1 Polysaccharides

Algal polysaccharides are economical, biodegradable and biocompatible natural non-toxic polymers found in abundance in nature. Figure 25.2 represents the life cycle of the virus and the general mechanism of action of antiviral polysaccharides derived from macroalgae.

2.1.1 Carrageenan

Carrageenan are natural anionic sulfated polysaccharides that are mostly found in the matrix of red algae such as *Gigartina*, *Chondrus*, *Hypnea*, and *Eucheuma.* They share structural and functional similarities with cellulose of higher order plants (Ahmadi et al. 2015).

They are classified based on the presence of 3,6-anhydrogalactopyranose and sulfated groups on the main structure into three types- namely- λ (*lambda*), κ (*kappa*), and ι (*iota*) carrageenan (Fig. 25.3). They selectively inhibit the binding of various enveloped and non-enveloped viruses on the host cells. They are an effective

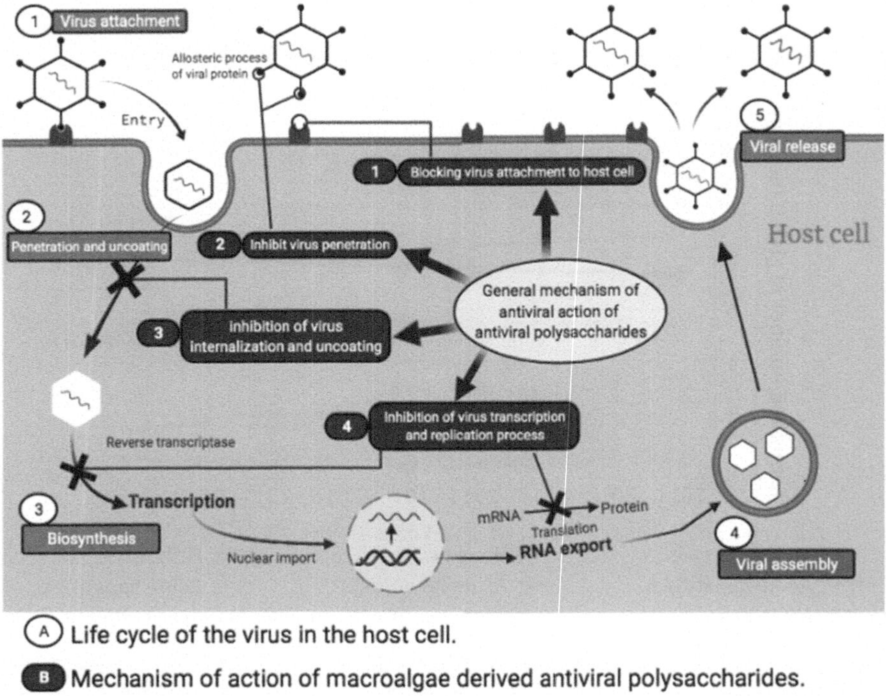

Fig. 25.2 (**a**) Life cycle of the virus in the host cell including (1) virus attachment, (2) penetration and uncoating, (3) biosynthesis (4) viral assembly and (5) viral release. (**b**) Mechanism of action of macroalgae derived antiviral polysaccharides which includes (1) blocking virus attachment to host cell, (2) inhibition of virus penetration (Hans et al. 2020)

inhibitor of human papilloma virus (HPV). λ carrageenan can lead to the inactivation of herpes simplex virus (HSV). 1T1 is a λ carrageenan isolated from *Gigartina skottsbegii* which showed antiviral activity against HSV-2. The activity was displayed in mice due to the interference with the virus attachment stage to the host cells (Carlucci et al. 2004). Carrageenan extracted from *Meristiella delirium* was found to be effective against HSV-2 and Dengue virus-2 (DENV-2) with no cytotoxicity on Vero cells (Paula et al. 2006). ι carrageenan inhibited human rhinovirus (HRV) replication during its primary phase. It also inhibited the replication of the dengue virus in mosquitoes by affecting cell proliferation and protein synthesis, whereas the Vero cell line studies have shown early inhibitory activity due to the presence of some primary receptors (Talarico et al. 2011). κ-carrageenan on sulfation and acetylation inhibits the influenza virus. Molecular weight along with sulfonation groups are linked to the antiviral properties of the acetylated carrageenan against HIV (Yamada et al. 1997). It also binds to Enterovirus 71 (EV 71) and forms carrageenan-viruses complexes, disrupting the virus-receptor interaction which makes it an ideal candidate for the development of anti-EV 71 agents (Chiu et al. 2012).

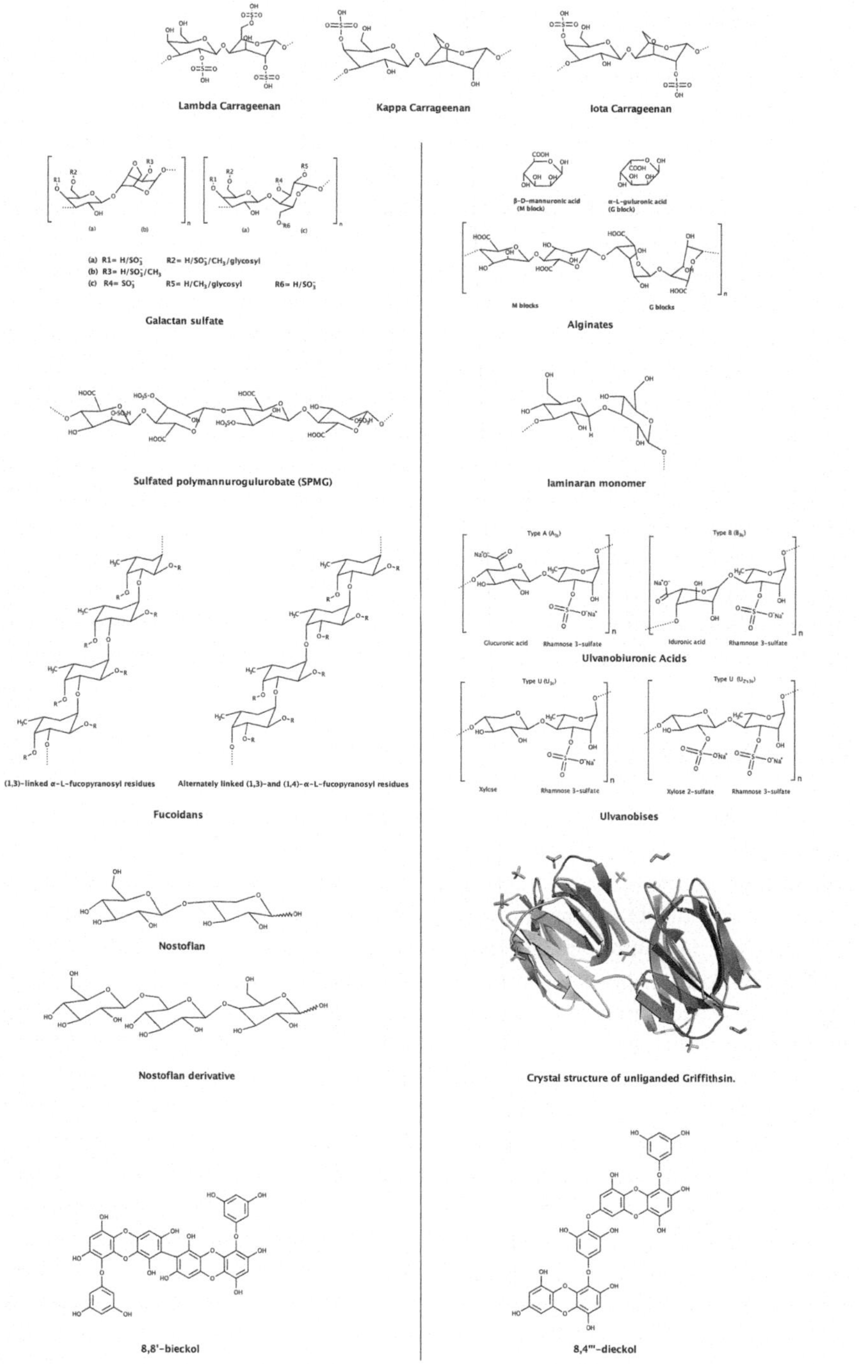

Fig. 25.3 Antiviral compounds from macroalgae. The structure of protein griffithsin was generated by using PyMOL software and protein data bank accession number 2GTY (Ziółkowska et al. 2006)

2.1.2 Galactan

Galactans, also known as sulfated galactans, are found in red algae as the main extracellular polysaccharide. The main structure comprises of linear chains of galactose but there are few exceptions to this. There is an alternate chain of 3-β-D-galactopyranose (G units) and 4-α-D-galactopyranose residues or 4-3,6-anhydrogalactopyranose residues form the structural backbone with the presence of D-unit in carrageenans and L-units in agarans. In some exceptional galactans the DL-hybrids that enclose the G unit is attached to both D and L units (Estevez et al. 2001). Figure 25.3 shows the structure of galactan sulfate derived from red macroalgae. These varied structural forms of galactans are found to be effective against many enveloped viruses, namely HSV-1, HSV-2, DENV, hepatitis A, HIV-1, and HIV-2 (Witvrouw et al. 1994). Antiviral action of these galactans extracted from *Callophyllis variegata* has shown inhibitory effect with less cytotoxicity when tested for HSV-1, HSV-2, and DENV-2 (Rodríguez et al. 2005). Galactan sulfate isolated from *Agardhiella tenera* inhibited adhesion of virus to the cells in HIV-1 and HIV-2. Galactan isolated from *Schizymenia binderi* is highly selective against HSV-1 and HSV-2. (Matsuhiro et al. 2005). Structural hybrids of D, L-galactan, CS2-3 extracted from *Cryptonemia crenulata* has shown inhibition in the multiplication of three clinical strains of DENV-2 virus in a Vero cell line. It gave activity by inhibiting the adsorption of the virus into the host cells (Talarico et al. 2007).

2.1.3 Alginates

Alginates are acid polysaccharides primarily found in brown algae such as *Ascophyllum nodusm*, *Macrocystis pyrifera*, *Laminaria hyperborea*, *Laminaria digitata*, and *Laminaria japonica*. They are anionic linear polysaccharides with a backbone of poly-D-glucuronic acid(G blocks) and poly-D-mannuronic acid (M blocks), with D-guluronic acid and D-mannuronic acid (GM blocks) (Wang et al. 2012). Figure 25.3 shows the structures of M and G blocks. Another drug 911 is an alginate polysaccharide and showed promising anti-HIV-1 and anti-hepatitis B (HBV) results. In the case of HIV-1, it inhibits replication of the virus by decreasing the activity of reverse transcriptase (RTase) which in turn decreases viral adhesion to the host cells. It is effective in both chronic infection of H9 cells and acute infection of MT4 cells *in vitro* as well as *in vivo* (Xianliang et al. 2000). In the case of HBV, 911 inhibits viral replication by decreasing the activity of DNA polymerase (Jiang et al. 2003). A sulfated form of alginate namely sulfated polymannuroguluronate (SPMG) (Fig. 25.3) can be an anti-AIDS drug candidate as it inhibits gp120 from attaching to CD4 molecules on the surface of T cells in case of HIV-1 infection. There is a correlation between the size of SPMG oligosaccharides and their inhibitory action. The SPMG fragment must at least be a hexasaccharide to interact and inhibit gp120 and this property increases with an increase in the size of SPGM fragment (Meiyu et al. 2003; Liu et al. 2005). *In vitro* and *in vivo* studies suggests the effectiveness of SPMG against HPV infection. It is understood that it blocks HPV binding and entry by interacting with capsid L1 protein (Wang et al. 2020).

2.1.4 Fucan and Fucoidan

These are sulfated polysaccharides with high molecular weight in the range of 100–1600 kDa. They are found in the mucilaginous matrix or intercellular tissues of brown algae and are classified into three major groups i.e. glycuronogalactofucans, fucoidans, and xylofucoglycuronans. The structure of fucan is diverse and depends on the source of brown algae species used for its extraction. Sulfated fucans from *Dictyota mertensii, Lobophora variegata, Fucus vesiculosus, and Spatoglossum schroederi* block the reverse transcriptase activity and prevent HIV infections (Queiroz et al. 2008). They are also obtained from *Cladosiphon okamuranus* having sulfated fucose units and glucuronic acid, which inhibits DENV-2 infection in the BHK-21 cell line (Adhikari et al. 2006). MC26 is a fucose polysaccharide obtained from brown algae *Sargassam piluliferm* found to be effective in influenza with less cytotoxicity (Akamatsu et al. 2003). Fucans obtained from brown algae *Cystoseira indica* is effective against HSV-1 and HSV-2, showing no cytotoxicity for Vero cell culture with a proposed reason that it inhibits virus adsorption to the host cells. (Mandal et al. 2007). Fucoidans consist of L-fucose and less than 10% of other monosaccharides (Fig. 25.3). There is a higher proportion of fucose in the extracellular matrix of brown algae such as mozuku, komby, limu moui, bladderwrack, wakame, sea cucumber, and hijiki. The structure of fucoidans is diverse and repeating alternate sequence of α-1,3-linked sulfated L-fucose with an α-1,4-glycosidic bond which forms the backbone of fucoidans (Tanna and Mishra 2019). They are generally sulfated and acetylated and may contain uronic acid (Berteau and Mulloy 2003; Cumashi et al. 2007; Pomin and Mourão 2008). They are found to be effective against a few human RNA and DNA viruses both *in vivo* and *in vitro* (Witvrouw and De Clercq 1997). They are effective anti-HSV-1 and HSV-2 agents with no cytotoxicity for Vero cell lines. Fucoidans mainly block the adhesion of the virus to the host cells and inhibit viral-induced syncytium formation (Hidari et al. 2008). They exhibit better antiviral potency as compared with ribavirin in the Newcastle disease virus in the Vero cell studies (Elizondo-Gonzalez et al. 2012). They are capable of alternating the proteins of the extracellular matrix and can induce cell apoptosis by affecting cell proliferation (Haroun-Bouhedja et al. 2000; Koyanagi et al. 2003; Aisa et al. 2005; Moon et al. 2008).

2.1.5 Laminaran

Laminaran is a glucan found widely in brown algae such as *Saccharina longicruris, F. vesiculosus,* and *Ascophyllum nodosum.* Laminaran is a linear polysaccharide and has β(1 → 3)-linked glucose as the central chain along with β(1 → 6)-linked side-chain branching (Peat et al. 1958). There are two types of laminaran; one with glucose residues (G-series) and other with terminal D-mannitol residues (M-series) (Nelson and Lewis 1974). Their composition may vary depending on the species and the other physical and biological factors. Laminaran exhibits antiviral properties with low cellular toxicity *in vivo*; they were found to be useful in HIV by inhibiting replication and proliferation of the virus (O'Doherty et al. 2010).

2.1.6 Ulvan

Ulvan is a gelling polysaccharide obtained from *Ulva* species, an edible green seaweed. Ulvans are polyanionic heteropolysaccharide and their sugar composition mainly consist of rhamnose, glucuronic acid, and xylose (Fig. 25.3). However, it contains a wide range of other monosaccharides. α- and β-(1,4)-linked monosaccharides with repeating disaccharide units form the backbone of the ulvan structure. Type A ulvanobiuronic acid and type B ulvanobiuronic acid are the major disaccharide repeating units found in ulvan whereas, ulvanobioses (type U) is the minor disaccharide present (Lahaye and Robic 2007).

2.1.7 Naviculan

It is a high molecular weight polysaccharide made of various sugar moieties like fructose, xylose, rhamnose, mannose, fucose, and sulfate groups. Naviculan is obtained from a diatom *Navicula directa.* It is found to be effective against HSV-1, HSV-2, and influenza virus and works by inhibiting the initial stages of virus replication. Studies suggest its effectiveness against enveloped viruses (Lee et al. 2006).

2.1.8 Calcium Spirulan (Ca-SP)

It is obtained from the marine blue-green algae *Arthrospira plantensis*, and is a novel sulfated polysaccharide. It comprises of ribose, fructose, mannose, glucose, xylose, galactose, rhamnose, galacturonic acid, glucuronic acid, calcium and sulfate. It was found to inhibit virus entry into the host cell during *in vitro* and Vero studies. It is a selective inhibitor of viruses like HSV-1, HCMV, influenza A, measles, HIV-1, polio, mumps, and Coxsackie virus. It also exhibits mild anticoagulant properties (Hayashi et al. 1996). Ca-SP can be a promising new anti-HIV drug candidate.

2.1.9 Nostoflan (NSF)

It is found in edible blue-green algae; *Nostoc flagelliforme.* On hydrolysis NSF yields two types of oligosaccharides namely PA-1 [β-D-GlcAp-(1 → 4)-Xyl-PA] and PA-2 [β-D-GlcAp-(1 → 6)-β-D-Glcp-(1 → 4)-Gal-PA] (Fig. 25.3). It is proposed to be effective against various enveloped viruses by inhibiting their binding with the host cells. It can be a great candidate for the newer antiherpes drug (Kanekiyo et al. 2005; Thuan et al. 2019).

2.1.10 Xylomannan Sulfate

Xylomannan is a novel anti-freeze agent, that was first isolated from an Alaskan beetle *Upis ceramboides*. Its structure consists of β-D-mannopyranosyl-(1 → 4)-β-D-xylopyranose-disaccharide-repeating units. Some seaweed-derived xylomannan has antiviral properties (Table 25.1). Sulphated polysaccharides can be an option for the development of an anti-COVID drug, since the early symptoms of COVID-19 manifests common cold and flu, with similarities in the mechanism of action of the virus. Carrageenans in the form of nasal spray is an effective treatment for the common cold in adults and children (Ron Eccles et al. 2010). *Iota* carrageenan co-administered with Zanamivir in the form of nasal spray relieves upper respiratory symptoms in the patients suffering from influenza A (R Eccles et al. 2015). Such polysaccharides can be evaluated for their effectiveness against SARS COVID-2 virus. The ability of various microalgal polysaccharides especially ulvans, fucoidans, and carrageenan to inhibit virus adhesion and replication can be tested for the current pandemic (Pereira and Critchley 2020).

2.2 *Lectin*

Lectins are proteins or glycoproteins that are found naturally in cells that bind reversibly to glycans of glycoproteins, glycolipids, and polysaccharides. They are responsible for cell-cell interaction and protein folding and poses bioactivity; they are used as probes to determine cell surface structure and function. Currently, they are used to develop chemotherapeutic and antiviral agents. They are currently found to be effective against HIV. Griffithsin (GRFT), cyanovirin (CV-N), and banana lectin (Banlec) are some of the promising lectins for the development of antiviral drugs (Lusvarghi and Bewley 2016).

2.2.1 Griffithsin (GRFT)

Griffithsin was first isolated from *Griffithsia* sp., a type of red algae. GRFT is a protein and shows no similarity with any other known protein. GRFT exists as a stable homodimer and every subunit consist of 121 amino acids. GRFT interacts with the terminal sugar moiety of oligosaccharides (Sanchez 2013) (Fig. 25.3). A study on mice infected with the SARS-CoV virus has shown a 100% recovery when administered with GRFT by intranasal route (Ishag et al. 2013). It also showed potential to be a good candidate for the development of topical antiviral agents.

Table 25.1 Antiviral compounds from macroalgae, their significant sources and effectiveness against various viral diseases

Antiviral compound	Significant source	Effective against	References
Carrageenan			
a) λ(*lambda*) carrageenan	*Gigartina skottsbegii* *Chondrus crispus* *Meristiella gelidium*	Herpes simplex Virus (HSV-1 and HSV-2) African swine fever virus (ASFV) Dengue virus (DENV	Carlucci et al. (2004), Paula et al. (2006) García-Villalón and Gil-Fernández (1991) Zhu et al. (2018) Piccini et al. (2020), Paula et al. (2006)
b) κ(*kappa*) carrageenan	*Kappaphycus alvarezii*	Human enterovirus 71 infections.	Chiu et al. (2012) Rudke et al. (2020)
c) ι(*iota*) carrageenan	*Eucheuma denticulatum* *Solieria filiformis*	Human rhinovirus (HRV)infection Herpes simplex virus (HSV-1)	Grassauer et al. (2008) Jönsson et al. (2020) Ana et al. (2021)
Galactan			
a) Sulfated galactan	*Callophyllis variegata*	HSV-1, HSV-2 and DENV-2	Rodríguez et al. (2005)
	Agardhiella tenera	HIV-1 and HIV-2	Witvrouw et al. (1994)
	Schizymenia binderi	HSV-1 and HSV-2	Matsuhiro et al. (2005)
	Cryptonemia crenulata	DENV-2	Talarico et al. (2007)
	Gymnogongrus griffithsiae, *Cryptonemia crenulata*	HSV-1 and HSV-2	Talarico et al. (2004)
	Gracilaria corticata	HSV-1 and HSV-2	Mazumder et al. (2002)
b) DL-galactan hybrid	*Gymnogongrus torulosus*	HSV-2 and DENV-2	Pujol et al. (2002)
Fucan and fucoidan			
a) Galactofucan	*Adenocystis utricularis*	HSV-1, HSV-2	Ponce et al. (2003)
	Dictyota dichotoma	HSV-1	Rabanal et al. (2014)
	Undaria pinnatifida	HSV-1, HSV-2, human cytomegalovirus (HCMV)	Hemmingson et al. (2006)
b) Glucuronic acid, sulfated fucose	*Cladosiphon okamuranus*	DENV-2	Hidari et al. (2008)h
c) Sulfated fucans	*Cystoseria indica*	HSV-1, HSV-2	Mandal et al. (2007)
d) Fucoidan	*Sargassum mcclurei*	HIV-1	Thuy et al. (2015)
	Fucus vesiculosus	Bovine viral diarrhea virus(BVDV)	Güven et al. (2020)
	Laminaria japonica	Avian influenza virus (H5N1)	Makarenkova et al. (2010)
	Sargassum trichophyllum	HSV-2	Lee et al. (2011)

(continued)

Table 25.1 (continued)

Antiviral compound	Significant source	Effective against	References
e) Xylan fucoidan	*Caulerpa brachypus*	HSV-1	Lee et al. (2004)
Ulvan			
	Enteromorpha compressa	HSV	Lopes et al. (2017)
	Ulva intestinalis	Measles virus	Morán-Santibañez et al. (2016)
	Ulva armoricana	HSV-1	Hardouin et al. (2016)
Xylomannan sulfate			
	Sebdenia polydactyla	HSV-1	Ghosh et al. (2009)
	Scinaia hatei	HSV-1 and HSV-2	Mandal et al. (2008)

(O'Keefe et al. 2009; Girard et al. 2018). GRFT binds with the glycoprotein enveloped by the virus and prevents CD4 and other antibodies from binding to the virus (Alexandre et al. 2010). Its immediate antiviral action is a great advantage compared to other antiviral agents which are being evaluated for the same purpose in HIV-1 (Emau et al. 2007). GRFT inhibits HCV, besides it can be effective against enveloped viruses such as the Japanese Encephalitis Virus (JEV), HSV-2, and HPV (Lusvarghi and Bewley 2016). It is also effective for the inhibition of different strains of coronavirus from replicating without cell proliferation. It can be a good candidate to test against respiratory infection for SARS COV-2 pandemic. The antiviral property of GRFT can be in synergism with other lectins (Ziółkowska et al. 2006; O'Keefe et al. 2010).

2.3 *Phlorotannins*

Phlorotannins are derivatives of a water-soluble polyphenolic compounds called tannins. Phlorotannins consist of polymer-forming phloroglucinol units which are biosynthesized by the acetate-malonate pathway. Phlorotannins are mainly obtained from brown and red algae (Nagayama et al. 2002; Kim et al. 2006). They are effective against the HIV-1 virus and exhibit inhibition of reverse transcriptase, protease, and integrase enzymes which play a vital role in virus replication inside the host cells. When obtained from various sources, they may elicit different inhibitory properties to these target enzymes (Kim and Karadeniz 2011). Phlorotannins derivatives obtained from brown alga *Ecklonia cava* demonstrated inhibition of protease and reverse transcriptase (RT) enzyme. Out of the four phlorotannins derivatives tested, 8,8 ′-bieckol and 8,4 ‴-dieckol (Fig. 25.3) inhibited HIV-1 RT efficiently while showing moderate inhibition towards HIV-1 protease enzyme (Ahn et al. 2004).

3 Conclusion and Future Perspective

Viral infection and re-infections are responsible for deadly diseases in human history and can be highly contagious causing an outbreak of epidemic. It is a lasting challenge for the healthcare sector, as with the rising human population and frequent travel throughout the globe has increased the contact between humans and animals. Countries with higher human-wildlife interactions are more prone to an enzootic and epizootic viral transmission that can be fatal for both humans and wildlife species. Moreover, viruses are a marvel of nature with a complex life cycle and ability to undergo mutations which makes it difficult for the development of antiviral drugs. Developing a vaccine for some old known viruses such as HIV or dengue has been challenging enough. The timely outbreaks of SARS, MERS, Ebola in recent years have called for immediate action for revolutionary discovery and development of antiviral drugs. Viral outbreaks are sudden as we can see in the case of the current SARS COV-2 pandemic; prior knowledge of viruses and the antiviral agents has helped for screening the candidates and for the development of vaccine against SARS-2. Viruses are diverse in their structure and function, thus there is a need for development of antivirals that can target a large group of viruses. Macroalgae are a great source of sulphated and halogenated polysaccharides, lectins, and phlorotannins which are promising candidates for the research and development of new antiviral agents. Most of these moieties are at early phases of development and more extensive investigations are required to develop an effective antiviral drug. The compounds belonging to these groups of metabolites can lead to groundbreaking discovery of new antivirals. They can be evaluated to be used as direct-acting antivirals or in a combination to enhance the effectiveness of already available antivirals. Moreover, they can be useful to cope up with the increasing cases of antiviral resistant strains and new viruses.

Acknowledgments We thank SVKM'S NMIMS for all the support.

References

Adhikari U, Mateu CG, Chattopadhyay K, Pujol CA, Damonte EB, Ray B (2006) Structure and antiviral activity of sulfated fucans from *Stoechospermum marginatum*. Phytochemistry. https://doi.org/10.1016/j.phytochem.2006.05.024

Ahmadi A, Moghadamtousi SZ, Abubakar S, Zandi K (2015) Antiviral potential of algae polysaccharides isolated from marine sources: a review. Biomed Res Int 2015. https://doi.org/10.1155/2015/825203

Ahn MJ, Yoon KD, Min SY, Lee JS, Kim JH, Kim TG, Kim SH, Kim NG, Huh H, Kim J (2004) Inhibition of HIV-1 reverse transcriptase and protease by phlorotannins from the brown alga *Ecklonia cava*. Biol Pharm Bull 27:544–547. https://doi.org/10.1248/bpb.27.544

Aisa Y, Miyakawa Y, Nakazato T, Shibata H, Saito K, Ikeda Y, Kizaki M (2005) Fucoidan induces apoptosis of human HS-sultan cells accompanied by activation of caspase-3 and down-regulation of ERK pathways. Am J Hematol 78:7–14. https://doi.org/10.1002/ajh.20182

Akamatsu E, Shimanaga M, Kamei Y (2003) Isolation of an anti-influenza virus substance, MC26 from a marine brown alga, *Sargassum piluliferum* and its antiviral activity against influenza virus. Coastal Bioenvironment - Saga University, Saga, Japan

Alexandre KB, Gray ES, Lambson BE, Moore PL, Choge IA, Mlisana K, Abdool Karim SS et al (2010) Mannose-rich glycosylation patterns on HIV-1 subtype C gp120 and sensitivity to the lectins, Griffithsin, Cyanovirin-N and Scytovirin. Virology 402:187–196. https://doi.org/10.1016/j.virol.2010.03.021

Ana P, Nathalie B, Gilles B, Daniel R, Tomás MS, Yolanda FP (2021) Anti-Herpes simplex virus (HSV-1) activity and antioxidant capacity of carrageenan-rich enzymatic extracts from *Solieria filiformis* (Gigartinales, Rhodophyta). Int J Biol Macromol 168:322–330. https://doi.org/10.1016/j.ijbiomac.2020.12.064

Anil S, Venkatesan J, Chalisserry EP, Nam SY, Kim S-K (2017) Applications of seaweed polysaccharides in dentistry. Seaweed Polysaccharides 331–340:12. https://doi.org/10.1016/B978-0-12-809816-5.00017-7

Berteau O, Mulloy B (2003) Sulfated fucans, fresh perspectives: structures, functions, and biological properties of sulfated fucans and an overview of enzymes active toward this class of polysaccharide. Glycobiology 13:29R–40R. https://doi.org/10.1093/glycob/cwg058

Carlucci MJ, Scolaro LA, Noseda MD, Cerezo AS, Damonte EB (2004) Protective effect of a natural carrageenan on genital herpes simplex virus infection in mice. Antivir Res 64:137–141. https://doi.org/10.1016/j.antiviral.2004.07.001

Charette M, Walter S (2010) The volume of earth' s ocean. J Oceanogr Soc 23:112–114

Chiu YH, Chan YL, Tsai LW, Li TL, Wu CJ (2012) Prevention of human enterovirus 71 infection by kappa carrageenan. Antiviral Res 95:128–134. https://doi.org/10.1016/j.antiviral.2012.05.009

Cumashi A, Ushakova NA, Preobrazhenskaya ME, D'Incecco A, Piccoli A, Totani L, Tinari N et al (2007) A comparative study of the anti-inflammatory, anticoagulant, antiangiogenic, and antiadhesive activities of nine different fucoidans from brown seaweeds. Glycobiology 17:541–552. https://doi.org/10.1093/glycob/cwm014

Eccles R, Meier C, Jawad M, Weinmüllner R, Grassauer A, Prieschl-Grassauer E (2010) Efficacy and safety of an antiviral iota-carrageenan nasal spray: a randomized, double-blind, placebo-controlled exploratory study in volunteers with early symptoms of the common cold. Respir Res 11:108. https://doi.org/10.1186/1465-9921-11-108

Eccles R, Winther B, Johnston SL, Robinson P, Trampisch M, Koelsch S (2015) Efficacy and safety of iota-carrageenan nasal spray versus placebo in early treatment of the common cold in adults: the ICICC trial. Respir Res 16:121. https://doi.org/10.1186/s12931-015-0281-8

Elizondo-Gonzalez R, Elizabeth Cruz-Suarez L, Ricque-Marie D, Mendoza-Gamboa E, Rodriguez-Padilla C, Trejo-Avila LM (2012) In vitro characterization of the antiviral activity of fucoidan from *Cladosiphon okamuranus* against Newcastle disease virus. Virol J 9:1–9. https://doi.org/10.1186/1743-422X-9-307

Emau P, Tian B, O'keefe BR, Mori T, McMahon JB, Palmer KE, Jiang Y, Bekele G, Tsai CC (2007) Griffithsin, a potent HIV entry inhibitor, is an excellent candidate for anti-HIV microbicide. J Med Primatol 36:244–253. https://doi.org/10.1111/j.1600-0684.2007.00242.x

Estevez JM, Ciancia M, Cerezo AS (2001) DL-Galactan hybrids and agarans from gametophytes of the red seaweed *Gymnogongrus torulosus*. Carbohydr Res 331:27–41. https://doi.org/10.1016/S0008-6215(01)00015-5

Gallimore W (2017) Marine metabolites: oceans of opportunity. Pharmacognosy: fundamentals, applications and strategy. Elsevier Inc. doi:https://doi.org/10.1016/B978-0-12-802104-0.00018-4

García-Villalón D, Gil-Fernández C (1991) Antiviral activity of sulfated polysaccharides against African swine fever virus. Antivir Res. https://doi.org/10.1016/0166-3542(91)90031-L

Ghosh T, Pujol CA, Damonte EB, Sinha S, Ray B (2009) Sulfated xylomannans from the red seaweed Sebdenia polydactyla: structural features, chemical modification and antiviral activity. Antivir Chem Chemother 19:235–242. https://doi.org/10.1177/095632020901900603

Girard L, Birse K, Holm JB, Gajer P, Humphrys MS, Garber D, Guenthner P et al (2018) Impact of the griffithsin anti-HIV microbicide and placebo gels on the rectal mucosal proteome and microbiome in non-human primates. Sci Rep 8:8059. https://doi.org/10.1038/s41598-018-26313-8

Grassauer A, Weinmuellner R, Meier C, Pretsch A, Prieschl-Grassauer E, Unger H (2008) Iota-carrageenan is a potent inhibitor of rhinovirus infection. Virol J 5:107. https://doi.org/10.1186/1743-422X-5-107

Güven KC, Coban B, Özdemir O (2020) Pharmacology of marine macroalgae. Encycl Mar Biotechnol 10:585–615. https://doi.org/10.1002/9781119143802.ch20

Hans N, Malik A, Naik S (2020) Antiviral activity of sulfated polysaccharides from marine algae and its application in combating COVID-19: mini review. Bioresour Technol Rep 13:100623. https://doi.org/10.1016/j.biteb.2020.100623

Hardouin K, Bedoux G, Burlot AS, Donnay-Moreno C, Bergé JP, Nyvall-Collén P, Bourgougnon N (2016) Enzyme-assisted extraction (EAE) for the production of antiviral and antioxidant extracts from the green seaweed Ulva armoricana (Ulvales, Ulvophyceae). Algal Res 16:233–239. https://doi.org/10.1016/j.algal.2016.03.013

Haroun-Bouhedja F, Ellouali M, Sinquin C, Boisson-Vidal C (2000) Relationship between sulfate groups and biological activities of Fucans. Thromb Res 100:453–459. https://doi.org/10.1016/S0049-3848(00)00338-8

Hayashi T, Hayashi K, Maeda M, Kojima I (1996) Calcium Spirulan, an inhibitor of enveloped virus replication, from a blue-green alga *Spirulina platensis*. J Nat Prod 59:83–87. https://doi.org/10.1021/np960017o

Hemmingson JA, Falshaw R, Furneaux RH, Thompson K (2006) Structure and antiviral activity of the Galactofucan sulfates extracted from *UndariaPinnatifida* (Phaeophyta). J Appl Phycol 18:185. https://doi.org/10.1007/s10811-006-9096-9

Hidari KIPJ, Takahashi N, Arihara M, Nagaoka M, Morita K, Suzuki T (2008) Structure and anti-dengue virus activity of sulfated polysaccharide from a marine alga. Biochem Biophys Res Commun 376:91–95. https://doi.org/10.1016/j.bbrc.2008.08.100

Institute for Health Metrics and Evaluation (2021) Global health data exchange. http://ghdx.healthdata.org. Accessed 12 Feb 2021

Ishag HZA, Li C, Huang L, Sun MX, Wang F, Ni B, Malik T, Chen PY, Mao X (2013) *Griffithsin* inhibits Japanese encephalitis virus infection in vitro and in vivo. Arch Virol 158:349–358. https://doi.org/10.1007/s00705-012-1489-2

Jiang B-f, Xiao-fei X, Li L, Yuan W (2003) Study on—911|| anti-HBV effect in HepG2. 2.15 cells culture. Modern Preventive Medicine 30:517–518

Jönsson M, Allahgholi L, Sardari RRR, Hreggviosson GO, Karlsson EN (2020) Extraction and modification of macroalgal polysaccharides for current and next-generation applications. Molecules 25. https://doi.org/10.3390/molecules25040930

Kandale A, Meena AK, Rao MM, Panda P, Mangal AK, Reddy G, Babu R (2011) Marine algae: An introduction, food value and medicinal uses. J Pharm Res 17

Kanekiyo K, Lee JB, Hayashi K, Takenaka H, Hayakawa Y, Endo S, Hayashi T (2005) Isolation of an antiviral polysaccharide, nostoflan, from a terrestrial cyanobacterium, Nostoc flagelliforme. J Nat Prod 68:1037–1041. https://doi.org/10.1021/np050056c

Kim SK, Karadeniz F (2011) Anti-HIV activity of extracts and compounds from marine algae. In: Advances in food and nutrition research, vol 64, 1st edn. Elsevier, London. https://doi.org/10.1016/B978-0-12-387669-0.00020-X

Kim M-M, Van Ta Q, Mendis E, Rajapakse N, Jung W-K, Byun H-G, Jeon Y-J, Kim S-K (2006) Phlorotannins in Ecklonia cava extract inhibit matrix metalloproteinase activity. Life Sci 79:1436–1443. https://doi.org/10.1016/j.lfs.2006.04.022

Kim SK, Vo TS, Ngo DH (2011) Potential application of marine algae as antiviral agents in medicinal foods. In: Advances in food and nutrition research, vol 64, 1st edn. Elsevier, London. https://doi.org/10.1016/B978-0-12-387669-0.00019-3

Koyanagi S, Tanigawa N, Nakagawa H, Soeda S, Shimeno H (2003) Oversulfation of fucoidan enhances its anti-angiogenic and antitumor activities. Biochem Pharmacol 65:173–179. https://doi.org/10.1016/S0006-2952(02)01478-8

Lahaye M, Robic A (2007) Structure and function properties of Ulvan, a polysaccharide from green seaweeds. Biomacromolecules 8:1765–1774. https://doi.org/10.1021/bm061185q

Lee JB, Hayashi K, Maeda M, Hayashi T (2004) Antiherpetic activities of sulfated polysaccharides from green algae. Planta Med 70:813–817. https://doi.org/10.1055/s-2004-827228

Lee JB, Hayashi K, Hirata M, Kuroda E, Suzuki E, Kubo Y, Hayashi T (2006) Antiviral sulfated polysaccharide from Navicula directa, a diatom collected from deep-sea water in Toyama Bay. Biol Pharm Bull 29:2135–2139. https://doi.org/10.1248/bpb.29.2135

Lee JB, Takeshita A, Hayashi K, Hayashi T (2011) Structures and antiviral activities of polysaccharides from *Sargassum trichophyllum*. Carbohydr Polym 86:995–999. https://doi.org/10.1016/j.carbpol.2011.05.059

Liu H, Geng M, Xin X, Li F, Zhang Z, Li J, Ding J (2005) Multiple and multivalent interactions of novel anti-AIDS drug candidates, sulfated polymannuronate (SPMG)-derived oligosaccharides, with gp120 and their anti-HIV activities. Glycobiology 15:501–510. https://doi.org/10.1093/glycob/cwi031

Lopes N, Ray S, Espada SF, Bomfim WA, Ray B, Faccin-Galhardi LC, Linhares REC, Nozawa C (2017) Green seaweed *Enteromorpha compressa* (Chlorophyta, Ulvaceae) derived sulphated polysaccharides inhibit herpes simplex virus. Int J Biol Macromol 102:605–612. https://doi.org/10.1016/j.ijbiomac.2017.04.043

Lusvarghi S, Bewley CA (2016) Griffithsin: An antiviral lectin with outstanding therapeutic potential. Viruses 8. https://doi.org/10.3390/v8100296

Makarenkova ID, Deriabin PG, Lvov DK, Zviagintseva TN, Besednova NN (2010) Antiviral activity of sulfated polysaccharide from the brown algae *Laminaria japonica* against avian influenza a (H5N1) virus infection in the cultured cells. Voprosy virusologii 55:41–45

Mandal P, Mateu CG, Chattopadhyay K, Pujol CA, Damonte EB, Ray B (2007) Structural features and antiviral activity of sulphated fucans from the brown seaweed *Cystoseira indica*. Antivir Chem Chemother 18:153–162. https://doi.org/10.1177/095632020701800305

Mandal P, Pujol CA, Carlucci MJ, Chattopadhyay K, Damonte EB, Ray B (2008) Anti-herpetic activity of a sulfated xylomannan from *Scinaia hatei*. Phytochemistry 69:2193–2199. https://doi.org/10.1016/j.phytochem.2008.05.004

Matsuhiro B, Conte AF, Damonte EB, Kolender AA, Matulewicz MC, Mejías EG, Pujol CA, Zúñiga EA (2005) Structural analysis and antiviral activity of a sulfated galactan from the red seaweed *Schizymenia binderi* (Gigartinales, Rhodophyta). Carbohydr Res 340:2392–2402. https://doi.org/10.1016/j.carres.2005.08.004

Mazumder S, Ghosal PK, Pujol CA, Carlucci MJ, Damonte EB, Ray B (2002) Isolation, chemical investigation and antiviral activity of polysaccharides from *Gracilaria corticata* (Gracilariaceae, Rhodophyta). Int J Biol Macromol 31:87–95. https://doi.org/10.1016/S0141-8130(02)00070-3

Meiyu G, Fuchuan L, Xianliang X, Jing L, Zuowei Y, Huashi G (2003) The potential molecular targets of marine sulfated polymannuroguluronate interfering with HIV-1 entry: interaction between SPMG and HIV-1 rgp120 and CD4 molecule. Antivir Res 59:127–135. https://doi.org/10.1016/S0166-3542(03)00068-8

Moon HJ, Lee SR, Shim SN, Jeong SH, Stonik VA, Rasskazov VA, Zvyagintseva T, Lee YH (2008) Fucoidan inhibits UVB-induced MMP-1 expression in human skin fibroblasts. Biol Pharm Bull 31:284–289. https://doi.org/10.1248/bpb.31.284

Morán-Santibañez K, Cruz-Suárez LE, Ricque-Marie D, Robledo D, Freile-Pelegrín Y, Peña-Hernández MA, Rodríguez-Padilla C, Trejo-Avila LM (2016) Synergistic effects of sulfated polysaccharides from Mexican seaweeds against measles virus. Biomed Res Int 2016. https://doi.org/10.1155/2016/8502123

Nagayama K, Iwamura Y, Shibata T, Hirayama I, Nakamura T (2002) Bactericidal activity of phlorotannins from the brown alga *Ecklonia kurome*. J Antimicrob Chemother 50:889–893. https://doi.org/10.1093/jac/dkf222

Nelson TE, Lewis BA (1974) Separation and characterization of the soluble and insoluble components of insoluble laminaran. Carbohydr Res 33:63–74. https://doi.org/10.1016/S0008-6215(00)82940-7

O'Doherty JV, Dillon S, Figat S, Callan JJ, Sweeney T (2010) The effects of lactose inclusion and seaweed extract derived from *Laminaria* spp. on performance, digestibility of diet components and microbial populations in newly weaned pigs. Anim Feed Sci Technol 157:173–180. https://doi.org/10.1016/j.anifeedsci.2010.03.004

O'Keefe BR, Vojdani F, Buffa V, Shattock RJ, Montefiori DC, Bakke J, Mirsalis J et al (2009) Scaleable manufacture of HIV-1 entry inhibitor griffithsin and validation of its safety and efficacy as a topical microbicide component. Proc Natl Acad Sci U S A 106:6099–6104. https://doi.org/10.1073/pnas.0901506106

O'Keefe BR, Giomarelli B, Barnard DL, Shenoy SR, Chan PKS, McMahon JB, Palmer KE et al (2010) Broad-Spectrum and Activity and Efficacy of the Antiviral Protein Griffithsin against Emerging Viruses of the Family Coronaviridae. J Virol 84:2511–2521. https://doi.org/10.1128/JVI.02322-09

Parrish CR, Holmes EC, Morens DM, Park E-C, Burke DS, Calisher CH, Laughlin CA, Saif LJ, Daszak P (2008) Cross-species virus transmission and the emergence of new epidemic diseases. Microbiol Mol Biol Rev 72:457–470. https://doi.org/10.1128/mmbr.00004-08

Paula PC, Talarico LB, Noseda MD, Silvia SM, Damonte EB, Maria ER, Duarte. (2006) Chemical structure and antiviral activity of carrageenans from *Meristiella gelidium* against herpes simplex and dengue virus. Carbohydr Polym 63:459–465. https://doi.org/10.1016/j.carbpol.2005.09.020

Peat S, Whelan WJ, Lawley HG (1958) The structure of laminarin. Part I. the main polymeric linkage. J Chem Soc 141:724–728. https://doi.org/10.1039/JR9580000724

Pereira L, Critchley AT (2020) The COVID 19 novel coronavirus pandemic 2020: seaweeds to the rescue? Why does substantial, supporting research about the antiviral properties of seaweed polysaccharides seem to go unrecognized by the pharmaceutical community in these desperate times? doi:https://doi.org/10.1007/s10811-020-02143-y/Published

Piccini LE, Carro AC, Quintana VM, Damonte EB (2020) Antibody-independent and dependent infection of human myeloid cells with dengue virus is inhibited by carrageenan. Virus Res 290. https://doi.org/10.1016/j.virusres.2020.198150

Pomin VH, Mourão PAS (2008) Structure, biology, evolution, and medical importance of sulfated fucans and galactans. Glycobiology 18:1016–1027. https://doi.org/10.1093/glycob/cwn085

Ponce NMA, Pujol CA, Damonte EB, Flores ML, Stortz CA (2003) Fucoidans from the brown seaweed *Adenocystis utricularis:* extraction methods, antiviral activity and structural studies. Carbohydr Res 338:153–165. https://doi.org/10.1016/S0008-6215(02)00403-2

Pujol CA, Estevez JM, Carlucci MJ, Ciancia M, Cerezo AS, Damonte EB (2002) Novel DL-galactan hybrids from the red seaweed *Gymnogongrus torulosus* are potent inhibitors of herpes simplex virus and dengue virus. Antivir Chem Chemother 13:83–89. https://doi.org/10.1177/095632020201300202

Queiroz KCS, Medeiros VP, Queiroz LS, Abreu LRD, Rocha HAO, Ferreira CV, Jucá MB, Aoyama H, Leite EL (2008) Inhibition of reverse transcriptase activity of HIV by polysaccharides of brown algae. Biomed Pharmacother 62:303–307. https://doi.org/10.1016/j.biopha.2008.03.006

Rabanal M, Ponce NMA, Navarro DA, Gómez RM, Stortz CA (2014) The system of fucoidans from the brown seaweed *Dictyota dichotoma*: chemical analysis and antiviral activity. Carbohydr Polym 101:804–811. https://doi.org/10.1016/j.carbpol.2013.10.019

Rodríguez MC, Merino ER, Pujol CA, Damonte EB, Cerezo AS, Matulewicz MC (2005) Galactans from cystocarpic plants of the red seaweed *Callophyllis variegata* (Kallymeniaceae, Gigartinales). Carbohydr Res 340:2742–2751. https://doi.org/10.1016/j.carres.2005.10.001

Rudke AR, de Andrade CJ, Ferreira SRS (2020) *Kappaphycus alvarezii* macroalgae: An unexplored and valuable biomass for green biorefinery conversion. Trends Food Sci Technol 103:214–224. https://doi.org/10.1016/j.tifs.2020.07.018

Sanchez G (2013) Las instituciones de ciencia y tecnología en los procesos de aprendizaje de la producción agroalimentaria en Argentina. *El sistema argentino de innovación: instituciones, empresas y redes*. El desafío de la creación y apropiación de conocimiento 670:661–670. https://doi.org/10.1002/prot

Talarico LB, Zibetti RGM, Faria PCS, Scolaro LA, Duarte MER, Noseda MD, Pujol CA, Damonte EB (2004) Anti-herpes simplex virus activity of sulfated galactans from the red seaweeds *Gymnogongrus griffithsiae* and *Cryptonemia crenulata*. Int J Biol Macromol 34:63–71. https://doi.org/10.1016/j.ijbiomac.2004.03.002

Talarico LB, Duarte MER, Zibetti RGM, Noseda MD, Damonte EB (2007) An algal-derived DL-galactan hybrid is an efficient preventing agent for in vitro dengue virus infection. Planta Med 73:1464–1468. https://doi.org/10.1055/s-2007-990241

Talarico LB, Noseda MD, Ducatti DRB, Duarte MER, Damonte EB (2011) Differential inhibition of dengue virus infection in mammalian and mosquito cells by iota-carrageenan. J Gen Virol 92:1332–1342. https://doi.org/10.1099/vir.0.028522-0

Tanna B, Mishra A (2019) Nutraceutical potential of seaweed polysaccharides: structure, bioactivity, safety, and toxicity. Compr Rev Food Sci Food Saf 18:817–831. https://doi.org/10.1111/1541-4337.12441

Thuan NH, An TT, Shrestha A, Canh NX, Sohng JK, Dhakal D (eds) (2019) Recent advances in exploration and biotechnological production of bioactive compounds in three cyanobacterial genera: *Nostoc*, *Lyngbya*, and *Microcystis*. Front Chem. https://doi.org/10.3389/fchem.2019.00604

Thuy TTT, Ly BM, Van TTT, Van Quang N, Tu HC, Zheng Y, Seguin-Devaux C, Mi B, Ai U (2015) Anti-HIV activity of fucoidans from three brown seaweed species. Carbohydr Polym 115:122–128. https://doi.org/10.1016/j.carbpol.2014.08.068

Wang W, Wang SX, Guan HS (2012) The antiviral activities and mechanisms of marine polysaccharides: An overview. Mar Drugs 10:2795–2816. https://doi.org/10.3390/md10122795

Wang S, Lu Z, Wang S, Liu W, Gao J, Tian L, Wang L et al (2020) The inhibitory effects and mechanisms of polymannuroguluronate sulfate against human papillomavirus infection in vitro and in vivo. Carbohydr Polym 241:116365. https://doi.org/10.1016/j.carbpol.2020.116365

Witvrouw M, De Clercq E (1997) Sulfated polysaccharides extracted from sea algae as potential antiviral drugs. Gen Pharmacol Vasc S 29:497–511. https://doi.org/10.1016/S0306-3623(96)00563-0

Witvrouw M, Este JA, Mateu MQ, Reymen D, Andrei G, Snoeck R, Ikeda S et al (1994) Activity of a sulfated polysaccharide extracted from the red seaweed *Aghardhiella tenera* against human immunodeficiency virus and other enveloped viruses. Antivir Chem Chemother 5:297–303. https://doi.org/10.1177/095632029400500503

Xianliang X, Meiyu G, Huashi G, Zelin L (2000) Study on the mechanism of inhibitory action of 911 on replication of HIV-1 in vitro. Chin J Marine Drugs 19:15–18

Yamada T, Ogamo A, Saito T, Watanabe J, Uchiyama H, Nakagawa Y (1997) Preparation and anti-HIV activity of low-molecular-weight carrageenans and their sulfated derivatives. Carbohydr Polym 32:51–55. https://doi.org/10.1016/S0144-8617(96)00128-2

Zhu B, Ni F, Sun Y, Zhu X, Yin H, Yao Z, Yuguang D (2018) Insight into carrageenases: major review of sources, category, property, purification method, structure, and applications. Crit Rev Biotechnol 38:1261–1276. https://doi.org/10.1080/07388551.2018.1472550

Ziółkowska NE, O'Keefe BR, Mori T, Zhu C, Giomarelli B, Vojdani F, Palmer KE, McMahon JB, Wlodawer A (2006) Domain-swapped structure of the potent antiviral protein Griffithsin and its mode of carbohydrate binding. Structure 14:1127–1135. https://doi.org/10.1016/j.str.2006.05.017

Chapter 26
Chemical Composition and Phytopharmaceuticals: An Overview of the *Caulerpa* and *Cystoseira* Genera

Gonçalo P. Rosa, Maria do Carmo Barreto, Ana M. L. Seca, and Diana C. G. A. Pinto

Abbreviations

ABTS	2,2′-Azino-bis(3-ethylbenzthiazoline-6-sulfonic acid)
AGS	Human gastric epithelial cell line
Aβ25-35	Amyloid-β-protein fragment 25-35
Bax	Bcl-2-associated X protein
Bcl-2	B-cell lymphoma 2
EGCG	Epigallocatechin gallate
HCT116	Human colon carcinoma cell line
HCT15	Human colon adenocarcinoma cell line
HeLa	Cervix adenocarcinoma cell line
HepG2	Human hepatocarcinoma cell line
IC50	Half maximal inhibitory concentration
KA3IT	Mouse embryonic fibroblast virally transformed cancer cell line
LOX	Lipoxygenase
LPS	Lipopolysaccharide

G. P. Rosa · A. M. L. Seca
LAQV-REQUIMTE, Department of Chemistry, University of Aveiro, Aveiro, Portugal

Faculty of Sciences and Technology, cE3c–Centre for Ecology, Evolution and Environmental Changes/Azorean Biodiversity Group and, University of Azores, Ponta Delgada, Portugal
e-mail: goncalo.p.rosa@uac.pt; ana.ml.seca@uac.pt

M. do Carmo Barreto
Faculty of Sciences and Technology, cE3c–Centre for Ecology, Evolution and Environmental Changes/Azorean Biodiversity Group and, University of Azores, Ponta Delgada, Portugal
e-mail: maria.cr.barreto@uac.pt

D. C. G. A. Pinto (✉)
LAQV-REQUIMTE, Department of Chemistry, University of Aveiro, Aveiro, Portugal
e-mail: diana@ua.pt

A. Ranga Rao, G. A. Ravishankar (eds.), *Sustainable Global Resources of Seaweeds Volume 2*, https://doi.org/10.1007/978-3-030-92174-3_26

MAPK	Mitogen activated protein kinases
NF-kB	Necrosis factor kB
NIH3T3	Mouse embryonic fibroblast cell line
PTP1B	Protein tyrosine phosphatase 1B
SH-SY5Y	Neuroblastoma cell line
SI	Selectivity index
SK13	Lysinibacillus sphaericus
THP-1	Human leukemic cell line
TNF-α	tumor necrosis factor α

1 Introduction

Natural products have, through time, contributed to the development of new drugs (Hassan et al. 2019) and remain a valuable source of bioactive compounds. In the last years, the focus of natural products research has shifted towards marine sources since they represent approximately 25% of the Earth's biodiversity (Jiménez 2018). Within these marine sources, researchers have focused mainly on marine macroalgae due to their vast biodiversity and the great variety of bioactive compounds isolated from this group of organisms (Shama et al. 2019). According to recent studies, about 72,500 different algae species are described worldwide, and marine macroalgae represent the larger group among them (Guiry 2012).

Marine macroalgae are subjected to harsh environmental conditions, such as salinity, pH, sunlight, and CO_2 supply, which can change dramatically during the day, namely for intertidal species (Martínez et al. 2012). Consequently, macroalgae developed various surviving strategies, which led to the production of numerous secondary metabolites with exclusive structural cores, namely alkaloids, cyclic peptides, diterpenoids, phlorotannins, polyketides, quinones, and sterols (Salehi et al. 2019). Macroalgae are an excellent supply of health-promoting metabolites featuring unique structures capable of acting on a diverse spectrum of diseases. These metabolites are excellent lead compounds in developing new drugs that attract the pharmaceutical industry.

Despite the remarkable increase in discovering new metabolites and biological activities from marine macroalgae seen in the last years, several unexplored compounds could serve as a source of novel added-value compounds. The following paragraphs will focus on studies regarding two genera of macroalgae, *Caulerpa* and *Cystoseira*, showing what is already known about their metabolite's biological activities and identifying the gaps and potential research opportunities within these two groups of macroalgae.

2 *Cystoseira* Genus

Cystoseira C. Agardh, 1820 is a genus of brown marine macroalgae of the Sargassaceae family. Algae from this genus are distributed along the Atlantic-Mediterranean coasts, and the genus includes around 40 species (Guiry and Guiry 2021). These species ensure food and shelter to numerous marine organisms, so they are considered essential for the marine forests' biogenic structure (Cheminée et al. 2013), and they also have economic value.

Extracts prepared from several *Cystoseira* species exhibited biological activities, such as antimycobacterial, antiviral, and antitumoral (Bruno De Sousa et al. 2017a). The results report the vast potential of *Cystoseira* genus species, indicating the opportunity to find new metabolites and further explore their bioactive potential. In this regard, the following paragraphs describe the most recent results on the bioactive compounds isolated from the *Cystoseira* genus.

2.1 *Diterpenoids*

The phytochemical analysis of *Cystoseira myrica* (S.G. Gmelin) C.Agardh led to the identification of 6 terpenoids, cystoseirol monoacetate (**1**), dictyol F monoacetate (**2**), dictyone (**3**), dictyone acetate (**4**), isodictytriol monoacetate (**5**), and pachydictyol (**6**) (Fig. 26.1). Dictyol F monoacetate was less cytotoxic against the KA3IT cell line (IC_{50} 10 μg/mL), whereas the rest showed an IC_{50} of 5 μg/mL. Unfortunately, their selectivity is reduced; they are also cytotoxic against the non-tumor cell line NIH3T3 (7.5–15 μg/mL) (Ayyad et al. 2003).

In opposition, isololiolide (**7**) (Fig. 26.1), a terpenoid isolated from *Cystoseira tamariscifolia* (Hudson) Papenfuss, showed both cytotoxic activity against HepG2 (IC_{50} = 13.15 μM), AGS (IC_{50} = 32.36 μM) and HCT15 (IC_{50} = 23.59 μM) cell lines and high selectivity, since its activity against non-tumor human fibroblasts (IC_{50} = 1131.76 μM) was low (Vizetto-Duarte et al. 2016a). The authors established that isololiolide (**7**) disrupts the normal cell cycle of HepG2 by altering the expression of proteins, namely by inducing the increased expression of p53 and decreasing the expression levels of procaspase-3 and Bcl-2 levels (Vizetto-Duarte et al. 2016a).

Compound **7** also showed antiparasitic activity, namely against *Trypanossoma cruzi*, the causing agent of Chaga's Disease (Lima et al. 2019). Both trypomastigote and intracellular amastigotes of *Trypanosoma cruzi* were affected by isoliolide (**7**) *in vitro*, with IC_{50} values of 32 μM and 40 μM, respectively, while no cytotoxicity against mammalian cells was described (> 200 μM). Parasite death was due to the disruption of the plasma membrane integrity and a strong depolarization of the mitochondrial membrane potential, induced by isoliolide (**7**) (Lima et al. 2019).

Other terpenoids, for which biological evaluations were not reported, have been isolated from *Cystoseira* spp., indicating a potential for further discoveries within this genus.

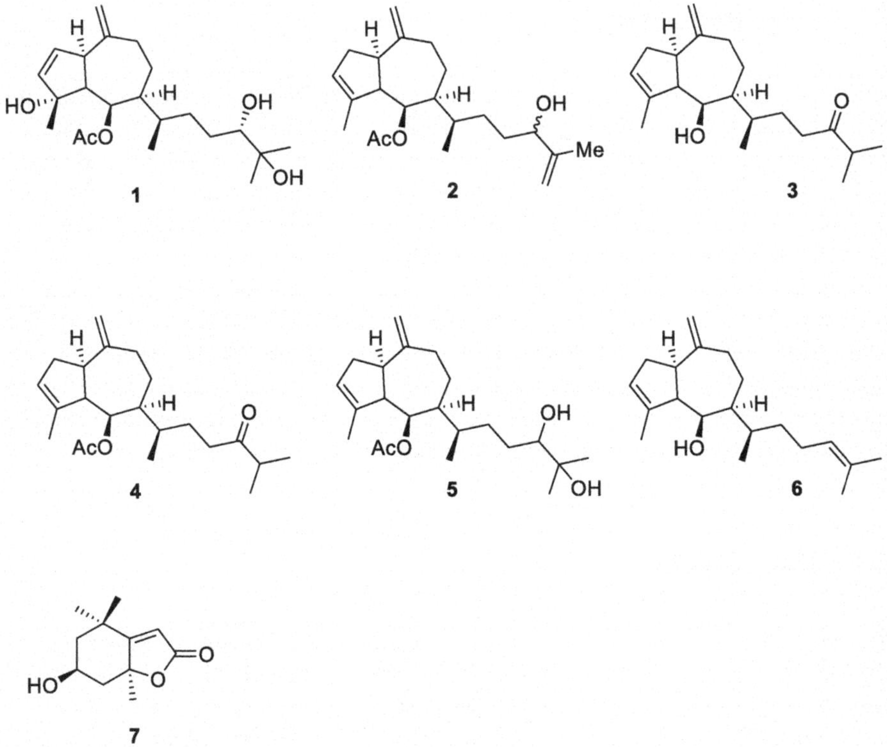

Fig. 26.1 Chemical structures of compounds 1–7

2.2 Meroterpenoids

Studies on the phytochemical composition of *Cystoseira abies -marina* (S.G.Gmelin) C.Agardh isolated four new compounds (Fig. 26.2): cystoazorol A (**8**), cystoazorol B (**9**), cystoazorone A (**10**), and cystoazorone B (**11**) (Gouveia et al. 2013). Compounds **8**, **10**, and **11** exhibited inhibitory activity against HeLa cells in both lag and log growth phases, with the best results obtained against cells in the log phase.

The most active compound was **8**, with an IC_{50} of 10.2 and 2.8 μg/mL against cells in lag and log phase, respectively. Interestingly, compound **8** showed a selectivity index (SI) of 1.64 and 2.46 for cells in lag and log phases of growth, respectively, which is higher than the observed for taxol, a drug currently used in chemotherapy (SI = 1.50 in lag phase and 0.50 in log phase) (Gouveia et al. 2013).

Demethoxy cystoketal chromane (**12**), a meroditerpenoid isolated from *Cystoseira tamariscifolia* (Vizetto-Duarte et al. 2016b) and *Cystoseira amentacea* var. *stricta* Montagne (Valls et al. 1996), significantly reduced the viability of HepG2 cells (IC_{50} = 14.77 μg/mL) while keeping a high selectivity towards the non-tumor S17 cell line (IC_{50} = 48.46 μg/mL, SI = 3.28) (Vizetto-Duarte et al. 2016b).

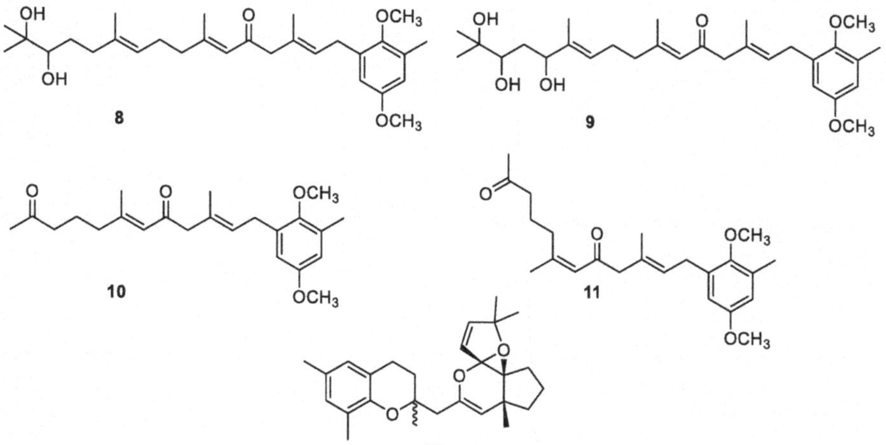

Fig. 26.2 Chemical structures of compounds **8–12**

A series of new tetraprenyltoluquinol (**13–18**), triprenyltoluquinol (**19–20**), and tetraprenyltoluquinone (**21–22**) derivatives were isolated from *Cystoseira crinita* Duby (Fig. 26.3) (Fisch et al. 2003). These compounds were tested for their antioxidant activities by different methods, which revealed that most of them were powerful antioxidants, with activity comparable to that of α-tocopherol, used as standard. 2-[(2′*E*,6′*E*,10′*E*)-5′,13′-Dioxo-3′,7′,11′,15′-tetramethyl-hexadeca-2′,6′,10′,14′-tetraenyl]-6-methylhydroquinone (**14**) and 5-oxo-isocystofuranoquinone (**22**) were the compounds that presented the higher antioxidant activities (Fisch et al. 2003).

The isomers (3*R*)- and (3*S*)-tetraprenyltoluquinols (**23**) and (**24**) were isolated from the hexane extract of *Cystoseira baccata* (S.G.Gmelin) P.C.Silva (Bruno De Sousa et al. 2017b) and demonstrated the ability to induce cytoplasmic vacuolization and disruption of the mitochondrial membrane potential of *Leishmania infantum* promastigotes and amastigotes, with an IC_{50} of 25.0 μM in the inhibition of the intracellular infection of this parasite. Moreover, they were not cytotoxic against mammalian macrophages, with a SI of 5.04, which indicates their potential to be safe anti-parasitic agents (Bruno De Sousa et al. 2017b).

Cystoseira tamariscifolia produces cystomethoxybifurcarenone (**25**), a compound that has displayed interesting antifungal activity against three tomato pathogenic fungi (*Botrytis cinerea*, *Fusarium oxysporum* sp. *lycopersici*, and *Verticillium alboatrum*) (Bennamara et al. 1999).

Mokrini et al. (2008) reported the antifouling activity of three meroditerpenoid derivatives (**26–28**) (Fig. 26.4), found in *Cystoseira baccata*. Compound **26** inhibited the settlement of *Sargassum muticum* and the phenoloxidase activity of mussels with IC_{50} of 2.5 and 1 μg/mL, respectively. Compound **27** was able to inhibit the settlement of *Sargassum muticum* and the settlement of *Ulva intestinalis* and the phenoloxidase activity, all with an IC_{50} of 1 μg/mL. Compound **28** inhibited the settlement of *Ulva intestinalis* and mussels' phenoloxidase activity (IC_{50} = 2.5 μg/mL). These compounds were nontoxic to oyster and sea urchin larvae, which shows

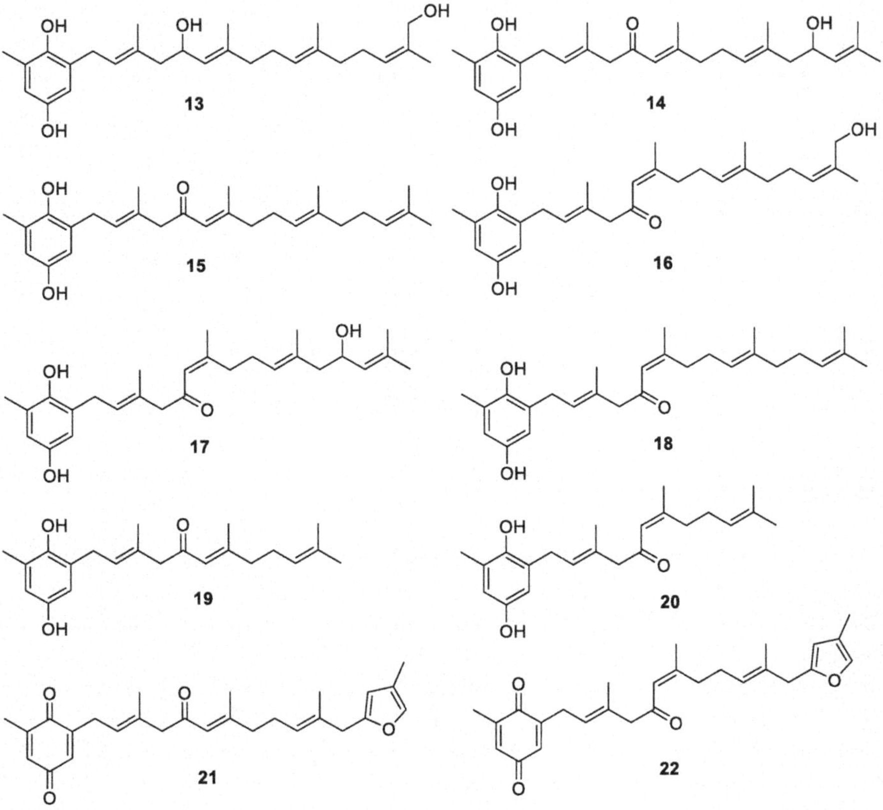

Fig. 26.3 Chemical structures of compounds **13–22**

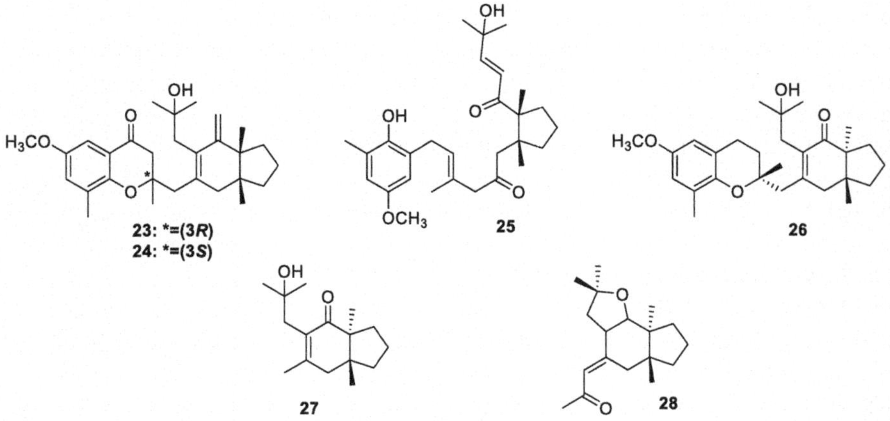

Fig. 26.4 Chemical structures of compounds **23–28**

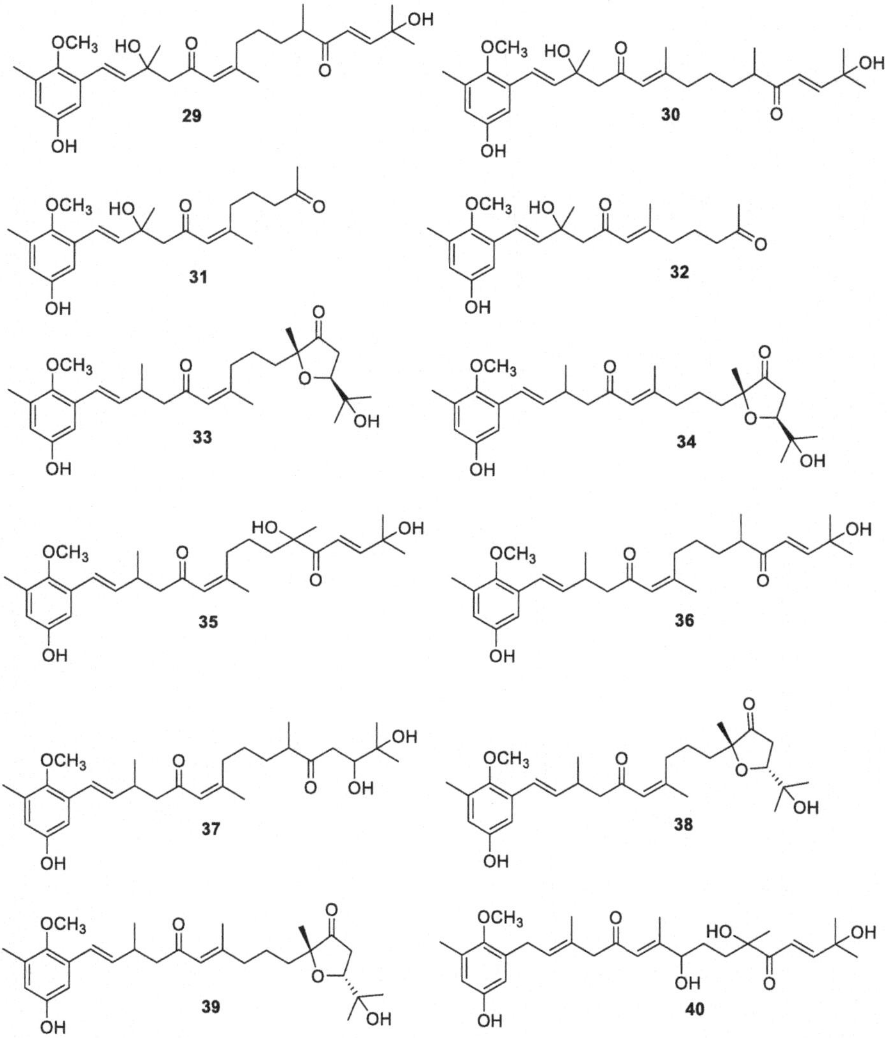

Fig. 26.5 Chemical structures of compounds **29–52**

their potential as natural antifouling agents for aquaculture systems (Mokrini et al. 2008).

Eighteen new meroterpenoids, cystodiones A–M (**29–40**) and cystones A–F (**41–46**), were isolated from *Cystoseira usneoides* (Linnaeus) M.Roberts, alongside the known meroterpenoids: cystomexicone A (**47**), cystomexicone B (**48**), usneoidone E (**49**), amentadione-1′-methyl ether (**50**), usneoidone Z (**51**) and 6-*cis*-amentadione-1′-methyl ether (**52**) (Fig. 26.5) (De Los Reyes et al. 2013, 2016). These compounds presented antioxidant activity in ABTS radical scavenging assay, with **29**, **30**, **35**, **36**, **50**, and **51** having an IC_{50} very close to Trolox's (about 26 μM).

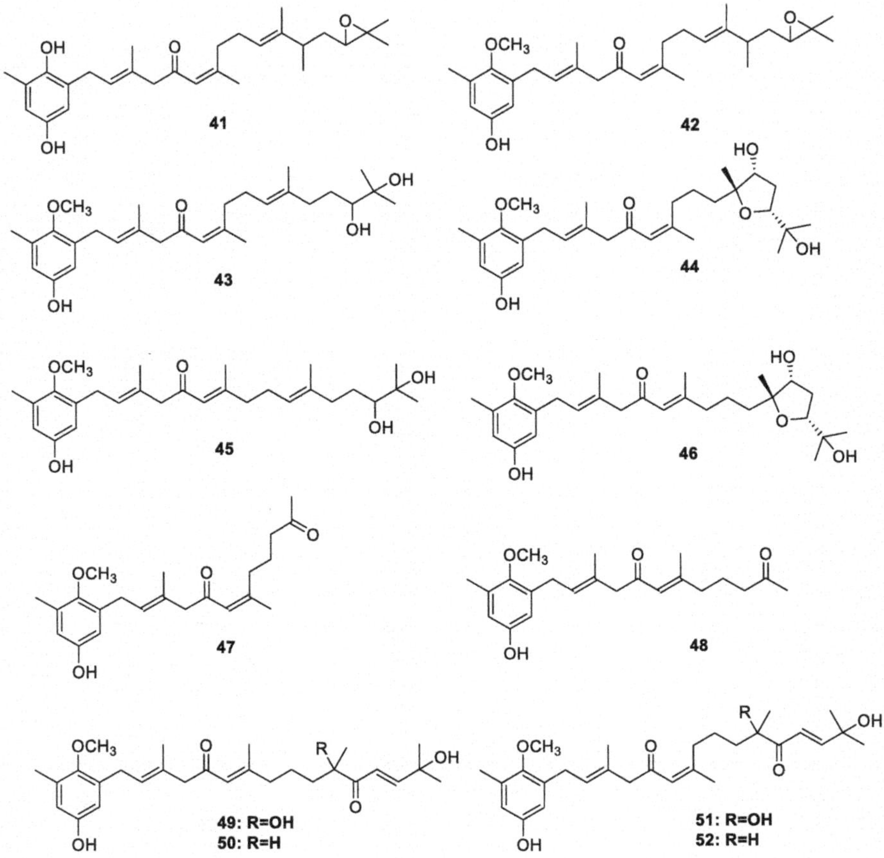

Fig. 26.5 (continued)

A concentration of 10 μM from **51** inhibited the production of TNF-α, a proinflammatory cytokine, in lipopolysaccharide (LPS)-stimulated THP-1 human macrophages by 73% when compared with untreated cells (De Los Reyes et al. 2013). Furthermore, the treatment of macrophages with cystodione G (**35**) at 10 μM and with cystodione M (**40**) at 8 μM caused significant inhibition of the production of TNF-α when compared to LPS-stimulated untreated cells (De Los Reyes et al. 2016).

2.3 *Sterols*

A wide array of sterols has been isolated from different *Cystoseira* species, namely *Cystoseira adriatica* Sauvageau (Kapetanović et al. 2005), *Cystoseira crinita*, *Cystoseira barbata* (Stackhouse) C.Agardh (Milkova et al. 1997), *Cystoseira nodicaulis* (Withering) M.Roberts, *Cystoseira tamariscifolia* and *Cystoseira usneoides*

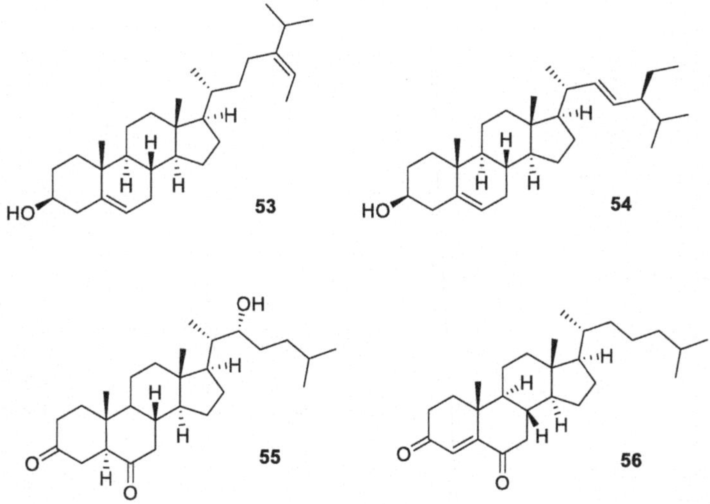

Fig. 26.6 Chemical structures of compounds **53–56**

(Andrade et al. 2013). However, only four of them (Fig. 26.6) have their biological activities studied, so further studies on *Cystoseira* sterols bioactivity are desirable.

Of the sterols found in *Cystoseira* spp., fucosterol (**53**), a characteristic metabolite of brown algae, is the most known. This compound's vast activities have been recently reviewed by Hannan et al. (2020).

Stigmasterol (**54**) is another bioactive sterol found in *Cystoseira* spp., able to inhibit the expression of several pro-inflammatory and matrix degradation mediators involved in osteoarthritis-induced cartilage degradation (Gabay et al. 2010). Furthermore, stigmasterol (**54**) also possesses strong apoptosis induction effects by increasing the expression of pro-apoptotic genes, like Bax and p53, and downregulating the expression of Bcl-2, an anti-apoptotic gene, in HepG2 cells (Kim et al. 2014), and vascular smooth muscle cells (Li et al. 2015). The compound **54** capacity to induce apoptosis in vascular smooth muscle cells shows its potential in preventing cardiac diseases.

Two less known steroids, 3-keto-22-epi-28-nor-cathasterone (**55**) and cholest-4-ene-3,6-di-one (**56**) were identified in *Cystoseira myrica* (Hamdy et al. 2009). Compound **55** was active against HEPG-2 and HCT116 cancer cells lines, with IC_{50} of 2.96 and 12.38 μM, respectively. Compound **56** was almost twelve times more potent against HCT116 (IC_{50} = 1.16 μM) (Hamdy et al. 2009).

2.4 Carotenoids

Fucoxanthin (**57**) (Fig. 26.7), a carotenoid of algal origin, was found in *Cystoseira brachycarpa* J.Agardh (Ragonese et al. 2014). The biological properties of this compound are widely studied and show its importance as a marine natural product.

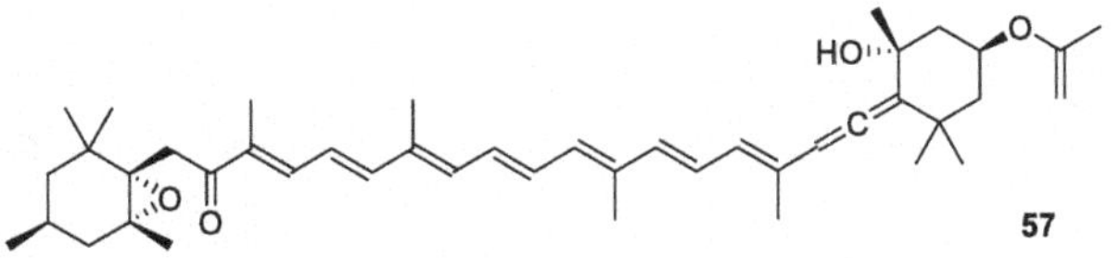

Fig. 26.7 Chemical structure of compound **57**

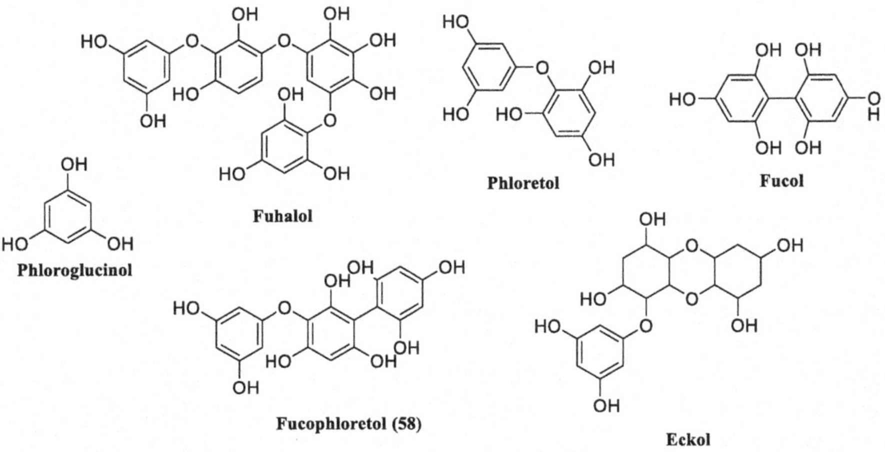

Fig. 26.8 Examples of different subclasses of phlorotannins and the structure of fucophloretol **58**

The most important activities displayed by fucoxanthin (**57**) are antioxidant, anti-inflammatory, anticancer, anti-obesity, antidiabetic, antimalarial, and antiangiogenic activities (Zhang et al. 2015). With its ability to treat life-style related diseases, fucoxanthin (**57**) is in high demand, so its presence should be studied in more species of *Cystoseira*, to determine which species produce higher amounts of fucoxanthin (**57**), increasing their value.

2.5 *Phlorotannins*

Phlorotannins are polymers of phloroglucinol (Fig. 26.8) with different degrees of polymerization and are among polyphenolic compounds produced by brown seaweed as secondary metabolites. They represent about 14% of brown algae's dry biomass (Machu et al. 2015). Phlorotannin classification is based on the types of linkages between the phloroglucinol units. There are four subclasses, namely, phlorotannins with ether linkages (*e.g.*, fuhalol and phloretol), those with phenyl linkages (*e.g.*, fucol), those with both ether and phenyl linkages (*e.g.*, fucophloretol), and those with a dibenzodioxin linkage (*e.g.*, eckol) (Singh and Sidana 2013) (Fig. 26.8).

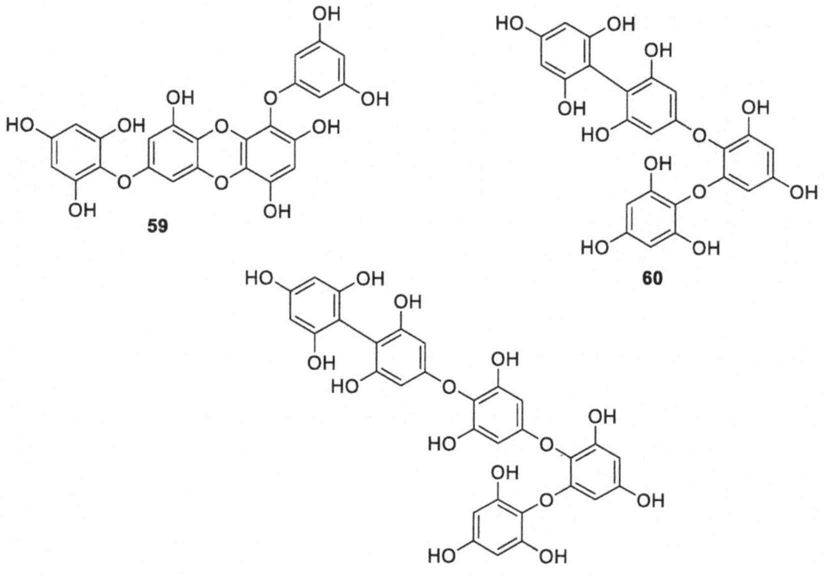

59 60 61

Fig. 26.9 Chemical structures of compounds **59–61**

In brown algae, several phlorotannins have been identified in *Cystoseira* spp. Different phlorotannins from the eckol and fucophloroetol groups were identified in *Cystoseira nodicaulis*, *Cystoseira tamariscifolia* and *Cystoseira usneoides* (Ferreres et al. 2012). Of all the phlorotannins identified, fucophloroetol (**58**), 7-phloroetckol (**59**), fucodiphloroetol (**60**) and fucotriphloroetol (**61**) (Fig. 26.9) are highlighted, since they were found on these species for the first time (Ferreres et al. 2012). *Cystoseira humilis* Schousboe ex Kützing also produces 7-phloreckol (**59**), while *Cystoseira baccata* and *Cystoseira nodicaulis* present traces of other phlorotannins such as fucols, phlorethols and fucophlorethols (Stiger-Pouvreau et al. 2014).

Several studies with *Cystoseira* spp. phlorotannins report the results of phlorotannin-rich extracts (Ferreres et al. 2012; Lopes et al. 2012, 2013; Stiger-Pouvreau et al. 2014) often composed of two or more different phlorotannins. This is due to the difficulty in separating them since the identification methods allow the determination of the different phlorotannins in the mixture but not their isolation. It is, therefore, difficult to attribute the bioactivities of a given extract to a specific component. Nonetheless, in general, phlorotannins are known to possess many bioactivities such as antioxidant, antitumor, anti-inflammatory, antidiabetic, antihypertensive, and antiallergic (Freitas et al. 2015; Rosa et al. 2020). Also, antibacterial and antifungal activities are particularly common for phlorotannins (Ford et al. 2020). A few phlorotannins have been found to inhibit tyrosinase, which is relevant in cosmetics since this activity is associated with the inhibition of melanogenesis, meaning that these compounds can be used as skin depigmentation agents (Wijesinghe and Jeon 2011).

3 *Caulerpa* Genus

The genus *Caulerpa* J.V. Lamouroux (1809) is a Chlorophyta (green algae) belonging to the Order Bryopsidales and the Family Caulerpaceae. There are currently described 97 species and over 100 varieties of *Caulerpa* (Guiry and Guiry 2021). The original distribution of *Caulerpa* spp. is the intertidal and intratidal zone of tropical and semitropical marine waters worldwide. Macroalgae of this genus, namely *Caulerpa lentillifera* J.Agardh and *Caulerpa racemosa* (ForssKål) J.Agardh, are widely consumed around the world, commonly named sea grapes, lelato, green caviar, and lai-lai (De Gaillande et al. 2017).

Many *Caulerpa* species were extensively utilized in aquariums because of their adaptability and eye-pleasing nature (Walters et al. 2006). The wide use of these species in aquariums worldwide led to their uncontrolled invasion of areas where they were not native, leading *Caulerpa* species to be listed as the world's worst invasive algal species (Walters et al. 2006). Actually, the genus gained much attention in recent decades mainly because of the invasive potential of species like *Caulerpa taxifolia* (M.Vahl) C.Agardh and *Caulerpa cylindracea* Sonder (Montefalcone et al. 2015).

Many reports have shown that extracts and compounds from *Caulerpa* spp. presented a diverse range of bioactivities like insecticidal, antimicrobial, antifouling, feeding deterrent, anti-inflammatory, cytotoxic, antiproliferative, and antimetastatic (Rushdi et al. 2020; Zubia et al. 2020). In this regard, the following paragraphs detail the findings of the compounds isolated from the *Cystoseira* genus in the last years.

3.1 Alkaloids

Two bisindole alkaloids, racemosins A (**62**) and B (**63**), were isolated for the first time from *Caulerpa racemosa* along with caulerpin (**64**) (Fig. 26.10).

These compounds were tested for their ability to protect neuronal cells (SH-SY5Y) against the neurotoxic effects of $A\beta_{25-35}$. Caulerpin (**64**) showed a level of neuroprotection similar to that of epigallocatechin gallate (EGCG), the positive control, increasing cell viability in 14.6% at 10 μM (16.57% for EGCG). On the other hand, racemosins A (**62**) and B (**63**) showed moderate/weak neuroprotective activity with 5.5% and 8.1% increase in cell viability (10 μM), respectively (Liu et al. 2013).

Caulerpin (**64**) is one of the best studied *Caulerpa* spp. compounds, exhibiting *in vitro* antitumor activity against a range of cell-lines, while showing low toxicity in mice models (Murugan and Iyer 2013; Li et al. 2018). The low toxicity observed is the reason it is still considered a potential antitumor drug lead, despite the low activity compared with taxol (Li et al. 2018). Furthermore, it showed antiviral activity against herpes simplex virus type 1, in pre-clinical assays with Vero cells, showing

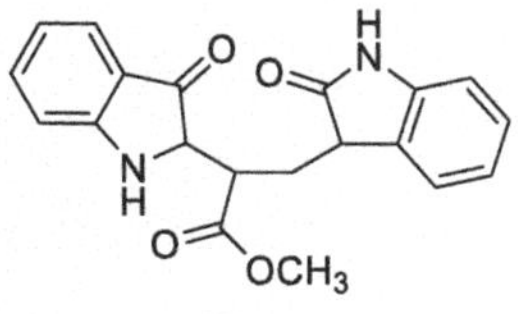

62

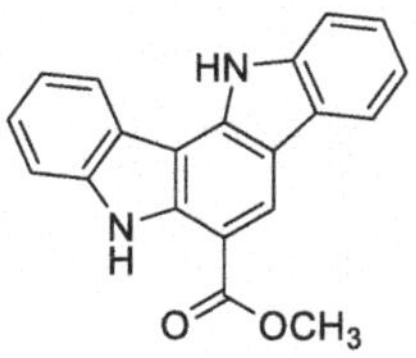

63

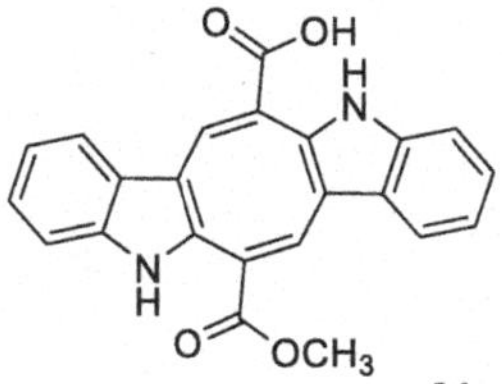

64

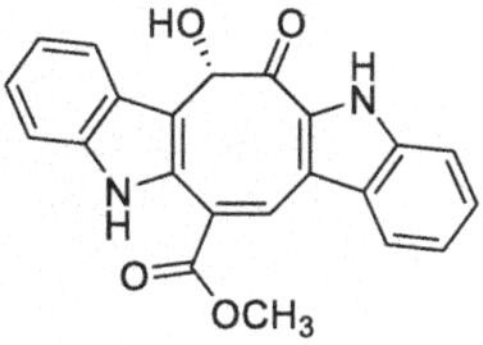

65

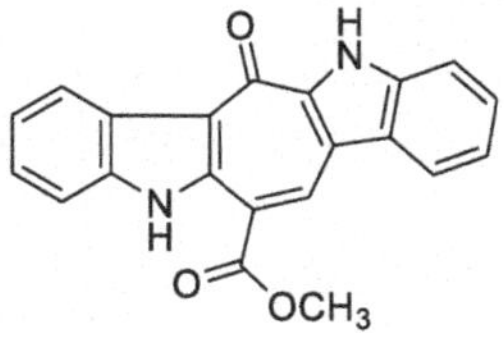

66

Fig. 26.10 Chemical structures of compounds **62–66**

an IC_{50} of 1.29 μM, similar to the value obtained for acyclovir, a commercial anti-herpetic drug (IC_{50} = 1.09 μM), presenting lower cytotoxicity than the one obtained with acyclovir (Macedo et al. 2012).

Other activities reported for compound **64** include antimicrobial (Vairappan 2004), antituberculosis (Chay et al. 2014), antispasmodic (Cavalcante-Silva et al. 2013), antinociceptive, anti-inflammatory (De Souza et al. 2009), and PTP1B inhibitory activities (Ornano et al. 2014).

Yang et al. (2014) isolated racemosin C (**65**), along with caulersin (**66**), from *Caulerpa racemosa* (Fig. 26.10). These compounds significantly inhibited tyrosine phosphatase B (PTP1B) activity, a negative regulator of insulin signaling, whose uncontrolled activity is associated with cancer development (Xu et al. 2019; Yu et al. 2019). The IC_{50} values for PTP1B were 5.86 and 7.14 μM for racemosin C (**65**) and caulersin (**66**), respectively, showing their potential as candidates for cancer therapy. Their activity was lower than caulerpin (**64**), whose IC_{50} was 3.77 μM (Yang et al. 2014).

3.2 Sesquiterpenes

Caulerpenyne (**67**) (Fig. 26.11) a sesquiterpene first isolated from *Caulerpa prolifera* (Forsskål) J.V.Lamouroux (Amico et al. 1978), showed to be an uncompetitive effective lipoxygenase (LOX) inhibitor *in vitro* (IC_{50} = 5.1 μM) (Cengiz et al. 2011). This compound also presents antibacterial, neurotoxic, phytotoxic, and anti-proliferative activities (Barbier et al. 2001; Mozzachiodi et al. 2001; Raniello et al. 2007). It is cytotoxic to eight different tumor cells of human origin, acting by

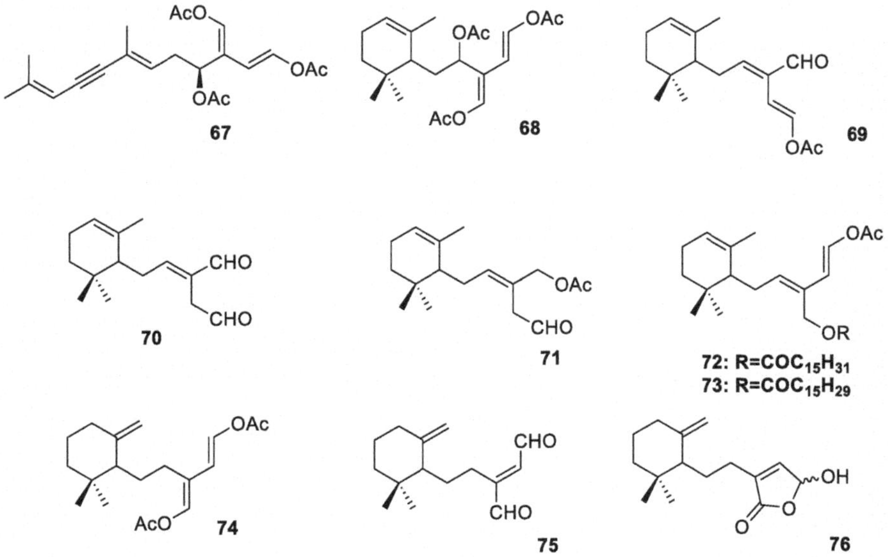

Fig. 26.11 Chemical structures of compounds **67–76**

modulating DNA synthesis and protein phosphorylation. It also blocks the cell cycle of sea urchin embryos at a metaphase like stage and inhibits the activation of MAPK proteins (Fischel et al. 1995).

Sesquiterpenes (**68–73**) (Fig. 26.11) were isolated from *Caulerpa ashmeadii* Harvey, and tested for field feeding preference, antimicrobial activity and ichthyotoxicity (Paul et al. 1987). All compounds except **72** and **73** showed antimicrobial activity against the fungus *Lagenidium callinectes*, and the bacteria *Vibrio leignathi*, *Vibrio phosphoreum*, and SK13, with compounds **70** and **71** being the most active. Also, these compounds were toxic to damselfish (*Pomacentrus phillipinus*) (Paul et al. 1987).

From *Caulerpa bikinensis* W.R.Taylor, were isolated the compounds **74–76** (Fig. 26.11) (Paul and Fenical 1982). The tests for the feeding deterrence of these compounds showed that compound **74** and **75** were toxic to *Pomacentrus phillipinus* at 5 μg/mL and were also cytotoxic to the fertilized egg of the Pacific sea urchin *Lytechinus pinctus*, (ED_{50} values of 2 and 1 μg/mL, respectively) (Paul and Fenical 1982), results that reinforce the possible roles of these compounds as agents of chemical defense.

The knowledge about the sesquiterpenes of *Caulerpa* spp. did not advance in recent years, which is surprising due to the bioactivities presented and the diverse structures found. They could be interesting leads for the discovery of new therapeutic agents.

3.3 Diterpenoids

Compound **77** (Fig. 26.12) was isolated from *Caulerpa brownii* (C.Agardh) Endlicher and exhibited antibacterial activity against bacteria *Escherichia coli*, *Staphylococcus aureus* and *Bacilus subtilis*. It also inhibits the growth of the marine bacteria *Vibrio harveyi*, *Vibrio leiognathid* and *Vibrio anguillarum* (Paul and Fenical 1985).

Caulerpa racemosa yielded 4,5-dehydrodiodictyonema A (**78**), racemobutenolids A and B, as a pair of epimers (**79 a** and **b**) and α-tocoxylenoxy (**80**) (Fig. 26.12) (Yang et al. 2015). Compound **78** inhibited PTP1B activity (IC_{50} = 2.30 μM), being one of the strongest inhibitors of this enzyme isolated from *Caulerpa* spp.

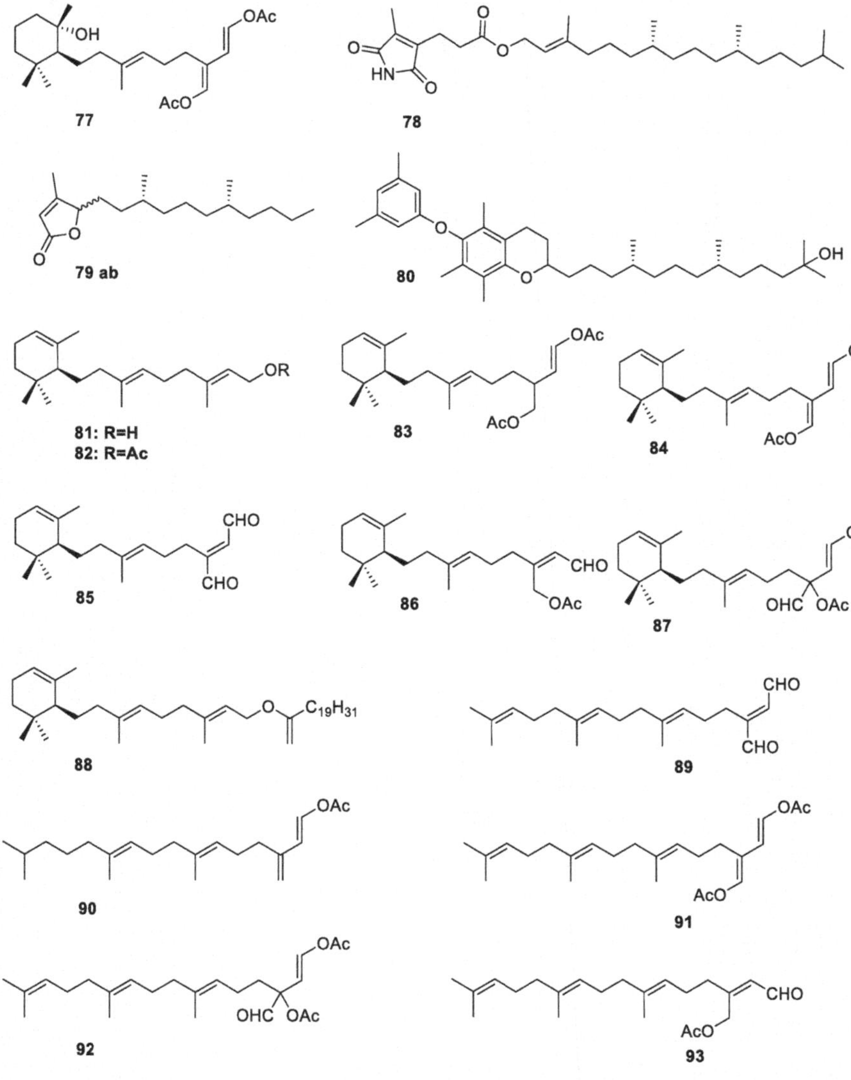

Fig. 26.12 Chemical structures of compounds **77–93**

The isolation of diterpenes (**81–93**) (Fig. 26.12) was reported by Handley and Blackman (2005) from both branched and unbranched specimens of *Caulerpa brownie*, however, the bioactivities of these compounds are not studied, which constitutes a gap to bridge.

3.4 Sterols

Further studies led to the isolation of sterols from *Caulerpa chemnitzia* (Esper) J.V.Lamouroux, *Caulerpa faridii* Nizamuddin, *Caulerpa manorensis* Nizamuddin, and *Caulerpa taxifolia*, including cholesterol (**94**), 24-methylcholesterol (**95**), 24-methylcholesta-7,22-diene-3β-ol (**96**), 4,24-dimethyl-cholesta-5,22-diene-3β-ol (**97**), and β-sitosterol (**98**) (Aliya and Shameel 2003) (Fig. 26.13).

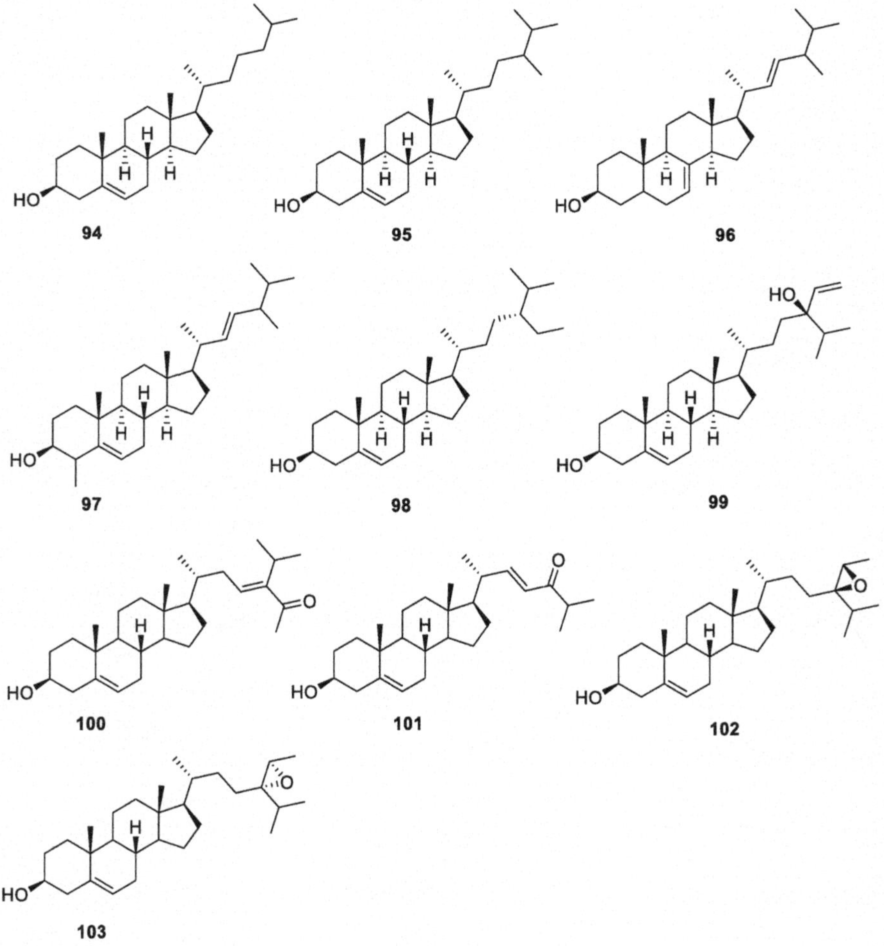

Fig. 26.13 Chemical structures of compounds **94–103**

In *Caulerpa racemosa*, fucosterol (**53**) (Fig. 26.6), saringosterol (**99**) and the oxidized sterols (**100–103**) (Fig. 26.13) were identified (Aliya and Shameel 2003; Yang et al. 2015), which were not found in any other *Caulerpa* species. Their PTP1B inhibitory activity was tested (IC_{50} 3.80–49.97 μM), compound **100** being the most potent inhibitor, like the positive control, oleanolic acid (IC_{50} = 3.02 μM). Compounds **100** and **101** presented neuroprotective activity, protecting the SH-SY5Y cell line against $A\beta_{25-35}$-induced damage, increasing cell viability in 11.31 and 15.98%, respectively, at 10 μM. The result obtained for compound **101** is very promising since it is very close to the one obtained with the positive control EGCG (16.57%) at the same concentration (Yang et al. 2015).

4 Conclusion and Future Perspectives

The data reviewed above show that the *Cystoseira* and *Caulerpa* genera are excellent sources for obtaining bioactive compounds. In *Cystoseira*, the terpenoids and meroterpenoids are the families with the highest number of compounds identified but steroids, phlorotannins and phenolic compounds are also present. These compounds have displayed very interesting biological activities like antioxidant, anti-inflammatory, cytotoxic, anticancer, cholinesterase inhibition, antidiabetic, anti-herpetic and antimicrobial activities.

The *Caulerpa* genus exhibits high chemodiversity, the most common families being alkaloids and linear and monocyclic terpenoids possessing aldehydic and enol-acetate functional groups. Compounds with terminal bis-enol acetate group are uniquely found in this genus. This functional group is represented by an acetylated bis-enol category of 1,4-dialdehyde group, imparting bioactivities to several species as discussed above. These compounds present a very wide array of interesting biological activities, such as neuroprotective, antitumor, anti-inflammatory, antibacterial, and antifungal.

It is very clear that compounds isolated from *Cystoseira* and *Caulerpa* genera present high medical potential, providing an extensive list of natural structures that could serve as scaffolds for designing novel leads for pharmacological purposes.

However, there are still species poorly explored and can be sources of new valuable compounds. In addition, many compounds only have one bioactivity reported and so could be assayed for other properties, to increase their array of possible applications and, consequently, their value.

Furthermore, the cosmetic potential of *Cystoseira* spp. and *Caulerpa* spp., has been neglected. This gap constitutes an opportunity directing the research to their cosmeceutical activities, which would increase these genera's economic value.

Acknowledgments Thanks are due to FCT - Fundação para a Ciência e a Tecnologia for supporting G.P.R.'s grant (SFRH/BD/144446/2019), through national and European funds and co-financed by the European Social Fund through the Regional Operational Programme Centro 2020, as well as to FCT, the European Union, QREN, FEDER, and COMPETE, through funding the cE3c center (UIDB/00329/2020) and the LAQV-REQUIMTE (UIDB/50006/2020).

References

Aliya R, Shameel M (2003) Marine natural products of *Caulerpa* (siphonocladophyceae). Pak J Bot 35:659–669

Amico V, Piattelli M, Tringali C, Fattorusso E, Magno S, Mayol L (1978) Caulerpenyne, an unusual sequiterpenoid from the green alga *Caulerpa prolifera*. Tetrahedron Lett 19:3593–3596

Andrade PB, Barbosa M, Matos RP, Lopes G, Vinholes J, Mouga T, Valentão P (2013) Valuable compounds in macroalgae extracts. Food Chem 138:1819–1828

Ayyad S-EN, Abdel-Halim OB, Shier WT, Hoye TR (2003) Cytotoxic hydroazulene diterpenes from the brown alga *Cystoseira myrica*. Zeitschrift für Naturforschung C 58:33–38

Barbier P, Guise S, Huitorel P, Amade P, Pesando D, Briand C, Peyrot V (2001) Caulerpenyne from *Caulerpa taxifolia* has an antiproliferative activity on tumor cell line SK-N-SH and modifies the microtubule network. Life Sci 70:415–429

Bennamara A, Abourriche A, Berrada M, Charrouf M, Chaib N, Boudouma M, Garneau FX (1999) Methoxybifurcarenone: an antifungal and antibacterial meroditerpenoid from the brown alga *Cystoseira tamariscifolia*. Phytochemistry 52:37–40

Bruno De Sousa C, Gangadhar KN, Macridachis J, Pavão M, Morais TR, Campino L, Varela J, Lago JHG (2017a) *Cystoseira* algae (Fucaceae): update on their chemical entities and biological activities. Tetrahedron Asymmetry 28:1486–1505

Bruno De Sousa C, Gangadhar KN, Morais TR, Conserva GAA, Vizetto-Duarte C, Pereira H, Laurenti MD, Campino L, Levy D, Uemi M, Barreira L, Custódio L, Passero LFD, Lago JHG, Varela J (2017b) Antileishmanial activity of meroditerpenoids from the macroalgae *Cystoseira baccata*. Exp Parasitol 174:1–9

Cavalcante-Silva L, de Carvalho Correia A, Barbosa-Filho J, da Silva B, de Oliveira Santos B, de Lira D, Sousa J, de Miranda G, de Andrade Cavalcante F, Alexandre-Moreira M (2013) Spasmolytic effect of caulerpine involves blockade of Ca^{2+} influx on Guinea pig ileum. Mar Drugs 11:1553–1564

Cengiz S, Cavas L, Yurdakoc K, Pohnert G (2011) The sesquiterpene caulerpenyne from *Caulerpa* spp. is a lipoxygenase inhibitor. Mar Biotechnol 13:321–326

Chay C, Cansino R, Pinzón C, Torres-Ochoa R, Martínez R (2014) Synthesis and anti-tuberculosis activity of the marine natural product caulerpin and its analogues. Mar Drugs 12:1757–1772

Cheminée A, Sala E, Pastor J, Bodilis P, Thiriet P, Mangialajo L, Cottalorda J-M, Francour P (2013) Nursery value of *Cystoseira* forests for Mediterranean rocky reef fishes. J Exp Mar Biol Ecol 442:70–79

De Gaillande C, Payri C, Remoissenet G, Zubia M (2017) *Caulerpa* consumption, nutritional value and farming in the indo-Pacific region. J Appl Phycol 29:2249–2266

De Los Reyes C, Zbakh H, Motilva V, Zubía E (2013) Antioxidant and anti-inflammatory meroterpenoids from the brown alga *Cystoseira usneoides*. J Nat Prod 76:621–629

De Los Reyes C, Ortega MJ, Zbakh H, Motilva V, Zubía E (2016) *Cystoseira usneoides*: a brown alga rich in antioxidant and anti-inflammatory meroditerpenoids. J Nat Prod 79:395–405

De Souza ÉT, Pereira de Lira D, Cavalcanti de Queiroz A, Costa da Silva DJ, Bezerra de Aquino A, Campessato Mella E, Prates Lorenzo V, de Miranda GE, de Araújo-Júnior JX, de Oliveira Chaves MC, Barbosa-Filho JM, Filgueiras de Athayde-Filho P, de Oliveira Santos BV, Alexandre-Moreira MS (2009) The antinociceptive and anti-inflammatory activities of caulerpin, a bisindole alkaloid isolated from seaweeds of the genus *Caulerpa*. Mar Drugs 7:689–704

Ferreres F, Lopes G, Gil-Izquierdo A, Andrade P, Sousa C, Mouga T, Valentão P (2012) Phlorotannin extracts from Fucales characterized by HPLC-DAD-ESI-MSn: approaches to hyaluronidase inhibitory capacity and antioxidant properties. Mar Drugs 10:2766–2781

Fisch KM, Böhm V, Wright AD, König GM (2003) Antioxidative meroterpenoids from the brown alga *Cystoseira crinita*. J Nat Prod 66:968–975

Fischel J, Lemee R, Formento P, Caldani C, Moll JL, Pesando D, Meinsez A, Grelier P, Pietra P, Guerriero A (1995) Cell growth inhibitory effects of caulerpenyne, a sesquiterpenoid from the marine algae *Caulerpa taxifolia*. Anticancer Res 15:2155–2160

Ford L, Stratakos AC, Theodoridou K, Dick JTA, Sheldrake GN, Linton M, Corcionivoschi N, Walsh PJ (2020) Polyphenols from brown seaweeds as a potential antimicrobial agent in animal feeds. ACS Omega 5:9093–9103

Freitas AC, Pereira L, Rodrigues D, Carvalho AP, Panteleitchouk T, Gomes AM, Duarte AC (2015) Marine functional foods. In: Kim S-K (ed) Springer handbook of marine biotechnology. Springer, Berlin, pp 969–994

Gabay O, Sanchez C, Salvat C, Chevy F, Breton M, Nourissat G, Wolf C, Jacques C, Berenbaum F (2010) Stigmasterol: a phytosterol with potential anti-osteoarthritic properties. Osteoarthr Cartil 18:106–116

Gouveia V, Seca AML, Barreto MC, Neto A, Kijjoa A, Silva AMS (2013) Cytotoxic meroterpenoids from *Cystoseira abies-marina*. Phytochem Lett 6:593–597

Guiry MD (2012) How many species of algae are there? J Phycol 48:1057–1063

Guiry MD, Guiry GM (2021) *AlgaeBase*. World-wide electronic publication. National University of Ireland, Galway. http://www.algaebase.org. Accessed 3 Jan 2021

Hamdy A-HA, Aboutabl EA, Sameer S, Hussein AA, Díaz-Marrero AR, Darias J, Cueto M (2009) 3-Keto-22-epi-28-nor-cathasterone, a brassinosteroid-related metabolite from *Cystoseira myrica*. Steroids 74:927–930

Handley JT, Blackman AJ (2005) Secondary metabolites from the marine alga *Caulerpa brownii* (Chlorophyta). Aust J Chem 58:39–46

Hannan MA, Sohag AAM, Dash R, Haque MN, Mohibbullah M, Oktaviani DF, Hossain MT, Choi HJ, Moon IS (2020) Phytosterols of marine algae: insights into the potential health benefits and molecular pharmacology. Phytomedicine 69:153201

Hassan SS, Jin HZ, Abu-Izneid T, Rauf A, Ishaq M, Suleria HAR (2019) Stress-driven discovery in the natural products: a gateway towards new drugs. Biomed Pharmacother 109:459–467

Jiménez C (2018) Marine natural products in medicinal chemistry. ACS Med Chem Lett 9:959–961

Kapetanović R, Sladić DM, Popov S, Zlatović MV, Kljajić Z, Gašić MJ (2005) Sterol composition of the Adriatic Sea algae *Ulva lactuca*, *Codium dichotomum*, *Cystoseira adriatica* and *Fucus virsoides*. J Serb Chem Soc 70:1395–1400

Kim Y-S, Li X-F, Kang K-H, Ryu B, Kim SK (2014) Stigmasterol isolated from marine microalgae *Navicula incerta* induces apoptosis in human hepatoma HepG2 cells. BMB Rep 47:433–438

Li C, Liu Y, Xie Z, Lu Q, Luo S (2015) Stigmasterol protects against Ang II-induced proliferation of the A7r5 aortic smooth muscle cell-line. Food Funct 6:2266–2272

Li H, Liao X, Sun Y, Zhou R, Long W, Li L, Gu L, Xu S (2018) An economical synthesis of caulerpin and evaluation of its new anticancer activities. ChemistrySelect 3:12406–12409

Lima ML, Romanelli MM, Borborema SET, Johns DM, Migotto AE, Lago JHG, Tempone AG (2019) Antitrypanosomal activity of isololiolide isolated from the marine hydroid *Macrorhynchia philippina* (Cnidaria, hydrozoa). Bioorg Chem 89:103002

Liu D-Q, Mao S-C, Zhang H-Y, Yu X-Q, Feng M-T, Wang B, Feng L-H, Guo Y-W (2013) Racemosins a and B, two novel bisindole alkaloids from the green alga *Caulerpa racemosa*. Fitoterapia 91:15–20

Lopes G, Sousa C, Silva LR, Pinto E, Andrade PB, Bernardo J, Mouga T, Valentão P (2012) Can phlorotannins purified extracts constitute a novel pharmacological alternative for microbial infections with associated inflammatory conditions? PLoS One 7:e31145

Lopes G, Pinto E, Andrade PB, Valentão P (2013) Antifungal activity of phlorotannins against dermatophytes and yeasts: approaches to the mechanism of action and influence on *Candida albicans* virulence factor. PLoS One 8:e72203

Macedo NRPV, Ribeiro MS, Villaça RC, Ferreira W, Pinto AM, Teixeira VL, Cirne-Santos C, Paixão ICNP, Giongo V (2012) Caulerpin as a potential antiviral drug against herpes simplex virus type 1. Rev Bras 22:861–867

Machu L, Misurcova L, Vavra Ambrozova J, Orsavova J, Mlcek J, Sochor J, Jurikova T (2015) Phenolic content and antioxidant capacity in algal food products. Molecules 20:1118–1133

Martínez B, Arenas F, Rubal M, Burgués S, Esteban R, García-Plazaola I, Figueroa FL, Pereira R, Saldaña L, Sousa-Pinto I, Trilla A, Viejo RM (2012) Physical factors driving intertidal macroalgae distribution: physiological stress of a dominant fucoid at its southern limit. Oecologia 170:341–353

Milkova T, Talev G, Christov R, Dimitrova-Konaklieva S, Popov S (1997) Sterols and volatiles in *Cystoseira barbata* and *Cystoseira crinita* from the black sea. Phytochemistry 45:93–95

Mokrini R, Mesaoud MB, Daoudi M, Hellio C, Maréchal J-P, El Hattab M, Ortalo-Magné A, Piovetti L, Culioli G (2008) Meroditerpenoids and derivatives from the brown alga *Cystoseira baccata* and their antifouling properties. J Nat Prod 71:1806–1811

Montefalcone M, Morri C, Parravicini V, Bianchi CN (2015) A tale of two invaders: divergent spreading kinetics of the alien green algae *Caulerpa taxifolia* and *Caulerpa cylindracea*. Biol Invasions 17:2717–2728

Mozzachiodi R, Scuri R, Roberto M, Brunelli M (2001) Caulerpenyne, a toxin from the seaweed *Caulerpa taxifolia*, depresses afterhyperpolarization in invertebrate neurons. Neuroscience 107:519–526

Murugan K, Iyer V (2013) Differential growth inhibition of cancer cell lines and antioxidant activity of extracts of red, brown, and green marine algae. *In Vitro Cellular & Developmental Biology*. Animal 49:324–334

Ornano L, Donno Y, Sanna C, Ballero M, Serafini M, Bianco A (2014) Phytochemical study of *Caulerpa racemosa* (Forsk.) J. Agarth, an invading alga in the habitat of La Maddalena archipelago. Nat Prod Res 28:1795–1799

Paul VJ, Fenical W (1982) Toxic feeding deterrents from the tropical marine alga *Caulerpa bikinensis* (chlorophyta). Tetrahedron Lett 23:5017–5020

Paul VJ, Fenical W (1985) Diterpenoid metabolites from pacific marine algae of the order Caulerpales (Chlorophyta). Phytochemistry 24:2239–2243

Paul VJ, Littler MM, Littler DS, Fenical W (1987) Evidence for chemical defense in tropical green alga *Caulerpa ashmeadii* (Caulerpaceae: Chlorophyta): isolation of new bioactive sesquiterpenoids. J Chem Ecol 13:1171–1185

Ragonese C, Tedone L, Beccaria M, Torre G, Cichello F, Cacciola F, Dugo P, Mondello L (2014) Characterization of lipid fraction of marine macroalgae by means of chromatography techniques coupled to mass spectrometry. Food Chem 145:932–940

Raniello R, Mollo E, Lorenti M, Gavagnin M, Buia MC (2007) Phytotoxic activity of caulerpenyne from the Mediterranean invasive variety of *Caulerpa racemosa*: a potential allelochemical. Biol Invasions 9:361–368

Rosa GP, Tavares WR, Sousa PMC, Pagès AK, Seca AML, Pinto DCGA (2020) Seaweed secondary metabolites with beneficial health effects: an overview of successes in *in vivo* studies and clinical trials. Mar Drugs 18:8

Rushdi MI, Abdel-Rahman IAM, Attia EZ, Abdelraheem WM, Saber H, Madkour HA, Amin E, Hassan HM, Abdelmohsen UR (2020) A review on the diversity, chemical and pharmacological potential of the green algae genus *Caulerpa*. S Afr J Bot 132:226–241

Salehi B, Sharifi-Rad J, Seca AML, Pinto DCGA, Michalak I, Trincone A, Mishra AP, Nigam M, Zam W, Martins N (2019) Current trends on seaweeds: looking at chemical composition, phytopharmacology, and cosmetic applications. Molecules 24:4182

Shama A, Joyce SG, Mari FD, Ranga Rao A, Ravishankar GA, Hudaa N (2019) Macroalgae and microalgae: novel sources of functional food and feed. In: Ravishankar GA, Rao AR (eds) Handbook of algal technologies and phytochemicals: Volume-I: food, health and nutraceutical applications. CRC, New York, pp 207–219

Singh IP, Sidana J (2013) Phlorotannins. In: Domínguez H (ed) Functional ingredients from algae for foods and nutraceuticals. Woodhead Publishing, Cambridge, pp 181–204

Stiger-Pouvreau V, Jegou C, Cerantola S, Guérard F, Le Lann K (2014) Phlorotannins in Sargassaceae species from Brittany (France): interesting molecules for ecophysiological and valorisation purposes. Adv Bot Res 71:379–411

Vairappan CS (2004) Antibacterial activity of major secondary metabolites found in four species of edible green macroalgae genus *Caulerpa*. Asian Journal of Microbiology Biotechnology and Environmental Sciences 6:197–201

Valls R, Mesguiche V, Piovetti L, Prost M, Peiffer G (1996) Meroditerpenes from the brown alga *Cystoseira amentacea* var. *stricta* collected off the French mediterranean coast. Phytochemistry 41:1367–1371

Vizetto-Duarte C, Custódio L, Gangadhar KN, Lago JHG, Dias C, Matos AM, Neng N, Nogueira JMF, Barreira L, Albericio F, Rauter AP, Varela J (2016a) Isololiolide, a carotenoid metabolite isolated from the brown alga *Cystoseira tamariscifolia*, is cytotoxic and able to induce apoptosis in hepatocarcinoma cells through caspase-3 activation, decreased Bcl-2 levels, increased p53 expression and PARP cleavage. Phytomedicine 23:550–557

Vizetto-Duarte C, Custódio L, Acosta G, Lago JH, Morais TR, Bruno de Sousa C, Gangadhar KN, Rodrigues MJ, Pereira H, Lima RT, Vasconcelos MH, Barreira L, Rauter AP, Albericio F, Varela J (2016b) Can macroalgae provide promising anti-tumoral compounds? A closer look at *Cystoseira tamariscifolia* as a source for antioxidant and anti-hepatocarcinoma compounds. PeerJ 4:41704

Walters LJ, Brown KR, Stam WT, Olsen JL (2006) E-commerce and *Caulerpa*: unregulated dispersal of invasive species. Front Ecol Environ 4:75–79

Wijesinghe WAJP, Jeon Y-J (2011) Biological activities and potential cosmeceutical applications of bioactive components from brown seaweeds: a review. Phytochem Rev 10:431–443

Xu Q, Wu N, Li X, Guo C, Li C, Jiang B, Wang H, Shi D (2019) Inhibition of PTP1B blocks pancreatic cancer progression by targeting the PKM2/AMPK/mTOC1 pathway. Cell Death Dis 10:874

Yang H, Liu D-Q, Liang T-J, Li J, Liu A-H, Yang P, Lin K, Yu X-Q, Guo Y-W, Mao S-C, Wang B (2014) Racemosin C, a novel minor bisindole alkaloid with protein tyrosine phosphatase-1B inhibitory activity from the green alga *Caulerpa racemosa*. J Asian Nat Prod Res 16:1158–1165

Yang P, Liu D-Q, Liang T-J, Li J, Zhang H-Y, Liu A-H, Guo Y-W, Mao S-C (2015) Bioactive constituents from the green alga *Caulerpa racemosa*. Bioorg Med Chem 23:38–45

Yu M, Liu Z, Liu Y, Zhou X, Sun F, Liu Y, Li L, Hua S, Zhao Y, Gao H, Zhu Z, Na M, Zhang Q, Yang R, Zhang J, Yao Y, Chen X (2019) PTP 1B markedly promotes breast cancer progression and is regulated by miR-193a-3p. FEBS J 286:1136–1153

Zhang H, Tang Y, Zhang Y, Zhang S, Qu J, Wang X, Kong R, Han C, Liu Z (2015) Fucoxanthin: a promising medicinal and nutritional ingredient. Evid Based Complement Alternat Med 2015:723515

Zubia M, Draisma SGA, Morrissey KL, Varela-Álvarez E, De Clerck O (2020) Concise review of the genus *Caulerpa* J.V. Lamouroux. J Appl Phycol 32:23–39

Chapter 27
Skin Whitening with Seaweeds: Looking into Emerging Products in the Natural Cosmeceutical Market.

Ayse Kose

1 Introduction

The pigmentation process is a complex metabolic flux determining the color of skin, hair, and eyes in mammalians (Vachtenheim and Borovanský 2010). The process starts with L-tyrosine or L-3,4-dihydroxyphenylalanine (L-DOPA) and ends to produce eumelanins (dark brown pigments) and pheomelanin (red-orange pigment) (D'Mello et al. 2016). Skin color has always been a fascinating biological process that humanity wanted to understand how and under what kind of circumstances it is developed.

Early observations on skin color were attributed to gender, ethnicity, and exposure to various environmental factors (Westerhof 2006). The underlying chemical under this fascinating phenotypic variation is melanin pigment which is named after the pioneering work of Jöns Jacop Berzelius in 1840. He extracted the pigment from the coroid layer of the eye and named it as "melanin", which is derived from "*melanos*" which means "*dark, black*". In the 15th and 16th centuries, pigmentation experiments were excessively investigated due to increased interaction with European and African populations. One of the pioneers in the extraction of melanin from the skin is Malpighi, who compared the skin of African and European individuals (Solano 2014). In his findings, he attributed that skin color to be a phenomenon developed on the upper layer because, in both individuals, the bottom layer of the skin was having the same pale pink color. The traces of melanin can be reached in fossils as well, which makes this pigment an enigmatic and extremely stable chemical (Zhang et al. 2010).

Melanin is produced in cells known as melanocytes, derived from neural crest cells in early development (Cheli et al. 2010). In melanocytes, melanin is produced in melanosome granules and transferred to surrounding keratinocytes (Smit et al.

A. Kose (✉)
Faculty of Engineering, Department of Bioengineering, Ege University, Izmir, Turkey

A. Ranga Rao, G. A. Ravishankar (eds.), *Sustainable Global Resources of Seaweeds Volume 2*, https://doi.org/10.1007/978-3-030-92174-3_27

2009). The molecular mechanism and transport of this pigment is a fascinating topic for understanding biological and biochemical machinery in the human body. Melanin is not only giving color to the skin but it is documented to protect skin from harmful chemicals, heavy metals, UV exposure and also oxidative reagents (Solano 2017). However, as any imbalance in biochemical machineries may cause a type of disturbance in the natural state of the system, excessive or decreased melanin production may cause pigmentation disorders in the skin (Lee et al. 2018). These disorders can be aesthetic or life-threatening. Although melanin is known to affect the visuals of humans, recent studies also found that mechanism underlying melanin production can have an interaction with severe neurological disorders such as Alzheimer and Parkinson (Tel et al. 2013; Bagherzadeh et al. 2015). These findings along with pigmentation disorders prove the importance of understanding molecular machinery for these biochemical disturbances.

Although pigmentation is a complex pathway of cell signaling molecular metabolisms, the first and rate limiting step is the conversion of L-tyrosine and/or L-DOPA by tyrosinase enzyme. It is an enzyme classified in polyphenol oxidases (Chang 2012). Melanin production in the skin as an enzymatic process was only discovered in by Bourquelet. This was a tremendous finding to understand enzyme kinetics and its effect on skin coloring. When tyrosinase enzyme is upregulated to produce an excessive amount of melanin, this issue is called as hyperpigmentation disorders (Smit et al. 2009). Hyperpigmentation problems are irregular dark patches, freckles, age spots or nevus bodies on skin. On the other hand, when tyrosinase enzyme activity is downregulated, subsequently melanin synthesis is decreased and white patches on skin could be observed as it is in vitiligo. However, due to an enormous number of hyperpigmentation related problems, tyrosinase inhibitors (also referred to as skin whitening molecules) are highly demanded in the cosmeceutical industry to reverse pigmentation issues and restore the evenly distributed skin color (Pillaiyar et al. 2018).

Seaweeds are also gaining attention due to promising tyrosinase inhibitory effects, easy in cultivation and as a sustainable source for bioactive molecules (Azam et al. 2017b). Recent studies clearly demonstrated that seaweeds are considered as future's organisms for tyrosinase inhibitors. Thus, in this chapter tyrosinase inhibitors from seaweeds discovered so far has been discussed.

2 State of the Art

Due to increasing demand on cosmeceutical products, investigation of potential sources for this purpose is inevitably high. It is estimated to have over 50 billion US dollars by 2027 and the market size is expected to grow rapidly. The cosmeceutical compounds mainly have effects as anti-aging, anti-cellulite, anti-wrinkle, antioxidant, anti-bacterial, anti-acne and antitumor to treat or sooth the (dermo)cosmetics problems (Thomas and Kim 2013). Tyrosinase inhibitors or in general skin whitening compounds are holding a major share in the cosmeceutical industry.

Since, hyperpigmentation disorders are worldwide aesthetic pigmentation issues (Hacker 2017); formulations to reverse darkened skin and restoring skin color integrity have a primary role in an individual's psychology. Thus the market for skin whitening products has a dramatic expansion by the beginning of 2000s and estimated to be over 30 billion US dollars (Pillaiyar et al. 2018). This amount is almost half of the estimations for global cosmeceuticals industry.

With respect to increase in consumer's conscious to use natural substances, skin whitening cosmeceuticals were excessively searched in natural sources such as plants, fungus, bacteria, dairy products and aquatic resources (Smit et al. 2009). Besides consumer consciousness, common tyrosinase inhibitors such as hydroquinone, kojic acid, ascorbic acid and arbutin face stability and toxicity issues (Abu et al. 2009). Hydroquinone is proved to be toxic and carcinogenic, thus European Committee banned hydroquinone from cosmeceuticals formulations. As a golden standard to work on pigmentation, kojic acid, an aromatic compound from *Aspergillus niger* was also found to be toxic on healthy cells thus the utilization in formulations was limited (Singh et al. 2016). Besides, these compounds also suffer from transdermal delivery issues (Feng et al. 2018). As a novel and sustainable resource, algae (referred as seaweeds in this chapter) were considered as natural and sustainable resources for cosmeceutical compounds (Wijesekara et al. 2011).

Seaweeds are classified into four major classes, the rhodophyceae (red algae), the phaeophyceae (brown algae), the cyanophyceae (blue-green algae), and the chlorophyceae (green algae) (Jesumani et al. 2019). They are highly diverse and can be found in many various aquatic systems. This diversity makes them unique sources for bioprospecting bio-functional molecules. Algae has been utilized in various aspects in biotechnology from biofuel to feed applications. Among them, when a bibliographic analysis is done, a great portion of algae studies are concentrated on cosmetics (Figure 27.1a). It is not quite surprising because algae has high amount of functional proteins, peptides, fatty acids, terpenoids, polysaccharides, phenolic compounds and pigments valuable as antioxidant, antibacterial, antifungal, anti-inflammatory and sunscreen bioactivities. These substances are known to be effective via various mechanism of action for cosmetics including skin whitening via antioxidant mechanisms, direct tyrosinase enzyme inhibition or pathway mediated downregulation of tyrosinase inhibition. In a database search obtained from Scopus, pioneering research on algae in cosmetics starts in 1990s and gradually increases up to date (Figure 27.1b). By the year 2010s, the annual number of publications were over 100. However, there is a rapid accumulation of utilization of seaweeds in tyrosinase inhibitors and considered as a reliable substance for regulation of mammalian melanogenesis.

When a bibliographic analysis on seaweed (Figure 27.1a) is done, it is clear that seaweeds contribution to tyrosinase inhibitors are high on brown algae species. Phenol derivatives from seaweeds also have a high impact on deciphering tyrosinase inhibitors and skin whitening compounds. The most co-occurred term is antioxidant activity. As most of the studies also suggests (Heo et al. 2005; Chang and Teo 2016), antioxidant activities are key determinators along with direct tyrosinase inhibitors.

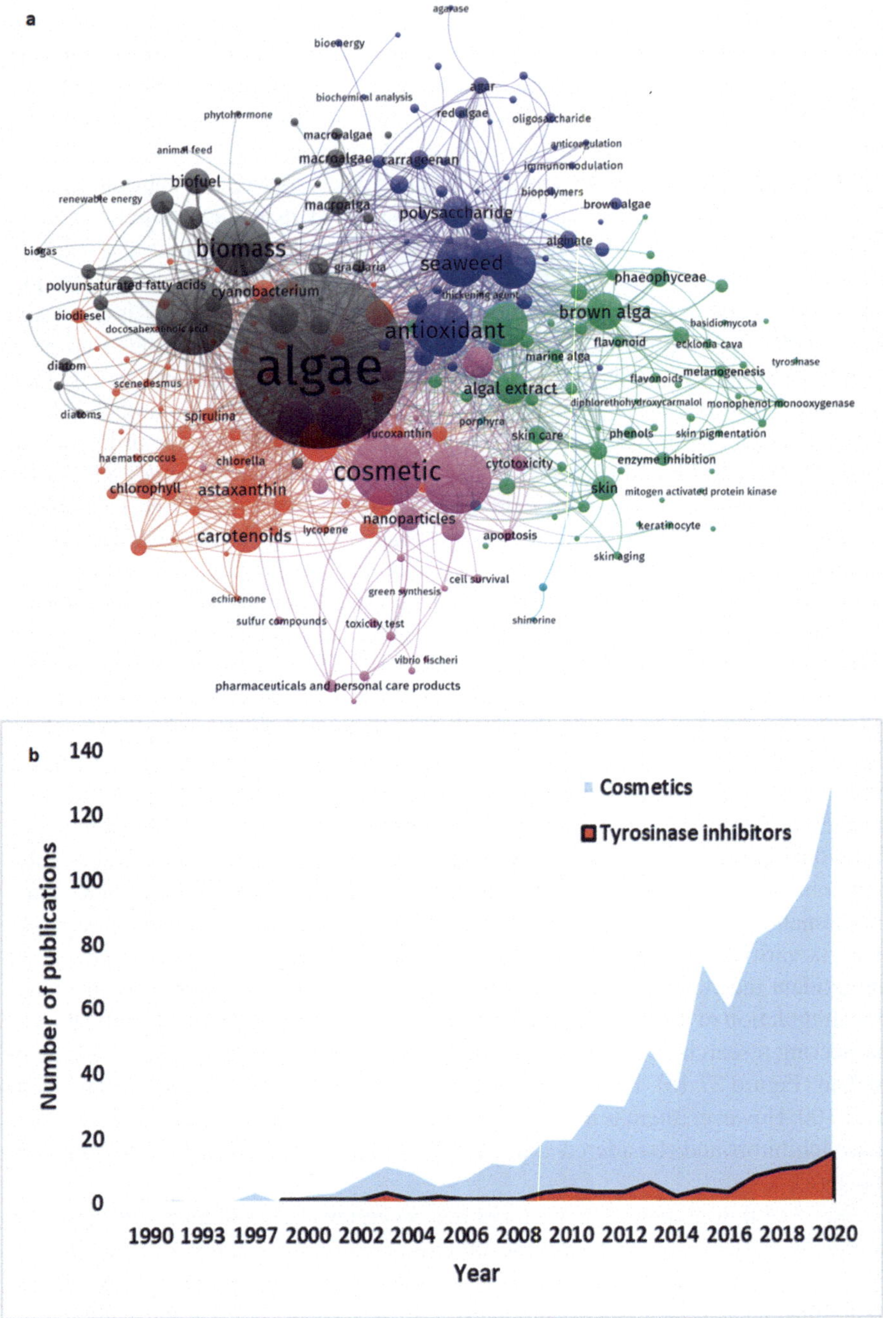

Fig. 27.1 Bibliographic overview of utilization of algae in biotechnology; (**a**) bibliographic data analysis via VOS viewer to capture algae utilization in biotechnology, (**b**) Number of publications between 1990–2020 about algae cosmetics and algae tyrosinase inhibitors based on Scopus data

3 Understanding the Molecular Mechanism of Skin Whitening

Tyrosinase enzyme is a polyphenol oxidase which can be found in plants, fungus, bacteria and animals (Pillaiyar et al. 2015). The enzyme can catalyze the oxidation of monophenols to diphenols or oxidation of phenols to quinones. Even though tyrosinase enzyme is found in a widespread life forms, the function of the enzyme may differ. In fungal and mammalian sources, enzyme is responsible for the color development (Park et al. 2004). However in plants tyrosinase enzyme acts as a part of the defense system to prevent damages from pathogen or insect invasion.

In humans, tyrosinase enzyme catalyzes the oxidative conversion of L-tyrosine to L-DOPA or Dopaquinone (DQ) (Kubo et al. 2000). This initial step is critical and rate limiting step for melanin synthesis. In order to catalyze tyrosinase mediated melanin synthesis, gene expression of tyrosinase related protein-1 (TRP-1) and TRP-2 is required (D'Mello et al. 2016). When there is free cysteine in the reaction environment, DQ reacts with cysteine to produce 3 or 5-cysteinyl DOPA which subsequently converts into pheomelanin (Meredith and Sarna 2006). If there is no cysteine in the environment, DQ autoxidize to dopachrome and finally eumelanin are synthesized (Różanowska et al. 1999). The synthesis of pheomelanin or eumelanin is highly dependent on the molecules existing in the reaction environment. There are several good review papers discussing the pathways, production and activity of melanin in humans (Kubo et al. 2000; T. S. Chang 2009; Cheli et al. 2010; D'Mello et al. 2016; Pillaiyar et al. 2017a, b).

Melanin synthesis is also regulatad by pivotal cell signaling pathways such as cyclic adenosine monophosphate (cAMP), mitogen-activated protein kinase (MAPK), phosphatidylinositol 3-kinase (PI3K)/Akt, Wnt/β-catenin, nitric oxide and autophagy related pathways (Shimomura et al. 2009; D'Mello et al. 2016; Pillaiyar et al. 2017a). It is also known that cAMP can crosstalk between ERK1/2 and other pathways to regulate melanogenesis (Kang et al. 2015).

Increase in α- melanin stimulating hormone (α-MSH) activates the tyrosinase enzyme expression via cAMP pathway (Azam et al. 2017a). When α-MSH binds to MC1R, intracelluar cAMP levels are increased which subsequently increases the expression of tyrosinase related genes, TRP-1, TRP-2 and TYR (D'Mello et al. 2016). The control of cAMP pathway on melanogenesis is resulted via CAMP/PKA to phosphorylate CREB (Azam et al. 2018) to induce the MITF expression. Increased MITF levels upregulates tyrosinase expression to produce melanin. Skin whitening compounds can interfere cAMP pathway to decrease melanin levels via binding PKA, deactivation of CREB, downregulation of MITF or expression of MC1R.

PI3K/Akt pathway regulates melanogenesis via inhibition of PI3K by cAMP, thus situmulates glycogen synthase kinase 3β (GSK3β) (Khaled et al. 2002; Nokinsee et al. 2015). Activation of GSK3β induces MITF phosphorylation and upregulates TYR related gene expression. Akt is activated by phosphorylation on Ser473 and Thr308 residues via activation of PI3K (K. C. Park et al. 2010; Azam et al. 2017a). Then Akt phosphorylates GSK3β to inactivate (Bellei et al. 2008; Kang et al. 2015). PI3K/Akt and cAMP pathway has a crosstalk inhibition or activation of tyrosinase related genes (Shimomura et al. 2009). PI3K/Akt can also be initiated with SCF/c-Kit interaction (D'Mello et al. 2016). Thus activation of PI3K/Akt via phosphorylation can be a target to decrease hyperpigmentation problems.

Wnt/β-catenin signaling is responsible of melanocyte differentiation and expansion through neural crest cells (Dissanayake et al. 2008; Liu et al. 2017). Thus the conribution of Wnt/ β-catenin pathway has a critical role to not only regulate melanogenesis but also directly effects the cell fate. When β-catenin accumulates in the cytosole it inhibits GSK3β and β-catenin is transferred to nucleus subsequently increase MITF expression resulting in the increase in melanin synthesis (Bellei et al. 2008; Mericli et al. 2017). Wnt/β-catenin signaling is known to interact with cAMP and Akt pathways to regulate melanogenesis. Thus inhibition of GSK3β and degredation of β-catenin can be another target to decrease melanin synthesis.

Melanin synthesis can be downregulated and controlled via several approaches. Chang has classified various mehanism of action to decrease melanin synthesis (Chang 2009). Since melanin synthesis is a combination of enzymatic and signaling metabolism it is not easy to clarify and classify molecules susceptible for melanin synthesis. Some molecules such as ascorbic acid can only interact with enzyme substrate such as DQ (Kubglomsong et al. 2018). Some of phenolic compounds as well known to interact with substrates to decrease avaible subatrate in the environment to initiate melanin synthesis. On the other hand acidic or basic compounds can degrade enzyme itself (Chang 2009). When enzyme is degraded, obviously melanin synthesis is diminished. Some molecules are known to be covalently linked with enzyme thus irreversibly inhibiting enzyme activity. However, according to Chang, true tyrosinase inhibitor can interact with enzyme reversibly to regulate catalytic acitivty via known enzymatic inhibition metabolism (direct, indirect or mixed inhibition).

To screen tyrosinase inhibition, usually mushroom tyrosinase enzyme is utilized as a cell free assay (Kubo et al. 2000). However some compounds, as also listed above, do not necessarily interact with enzyme. When these compounds are investigated in cell cultures, they can downregulate melanin synthesis and tyrosinase gene expression via various pathways. One of good examples to this type of inhibition is with docosahexaenoic acid (DHA) (Balcos et al. 2014). DHA is a well-known omega-3 fatty acid which is also abundant in algal resources (Isleten-Hosoglu and Elibol 2017). Balcos et al. (2014) investigated DHA as a potential tyrosinase inhibitory compound because fatty acids (based on their saturation levels) are known to either increase or decrease tyrosinase enzyme activity via direct enzymatic inhibition or proteasomal degradation (Ando et al. 2004). However DHA did not inhibit mushroom tyrosinase but when B16 cells cultured with various amount of DHA, significant tyrosinase inhibition and melanin decrease was observed. The tyrosinase inhibition is known to due to proteasomal degradation, Akt, ERK, CREB or MITF related inhibition of tyrosinase activity was not found either. This suggest that tyrosinase enzyme inhibition is related to other pathways causing tyrosinase degradation.

Besides these mechanisms of actions, there are several other ways to decrease melanin synthesis. Before discussing other mechanisms, it is crucial to point the differences between tyrosinase inhibitors and melanogenesis inhibitors. It is clear that when tyrosinase inhibition occurs via direct enzyme inhibition or pathway-mediated inhibition, melanin production is downregulated. However, it is not necessarily required to decrease melanin accumulation via actions on the tyrosinase enzyme. There are a few routes known to decrease the final melanin content on the skin via inhibiting melanosome transfer (Pillaiyar et al. 2017a). After melanin is synthesized, melanosome granules are responsible for transfer of melanin from melanocyte to surrounding keratinocytes (Smit et al. 2009; Vachtenheim and Borovanský 2010). Although several mechanisms are proposed for melanosome transfer, the overall metabolism is not fully understood. Another group of compounds are antioxidant molecules which can contribute to decrease in hyper-accumulation of melanin pigment (Rangkadilok et al. 2007). Since UV exposure is one of the main triggers to upregulate melanin synthesis, UV exposure can also elevate the levels of reactive oxygen species (Kumar and Mandal 2019). When antioxidant molecules are utilized, with respect to decrease of reactive oxygen species, a significant decrease in melanin synthesis can occur as well.

4 Bioprospecting Seaweeds for Skin Whitening

There are reliable numbers of studies considering seaweeds as a feedstock for skin whitening compounds (Table 27.1). The concept of using seaweeds in cosmetics is not a novel idea however, deciphering pure bioactive biochemicals in seaweeds for tyrosinase inhibition and skin whitening is a promising field of interest to increase the utilization of natural substances for topical cosmeceuticals.

Polyphenols in seaweeds are the mostly studied compounds in terms of skin whitening effect (Azam et al. 2017b). Especially phlorotannins are known to have beneficial cosmeceutical effect including anti-tyrosinase, anti-aging, antioxidant and anti-inflammatory effects (Thomas and Kim 2013). Among seaweed phlorotannins, dieckol, eckol, octaphlorethol A and phloroglucinol are reported to inhibit tyrosinase activity. Besides; compounds such as fucoidan (Kang et al. 2015), fucoxanthin (Shimoda et al. 2010), some polysaccharides (Pratoomthai et al. 2018) and bromophenol compounds (Paudel et al. 2019) are also found to have significant amount of tyrosinase inhibition.

Kang et al. (2004) used 17 different seaweed species (2 of Chlorophyta, 5 of Phaeophyta, and 10 of Rhodophyta) using L-tyrosine as a substrate for mushroom tyrosinase activity. *Ecklonia stolonifera* OKAMURA (*Laminariaceae*, brown algae) showed an inhibitory activity with an IC_{50} value of 345 μg/mL. Ethyl acetate fraction of methanolic extracts showed the highest tyrosinase inhibition (IC_{50} value of 67 μg/mL) which was abundant in phlorotannins. Among them, dieckol was the most potent (IC_{50} value of 2.12 μg/mL) followed by eckol (IC_{50} value of 33.2 μg/mL). Kojic acid, as a reference had IC_{50} value of 6.32 μg/mL which makes dieckol

Table 27.1 Tyrosinase inhibitory compounds from several seaweed sources

Seaweed	Compund	Action mechanism	Tyrosinase assay	Reference
S.plagyophyllum(SP) *E. cottonii (EC)*	Methanolic extracts	**IC_{50} values** ***Monophenolase activity*** **SP:** 2195.206 µg/mL **EC:** 2691.478 µg/mL ***Diphenolase activity*** **SP:** 1769.336 µg/mL **EC:** 2631.648 µg/mL	Mushroom	Dolorosa et al. (2019)
Ecklonia cava	Triphlorethol A (1) Eckol (2) 2-phloroeckol (3) Phlorofucofuroeckol A (4) 2-O-(2,4,6trihydroxyphenyl)-6,6′-bieckol (5) 6,8′-bieckol (6) 8,8′-bieckol (7)	• **IC_{50} values** Compound 3: 7.0 ± 0.2 µM Compound 5: 8.8 ± 0.1 µM • Competitive inhibition • Interaction with His85 and Asn260 at the active site.	Mushroom Molecular docking	Yang et al. (2019)
Ecklonia stolonifera	974-A	**IC_{50} values** ***Monophenolase activity*** 1.57 ± 0.08 µM ***Diphenolase activity*** 3.56 ± 0.22 µM • Competitive inhibition • Downregulated the expression of tyrosinase, tyrosinase-related protein (TRP)-1, and TRP-2 in B16F10 • Antioxidant	Mushroom B16/F10 Molecular docking	Manandhar et al. (2019)

Seaweed	Compund	Action mechanism	Tyrosinase assay	Reference
E. cava	Eckol	• Non-competitive inhibitor to mushroom tyrosinase • Reduced α-MSH-induced expression of tyrosinase, TRP1, and TRP2	Mushroom Molecular docking on *bacillus megaterium* tyrosinase B16-F10	Lee et al. (2015)
E. cava	Dieckol	• IC_{50} value 20 μM • Non-competitive for mushroom • Inhibition in B16-F10 tyrosinase	Mushroom B16-F10	Kang et al. (2012)
E. stolonifera	Dioxinodehydroeckol	• PI3K/Akt-mediated downregulation of MITF in B16-F10 cells	B16F10	Lee et al. (2018)
Dictyota coriacea	1,9-dihydroxycrenulide Epiloliolide	• 1,9-dihydroxycrenulide and Epiloliolide at 30 μg/ml 27.8 and 22.6% inhibition respectively	Mushroom	Kyeong et al. (2013)
Myagropsis myagroides	Sargachromanol G Sargachromanol I Mojabanchromanol b	Tyrosinase inhibition (%) • Sargachromanol G 41.27 ± 4.70 • Sargachromanol I 19.92 ± 8.59 • Mojabanchromanol b 48.58 ± 2.42	Mushroom	Kim et al. (2013b)

(continued)

Table 27.1 (continued)

Seaweed	Compund	Action mechanism	Tyrosinase assay	Reference
A. nodosum *L. japónica* *L. Trabeculate* *L. Nigrecen*	Extracts	• L. Trabeculate inhibited tyrosinase activity, with an MAE fraction of 33.73% • Not effective compared to kojic acid	Mushroom	Yuan et al. (2018)
Gracilaria fisheri	Sulfated galactans	• Did not have a direct inhibition against either mushroom tyrosinase activity or B16F10 cells-free tyrosinase activity • Decreased the expression of TYR, TRP-1, TRP-2, and MITF genes in B16-F10	Mushroom B16-F10	Pratoomthai et al. (2018)
Gracilaria arcuata (Zanardini)	Methanol extracts	• IC_{50} value 0.33 mg/ml	Mushroom	Layse et al. (2011)
Hizikia Fusiformis	4-hydroxyphenethyl alcohol	• Mushroom tyrosinase inhibition (0.1 mg/ml 75% inhibition). • 10 μg/mL resulted in a melanin decrease of approximately 27% in B16-F10 • Reduction of pigmented spots	Mushroom B16-F10 Guinea pigs	Jang et al. (2014)
S. latiuscula	Bromophenol compounds 1,2,3	• 1–3 competitive inhibitor • IC_{50} of 3 2.92 ug/ml for L-tyrosine • 1–3 significantly toxic to B16-F10 • Bromophenols 1 and 3 significantly reduced the expression levels of tyrosinase	Mushroom B16-F10	Paudel et al. (2019)
Sargassum fusiforme	Polysaccharide	• Inhibitor for monophenolase and diphenolase activity • Competitive uncompetitive mixed type inhibitor	Mushroom	Chen et al. (2016)

Seaweed	Compund	Action mechanism	Tyrosinase assay	Reference
Sargassum serratifolium	Sargaquinoic acid	• Inhibits the expression of TYR, TRP1 and TRP2. • Inhibits cellular TYR activity dose-dependently • Reduced intracellular cAMP accumulation, suppressed phosphorylation CREB, downregulated MITF • Increased the phosphorylation of ERK1/2 and MITF (Ser73), inducing proteasomal degradation of MITF • Suppressed melanin production through the cAMP/CREB- and ERK1/2-mediated downregulation of MITF	B16-F10	Azam et al. (2018) Azam et al. (2017a, b)
Eucheuma cottonii	Methanolic extracts	• IC_{50} value 234.33 μg/mL	Mushroom	Chang and Teo (2016)
Ishige okamurae	Diphlorethohydroxycarmalol	• Inhibition of mushroom TYR and melanin synthesis • Reduction of UV-B induced ROS levels	Mushroom B16-F10	Heo et al. (2009a, b)

(continued)

Table 27.1 (continued)

Seaweed	Compund	Action mechanism	Tyrosinase assay	Reference
Ishige foliacea	Octaphlorethol A	• ERK1/2-mediated downregulation of MITF, TYR TRP-1 and TRP-2 in B16-F10. • Reduces p38 MAPK protein levels and activates extracellular signal-regulated kinase (ERK) and c-Jun N-terminal kinases (JNKs) protein expressions in B16F10 cells.* • More than 90% of subject embryos survived upon exposure to concentrations below 25 μM • Inhibition of *in vivo* TYR embryo activity and melanin synthesis	B16F10 zebra fish	Kim et al. (2015) Kim et al. (2013a, b)*
Odonthalia corymbifera	2,3-dibromo-4,5-dihydroxylbenzyl moieties	• Compound 6 showed the most potent inhibitor (IC50 = 1.0 μM) • Compound 1 was examined for kinetic analysis to be non-competitive inhibition (Ki; 2.4 μM)	Mushroom	Islam et al. (2017)
Ecklonia cava	Dieckol	• Dieckol decreased tyrosinase activity 92.7% at the concentration of 100 μM. • Dieckol effectively decreased the generated ROS (%234) to 100.7% at 250 μM • Strong protective properties against UV-B radiation-induced DNA damage	B16-F10 Fibroblasts	Heo et al. (2009a)

Seaweed	Compund	Action mechanism	Tyrosinase assay	Reference
Ecklonia cava	7-phloroeckol	• IC_{50} value 0.85 μM (arbutin, 243.16 μM) and kojic acid 40.28 μM) • Non-competitive inhibitor • Inhibitor for 3-isobutyl-1-methylxanthine (IBMX)-induced melanin formation	B16-F10 Mushroom	Yoon et al. (2009)
Ecklonia cava cultured with magma seawater of Jeju	MSWE extract	• Inhibited melanin synthesis and decreased the expression of melanogenesis-related protein in α-MSH B16F10 • MSWE is non-toxic to B16-F10	B16-F10	Ding et al. (2019)
Ecklonia cava	EC extracts	• Tyrosinase inhibition	Mushroom	Kim et al. (2013b)
Endarachne binghamiae	Ethyl acetate extracts	• Non inhibition on mushroom tyrosinase or cell extracted tyrosinase • Inhibition of melanin synthesis via inhibition of α-glucosidase-dependent glycosylation of tyrosinase	Melan-a cells	Jeon et al. (2013)

(continued)

Table 27.1 (continued)

Seaweed	Compund	Action mechanism	Tyrosinase assay	Reference
Fucus vesiculosus	Fucoidan	• Fucoidan inhibited melanin synthesis by down- regulating MITF and tyrosinase protein expression. • Induced phosphorylation of ERK, but not Akt	Mel-ab	Song et al. (2015)
Kelp	Fucoidan	• Competitive inhibitor for monophenolase activity • Ki; 0.9907 mg/mL	Mushroom	Yu and Sun (2014)
Fucus vesiculosus	Fucoidan	• IC*50* value 550 ± 4.3 μg/mL • Decrease melanin synthesis • Upregulated the expression of tyrosinase and MITF • IC_{50} was 530 ± 3.32 μg/ml for cell viability	B16-F10	Wang et al. (2017)
Prasiola japonica	Loliolide and Pj-EE	• Decreased expression of MITF and tyrosinase in B16F10 treated with α-MSH • Reduced melanin secretion	B16-F10	Park et al. (2018)
Padina boryana	Ethanol extracts (PBE)	• Downregulation of MITF, TRP-1 and TRP-2. • The phosphorylation of ERK was sustained via PBE and hence declined the ultimate melanin synthesis.	B16-F10	Jayawardena et al. (2020)
Grateloupia lancifolia	Diethyl ether extracts	• IC_{50} values of 47.8 μg/ml	Mushroom	Han (2012)

Seaweed	Compund	Action mechanism	Tyrosinase assay	Reference
Digenea simplex, *Laurencia papillosa* *Laurencia paniculata*	Methanolic extracts	• Effective mushroom inhibiton on monophenolase and diphenolase activity • *D. simplex* extract inhibited tyrosinase activity by 43.18% in zebrafish model • Decreased total melanin content of zebrafish by 47.27%	Mushroom Zebrafish	Namjoyan et al. (2019)
Laminaria japónica	Fucoxanthin	• Reduced TYR activity and melanin content in B16-F10 • Suppress PGE2, MSH, TRP1, NTR, EP1 and MC1R	B16-F10 Guinea pigs UV induction	Shimoda et al. (2010)

as 3 times more potent than kojic acid. Inhibitory metabolism of action for the compounds were competitive for phloroglucinol and eckstolonol; non-competitve for dieckol, eckol and Phlorofucofuroeckol A. for phloroglucinol derived compounds antioxidant mediated tyrosinase inhibition is also considered as an effective inhibitory pathway (Kang et al. 2004).

Lee et al. purified eckol from *Ecklonia cava* which is a common type of edible seaweed in Asian coasts. Based on their previous work they only focus on eckol as a tyrosinase inhibitory compound. Their study is an important aspect on utilization of molecular dynamics simulation and molecular modelling to predict binding efficiency of ligand-substrate. These sort of computational experiments (Kang et al. 2012; Wang et al. 2012; Paudel et al. 2019) are critical for rapid screening of a large bioactive substance library. Based on enzyme inhibition kinetics, experiments on mushroom tyrosinase, eckol is a non-competitive inhibitor for tyrosinase. Molecular docking studies revealed that eckol is binding to Asn205, His208, and Arg20 residues meanwhile arbutin is binding Asn205, His208, and Gly216. From the results we can assume Asn205 and His208 residues can be critical for deciphering potential tyrosinase inhibitor molecules. Another study done by Kang et al. revealed the potential of dieckol as a high potential tyrosinase inhibitor which is also confirmed by molecular dynamics studies by computational models using tyrosinase from *Bacillus megaterium* tyrosinase (PDB ID: 3NM8). Dieckol inhibited mushroom tyrosinase non-competitively with an IC_{50} of 20 μM and was more effective as than arbutin. Dieckol was binding tyrosinase enzyme via His208, Met215, and Gly46 residues. Lee et al., also find His208 as a residue important for tyrosinase inhibition (Kang et al. 2012).

Up to date, it is known that algae species such as *Ecklonia stolonifera, E. cava, Sargassum, Laminaria japonica, Ishige okamurae, I. foliacea, Hizikia fusiformis, Fucus vesiculosus, S. serratifolium, S. polycystum, S. plagyophyllum E. cottonii* and *Gracilaria fisheri* species were extensively investigated for their potential on tyrosinase inhibition. There is one product in the market, known as Whitanyl. The active ingredient is a mixture of oligosaccharides from red algae *Palmaria palmata*. The active ingredient is known to whiten sun exposure derived dark spots via blocking melanosome transfer (Pereira 2018).

Apart from biochemicals in seaweed, pigments such as astaxanthin from *Haematococcus pluvialis* (Rao et al. 2013) and Zeaxanthin from *Nannochloropsis* are known to decrease melanin synthesis (Shen et al. 2011). Both of these pigments are elements in photosynthetic metabolism in algae and are known to be highly effective antioxidant molecules (Morone et al. 2019). Also fucoxanthin from brown seaweed is known to be a tyrosinase inhibitor (Shimoda et al. 2010) but diatoms such as *Phaeodactylum tricornutum* are high producers for fucoxanthin as well. However there is not enough research on the effects of fucoxanthin from diatoms for tyrosinase inhibition. Not only seaweed itself but microalgae species are susceptible to tyrosinase inhibitor studies. Thus a more extensive screening of algal metabolites via computational and experimental procedures are emerging. Even though there are approaches to utilize cosmetics from microalgae, there is not enough evidence on these substances for tyrosinase inhibition.

Recently peptides, proteins and protein hydrolysates are considered as non-toxic and high potential compounds to decrease tyrosinase enzyme activity both in mushroom and intracellular tyrosinase activity (Schurink et al. 2007). Certain amino acid combinations such as arginine or phenylalanine residues along with non-polar amino acids; valine, alanine and leucine known to strongly bind and decrease tyrosinase activity via competitive and/or noncompetitive action (Schurink et al. 2007; Nie et al. 2017). Also C-terminal amino acid sequence is a strong evidence to inhibit tyrosinase (Ochiai et al. 2016). The location and combination of tyrosine and phenylalanine is more critical because phenylalanine is structural analog to tyrosine, the substrate of tyrosinase enzyme. Thus, position of these residues can either be inhibitor or activator.

There is not much data on tyrosinase inhibitory peptides and proteins from marine resources. Only a few of studies indicates that peptides such as non-ribosomal peptides from cyanobacteria can be considered as strong tyrosinase inhibitors. Due to their un-natural amino acid combinations and cyclic orientations these peptides are highly susceptible to be considered for tyrosinase inhibition. For example Oscillapeptin G from toxic algae *Oscillatoria agardhii* is a strong inhibitor for mushroom tyrosinase (Sano and Kaya 1996). Also microcystins from *Microcystis viridis* showed tyrosinase inhibitory effect on mushroom tyrosinase as well (Morone et al. 2019). Another protein from algal resources was phycocyanin from cyanobacteria. Although phycocyanin was not a direct inhibitor on tyrosinase enzyme, it decreased melanin synthesis via upregulation of MAP/ERK dependent downregulation of MITF synthesis and downregulation of p38 MAPK regulated down regulation of CREB synthesis. Phycocyanin also decreased melanin via GRB2-ERK 1/2 pathway (Wu et al. 2011; Faccio et al. 2014).

As it is stated in here, algae is a valuable resource for tyrosinase inhibitory molecules. Combination of computational tools to simulate molecular dynamics (Lee et al. 2015) and interaction with tyrosinase enzyme residues is a critical approach to decrease trial and error approach to decipher inhibitory molecules. On the other hand, computational tools can give a rapid screening strategy via obtaining already defined chemical structures from algae through natural molecule database search (Musuamba et al. 2020). However, most of the molecular dynamics analyses are done on mushroom or microbial tyrosinase enzyme. It is known that when human cells are introduced, the relation with molecules to decrease melanin synthesis is variable. Thus obtaining of crystal structure of human tyrosinase to work on molecular dynamics will pave and increase the pace of the ongoing search on the discovery of potent tyrosinase inhibitors from algal resources.

5 Conclusion and Future Perspective

Skin pigmentation disorders are the undeniable reality of society. Compounds inhibiting tyrosinase enzyme activity or resulting in the decreased accumulation of melanin have a critical role of designing novel functional topical cosmetics products. Due to the high demand in the cosmetics industry for skin whitening products, semi-synthetic and synthetic compounds are started to be replaced by natural substances for a conscious consumption ideology. However, even though we have an enormous number of natural substances from various sources (plants, algae, bacteria and fungus etc.), the commercial availability of these compounds are lacking. Seaweeds are ancient formulas for beauty and healthy skin, thus renovating the idea of using seaweed as tyrosinase inhibitors are important. Up to date, the compounds such as fucoxanthin and phlorotannins are proven to be efficient tyrosinase inhibitor with low or no cytotoxicity. The future of seaweed based tyrosinase inhibitor thus should focus on the downstream process and efficient extraction of bioactive compounds in high quantity to reach commercial scale applications. Another point is to design skin whitening products to see the stability and effect of the final formula in terms of tyrosinase inhibition. When these challenges are met, the future is bright to use seaweed as skin whitening feedstocks.

References

Abu U, Anan LZ, Wang Y, Hantash BM (2009) Short-sequence oligopeptides with inhibitory activity against mushroom and human tyrosinase. J Invest Dermatol 129:2242–2249. https://doi.org/10.1038/jid.2009.124

Ando H, Watabe H, Valencia JC, Yasumoto K-i, Furumura M, Funasaka Y, Oka M, Ichihashi M, Hearing VJ (2004) Fatty acids regulate pigmentation via proteasomal degradation of Tyrosinase. J Biol Chem 279:15427–15433. https://doi.org/10.1074/jbc.m313701200

Azam MS, Joung EJ, Choi J, Kim HR (2017a) Ethanolic extract from *Sargassum serratifolium* attenuates hyperpigmentation through CREB/ERK signaling pathways in α-MSH-stimulated B16F10 melanoma cells. J Appl Phycol 29:2089–2096

Azam MS, Choi J, Lee MS, Kim HR (2017b) Hypopigmenting effects of brown algae-derived phytochemicals: a review on molecular mechanisms. Mar Drugs 15. https://doi.org/10.3390/md15100297

Azam MS, Kwon M, Choi J, Kim HR (2018) Sargaquinoic acid ameliorates hyperpigmentation through cAMP and ERK-mediated downregulation of MITF in α-MSH-stimulated B16F10 cells. Biomed Pharmacother 104:582–589. https://doi.org/10.1016/j.biopha.2018.05.083

Bagherzadeh K, Talari FS, Sharifi A, Ganjali MR, Saboury AA, Amanlou M (2015) A new insight into mushroom tyrosinase inhibitors: Docking, pharmacophore-based virtual screening, and molecular modeling studies. J Biomol Struct Dynam 33:487–501. https://doi.org/10.1080/07391102.2014.893203

Balcos MC, Kim SY, Jeong H-s, Yun H-y, Baek KJ, Kwon NS, Kim D-s (2014) Docosahexaenoic acid inhibits melanin synthesis in murine melanoma cells in vitro through increasing tyrosinase degradation. Nature Publishing Group 11:489–495. https://doi.org/10.1038/aps.2013.174

Bellei B, Flori E, Izzo E, Maresca V, Picardo M (2008) GSK3β inhibition promotes melanogenesis in mouse B16 melanoma cells and normal human melanocytes. Cell Signal 20:1750–1761. https://doi.org/10.1016/j.cellsig.2008.06.001

Chang TS (2009) An updated review of tyrosinase inhibitors. Int J Mol Sci 10:2440–2475. https://doi.org/10.3390/ijms10062440

Chang TS (2012) Natural melanogenesis inhibitors acting through the down-regulation of tyrosinase activity. Materials 5:1661–1685. https://doi.org/10.3390/ma5091661

Chang VS, Teo SS (2016) Evaluation of heavy metal, antioxidant and anti-tyrosinase activities of red seaweed (*Eucheuma cottonii*). Int Food Res J 23:2370–2374

Cheli Y, Ohanna M, Ballotti R, Bertolotto C (2010) Fifteen-year quest for microphthalmia-associated transcription factor target genes. Pigment Cell Melan Res 23:27–40. https://doi.org/10.1111/j.1755-148X.2009.00653.x

Chen BJ, Shi MJ, Cui S, Hao SX, Hider RC, Zhou T (2016) Improved antioxidant and anti-tyrosinase activity of polysaccharide from *Sargassum fusiforme* by degradation. Int J Biol Macromol 92:715–722

D'Mello SAN, Finlay GJ, Baguley BC, Askarian-Amiri ME (2016) Signaling pathways in melanogenesis. Int J Mol Sci 17:1–18. https://doi.org/10.3390/ijms17071144

Ding Y, Kim SH, Lee JJ, Hong JT, Kim EA, Kang DH, Heo SJ, Lee SH (2019) Anti-melanogenesis activity of *Ecklonia cava* extract cultured in tanks with magma seawater of Jeju Island. Algae 34:163–175

Dissanayake SK, Olkhanud PB, O'Connell MP, Carter A, French AD, Camilli TC, Emeche CD et al (2008) Wnt5A regulates expression of tumor-associated antigens in melanoma via changes in signal transducers and activators of transcription 3 phosphorylation. Cancer Res 68:10205–10214. https://doi.org/10.1158/0008-5472.CAN-08-2149

Dolorosa MT, Nurjanah, Purwaningsih S, Anwar E, Hidayat T (2019) Tyrosinase inhibitory activity of *Sargassum plagyophyllum*. IOP Conf. Series: Earth Environ Sci 278:012020. https://doi.org/10.1088/1755-1315/278/1/012020

Faccio G, Kämpf MM, Piatti C, Thöny-Meyer L, Richter M (2014) Tyrosinase-catalyzed site-specific immobilization of engineered C-phycocyanin to surface. Sci Rep 4:1–8. https://doi.org/10.1038/srep05370

Feng L, Shi N, Cai S, Qiao X, Chu P, Wang H, Long F et al (2018) De novo molecular design of a novel octapeptide that inhibits in vivo melanogenesis and has great transdermal ability. J Med Chem 61:6846–6857. https://doi.org/10.1021/acs.jmedchem.8b00737

Hacker SM (2017) Common disorders of pigmentation when are more than cosmetic cover-ups required ? Postgrad Med 125:5481. https://doi.org/10.1080/00325481.1996.11946143

Han N (2012) Antioxidative, anticholinesterase and antityrosinase activities of the red alga *Grateloupia lancifolia* extracts. Afr J Biotechnol 11:9457–9467

Heo SJ, Park EJ, Lee KW, Jeon YJ (2005) Antioxidant activities of enzymatic extracts from brown seaweeds. Bioresour Technol 96:1613–1623. https://doi.org/10.1016/j.biortech.2004.07.013

Heo SJ, Ko SC, Cha SH, Kang DH, Park HS, Choi YU, Kim D, Jung WK, Jeon YJ (2009a) Effect of phlorotannins isolated from *Ecklonia cava* on melanogenesis and their protective effect against photo-oxidative stress induced by UV-B radiation. Toxicol In Vitro 23(6):1123–1130. https://doi.org/10.1016/j.tiv.2009.05.013

Heo S-J, Ko S-C, Kang S-M, Cha S-H, Lee S-H, Kang D-H, Jung W-K, Affan A, Chulhong O, Jeon Y-J (2009b) Inhibitory effect of diphlorethohydroxycarmalol on melanogenesis and its protective effect against UV-B radiation-induced cell damage. Food Chem Toxicol 48:1355–1361

Islam R, Mikami D, Kurihara H (2017) Tyrosinase inhibitory and antioxidant activity by bromophenols from the alga *Odonthalia corymbifera*. Nat Prod Ind J 13:1–12

Isleten-Hosoglu M, Elibol M (2017) Improvement of medium composition and cultivation conditions for growth and lipid production by *Crypthecodinium cohnii*. Roman Biotechnol Lett 22:13086–13095

Jang MS, Park HY, Nam KH (2014) Whitening effects of 4-hydroxyphenethyl alcohol isolated from water boiled with *Hizikia fusiformis*. Food Sci Biotechnol 23:555–560

Jayawardena TU, Asanka Sanjeewa KK, Kim H-S, Lee HG, Wang L, Lee D-S, Jeon J (2020) *Padina boryana*, a brown alga from the Maldives: inhibition of α-MSH-stimulated melanogenesis via the activation of ERK in B16F10 cells. Fish Aqua Sci 23:118

Jeon Y, Jung Y, Youm J-K, Kim YK, Kim S-N (2013) Inhibitory effect of Endarachne binghamiae extract on melanin synthesis. Korean J Plant Resour 26:526–532

Jesumani V, Hong D, Aslam M, Pei P, Huang N (2019) Potential use of seaweed bioactive compounds in skincare—a review. Mar Drugs 17:1–19. https://doi.org/10.3390/md17120688

Kang HS, Kim HR, Byun DS, Son BW, Nam TJ, Choi JS (2004) Tyrosinase inhibitors isolated from the edible brown alga *Ecklonia stolonifera*. Arch Pharm Res 27:1226–1232. https://doi.org/10.1007/BF02975886

Kang SM, Heo SJ, Kim KN, Lee SH, Yang HM, Kim AD, Jeon YJ (2012) Molecular docking studies of a phlorotannin, dieckol isolated from Ecklonia cava with tyrosinase inhibitory activity. Bioorg Med Chem 20:311–316. https://doi.org/10.1016/j.bmc.2011.10.078

Kang SJ, Choi BR, Lee EK, Kim SH, Yi HY, Park HR, Song CH, Lee YJ, Sae Kwang K (2015) Inhibitory effect of dried pomegranate concentration powder on melanogenesis in B16F10 melanoma cells; involvement of p38 and PKA signaling pathways. Int J Mol Sci 16:24219–24242. https://doi.org/10.3390/ijms161024219

Khaled M, Larribere L, Bille K, Aberdam E, Ortonne JP, Ballotti R, Bertolotto C (2002) Glycogen synthase kinase 3β is activated by cAMP and plays an active role in the regulation of melanogenesis. J Biol Chem 277:33690–33697. https://doi.org/10.1074/jbc.M202939200

Kim KN, Yang HM, Kang SM, Kim D, Ahn G, Jeon YJ (2013a) Octaphlorethol A isolated from *Ishige foliacea* inhibits α-MSH-stimulated induced melanogenesis via ERK pathway in B16F10 melanoma cells. Food Chem Toxicol 59:521–526

Kim SJ, Kim D, Park J, Lee T-k (2013b) Activities of *Ecklonia cava* extracted with the ultrasonic wave method. J Life Sci 23(7):913–918

Kim KN, Yang HM, Kang SM, Ahn G, Roh SW, Lee WW, Kim D, Jeon YJ (2015) Whitening effect of octaphlorethol a isolated from ishige foliacea in an in vivo zebrafish model. J Microbiol Biotechnol 25:448–451

Kubglomsong S, Theerakulkait C, Reed RL, Yang L, Maier CS, Stevens JF (2018) Isolation and identification of Tyrosinase-inhibitory and copper-chelating peptides from hydrolyzed Rice-bran-derived albumin. J Agric Food Chem 66:8346–8354. https://doi.org/10.1021/acs.jafc.8b01849

Kubo I, Kinst-Hori I, Kubo Y, Yamagiwa Y, Kamikawa T, Haraguchi H (2000) Molecular design of antibrowning agents. J Agric Food Chem 48:1393–1399. https://doi.org/10.1021/jf990926u

Kumar JP, Mandal BB (2019) The inhibitory effect of silk sericin against ultraviolet-induced melanogenesis and its potential use in cosmeceutics as an anti-hyperpigmentation compound. Photochemical and Photobiological sciences 18. Royal Society of. Chemistry 2497–2508. https://doi.org/10.1039/c9pp00059c

Kyeong K, Ryeo M-CK, Kim SS, Tae Heon O, Kim G-O, Hyun C-G, Hyun JW, Lee NH (2013) Anti-melanogenesis Constituents from the Seaweed *Dictyota coriacea*. Nat Prod Commun 14:8

Layse C, De Almeida F, Heloina De S, Falcão GR, De ML, Montenegro CDA, Lira NS et al (2011) Bioactivities from marine algae of the genus Gracilaria. Int J Mol Sci 12:4550–4573

Lee SH, Kang SM, Sok CH, Hong JT, Jae Young O, Jeon YJ (2015) Cellular activities and docking studies of eckol isolated from *Ecklonia cava* (Laminariales, Phaeophyceae) as potential tyrosinase inhibitor. Algae 30:163–170. https://doi.org/10.4490/algae.2015.30.2.163

Lee A, Kim JY, Heo J, Cho DH, Kim HS, An IS, An S, Bae S (2018) The inhibition of melanogenesis via the PKA and ERK signaling pathways by *Chlamydomonas reinhardtii* extract in B16f10 melanoma cells and artificial human skin equivalents. J Microbiol Biotechnol 28:2121–2132. https://doi.org/10.4014/jmb.1810.10008

Liu YY, Xiu Rong S, Liu SS, Yang SS, Jiang CY, Zhang Y, Zhang S (2017) Zebrafish phosvitin-derived peptide Pt5 inhibits melanogenesis via cAMP pathway. Fish Physiol Biochem 43:517–525. https://doi.org/10.1007/s10695-016-0306-3

Manandhar B, Wagle A, Seong SH, Paudel P, Kim H-R, Jung HA, Choi JS (2019) Phlorotannins with Potential Anti-Tyrosinase and Antioxidant Activity Isolated from the Marine Seaweed *Ecklonia stolonifera*. Antioxidants (Basel) 8(8):240

Meredith P, Sarna T. 2006. The physical and chemical properties of eumelanin. doi:https://doi.org/10.1111/j.1600-0749.2006.00345.x

Mericli F, Becer E, Kabadayi H, Hanoglu A, Hanoglu DY, Yavuz DO, Ozek T, Vatansever S (2017) Fatty acid composition and anticancer activity in colon carcinoma cell lines of *Prunus dulcis* seed oil. Pharmaceutical Biology 55:1239–1248. https://doi.org/10.1080/13880209.2017.1296003

Morone J, Alfeus A, Vasconcelos V, Martins R (2019) Revealing the potential of cyanobacteria in cosmetics and cosmeceuticals — a new bioactive approach. Algal Res 41. Elsevier:101541. https://doi.org/10.1016/J.ALGAL.2019.101541

Musuamba FT, Bursi R, Manolis E, Karlsson K, Kulesza A, Courcelles E, Boissel JP et al (2020) Verifying and validating quantitative systems pharmacology and in silico models in drug development: current needs, gaps, and challenges. CPT: Pharmacometrics and systems. Pharmacology 9:195–197. https://doi.org/10.1002/psp4.12504

Namjoyan F, Farasat M, Alishahi M, Jahangiri A, Mousavi H (2019) The anti-melanogenesis activities of some selected red macroalgae from northern coasts of the Persian Gulf. Iranian Journal of Pharmaceutical Research 18

Nie H, Lin L, Yang H, Guo H, Liu X, Tan Y, Wang W, Quan J, Zhu L (2017) A novel Heptapeptide with Tyrosinase inhibitory activity identified from a phage display library. Applied biochemistry and biotechnology 181. Applied biochemistry and biotechnology: 219–232. https://doi.org/10.1007/s12010-016-2208-3

Nokinsee D, Shank L, Lee VS, Nimmanpipug P (2015) Estimation of inhibitory effect against tyrosinase activity through homology modeling and molecular docking. Enzyme Research 2015. https://doi.org/10.1155/2015/262364

Ochiai A, Tanaka S, Tanaka T, Taniguchi M (2016) Rice bran protein as a potent source of anti-melanogenic peptides with tyrosinase inhibitory activity. J Nat Prod 79:2545–2551. https://doi.org/10.1021/acs.jnatprod.6b00449

Park SH, Kim DS, Kim WG, Ryoo IJ, Lee DH, Huh CH, Youn SW, Yoo ID, Park KC (2004) Terrein: a new melanogenesis inhibitor and its mechanism. Cell Mol Life Sci 61:2878–2885. https://doi.org/10.1007/s00018-004-4341-3

Park KC, Huh SY, Choi HR, Kim DS (2010) Biology of melanogenesis and the search for hypopigmenting agents. Dermatologica Sinica 28:53–58. https://doi.org/10.1016/S1027-8117(10)60011-0

Park SH, Choi E, Kim S, Kim DS, Kim JH, Chang S, Choi JS et al (2018) Molecular sciences oxidative stress-protective and anti-melanogenic effects of Loliolide and ethanol extract from fresh water green algae. *Prasiola japonica* Int J Mol Sci 19(9):2825. https://doi.org/10.3390/ijms19092825

Paudel P, Aditi Wagle S, Seong H, Park HJ, Jung HA, Choi JS (eds) (2019) A new Tyrosinase inhibitor from the red alga *Symphyocladia latiuscula* (Harvey), Yamada (Rhodomelaceae)

Pereira L (2018) Seaweeds as source of bioactive substances and skin care therapy—cosmeceuticals, Algotheraphy, and thalassotherapy. Cosmetics 5. https://doi.org/10.3390/cosmetics5040068

Pillaiyar, Thanigaimalai, Manoj Manickam, and Sang-Hun Jung. 2015. Inhibitors of melanogenesis: a patent review (2009–2014). Expert Opin Ther Pat 25: 775–788. doi:https://doi.org/10.1517/13543776.2015.1039985

Pillaiyar T, Manickam M, Jung S-H (2017a) Downregulation of melanogenesis: drug discovery and therapeutic options. Drug Discov Tod Trends 22:282–298. https://doi.org/10.1016/J.DRUDIS.2016.09.016

Pillaiyar T, Manickam M, Jung SH (2017b) Recent development of signaling pathways inhibitors of melanogenesis. Cell Signal 40:99–115. https://doi.org/10.1016/j.cellsig.2017.09.004

Pillaiyar T, Namasivayam V, Manickam M, Jung SH (2018) Inhibitors of Melanogenesis: An updated review. J Med Chem 61:7395–7418. https://doi.org/10.1021/acs.jmedchem.7b00967

Pratoomthai B, Songtavisin T, Gangnonngiw W, Wongprasert K (2018) In vitro inhibitory effect of sulfated galactans isolated from red alga *Gracilaria fisheri* on melanogenesis in B16F10 melanoma cells. J Appl Phycol 30:2611–2618

Rangkadilok N, Sitthimonchai S, Worasuttayangkurn L, Mahidol C, Ruchirawat M, Satayavivad J (2007) Evaluation of free radical scavenging and antityrosinase activities of standardized longan fruit extract. Food Chem Toxicol. https://doi.org/10.1016/j.fct.2006.08.022

Rao AR, Sindhuja HN, Dharmesh SM, Sankar KU, Sarada R, Ravishankar GA (2013) Effective inhibition of skin cancer, tyrosinase, and antioxidative properties by astaxanthin and astaxanthin esters from the green alga *haematococcus pluvialis*. J Agric Food Chem 61:3842–3851. https://doi.org/10.1021/jf304609j

Różanowska M, Sarna T, Land EJ, George Truscott T (1999) Free radical scavenging properties of melanin: interaction of eu- and pheo-melanin models with reducing and oxidising radicals. Free Radic Biol Med 26:Pergamon: 518–525. https://doi.org/10.1016/S0891-5849(98)00234-2

Sano T, Kaya K (1996) Oscillapeptin G, a tyrosinase inhibitor from toxic *Oscillatoria agardhii*. J Nat Prod 59:90–92. https://doi.org/10.1021/np9600210

Schurink M, van Berkel WJH, Wichers HJ, Boeriu CG (2007) Novel peptides with tyrosinase inhibitory activity. Peptides 28(3):485–495. https://doi.org/10.1016/j.peptides.2006.11.023

Shen CT, Chen PY, Wu JJ, Lee TM, Hsu SL, Chang CMJ, Young CC, Shieh CJ (2011) Purification of algal anti-tyrosinase zeaxanthin from *Nannochloropsis oculata* using supercritical anti-solvent precipitation. J Supercrit Fluids 55:955–962. https://doi.org/10.1016/j.supflu.2010.10.003

Shimoda H, Tanaka J, Shan SJ, Maoka T (2010) Anti-pigmentary activity of fucoxanthin and its influence on skin mRNA expression of melanogenic molecules. J Pharm Pharmacol 62:1137–1145. https://doi.org/10.1111/j.2042-7158.2010.01139.x

Shimomura Y, Wajid M, Ishii Y (2009) Dioxinodehydroeckol inhibits melanin synthesis through PI3K/Akt signalling pathway in a-melanocyte-stimulating hormone-treated B16F10 cells. Proc Natl Acad Sci 40:653–668

Singh BK, Park SH, Lee HB, Goo YA, Kim HS, Cho SH, Lee JH et al (2016) Kojic acid peptide: a new compound with anti-tyrosinase potential. Ann Dermatol 28:555–561. https://doi.org/10.5021/ad.2016.28.5.555

Smit N, Vicanova J, Pavel S (2009) The hunt for natural skin whitening agents. Int J Mol Sci 10:5326–5349. https://doi.org/10.3390/ijms10125326

Solano F (2014) Melanins: skin pigments and much more—types, structural models, biological functions, and formation routes. New J Sci lArticle ID 498276. https://doi.org/10.1155/2014/498276

Solano F (2017) Melanin and melanin-related polymers as materials with biomedical and biotechnological applications—cuttlefish ink and mussel foot proteins as inspired biomolecules. Int J Mol Sci 18(7):1561. https://doi.org/10.3390/ijms18071561

Song YS, Balcos MC, Yun HY, Baek KJ, Kwon NS, Kim MK, Kim DS (2015) ERK activation by fucoidan leads to inhibition of melanogenesis in Mel-ab cells. Korean J Physiol Pharmacol 19:29–34. https://doi.org/10.4196/kjpp.2015.19.1.29

Tel G, Öztürk M, Duru ME, Doğan B, Harmandar M (2013) Fatty acid composition, antioxidant, anticholinesterase and tyrosinase inhibitory activities of four *Serratula* species from anatolia. Rec Nat Prod 7:86–95

Thomas NV, Kim SK (2013) Beneficial effects of marine algal compounds in cosmeceuticals. Mar Drugs 11:146–164. https://doi.org/10.3390/md11010146

Vachtenheim J, Borovanský J (2010) "Transcription physiology" of pigment formation in melanocytes: central role of MITF. Exp Dermatol 19:617–627. https://doi.org/10.1111/j.1600-0625.2009.01053.x

Wang ZJ, Si YX, Sangho O, Yang JM, Yin SJ, Park YD, Leeb J, Qian GY (2012) The effect of fucoidan on tyrosinase: computational molecular dynamics integrating inhibition kinetics. J Biomol Struct Dyn 30:460–473. https://doi.org/10.1080/07391102.2012.682211

Wang ZJ, Wei X, Liang JW, Wang CS, Kang Y (2017) Effect of Fucoidan on B16 murine melanoma cell melanin formation and apoptosis. Afr J Trad Complem Altern Med 14:149–155

Westerhof W (2006) The discovery of the human melanocyte, pp 183–193. https://doi.org/10.1111/j.1600-0749.2006.00313.x

Wijesekara I, Kim SK, Li Y, Li YX (2011) Phlorotannins as bioactive agents from brown algae. Process Biochemistry Elsevier. https://doi.org/10.1016/j.procbio.2011.09.015

Wu LC, Lin YY, Yang SY, Weng YT, Tsai YT (2011) Antimelanogenic effect of c-phycocyanin through modulation of tyrosinase expression by upregulation of ERK and downregulation of p38 MAPK signaling pathways. J Biomed Sci 18:1–11. https://doi.org/10.1186/1423-0127-18-74

Yang SY, Kim JH, Lee S, Park S, Park JS, Kim YH (2019) Slow-binding inhibition of Tyrosinase by *Ecklonia cava* Phlorotannins. Mar Drugs 17:2–11

Yoon NY, Eom TK, Kim MM, Kim SK (2009) Inhibitory effect of phlorotannins isolated from *Ecklonia cava* on mushroom tyrosinase activity and melanin formation in mouse B16F10 melanoma cells. J Agric Food Chem 57:4124–4129

Yu P, Sun H (2014) Purification of a fucoidan from kelp polysaccharide and its inhibitory kinetics for tyrosinase. Carbohydr Polym 99:278–283

Yuan Y, Zhang J, Fan J, Clark J, Shen P, Li Y, Zhang C (2018) Microwave assisted extraction of phenolic compounds from four economic brown macroalgae species and evaluation of their antioxidant activities and inhibitory effects on α-amylase, α-glucosidase, pancreatic lipase and tyrosinase. Food Res Int 113:288–297

Zhang F, Kearns SL, Orr PJ, Benton MJ, Zhou Z, Johnson D, Xing X, Wang X (2010) Fossilized melanosomes and the colour of Cretaceous dinosaurs and birds. Nature 1:4–7. https://doi.org/10.1038/nature08740

Chapter 28
Current Trends and Future Prospective of Anti-biofilm Compounds from Marine Macroalgae: An Overview

Nadarajan Viju, Stanislaus Mary Josephine Punitha, Ambati Ranga Rao, Gokare A. Ravishankar, and Sathianeson Satheesh

Abbreviations

Ag^2+	Silver ion
AHL	Acylated homoserine lactone
Anti-QS	Anti-quorum sensing
Fe_3O_4	Iron (II,III) oxide
QS	Quorum sensing
QSI	Quorum sensing inhibition
TiO_2	Titanium dioxide
ZnO	Zinc oxide

N. Viju · S. M. J. Punitha
Centre for Marine Science and Technology, Manonmaniam Sundaranar University, Kanyakumari, Tamil Nadu, India

A. Ranga Rao
Department of Biotechnology, Vignan's Foundation of Science, Technology and Research Deemed to be University, Vadlamudi, Guntur, Andhra Pradesh, India

G. A. Ravishankar
Dr. C. D. Sagar Centre for Life Sciences, Dayananda Sagar College of Engineering, Dayananda Sagar Institutions, Bengaluru, Karnataka, India

S. Satheesh (✉)
Department of Marine Biology, Faculty of Marine Sciences, King Abdulaziz University, Jeddah, Saudi Arabia
e-mail: ssathianeson@kau.edu.sa

A. Ranga Rao, G. A. Ravishankar (eds.), *Sustainable Global Resources of Seaweeds Volume 2*, https://doi.org/10.1007/978-3-030-92174-3_28

1 Introduction

The negative impact of biofilms has posed serious problems to public health, medicine, industries and marine transportation which led to the loss of the world's economy (Epstein et al. 2011). Consequently, the application of anti-microbial agents has been used to get rid of biofilm-associated problems (Danese 2002). However, the widespread use of anti-biofilm agents has been responsible for the development of numerous problems including the emergence of drug-resistant among biofilm microbes and induced toxicity against non-target organisms. In this circumstance, biologically active compounds isolated from natural sources are considered as highly valuable anti-biofilm agents, as they are less toxic and exhibit sufficient bioactive properties (Dobretsov et al. 2006). Consequently, the search for eco-friendly anti-biofilm agents has been triggered among the researchers working on anti-biofilm studies.

The anti-biofilm properties of naturally derived compounds have been expounded and documented extensively (Nithyanand et al. 2010). Among the natural sources, marine macroalgae produce a variety of bioactive compounds which include phenolics, terpenes, sterols, lipids and acrylic acids (Arunkumar et al. 2010). The rationale for the production of the compounds that possess bioactivities by marine macroalgae is clarified as a natural chemical defence strategy thereby macroalgae defend microbial pathogens, bio-foulers and grazers (Pereira and da Gama 2008). Notably, a bountiful of anti-biofilm agents obtained from marine macroalgae have been documented and related investigations are being performed across the world in order to obtain a novel promising anti-biofilm agent (Gadhi et al. 2018). The chemical defence mechanism of marine macroalgae is the key factor that attracts researchers to pick them up as a resource of biologically active components. Therefore, it is indispensable to analyse the anti-biofilm properties of marine macroalgae in order to use macroalgae as the potential resource of unidentified anti-biofilm agents.

2 Biofilms

Microorganisms generally occur in the environment as an assemblage called biofilm that exists as a single layer or three-dimensional structures on abiotic and biotic surfaces (Kumar and Prasad 2006). A biofilm community can be formed by a distinct kind of microorganisms or consortium of many microbial species such as bacteria, fungi etc. However, bacteria have been reported as the dominant group that is responsible for the formation of most of the biofilms (Costerton et al. 1978, 1999). When it comes to the medical sector, the bacterial genus *Enterococcus*, *Staphylococcus*, *Streptococcus*, *Escherichia*, *Klebsiella, Proteus* and *Pseudomonas* are predominantly involved in biofilm formation (Khatoon et al. 2018). Whereas *Pseudomonas, Bacillus, Escherichia, Listeria, Salmonella, Staphylococcus* and *Yersinia* are found in the biofilm related to food industries (Ruan et al. 2015;

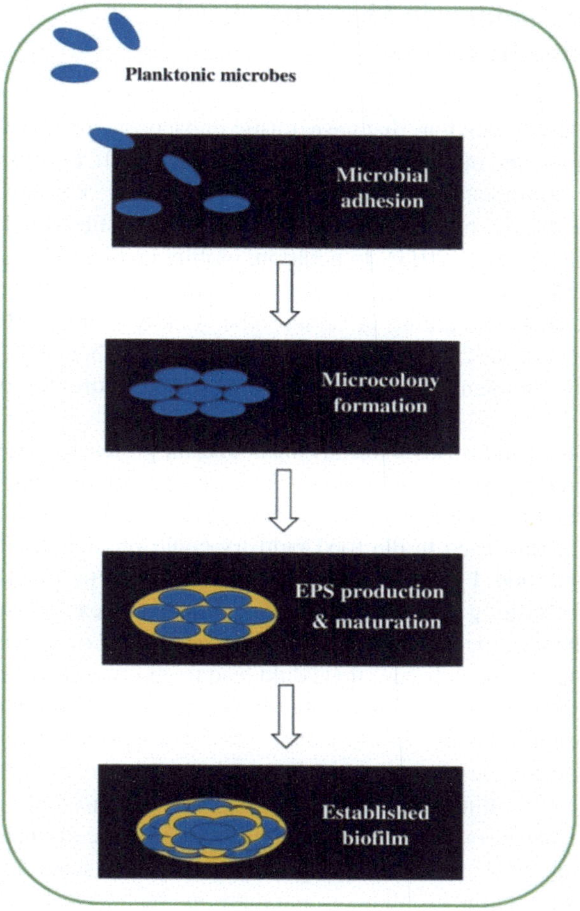

Fig. 28.1 Steps involved in biofilm formation in an aquatic environment

Rothrock et al. 2017). The bacterial species such as *Roseobacter, Alteromonas, Pseudoalteromonas*, *Pseudomonas*, *Enterobacter*, *Aeromonas*, *Cytophaga, Flavobacterium, Micrococcus, Bacillus* and *Vibrio* have been identified in the biofilm community in the structures submerged in the marine environment (Caruso 2020).

Biofilm formation is a step by step microbial action (Fig. 28.1) that starts with the reversible attachment of bacteria to a substratum (Cheng et al. 2007). Subsequently, the adherent cells would develop a microcolony by producing extracellular polymeric substances which enhance the formation and structure of mature biofilm (Leid et al. 2005; De Carvalho 2007). Generally, biofilms allow organisms to survive and thrive in unfriendly environmental conditions (Costerton et al. 1987; Hall-Stoodley et al. 2004). However, biofilms have beneficial role in food chain, agriculture, sewage treatment plants, power plants, petroleum refinery and mining sectors (Coulon et al. 2012),whereas, a negative impact of biofilms on the society has been reported (Davies 2003; Epstein et al. 2011).

3 Impact of Biofilms in Medicinal, Food and Marine Industries

Typically, biofilms are noxious that have drastic impacts on our society that ranging from medical, food and marine industries (Lopez et al. 2010). In medical industries, biofilm causes life-threatening diseases including chronic osteomyelitis, cystitis, prostatitis, otitis media, pneumonia in patients with cystic fibrosis (Francolini and Donelli 2010; Hoiby et al. 2011). In addition, biofilm causes various infections on the biomaterials such as vascular and urinary catheters, orthopaedic appliances, disposable lenses, artificial cardiac valve and other sundry implants (Leunisse et al. 2001; Campoccia et al. 2006). It has been reported that about 80% of microbial infections in the human body are caused by biofilm-forming bacteria (Mah et al. 2003).

In terms of food industries, biofilms are found on pasteurizer, reverse osmosis membranes, liquid pipelines and utensils used for the preparation, package and storage of materials (Abdallah et al. 2014; Colagiorgi et al. 2017). The biofilm developed on the materials used in the food industry could produce toxins that can be contaminated the food. For instance, the diarrheal and emetic toxins produced by the food biofilm forming *Bacillus cereus* can cause diarrhoea and vomiting (Galie et al. 2018), whereas, the food-biofilm forming bacterium *Listeria monocytogenes* causes listeriosis, a critical disease that could lead to abortion (Ferreira et al. 2014).

Notably, in the marine environment, biofilm caused by microorganisms leads to the development of bio-fouling. It has been stated that marine biofilms act as preferential substrate for the settlement of the larvae and spores of higher organisms whereby drastic biofouling is developed (Hadfield 2011). Bio-fouling is caused by more than 4000 marine organisms and it varied from one type of microorganisms to other (Yebra et al. 2004). Bio-fouling represents a major nuisance and causes huge economic losses to marine installations and maritime sectors such as offshore pipelines, desalination plants, marine vessels etc. (Schultz et al. 2011). Moreover, biofilm plays an important role in bio-corrosion that causes economic loss by degrading the metals used in industries (Javaherdashti 1999; Narenkumar et al. 2019). The governments and private industries spend a lot of money to prevent/control biofilm-associated problems.

4 Current Anti-biofilm Strategies

To overcome the problems connected with biofilm, several strategies are being carried out. Typically, the anti-biofilm strategy is classified as biofilm prevention methods and biofilm destruction methods. The biofilm prevention method is further divided into anti-adhesion therapy and anti-quorum sensing therapy, whereas, biofilm destruction method is divided into two subgroups such as combination therapy and biofilm disassociation therapy (Fig. 28.2). The anti-adhesion therapy is defined

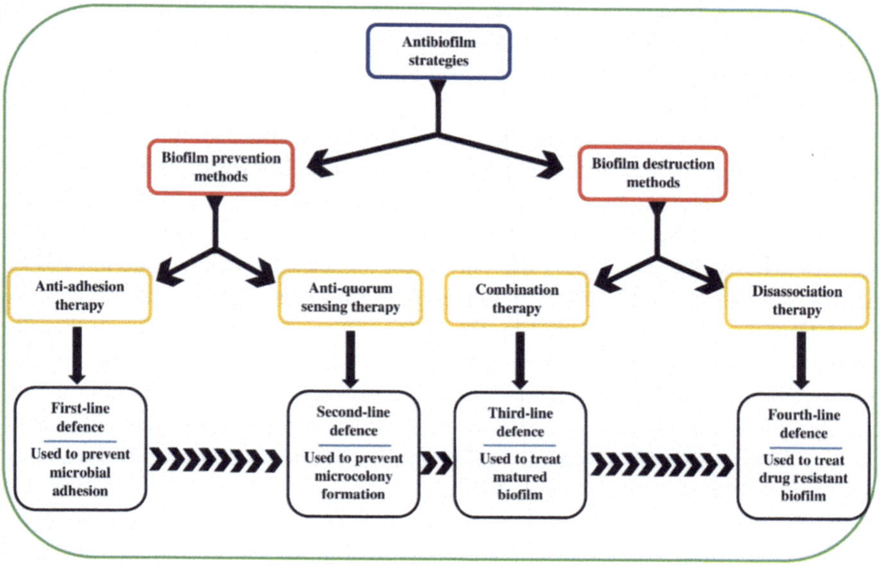

Fig. 28.2 Anti-biofilm strategies used in medical, food and maritime sectors

as the first line of defence that prevents microbial adhesion to different surfaces. Anti-adhesion methods are applied to fend off biofilm formation on medical implants/devices either by coating the surfaces or by modifying the surfaces with anti-biofilm agents (Narayana and Srihari 2020; Paluch et al. 2020). Generally, anti-microbial peptides, anti-biotics, enzymes, nanoparticles and quaternary ammonium compounds are used as the active component in anti-adhesion therapy.

In the surface modification method, the medical devices are impregnated with the anti-biotics such as nitrofurazone, gentamicin and norfloxacin (Francolini et al. 2010). Contrarily, antibiotics/antimicrobials are incorporated into a polymer matrix and applied on medical devices. For example, the polyurethane polymer incorporated with 1,2,3 triazole and palladium nanoparticles is used to keep off *Pseudomonas aeruginosa* biofilm (Hong et al. 2016). Similar surface coating methods are used to fight off the biofilm development in food, marine and other industries. For instance, steel coating developed with nanoparticles (e.g. Ag2+, Fe3O4, TiO2, ZnO) is used to inhibit the biofilm formation in food industries (Beyth et al. 2015). Whereas, in the marine environment biocide based surface coatings are used to control marine biofilm related problems (Thomas et al. 2001).

The anti-quorum sensing (anti-QS) strategy is one of the promising method of biofilm control and considered as the second line of defence, as used to treat the microbes managed to escape from the first line of defence (anti-adhesion). Bacterial quorum sensing (QS) regulate the biofilm development by increasing the bacterial density by way of the expression of specific sets of genes (Costerton et al. 2007). The QS can be terminated by inhibiting the signal molecules from binding to the receptor (Gonzalez and Keshavan 2006). The anti-QS mode of biofilm inhibition has been expounded and documented by many researchers. For instance, Francolini et al. (2004) reported the biofilm inhibition of usnic acid by interfering with the signalling pathways of *Pseudomonas aeruginosa*.

On the other hand, combination therapy is defined as the third line of defence strategy that is used to eradicate the established mono-species biofilm and multispecies biofilm. The multispecies biofilm display a high level of anti-biotic resistance, thus eradication of multispecies biofilm is a challenging task. Therefore, antibiotics with different action mechanisms are combined to enhance the anti-biofilm activity by attacking different targets/species (Brackman et al. 2011). Another study, Raad et al. (2007) combined the antibiotics rifampicin and vancomycin to treat the biofilm developed by *Staphylococcus aureus* and found out that the antibiotic combo significantly inhibited the biofilm developed by *Staphylococcus aureus*. Similar combined anti-biotic formulations such as fosfomycin/tobramycin, clarithromycin/tobramycin, clarithromycin/erythromycin, clarithromycin/vancomycin, clarithromycin/daptomycin and ceftaroline/daptomycin have been used for anti-biofilm treatment (Ciofu et al. 2017).

The biofilm disassociation therapy is the fourth line of defence, used to treat the biofilms that acquired resistance against anti-biofilm agents. In this method, anti-biofilm agents are combined with extracellular matrix degrading substances (enzymes) such as α-amylase, glycosidases, proteases, DNases, dispersin B, and lysostaphin (Ammar et al. 2017). The action mechanism of the therapy is that the matrix degrading substances would destroy the physical integrity of the biofilm matrix thereby the microbes embedded in the biofilm would expose to antibiotics. Subsequently, the microbes exposed to antibiotics would be inhibited/killed and resulting in biofilm elimination (Kaplan 2009). The combination of the enzyme lysostaphin and the antibiotic nafcillin effectively eliminated the biofilm developed by multidrug-resistant *Staphylococcus aureus* on a medical device (John et al. 2009).

5 Need for Alternative Anti-biofilm Agents

There are two key factors that have been addressed for the need for novel anti-biofilm agents, they are (1) induced toxicity to untargeted organisms and (2) induced antimicrobial resistance among the biofilm microbes. Figure 28.3 explains the need for alternative anti-biofilm agents for biofilm control. The administration of antimicrobial agents has been an effective strategy to throw out biofilms (Danese 2002). However, continual application of such antimicrobials in the environment can trigger toxicity to unfocused cells/organisms (Langsrud et al. 2003; Simoes and Vieira 2009). For instance, the antibiofilm agent triclosan has toxic effects such as modification in mitochondrial function, calcium signalling, homeostasis and immunological parameters (Weatherly and Gosse 2017). Similarly, heavy metals like mercury, silver, copper, zinc and nickel previously used as anti-biofilm agents are not in use today owing to their consequential danger to health (Teitzel and Parsek 2003). Comparably, the organotin based surface coating used to prevent marine biofilm/

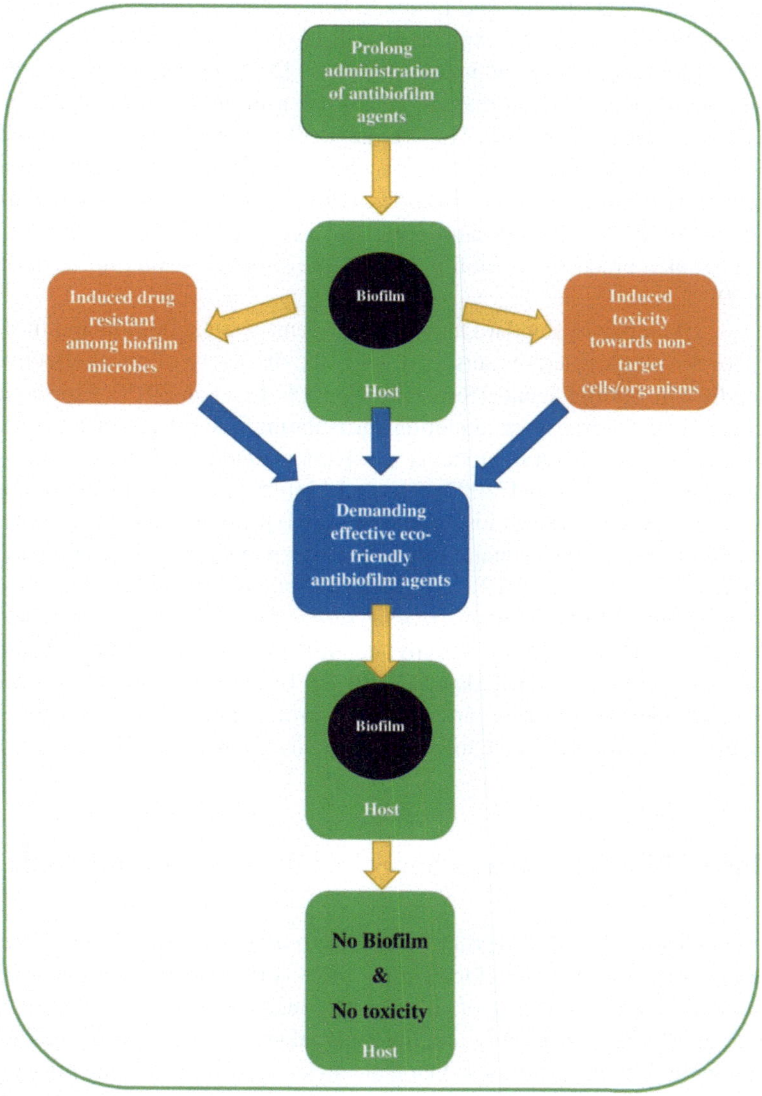

Fig. 28.3 Graphical explanation of the need of alternative non-toxic anti-biofilm agents

bio-fouling caused a deleterious effect on mussel, snail and oyster; thereby the application of them in the marine environment was banned (Thomas and Brooks 2010).

In terms of resistance, bacteria embedded in biofilm matrix acquire the long-suffering characteristics against anti-biofilm agents and host immune responses (Abebe 2020). Typically, biofilms are 10–1000 fold more resistant to antimicrobials compared to planktonic cells which make the eradication of biofilm a complicated process (Simoes et al. 2006). The rationale behind the increased antibiotic resistance of biofilm has not been explained well. However, few antibiotic resistance practices of biofilm have been addressed, they are (1) the production of extracellular polymeric substances that can restrict the diffusion of antibiotics into the biofilm (Wajiha et al. 2010), (2) biofilm bacteria generally grow slower and the reduced metabolic activity provide them less susceptible against most of the antibiotics (Estrela and Abraham 2010) and (iii), the biofilm bacteria produce the compound that could detoxify the effect of anti-biotic (Mah et al. 2003). These complex antibiotics resistance practices of biofilm lead to the survival of biofilm bacteria (Ito et al. 2009).

The low efficiency of standard anti-biofilm agents against biofilm and increased toxicity towards untargeted organisms triggers the discovery of novel effective and eco-friendly anti-biofilm agents (Simoes et al. 2006; Lu et al. 2019). Consequently, researchers have focused their attention on isolating natural product based anti-biofilm agents, since natural products are biodegradable and possess non-toxic (Sipkema et al. 2005; You et al. 2007). The anti-biofilm activities of natural products belong to phenolics, terpenoids, alkaloids, polypeptides, and polyacetylenes obtained from various biological sources have been documented (Bakkiyaraj and Pandian 2010; Yong et al. 2019). For instance, the compounds styrylpyrone and neolignans obtained from the plant *Helichrysum italicum* (Family - Asteracea) displayed significant anti-biofilm activity against the biofilm-forming bacterium *Pseudomonas aeruginosa* (Brigida et al. (2013). Besides, the compounds such as berberine, tetrandrine, piperine, artocarpin, malvin, isosteviol, eugenol have also been reported as natural product anti-biofilm agents (Xun et al. 2017).

6 Marine Macroalgae as a Source of Bioactive Compounds

Marine organisms have idiosyncratic strategies to fight off invaders, predators and other competitors (Wahl 1989). One among them is the chemical defence strategy that accomplished by producing of certain compounds against the target organisms (Selvin et al. 2010). Remarkably, marine macroalgae produce lots of biologically active compounds that can act as a chemical barrier against pathogenic microorganisms, herbivores and biofoulers (Charles and Victoria 2005; Paul et al. 2006a, 2006b). It has been hypothesized that the chemical defence is the major defence strategy of marine macroalgae (Paul et al. 2001; Pereira and da Gama 2008). Figure 28.4 explains the chemical defence strategies of marine macroalgae. The marine macroalgae belong to the class rhodophyta (red algae) are known to produce halogenated compounds that involve in the ecological role of chemical defence (Paul et al. 1987). The halogenated terpenes and sesquiterpenes have an important role in the macroalgal chemical defence by defending the herbivores and bio-fouling organisms (Da Gama et al. 2003; Paradas et al. 2010). Besides red algae, the chemical defence strategy of green (Chlorophyte) and brown (Phaeophyta) algae have also been reported extensively (Pereira and da Gama 2008). The diterpenoid alcohol

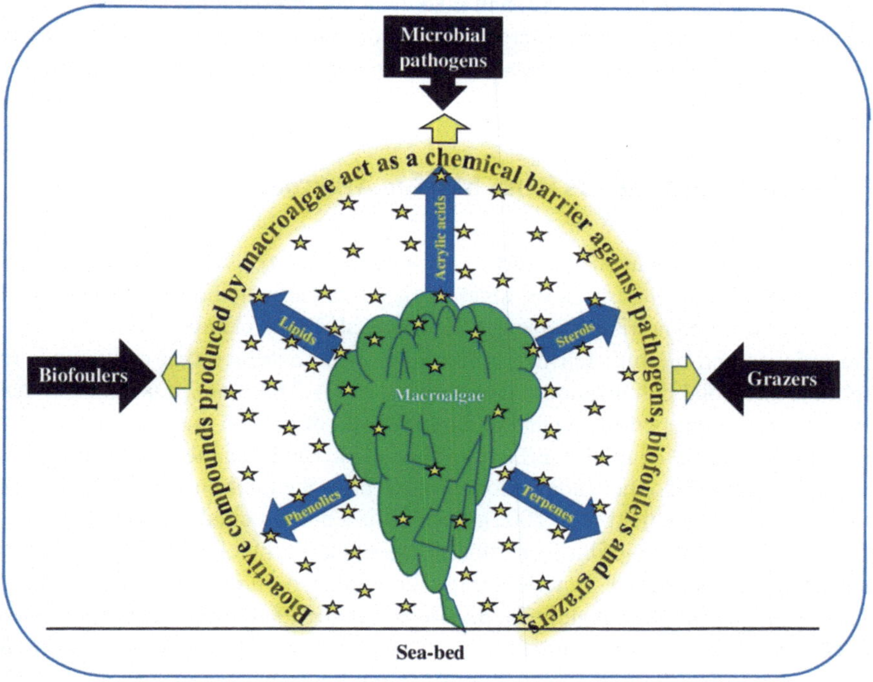

Fig. 28.4 Marine macroalgal chemical defence mechanism against microbial pathogens, biofoulers and grazers

like pachydictyol-A produced by the brown alga *Dictyota dichotoma* has been reported to defend fish like grazers (Hay et al. 1987).

Over 3000 marine macroalgal derived natural products mainly belong to polysaccharides, lipids, fatty acids, sterols, phenolics and carotenoids have been documented (Perez et al. 2016a, 2016b). These macroalgae derived products have potential application in food (human and animal, Shama et al. 2019) and pharmaceutical industries (Leandro et al. 2019). The cytotoxic kahalalide F produced by the green algae *Bryopsis pennata* exhibited anti-cancer activity (Wang et al. 2015). Similarly, the phenolic compounds (flavonoids, chlorophyll, and carotenoids) produced by marine algae possess antioxidant property (Hwang and Thi 2014). In an investigation, Chia et al. (2015) observed the antioxidant activity of the macroalgae *Padina tetrastromatica*, *Caulerpa racemosa* and *Turbinaria ornate*. *Bonnemaisonia hamifera* has shown antibacterial and antifouling activities by producing the bioactive component 1,1,3,3-tetrabromo-2-heptanone (Persson et al. 2011). Comparably, Kathryn et al. (2014) isolated the compound Dopamine which showed antifouling activity from the macroalgae Ulvaria obscura. Anti-fouling property is the natural chemical defence strategy of marine macroalgae and production of many antifouling compounds such as Omaezallene, Chromanols, Bromosphaerol etc. have been reported (Piazza et al. 2011; Cho 2013). Besides, marine macroalgae produce the bioactive compounds that possess anti-coagulant, anti-thrombotic, anti-aging, anti-depression, anti-fatigue, anti-allergic anti-cancer, anti-inflammatory, anti-fungal, anti-viral and anti-biofilm activities have been reported (Biris-Dorhoi et al. 2020).

7 Anti-biofilm Agents from Marine Macroalgae

Microbial colonization on the surface of marine macroalgae can cause serious health problems such as diseases, algal tissue degradation and decrease the rate of photosynthesis and nutrient exchange (Lachnit et al. 2013). Competitively, macroalgae produce anti-quorum sensing, anti-adhesive and antimicrobial compounds to fight off microbial colonization (Goecke et al. 2010). Therefore, it is believed that the anti-biofilm property is a general characteristic of marine macroalgae, thereby microbial colonization on the surfaces is prevented. The furanones produced by the genus *Delisea* defend the colonization of bacteria by the quorum sensing inhibition (QSI) method (Manefield et al. 2002). The QSI is a non-biocidal method of antibiofilm strategy and the biofilm formation is prevented by blocking the QS-inducers like acylated homoserine lactone (AHL), autoinducers, and autoinducers type 2 (Ciric et al. 2019). Correspondingly, Skindersoe et al. (2008) reported the QSI of the macroalgal (*Caulerpa* sp.) extract by interfering with the AHL regulatory system. It has been reported that the production of QS-inhibitors by macroalgae could be a natural biofilm preventive mechanism (Jha et al. 2013). Evidently, a research by Cumberbatch (2002) analysed the anti-QS effect of the 30 different macroalgal species collected from South Florida and observed the production of anti-quorum sensing components by 73% (22 out of 30) tested macroalgae. Like QSI, several non-biocidal strategies that control biofilm formation have been proposed which include the anti-adhesion strategy (Travier et al. 2013). For instance, Viano et al. (2009) described the anti-adhesion property of the terpenoids (isolated from *Dictyota* sp.) against the biofilm formation by marine *Pseudoalteromonas* sp. Some of the reported macroalgae that produced anti-quorum sensing and anti-adhesion compounds are given in Table 28.1.

The compounds with biocidal (antimicrobial) mode of biofilm inhibition derived from macroalgae have been reported extensively (Perez et al. 2016a, b; Li et al. 2018). Vellaisamy et al. (2011) reported the antibacterial activity of different macroalgal species (*Ulva lactuca, Caulerpa scalpelliformis, Padina boergesenii, Caulerpa* sp. and *Chaetomorpha linoides*) against the biofilm forming bacteria. Likewise, hundreds of macroalgae-derived compounds that exhibit antimicrobial activity against various clinical (e.g. *Staphylococcus* sp., *Escherichia* sp., *Bacillus* sp., *Klebsiella* sp., *Streptococcus* sp., *Streptococcus* sp. etc.) and marine (e.g. *Flavobacterium* sp., *Enterobacter* sp. *Bacillus* sp., *Shewanella* sp., *Pseudoalteromonas* sp., *Cobetia* sp., *Marinobacterium* sp., *Vibrio* sp. etc.) pathogenic/biofilm-forming bacteria have been documented (Culioli et al. 2008; Shanmughapriya et al. 2008). The details of some of the macroalgae that displayed

Table 28.1 Non-biocidal mode of anti-biofilm/anti-bacterial activity of marine macroalgae

S. No.	Marine Macroalgae	Responsible compound/ extract	Bioactivity	Reference
1	*Gloiopeltis furcate*	Polysaccharide	Antiadhesion	Saeki (1994)
2	*Gloiopeltis furcate*	Funoran	Antiadhesion	Sato et al. (1998)
3	*Delisea pulchra*	Halogenated furanones	Antiquorum sensing	Manefield et al. (1999)
4	*Laminaria digitate*	Halogens	Antiquorum sensing	Borchardt et al. (2001)
5	*Chlamydomonas reinhardtii*	Lumichrome	Antiquorum sensing	Teplitski et al. (2004)
6	*Ahnfeltiopsis flabelliformis*	Betonicine	Antiquorum sensing	Kim et al. (2007)
7	*Laurencia* sp.	Algal extract	Antiquorum sensing	Skindersoe et al. (2008)
8	*Galaxauraceae*	Algal extract	Antiquorum sensing	Skindersoe et al. (2008)
9	*Dicytosphaeria ocellata*	Algal extract	Antiadhesion	Sneed and Pohnert (2011)
10	*Bonnemaisonia hamifera*	1,1,3,3-Tetrabromo-2-heptanone	Antiadhesion	Persson et al. (2011)
11	*Fucus vesiculosus*	Proline	Antiadhesion	Saha et al. (2012)
12	*Chondrus crispus*	Algal extract	Antiquorum sensing	Salta et al. (2013)
13	*Asparagopsis taxiformis*	Algal extract	Antiquorum sensing	Jha et al. (2013)
14	*Ulva fasciata*	Algal extract	Antiquorum sensing	Batista et al. (2014)
15	*Taonia atomaria*	Sesquiterpenes	Antiadhesion	Othmani et al. (2015)
16	*Canistrocarpus cervicornis*	Algal extract	Antiquorum sensing	Carvalho et al. (2017)
17	*Symbiodinium* sp.	Algal extract	Antiquorum sensing	Zea-Obando et al. (2018)
18	*Hizikia fusiforme*	Phlorotannins	Antiquorum sensing	Tang et al. (2020)

antimicrobial activity against various pathogenic and biofilm-forming bacteria are given in Table 28.2.

8 Future Perspectives

Marine macroalgae have copious amounts of bioactive compounds that have diverse health properties (Rajapakse and Kim 2011; Skrovankova 2011). Many studies have been carried out for the isolation of novel bioactive compounds using marine

Table 28.2 Biocidal mode of anti-biofilm/anti-bacterial activity of marine macroalgae

S. No.	Marine Macroalgae	Responsible compound/ extract	Bioactivity	Reference
1	*Sargassum* sp.	Polyphenols	Antibacterial	Sieburth and Conover (1965)
2	*Caulerpa* sp.	Sesquiterpenoids	Antibacterial	Paul et al. (1987)
3	*Dilophus guineensis*	Dilophic acid	Antibacterial	Schlenk and Gerwick (1987)
4	*Cystoseira tamariscifolia*	Methoxybifurcarenone	Antibacterial	Bennamara et al. (1999)
5	*Laurencia pannosa*	Pannosanol, pannosane	Antibacterial	Suzuki et al. (2001)
6	*Sphaerococcus coronopifolius*	Bromosphaerone	Antibacterial	Etahiri et al. (2001)
7	*Codium iyengarii*	Iyengaroside-A	Antibacterial	Ali et al. (2002)
8	*Rhodomela confervoides*	Bromophenols	Antibacterial	Xu et al. (2003)
9	*Fucus vesiculosus*	Polyhydroxylated fucophlorethol	Antibacterial	Sandsdalen et al. (2003)
10	*Laurencia* sp.	Sesquiterpenes	Antibacterial	Bansemir et al. (2004)
11	*Asparagopsis armata*	Halomethanes, haloether, haloacetales	Antibacterial	Paul et al. (2006)
12	*Osmundaria serrata*	Lanosol ethyl ether	Antibacterial	Barreto and Meyer (2006)
13	*Ulva fasciata*	Diterpenoids	Antibacterial	Chakraborty et al. (2010)
14	*Laurencia chilensis*	Sesquiterpenes	Antibacterial	Vairappan et al. (2010)
15	*Sargassum horneri*	Chromanols	Antibacterial	Cho (2013)
16	*Gracilaria fisheri*	Algal extract	Antibacterial	Srikong et al. (2015)
17	*Halidrys siliquosa*	Algal extract	Antibacterial	Busetti et al. (2015)
18	*Dictyosphaeria cavernosa*	Algal extract	Antibacterial	Deepa et al. (2017)

macroalgae (Carpena et al. 2021; Habeebullah et al. 2020). Jha et al. (2013) reported the presence of QS-inhibitors in the methanol extract of the macroalgae *Asparagopsis taxiformis*. Similarly, the bioactivity of macroalgal extracts obtained from various red algae (e.g. *Alsidium corallinum, Ceramium rubrum, Chondracanthus canaliculatus, Ganonema farinosum, Gelidium pusillum* etc.), green algae (e.g. *Bryopsis pennata, Caulerpa racemose, Cladophora glomerata, Enteromorpha linza, Ulva prolifera* etc.) and brown algae (e.g. *Cladophora rupestris, Colpomenia tuberculate, Dictyopteris delicatula, Eisenia bicyclis, Padina mexicana* etc.) have been documented liberally (Perez et al. 2016a, 2016b; Silva et al. 2020). Analysing the previous dataset, we found out that the extraction and isolation of the compounds responsible for the bioactivity is the bottleneck that hinders the discovery of novel bioactive agents from marine macroalgae.

A natural source could produce thousands compounds that may possess or may not possess bioactive properties. To analyse the bioactive (e.g. anti-cancer, anti-biofilm etc.) properties, the compounds must be separated from the source (raw material). Generally, the compounds produced by the natural sources are extracted using conventional (e.g. soxhlet extraction, maceration etc.) and contemporary (e.g. ultrasound assisted extraction, pulsed electric field extraction etc.) methods (Rasul 2018). Each method used for the extraction of a particular group of compounds is mainly based on the polar, non-polar and mid-polar of the compounds. The extraction method-decoction is applied to extract polar compounds, while this method is not suitable to extract thermo-labile components (Zhang et al. 2018). Therefore, distinctly most efficient extraction methods are adopted.

On the other hand, purification of the desired compound from the extract is a highly skilled process. Typically, chromatography methods used for the separation of natural products include adsorption column chromatography, partition chromatography, counter-current chromatography, gel filtration chromatography, ion-exchange chromatography, preparative gas chromatography, supercritical fluid chromatography, multi-dimensional chromatography and other chromatographic techniques (Lau et al. 2015; Zhang et al. 2018). In chromatography, the separation of natural products is carried out based on the properties such as adsorption, partition coefficient, molecular size and ionic strength of the product (Coskun 2016). Hence, thorough knowledge in chromatography analysis would be very handy to isolate the desired compounds from the crude extract.

9 Conclusion

Nature has offered thousands of marine macroalgal species that can produce plenty of bioactive compounds. Therefore, screening marine macroalgae for anti-biofilm compounds would lead to the identification of many more novel compounds as anti-biofilm agents. Macroalgal biomass harvesting from the marine environment is also one of the issues from the viewpoint of biodiversity conservation. However, most of the algal species are cultivable hence their supply may not be a limitation.

Acknowledgments NV and SMJP acknowledge the Centre for Marine Science and Technology, Manonmaniam Sundaranar University for providing necessary facility to prepare this chapter. ARR acknowledges support and resources from the Improvement of Science & Technology Infrastructure in Higher Educational Institutions (FIST Project no: LSI-576/2013) from the Department of Science and Technology, Government of India and Centre of Excellence, Department of Biotechnology, Vignan's Foundation for Science, Technology and Research. SS acknowledges the Faculty of Marine Sciences, King Abdulaziz University, Saudi Arabia for providing the facility to prepare this chapter.

References

Abdallah M, Benoliel C, Drider D, Dhulster P, Chihib NE (2014) Biofilm formation and persistence on abiotic surfaces in the context of food and medical environments. Arch Microbiol 196(7):453–472. https://doi.org/10.1007/s00203-014-0983-1

Abebe GM (2020) The Role of Bacterial Biofilm in antibiotic resistance and food contamination. Int J Microbiol. https://doi.org/10.1155/2020/1705814

Ali MS, Saleem M, Yamdagni R, Ali MA (2002) Steroid and antibacterial steroidal glycosides from marine green alga *Codium iyengarii*. Nat Prod Lett 16:407–413

Arunkumar K, Sivakumar SR, Rengasamy R (2010) Review on bioactive potential in seaweeds (Marine Macroalgae): a special emphasis on bioactivity of seaweeds against plant pathogens. Asian J Plant Sci 9:227–240. https://doi.org/10.3923/ajps.2010.227.240

Bakkiyaraj D, Pandian SK (2010) In vitro and in vivo antibiofilm activity of a coral associated actinomycete against drug resistant *Staphylococcus aureus* biofilms. Biofouling 26(6):711–717. https://doi.org/10.1080/08927014.2010.511200

Bansemir A, Just N, Michalik M, Lindequist U, Lalk M (2004) Extracts and sesquiterpene derivatives from the red alga *Laurencia chondrioides* with antibacterial activity against fish and human pathogenic bacteria. Chem Biodivers 1:463–467

Barreto M, Meyer JJM (2006) Isolation and antimicrobial activity of a lanosol derivative from *Osmundaria serrata* (Rhodophyta) and a visual exploration of its biofilm covering. S Afr J Bot 72(4):521–528. https://doi.org/10.1016/j.sajb.2006.01.006

Batista D, Carvalho A, Costa R, Coutinho R, Dobretsov S (2014) Extracts of macroalgae from the Brazilian coast inhibit bacterial quorum sensing. Bot Mar 57:441–447

Bennamara A, Abourriche A, Berrada M, Charrouf M, Chaib N, Boudouma M, Garneau FX (1999) Methoxybifurcarenone: an antifungal and antibacterial meroditerpenoid from the brown alga *Cystoseira tamariscifolia*. Phytochemistry 52(1):37–40. https://doi.org/10.1016/s0031-9422(99)00040-0

Beyth N, Houri-Haddad Y, Domb A, Khan W, Hazan R (2015) Alternative antimicrobial approach: nano-Antimicrobial materials. Evid Based Complement Alternat Med 2015:246012. https://doi.org/10.1155/2015/246012

Biris-Dorhoi ES, Michiu D, Pop CR, Rotar AM, Tofana M, Pop OL, Socaci SA, Farcas AC (2020) Macroalgae-A sustainable source of chemical compounds with biological activities. Nutrients 12(10):3085. https://doi.org/10.3390/nu12103085

Borchardt SA, Allain EJ, Michels JJ, Stearns GW, Kelly RF, McCoy WF (2001) Reaction of acylated homoserine lactone bacterial signaling molecules with oxidized halogen antimicrobials. Appl Environ Microbiol 67:3174–3179

Brackman G, Cos P, Maes L, Nelis HJ, Coenye T (2011) Quorum sensing inhibitors increase the susceptibility of bacterial biofilms to antibiotics in vitro and in vivo. Antimicrob Agents Chemother 55(6):2655–2661. https://doi.org/10.1128/AAC.00045-11

Brigida D, Elisabetta B, Grazia D, Lorena C, Monica S, Valeria S, Severina P, Giovanna D, Antonio F (2013) Spectroscopic identification and anti-biofilm properties of polar metabolites from the medicinal plant *Helichrysum italicum* against *Pseudomonas aeruginosa*. Bioorg Med Chem 21:7038–7046. https://doi.org/10.1016/j.bmc.2013.09.019

Busetti A, Thompson TP, Tegazzini D, Megaw J, Maggs CA, Gilmore BF (2015) Antibiofilm activity of the brown alga *Halidrys siliquosa* against clinically relevant human pathogens. Mar Drugs 13(6):3581–3605. https://doi.org/10.3390/md13063581

Campoccia D, Montanaro L, Arciola CR (2006) The significance of infection related to orthopedic devices and issues of antibiotic resistance. Biomaterials 27:2331–2339

Carpena M, Caleja C, Pereira E, Pereira C, Ciric A, Sokovic M, Soria-Lopez A, Fraga-Corral M, Simal-Gandara J, Ferreira ICFR et al (2021) Red seaweeds as a source of nutrients and bioactive compounds: optimization of the extraction. Chemosensors 9:132. https://doi.org/10.3390/chemosensors9060132

Caruso G (2020) Microbial colonization in marine environments: overview of current knowledge and emerging research topics. J Mar Sci Eng 8:78. https://doi.org/10.3390/jmse8020078

Carvalho AP, Batista D, Dobretsov S (2017) Extracts of seaweeds as potential inhibitors of quorum sensing and bacterial growth. J Appl Phycol 29:789–797. https://doi.org/10.1007/s10811-016-1014-1

Chakraborty K, Lipton AP, Raj RP, Vijayan KK (2010) Antibacterial labdane diterpenoids of *Ulva fasciata* Delile from southwestern coast of the Indian Peninsula. Food Chem 119:1399–1408

Charles A, Victoria F (2005) Defensive and sensory chemical ecology of brown algae. Adv Bot Res 43:1–91. https://doi.org/10.1016/S0065-2296(05)43001-3

Cheng G, Zhang Z, Chen S, Bryers JD, Jiang S (2007) Inhibition of bacterial adhesion and biofilm formation on zwitterionic surfaces. Biomaterials 28(29):4192–4199. https://doi.org/10.1016/j.biomaterials.2007.05.041

Chia YY, Kanthimathi MS, Khoo KS, Rajarajeswaran J, Cheng HM, Yap WS (2015) Antioxidant and cytotoxic activities of three species of tropical seaweeds. BMC Complement Altern Med 15:339. https://doi.org/10.1186/s12906-015-0867-1

Cho J (2013) Antifouling chromanols isolated from brown alga *Sargassum horneri*. J Appl Phycol 25:299–309

Ciofu O, Rojo-Molinero E, Macià MD, Oliver A (2017) Antibiotic treatment of biofilm infections. APMIS 125(4):304–319

Ciric AD, Petrovic JD, Glamoclija JM, Smiljkovic MS, Nikolic MM, Stojkovic DS et al (2019) Natural products as biofilm formation antagonists and regulators of quorum sensing functions: a comprehensive review update and future trends. South Afr J Bot 120:65–80. https://doi.org/10.1016/j.sajb.2018.09.010

Colagiorgi A, Bruini I, Di Ciccio PA, Zanardi E, Ghidini S, Ianieri A (2017) *Listeria monocytogenes* biofilms in the wonderland of food industry. Pathogens 6(3):41. https://doi.org/10.3390/pathogens6030041

Coskun O (2016) Separation techniques: chromatography. North Clin Istanb 3(2):156–160. https://doi.org/10.14744/nci.2016.32757

Costerton JW, Geesey GG, Cheng KJ (1978) How bacteria stick. Sci Am Jan 238(1):86–95. https://doi.org/10.1038/scientificamerican0178-86

Costerton JW, Cheng KJ, Geesey GG, Ladd TI, Nickel JC, Dasgupta M, Marrie TJ (1987) Bacterial biofilms in nature and disease. Annu Rev Microbiol 41:435–464

Costerton JW, Stewart PS, Greenberg EP (1999) Bacterial biofilms: a common cause of persistent infections. Science 284(5418):1318–1322. https://doi.org/10.1126/science.284.5418.1318

Costerton J, Montanaro L, Arciola CR (2007) Bacterial communications in implant infections: a target for an intelligence war. Int J Artif Organs 30:757–763. https://doi.org/10.1177/039139880703000903

Coulon F, Chronopoulou PM, Fahy A, Paisse S, Goni-Urriza M, Peperzak L, Alvarez LA, McKew BA, Brussaard CP, Underwood GJ (2012) Central role of dynamic tidal biofilms dominated by aerobic hydrocarbonoclastic bacteria and diatoms in the biodegradation of hydrocarbons in coastal mudflats. Appl Environ Microbiol 78:3638–3648

Culioli G, Ortalo-Magné A, Valls R, Hellio C, Clare AS, Piovetti L (2008) Antifouling activity of meroditerpenoids from the marine brown alga *Halidrys siliquosa*. J Nat Prod 71:1121–1126

Cumberbatch A (2002) Characterization of the anti-quorum sensing activity exhibited by marine macroalgae of South Florida. Miami: Department of Biological Sciences, Florida International University

Da Gama BAP, Pereira RC, Soares AR, Teixeira VL, Yoneshigue-Valentin Y (2003) Is the mussel test a good indicator of antifouling activity? A comparison between laboratory and field tests. Biofouling 19:161–169

Danese PN (2002) Antibiofilm approaches: prevention of catheter colonization. Chem Biol 9(8):873–880. https://doi.org/10.1016/s1074-5521(02)00192-8

Davies D (2003) Understanding biofilm resistance to antibacterial agents. Nat Rev Drug Discov 2(2):114–122. https://doi.org/10.1038/nrd1008

De Carvalho CC (2007) Biofilms: recent developments on an old battle. Recent Pat Biotechnol 1:49–57
Deepa S, Venkateshwaran P, Vinithkumar NV, Kirubagaran R (2017) Bioactive propensity of macroalgae from the Andaman & Nicobar Islands. Pharm J 9(6):815–820
Dobretsov S, Dahms HU, Qian PY (2006) Inhibition of biofouling by marine microorganisms and their metabolites. Biofouling 22:43–54
Epstein AK, Pokroy B, Seminara A, Aizenberg J (2011) Bacterial biofilm shows persistent resistance to liquid wetting and gas penetration. Proc Natl Acad Sci U S A 108:995–1000
Estrela AB, Abraham WR (2010) Combining biofilm-controlling compounds and antibiotics as a promising new way to control biofilm infections. Pharmaceuticals 3(5):1374–1393. https://doi.org/10.3390/ph3051374
Etahiri S, Bultel-Ponce V, Caux C, Guyot M (2001) New bromoditerpenes from the red alga *Sphaerococcus coronopifolius*. J Nat Prod 64:1024–1027
Ferreira V, Wiedmann M, Teixeira P, Stasiewicz MJ (2014) *Listeria monocytogenes* persistence in food-associated environments: epidemiology, strain characteristics, and implications for public health. J Food Prot 77(1):150–170. https://doi.org/10.4315/0362-028X.JFP-13-150
Francolini I, Donelli G (2010) Prevention and control of biofilm-based medical-device-related infections. FEMS Immunol Med Microbiol 59:227–238
Francolini I, Norris P, Piozzi A, Donelli G, Stoodley P (2004) Usnic acid, a natural antimicrobial agent able to inhibit bacterial biofilm formation on polymer surfaces. Antimicrob Agents Ch 48:4360–4365
Francolini I, D'Ilario L, Guaglianone E, Donelli G, Martinelli A, Piozzi A (2010) Polyurethane anionomers containing metal ions with antimicrobial properties: thermal, mechanical and biological characterization. Acta Biomater. https://doi.org/10.1016/j.actbio.2010.03.042
Gadhi AAA, El-Sherbiny MM, Al-Sofyani AMA, BaAkdah MA, Satheesh S (2018) Antibiofilm activities of extracts of the macroalga *Halimeda* sp. from the Red Sea. J Mar Sci Tech 26(6):838–846
Galie S, Garcia-Gutiérrez C, Miguélez EM, Villar CJ, Lombó F (2018) Biofilms in the food industry: health aspects and control methods. Front Microbiol 9:898. https://doi.org/10.3389/fmicb.2018.00898
Goecke F, Labes A, Wiese J, Imhoff JF (2010) Chemical interactions between marine macroalgae and bacteria. Mar Ecol Prog Ser 409:267–299
Gonzalez JE, Keshavan ND (2006) Messing with bacterial Quorum Sensing. Microbiol Mol Biol Rev 70(4):859–875
Habeebullah K, Alagarsamy S, Sattari S et al (2020) Enzyme-assisted extraction of bioactive compounds from brown seaweeds and characterization. J Appl Phycol 32:615–629. https://doi.org/10.1007/s10811-019-01906-6
Hadfield MG (2011) Biofilms and marine invertebrate larvae: what bacteria produce that larvae use to choose settlement sites. Annu Rev Mar Sci 3:453–470
Hall-Stoodley L, Costerton WJ, Stoodley P (2004) Bacterial biofilms: from the natural environment to infectious diseases. Nat Rev Microbiol 2:95–108
Hay ME, Fenical W, Gustafson K (1987) Chemical defense against diverse coral reef herbivores. Ecology 68:1581–1591
Hoiby N, Ciofu O, Johansen HK, Song ZJ, Moser C, Jensen PO (2011) The clinical impact of bacterial biofilms. Int J Oral Sci 3:55–65
Hong C, Yihui X, Francisco VL, Liyan S, Klaus-Viktor P, Suzana N, Peiying H (2016) Antibiofilm effect enhanced by modification of 1,2,3-triazole and palladium nanoparticles on polysulfone membranes. Sci Rep 6:24289. https://doi.org/10.1038/srep24289
Hwang ES, Thi ND (2014) Effects of extraction and processing methods on antioxidant compound contents and radical scavenging activities of laver (*Porphyra tenera*). Prev Nutr Food Sci 19(1):40–48. https://doi.org/10.3746/pnf.2014.19.1.040

Ito A, Taniuchi A, May T, Kawata K, Okabe S (2009) Increased antibiotic resistance of *Escherichia coli* in mature biofilms. Appl Environ Microbiol 75(12):4093–4100. https://doi.org/10.1128/AEM.02949-08

Javaherdashti RA (1999) A review of some characteristics of MIC caused by sulfate-reducing bacteria: past, present and future. Anti-Corr Method Mater 46:173–180. https://doi.org/10.1108/00035599910273142

Jha B, Kavita K, Westphal J, Hartmann A, Schmitt-Kopplin P (2013) Quorum sensing inhibition by *Asparagopsis taxiformis*, a marine macro alga: separation of the compound that interrupts bacterial communication. Mar Drugs 11:253–265. https://doi.org/10.3390/md11010253

John KK, Tanya C, James M (2009) Lysostaphin established *Staphylococcus aureus* biofilms in jugular vein catheterized mice. J Antimicrob Chemother 64:94–100. https://doi.org/10.1093/jac/dkp145

Kaplan JB (2009) Therapeutic potential of biofilm-dispersing enzymes. Int J Artif Organs 32(9):545–554. https://doi.org/10.1177/039139880903200903

Kathryn VA, Elizabeth H, Marianne C (2014) Effects of dopamine, a compound released by the green-tide macroalga *Ulvaria obscura* (Chlorophyta), on marine algae and invertebrate larvae and juveniles. Phycologia 53:195–202. https://doi.org/10.2216/13-237.1

Khatoon Z, McTiernan CD, Suuronen EJ, Mah TF, Alarcon EI (2018) Bacterial biofilm formation on implantable devices and approaches to its treatment and prevention. Heliyon 4(12):e01067. https://doi.org/10.1016/j.heliyon.2018.e01067

Kim JS, Kim YH, Seo YW (2007) Quorum sensing inhibitors from the red alga, *Ahnfeltiopsis flabelliformis*. Biotechnol Bioprocess Eng 12:308. https://doi.org/10.1007/BF02931109

Kumar A, Prasad R (2006) Biofilm. JK Sci 8:14–17

Lachnit T, Fischer M, Künzel S, Baines JF, Harder T (2013) Compounds associated with algal surfaces mediate epiphytic colonization of the marine macroalga *Fucus vesiculosus*. FEMS Microbiol Ecol 84(2):411–420. https://doi.org/10.1111/1574-6941.12071

Langsrud S, Sidhu MS, Heir E, Holck AL (2003) Bacterial disinfectant resistance – a challenge for the food industry. Int Biodeterior Biodegr 51:283–290

Lau EC, Mason DJ, Eichhorst N, Engelder P, Mesa C, Kithsiri Wijeratne EM, Gunaherath GM, Gunatilaka AA, La Clair JJ, Chapman E (2015) Functional chromatographic technique for natural product isolation. Org Biomol Chem 13(8):2255–2259. https://doi.org/10.1039/c4ob02292k

Leandro A, Pereira L, Gonçalves AMM (2019) Diverse applications of marine macroalgae. Mar Drugs 18(1):17. https://doi.org/10.3390/md18010017

Leid JG, Willson CJ, Shirtliff ME, Hassett DJ, Parsek MR, Jeffers AK (2005) The exopolysaccharide alginate protects *Pseudomonas aeruginosa* biofilm bacteria from IFN-gamma-mediated macrophage killing. J Immunol 175:7512–7518

Leunisse C, Van Weissenbruch R, Busscher JH, Van Der Mei HC, Dijk F, Albers FW (2001) Biofilm formation and design features of indwelling silicone rubber tracheoesophageal voice prostheses–an electron microscopical study. J Biomed Mater Res 58(B):556–563

Li Y, Sun S, Pu X, Yang Y, Zhu F, Zhang S, Xu N (2018) Evaluation of antimicrobial activities of seaweed resources from Zhejiang Coast, China. Sustainability 10(7):2158. https://doi.org/10.3390/su10072158

Lopez D, Vlamakis H, Kolter R (2010) Biofilms. Cold Spring Harb Perspect Biol 2:1–11

Mah TF, Pitts B, Pellock B, Walker GC, Stewart PS, O'Toole GA (2003) A genetic basis for *Pseudomonas aeruginosa* biofilm antibiotic resistance. Nature 426(6964):306–310. https://doi.org/10.1038/nature02122

Manefield M, de Nys R, Kumar N, Read R, Givskov M, Steinberg P, Kjelleberg S (1999) Evidence that halogenated furanones from *Delisea pulchra* inhibit acylated homoserine lactone (AHL)-mediated gene expression by displacing the AHL signal from its receptor protein. Microbiology 145:283–291

Manefield M, Rasmussen TB, Henzter M, Andersen JB, Steinberg P, Kjelleberg S, Givskov M (2002) Halogenated furanones inhibit quorum sensing through accelerated LuxR turnover. Microbiology 148(4):1119–1127. https://doi.org/10.1099/00221287-148-4-1119

Narayana P, Srihari P (2020) A Review on surface modifications and coatings on implants to prevent biofilm. Regen Eng Transl Med 6:330–346. https://doi.org/10.1007/s40883-019-00116-3

Narenkumar J, AlSalhi MS, Arul Prakash A, Abilaji S, Devanesan S, Rajasekar A, Alfuraydi AA (2019) Impact and role of bacterial communities on biocorrosion of metals used in the processing industry. ACS Omega 4(25):21353–21360. https://doi.org/10.1021/acsomega.9b02954

Nithyanand P, Thenmozhi R, Rathna J, Pandian SK (2010) Inhibition of *Streptococcus pyogenes* biofilm formation by coral-associated actinomycetes. Curr Microbiol 60(6):454–460. https://doi.org/10.1007/s00284-009-9564-y

Othmani A, Bunet R, Bonnefont J, Briand JF, Culioli G (2015) Settlement inhibition of marine biofilm bacteria and barnacle larvae by compounds isolated from the Mediterranean brown alga *Taonia atomaria*. J Appl Phycol 28:1975–1986

Paluch E, Rewak-Soroczyńska J, Jędrusik I (2020) Prevention of biofilm formation by quorum quenching. Appl Microbiol Biotechnol 104:1871–1881. https://doi.org/10.1007/s00253-020-10349-w

Paradas WC, Salgado LT, Sudatti DB, Crapez MA, Fujii MT, Coutinho R, Pereira RC, Amado Filho GM (2010) Induction of halogenated vesicle transport in cells of the red seaweed *Laurencia obtusa*. Biofouling 26(3):277–286. https://doi.org/10.1080/08927010903515122

Paul VJ, Littler MM, Littler DS, Fenical W (1987) Evidence for chemical defense in tropical green alga *Caulerpa ashmeadii* (Caulerpaceae: Chlorophyta): isolation of new bioactive sesquiterpenoids. J Chem Ecol 13:1171–1185

Paul VJ, Cruz-Rivera E, Thacker RW (2001) Chemical mediation of macroalgal-herbivore interactions: ecological and evolutionary perspectives. In: McClintock JB, Baker BJ (eds) Marine chemical ecology. CRC, pp 227–265

Paul NA, de Nys R, Steinberg PD (2006a) Chemical defense against bacteria in the red alga Asparagopsis armata: linking structure with function. Mar Ecol Prog Ser 306:87–101

Paul VJ, Puglisi MP, Ritson-Williams R (2006b) Marine chemical ecology. Nat Prod Rep 23:153–180

Pereira RC, da Gama BAP (2008) Macroalgal chemical defenses and their roles in structuring tropical marine communities. In: Amsler CD (ed) Algal chemical ecology. Springer, Berlin, Heidelberg. https://doi.org/10.1007/978-3-540-74181-7_2

Perez MJ, Falque E, Domínguez H (2016a) Antimicrobial action of compounds from marine seaweed. Mar Drugs 14(3):52. https://doi.org/10.3390/md14030052

Perez MJ, Falqué E, Domínguez H (2016b) Antimicrobial action of compounds from marine seaweed. Mar Drugs 14(3):52. https://doi.org/10.3390/md14030052

Persson F, Svensson R, Nylund GM, Fredriksson NJ, Pavia H, Hermansson M (2011) Ecological role of a seaweed secondary metabolite for a colonizing bacterial community. Biofouling 27:579–588

Piazza V, Roussis V, Garaventa F, Greco G, Smyrniotopoulos V, Vagias C et al (2011) Terpenes from the red alga *Sphaerococcus coronopifolius* inhibit the settlement of barnacles. Mar Biotechnol 13:764–772. https://doi.org/10.1007/s10126-010-9337-4

Raad I, Hanna H, Dvorak T, Chaiban G, Hachem R (2007) Optimal antimicrobial catheter lock solution, using different combinations of minocycline, EDTA, and 25-percent ethanol, rapidly eradicates organisms embedded in biofilm. Antimicrob Agents Chemother 51:78–83

Rajapakse N, Kim SK (2011) Nutritional and digestive health benefits of seaweed. Adv Food Nutr Res 64:17–28. https://doi.org/10.1016/B978-0-12-387669-0.00002-8

Rasul MG (2018) Extraction, isolation and characterization of natural products from medicinal plants. Int J Basic Sci Appl Comput 2(6):2394–2367

Rothrock MJJ, Davis ML, Locatelli A, Bodie A, McIntosh TG, Donaldson JR, Ricke SC (2017) Listeria occurrence in poultry flocks: detection and potential implications. Front Vet Sci 4:125. https://doi.org/10.3389/fvets.2017.00125

Ruan L, Crickmore N, Peng D, Sun M (2015) Are nematodes a missing link in the confounded ecology of the entomopathogen *Bacillus thuringiensis*? Trends Microbiol 23(6):341–346. https://doi.org/10.1016/j.tim.2015.02.011

Saeki Y (1994) Effect of seaweed extracts on Streptococcus sobrinus adsorption to saliva-coated hydroxyapatite. Bull Tokyo Dent Coll 35:9–15

Saha M, Rempt M, Gebser B, Grueneberg J, Pohnert G, Weinberger F (2012) Dimethylsulphopropionate (DMSP) and proline from the surface of the brown alga *Fucus vesiculosus* inhibit bacterial attachment. Biofouling 28:37–41

Salta M, Wharton JA, Dennington SP, Stoodley P, Stokes KR (2013) Anti-biofilm performance of three natural products against initial bacterial attachment. Int J Mol Sci 14(11):21757–21780

Sandsdalen E, Haug T, Stensvåg K et al (2003) The antibacterial effect of a polyhydroxylated fucophlorethol from the marine brown alga, *Fucus vesiculosus*. World J Microbiol Biotechnol 19:777–782. https://doi.org/10.1023/A:1026052715260

Sato S, Yoshinuma N, Ito K, Tokumoto T, Takiguchi T, Suzuki Y, Murai S (1998) The inhibitory effect of funoran and eucalyptus extract-containing chewing gum on plaque formation. J Oral Sci 40:115–117

Schlenk D, Gerwick WH (1987) Dilophic acid a diterpenoid from the tropical brown seaweed *Dilophus guineensis*. Phytochemistry 26:1081–1084

Schultz MP, Bendick JA, Holm ER, Hertel WM (2011) Economic impact of biofouling on a naval surface ship. Biofouling 27(1):87–98

Selvin J, Ninawe AS, Kiran GS, Lipton AP (2010) Sponge-microbial interactions: ecological implications and bioprospecting avenues. Crit Rev Microbiol 36:82–90

Shama A, Joyce SG, Mari FD, Ranga Rao A, Ravishankar GA, Hudaa N (2019) Macroalgae and microalgae: novel sources of functional food and feed. In: Ravishankar GA, Ranga Rao A (eds) Handbook of algal technologies and phytochemicals: volume-I: food, health and nutraceutical applications. CRC Press, pp 207–219

Shanmughapriya S, Manilal A, Sujith S et al (2008) Antimicrobial activity of seaweeds extracts against multiresistant pathogens. Ann Microbiol 58:535–541. https://doi.org/10.1007/BF03175554

Sieburth JM, Conover JT (1965) Sargassum tannin, an antibiotic which retards fouling. Nature 208:52–53

Silva A, Silva SA, Carpena M, Garcia-Oliveira P, Gullón P, Barroso MF, Prieto MA, Simal-Gandara J (2020) Macroalgae as a source of valuable antimicrobial compounds: extraction and applications. Antibiotics 9(10):642. https://doi.org/10.3390/antibiotics9100642

Simoes M, Vieira MJ (2009) Persister cells in Pseudomonas fluorescens biofilms treated with a biocide. In: Proceedings of the international conference processes in biofilms: fundamentals to applications, Davis, CA, pp 58–62

Simoes M, Simões LC, Machado I, Pereira MO, Vieira MJ (2006) Control of flow-generated biofilms using surfactants - evidence of resistance and recovery. Food Bioprod Process 84:338–345

Sipkema D, Franssen MCR, Osinga R, Tramper J, Wijffels RH (2005) Marine sponges as pharmacy. Mar Biotechnol 7:142–162

Skindersoe ME, Ettinger-Epstein P, Rasmussen TB, Bjarnsholt T, de Nys R, Givskov M (2008) Quorum sensing antagonism from marine organisms. Mar Biotechnol 10:56–63

Skrovankova S (2011) Seaweed vitamins as nutraceuticals. Adv Food Nutr Res 64:357–369. https://doi.org/10.1016/B978-0-12-387669-0.00028-4

Sneed JM, Pohnert G (2011) The green alga *Dicytosphaeria ocellata* and its organic extracts alter natural bacterial biofilm communities. Biofouling 27:347–356

Srikong W, Mittraparp-arthorn P, Rattanaporn O, Bovornreungroj N, Bovornreungroj P (2015) Antimicrobial activity of seaweed extracts from Pattani, Southeast coast of Thailand. Food Appl Biosci J 3:39–49

Suzuki M, Daitoh M, Vairappan CS, Abe T, Masuda M (2001) Novel halogenated metabolites from the Malaysian *Laurencia pannosa*. J Nat Prod 64:597–602

Tang J, Wang W, Chu W (2020) Antimicrobial and anti-quorum sensing activities of phlorotannins from seaweed (*Hizikia fusiforme*). Front Cell Infect Microbiol 10:586750. https://doi.org/10.3389/fcimb.2020.586750

Teitzel GM, Parsek MR (2003) Heavy metal resistance of biofilm and planktonic *Pseudomonas aeruginosa*. Appl Environ Microbiol 69(4):2313–2320. https://doi.org/10.1128/aem.69.4.2313-2320.2003

Teplitski M, Chen H, Rajamani S, Gao M et al (2004) *Chlamydomonas reinhardtii* secretes compounds that mimic bacterial signals and interfere with quorum sensing regulation in bacteria. Plant Physiol 134:137–146

Thomas KV, Fileman TW, Readman JW, Waldock MJ (2001) Antifouling paint booster biocides in the UK coastal environment and potential risks of biological effects. Mar Pollut Bull 42:677–688

Travier L, Rendueles O, Ferrières L, Herry JM, Ghigo JM (2013) *Escherichia coli* resistance to nonbiocidal antibiofilm polysaccharides is rare and mediated by multiple mutations leading to surface physicochemical modifications. Antimicrob Agents Chemother 57(8):3960–3968. https://doi.org/10.1128/AAC.02606-12

Vairappan CS, Anangdan SP, Tan KL, Matsunaga S (2010) Role of secondary metabolites as defense chemicals against ice-ice disease bacteria in biofouler at carrageenophyte farms. J Appl Phycol 22:305–311

Vellaisamy B, Palanichamy S, Rajaram R (2011) Effects of certain seaweed extracts on the primary bio-film forming bacteria. J Mar Biol Ass India 53(1):104–110

Viano Y, Bonhomme D, Camps M, Briand JF, Ortalo-Magné A, Blache Y, Piovetti L, Culioli G (2009) Diterpenoids from the Mediterranean brown alga *Dictyota* sp. evaluated as antifouling substances against a marine bacterial biofilm. J Nat Prod 72(7):1299–1304. https://doi.org/10.1021/np900102f

Wahl M (1989) Marine epibiosis.1. Fouling and antifouling - some basic aspects. Mar Ecol Prog Ser 58(1–2):175–189

Wajiha K, Steve B, Sherry K, John H, Fariha H, George O (2010) Aminoglycoside resistance of *Pseudomonas aeruginosa* biofilms modulated by extracellular polysaccharide. Int Microbiol 13:207–212. https://doi.org/10.2436/20.1501.01.127

Wang B, Waters AL, Valeriote FA, Hamann MT (2015) An efficient and cost-effective approach to kahalalide F N-terminal modifications using a nuisance algal bloom of *Bryopsis pennata*. Biochim Biophys Acta Gen Subj 1850:1849–1854

Weatherly LM, Gosse JA (2017) Triclosan exposure, transformation, and human health effects. J Toxicol Environ Health B Crit Rev 20(8):447–469. https://doi.org/10.1080/10937404.2017.1399306

Xu N, Fan X, Yan X, Li X, Niu R, Tseng CK (2003) Antibacterial bromophenols from the marine red alga *Rhodomela confervoides*. Phytochemistry 62:1221–1224

Xun S, Yi-Xuan X, Zhen H, Hongjie Z (2017) A review of natural products with anti-biofilm activity. Curr Org Chem 21. https://doi.org/10.2174/1385272821666170620110041

Yebra DM, Kiil S, Dam-Johansen K (2004) Review. Antifouling technology-past, present and future steps towards efficient and environmentally friendly antifouling coatings. Prog Org Coat 50:75–104

Yong YY, Dykes GA, Choo WS (2019) Biofilm formation by staphylococci in health-related environments and recent reports on their control using natural compounds. Crit Rev Microbiol 45:201–222. https://doi.org/10.1080/1040841X.2019.1573802

You J, Xue X, Cao L, Lu X, Wang J, Zhang L, Zhou S (2007) Inhibition of Vibrio biofilm formation by a marine actinomycete strain A66. Appl Microbiol Biotechnol 76:1137–1144

Zea-Obando C, Tunin-Ley A, Turquet J, Culioli G, Briand JF, Bazire A, Réhel K, Faÿ F, Linossier I (2018) Anti-bacterial adhesion activity of tropical microalgae extracts. Molecules 23(9):2180. https://doi.org/10.3390/molecules23092180

Zhang QW, Lin LG, Ye WC (2018) Techniques for extraction and isolation of natural products: a comprehensive review. Chin Med 13:20. https://doi.org/10.1186/s13020-018-0177-x

Chapter 29
Biological Activities and Health Benefits of Seaweed Carotenoids with Special Reference to Fucoxanthin

Rangaswamy Lakshminarayana, Kariyappa Vijay, Rudrappa Ambedkar, Ambati Ranga Rao, and Gokare A. Ravishankar

Abbreviations

ACOX1	Acyl-CoA oxidase-1
AMD	Age-related macular degeneration
AP-I	Activator protein-1
CCDs	Carotenoid cleavage dioxygenases
CDK	Cyclin-dependent kinase
CoA	Coenzyme A
CPT1	Carnitine palmitoyl-transferase
CVD	Cardiovascular diseases
DMAP	Dimethylallyl diphosphate
EBPα	Enhancer-binding protein-α
FFA	Free fatty acid
GGPP	Geranyl-geranyl diphosphate
HIF-1α	Hypoxia-inducible factor-1α
IL	Interleukin
iNOS	Indusible nitric oxide synthase
IPP	Isopentenyl diphosphate
LDL-R	Low-density lipoprotein receptor

R. Lakshminarayana (✉) · K. Vijay · R. Ambedkar
Department of Microbiology and Biotechnology, Bangalore University, Bengaluru, Karnataka, India

A. Ranga Rao
Department of Biotechnology, Vignan's Foundation of Science, Technology and Research Deemed to be University, Vadlamudi, Guntur, Andhra Pradesh, India

G. A. Ravishankar
Dr. C. D. Sagar Centre for Life Sciences, Dayananda Sagar College of Engineering, Dayananda Sagar Institutions, Bengaluru, Karnataka, India

A. Ranga Rao, G. A. Ravishankar (eds.), *Sustainable Global Resources of Seaweeds Volume 2*, https://doi.org/10.1007/978-3-030-92174-3_29

MDA	Malondialdehyde
MEP	Methylerythritol phosphate
MMP-2	Matrix metalloproteinases-2
mRNA	Messenger ribonucleic acid
MVA	Mevalonate
NADPH	Nicotinamide adenine dinucleotide phosphate
NF-kB	Nuclear factor kappa-B
Nrf2	Nuclear factor-erythroid 2 related factor-2
NSIDSs	Non-steroidal anti-inflammatory drugs
PDS	Phytoene desaturase synthase
PGE2	Prostaglandin E2
PI3K	Phosphatidylinositol-3-kinase
PPARα	Peroxisome proliferators activated receptor-α
RNA	Ribonucleic acid
RNS	Reactive nitrogen species
ROS	Reactive oxygen species
SCD-1	Stearoyl-coenzyme A desaturase-1
STAT3	Signal transducer and activator of transcription 3
TNF-α	Tumor necrosis factor α
UV	Ultra-Violet
VAD	Vitamin A deficiency
ZDS	ζ--carotene desaturase

1 Introduction

Carotenoids are natural pigments belonging to the category of tetraterpenoids (C_{40} atoms) consisting of a long polyene chain with alternative conjugated double bonds (Britton 1995). There are over 750 known carotenoids found in nature, and are categorized into two major classes, xanthophylls and carotenes. Further, carotenoids are also classified based on the presence and absence of functional groups, chemical structure, and biological activity (Fig. 29.1). Among carotenoids, β- carotene, lycopene, lutein, astaxanthin, canthaxanthin, and fucoxanthin, are recognized as superior antioxidants and considered as bioactive nutrients. Humans and animals are not able to synthesize carotenoids *de novo* and they need to obtain them through diet. Carotenoids exhibit vibrant colors in nature like yellow, orange, red, brown, etc. They are found in a large number of fruits and vegetables, animal products (eggs, butter, and milk), certain microorganisms, and seafood (salmon, crustaceans, trout, k mollusk, etc.) (Britton et al. 2009). Carotenoids are a large class of plant isoprenoids, that participate in essential processes such as respiration (ubiquinone), photosynthesis (carotenoids), regulation of growth and development (strigolactones, cytokinins, brassinosteroids, gibberellins, and abscisic acid), and as attractants for other organisms (pollinating insects, and seed-distributing herbivores) (Arathi et al. 2017). Secondary isoprenoid metabolites have commercial value as flavors, pigments, polymers, or drugs. However, only a limited supply of these compounds is

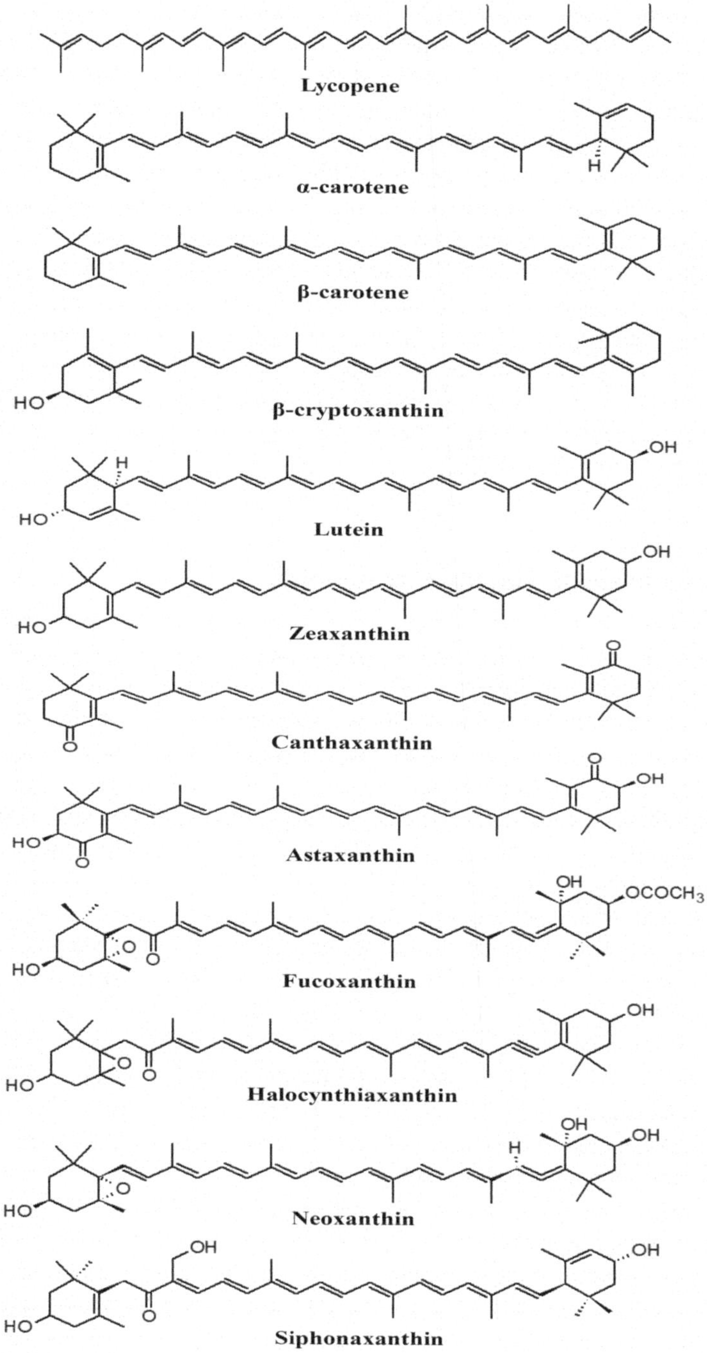

Fig. 29.1 Molecular structure of carotenoids

usually available from natural plant sources. Majorly, plants provide a primary source of essential vitamins and carotenoids for human diets. Several epidemiological and clinical trials have well correlated the consumption of dietary carotenoids with decreased risk of vitamin A deficiency (VAD), cancer, cardiovascular diseases (CVD), and age-related macular degeneration (AMD) (Britton et al. 2009). The α-carotene, β-carotene, and β- cryptoxanthin are known precursors of vitamin A. The non-pro vitamin A carotenoid, such as lycopene has gained special attention for its role in the efficient singlet oxygen quenching and reduction of CVD and prostate cancers (Arathi et al. 2015). Lutein and zeaxanthin are considered as vital macular pigment selectively accumulated in the central region of the retina, acting as a filter of high-energy blue light, quenchers of singlet oxygen, chain-breaking antioxidants, and are involved in the reduction of phototoxic damage to the eye (Arathi et al. 2015). Fucoxanthin is another promising marine carotenoid with an anticipated role against diabetes, obesity, and angiogenesis (Hussein et al. 2006; Woo et al. 2009; Arathi et al. 2015).

2 Biosynthesis of Marine Carotenoids

Carotenoids are widely distributed in nature, and synthesized in photosynthetic (cyanobacteria, algae, and plants) and non-photosynthetic organisms (certain fungi and bacteria) (Botella-Pavía and Rodríguez-Concepción 2006; Ranga Rao et al. 2019). In nature, two distinct pathways have evolved for the synthesis of the universal precursors of all isoprenoid products, namely a 5- carbon isoprene unit, isopentenyl diphosphate (IPP), and its allylic isomer dimethylallyl diphosphate (DMAPP). Since, the discovery of the mevalonate (MVA) pathway in the 1950s, it was assumed that IPP was synthesized from acetyl-CoA via mevalonate (Fraser and Bramley 2004). Although MVA pathway contributes to carotenoid synthesis in some cases, such as in etiolated seedlings (Rodriguez-Concepcion et al. 2004), plant carotenoids are mainly produced through the methylerythritol phosphate (MEP)-derived pathway in phototrophic plants (Botella-Pavía and Rodríguez-Concepción 2006). The first step in carotenoid biosynthesis is the condensation of two geranyl-geranyl diphosphate (GGPP; C_{20}) molecules to form phytoene (C_{40})· Phytoene is a colorless carotenoid produced as a 15-cis isomer by the enzyme phytoene synthase (Fig. 29.2). A detailed biosynthesis pathway of carotenoids and their derivatives in various natural sources are shown in Fig. 29.2.

Many marine organisms are of color mainly due to the accumulation of carotenoids through consumption. However, they play an important role in determining the quality of seafood, such as shrimp, lobsters, crabs, salmon, and tuna. Among marine carotenoids, fucoxanthin (the dominant carotenoid in brown seaweed) is the most abundant and contributes about 10% to the total worldwide production of natural carotenoids (Matsuno 2001).

Among carotenoids, fucoxanthin is considered as major xanthophyll carotenoid abundantly found in brown seaweeds, and certain microalgae (Peng et al. 2011).

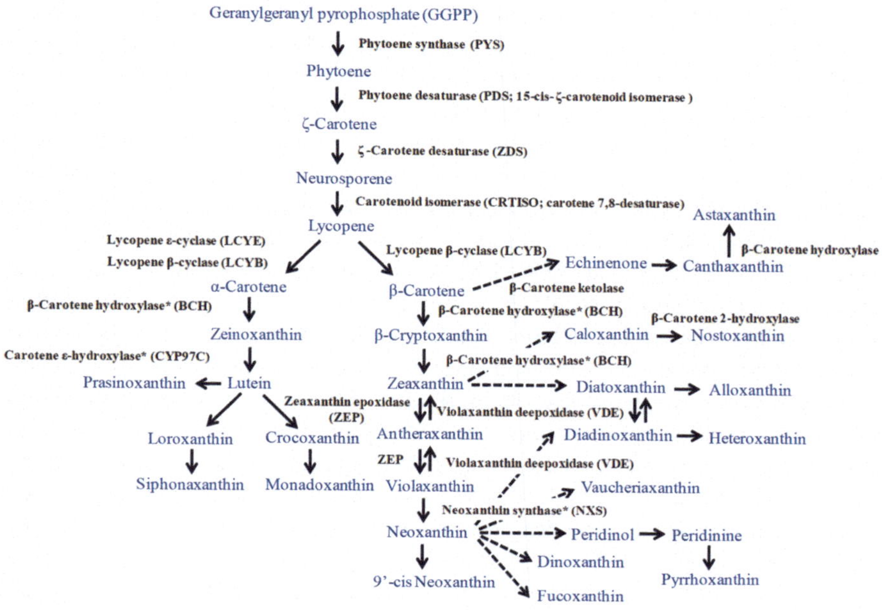

Fig. 29.2 The scheme of biosynthesis of marine carotenoids

Padina tetrastromatica, *Sargassum fusiforme, Undaria pinnatifida*, and *Laminaria japonica*, are the common edible brown algae consumed in European and South-East Asian countries. (Sangeetha et al. 2010). These macroalgae are recognised as rich sources of fucoxanthin (Kim et al. 2012). In the food industry, *Phaeodactylum tricornutum*, (diatoms) are preferentially used as a source of fucoxanthin due to higher content and extraction efficiency than from other macroalgae (Wang et al. 2018). Several studies have suggested that fucoxanthin has an immense potential to prevent various chronic diseases including obesity, diabetes, and cancers.

3 Antioxidant Activity of Carotenoids

Many nutrition and health-associated clinical trials have recommended regular dietary consumption of carotenoids from rich sources thereby decreasing the incidence of several chronic diseases. These chronic diseases are allied with higher oxidative stress and elevated circulating systemic chronic inflammatory markers or components such as cytokines, prostaglandin E2 (PGE2), tumor necrosis factor α (TNF-α), C- reactive protein. Also, immune cells are associated with inflammatory responses prostaglandins, isoprostanes, oxidized cholesterol, and malondialdehyde (MDA) levels. These conditions may have resulted in tissue damage and eventually causes or aggravate the disease. The extended electrophilic conjugated double bonds are considered an important feature of carotenoids since it assists in stabilizing unpaired electrons after quenching of free radicals. As of this conjugated double-bond structure, carotenoids are potentially involved in scavenging ROS such as singlet oxygen ($^1O_2$) and peroxy radicals (Britton 1995). Possibly, they react by physical quenching, electron gaining or donation, and adduct formation. The scavenging of $^1O_2$ depends on the number of conjugated double bonds. Carotenoids as lipophilic molecules; they are typically located inside the cell membranes. Xanthophylls, which are less hydrophobic than carotenes, are found in cellular membranes at the lipid/aqueous interface, and they can scavenge lipid and aqueous phase radicals (Agamey et al. 2004). β-carotene or lycopene scavenge radicals in the lipid phase, hence these molecules are arranged-exclusively within the inner part of the lipid bilayer (Gruszecki and Strzalka 2005). The incorporation of a specific type of carotenoids and their chemical nature and structure may greatly affect membranes properties (mechanical strength, rigidity, fluidity), which are considered to be crucial for effective functioning, especially protection against oxidative stress and reactive oxygen species (Gruszecki 2010). Apart from these, carotenoids are shown pro-oxidants activity under higher oxygen tension in the cellular environment. In addition, others have demonstrated that carotenoids including marine origin involved in the inhibition of cancer cells by exhibiting pro-oxidant effects at higher cellular oxygen (O_2) tension. Burton and Ingold (1984) and Palozza et al. (2004) have demonstrated the pro-$^1O_2$ oxidant activity of carotenoids and their mediated cytotoxicity at an increased concentration under higher oxygen tension. Furthermore, Gorrini et al. (2013) reported that regulation of oxidative stress is considered as a critical factor in both tumor development and responses to anticancer therapies. In this regard, many signaling pathways of chronic diseases are linked to ROS-mediated mechanisms (Fig. 29.3). Since carotenoids are potent antioxidants, the formation of carotenoid radicals resulting from ROS detoxification and subsequent addition of oxygen to carotenoid radicals may generate peroxy radicals and hydroperoxides at higher oxygen tension (Gamey et al. 2004; Lakshminarayana et al. 2013; Sowmya et al. 2017). The antioxidant or pro-oxidant actions of carotenoids in the biological system depend on their structure, concentration, and oxygen tension as well as the presence and absence of other antioxidants in the cells. The pro-oxidant activity of carotenoids is predominate over the antioxidant mechanism, resulting in the association of chain-carrying peroxyl radical generation with higher partial pressures of oxygen (Burton and Ingold 1984; Arathi et al. 2016), which may lead to redox imbalance in cells, especially in chronic conditions. In contrast, carotenoids optimized the redox status as a defense mechanism against peroxyl radicals or its mediated oxidative stress in normal metabolic processes versus accelerated metabolism in diseased or cancer cells. This circuitous path may include interactions with cellular signaling cascades directed to endogenous defense systems and anti-inflammation is of biological relevance.

In contrast to other carotenoids, fucoxanthin consists of unique structural properties that comprise an allenic bond, epoxy, hydroxyl, carbonyl, and carboxyl groups (Fig. 29.1). Among carotenoids, only about specific carotenoids contain an allenic bond (Peng et al. 2011). It has known that fucoxanthin considers having a robust antioxidant capacity by scavenging singlet oxygen and free radicals (Sachindra

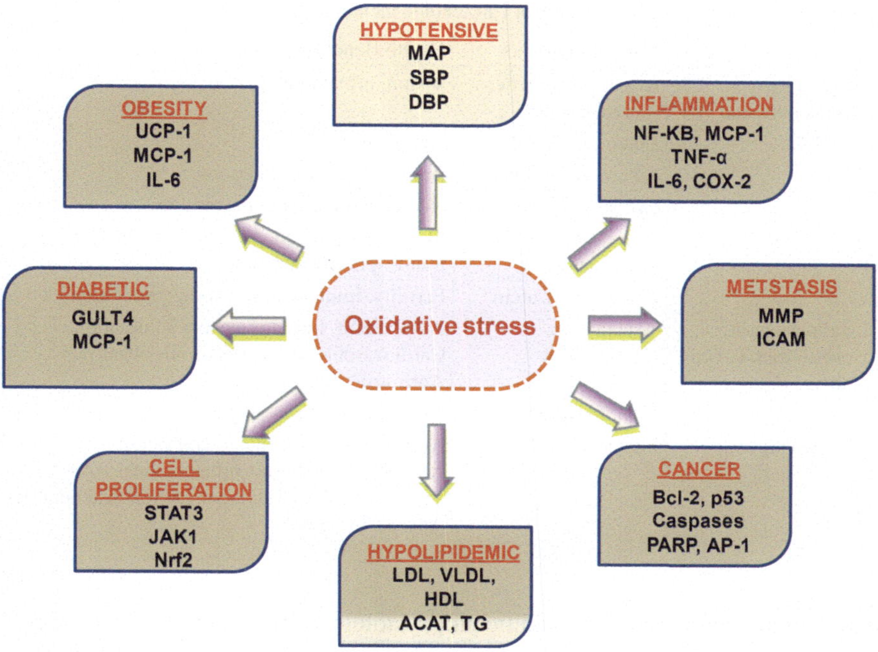

Fig. 29.3 Schematic representation of major transcription factors modulated by reactive oxygen species

et al. 2007). However, compared to other carotenoids, fucoxanthin is believed to be slightly stable to extreme conditions (light, heat, and atmospheric oxygen), as its exposure can degrade quickly (Zhao et al. 2014). As carotenoids are good scavengers of free radicals, however, their capability depends on the presence or absence of specific functional groups in their terminal rings (Miller et al. 1996). Liu et al. (2011) investigated the fucoxanthin's role in reducing the intracellular levels of ROS, lipid peroxides, and protein carbonyl in ferric nitrilotriacetate induced murine embryonic hepatic cells. Zeng et al. (2018) demonstrated the protective effect of fucoxanthin and its metabolite(s) on lowered lipid peroxidation levels in HepG2 cells treated with tributyltin. Besides, fucoxanthin is shown to enhance the endogenous antioxidant defense mechanism by increasing enzymes involved in synthesizing glutathione in the cells (Zheng et al. 2014).

4 Health Benefits of Carotenoids Including Fucoxanthin

The potential contribution of carotenoids on human health benefits is summarized in Table 29.1. Generally, the primary health benefits of carotenoids are explained by their antioxidant mechanism. Apart from these, specific carotenoids may perform a

Table 29.1 General health benefits of carotenoids including fucoxanthin

Sources	Natural pigments	Health Benefits
Green leafy vegetables and fruits	α- and β-Carotene, β-cryptoxanthin	Vision, provitamin A
	Lutein, Zeaxanthin	Eye health
	Lutein, β-carotene	Brain – cognitive functions
	Lycopene	Heart health, Cancer prevention
	Lycopene, β- carotene	Skin – UV protection, Genomic effects on transcription/translation
	β-carotene, Lutein	Fertility, Immune modulation/stimulation
Salmon, trout, crustacean, Green algae	Astaxanthin, Cantaxanthin	Antioxidant, Immune system stimulation, Cardiovascular protective, Anti- obesity, Anti-proliferative
Green algae	Siphonaxanthin	Anticancer, Anti-angiogenic
Brown algae	Fucoxanthin	Antioxidant, Anticancer, Anti-obesity, Neuroprotective, Anti-inflammatory. Anti-angiogenic, Phtoprotective, Prevent osteoporosis Anti-proliferative

unique biological function by additional physiological mechanisms. For example, β-carotene has added benefits due to biological ability to convert into vitamin A molecule, while lutein and zeaxanthin absorb specific wavelengths of light which could help to protect the eyes from photo-toxicity (Barker et al. 2011). Carotenoids may guard against certain types of cancer by limiting the abnormal growth of cells by enhancing gap-junction communication. In addition, carotenoids are involved in preventing coronary heart disease by the inhibition of the formation and oxidation of low-density lipoprotein (Iwamoto et al. 2000). In this regard, the EAT-Lancet Commission is planning to implement a feasible global food system that can deliver a healthy sustainable diet to a world population and is targeted to reach 9 billion by 2050 (Hirvonen et al. 2020). Consequently, identification of such compounds from the edible sources, which are performing a stable and superior activity in the cellular vicinity may support further development of a potent functional ingredient or specialty -foods.

4.1 Anti-inflammatory Effect of Fucoxanthin

The sources of inflammation are widespread and may be due to various factors including microbial and viral infections, exposure to radiation, sensitivity to allergens, bio-magnifications of toxic chemicals, autoimmune diseases, obesity, smoking, consumption of alcohol and high-calorie dietary habits (Schetter et al. 2010). In general, two stages of inflammation exist, such as acute and chronic inflammation. Acute inflammation (innate immunity) is a preliminary stage of inflammation, which is mediated by the activation of the immune system. This kind of inflammation persists only for a short period and is beneficial for the host. If the inflammation resides for a longer period, it may lead to various illnesses of the host. Chronic inflammation can lead to diabetes, cardiovascular, pulmonary, neurological, age-related degenerative diseases and cancers (Lin and Karin 2007).

Alternatively, inflammatory cells also produce soluble mediators, such as metabolites of arachidonic acid, cytokines, and chemokines, which act by further recruiting inflammatory cells to the site of damage and producing more reactive species. These key factors or mediators, activate signal transduction cascades and inducing changes in transcription factors, such as nuclear factor κB (NF-κB), hypoxia-inducible factor-1α (HIF-1α), signal transducer and activator of transcription 3 (STAT3), activator protein-1 (AP-1), nuclear factor of activated T cells, and NF-E2 related factor-2 (Nrf2). Induction of cyclooxygenase-2 and inducible nitric oxide synthase (iNOS), expression of tumor necrosis factor-α (TNF-α), interleukin (IL-1, IL-6, IL-8) and chemokines, and alterations in the specific micro RNAs expression are documented in oxidative stress-induced inflammation (Sedgera and McDermott 2014) (Figs. 29.3 and 29.4). This sustained inflammatory and oxidative environment lead to a vicious condition, further which can damage healthy neighboring epithelial and stromal cells, and consequently may responsible for carcinogenesis (Federico et al. 2007). Control and development of cancers are associated with the expression of pro-inflammatory gene products mainly regulated by transcription factor NF-kB, which is constitutively active in most tumors and is induced by carcinogens (such as cigarette smoke), tumor promoters, carcinogenic viral proteins, chemotherapeutic agents, and γ-irradiation. Therefore, regulation of NF-κB or NF-κB expression by anti-inflammatory agents is considered to be important in both the prevention and treatment of cancer.

Overall available research report provides the strong link between chronic inflammation and the development of chronic diseases (Fig. 29.4). Further, inflammatory biomarkers as described can be used to monitor or diagnosis the progression of the disease. Also, these biomarkers can be exploited to develop new anti-inflammatory drugs either synthetic or natural compounds including dietary components to prevent the onset of several chronic diseases, particularly cancer. Currently, drugs used as an adjuvant in chemotherapy and radiotherapy, by themselves activate NF-κB and mediate resistance (Ahmed et al. 2006). Numerous natural anti-inflammatory agents have been shown to demonstrate chemoprevention actions (Aggarwal and Shishodia 2006; Aggarwal et al. 2006). Therefore, compounds explored from natural origin can be used not only for prevention but also to treat cancer. The lack of toxicity associated with the natural agents vs. synthetic anti-inflammatory drugs with secondary complications targeting anti-inflammation and antiproliferation of cancer cells is off our current interest to explore additional advantages. Therefore, regulation of NF-κB or NF-κB expression by anti-inflammatory agents is considered to be important in both the prevention and treatment of cancer.

Among natural products or compounds, the influence of carotenoids was extensively correlated against inflammatory disease from the past two decades which is mainly due to accumulation of carotenoids in human tissues and circulated in the

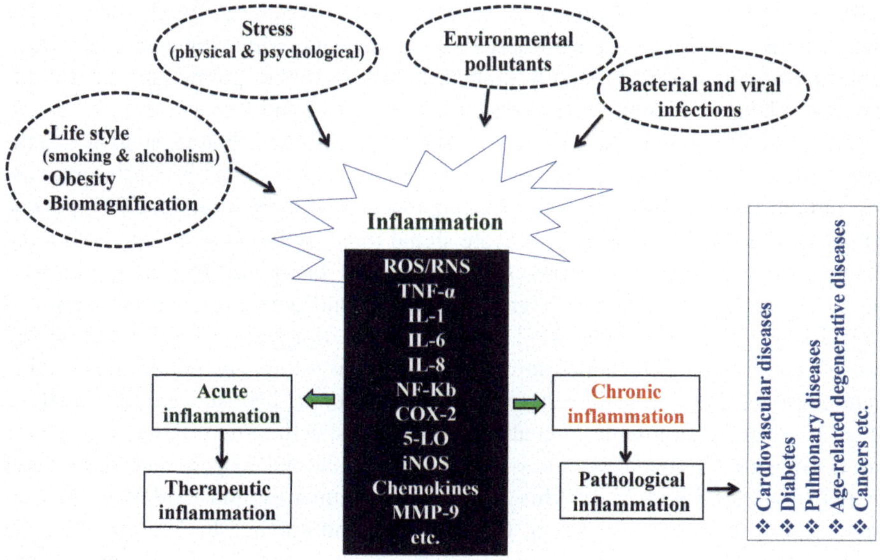

Fig. 29.4 Schematic representation of different factors involved in inflammation

blood via dietary consumption of fruits, vegetables, and algal sources (Khachik et al. 1997; Scarmo et al. 2010; Eggersdorfer and Wyss 2018). Epidemiological and clinical studies have demonstrated that regular dietary intake of carotenoid-rich food reduced the risk of atherosclerosis, cancers, and age-related degenerative diseases. In this context, many notable studies have correlated the plasma carotenoids levels and decreased risk of oxidative stress and mediated inflammatory responses (Thomson et al. 2007; Hozawa et al. 2007; Butalla et al. 2012). Liu et al. (2020) summarized the protective effect of fucoxanthin and its related molecular mechanism as a promising compound for inflammation-allied diseases in humans.

4.2 *Anti-cancer Effect of Fucoxanthin*

Cancer is the second leading cause of death occurring globally. Moreover, the most common among them is lung cancer, followed by breast, colorectal, stomach, and liver cancer (Bray et al. 2018). The diagnosis of cancer at the early stage is essential to reduce the incidence of cancer-associated death. Among natural compounds, carotenoids are the most abundantly found in nature and have been explored due to possess vast potential as nutraceuticals (Ranga Rao et al. 2019). Moreover, there is a wealth of information relevant to anticancer research in natural compounds (Jansen et al. 2000), particularly the apoptosis-inducing properties of carotenoids. Hence, exploration of carotenoids from new sources would be an important approach to chemo-prevention and/or chemotherapy.

Currently, natural compounds are broadly used as an alternative strategy to treat various cancers, owing to fewer side effects. Though various natural compounds have been shown to possess anticancer activities, treatment is still being practiced with synthetic anticancer drugs due to a lack of evidence on the efficiency of natural compounds compared to chemically derived drugs (Chegaev et al. 2013). Therefore, exploration of potent natural compounds on efficient control of tumor proliferation is an active area of research among them are carotenoids (Sowmya et al. 2017). Epidemiological and clinical trials have suggested that carotenoids present in rich green leafy vegetables, fruits, edible marine seaweeds, and their supplements are associated with reduced risk of certain cancer (Amin et al. 2009; Eliassen et al. 2015; Fergusona et al. 2015; Bakker et al. 2016; Qiu et al. 2020). Moreover, studies have revealed the role of terrestrial carotenoids such as β-carotene (for breast cancer) (Gloria et al. 2014), lycopene (for prostate cancer) (Arathi et al. 2016), and lutein (mammary gland cancer) (Chew et al. 2003) in their inhibition of specific cancer cell proliferation through various mechanisms (Fig. 29.5).

Recently, fucoxanthin and astaxanthin have been demonstrated to possess bioactivity against acute inflammation and tumor growth (Islam et al. 2013; Mei et al. 2017; Méresse et al. 2020). The chemoprevention of cancer by carotenoids may occur through modulation of intercellular oxidative status, signal transduction and enhancement of gap-junction communication (Siems et al. 2002; Palozza et al. 2004). In fact, studies have found that fucoxanthin and its metabolites, neoxanthin, canthaxanthin, and peridinin from marine sources induce apoptosis in cancer cells (Hosokawa et al. 2004; Konishi et al. 2006; Sugawara et al. 2007).

Anti-cancer effects of fucoxanthin are also mediated by cell cycle arrest, inducing apoptosis, and inhibiting metastasis. Fucoxanthin is demonstrated to reduce cancer cell proliferation in various cell models. The antiproliferation activity of fucoxanthin on cancer cell proliferation was controlled by increased p21, a cyclin-dependent kinase (CDK)-inhibitory protein, and decreased CDK-2/CDK-4, cyclin D1. Also, fucoxanthin induces apoptosis of cells by progressive expression of Bax with concomitant decreasing Bcl-2 expression and increased caspase-3 activity. Further, the influence of fucoxanthin on the induction of cancer cell death mediated through inhibition of phosphatidylinositol-3-kinase (PI3K)/Akt pathways in cancer cells (Xu et al. 2014; Liu et al. 2016; Yang et al. 2018). In addition, fucoxanthin induces apoptosis by reducing protein expression of an anti-apoptotic protein myeloid cell leukemia 1, and a transcription factor involved in apoptosis, signal transducer, and activator of transcription 3 (STAT3 in addition, the role of fucoxanthin on cancer cell death demonstrated in vivo models. Fucoxanthin dose of mice inhibited polyp formation and increased anoikis-like cells in colonic mucosa treated with colorectal tumor inducers (Terasaki et al. 2019). Additionally, the anti-cancer effect of fucoxanthin is mediated by inhibiting cancer cell migration or invasion of U87 and U251 cells *in vitro*, as measured by the scratch wound-healing assay. In these cells, fucoxanthin efficiently decreased matrix metalloproteinases 2 (MMP-2) and MMP-9 levels (Chung et al. 2013; Liu et al. 2016), as they play a vital role in tumor invasion and metastasis.

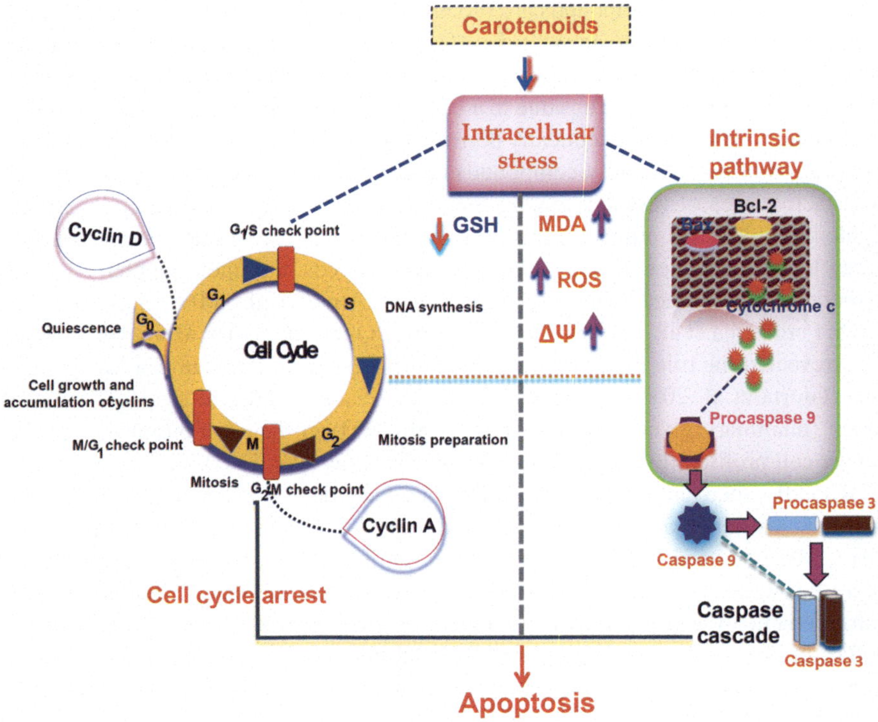

Fig. 29.5 Schematic representation of various mechanisms in apoptosisir

4.3 Anti-obesity Effect of Fucoxanthin

Fucoxanthin is a major marine carotenoid, which is distinct from other carotenoids consisting of epoxide functional groups. Dietary fucoxanthin metabolically converted to amarouciaxanthin A through fucoxanthinol in the gastrointestinal tract due to reaction with digestive enzymes (lipase and cholesterol esterase) and converted to amarouciaxanthin A in the liver by dehydrogenation/isomerization (Asai et al. 2004). The major metabolic products of fucoxanthin are shown in Fig. 29.6.

Amarouciaxanthin A is preferentially accumulate in the adipose tissue, whereas fucoxanthinol mounts up in the other organs and tissues (Hashimoto et al. 2009). Sangeetha et al. (2010) have reported various fucoxanthin metabolites besides the major metabolic products- fucoxanthinol and amarouciaxanthin A in rats. Further, they proposed a possible metabolic conversion pathway of fucoxanthin in the plasma and liver tissue samples. They hypothesized that these fucoxanthin metabolites are formed due to isomerization, demethylation, deacetylation, dehydrogenation, and oxidation. Therefore, it is assumed that fucoxanthin is believed to be the active form exerting various physiological functions. Amarouciaxanthin A is mainly stored in abdominal white adipose tissue, while fucoxanthinol circulates through the bloodstream and accumulates in erythrocytes, liver, lung, heart, spleen, adipose

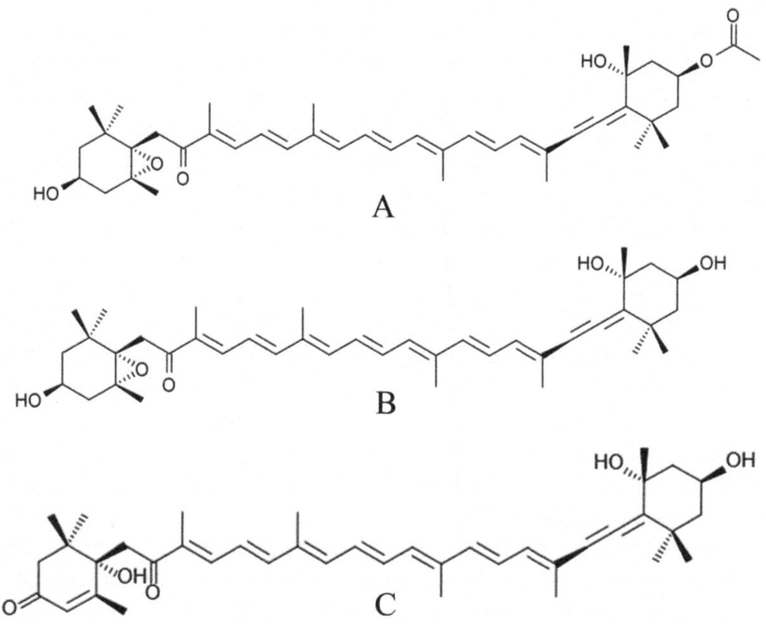

Fig. 29.6 Fucoxanthin (**a**) and its major metabolites; fucoxanthinol (**b**) and amarouciaxanthin A (**c**) found *in vivo* studies

tissue, and kidney. Epoxy-carotenoid-like fucoxanthin absorption rate is generally influenced by the dietary food matrix. Long-term inadequate diet and lifestyle may modify the lipid metabolism and increase visceral fat accumulation, thus resulting in obesity and associated metabolic disorders (diabetes mellitus, dyslipidemia, hypertension, and cardiovascular diseases).

Fucoxanthin appreciably minimizes the accumulation of triglyceride concentrations in plasma and liver plasma and hepatic tissues by favorably influencing cholesterol-regulatory enzymes (3-hydroxy-3-methylglutaryl coenzyme A reductase and acyl-coenzyme A) (Hu et al. 2012). Also, fucoxanthin is constructively affected by gene expression associated with lipid metabolism. In rats, supplementation of fucoxanthin decreased mRNA expression of hepatic Acetyl-CoA carboxylase that catalyzes the irreversible carboxylation of acetyl-CoA to produce malonyl-CoA. The increased production of this enzyme up-regulate the metabolism of the fatty acid.

Besides, fucoxanthin increased high-density lipoprotein in KK-Ay mice (a knock-out mouse model for type 2 diabetics that exhibits dyslipidemia, marked obesity, and hypertension) by inducing sterol regulatory element-binding protein expression and reduced cholesterol uptake in the liver via down-regulation of low-density lipoprotein -receptor (LDL-R) and scavenger receptor class B1 protein. Beppu et al. (2012) suggested that dietary fucoxanthin modulates protein convert as a subtilisin expression, which stimulates the degradation of LDL-R intracellularly in lysosomes. Also, supplementation of fucoxanthin decreased expression of fatty acid synthase; a multi-enzyme protein catalyzes fatty acid synthesis, mainly a synthesis of palmitate into long-chain saturated fatty acids from acetyl malonyl-CoA. Subsequently, in another study, they demonstrated down-regulation of stearoyl-coenzyme A desaturase-1 (SCD1) implicated in reducing obesity and related health problems. Also, leptin levels in serum significantly decreased in hyperleptinemia KK-A(y) mice fed with fucoxanthin; however, its effect was suppressive on hepatic SCD1 and gain in body weight (Beppu et al. 2013).

Further, fucoxanthin activates the enzyme glucose-6-phosphate dehydrogenase in the pentose-phosphate pathway, which supplies energy to the cells with maintaining the nicotinamide adenine dinucleotide phosphate (NADPH) level. This helps retain appropriate glutathione levels in the cells and supports the prevention of oxidative insult (Aster et al. 2010). Ferre and Foufelle (2010) revealed that nutrients like fucoxanthin influenced Acyl-CoA cholesterol acyltransferase and hydroxy-3-methylglutaryl-coenzyme A, as well as sterol regulatory element-binding transcription factors related to synthesis or production of lipid and cholesterol. Likewise, the expression of lecithin-cholesterol acyltransferase (an enzyme that converts free cholesterol into cholesteryl ester) and carnitinepalmitoyl-transferase (CPT1) considerably increased, followed by the fucoxanthin supplement. The synthesized cholesteryl ester is then confiscating into the core of a lipoprotein, high-density lipoprotein (DeVries et al. 2004). In these metabolic disorders, the level of free fatty acid gets elevated and accumulated in skeletal muscle tissue; this leads to a decrease in the ability of muscles to oxidize fatty acids. The elevated levels of malonyl-CoA caused by hyperglycemia, inhibited the CPT1. This situation decreases long-chain fatty acids transport into muscle and heart mitochondria, thus declining fatty acid oxidation thereby increasing FFA levels and the accumulation of fat in skeletal muscle. In this regard, up-regulation of CPT1 by fucoxanthin plays a vital role in reducing the incidence of these symptoms (Rasmussen et al. 2002).

In addition, fucoxanthin metabolite down-regulate PPARγ and exhibited suppress adipocyte differentiation efficiently than its parent molecule (Maeda et al. 2009). Likewise, others have shown that amarouciaxanthin A suppress PPARγ and C/EBPα expression during adipocyte differentiation, as compared to fucoxanthinol, amarouciaxanthin A markedly down-regulated the expression of adipocyte fatty acid-binding protein, lipoprotein lipase, and glucose-transporter 4 (Yim et al. 2011). Besides, many studies attempted to elucidate the influence of various factors on fucoxanthin on obesity-related health problems and their consequences.

4.4 Anti-diabetic Effect of fucoxanthin

Diabetes mellitus is an intricate metabolic disorder closely allied to poor dietary adaptation, altered lifestyle, and obesity, characterized by elevated blood glucose levels, primarily due to insulin resistance developed by excessive obesity (Zhang et al. 2015). Several chronic health complications (cardiovascular diseases, kidney disease, and blindness) are closely associated with the person due to Diabetes mellitus condition (Forbes and Cooper 2013). Therefore, dietary modification is considered crucial to prevent or treat diabetes complications. Maeda et al. (2015) have shown the anti-diabetic effect of fucoxanthin by regulating the insulin signaling pathway in obese/diabetic KK-Ay mice model. Also, the consumption of fucoxanthin showed significantly reduced blood glucose levels and concomitant with decreased expression of pro-inflammatory genes (TNFα and MCP-1) in adipose tissue. Further, this anti-diabetic effect of fucoxanthin was correlated with improved translocation of glucose transporter 4 to the cell membrane in muscle, indicated the insulin sensitivity against fucoxanthin treatment in mice. Nishikawa et al. (2012) reveal the influence of fucoxanthin on insulin receptor expression and Akt phosphorylation in muscles. In addition, Maeda et al. (2007) showed that adequate intake of fucoxanthin for four weeks reduced the leptin and tumor necrosis factor α levels in white adipose tissue, which further supports increasing insulin resistance depletion of glucose levels in the blood. Zhang et al. (2018) revealed that supplements of fucoxanthin reduced the fasting blood glucose level and improved intraperitoneal glucose-insulin tolerances. Woo et al. (2010) observed the role of fucoxanthin on insulin receptor substrate 1/PI3K/Akt and AMPK signaling pathways in the liver and skeletal muscle. Likewise, fucoxanthin markedly reduced the blood glucose and hemoglobin levels allied with a decrease in insulin and resistin levels in plasma samples of mice model with diet-induced obesity. Recently, many studies have suggested the health benefits of brown seaweeds and its pigment constituents as right functional food or nutraceuticals for natural diabetes therapy (Miyashita et al. 2020; Oliyaei et al. 2021). In addition, others have investigated the effect of fucoxanthin-rich wakame lipid extract on high-fat diet-induced mice models. This study resulted in markedly decreased insulin and glucose levels in plasma with concomitant increase in expression of glucose transporter 4 and β3-adrenergic receptor; however, monocyte chemo attractant protein-1 expression decreased (Maeda et al. 2009). These notable studies have strongly suggested the anti-diabetic activity of fucoxanthin.

5 Conclusion

Overall several studies have positively illustrated the potential health benefits of marine carotenoid- fucoxanthin on the reduction of chronic diseases, such as cancer, obesity, and diabetes mellitus. Therefore, carotenoids, including fucoxanthin and astaxanthin, are exploited as potential therapeutic agents for the treatment of diseases. In these contexts, experimental results have strongly confirmed the role of fucoxanthin on various biochemical and molecular mechanisms involved in the reduction of chronic diseases in various cells and animal models. However, studies on the safety of fucoxanthin intake in humans are still not well detailed. Furthermore, epidemiological and clinical trials are needed to exploit its potential health benefits.

Acknowledgments RL, KV and RA thanks Department of Microbiology and Biotechnology, Jnana Bharathi Campus, Bangalore University, Bengaluru for their constant support and also providing facilities. ARR acknowledge the FIST Project No. LSI-576/2013 from the Department of Science and Technology, Government of India and Centre of Excellence, Department of Biotechnology, Vignan's Foundation for Science, Technology and Research for the facilities and also support. GAR thanks Dr. Prema Chandra Sagar, Vice Chairman, Dayananda Sagar Institutions, and Pro-Chancellor, Dayananda Sagar University, Bengaluru, India, for his encouragement and support.

References

Agamey AE, Lowe GM, McGarveya DJ, Mortensenc A, Phillip DM, Truscott TG, Young AJ (2004) Carotenoid radical chemistry and antioxidant/pro-oxidant properties. Arch Biochem Biophys 430:37–48

Aggarwal BB, Shishodia S (2006) Molecular targets of dietary agents for prevention and therapy of cancer. Biochem Pharmacol 71:1397–1421

Aggarwal BB, Shishodia S, Sandur SK, Pandey MK, Sethi G (2006) Inflammation and cancer: How hot is the link? Biochem Pharmacol 72:1605–1621

Ahmed KM, Cao N, Li JJ (2006) HER-2 and NF-κB as the targets for therapy-resistant breast cancer. Anticancer Res 26:4235–4243

Amin ARM, Kucuk O, Kuhari FR (2009) Perspectives for cancer prevention with natural compounds. J Clin Oncol 27:2712–2725

Arathi BP, Sowmya PR, Vijay K, Baskaran V, Lakshminarayana R (2015) Metabolomics of carotenoids: The challenges and prospects-A review. Trends Food Sci Tech 45:105–117

Arathi BP, Sowmya PR, Kuriakose GC, Vijay K, Baskaran V, Jayabaskaran and C, Lakshminarayana R. (2016) Enhanced cytotoxic and apoptosis inducing activity of lycopene oxidation products in different cancer cell lines. Food Chem Toxicol 97:265–276

Arathi BP, Sowmya PR, Vijay K, Baskaran V, Lakshminarayana R (2017) Progress in enrichment and metabolic profiling of diverse carotenoids in tropical fruits: Importance of hyphenated techniques. In: Noureddine B (ed) Modern Biotechnologies and phytonutritional improvement of crops. Wiley, pp 271–307

Asai A, Sugawara T, Ono H, Nagao A (2004) Biotransformation of fucoxanthinol into amarouciaxanthin A in mice and HepG2 cells: formation and cytotoxicity of fucoxanthin metabolites. Drug Metab Dispos 32:205–211

Aster J, Kumar V, Robbins SL, Abbas AK, Fausto N, Cotran RS (2010) Robbins and cotran pathologic basis of disease, vol 33, 8th edn. Saunders/Elsevier, Philadelphia, PA, pp 340–341

Bakker MF, Peeters PH, Klaasen VM et al (2016) Plasma carotenoids, vitamin C, tocopherols, and retinol and the risk of breast cancer in the European prospective investigation into cancer and nutrition cohort. Am J Clin Nutr 103(2):454–464

Barker FM, Snodderly DM, Johnson EJ, Schalch W, Koepcke W, Gerss J, Neuringer M (2011) Nutritional manipulation of primate retinas, V: effects of lutein, zeaxanthin, and n-3 fatty acids on retinal sensitivity to blue-light-induced damage. Invest Ophthalmol Vis Sci 52:3934–3942

Beppu F, Hosokawa M, Niwano Y, Miyashita K (2012) Effects of dietary fucoxanthin on cholesterol metabolism in diabetic/obese KK-A(y) mice. Lipids Health Dis 11:112

Beppu F, Hosokawa M, Yim MJ, Shinoda T, Miyashita K (2013) Down-regulation of hepatic stearoyl-CoA desaturase-1 expression by Fucoxanthin via leptin signaling in diabetic/obese KK-Ay mice. Lipids 48:449–455

Botella-Pavía P, Rodríguez-Concepción M (2006) Carotenoid biotechnology in plants for nutritionally improved foods. Physiol Plant 126:369–381

Bray F, Ferlay J, Soerjomataram I, Siegel RL, Torre LA, Jemal A (2018) Global cancer statistics 2018: GLOBOCAN estimates of incidence and mortality worldwide for 36 cancers in 185 countries. CA Cancer J Clin 68:394–424

Britton G (1995) Structure and properties of carotenoids in relation to function. FASEB J 9:1551–1558

Britton G, Liaaen-Jensen S, Pfander H (2009) Carotenoids volume-5: nutritionand health. Birkhauser Basel, Springer Nature, Cham, 431 p

Burton GW, Ingold KU (1984) β-carotene: an unusual type of lipid antioxidant. Science 224:569–573

Butalla AC, Crane TE, Patil B, Wertheim BC, Thompson P, Thomson CA (2012) Effects of a carrot juice intervention on plasma carotenoids, oxidative stress, and inflammation in overweight breast cancer survivors. Nutr Cancer 64:331–341

Chegaev K, Rignati C, Rolando B, Lazzarato L, Gazzano E, Guglielmo S, Ghigo D, Fruttero R, Gascoa A (2013) Doxorubicin-antioxidant co-drugs. Bioorganic Med Chem Lett 23:5307–5310

Chew BP, Brown CM, Park JS, Mixter PF (2003) Dietary lutein inhibits mouse mammary tumor growth by regulating angiogenesis and apoptosis. Anticancer Res 23:3333–3339

Chung TW, Choi HJ, Lee JY, Jeong HS, Kim CH, Joo M, Choi JY, Han CW, Kim SY, Choi JS, Ha KT (2013) Marine algal fucoxanthin inhibits the metastatic potential of cancer cells. Biochem Biophys Res Commun 439:580–585

DeVries R, Borggreve SE, Dullaart RP (2004) Role of lipases, lecithin: Cholesterol acyltransferase and cholesteryl ester transfer protein in abnormal high density lipoprotein metabolism in insulin resistance and type 2 diabetes mellitus. Clin Lab 49:601–613

Eggersdorfer M, Wyss A (2018) Carotenoids in human nutrition and health. Arch Biochem Biophys 15:18–26

Eliassen AH, Liao X, Rosner B, Tamimi RM, Tworoger SS, Hankinson SE (2015) Plasma carotenoids and risk of breast cancer over 20 y of follow-up. Am J Clin Nutr 101:1197–1205

Federico A, Morgillo F, Tuccillo C, Ciardiello F, Loguercio C (2007) Chronic inflammation and oxidative stress in human carcinogenesis. Int J Cancer 121:2381–2386

Fergusona LR, Chenb H, Collins AR et al (2015) Genomic instability in human cancer: molecular insights and opportunities for therapeutic attack and prevention through diet and nutrition. Semin Cancer Biol 35:S5–S24

Ferre P, Foufelle F (2010) Hepatic steatosis: A role for *de novo* lipogenesis and the transcription factor SREBP-1c. Diabetes Obes Metab 12:83–92

Forbes JM, Cooper ME (2013) Mechanisms of diabetic complications. Physiol Rev 93:137–188

Fraser PD, Bramley PM (2004) The biosynthesis and nutritional uses of carotenoids. Prog Lipid Res 43:228–265

Gloria NF, Soares N, Brand C, Oliveira FL, Borojevic R, Teodoro AJ (2014) Lycopene and β-carotene induce cell cycle arrest and apoptosis in human breast cancer cell lines. Anticancer Res 34:1377–1386

Gorrini C, Harris IS, Mak TW (2013) Modulation of oxidative stress as an anticancerstrategy. Nat Rev Drug Discov 12(12):931–947

Gruszecki WI (2010) Carotenoids in lipid membranes. In: Landrum JT (ed) Carotenoids, physical, chemical, and biological functions and properties. Taylor and Francis Group, New York, pp 19–30

Gruszecki WI, Strzalka K (2005) Carotenoids as modulators of lipid membrane physical properties. Biochim Biophys Acta 1740:108–115

Hashimoto T, Ozaki Y, Taminato M, Das SK, Mizuno M, Yoshimura K, Maoka T, Kanazawa K (2009) The distribution and accumulation of fucoxanthin and its metabolites after oral administration in mice. Br J Nutr 102:242–248

Hirvonen K, Bai Y, Headey D, Masters WA (2020) Affordability of the EAT–Lancet reference diet: a global analysis. Lancet 8:e59–e66

Hosokawa M, Kudo M, Maeda H, Kohno H, Tanaka T, Miyashita K (2004) Fucoxanthin induces apoptosis and enhances the antiproliferative effect of the PPARγ ligand, troglitazone, on colon cancer cells. Biochim Biophys Acta 1675:113–119

Hozawa A, Jacobs DR, JrSteffes MW, Gross MD, Steffen LM, Lee DH (2007) Relationships of circulating carotenoid concentrations with several markers of inflammation, oxidative stress, and endothelial dysfunction: The coronary artery risk development in young adults (CARDIA)/ young adult longitudinal trends in antioxidants (YALTA) study. Clin Chem 53:447–455

Hu X, Li Y, Li C, Fu Y, Cai F, Chen Q, Li D (2012) Combination of fucoxanthin and conjugated linoleic acid attenuates body weight gain and improves lipid metabolism in high-fat diet-induced obese rats. Arch Biochem Biophys 519:59–65

Hussein G, Sankawa U, Goto H, Matsumoto K, Watanabe H (2006) Astaxanthin, a carotenoid with potential in human health and nutrition. J Nat Product 69:443–449

Islam MD, Ishita IJ, Jin SE, Choi RJ, Lee CM, Kim YS, Jung HA, Choi SJ (2013) Anti- inflammatory activity of edible brown alga *Saccharina japonica* and its constituents pheophorbide a and pheophytina in LPS-stimulated RAW 264.7 macrophage cells. Food Chem Tox 55:541–548

Iwamoto T, Hosoda K, Hirano R, Kurata H, Matsumoto A, Miki W, Kamiyama M, Itakura H, Yamamoto S, Kondo K (2000) Inhibition of low-density lipoprotein oxidation by astaxanthin. J Atheroscler Thromb 7:216–222

Jansen B, Wacheck V, Heere-Ress E, Schlagbauer-Wadl H, Hoeller C, Lucas T, Hoermann M, Hollenstein U, Wolff K, Pehamberger H (2000) Chemosensitisation of malignant melanoma by Bcl-2 antisense therapy. Lancet 356:1728–1733

Kaulmann A, Bohn T (2014) Carotenoids, inflammation, and oxidative stress-implications of cellular signaling pathways and relation to chronic disease prevention. Nutr Res 34:907–929

Khachik F, Spangler CJ, Smith JC, Canfield LM, Steck A, Pfander H (1997) Identification, quantification, and relative concentrations of carotenoids and their metabolites in human milk and serum. Anal Chem 69:1873–1881

Kim SM, Jung YJ, Kwon ON, Cha KH, Um BH, Chung D, Pan CH (2012) A potential commercial source of fucoxanthin extracted from the microalga *Phaeodactylumtricornutum*. Appl Biochem Biotechnol 166:1843–1855

Konishi I, Hosokawa M, Sashima T, Kobayashi H, Miyashita K (2006) Halocynthiaxanthin and fucoxanthinol isolated from *Halocynthiaroretzii*induce apoptosis in human leukemia, breast and colon cancer cells. Comp Biochem Physiol C 142:53–59

Lakshminarayana R, Aruna G, Sathish V, Shylaja MD, Baskaran V (2013) Structural elucidation of possible lutein oxidation products mediated through peroxyl radical inducer 2,2′-Azobis (2 methylpropionamidine) dihydrochloride: antioxidant and cytotoxic influence of oxidized lutein in HeLa cells. Chem Biol Interact 203:448–455

Lin WW, Karin MA (2007) Cytokine-mediated link between innate immunity, inflammation, and cancer. J Clin Invest 117:1175–1183

Liu CL, Chiu YT, Hu ML (2011) Fucoxanthin enhances HO-1 and NQO1 expression in murine hepatic BNL CL.2 cells through activation of the Nrf2/ARE system partially by its pro-oxidant activity. J Agric Food Chem 59:11344–11351

Liu Y, Zheng J, Zhang Y, Wang Z, Yang Y, Bai M, Dai Y (2016) Fucoxanthin activates apoptosis via inhibition of PI3K/Akt/mTOR pathway and suppresses invasion and migration by restriction of p38-MMP-2/9 pathway in human glioblastoma cells. Neurochem Res 41:2728–2751

Liu M, Li W, Chen Y, Wan X, Wang J (2020) Fucoxanthin: A promising compound for human inflammation-related diseases. Life Sci 255:117850

Maeda H, Hosokawa M, Sashima T, Miyashita K (2007) Dietary combination of fucoxanthin and fish oil attenuates the weight gain of white adipose tissue and decreases blood glucose in obese/ diabetic KK-Ay mice. J Agric Food Chem 55:7701–7706

Maeda H, Hosokawa M, Sashima T, Murakami-Funayama K, Miyashita K (2009) Anti-obesity and anti-diabetic effects of fucoxanthin on diet-induced obesity conditionsin a murine model. Mol Med Rep 2:897–902

Maeda H, Kanno S, Kodate M, Hosokawa M, Miyashita K (2015) Fucoxanthinol, metabolite of fucoxanthin, improves obesity-induced inflammation in adipocyte cells. Mar Drugs 13:4799–4813

Matsuno T (2001) Aquatic animal carotenoids. Fisher Sci 67:771–783

Mei CH, Zhou SC, Zhu L, Ming JX, Zeng FD, Xu R (2017) Antitumor effects of laminaria extract fucoxanthin on Lung Cancer. Mar Drugs 15(2):39

Méresse S, Fodil M, Fleury F, Chénais B (2020) Fucoxanthin, a marine-derived carotenoid from brown seaweeds and microalgae: A promising bioactive compound for cancer therapy. Int J Mol Sci 21:9273

Miller NJ, Sampson J, Candeias LP, Bramley PM, Rice-Evans CA (1996) Antioxidant activities of carotenes and xanthophylls. FEBS Lett 384:240–242

Miyashita K, Beppu F, Hosokawa M, Liu X, Wang S (2020) Nutraceutical characteristics of the brown seaweed carotenoid fucoxanthin. Arch Biochem Biophy 686:108364

Nishikawa S, Hosokawa M, Miyashita K (2012) Fucoxanthin promotes translocation and induction of glucose transporter 4 in skeletal muscles of diabetic/obese KK-Ay mice. Phytomedicine 19:389–394

Oliyaei N, Moosavi-Nasab M, Tamaddon AM, Tanideh N (2021) Antidiabetic effect of fucoxanthin extracted from *Sargassumangustifolium* on streptozotocin-nicotinamide-induced type 2 diabetic mice. Food Sci Nutr 9:3521–3529

Palozza P, Serini S, Di Nicuolo F, Calviello G (2004) Modulation of apoptotic signalingbycarotenoidsincancercells. Arch Biochem Biophys 430:104–109

Peng J, Yuan JP, Wu CF, Wang JH (2011) Fucoxanthin, a marine carotenoid present in brown seaweeds and diatoms: metabolism and bioactivities relevant to human health. Mar Drugs 9:1806–1828

Qiu Y, Yu H, Zeng R, Guo S, Daniyal M, Deng Z, Wang W (2020) Recent development on anti-obesity compounds and their mechanisms of action: a review. Curr Med Chem 27:3577–3597

Ranga Rao A, Deepika G, Ravishankar GA, Sarada R, Narasimharao BP, Bo L, Su Y (2019) Industrial potential of carotenoid pigments from microalgae: Current trends and future prospects. Crit Rev Food Sci Nutr 59(12):1880–1902

Rasmussen BB, Holmbäck UC, Volpi E, Morio-Liondore B, Paddon-Jones D, Wolfe RR (2002) Malonyl coenzyme A and the regulation of functional carnitine palmitoyltransferase-1 activity and fat oxidation in human skeletal muscle. J Clin Invest 110:1687–1693

Rodriguez-Concepcion M, Forés O, Martinez-Garcia JF, Gonzalez V, Phillips MA, Ferrer A, Boronat A (2004) Distinct light-mediated pathways regulate the biosynthesis and exchange of isoprenoid precursors during Arabidopsis seedling development. Plant Cell 16:144–156

Sachindra NM, Sato E, Maeda H, Hosokawa M, Niwano Y, Kohno M, Miyashita K (2007) Radical scavenging and singlet oxygen quenching activity of marine carotenoid fucoxanthin and its metabolites. J Agric Food Chem 55:8516–8522

Sangeetha RK, Bhaskar N, Divakar S, Baskaran V (2010) Bioavailability and metabolism of fucoxanthin in rats: structural characterization of metabolites by LC-MS (APCI). Mol Cell Biochem 333:299–310

Scarmo S, Cartmel B, Lin H, Leffell DJ, Welch E, Bhosale P, Bernstein PS, Mayne ST (2010) Significant correlations of dermal total carotenoids and dermal lycopene with their respective plasma levels in healthy adults. Arch Biochem Biophys 504:34–39

Schetter A, Heegaard NH, Harris CC (2010) Inflammation and cancer: interweaving microRNA, free radical, cytokine and p53 pathways. Carcinogenesis 31:37–49

Sedgera LM, McDermott MF (2014) TNF and TNF-receptors: From mediators of cell death and inflammation to therapeutic giants - past, present and future. Cytokine Growth Factor Rev 25:453–472

Siems W, Sommerburg O, Schild L, Augustin W, Langhans CD, Wiswedel I (2002) β- carotene cleavage products induce oxidative stress *in vitro* by impairing mitochondrial respiration. FASEB J 16:1289–1291

Sowmya PR, Arathi BP, Vijay K, Baskaran V, Lakshminarayana R (2017) Astaxanthin from shrimp efficiently modulates oxidative stress and allied cell death progression in MCF-7 cells treated synergistically with β-carotene and lutein from greens. Food Chem Toxicol 106:58–69

Sugawara T, Yamashita K, Sakai S, Asai A, Nagao A, Shiraishi T, Imai I, Hirata T (2007) Induction of apoptosis in DLD-1 human colon cancer cells by peridinin isolated from the *dinoflagellate, Heterocapsatriguetra*. Biosci Biotechnol Biochem 71:1069–1072

Terasaki M, Iida T, Kikuchi F et al (2019) Fucoxanthin potentiates anoikis in colon mucosa and prevents carcinogenesis in AOM/DSS model mice. J Nutr Biochem 64:198–205

Thomson CA, Stendell-Hollis NR, Rock CL, Cussler EC, Flatt SW, Pierce JP (2007) Plasma and dietary carotenoids are associated with reduced oxidative stress in women previously treated for breast cancer. Cancer Epidemiol Biomark Prev 16:2008–2015

Wang LJ, Fan Y, Parsons RL, Hu GR, Zhang PY, Li FL (2018) A rapid method for the determination of fucoxanthin in diatom. Mar Drugs 16:33

Woo MN, Jeon SM, Shin YC, Lee MK, Kang MA, Choi MS (2009) Anti-obese property of fucoxanthin is partly mediated by altering lipid-regulating enzymes and uncoupling proteins of visceral adipose tissue in mice. Mol Nutr Food Res 53:1603–1611

Woo MN, Jeon SM, Kim HJ, Lee MK, Shin SK, Shin YC, Park YB, Choi MS (2010) Fucoxanthin supplementation improves plasma and hepatic lipid metabolism and blood glucose concentration in high-fat fed C57BL/6N mice. Chem Biol Interact 186:316–322

Xu C, Wang Q, Feng X, Bo Y (2014) Effect of evodiagenine mediates photocytotoxicity on human breast cancer cells MDA-MB-231 through inhibition of PI3K/AKT/mTOR and activation of p38 pathways. Fitoterapia 99:292–299

Yang J, Pi C, Wang G (2018) Inhibition of PI3K/Akt/mTOR pathway by apigenin induces apoptosis and autophagy in hepatocellular carcinoma cells. Biomed Pharmacother 103:699–707

Yim MJ, Hosokawa M, Mizushina Y, Yoshida H, Saito Y, Miyashita K (2011) Suppressive effects of amarouciaxanthin A on 3T3-L1 adipocyte differentiation through down-regulation of PPAR-γ and C/EBPr mRNA expression. J Agric Food Chem 59:1646–1652

Zeng J, Zhang Y, Ruan J, Yang Z, Wang C, Hong Z, Zuo Z (2018) Protective effects of fucoxanthin and fucoxanthinol against tributyltin-induced oxidative stress in HepG2 cells. Environ Sci Pollut Res Int 25:5582–5589

Zhang H, Tang Y, Zhang Y, Zhang Z, Qu J, Wang X, Kong R, Han and Liu Z. (2015) Fucoxanthin: a promising medicinal and nutritional ingredient. Evid Based Compl Alter Med 2015:723515

Zhang Y, Xu W, Huang X, Zhao Y, Ren Q, Hong Z, Huang, and Xing X. (2018) Fucoxanthin ameliorates hyperglycemia, hyperlipidemia and insulin resistance in diabetic mice partially through irs-1/pi3k/akt and ampk pathways. J Funct Foods 48:515–524

Zhao D, Kim SM, Pan CH, Chung D (2014) Effects of heating, aerial exposure and illumination on stability of fucoxanthin in canola oil. Food Chem 145:505–513

Zheng J, Piao MJ, Kim KC, Yao CW, Cha JW, Hyun JW (2014) Fucoxanthin enhances the level of reduced glutathione via the Nrf2-mediated pathway in human keratinocytes. Mar Drugs 12:4214–4230

Chapter 30
Cosmeceuticals from Macrophyte Algae

Tatiana V. Puchkova, Sofia A. Khapchaeva, Vasily S. Zotov, Alexandr A. Lukyanov, Svetlana G. Vasilieva, Ambati Ranga Rao, Gokare A. Ravishankar, and Alexei E. Solovchenko

Abbreviations

DNA	Deoxyribonucleic acid
DW	dry weight
kDa	Kilodaltons
ROS	Reactive oxygen species
MAA	mycosporine-like amino acids
UV B	Ultraviolet B
UV A	Ultraviolet A

T. V. Puchkova · S. A. Khapchaeva · V. S. Zotov
Federal Research Centre "Fundamentals of Biotechnology" RAS, Moscow, Russia

A. A. Lukyanov · S. G. Vasilieva
Lomonosov Moscow State University, Moscow, Russia
e-mail: lukyanov@mail.bio.msu.ru

A. Ranga Rao
Department of Biotechnology, Vignan's Foundation of Science, Technology and Research Deemed to be University, Vadlamudi, Guntur, Andhra Pradesh, India

G. A. Ravishankar
Dr. C. D. Sagar Centre for Life Sciences, Dayananda Sagar College of Engineering, Dayananda Sagar Institutions, Bengaluru, Karnataka, India

A. E. Solovchenko (✉)
Lomonosov Moscow State University, Moscow, Russia

Derzhavin Tambov State University, Tambov, Russia
e-mail: solovchenko@mail.bio.msu.ru

A. Ranga Rao, G. A. Ravishankar (eds.), *Sustainable Global Resources of Seaweeds Volume 2*, https://doi.org/10.1007/978-3-030-92174-3_30

1 Introduction

Algae are a large group of oxygenic phototrophic microorganisms. They dwell in marine and freshwater environments serving an important ecosystem function as the primary producers (Richmond 2008; Masojídek et al. 2013; Puchkova et al. 2021). Macroalgae (seaweed) and microalgae possess flexible metabolic pathways yielding a plethora of bioactive molecules (Ravishankar and Ranga Rao 2019a, 2019b). Frequently, these molecules exert assorted beneficial effects on human health, particularly on skin condition (Rastogi et al. 2017; Thomas and Kim 2013; Kim 2011). Many metabolites synthesized by macrophyte algae exert anti-viral, anti-fungal, or anti-biotic effects (Masojídek et al. 2013; Rastogi et al. 2017; Mu et al. 2019; Coates et al. 2013). Considerable attention is drawn to the research on new natural biologically active substances from marine microalgae. This effort is also supported by the ever-increasing demand for "green" raw materials for pharmaceuticals and cosmetology (García et al. 2017; Kim 2011; Rastogi et al. 2017).

The biodiversity of algae is enormous, with more than 50,000 different species, and only a small part of these organisms has been studied (Richmond 2008; Barbosa and Roque 2019). Macroalgae represent an ample source of commercially relevant substances (Kijjoa and Sawangwong 2004), most of those are of direct interest for the production of cosmetics (Table 30.1). Those substances such as crude or purified pigments, polysaccharides, together with fatty acids are used in modern cosmetics formulations (Thomas and Kim 2013; Borowitzka 2013). Most of the bioactive substances of algae are represented by secondary metabolites that are induced to accumulate in algal cells under harsh environmental conditions (Mulders et al. 2014; Solovchenko 2013). Frequently, these metabolites feature a unique chemical structure that is not found in terrestrial organisms and hardly can be reproduced by chemical synthesis.

The world scientific community has rapidly accumulated extensive information on the chemistry and diverse effects of substances and metabolites of microalgae (Levine 2018; García et al. 2017; Coates et al. 2013). Many substances of microalgal origin found extensive use in the cosmetic industry, the list of extracts and individual chemicals isolated from them thoroughly tested for safety and effectiveness is not yet very large (Scott 2015). To complement recent reviews of individual cosmeceutical groups (Gong and Bassi 2016; Fox and Zimba 2018; Julius 2018; Mimouni et al. 2018; Morançais et al. 2018; Novoveska et al. 2019; Puchkova et al. 2021; Pangestuti et al. 2021), we covered certain important classes of such compounds from seaweed important for cosmeceutical use.

2 Polysaccharides

The bulk of the carbohydrates that make up algae are polysaccharides constituting up to 55% of their biomass dry weight (DW) (Rastogi et al. 2017; Briand 2003). Structural polysaccharides are found in seaweed in different combinations (Boney 1965). Most often, these are homopolysaccharides from D-mannose, D-galactose,

Table 30.1 Seaweed metabolites for cosmetics applications

Seaweeds	Metabolites	Properties	Reference
Saccharina longicruris	Galactofucan	Fibroblasts growth rate, synthesis of matrix Metalloproteinase and collagen-I	Rioux et al. (2013)
Ecklonia cava	Phlorotannins	Adipogenesis inhibitory effect	Ko et al. (2013)
Sargassum polycystum	Saponins, flavonoids, tannins, terpenoids, phenols, amino acids and amines	Anti-melanogenesis, skin-whitening effect	Song et al. (2009), Chan et al. (2011)
Corallina officinalis	Sulphated	Anti-oxidant	Yang et al. (2011)
Palmaria palmate	MAAs	Anti-UV	Yuan et al. (2009)
Porphyra haitanensis	Sulphated galactans	Anti-oxidant activity	Zhang et al. (2004)
Porphyra umbilicalis	MAAs	Anti-UVA	Carreto and Carignan (2011)
Ulva lactuca	Clay and polysaccharide, Lipopeptide, fatty acids, Tripeptide	Anti-elastase Collagen synthesis In vitro and in vivo Nrf2-ARE activation	Demais et al. (2007), Delaunay and Volle (2011), Wang et al. (2013), Guglielmo and Montanari (2008)
Fucus vesiculosus	Fucoidan	Anticoagulant	Ruperez et al. (2002)
Kjellmaniella crassifolia	Fucoidan	Antiaging, antiwrinkle	Mizutani et al. (2010)
Pelvetia canaliculata	Flavonoids, amino acids, fucoidans, polyols	Collagen synthesis, proteoglycans synthesis stimulation	Hupel et al. (2011)
Hijikia fusiformis	Fucoxanthin	*In vivo* inducer of the Nrf2-ARE	Liu et al. (2011)

and D-xylose forming branched and/or linear chains. They are part of both cell walls and mucus, as well as intercellular and intracellular colloids. Algal polysaccharides have unique physical and chemical properties and are widely used in the cosmetics industry (Rastogi et al. 2017; Scott 2015; Thibodeau and Lacasse 2006).

2.1 Alginic Acids

Alginic acids of algae, as well as alginates, are the main structural polysaccharides, are part of almost all brown algae. They are especially ample in various species of the genera *Laminaria*, *Macrocystis*, and *Fucus*—40%-50% of the dry weight

(Podkorytova et al. 2007; Mišurcová 2012). Usually, the content of alginic acids ranges from 10–40% DW, but in some species, it can be higher, depending on the type of algae, its growing conditions, and the season of harvest. Alginic acid is an irregular heteropolysaccharide with a molecular weight of 90 to 200 thousand, formed by two monomers of D-mannuronic and L-guluronic acids connected by β-1,4 bonds (Podkorytova et al. 2007). It is insoluble in water. The structure is variable, and the proportions of these acids can vary even within the same species. The degree of polymerization of native polymers is quite large and the number of monosaccharide residues can be from 1000 to 10,000. When interacting with cations, they form salts (alginates).

2.2 Alginates

Alginates are the predominant industrial product derived from seaweed. In the food industry, they are used as emulsifiers, stabilizers, gelling and moisture-retaining components. The global food industry consumes about 30% of the produced alginates (Bedoux et al. 2014). As an eco-friendly and renewable resource, alginates are becoming increasingly important. These are non-toxic compounds with colloidal properties, which is why they are increasingly demanded by the cosmetics, food, pharmaceutical, and textile industries 4, 32. Alginates are obtained from more than 300 species of brown algae. The main sources are large brown algae-representatives of Laminariaceae and Fucaceae: *Laminaria hyperborea*, *Macrocystis pyrifera* and *Ascophyllum nodosum*; to a lesser extent, *L. digitata*, *L. japonica*, *Ecklonia maxima*, *Lessonia nigrescens* and some *Sargassum* species (Mišurcová 2012; Rastogi et al. 2017).

2.3 Laminarin

Laminarin (laminaran) or algal starch is a β-glucan, a reserve substance of brown algae with a molecular weight of 3500–5000. It is a linear or weakly branched polysaccharide consisting of D-glucopyranose residues (Rastogi et al. 2017). In linear chains, 1–3 glycoside bonds are most common, branched chains contain the 1–6 bonds. Both the linear and branched forms of laminarin contain 2.4–3.7% mannitol and a small amount of D-mannose. Laminarin from different types of algae differs in its degree of branching, the content of mannitol, the number and the pattern of 1–6-bonds, as well as in molecular weight (Rastogi et al. 2017).

Laminarin is abundant (10–20% DW) in various species of *Laminaria*: *Laminaria japonica*, *L. fragilis*, *L. religiosa*, *L. angustata* and *Alaria*. In *Fucus*, the laminarin content is usually below 6–10% DW. The common source of "soluble" laminarin is *Laminaria digitata*, and "insoluble" form of laminarin is found in *Laminaria hyperborea* (Rastogi et al. 2017; Podkorytova et al. 2007). Laminarin is also found in diatoms (where it was termed as chrysolaminarin or leucosin), green algae, and euglena algae (termed paramylon) (Rastogi et al. 2017; Podkorytova et al. 2007). Its content in algae is subject to large seasonal fluctuations. Laminarin is obtained from algae by extraction with dilute acids, followed by its precipitation with alcohol.

Polysaccharides commonly named laminarin are colorless amorphous substances lacking smell and taste. They are weak reducing agents, relatively easily hydrolyzed by acids to glucose. The latter reaction is the basis of the method of laminarin assay in raw materials (Rastogi et al. 2017; Podkorytova et al. 2007). The 1,3- and 1,6-glucans including laminarins, are directly involved in maintaining the immunity of animals and plants. Numerous studies have shown that laminarins and their derivatives are efficient immunostimulants, radio- and cryoprotectors (Rastogi et al. 2017; Podkorytova et al. 2007). Beta-glucans of algae, in particular laminarin exhibit moisturizing, antioxidant, and anti-inflammatory properties and have become very popular in skin care products over the past decade (Ozanne et al. 2020).

2.4 *Fucans and Fucoidans*

Fucans and fucoidans were found in all known brown algae. They are sulfated linear heteropolysaccharides consisting of α-L-fucose residues esterified with sulfuric acid (Li et al. 2008; Podkorytova et al. 2007). The largest amount of these polysaccharides, up to 20% DW, is found in *Fucus* plants. In Laminariaceae, their content usually does not exceed 5% DW (Li et al. 2008; Foley et al. 2011). The molecular weight of fucoidans varies from 10,000 to 900,000 Da. Brown algae contain a mixture of fucoidan fractions of different molecular weights. Fractions of fucoidans with molecular weights of 18, 50, 130, and 850 kDa were isolated from *Ascophyllum nodosum*; 40, 50, 80, and 850 kDa fractions were found in *Pelvetia canaliculata*; the 11.45 and 450 kDa fractions are typical of *Fucus vesiculosus* (Li et al. 2008; Foley et al. 2011).

At present, the structure of individual fucoidans from brown algae *Chorda filum*, *Laminaria saccharina*, *Lessonia valdosa*, *Ascofillum nodosum* and several other species have been established. The well-known Japanese algae "wakame" (*Undaria pinnatifida*) contains fucoidan enriched in galactose and acetyl groups. Although very similar in composition, fucoidans of different origins differ significantly in their bioactivity, and no reliable correlations between their structure and biological activity have been established. The main reason is the complexity of mixtures of sulfated polysaccharides of different chemical nature and the irregular distribution of individual structural elements along the polymer chain (Li et al. 2008; Foley et al. 2011). Nevertheless, fucoidans are considered very promising polymers, almost devoid of toxic properties. In cosmetics, fucoidans are used as ingredients with a good moisturizing effect, a pronounced anti-inflammatory and regenerating effect. In recent years, fucoidans have also been used as a skin-lightening agent (Foley et al. 2011; Barel et al. 2014).

2.5 Agar-Agar

Agar-agar is a natural gelling agent obtained from various types of red seaweed. It consists of a mixture of agarose polysaccharides (a linear polymer constructed from galactose residues) and agaropectin, in which galactose residues form ether bonds with sulfuric acid residues. Hot solutions form dense gels after cooling. It has gelling and thickening properties, high moisture-absorbing ability. In cosmetics, it is used as an emulsifying, softening, and thickening component in the production of creams, gels, toothpaste, and in products for oily skin (Barel et al. 2014).

2.6 Carrageenan

Carrageenan is a linear sulfated polysaccharide of red algae formed by the monomers of D-galactose connected by regularly alternating β-1-4 and α-1-3 glycosidic bonds (Kim 2014). The hydroxyl groups are partially esterified with sulfuric acid, and the content of sulfuric acid residues is high, on average 20–30%. The chemical composition of carrageenan and the ratio between its forms depending on the type of algae. The molecular weight ranges from 100 to 1000 kDa. Different types of carrageenan can be obtained from different types of macroalgae depending on the growing site and conditions, as well as on the extraction technology (Kim 2014). It is discovered in the thalli of red seaweed (Rhodophyceae) from the genera *Chondrus*, *Gigartina*, *Mastocarpus,* and other gigartine algae. The carrageenan content can be as high as 50% DW.

Carrageenan is a sulfated polysaccharide extracted from red seaweed—*Chondrus crispus* or *Gigartina mamillosa*, otherwise called Irish moss. It is a gelatinous mass, well soluble in water; 2% solution of carrageenan in hot water is a rather viscous, slimy colloid. If the concentration of carrageenan is increased to 3–5%, the solution thickens, and a gel is formed. The solutions have a pH of 7.0–8.5. The composition of carrageenan includes a fraction of inorganic compounds that are soluble in cold water and a gel-forming fraction that is soluble in hot water and thickens when cooled. The gel-forming fraction consists of calcium salts of sulfonated 200-kDa polysaccharides consisting of galactose and 3,6-anhydrogalactose. The viscosity of the resulting gels depends on the pH of the medium, molecular weight, concentration, temperature, and the electrolytes in the medium. In the presence of sodium chloride, a fairly hard, dense gel is formed. In the presence of calcium salts, the gel begins to absorb moisture more strongly. It is compatible with many natural and synthetic types of cosmetic raw materials; it is well dispersed in glycerin. In solutions with a high ethanol content, carrageenan precipitates (Barel et al. 2014).

In an acidic environment, carrageenan begins to depolymerize, and the higher the temperature of the mixture, the faster the depolymerization occurs. It is included in the formulations of creams, cosmetic milk, toothpaste. It is used as a gel-forming agent, thickener, softening and emulsifying component, which ensures the stability of the emulsion and suspension. As a foam stabilizer, it is used in shampoos, shaving creams. Toothpaste with carrageenans has an attractive appearance and is stable during storage. Carrageenan sodium and calcium salts, carrageenates, are fine powders of white or cream color without smell and taste. Carrageenan extract contains vitamins A, B1, B2, C, E, D, fucoxanthin, iodine, sulfo amino acids. It stimulates the regeneration of skin cells, has a softening and light bactericidal effect, effectively moisturizes the skin, and retains moisture due to the high content of polysaccharides, organic acids, mineral salts. In addition to this, carrageenate promotes the removal of toxins, has an antioxidant effect, nourishes the skin (Barel et al. 2014).

2.7 Mannitol

Mannitol, the hexacarbon hexatomic alcohol is a component of brown algae. This is one of the first products of photosynthesis serving as a reserve substance and cryoprotector in algae (Podkorytova et al. 2007; Li et al. 2008). The content of mannitol varies from 1% to 28% DW depending on the species, thallus part, and the season of harvest. In algae, only D-mannitol is found, representatives of the genus *Laminaria* contain the highest amounts of this compound during the periods of maximum photosynthetic activity. Obtaining mannitol from brown algae is several times cheaper than its chemical synthesis. Large amounts of mannitol (about 20% DW) were found in representatives of the genera *Laminaria*, *Alaria*, and *Arthrothamnus*; the highest mannitol content was recorded in the latter species (up to 26% DW) (Podkorytova et al. 2007; Rastogi et al. 2017). Since *Arthrothamnus bifidus* is also rich in alginic acids, it is one of the most promising sources for obtaining both ingredients. In cosmetics, it is used as a moisturizing component of creams and lotions, hair conditioners, as well as a sweetener in toothpastes. It also has emulsifying and antioxidant properties.

An illustrious example of cosmetic formulations with algal polysaccharides is Fucogel. It is an opalescent viscous liquid, an aqueous solution of an anionic polysaccharide with a linear structure consisting of molecules of galacturonic acid, L-fucose, and D-galactose. The multifunctional active ingredient, suitable for all types of cosmetic products and age groups. Fucogel regulates cellular interactions by binding membrane receptors to specific sugars, which leads to a decrease in skin reactivity and a calming effect. The product has an anti-aging effect. As a polysaccharide with film-forming properties, Fucogel produced by the company "Solabia Group" has a moisturizing effect on the skin, enhances the shine of the hair, improves the sensory properties of cosmetic products.

3 Lipids and Fatty Acids

Marine macrophytes are not particularly rich in lipids, but the huge reserves of seaweed biomass allow one to consider them as a potentially commercially relevant source of lipids and fatty acids. The content of lipids in algae varies significantly depending on the habitat conditions and season, the age of the thallus and the growth phase. The higher thalli grow on the littoral, the higher its lipid content ranging from 1 to 3% DW (Rastogi et al. 2017) The lipid composition of seaweed is often dominated by neutral lipids—triglycerides containing unsaturated and saturated fatty acids. In seaweed, they make up the bulk of cell lipids. In brown algae, triglycerides account for up to 83% of all cell lipids. As an example, the total lipid content of *Fucus vesiculosus* is 1.8–5%; *Enteromorpha intestinalis*—18.5%; *Undaria pinnatifida*—36.1%; *Porphyra umbilicalis*—37.7%.

The nutraceutical value of algae raw materials is largely determined by the fatty acid composition and the percentage of certain fatty acids. Usually, the content of fatty acids in macrophyte algae does not exceed 1% DW. The main groups of marine macroalgae differ in their composition of polyunsaturated fatty acids. According to their content, brown algae occupy an intermediate position between red and green algae. The C_{16}-C_{18} fatty acids predominate in green algae whereas very-long-chain (C_{20}+) polyene fatty acids are more abundant in red algae (Rastogi et al. 2017; Khozin-Goldberg and Cohen 2011).

Among brown algae, fatty acids composition is known mostly for the species of the orders Laminariales and Fucales. Brown algae have the same composition of fatty acids but differ in their proportions. The main fatty acids are palmitic and oleic acids from the C_{16} family and polyunsaturated fatty acids from C_{18} (18:2, 18:3, 18:4) and C_{20} (20:4, 20:5), the content of the latter varies from 29% to 69% of the total fatty acids. This feature of fatty acid composition distinguishes brown algae from red and green algae. Polyunsaturated fatty acids in macrophytes may account for up to 77% of all fatty acids and reach 6.7 mg/g DW in *Porphyra* and 0.5 mg/g in *Fucus vesiculosus*.

Seaweeds are an important sources of rare biologically active fatty acids, such as gamma-linolenic acid, arachidonic acid, and eicosapentaenoic acid (Malcata et al. 2018). The highest content of polyunsaturated fatty acids (arachidonic and eicosapentaenoic) is a feature of red algae. Eicosapentaenoic acid usually dominates their fatty acid profile reaching 50% of the total fatty acids (Ward and Singh 2005). Brown algae fatty acid profile is normally dominated by linoleic and α-linolenic fatty acids. In almost all species of the genus *Laminaria*, the proportion of polyunsaturated fatty acids of the (n-3) series is higher than that of the (n-6) series. Representatives of the Fucaceae family have a low level of polyunsaturated fatty acids (n-3) of the series (18:3, 18:4, and 20:5), and a higher content of myristic acid and, especially, oleic acid than in other species (Ward and Singh 2005).

There are reports on the antimicrobial, antitumor, and anti-inflammatory activity of brown algae lipids. Thus, the consumption of polyene fatty acids, such as docosahexaenoic acid (22:6n-3) and eicosapentaenoic acid (20:5n-3), reduces the risk of cardiovascular and inflammatory diseases (Khozin-Goldberg and Cohen 2011). So-called "algae oil", a concentrate of essential for humans and higher animals ω-3,6 polyunsaturated fatty acids obtained from the algae Laminariales and Fucales are becoming widespread in the cosmetics industry.

4 Carotenoid Fucoxanthin

Fucoxanthin is one of the most common carotenoids, constituting more than 10% of natural carotenoids, especially in the marine environment. It is an auxiliary light-harvesting pigment in the chloroplasts of brown seaweed and diatoms giving them a brown or olive-green color. It is found particularly in *Cystoseira* sp. Structurally, fucoxanthin has an unusual double allyl carbon and two hydroxyl groups, which is thought to be due to its high energy transfer efficiency (80%). Like other carotenoids, it is a good antioxidant due to its ability to quench singlet oxygen and scavenge free radicals. Fucoxanthin does not exhibit toxicity and mutagenicity under experimental conditions, it increases the level of circulating cholesterol in rodent models (Novoveska et al. 2019; Kijjoa and Sawangwong 2004). The content of fucoxanthin in seaweed varies depending on the season and the stage of development. Studies of the biological activity of fucoxanthin established its antioxidant, antitumor, antidiabetic and other properties were established. In cosmetics, it is used to whiten and improve the condition of the skin, also as a natural antioxidant, lipolytic agent (Spagolla Napoleão Tavares et al. 2020).

5 Mycosporine-Like Amino Acids

Mycosporine-like amino acids (MAA) are secondary metabolites found in marine and freshwater organisms dwelling in habitats with high fluxes of solar UV radiation. MAAs are produced by those organisms to protect against solar radiation (Řezanka et al. 2004; Shick and Dunlap 2002). MAAs are compounds of low molecular weight, usually <400 Da, colorless, uncharged, water-soluble ampholites. They have a similar chemical structure but differ in functional groups and/or conjugated amino acids. They consist of cyclohexenone or cyclohexenimine conjugated with an amino alcohol group or a nitrogen subgroup of an amino acid (Řezanka et al. 2004; Shick and Dunlap 2002).

Since the bulk biomass of macroalgae is produced along the coastline or on continental shelves and is subjected to UV stress, especially in southern latitudes, they also contain ample MAAs, the concentration of which varies greatly between classes of algae. The bulk of supralittoral algae exposed to maximum UV-B light produces large amounts of MAAs (e.g. *Porphyra umbilicalis*). In some species of supralittoral algae, the production of MAAs can potentially be induced by sunlight (*Palmaria decipiens*).

Due to the high photoprotective properties of MAAs, they began to be actively used as UV filters in cosmetics. Since the red alga *Porphyra umbilicalis* contains a large amount of MAAs, its extracts have already become a popular cosmetic ingredient that protects the skin in the UV-A range. Such ingredients are especially interesting for the category of natural cosmetics, that is, MAAs act as an antioxidant, preventing the formation of reactive oxygen species and interrupt the chain of peroxidation reactions. Below, we consider two algal MAA-based cosmetical formulations.

Helioguard®365 is a formulation developed by Schmid et al., containing liposomal MAA, is manufactured by MIbelle Biochemistry, and is currently available on the global market (Schmid et al. 2006). In addition to its high anti-aging activity, the formulation protects cells against UV-A-induced DNA damage. In vitro studies have shown that Helioguard 365, added at concentrations of 0.125% and 0.25% to human HaCaT keratinocytes exposed to 10-minute UV-A radiation, improved their viability in a dose-dependent manner; cell viability in the presence of 0.25% Helioguard 365 was 97.8%. The addition of 3% and 5% Helioguard 365 to human IMR-90 fibroblasts irradiated by UV-A significantly reduced DNA damage, depending on the concentration. Moreover, Schmid et al. the MAAs content in the preparation was found to be stable at 4 °CC and room temperature for at least 3 months, and at 37 °C shelf life decreases by 20% [183]. In addition, three-month exposure to the composition of the drug by simultaneous exposure to UV-A and different temperatures does not affect the stability of MAAs. Thus, Helioguard 365 has high preventive effectiveness against human skin damage caused by UV-A.

Another example is constituted by Helionori® containing MAA sunscreens extracted from *Porfira umbilicalis* as active ingredients. Produced by Gelima company, the product is resistant to sunlight for 6 hours and up to 120 °C for 30 min, stored at a temperature of 15 to 25 °C and stable for at least 18 months. The use of the cream showed effective protection of the metabolism of fibroblasts and keratinocytes exposed to UV-A against oxidative stress. After 24 h of irradiation in the presence of 2% of the extract, the protection of keratinocytes increased by 57%, and fibroblasts by 135%. The extract protects cellular components from UV-A. Thus, 2% Helionori protects keratinocyte membrane lipids by 139% and fibroblasts by 134%, and also provides maximum DNA protection (Singh et al. 2021).

6 Commercial Application of Macroalgal Products

In the current scenario, the production of bioactive compounds from macroalgae has become one of the most successful activities in biotechnology. Various bioactive compounds from macroalgae such as phlorotannins, saponins, flavonoids,

Fig. 30.1 Seaweed cosmetic products available in the market (Source from web sources). 1. Seaweed soap, 2 and 3. Oil containing gel cream; 4. Gel for body wash; 5. Skin whitening, moisturizing and tightening body lotion; 6. Seaweed powder; 7. Hydrating and moisturizing; 8. Provides sun protection-smoother, younger and facial skin; 9. Seaweed eye cream, and 10. Promote skin elasticity, firming, protection and regeneration

terpenoids, phenols, amino acids, galactofucan, sulphated polysaccharides, MAAs, Peptides, fucoidan, and fucoxanthin have a great demand in cosmeceutical applications (Table 30.1). Because of their use in various applications, macroalgal production technologies are increasing globally. Presently, various cosmetic products prepared from macroalgae are available in the market (Fig. 30.1). They include products such as sunscreen, shampoos, oils, gels, lotions, ointments, and creams. Some of the seaweed formulations are also prepared with combination of other ingredients such as vitamins and amino acids etc. Seaweed bioactive compounds such as extracts, polysaccharides, proteins, fucoidan, fucose, rhamnose are improved appearance of the skin, skin resilience, and reduce wrinkles, improved the moisture to the skin and hair, showed beneficial effects on skin ailments, and acted as antioxidant and anti-aging properties (Table 30.2.) Patent applications are available in the literature on the use of seaweeds in cosmetic applications(Table 30.3).

Table 30.2 List of some cosmeceutical products produced from macro algal forms by industries (Calado et al. 2019; Pimentel et al. 2018)

Products	Seaweeds	Industry	Cosmeceutical use
3 M3.Whiterig®	Concentrate of *dictyopteris membranacea*	ProTec Ingredia, UK	Anti-ageing, anti-dullness, brightener, whitening
ACB wakame bioferment advanced	*Undaria pinnatifida* cell culture ferment extract	Active Concepts LLC, USA	Anti-ageing, anti-oxidant, mitochondrial metabolism activator
Alariane ad®	*Alaria esculenta* extract	SePPIC, USA	Hair care
Ambre Oceane®	*Pelvetia canaliculata* extract	SePPIC, USA	Anti-ageing, Plumping
Antileukine™ 6	*Laminaria ochroleuca* extract	SePPIC, USA	Photoageing protection, suncare
Bioenergizer P BG PF®	*Pelvetia canaliculata* with *Laminaria digitata* extracts	SePPIC, USA	Hair care
BIORESTORER PF®	*Hypnea musciformis* extract	SePPIC, USA	Hair care
Cellynkage®	*Halomonas eurihalina* EPS	Lipotec SAU Spain	Menopausal rejuvenator, collagen inducer
Chondrus Crispus Flakes®	*Chondrus crispus* powder	SePPIC, USA	Spa treatments
Codiavelane® BG PF	*Codium tomentosum* extract	SePPIC, USA	Moisturizing, hydration, skin barrier
Coraline Concentrate®	*Corallina officinalis* extract	ERICSON LABORATOIRE, France	Anti-hunger, slimming
Dictyopteris oil®	*Dictyopteris membranacea* supercritical extracted oil	CODIF Technologie naturelle, France	Plumping, filler
Earlyboost®	*Jania rubens* taurine	ProTec Ingredia, UK	Anti-ageing, anti-pollution
Ephemer®	*Undaria pinnatifida* extract	SePPIC, USA	Anti-ageing, colourant and suncare
Esculane®	*Laminaria digitata* extract	SePPIC, USA	Hair care
Gelalg®	*Chondrus crispus* extract	SePPIC, USA	Tightening, anti-ageing, marine silicone, structural component, emulsifier
Homeo-shield ™	*Fucus serratus* extract	COCO skin clinic, USA	Moisturizing, hydrating, skin barrier
Homeo-Soothe™	*Ascophyllum nodosum* extract	Cosmetic Solutions, USA	Shooting, sensitive skin
Hydranov®	*Furcellaria lumbricalis* oligofurcellaran	ProTec Ingredia, UK	Hydration

(continued)

Table 30.2 (continued)

Products	Seaweeds	Industry	Cosmeceutical use
Juvenessence ad®	*Alaria esculenta* extract	SePPIC, USA	Anti-ageing, antipollution
Kalpariane ad®	*Alaria esculenta* extract	KALPARIANE AD®	Plumping, anti-ageing
Matrigenics 14G®	Fertile bases of *Undaria pinnatifida*	CODIF Technologie naturelle, France	Anti-ageing, antipollution
Neuroguard®	*Laminaria hyperborean* plus *Lessonia nigrescens* oligosaccharides	CODIF Technologie naturelle, France	Anti-ageing (neuro-ageing), anti-pollution
Oligophycocorail spe®	*Corallina officinalis* extract	SePPIC, USA	Energizing, anti-ageing
Pheohydrane®	Polysaccharide from *Laminaria digitata* plus amino acids from *Chlorella vulgaris*	CODIF Technologie naturelle, France	Hydration
Phycoboreane® 2C	*Laminaria hyperborea* extract	SePPIC, USA	Body shape, slimming
Phycocorail™	*Lithothamnion calcareum* extract	SePPIC, USA	Photoageing protection, suncare
Phycojuvenine®	*Laminaria digitata* extract	ProTec Ingredia, UK	Anti-ageing, antipollution
Phycosaccharide AC®	*Laminaria sp.* symbiotic microorganism oligosaccharide	CODIF Technologie naturelle, France	Healing, antiacne, anti-dullness
Phycosaccharide AI®	Oligosaccharide derived from *Laminaria digitata*	CODIF Technologie naturelle, France	Healing, anti-wrinkles and anti-dullness
Rhodofiltrat® Palmaria	*Palmaria palmata* carrageen concentrate	CODIF Technologie naturelle, France	Slimming, glow
Rhodysterol™ S	*Gelidium cartilagineum* extract	SePPIC, USA	Spa treatments
Scopariane®	*Sphacelaria scoparia* concentrate	ProTec Ingredia, UK	Firming, slimming
Seashine™	*Alaria esculenta* plus *Undaria pinnatifida* extract	SePPIC, USA	Moisturizing, whitening
Wakamine™ Wakamine 1% (Peptidic extract) Wakamine XP	*Undaria pinnatifida* extract	Givaudan Active Beauty, Switzerland	Moisturizing; Whitening agent Lightening agent
Xcell-30®	*Halymenia durvillei* extract	Derma elements, Hongkong	Anti-ageing
Helionori®	*Porphyra umbilicalis MAAs*	Gelyma, French Biosoil Technologies, Inc. USA	Photoprotective (UV-A) DNA Protection Prevention of sunburn

(continued)

Table 30.2 (continued)

Products	Seaweeds	Industry	Cosmeceutical use
Helioguard365	Porphyra-334 and Shinorine Porphyra-334 and Shinorine	Mibelle Biochemistry, Switzerland	Photo-protective (UV-A)
Algae gorria, Alga marris	*Porphyra Umbilicalis*	Laboratoires de Biarritz, French	Photoprotective (UV-A)
Fucorich	*Undaria Pinnatifida (Fucoidan)*	Marinova, Australia	Anti-aging
Maritech reverse	*Fucus vesiculosus* (Fucoidan)	Marinova, Australia	Anti-aging; antioxidant; anti-inflammation
Maritech synergy	*Fucus vesiculosus* (Fucoidan and polyphenol complex)	Marinova, Australia	Anti-aging; antioxidant; anti-inflammation
Gelcarin® PC 379 Gelcarin® PC812	*Chondrus crispus*	Givaudan Active Beauty, Switzerland	Decorative cosmetic care applications Lipsticks and deodorants
Akomarine® Fucus	*Fucus vesiculosus*	Akkot, Italy	Slimming and anti-cellulitis cosmetic, Skin softness and elasticity
Gracilaria Hydrogel	*Gracilaria conferta*	Sealaria LTD, Israel	Skincare products
Hijiki Extract	*Hizikia Fusiforme*	Elma Skin Care, Canada	Whitening agent, whitening preparations
Chlorofiltrat® Ulva HG	*Ulva lactuca*	CODIF Technologie naturelle, France	Skin care products Moisturizing and anti-inflammatory agent
AT UV Protector P	*Porphyra tenera*	Athena cosmetics Inc., USA	Photo-protection, Skin and sun care
Xylishine™	*Pelvetia canaliculata*	SePPIC, USA	Hair moisturizer, hair formulations

7 Conclusions and Future Perspectives

Macroalgae synthesizing a plethora of bioactive and antistress compounds are admirably adapted to coping with diverse stresses. In human body, these compounds demonstrate many beneficial effects which are well-documented in the literature. This is not surprising because many damages induced by environmental and other stress in the human cells are implemented via the same mechanisms such as free-radical attacks and lipid peroxidation as in algal cells. Therefore, macrophyte algae are a rich source of substances showing great potential for mitigating risks associated with the stress effects on human skin on a day-to-day basis. In many cases these compounds appear to be less toxic, allergenic, and, in general, more "biocompatible" than most of their synthetic counterparts. At the same time, the large- scale

Table 30.3 [a]Patent applications of cosmetics derived from macro algal forms

Patent No	Title	Purpose	Reference
US20210093540A1	Seaweed-derived cosmetic compositions	The appearance of the skin is improved	Athwal (2021)
US20210161980A1	Seaweed extracts, isolated compounds and methods of treatment	Treated various diseases-Alzheimer's, stroke, and disorder of aging	Luesch and Bousquet (2021)
US10493007B2	Microalgae derived compositions for improving the health and appearance of skin	Improved the health and appearance of skin	Dillon et al. (2019)
US9717932B2	Marine extracts and biofermentations for use in cosmetics	A skin care active ingredient for anti-aging cosmetic applications	Ceccoli et al. (2017)
US20160228352A1	Marine extract compositions and methods of use	Showed beneficial effects on skin ailments	Lewis (2016)
CN105777933A	Preparation of algal polysaccharide and application of algal polysaccharide in cosmetics	Moisturizing agent reduce skin wrinkles and delaying ageing.	Liu et al. (2016)
PCT/ KR2011/008910	Cosmetic composition containing gulfweed extract sea staghorn extract and brown seaweed extract.	Improved skin resilience, and reduces wrinkles.	Kim et al. (2014)
WO2012011907A1,	*Laminaria Saccharina* extract and vitamin B3 as whitening agents. WIPO (PCT)	Appearance of the hyperpigmented spot; Improved skin tone	Hazozaki et al. (2012)
EP1433463B1	Use of algal proteins in cosmetics.	Effects on skin and hair	Hagino and Saito (2010)
US7678368B2	Fucoidan containing cosmetics.	Fucoidan act as effective ingredient in cosmetics	Mizutani et al. (2010)
TW200914061A	Method for using green algae extract to retard aging of skin cells and cosmetic composition containing green algae extract	Performs the effect of retarding extrinsic skin aging	Shih and Shih (2009)
US20060115443A1	Cosmetic composition of two polysaccharides based on fucose and rhamnose	Fucose and rhamnose act on cutaneous, epithelial and conjunctive tissue	Gesztesi et al. (2006)
FR2822701B1	Use of algae *Phaeodactylum* extract as a cosmetic agent promoting proteasome activity of skin cells and cosmetic composition containing same	Promote proteasome activity of skin cells	Nizard et al. (2005)

[a]Representative list only

production of the algal biomass enriched with the cosmeceutical compounds seems to be economically viable. Even such a short consideration clearly shows the potential of marine algae for the production of "green" cosmeceuticals. Still, a large part of the chemo- and biodiversity of macroalgae remains underexplored. Therefore, ecological and biochemical characterization of new species, especially those from places with harsh environmental conditions constitutes the priority directions for further research and development, along with the development of cost-effective and sustainable methods of algal biomass production, harvesting, extraction, and purification of the valuable cosmeceuticals.

Acknowledgments ARR acknowledges the FIST Project no: LSI-576/2013 by the Department of Science and Technology, Government of India and Centre of Excellence, Department of Biotechnology, Vignan's Foundation for Science, Technology and Research for providing the facilities to prepare this manuscript. GAR thanks Dr. Premachandra Sagar, Vice Chairman, Dayananda Sagar Institutions, and Pro-Chancellor, Dayananda Sagar University, Bengaluru, India, for his encouragement and support. AS acknowledges the support of the Russian Science Foundation (grant # 21-74-20004).

References

Athwal G (2021) Seaweed derived cosmetic compositions. Patent US20210093540A1, 01 April 2021

Barbosa AJ, Roque AC (2019) Free marine natural products databases for biotechnology and bioengineering. Biotechnol J 14(11):1800607

Barel AO, Paye M, Maibach HI (2014) Handbook of cosmetic science and technology. CRC Press

Bedoux G, Hardouin K, Burlot AS, Bourgougnon N (2014) Bioactive components from seaweeds: Cosmetic applications and future development. Adv Bot Res 71:345–378

Boney A (1965) Aspects of the biology of the seaweeds of economic importance. In: Advances in marine biology, vol 3. Elsevier, pp 105–253

Borowitzka MA (2013) High-value products from microalgae—their development and commercialisation. J Appl Phycol 25(3):743–756

Briand X (2003) Algal active substances. Cosmet Toiletr 118(2):55–66

Calado R, Costa Leal M, Gaspar H, Santos S, Marques A, Leonor Nunes M, Vieira H (2019) How to succeed in marketing marine natural products for nutraceutical, pharmaceutical and cosmeceutical markets. In: Rampelotto PH, Trincone A (eds) Grand challenges in marine biology and biotechnology. Springer, pp 317–404

Carreto JI, Carignan MO (2011) Mycosporine-like amino acids: relevant secondary metabolites. Chemical ecological aspects. Mar Drugs 9:387–446

Ceccoli JD, Costello B, Pillai S (2017) Marine extracts and biofermentations for use in cosmetics. Patent US9717932B2, 1 Aug 2017

Chan YY, Kim KH, Cheah SH (2011) Inhibitory effects of *Sargassum polycystum* on tyrosinase activity and melanin formation in B16F10 murine melanoma cells. J Ethnopharmacol 137(3):1183–1188

Coates RC, Trentacoste E, Gerwick WH (2013) Bioactive and novel chemicals from microalgae. In: Handbook of microalgal culture, pp 504–531

Delaunay D, and Volle I (2011) Composition dermatologique et/ou cosmétique utilisée pour la régénération de la peau. WO 2011051591 A1

Demais H, Brendle J, Le Deit H, Laza Anca L, Lurton L, Brault D (2007) Argiles intercalées. EP 1786862 A1

Dillon HF, Zaman A, Day AG, Coragliotti A (2019) Microalgae derived compositions for improving the health and appearance of skin. Patent US10493007B2, 3 Dec 2019

Foley SA, Szegezdi E, Mulloy B, Samali A, Tuohy MG (2011) An unfractionated fucoidan from Ascophyllum nodosum: extraction, characterization, and apoptotic effects in vitro. J Nat Prod 74(9):1851–1861

Fox JM, Zimba PV (2018) Chapter 8 - Minerals and trace elements in microalgae. In: Levine IA, Fleurence J (eds) Microalgae in health and disease prevention. Academic, pp 177–193. https://doi.org/10.1016/B978-0-12-811405-6.00008-6

García JL, de Vicente M, Galán B (2017) Microalgae, old sustainable food and fashion nutraceuticals. Microb Biotechnol 10(5):1017–1024

Gesztesi J, Silva LVN, Robert L, Robert A (2006) Cosmetic composition of two polysaccharides based on fucose and rhamnose. Patent S20060115443A1, 1 June 2006

Gong M, Bassi A (2016) Carotenoids from microalgae: a review of recent developments. Biotechnol Adv. https://doi.org/10.1016/j.biotechadv.2016.10.005

Guglielmo M, Montanari D (2008) Cosmetic composition with a lifting effect for sustaining relaxed tissues. Patent WO 2008146116 A2

Hagino H, Saito M (2010) Use of algal proteins in cosmetics. European patent EP1433463B1, 18 Dec 2010

Hazozaki T, Laughlin LT, Swanson CL (2012) *Laminaria Saccharina* extract and vitamin B3 as whitening agents. WIPO (PCT) Patent WO2012011907A1, 26 Jan 2012

Hupel M, Lecointre C, Meudec A, Poupart N, Ar Gall E (2011) Comparison of photoprotective responses to UV radiation in the brown seaweed *Pelvetia canaliculata* and the marine angiosperm *Salicornia ramosissima*. J Exp Mar Biol Ecol 401:36–47

Julius ML (2018) Chapter 6 - Carbohydrate diversity in microalgae: a phylogenetically arranged presentation. In: Levine IA, Fleurence J (eds) Microalgae in health and disease prevention. Academic, pp 133–144. https://doi.org/10.1016/B978-0-12-811405-6.00006-2

Khozin-Goldberg I, Cohen Z (2011) Unraveling algal lipid metabolism: Recent advances in gene identification. Biochimie 93(1):91–100. https://doi.org/10.1016/j.biochi.2010.07.020

Kijjoa A, Sawangwong P (2004) Drugs and cosmetics from the sea. Mar Drugs 2(2):73–82

Kim S-K (2011) Marine cosmeceuticals: trends and prospects. CRC Press

Kim SK (2014) Marine cosmeceuticals. J Cosmet Dermatol 13(1):56–67

Kim HC, Hong YJ, Kim YJ Han SH (2014) Cosmetic composition containing gulfweed extract sea staghorn extract and brown seaweed extract. Application no. 4053/DELNP/2013, PCT International Application no: PCT/KR2011/008910, 21 Nov 2014

Ko SC, Myoungsook L, Ji-Hyeok L, Seung-Hong L, Yunsook L, Jeon YJ (2013) Dieckol, a phlorotannin isolated from a brown seaweed, *Ecklonia cava*, inhibits adipogenesis through AMP-activated protein kinase (AMPK) activation in 3T3-L1 preadipocytes. Environ Toxicol Pharmacol 36:1253–1260

Levine IA (2018) Chapter 1 - Algae: a way of life and health. In: Levine IA, Fleurence J (eds) Microalgae in health and disease prevention. Academic, pp 1–10. https://doi.org/10.1016/B978-0-12-811405-6.00001-3

Lewis E (2016) Marine extract compositions and methods of use. Patent US20160228352A1, 11 Aug 2016

Li B, Lu F, Wei X, Zhao R (2008) Fucoidan: structure and bioactivity. Molecules 13(8):1671–1695

Liu CL, Chiu YT, Hu ML (2011) Fucoxanthin enhances HO-1 and NQO1 expression in murine hepatic BNL CL2 cells through activation of the Nrf2/ARE system partially by its pro-oxidant activity. J Agric Food Chem 59:11344–11351

Liu X, Gao H, Chen X (2016) Preparation of algal polysaccharide and application of algal polysaccharide in cosmetics. Patent CN-105777933, 2 Mar 2016

Luesch H, Bousquet MS (2021) Seaweed extracts, isolated compounds and methods of treatment. Patent US20210161980A1, 3 June 2021

Malcata FX, Pinto IS, Guedes AC (2018) Marine macro-and microalgae: An overview. CRC Press, Boca Raton, FL

Masojídek J, Torzillo G, Koblízek M (2013) Photosynthesis in microalgae. In: Richmond A, Hu Q (eds) Handbook of microalgal culture: applied phycology and biotechnology, 2nd edn. Wiley, Chichester, pp 21–35

Mimouni V, Couzinet-Mossion A, Ulmann L, Wielgosz-Collin G (2018) Chapter 5 - Lipids from microalgae. In: Levine IA, Fleurence J (eds) Microalgae in health and disease prevention. Academic, pp 109–131. https://doi.org/10.1016/B978-0-12-811405-6.00005-0

Mišurcová L (2012) Chemical composition of seaweeds. Wiley Online Library

Mizutani S, Deguchi S, Kobayashi E, Nishiyama E, Sagawa H, Kato I. (2010) Fucoidan containing cosmetics. Patent US7678368B2, 16 Mar 2010

Morançais M, Mouget J-L, Dumay J (2018) Chapter 7 - Proteins and pigments. In: Levine IA, Fleurence J (eds) Microalgae in health and disease prevention. Academic, pp 145–175. https://doi.org/10.1016/B978-0-12-811405-6.00007-4

Mu N, Mehar JG, Mudliar SN, Shekh AY (2019) Recent advances in microalgal bioactives for food, feed, and healthcare products: commercial potential, market space, and sustainability. Compr Rev Food Sci Food Saf 18(6):1882–1897

Mulders KJM, Lamers PP, Martens DE, Wijffels RH (2014) Phototrophic pigment production with microalgae: biological constraints and opportunities. J Phycol 50(2):229–242. https://doi.org/10.1111/jpy.12173

Nizard C, Friguet B, Moreau M, Bulteau AL Saunois A (2005) Use of algae *Phaeodactylum* extract as a cosmetic agent promoting proteasome activity of skin cells and cosmetic composition containing same. Patent FR2822701B1, 18 Mar 2005

Novoveska L, Ross ME, Stanley MS, Pradelles R, Wasiolek V, Sassi JF (2019) Microalgal carotenoids: a review of production, current markets, regulations, and future direction. Mar Drugs 17(11):640. https://doi.org/10.3390/md17110640

Ozanne H, Toumi H, Roubinet B, Landemarre L, Lespessailles E, Daniellou R, Cesaro A (2020) Laminarin effects, a β-(1,3)-glucan, on skin cell inflammation and oxidation. Cosmetics 7:66

Pangestuti R, Shin K-H, Kim S-K (2021) Anti-photoaging and potential skin health benefits of seaweeds. Mar Drugs 19:172

Pimentel FB, Alves RC, Rodrigues F, Oliveira MBPP (2018) Macroalgae-derived ingredients for cosmetic industry-an update. Cosmetics 5:2

Podkorytova A, Vafina L, Kovaleva E, Mikhailov V (2007) Production of algal gels from the brown alga, Laminaria japonica Aresch., and their biotechnological applications. J Appl Phycol 19(6):827–830

Puchkova T, Khapchaeva S, Zotov V, Lukyanov A, Solovchenko A (2021) Microalgae as a sustainable source of cosmeceuticals. Mar Biol J 6(1):67–81. https://doi.org/10.21072/mbj.2021.06.1.06

Rastogi RP, Madamwar D, Pandey A (2017) Algal green chemistry: recent progress in biotechnology. Elsevier

Ravishankar GA, Ranga Rao A (eds) (2019a) Handbook of algal technolgies and phytochemicals: volume-I food, health and nutraceutical applications. CRC Press., ISBN: 9780429054242, p 322

Ravishankar GA, Ranga Rao A (eds) (2019b) Handbook of algal technologies and phytochemicals: volume II phycoremediation, biofuels and global biomass production. CRC Press, p 295. ISBN: 9780367178192

Řezanka T, Temina M, Tolstikov A, Dembitsky V (2004) Natural microbial UV radiation filters—Mycosporine-like amino acids. Folia Microbiol 49(4):339–352

Richmond A (2008) Handbook of microalgal culture: biotechnology and applied phycology. Wiley-Blackwell

Rioux LE, Moulin V, Beaulieu M, Turgeon SL (2013) Human skin fibroblast response I sdifferentially regulated by galactofucan and low molecualr weight galactofucan. Bioact Carbohydr Diet Fibre 1(2):105–110

Ruperez P, Ahrazem O, Leal JA (2002) Potential antioxydant capacity of sulfated polysaccharides from the edible marine brown seaweed *Fucus vesiculosus*. J Agric Food Chem 50:840–845

Schmid D, Schürch C, Zülli F (2006) Mycosporine-like amino acids from red algae protect against premature skin-aging. Euro Cosmet 9:1–4

Scott R (2015) Marine ingredients: latest actives from the deep. Personal Care 4:43–44

Shick J, Dunlap W (2002) Mycosporine-like amino acids and related gadusols: biosynthesis, accumulation, and UV-protective functions in aquatic organisms. Annu Rev Physiol 64(1):223–262

Shih MF, Shih MH (2009) Method for using algae extract to retard aging of skin cells and cosmetic composition containing green algae extract. Patent TW 200914061A, 01 Apr 2009

Singh A, Čížková M, Bišová K, Vítová M (2021) Exploring mycosporine-like amino acids (MAAs) as safe and natural protective agents against UV-induced skin damage. Antioxidants 10(5):683

Solovchenko AE (2013) Physiology and adaptive significance of secondary carotenogenesis in green microalgae. Russ J Plant Physiol 60(1):1–13. https://doi.org/10.1134/s1021443713010081

Song TY, Chen CH, Yang NC, Fu CS, Chang YT, Chen CL (2009) The correlation of in vitro mushroom tyrosinase activity with cellular tyrosinase activity and melanin formation in melanoma cells A2058. J Food Drug Anal 17:156–162

Spagolla Napoleão Tavares R, Maria-Engler SS, Colepicolo P, Debonsi HM, Schäfer-Korting M, Marx U, Gaspar LR, Zoschke C (2020) Skin irritation testing beyond tissue viability: Fucoxanthin effects on inflammation, homeostasis, and metabolism. Pharmaceutics 12:136

Thibodeau A, Lacasse I (2006) Actif issu de polysaccharides d'algues. Parfums cosmétiques actualités 188

Thomas NV, Kim S-K (2013) Beneficial effects of marine algal compounds in cosmeceuticals. Mar Drugs 11(1):146–164

Wang R, Paul VJ, Luesch H (2013) Seaweed extracts and unsaturated fatty acid constituents from the green alga *Ulva lactuca* as activators of the cytoprotective Nrf2-ARE pathway. Free Rad Biol Med 57:141–153

Ward OP, Singh A (2005) Omega-3/6 fatty acids: alternative sources of production. Process Biochem 40(12):3627–3652

Yang Y, Liu D, Wu J, Chen Y, Wang S (2011) *In vitro* antioxidant activities of sulfated polysaccharide fractions extracted from *Corallina officinalis*. Int J Biol Macromol 49:1031–1037

Yuan YV, Westcott ND, Hu C, Kitts DD (2009) Mycosporine-like amino acid composition of edible red alga *Palmaria palmata* (dulse) harvested from the west and east coasts of Grand Manan Island, New Brunswick. Food Chem 112:321–328

Zhang Q, Li N, Liu X, Zhao Z, Li Z, Xu Z (2004) The structure of a sulfated galactan from *Porphyra haitanensis* and its in vivo antioxidant activity. Carbohydr Res 339:105–111

Chapter 31
Anti-Diabetic Properties of Fucoidan from Different *Fucus* Species

Irina G. Danilova, Saied A. Aboushanab, Ksenia V. Sokolova, Gokare A. Ravishankar, Ambati Ranga Rao, and Elena G. Kovaleva

Abbreviations

ALP	Alkaline phosphatase
ALT	Alanine aminotransferase
ApoC-II	Apolipoprotein C-II
AST	Aspartate aminotransferase
C/ebp	Enhancer-binding protein
cAMP	Adenosine 3′,5′-cyclic monophosphate
CVD	Cardiovascular disease
DA	Malondialdehyde
DN	Diabetic nephropathy
DR	Diabetic retinopathy
ERK	Extracellular signaling regulatory kinases
FBG	Fasting blood glucose

I. G. Danilova
Institute of Natural Sciences and Mathematics, Ural Federal University Named After the First President of Russia B. N. Yeltsin, Yekaterinburg, Russia

Institute of Immunology and Physiology, Ural Branch of the Russian Academy of Sciences, Yekaterinburg, Russia

S. A. Aboushanab · E. G. Kovaleva (✉)
Institute of Chemical Engineering, Ural Federal University Named After the First President of Russia B. N. Yeltsin, Yekaterinburg, Russia
e-mail: sabushanab@urfu.ru; e.g.kovaleva@urfu.ru

K. V. Sokolova
Institute of Immunology and Physiology, Ural Branch of the Russian Academy of Sciences, Yekaterinburg, Russia

G. A. Ravishankar
Dr. C. D. Sagar Centre for Life Sciences, Dayananda Sagar College of Engineering, Dayananda Sagar Institutions, Bengaluru, Karnataka, India

A. Ranga Rao
Department of Biotechnology, Vignan's Foundation of Science, Technology and Research Deemed to be University, Vadlamudi, Guntur, Andhra Pradesh, India

A. Ranga Rao, G. A. Ravishankar (eds.), *Sustainable Global Resources of Seaweeds Volume 2*, https://doi.org/10.1007/978-3-030-92174-3_31

JNK	JunN-terminal kinase
LPL	Lipoprotein lipase
LPO	Lipid peroxidation
MAPK	Mitogen-activated protein kinases
NF-κB	Nuclear factor kappa
OMM-1	Uveal melanoma cell line
PPAR	Peroxisome proliferator-activated receptor
PPG	postprandial glucose
ROS	Reactive oxygen species
RPE	Retinal pigment epithelium
T1DM	Type 1 diabetes mellitus
T2DM	Type 2 diabetes mellitus
TGF-β1	Transforming growth factor-β1
TLR	Toll-like receptor
UCP-1	Uncoupling Protein 1
VEGF	Vascular endothelial growth factor
WHO	World Health Organization

1 Introduction

Diabetes mellitus (DM) is a heterogeneous metabolic disorder that affects over 400 million worldwide (WHO) out of which approximately 87–91% of the population suffering from type-2 diabetes mellitus (T2DM) (Oza and Kulkarni 2018, 2020). The life expectancy of an individual suffering from T2DM reduces by 15 approximately years primarily due to the development of vascular complications (Besednova et al. 2019; Mabate et al. 2021; Wu et al. 2021a, b). Pathogenesis of T2DM is associated with prominent insulin resistance and insufficient insulin secretion. The impaired function of pancreatic β cells leads to further complications in the kidney, heart, nervous system, and eyes (Aleissa et al. 2020).

The urgent need to develop a novel therapy for diabetes treatment is a challenge-led approach. In that sense, marine products are a valuable food source that makes a substantial contribution to preventing T2DM with less or negligible adverse effects (Bocanegra et al. 2021). The natural products originated from seaweeds have attracted attention recently, among them naturally occurring anti-diabetic ingredients are abundant in brown algae (Gunathilaka et al. 2020). Fucoidan, as a natural compound, has attracted interest due to its outstanding biological activity (Mabate et al. 2021). In recent years, encouraging results have been published in a considerable amount of literature on the relevant potential role of some *Fucus* species in the treatment of diabetes (Table 31.1).

Brown algae potentially comprise a crucial group of bioactive compounds named fucoidans (Li et al. 2008). The main chain of fucoidan primarily consists of α-(1,3)- and/or β-(1,4)-linked fucose, and C-2, C-4 and/or C-3 positions on L-fucose residue are simply substituted by sulfate (Fig. 31.1). With its complex structure and health-promoting properties, fucoidan exhibits various therapeutic functions such as

Table 31.1 The potential role of fucoidan from some *Fucus* species in the treatment of diabetes

Fucus species	Model/extract concentrations	Anti-diabetic properties	Reference
F. distichus	*In vitro* biochemical analysis/20 or 35μl of extract	↓ α-glucosidase and α-amylase ↓ absorption of digested carbohydrates ↓ post-prandial hyperglycemia	Kellogg et al. (2014)
F. distichus	*In vitro* assays (RAW 264.7 and 3T3-L1 cell lines)/12.5–50μg/ml of extract	↓ lipid accumulation in 3T3-L1 adipocytes ↓ TLR4 and TLR9 ↑ adiponectin and UCP-1 ↓ leptin mRNA expression	Kellogg et al. (2015)
F. guiryi, F. serratus, F. spiralis and *F. vesiculosus*	*In vitro* biochemical analysis/20, 50 and 100μl of extract	↓ α-amylase, α-glucosidase and xanthine oxidase	Lopes et al. (2019)
F. vesiculosus	*In vitro* α-Amylase inhibition activity/1 mg fucoidan/ml	No ↓ in α-amylase activity	Kim et al. (2015)
F. vesiculosus, F. distichus subsp. evanescens and *F. serratus*	*In vitro* assays (OMM-1 and RPE cell line ARPE19 cell lines)/1–100μg/ml of fucoidan extract	↓ OMM-1 oxidative stress ↓ VEGF secretion in ARPE19	Dorschmann et al. (2019)
F. vesiculosus	*In vitro* (3T3-L1 3T3-L1 preadipocyte cell lines)/ fucoidan fractions (F0.5, 0.9, 1.0, 2.0) 100–1000μg/ml	(F0.5 and 0.9)-↑ lipid in 3T3-L1 adipocytes (up-regulation of C/ebpα, C/ebpβ, and PPAR gamma) (F1.0 and 2.0)-↓ adipogenic activity	Oliveira et al. (2018)
F. vesiculosus	*In vivo* (male Wistar rats)/7.5 mg/kg·bw daily for 5 weeks	↓ microvesicular steatosis ↓ ALT, AST, ALP, conjugated bilirubin and triglycerides ↓ postprandial glycemia	Oliveira et al. (2018)
F. vesiculosus	Clinical trial / 6 months administration (*Ascophyllum Nodosum* and *F. Vesiculosus* extract + chromium picolinate)	↓ HbA1c, ↓ FBG and PPG No change in lipid profile	Derosa et al. (2019)

reducing cholesterol levels, cancer prevention, and other biological activities e.g. anti-thrombotic, anti-viral, and anti-inflammatory (Fitton et al. 2015).

Clinical trials have recommended a daily dose of fucoidan of 1 g supplementation/day 12 weeks (Myers et al. 2010) or 3 g on a daily basis for 12 days (Irhimeh et al. 2009). Moreover, fucoidan was reported to have anti-hyperglycemic effect, improve lipid metabolism and reduce insulin resistance (Yang et al. 2017). This book chapter provides the anti-diabetic potential of fucoidan and its role in the treatment of diabetic complications. Further, its mechanisms of action that scrutinize the future recommendations of fucoidan in the treatment of diabetes are discussed.

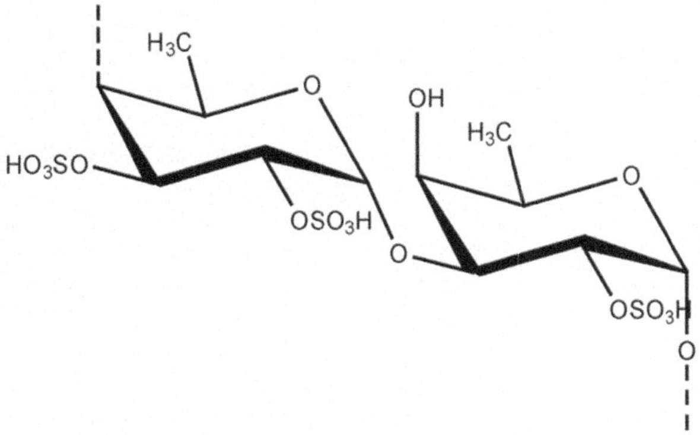

Fig. 31.1 Chemical structure of fucoidan as sulfated polysaccharide

2 Anti-diabetic Properties of Fucoidan

Diabetes mellitus type 2 (T2DM) is a global catastrophe and a life-threatening metabolic disorder that possibly leads to serious health conditions (Zhao et al. 2021). Numerous complications (nephropathy, neuropathy, retinopathy, cardiovascular diseases, etc.) may significantly reduce the life quality and increase disability and mortality among diabetic patients (Saeedi et al. 2019). In this regard, there is an urgent need to develop new therapeutic strategies and approaches for producing novel harmless drugs with multi-functional properties. Fucoidans are heteropolysaccharides derived from the cell walls and intercellular spaces of the brown seaweed (*Phaeophyceae*). Fucoidans are considered water-soluble sulfated heteropolysaccharides with L-fucose-4-sulfate as the main monosaccharide component (Wang et al. 2019). These heteropolysaccharides contain other compounds such as galactose, xylose, mannose, rhamnose, glucose, and arabinose (Jiao et al. 2011). Recent publications have shown that fucoidans from different brown algae vary significantly in their structure, monosaccharides components, and sulfate content (Ponce and Stortz 2020). Based on this fact, it was found that the degree of sulfation significantly affects the biological properties of fucoidans (Ale et al. 2011). These substances are of extremely high interest due to their potential biological activities and are safe and non-toxic for human consumption (Wang et al. 2019; Luthuli et al. 2019; Pradhan et al. 2020).

3 Fucoidan, Diabetic Hyperglycemia and Insulin Secretion

Insulin is an anabolic hormone that reduces blood sugar levels and regulates glucose homeostasis in the cell. It is well known that insulin dysfunction (concerning secretion or action) is always associated with the development and progression of diabetes (Wu et al. 2021a, b). The pathogenesis of type-1 (T1DM) and type-2 diabetes

(T2DM) is largely different. T1DM is mainly associated with impaired humoral and cellular immunity that results in the destruction of pancreatic β-cell. In turn, T2DM is characterized by insulin resistance which is basically developed due to insulin receptor insensitivity (Andoh 2016). Both types of diabetes mellitus are characterized by the inability of the endocrine cells (β-cells) of the pancreatic islets to satisfy the body's need for insulin due to either complete loss (T1DM) or functional deficiency in β-cells (T2DM). Thus, hindering the destruction of β-cells in the pancreatic islets of Langerhans, as well as improving insulin sensitivity, are essential to alleviate the progression of diabetes (Moin and Butler 2019).

In this regard, fucoidans strongly manifest pronounced anti-diabetic potentials. Among them, fucoidan of low molecular weight and a high degree of sulfation was found to display better anti-diabetic properties (Kim et al. 2014a; Jeong et al. 2013). In *in vitro* experiments using RIN-5F cell culture derived from pancreatic β-cell lines of diabetic rats, heteropolysaccharides from algae extract were documented to stimulate insulin synthesis depending on their structural characteristics (Zhang et al. 2008).

Similarly, administering fucoidan extracted from *Fucus vesiculosis* was revealed to demonstrate a time- and dose-dependent stimulation of insulin secretion mediated via the regulation of the cAMP signaling pathway (Jiang et al. 2015). Numerous animal studies have also noticed the evidence of the anti-hyperglycemic effect of fucoidans and their fundamental role in promoting insulin secretion (Mabate et al. 2021). Additionally, fucoidan was recorded to prevent the development of hyperglycemia in intact animals and reduce glycemic parameters in C57Bl/KSJ m +/+ db, C57Bl/KSJ db/db mice, and Goto-Kakizaki rats (Jiang et al. 2015; Kim et al. 2014a).

A few studies suggested that low molecular weight compounds would exhibit greater anti-diabetic potentials. For instance, low molecular weight fucoidan was investigated to substantially reduce the serum glucose levels and ameliorate the glucose homeostasis via elevating the glucose tolerance in leptin receptor-deficient db/db mice. Importantly, the main mechanism of fucoidan effect is associated with activation of AMP-activated protein kinase, which may contribute to decreasing insulin resistance (Jeong et al. 2013).

It is known that β-cells of the islets of Langerhans have a low level of antioxidant protection. This possibly may make the β-cells more vulnerable to the oxidative actions of reactive oxygen species. Fucoidans extracted from *Laminaria japonica* and *Undaria pinnatifida* exhibit significant antioxidant activity. Accordingly, it can be potentially used to treat diseases caused by free radical damage such as diabetes mellitus (Rodriguez-Jasso et al. 2011; Mak et al. 2013). Hence, fucoidan was outlined to reduce the accumulation of amyloid-β and reactive oxygen species against the amyloid-β-induced toxicity (Wang et al. 2018b).

4 Inhibition of α-Glucosidase and α-Amylase

Inhibition of α-glucosidase and α-amylase is an effective therapeutic approach for type-2 diabetes mellitus, which helps largely in the reduction of postprandial hyperglycemia. Since only a few drugs were reported to inhibit the activity of these enzymes, a search for naturally occurring compounds that inhibit α-glucosidases, with low toxicity and negligible adverse effect is an urgent problem of modern diabetology.

The level of the inhibitory effect of polysaccharides-derived-α-amylase depends on the seaweeds. It was found that fucoidan derived from the algae *Ascophyllum nodosum* was demonstrated to reduce the α-amylase activity, while the heteropolysaccharide derived from *Fucus vesiculosis* was recorded with no effect on the activity of this enzyme (Kim et al. 2014a). When evaluating the inhibitory effect of α-glucosidases and α-amylase of 11 fucoidans extracted from different species of brown algae, fucoidans derived from *Fucus vesiculosus* showed the highest inhibitory activity (Gheda et al. 2021). Accordingly, fucoidans were potentially stated to be promising inhibitors of α-glucosidase that in turn could be an efficient treatment of type-2 diabetes mellitus (Shan et al. 2016).

One of the possible mechanisms of fucoidan effect on postprandial glycemia through suppressing the glucose uptake is the increase in viscosity of intestine and small intestine contents. Accordingly, fucoidans were assumed to potentially stimulate the expression of intestinal proglucagon and the secretion of glucagon-like peptides (Mansour et al. 2013).

In the present therapeutic context, it is speculated that fucoidans may play a vital role as a prebiotic. This built a baseline for understanding the potential ability of fucoidan to normalize the ecosystem and the gut microbiota as well as maintain the homeostatic microbial diversity. Recent studies have noted that fucoidans as prebiotics may possibly regulate the blood glucose levels and blood glucose metabolism and create a favorable condition for the stimulation and growth of probiotics (Chen et al. 2019; Parnell and Reimer 2012).

5 Antioxidant Activity of Fucoidan

Diabetes mellitus (DM) is a global metabolic disorder that determines its course and prognosis. Chronic hyperglycemia and dyslipoproteinemia are accompanied by the development of oxidative stress in DM. At the same time, there is a significant increase in lipid peroxidation (LPO) and the activity of antioxidant enzymes such as superoxide dismutase, glutathione peroxidase, glutathione reductase, and catalase. Besides, a decrease in the concentration of endogenous non-enzymatic antioxidant molecules (for example, vitamins A, C and E, glutathione) and β-carotene, as well as increased concentration of malondialdehyde (MDA) and accumulation of oxidation products, are common features of diabetes mellitus (Tiwari et al. 2013).

It has been reported that oxidative stress affects both insulin secretion (through β-cell dysfunction caused by autoimmune reactions, cytokines and inflammatory proteins) (Jiang et al. 2015) and insulin function (through insulin resistance resulting from interference of Reactive Oxygen Species (ROS) with insulin signal transduction (Hurrle and Hsu 2017).

Antioxidants are chemical or biological agents that are capable of neutralizing the potentially damaging effects of free radicals through a variety of mechanisms, including catalytic systems to neutralize or reject ROS, and binding or inactivating metal ions to prevent ROS formation (Weng et al. 2019).

The identification of non-toxic antioxidant compounds derived from brown algae is a field of intensive research. The antioxidant properties of fucoidan are determined by its structure and are associated with the active absorption of ROS (Begum et al. 2021). Fucoidan is a well-known ROS scavenger. For instance, the heteropolysaccharides extracted from Japanese kelp (*L. japonica*) were utilized to treat diseases caused by free radical damage and showed its antioxidant properties (Rodriguez-Jasso et al. 2011; Wang et al. 2008). Likewise, fucoidan extracted from *Undaria pinnatifida* exhibits a significant antioxidant activity (Mak et al. 2013). Superoxide dismutase and glutathione activities were also induced after fucoidan treatment (Wei et al. 2017). Several factors determine the antioxidant activity of fucoidan, including concentration, degree of sulfation, type of sugar, and glycosidation branching (Melo et al. 2002; Zhao et al. 2018; Jin et al. 2018). Interestingly, the factors that determine the antioxidant activity of fucoidan are complex. Thus, the antioxidant mechanism of fucoidan activity is not fully understood. Chemical modifications of fucoidan can improve its antioxidant activity, which makes it a promising medication as an antioxidant in the context of oxidative stress developed in type-2 diabetes.

6 Role of Fucoidan in Diabetic Complications

6.1 Fucoidan and Metabolic Syndrome

Metabolic syndrome is the pathological prerequisite of cardiovascular and cerebrovascular diseases in type-2 diabetes. In addition, insulin resistance and/or type-2 diabetes mellitus, abdominal obesity, arterial hypertension, and dyslipidemia are considered mandatory components of metabolic syndrome. The prevalence of metabolic syndrome is substantially increasing catastrophically and, along with diabetes, is classified as a pandemic of the twenty-first century (Gheita et al. 2012).

Marine polysaccharides including plant polysaccharides can alleviate metabolic syndrome through various regulatory mechanisms (Wang et al. 2018a). Clinically, fucoidan may potentially play an important role in regulating the functions of body organs of diabetic patients thereby inhibiting complications of diabetes (Fig. 31.2). Obesity, in this regard, is deemed to be one of the manifestations of metabolic

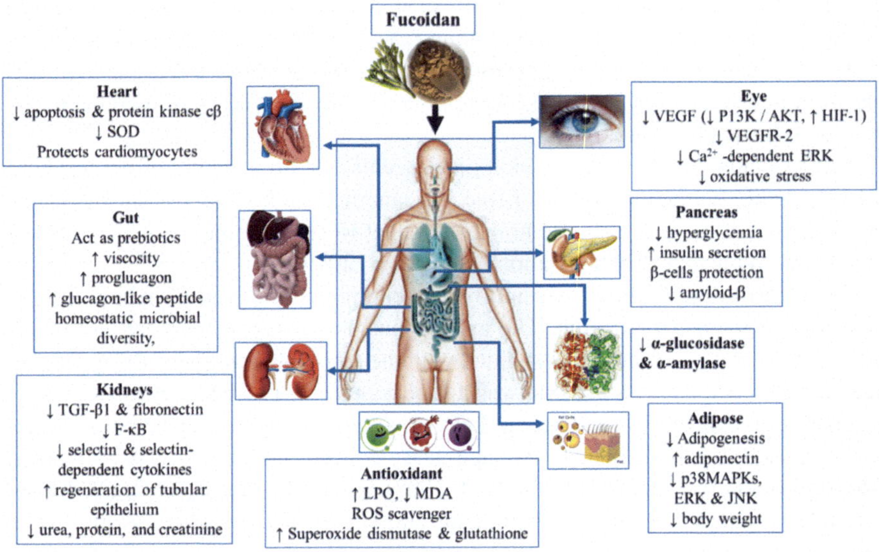

Fig. 31.2 The possible regulating role of fucoidan in alleviating the complications of type 2 DM in some major organs of the body

syndrome. It has been reported that the use of low-molecular-weight fucoidan (250–500 mg/kg daily *per os* for 6 weeks) in db/db mice may reduce the mass of white adipose tissue in the body of animals within 14–42 days after the beginning of treatment (Jeong et al. 2013). At the same time, a decrease in the level of triglycerides, low-density lipoproteins, and total cholesterol in the blood was observed. The authors also observed a decrease in glucose tolerance which could be explained by the increase in serum adiponectin (Jeong et al. 2013). Weight loss after the administration of fucoidan apparently could be associated with inhibition of adipogenesis. The treatment of the 3T3-L1 cell culture of preadipocytes with fucoidan at various doses led to a reduction of adipogenesis and a decrease in the expression of adipogenic genes aP2, ACC, PPARγ (Kim et al. 2009).

There are a number of drugs used in the treatment of T2DM and metabolic syndrome that are agonists of the nuclear receptor PPARγ, which belongs to the PPAR family (Kim et al. 2009). In addition, these authors proved that the suppression of lipogenesis by fucoidan occurs through the MAPK signaling pathway since the heteropolysaccharide inhibits the early activation of p38 mitogen-activated protein kinases (p38MAPKs), extracellular signaling regulatory kinases (ERK), and JunN-terminal kinase (JNK) (Kim et al. 2012).

The inhibition of lipogenesis and enhancement of lipolysis make fucoidan an attractive therapy for obesity treatment. Fucoidan can also make a substantial contribution to down-regulating the epididymal expression in adipose tissue in HFD mice, thereby improving obesity (Kim et al. 2014b). In this case, adipogenic markers and validation-associated cytokines including adipocyte protein 2, peroxisome proliferator-activated receptor-γ and CCAAR/enhancer-binding proteins-α may be reduced (Kim and Lee 2012). Thus, fucoidan, like many other polysaccharides isolated from seaweed, could be utilized for a long time as a functional nutrient which may potentially assist in a gradual decrease of the body weight (Wan-Loy and Siew-Moi 2016).

Lipid metabolism disorders easily diagnosed by the main biochemical parameters of blood plasma may cause metabolic syndrome (Parhofer 2016). With its development, lipid metabolism disorder may lead to an increase in the content of triglycerides, low-density lipoproteins, as well as the atherogenic index (Sascau et al. 2021). This complexity of biochemical disorders ultimately could result in increasing the lipid peroxidation, formation of oxidized-modified LDL, which may contribute to the formation of atherosclerotic plaques, and the development of coronary sclerosis (Sascau et al. 2021). In this context, heteropolysaccharides from seaweed, particularly fucoidan, meaningfully contribute to the reduction of the total blood cholesterol level (Borai et al. 2015). The authors pointed out that the decrease in total cholesterol level occurs due to the binding of bile acids with heteropolysaccharides, which in turn lead to an appropriate decrease in the level of bile acids in the intestine. This additionally promotes an increased synthesis of bile acids in the liver, the substrate which induces the secretion of the LPL cofactor, apolipoprotein C-II (ApoC-II) (Jakobsdottir et al. 2014). These results proposed that fucoidan acts as heparin, releasing LPL in addition to its role to increase intracellular transport and decrease LPL degradation in the medium. In addition, fucoidan-induced LPL and ApoC-II secretion may be involved in the regulation of plasma triglyceride clearance (Yokota et al. 2009). Fucoidan increased LPL activity in Apo E deficient mice, while simultaneously decreasing the intensity of the inflammatory process and the indicators of oxidative stress (Yokota et al. 2016). Furthermore, fucoidan is also able to increase the activity of the enzyme lecithin cholesterol acyltransferase, which is an indicator of a decreased risk for the incidence of atherosclerosis (Huang et al. 2010).

Despite a significant number of studies devoted to the effect of fucoidan on metabolic syndrome and fat metabolism, information on the mechanism of fucoidan activity is still fragmentary and not systematized. First, this is because the lipid-lowering effect of polysaccharides depends on the molecular weight, degree of sulfation, extraction method, and mode of administration (Qi et al. 2010; Cuong et al. 2015). In conclusion, fucoidans are promising compounds for the correction of metabolic syndrome and lipid metabolism disorders that necessarily accompany the development and progression of diabetes mellitus.

6.2 *Fucoidan and Diabetic Nephropathy*

Diabetic nephropathy (DN) is one of the most common microvascular complications of diabetes (Shahcheraghi et al. 2021). It is the leading cause of end-stage renal disease worldwide (Elmarakby and Sullivan 2012). DN is characterized by morphological and ultrastructural changes in the kidney including renal hypertrophy, changes in the structure and functions of glomeruli and tubules, accumulation of extracellular matrix components, interstitial fibrosis, and glomerulosclerosis. Hyperglycemia is recognized as a pivotal factor in the initiation and progression of DN through increased oxidative stress, inflammation, renal polyol formation, accumulation of advanced glycation end products. These end products may ultimately lead to altered hemodynamics such as systemic hypertension and increased intraglomerular pressure.

In vitro and *in vivo* studies were carried out to ascertain the protective effect of fucoidan against diabetic nephropathy (Wang et al. 2015). However, there is no general agreement on the mechanisms of fucoidan action. It is believed that fucoidan and its fractions can prevent or inhibit the progression of diabetic renal complications (Wang et al. 2014). This mechanism of action is primarily associated with its antioxidant activity, which is manifested in both decreasing the lipid peroxidation level and an increase in the activity of antioxidant enzymes (Cohen et al. 1996). On the other hand, the authors conducted experiments with modeling streptozotocin diabetes in rats suggested that due to the characteristic structure and the presence of sulfo groups, fucoidan can reduce heparinase activity in rats with DN. This could potentially contribute to a decrease in the cleavage of heparan sulfate, the main polysaccharide of the basement membrane of the renal glomerulus (Naito et al. 2004).

The experiments on Goto-Kakizaki rats with spontaneous diabetes demonstrated that administration of fucoidan for 13 weeks significantly reduced hyperglycemia (Jiang et al. 2015). In the kidney tissue, regeneration of cells of the tubular epithelium was observed and the degree of lymphocytic infiltration of the organ decreased.

The main factors, which regulate fibrosis, such as Collagen IV level, expression levels of transforming growth factor-β1 (TGF-β1) and fibronectin had also decreased in the renal cortex. Fucoidan also reduces urea, protein, and creatinine in the urine. The authors believed that the main mechanism of fucoidan action, preventing or inhibiting the development of DN, is the decrease in the activation of the NF-κB signaling pathway (Wang et al. 2015). Some authors argued that the prospect of using low molecular weight fucoidans extracted from *Saccharina japonica* in diabetic nephropathy in animals is associated with the ability of heteropolysaccharides to inhibit the level of P-selectin and selectin-dependent cytokines (Guan et al. 2020). This, to a large extent, helps to reduce the inflammatory response in the kidneys and preserves their structure (Xu et al. 2016). Thus, there is no doubt that fucoidan is a potential promising therapeutic agent in the treatment of diabetic nephropathy. However, the exact mechanism of its action requires further study.

6.3 Fucoidan and Diabetic Retinopathy

Diabetic retinopathy (DR) is a neuro-microvascular complication of diabetes, which remains a leading cause of vision loss globally (Stitt et al. 2016). It is estimated to result in blindness of over 10,000 diabetic patients annually and affects nearly all

T1DM and > 60% of T2DM patients during the first two decades of diabetes (Cheung et al. 2010). Diabetic retinopathy is characterized by progressive retinal vascular injury, microaneurysms, intraocular neovascularization, bleeding, and retinal edema. Violation of the blood supply leads to pathological changes in the main cells of the retina, including the pigment layer. The main causes of DR are hyperglycemia, accumulation of glycation products, inflammation, and the production of pro-inflammatory cytokines and growth factors (Wan et al. 2015). While studying the action of fucoidan *in vivo*, it was shown that fucoidan is able to reduce VEGF-stimulated angiogenesis by inhibiting the P13K/AKT signaling pathway and increasing the expression of the angiogenesis regulator HIF-1 (Narazaki et al. 2008).

In vitro experiments have shown that the introduction of fucoidan reduces the expression of VEGF receptors, and the level of VEGF in the pigment layer of the retina, which leads to a decrease in angiogenesis and neovascularization of the retina (Dorschmann et al. 2019). The pathways by which fucoidan reduces the expression and secretion of VEGF are unknown, however, it is suggested that fucoidan may inhibit autocrine VEGFR-2 signaling. The data obtained suggest the possibility of using fucoidan preparations as a VEGF antagonist or as an additional therapy in the treatment of DR (Narazaki et al. 2008; Dithmer et al. 2014). The positive effect of fucoidan in the treatment of diabetic retinopathy may be due to its antioxidant properties, which directly depend on the sulfate/fucose ratio in the sample (Wang et al. 2008).

In *in vitro* experiments on cell cultures of the retinal pigment epithelium, showed that the oxidative stress caused by hyperglycemia is reduced with the introduction of fucoidan. The mechanism of fucoidan action, in this case, is the inhibition of ROS formation through the Ca^{2+} – dependent ERK signaling pathway (Li et al. 2015). This study confirms the possibility of using fucoidans to develop new clinical treatments for diabetic retinopathy.

6.4 Fucoidan and Cardiovascular Dysfunction in Diabetes

Cardiovascular disease (CVD) remains the leading cause of morbidity and mortality among people with diabetes (Einarson et al. 2018). It accounts for up to two-thirds of deaths among people with diabetes and results in a cost of more than $ 37.3 billion in diabetes-associated expenditure per year (Low Wang et al. 2016).

Atherosclerotic CVD, including coronary heart disease, cerebrovascular disease, and peripheral arterial disease, and heart failure are the prominent CVD manifested in diabetes (Almourani et al. 2019). Coronary artery disease, cerebrovascular disease, peripheral arterial disease and heart failure are the most common cardiovascular diseases associated with diabetes (Fu et al. 2004). In experiments carried out on the Goto-Kakizaki rat line (spontaneous diabetes), it was shown that when fucoidan is administered for 3 months at various doses, the activity of superoxide dismutase in cardiomyocytes is inhibited, the levels of apoptosis and protein kinase cβ activity decrease. Thus, the authors argued that fucoidan has a protective effect on cardiomyocytes (Yu et al. 2014).

The recent articles contain mainly experimental materials, concerning the anti-diabetic action of fucoidans, therefore, the clinical trial data are of great value. Hernández-Corona et al. (2014) discussed the results of a double-blind, randomized, placebo-controlled clinical trial of the efficacy of fucoidan, obtained on 13 obese or overweight diabetic volunteers who received fucoidan *per os* daily for three months. All participants in the study stated that there has been an increase in the insulin secretion and a decrease in insulin resistance. During the basic therapy with fucol (dietary supplement based on fucoidan from *F. evanescens*), a decrease in the level of glycemia was observed in patients with dyslipoproteinemia (Wei et al. 2017).

7 Conclusion and Future Recommendations

The results presented in this chapter proved the promise of using fucoidans preparations as a potential anti-diabetic agent. It was noticed that relevant research about fucoidan is still ongoing around the world, since its mechanisms of action are yet not completely investigated. Although fucoidans have shown promising positive results in the treatment of a wide range of diseases *in vivo* and *in vitro* experiments, its potential as a dietary supplement has been proven. Some studies indicated that oral administration of fucoidans is preferred, however, its effectiveness in therapy is not well understood (Cunha and Grenha 2016). Since fucoidans are not broken down by digestive enzymes, when administered orally, the possibility of its synergistic action with short-chain fatty acids, which create a favorable environment for the action of heteropolysaccharides, cannot be excluded (Tran et al. 2021). However, the way in which these polysaccharides retain their functional properties at various stages of processing in industrial application remains unclear and needs further demonstrative studies. The identification of potential mechanisms of the anti-diabetic action of fucoidans is also suggested to be assessed on the liver, pancreas, and intestines.

Acknowledgments The work was carried out with financial support from the Ministry of Science and Education of the Russian Federation, State Contract no. FEUZ-2020-0058 (H687.42B.223/20).

References

Ale MT, Mikkelsen JD, Meyer AS (2011) Important determinants for fucoidan bioactivity: a critical review of structure-function relations and extraction methods for fucose-containing sulfated polysaccharides from brown seaweeds. Mar Drugs 9(10):2106–2130. https://doi.org/10.3390/md9102106

Aleissa MS, Alkahtani S, Abd Eldaim MA, Ahmed AM, Bungău SG, Almutairi B, Bin-Jumah M, Alkahtane AA, Alyousif MS, Abdel-Daim MM (2020) Fucoidan ameliorates oxidative stress, inflammation, DNA damage, and hepatorenal injuries in diabetic rats intoxicated with aflatoxin B_1. Oxidative Med Cell Longev. https://doi.org/10.1155/2020/9316751

Almourani R, Chinnakotla B, Patel R, Kurukulasuriya LR, Sowers J (2019) Diabetes and cardiovascular disease: an update. Curr Diab Rep 19(12). https://doi.org/10.1007/s11892-019-1239-x

Andoh T (2016) Subchapter 19A - Insulin. In: Takei Y, Ando H, Tsutsui K (eds) Handbook of hormones. Academic, San Diego, pp 157–153. https://doi.org/10.1016/B978-0-12-801028-0.00148-3

Begum R, Howlader S, Mamun-Or-Rashid ANM, Rafiquzzamam SM, Ashraf GM, Albadrani GM, Sayed AA, Peluso I, Abdel-Daim MM, Uddin MS (2021) Antioxidant and signal modulating effects of brown seaweed-derived compounds against oxidative stress associated pathology. Oxidative Med Cell Longev 2021:9974890, 22 p. https://doi.org/10.1155/2021/9974890

Besednova NN, Ermakova SP, Kuznetsova TA, Makarenkova ID, Krizhanovsky SP, Andryukov BG, Saporozhets TS (2019) Algae and type 2 diabetes: new treatment strategies. Antibiotiki i Khimioterapiya 64(11–12):54–67. https://doi.org/10.1016/0235-2990-2019-64-11-12-54-67

Bocanegra A, Macho-Gonzalez A, Garcimartín A, Benedí J, Sanchez-Muniz FJ (2021) Whole alga, algal extracts, and compounds as ingredients of functional foods: composition and action mechanism relationships in the prevention and treatment of type-2 diabetes mellitus. Int J Mol Sci 22:3816. https://doi.org/10.3390/ijms22083816

Borai IH, Ezz MK, Rizk MZ, Matloub A, Aly H, El A, Farrag R, Fouad GI (2015) Hypolipidemic and anti-atherogenic effect of sulphated polysaccharides from the green alga *Ulva fasciata*. Int J Pharm Sci Rev Res 31(1):1–12

Chen Q, Liu M, Zhang P, Fan S, Huang J, Yu S, Zhang C, Li H (2019) Fucoidan and galactooligosaccharides ameliorate high-fat diet–induced dyslipidemia in rats by modulating the gut microbiota and bile acid metabolism. Nutrition 65:50–59. https://doi.org/10.1016/j.nut.2019.03.001

Cheung N, Mitchell P, Wong TY (2010) Diabetic retinopathy. Lancet 376(9735):124–136. https://doi.org/10.1016/S0140-6736(09)62124-3

Cohen MP, Clements RS, Cohen JA, Shearman CW (1996) Prevention of decline in renal function in the diabetic db/db mouse. Diabetologia 39(3):270–274. https://doi.org/10.1007/BF00418341

Cunha L, Grenha A (2016) Sulfated seaweed polysaccharides as multifunctional materials in drug delivery applications. Mar Drugs 14(3). https://doi.org/10.3390/md14030042

Cuong HD, Thuy TTT, Huong TT, Ly BM, Van TTT (2015) Structure and hypolipidaemic activity of fucoidan extracted from brown seaweed *Sargassum henslowianum*. Nat Prod Res 29(5):411–415. https://doi.org/10.1080/14786419.2014.948436

Derosa G, Pascuzzo MD, D'Angelo A, Maffioli P (2019) Ascophyllum nodosum, fucus vesiculosus and chromium picolinate nutraceutical composition can help to treat type 2 diabetic patients. Diabet Metab Synd Obesit Targ Ther 12:1861–1865. https://doi.org/10.2147/DMSO.S212429

Dithmer M, Fuchs S, Shi Y, Schmidt H, Richert E, Roider J, Klettner A (2014) Fucoidan reduces secretion and expression of vascular endothelial growth factor in the retinal pigment epithelium and reduces angiogenesis in vitro. PLoS One 9(2). https://doi.org/10.1371/journal.pone.0089150

Dorschmann P, Bittkau KS, Neupane S, Roider J, Alban S, Klettner A (2019) Effects of fucoidans from five different brown algae on oxidative stress and VEGF interference in ocular cells. Mar Drugs 17(5). https://doi.org/10.3390/md17050258

Einarson TR, Acs A, Ludwig C, Panton UH (2018) Prevalence of cardiovascular disease in type 2 diabetes: a systematic literature review of scientific evidence from across the world in 2007-2017. Cardiovasc Diabetol 17(1):83. https://doi.org/10.1186/s12933-018-0728-6

Elmarakby AA, Sullivan JC (2012) Relationship between oxidative stress and inflammatory cytokines in diabetic nephropathy. Cardiovasc Ther 30(1):49–59. https://doi.org/10.1111/j.1755-5922.2010.00218.x

Fitton JH, Stringer DN, Karpiniec SS (2015) Therapies from fucoidan: an update. Mar Drugs 13(9):5920–5946. https://doi.org/10.3390/md13095920

Fu X, Xue C, Ning Y, Li Z, Xu J (2004) Acute antihypertensive effects of fucoidan oligosaccharides prepared from *Laminaria japonica* on renovascular hypertensive rats. J Ocean Univ China (Nat Sci) 34(4):560–564

Gheda S, Naby MA, Mohamed T, Pereira L, Khamis A (2021) Antidiabetic and antioxidant activity of phlorotannins extracted from the brown seaweed *Cystoseira compressa* in streptozotocin-induced diabetic rats. Environ Sci Pollut Res 28(18):22886–22901. https://doi.org/10.1007/s11356-021-12347-5

Gheita TA, El-Fishawy HS, Nasrallah MM, Hussein H (2012) Insulin resistance and metabolic syndrome in primary gout: relation to punched-out erosions. Int J Rheum Dis 15(6):521–525. https://doi.org/10.1111/1756-185X.12007

Guan Z, Shi L, Wang T, Xu Y, Xu T (2020) Low molecular weight fucoidan from *saccharina japonica* ameliorates the antioxidant capacity and reduces plaque areas in aorta in apoe-deficient mice with atherosclerosis. Pharm Chem J 54:804–810. https://doi.org/10.1007/s11094-020-02278-9

Gunathilaka TL, Samarakoon K, Ranasinghe P, Peiris LDC (2020) Antidiabetic potential of marine brown algae—a mini review. J Diabetes Res 2020:1230218, 13 p. https://doi.org/10.1155/2020/1230218

Hernández-Corona DM, Martínez-Abundis E, González-Ortiz M (2014) Effect of Fucoidan administration on insulin secretion and insulin resistance in overweight or obese adults. J Med Food 17(7):830–832. https://doi.org/10.1089/jmf.2013.0053

Huang L, Wen K, Gao X, Liu Y (2010) Hypolipidemic effect of fucoidan from *Laminaria japonica* in hyperlipidemic rats. Pharm Biol 48(4):422–426. https://doi.org/10.3109/13880200903150435

Hurrle S, Hsu WH (2017) The etiology of oxidative stress in insulin resistance. Biom J 40(5):257–262. https://doi.org/10.1016/j.bj.2017.06.007

Irhimeh MR, Fitton JH, Lowenthal RM (2009) Pilot clinical study to evaluate the anticoagulant activity of fucoidan. Blood Coagul Fibrinolysis 20(7):607–610. https://doi.org/10.1097/MBC.0b013e32833135fe

Jakobsdottir G, Nilsson U, Blanco N, Sterner O, Nyman M (2014) Effects of soluble and insoluble fractions from bilberries, black currants, and raspberries on short-chain fatty acid formation, anthocyanin excretion, and cholesterol in rats. J Agric Food Chem 62(19):4359–4368. https://doi.org/10.1021/jf5007566

Jeong YT, Kim YD, Jung YM, Park DC, Lee DS, Ku SK, Li X, Lu Y, Chao GH, Kim KJ, Lee JY, Baek MC, Kang W, Hwang SL, Chang HW (2013) Low molecular weight fucoidan improves endoplasmic reticulum stress-reduced insulin sensitivity through ampactivated protein kinase activation in l6 myotubes and restores lipid homeostasis in a mouse model of type 2 diabetess. Mol Pharmacol 84(1):147–157. https://doi.org/10.1124/mol.113.085100

Jiang X, Yu J, Ma Z, Zhang H, Xie F (2015) Effects of fucoidan on insulin stimulation and pancreatic protection via the cAMP signaling pathway in vivo and in vitro. Mol Med Rep 12(3):4501–4507. https://doi.org/10.3892/mmr.2015.3989

Jiao G, Yu G, Zhang J, Ewart HS (2011) Chemical structures and bioactivities of sulfated polysaccharides from marine algae. Mar Drugs 9(2):196–233. https://doi.org/10.3390/md9020196

Jin W, Ren L, Liu B, Zhang Q, Zhong W (2018) Structural features of sulfated glucuronomannan oligosaccharides and their antioxidant activity. Mar Drugs 16(9). https://doi.org/10.3390/md16090291

Kellogg J, Grace MH, Lila MA (2014) Phlorotannins from alaskan seaweed inhibit carbolytic enzyme activity. Mar Drugs 12(10):5277–5294. https://doi.org/10.3390/md12105277

Kellogg J, Esposito D, Grace MH, Komarnytsky S, Lila MA (2015) Alaskan seaweeds lower inflammation in RAW 264.7 macrophages and decrease lipid accumulation in 3T3-L1 adipocytes. J Funct Foods 15:396–407. https://doi.org/10.1016/j.jff.2015.03.049

Kim KJ, Lee BY (2012) Fucoidan from the sporophyll of *Undaria pinnatifida* suppresses adipocyte differentiation by inhibition of inflammation-related cytokines in 3T3-L1 cells. Nutr Res 32(6):439–447. https://doi.org/10.1016/j.nutres.2012.04.003

Kim MJ, Chang UJ, Lee JS (2009) Inhibitory effects of fucoidan in 3T3-L1 adipocyte differentiation. Mar Biotechnol 11(5):557–562. https://doi.org/10.1007/s10126-008-9170-1

Kim KJ, Yoon KY, Lee BY (2012) Fucoidan regulate blood glucose homeostasis in C57BL/KSJ m+/+db and C57BL/KSJ db/db mice. Fitoterapia 83(6):1105–1109. https://doi.org/10.1016/j.fitote.2012.04.027

Kim KT, Rioux LE, Turgeon SL (2014a) Alpha-amylase and alpha-glucosidase inhibition is differentially modulated by fucoidan obtained from *Fucus vesiculosus* and *Ascophyllum nodosum*. Phytochemistry 98:27–33. https://doi.org/10.1016/j.phytochem.2013.12.003

Kim MJ, Jeon J, Lee JS (2014b) Fucoidan prevents high-fat diet-induced obesity in animals by suppression of fat accumulation. Phytother Res 28(1):137–143. https://doi.org/10.1002/ptr.4965

Kim KT, Rioux LE, Turgeon SL (2015) Molecular weight and sulfate content modulate the inhibition of α-amylase by fucoidan relevant for type 2 diabetes management. Pharm Nutrit 3(3):108–114. https://doi.org/10.1016/j.phanu.2015.02.001

Li B, Lu F, Wei X, Zhao R (2008) Fucoidan: structure and bioactivity. Molecules 13(8):1671–1695. https://doi.org/10.3390/molecules13081671

Li X, Zhao H, Wang Q, Liang H, Jiang X (2015) Fucoidan protects ARPE-19 cells from oxidative stress via normalization of reactive oxygen species generation through the Ca2+–dependent ERK signaling pathway. Mol Med Rep 11(5):3746–3752. https://doi.org/10.3892/mmr.2015.3224

Lopes G, Barbosa M, Andrade PB, Valentão P (2019) Phlorotannins from Fucales: potential to control hyperglycemia and diabetes-related vascular complications. J Appl Phycol 31(5):3143–3152. https://doi.org/10.1007/s10811-019-01816-7

Low Wang CC, Hess CN, Hiatt WR, Goldfine AB (2016) Clinical update: cardiovascular disease in diabetes mellitus. Circulation 133(24):2459–2502. https://doi.org/10.1161/CIRCULATIONAHA.116.022194

Luthuli S, Wu S, Cheng Y, Zheng X, Wu M, Tong H (2019) Therapeutic effects of fucoidan: a review on recent studies. Mar Drugs 17(9). https://doi.org/10.3390/md17090487

Mabate B, Daub CD, Malgas S, Edkins AL, Pletschke BI (2021) Fucoidan structure and its impact on glucose metabolism: implications for diabetes and cancer therapy. Mar Drugs 19(1). https://doi.org/10.3390/md19010030

Mak W, Hamid N, Liu T, Lu J, White WL (2013) Fucoidan from New Zealand *Undaria pinnatifida:* monthly variations and determination of antioxidant activities. Carbohydr Polym 95(1):606–614. https://doi.org/10.1016/j.carbpol.2013.02.047

Mansour A, Hosseini S, Larijani B, Pajouhi M, Mohajeri-Tehrani MR (2013) Nutrients related to GLP1 secretory responses. Nutrition 29(6):813–820. https://doi.org/10.1016/j.nut.2012.11.015

Melo MRS, Feitosa JPA, Freitas ALP, De Paula RCM (2002) Isolation and characterization of soluble sulfated polysaccharide from the red seaweed *Gracilaria cornea*. Carbohydr Polym 49(4):491–498. https://doi.org/10.1016/S0144-8617(02)00006-1

Moin ASM, Butler AE (2019) Alterations in beta cell identity in type 1 and type 2 diabetes. Curr Diab Rep 19(9). https://doi.org/10.1007/s11892-019-1194-6

Myers SP, O'Connor J, Fitton JH, Brooks L, Rolfe M, Connellan P, Wohlmuth H, Cheras PA, Morris C (2010) A combined phase I and II open-label study on the effects of a seaweed extract nutrient complex on osteoarthritis. Biol Targ Ther 4:33–44. https://doi.org/10.2147/btt.s8354

Naito Y, Uchiyama K, Aoi W, Hasegawa G, Nakamura N, Yoshida N, Maoka T, Takahashi J, Yoshikawa T (2004) Prevention of diabetic nephropathy by treatment with astaxanthin in diabetic db/db mice. Bio Factors 20(1):49–59. https://doi.org/10.1002/biof.5520200105

Narazaki M, Segarra M, Tosato G (2008) Sulfated polysaccharides identified as inducers of neuropilin-1 internalization and functional inhibition of VEGF 165 and semaphorin3A. Blood 111(8):4126–4136. https://doi.org/10.1182/blood-2007-09-112474

Oliveira RM, Câmara RBG, Monte JFS, Viana RLS, Melo KRT, Queiroz MF, Filgueira LGA, Oyama LM, Rocha HAO (2018) Commercial fucoidans from *Fucus vesiculosus* can be grouped into antiadipogenic and adipogenic agents. Mar Drugs 16(6). https://doi.org/10.3390/md16060193

Oza MJ, Kulkarni YA (2018) Formononetin treatment in type 2 diabetic rats reduces insulin resistance and hyperglycemia. Front Pharmacol 9:739. https://doi.org/10.3389/fphar.2018.00739

Oza MJ, Kulkarni YA (2020) Trifolium pratense (Red Clover) Improve SIRT1 Expression and Glycogen Content in High Fat Diet-Streptozotocin Induced Type 2 Diabetes in Rats. Chem Biodivers 17(4). https://doi.org/10.1002/cbdv.202000019

Parhofer KG (2016) The treatment of disorders of lipid metabolism. Deutsch Arztebt Int 113(15):261–268. https://doi.org/10.3238/arztebl.2016.0261

Parnell JA, Reimer RA (2012) Prebiotic fiber modulation of the gut microbiota improves risk factors for obesity and the metabolic syndrome. Gut Microbes 3(1):29–34. https://doi.org/10.4161/gmic.19246

Ponce NMA, Stortz CA (2020) A comprehensive and comparative analysis of the fucoidan compositional data across the *phaeophyceae*. Front Plant Sci 11:556312. https://doi.org/10.3389/fpls.2020.556312

Pradhan B, Patra S, Nayak R, Behera C, Dash SR, Nayak S, Sahu BB, Bhutia SK, Jena M (2020) Multifunctional role of fucoidan, sulfated polysaccharides in human health and disease: a journey under the sea in pursuit of potent therapeutic agents. Int J Biol Macromol 164:4263–4278. https://doi.org/10.1016/j.ijbiomac.2020.09.019

Qi H, Liu X, Ma J, Zhang Q, Li Z (2010) In vitro antioxidant activity of acetylated derivatives of polysaccharide extracted from *Ulva pertusa* (Cholorophta). J Med Plants Res 4(23):2445–2451. https://doi.org/10.5897/jmpr10.019

Rodriguez-Jasso RM, Mussatto SI, Pastrana L, Aguilar CN, Teixeira JA (2011) Microwave-assisted extraction of sulfated polysaccharides (fucoidan) from brown seaweed. Carbohydr Polym 86(3):1137–1144. https://doi.org/10.1016/j.carbpol.2011.06.006

Saeedi P, Petersohn I, Salpea P, Malanda B, Karuranga S, Unwin N, Colagiuri S, Guariguata L, Motala AA, Ogurtsova K, Shaw JE, Bright D, Williams R (2019) Global and regional diabetes prevalence estimates for 2019 and projections for 2030 and 2045: results from the International Diabetes Federation Diabetes Atlas, 9th edition. Diabetes Res Clin Pract 157:107843. https://doi.org/10.1016/j.diabres.2019.107843

Sascau R, Clement A, Radu R, Prisacariu C, Statescu C (2021) Triglyceride-rich lipoproteins and their remnants as silent promoters of atherosclerotic cardiovascular disease and other metabolic disorders: a review. Nutrients 13(6):1774. https://doi.org/10.3390/nu13061774

Shahcheraghi SH, Aljabali AAA, Al Zoubi MS, Mishra V, Charbe NB, Haggag YA, Shrivastava G, Almutary AG, Alnuqaydan AM, Barh D, Dua K, Chellappan DK, Gupta G, Lotfi M, Serrano-Aroca Á, Bahar B, Mishra YK, Takayama K, Panda PK, Bakshi HA, Tambuwala MM (2021) Overview of key molecular and pharmacological targets for diabetes and associated diseases. Life Sci 278:119632. https://doi.org/10.1016/j.lfs.2021.119632

Shan X, Liu X, Hao J, Cai C, Fan F, Dun Y, Zhao X, Liu X, Li C, Yu G (2016) In vitro and in vivo hypoglycemic effects of brown algal fucoidans. Int J Biol Macromol 82:249–255. https://doi.org/10.1016/j.ijbiomac.2015.11.036

Stitt AW, Curtis TM, Chen M, Medina RJ, McKay GJ, Jenkins A, Gardiner TA, Lyons TJ, Hammes HP, Simó R, Lois N (2016) The progress in understanding and treatment of diabetic retinopathy. Prog Retin Eye Res 51:156–186. https://doi.org/10.1016/j.preteyeres.2015.08.001

Tiwari BK, Pandey KB, Abidi A, Rizvi SI (2013) Markers of oxidative stress during diabetes mellitus. J Biomark 2013:378790. https://doi.org/10.1155/2013/378790

Tran PHL, Lee BJ, Tran TTD (2021) Current developments in the oral drug delivery of fucoidan. Int J Pharm 598:120371. https://doi.org/10.1016/j.ijpharm.2021.120371

Wan TT, Li XF, Sun YM, Li YB, Su Y (2015) Recent advances in understanding the biochemical and molecular mechanism of diabetic retinopathy. Biomed Pharmacother 74:145–147. https://doi.org/10.1016/j.biopha.2015.08.002

Wang J, Zhang Q, Zhang Z, Li Z (2008) Antioxidant activity of sulfated polysaccharide fractions extracted from *Laminaria japonica*. Int J Biol Macromol 42(2):127–132. https://doi.org/10.1016/j.ijbiomac.2007.10.003

Wang J, Liu H, Li N, Zhang Q, Zhang H (2014) The protective effect of fucoidan in rats with streptozotocin-induced diabetic nephropathy. Mar Drugs 12(6):3292–3306. https://doi.org/10.3390/md12063292

Wang Y, Nie M, Lu Y, Wang R, Li J, Yang B, Xia M, Zhang H, Li X (2015) Fucoidan exerts protective effects against diabetic nephropathy related to spontaneous diabetes through the NF-κB signaling pathway *in vivo* and *in vitro*. Int J Mol Med 35(4):1067–1073. https://doi.org/10.3892/ijmm.2015.2095

Wang X, Wang X, Jiang H, Cai C, Li G, Hao J, Yu G (2018a) Marine polysaccharides attenuate metabolic syndrome by fermentation products and altering gut microbiota: an overview. Carbohydr Polym 195:601–612. https://doi.org/10.1016/j.carbpol.2018.05.003

Wang X, Yi K, Zhao Y (2018b) Fucoidan inhibits amyloid-β-induced toxicity in transgenic: *Caenorhabditis elegans* by reducing the accumulation of amyloid-β and decreasing the production of reactive oxygen species. Food Funct 9(1):552–560. https://doi.org/10.1039/c7fo00662d

Wang Y, Xing M, Cao Q, Ji A, Liang H, Song S (2019) Biological activities of fucoidan and the factors mediating its therapeutic effects: a review of recent studies. Mar Drugs 17(3). https://doi.org/10.3390/md17030183

Wan-Loy C, Siew-Moi P (2016) Marine algae as a potential source for anti-obesity agents. Mar Drugs 14(12). https://doi.org/10.3390/md14120222

Wei H, Gao Z, Zheng L, Zhang C, Liu Z, Yang Y, Teng H, Hou L, Yin Y, Zou X (2017) Protective effects of fucoidan on Aβ25-35 and D-gal-induced neurotoxicity in PC12 cells and D-gal-induced cognitive dysfunction in mice. Mar Drugs 15(3). https://doi.org/10.3390/md15030077

Weng L, Zhang F, Wang R, Ma W, Song Y (2019) A review on protective role of genistein against oxidative stress in diabetes and related complications. Chem Biol Interact 310:108665. https://doi.org/10.1016/j.cbi.2019.05.031

Wu J, Morrison F, Zhao Z, Haynes G, He X, Ali AK, Shubina M, Malmasi S, Ge W, Peng X, Turchin A (2021a) Reasons for discontinuing insulin and factors associated with insulin discontinuation in patients with type 2 diabetes mellitus: a real-world evidence study. Clin Diabet Endocrinol 7(1):1. https://doi.org/10.1186/s40842-020-00115-2

Wu Q, Wu S, Cheng Y, Zhang Z, Mao G, Li S, Yang Y, Zhang X, Wu M, Tong H (2021b) Sargassum fusiforme fucoidan modifies gut microbiota and intestinal metabolites during alleviation of hyperglycemia in type 2 diabetic mice. Food Funct 12(8):3572–3585. https://doi.org/10.1039/d0fo03329d

Xu Y, Zhang Q, Luo D, Wang J, Duan D (2016) Low molecular weight fucoidan modulates P-selectin and alleviates diabetic nephropathy. Int J Biol Macromol 91:233–240. https://doi.org/10.1016/j.ijbiomac.2016.05.081

Yang XD, Liu CG, Tian YJ, Gao DH, Li WS, Ma HL (2017) Inhibitory effect of fucoidan on hypoglycemia in diabetes mellitus anim. Int J Clin Exp Med 10(5):8529–8534

Yokota T, Nagashima M, Ghazizadeh M, Kawanami O (2009) Increased effect of fucoidan on lipoprotein lipase secretion in adipocytes. Life Sci 84(15–16):523–529. https://doi.org/10.1016/j.lfs.2009.01.020

Yokota T, Nomura K, Nagashima M, Kamimura N (2016) Fucoidan alleviates high-fat diet-induced dyslipidemia and atherosclerosis in ApoEshl mice deficient in apolipoprotein E expression. J Nutrit Biochem 32:46–54. https://doi.org/10.1016/j.jnutbio.2016.01.011

Yu X, Zhang Q, Cui W, Zeng Z, Yang W, Zhang C, Zhao H, Gao W, Wang X, Luo D (2014) Low molecular weight fucoidan alleviates cardiac dysfunction in diabetic goto-kakizaki rats by reducing oxidative stress and cardiomyocyte apoptosis. J Diabetes Res 2014:420929. https://doi.org/10.1155/2014/420929

Zhang D, Fujii I, Lin C, Ito K, Guan H, Zhao J, Shinohara M, Matsukura M (2008) The stimulatory activities of polysaccharide compounds derived from algae extracts on insulin secretion in vitro. Biol Pharm Bull 31(5):921–924. https://doi.org/10.1248/bpb.31.921

Zhao D, Xu J, Xu X (2018) Bioactivity of fucoidan extracted from *Laminaria japonica* using a novel procedure with high yield. Food Chem 245:911–918. https://doi.org/10.1016/j.foodchem.2017.11.083

Zhao G, Bhatia D, Jung F, Lipscombe L (2021) Risk of type 2 diabetes mellitus in women with prior hypertensive disorders of pregnancy: a systematic review and meta-analysis. Diabetologia 64(3):491–503. https://doi.org/10.1007/s00125-020-05343-w

Chapter 32
Pharmacological Importance of Bioactive Molecules of Seaweeds

Naveen Jayapala, Madan Kumar Perumal, Revathy Baskaran, and Baskaran Vallikannan

Abbreviation

DHA	Docosahexaenoic acid
EPA	Eicosapentaenoic acid
Fxn	Fucoxanthin
HFD	High fat diet
WHO	World Health Organization

1 Introduction

"Nutritive Food is the key to Health"- dietary food habits and rich nutritive food intake improves human health. Hence, the Greek physician Hippocrates mentioned the quote **"Let food be thy medicine, and medicine be thy food"**. Nutrition imbalance will lead to various health problems such as diabetes, obesity, cardiovascular,

N. Jayapala
Department of Biochemistry, CSIR-Central Food Technological Research Institute, Mysore, Karnataka, India

M. K. Perumal · B. Vallikannan (✉)
Department of Biochemistry, CSIR-Central Food Technological Research Institute, Mysore, Karnataka, India

Academy of Scientific and Innovative Research (AcSIR), Ghaziabad, Uttar Pradesh, India
e-mail: baskaranv@cftri.res.in

R. Baskaran
Academy of Scientific and Innovative Research (AcSIR), Ghaziabad, Uttar Pradesh, India

Department of Fruit and Vegetable Technology, CSIR-Central Food Technological Research Institute, Mysore, Karnataka, India

A. Ranga Rao, G. A. Ravishankar (eds.), *Sustainable Global Resources of Seaweeds Volume 2*, https://doi.org/10.1007/978-3-030-92174-3_32

and cancer. According to the World Health Organization (WHO) report, lower intake of vegetables and fruits in the diet has accounted for about 2.7 million deaths including 14% of gastrointestinal cancer deaths, 11% of ischemic heart disease deaths, and 9% of deaths due to stroke (WHO Fact Sheet 2018). The leafy vegetables, vegetables, and fruits are rich sources of nutrition and play a vital role in the prevention of metabolic diseases (Naveen and Baskaran 2018; Naveen et al. 2021a; Vani et al. 2021; Venkateish et al. 2021). Like terrestrial plant sources, marine sources especially seaweeds are gaining more importance in recent years.

Consumption of marine algae as food is practiced in Asian countries like China, Korea, and Japan. For example, approximately one-fifth of Japanese meals contained seaweed (MacArtain et al. 2007; Rebours et al. 2014). The current surge of research interest is on the pharmacologically important bioactives of seaweed. Isolation, characterization, and functional role of bioactives have potential applications in the functional food, pharmaceutical, and nutraceutical industries, with motivation towards modulating the metabolic diseases (Collins et al. 2016). Hence, the present chapter will focus on discussing the health benefits of various seaweed bioactives.

2 Seaweed Bioactives

Marine algal biomass are always a rich source of novel bioactives with immense applications such as nutraceutical, pharmaceutical, agrochemical, and cosmeceutical (Ravishankar and Ranga Rao 2019). The major pharmacological bioactives of seaweed are carotenoids, polysaccharides, polyphenols, lipids, and fatty acids. Various bioactives and their pharmacological activities of selected seaweeds are given in Table 32.1. The bioactives from seaweeds possess a wide range of pharmacological applications including antioxidant, anti-microbial, anti-diabetic, anti-obese, anti-inflammatory, anti-cancer, skin protective and neuroprotective effects (Fig. 32.1).

2.1 Carotenoids

Carotenoids are chemically and structurally diverse groups of yellow to red-colored compounds (>700 different compounds have been identified) comprising of 3–13 conjugated with double bonds and in some cases, six carbon hydroxylated ring structures at one or both ends of the molecule. Due to the increased application of carotenoids in the food and pharma industries, the demand for their production and screening from novel sources has increased (Craft et al. 2012). Therefore, the biotechnological production of carotenoids naturally from seaweeds has increased mainly because seaweeds are cost-effective and abundant in nature compared to terrestrial plants (Galasso et al. 2017). These seaweed carotenoids display numerous biological activities such as antioxidant, anti-inflammatory, anti-diabetic, anti-obesity, anti-cancer, and anti-angiogenic (Pangestuti and Siahaan 2018).

Table 32.1 Bioactives and their pharmacological activities of selected seaweeds

Seaweed name	Bioactive name	Biological activity	Reference
Undaria pinnatifida	Fucoxanthin	Antioxidant, Anti-mutagenic	Yan et al. (1999), Cho and Rhee (1997)
Porphyra sp.	Phycoerythrobilin	Antioxidant	Yabuta et al. (2010)
Phaeophyceae	Sulfated fucoidans	Anticoagulant	Chevolot and Foucault (1999)
Rhodophyceae	Sulfated galactans	Anticoagulant	Kolender et al. (1997)
Codium fragile	Xyloarabinogalactans	Anticoagulant	Jurd and Rogers (1995)
Sargassum thunbergii	Phlorotannins	Anti-thrombotic	Bae (2011)
A. nodosum and *Alaria esculenta*	Phlorotannins	Anti-proliferative	Nwosu et al. (2011)
H. fusiforme	Phlorotannins	Anti-microbial	Tang et al. (2020)
Symphyocladia latiuscula	Bromophenols	Antioxidant	Duan et al. (2007)
Saccharina japonica	Fucoidans	Immunomodulatory	Mayer and Lehmann (2000), Shibata et al. (2002), Kim and Joo (2008)
Acanthophora spicifera	Phloroglucinol	Tumoricidal	Vasanthi and Rajamanickam (2004)
Corallina pilulifera	Ethanolic extract	Anti-inflammatary	Yang and Zhang (2009)
Schizymenia dubyi	Sulfated glucuronogalactan	Anticoagulant	Bourgougnon and Lahaye (1996)
Lobophora variegate	Fucans	Anti-inflammatary	Jiao and Yu (2011)
Sargassum hemiphyllum	Fucoidan	Anti-cancer	Yan et al. (2015)
Ecklonia cava	Phlorotannin 6,6′-bieckol	Anti-inflammatary	Kazłowska et al. 2010
Palmaria palmata	Mycosporine-like amino acids	Antioxidant	Nishida et al. (2020)
Porphyria dentate	Catechol, rutin and hesperidin, MGDG, DGDG, SQDG	Anti-viral	de Souza et al. (2012), Saha et al. (2012)
Sargassum vulgare, Caulerpa racemosa Stypopodium zonale	Merodíterpenoids, aromatic acid, epitaondiol and peroxylactone	Anti-viral	Soares et al. (2007), Wang et al. (2008), Koishi et al. (2012)
Padina gymnospora, Palisada perforate, Caulerpa racemose	Crude extract	Antibacterial	Alghazeer et al. (2013)
Sargassum wightii, Cystoseira barbata	Alkaloids	Antibiotic	Marudhupandi and Kumar (2013)

(continued)

Table 32.1 (continued)

Seaweed name	Bioactive name	Biological activity	Reference
Ulva armoricana and Solieria chordalis	Lipid fractions	Anti-proliferative	Kendel et al. (2015)
Solieria chordalis and Sargassum muticum	Lipid fractions	Free radical scavening	Terme et al. (2018)
Ecklonia kurome	Fucan sulfate	Anticoagulant	Nishino and Nagumo (1991)
Turbinaria ornata, Dictyo pterisdelicatula	Sulphated polysaccharides	Anticoagulant	Arivuselvan et al. (2011)
Codium divaricatum, C. adhaerence, C. latum, C. fragile		Anticoagulant, Anti-tumor	Magalhaes et al. (2011)
Padina gymnospora	Heterofucan	Anticoagulant	Silva et al. (2005)
Sargassum polycystum	Crude extract	Anti-diabetic	Motshakeri et al. (2014)

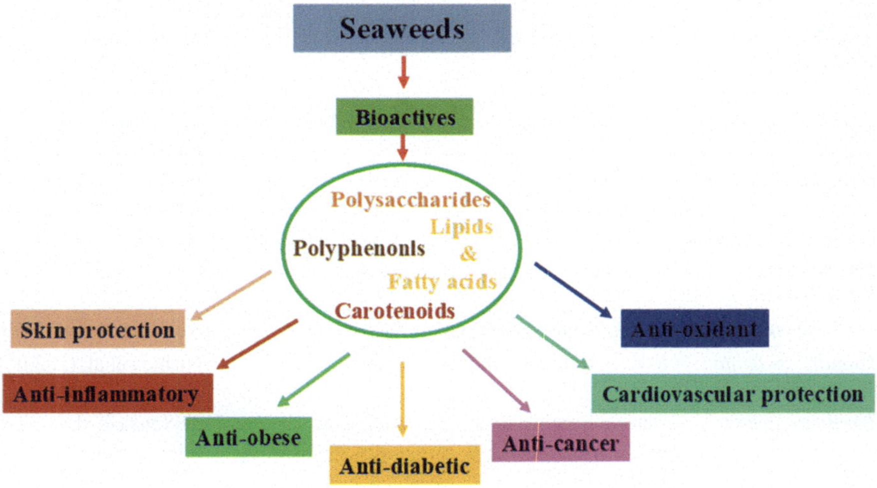

Fig. 32.1 Pharmacological activities of seaweed bioactives

Fucoxanthin (Fxn; Fig. 32.2) is the most extensively studied seaweed carotenoid found abundantly in brown seaweeds (Pereira et al. 2021). Fxn is commonly present in the edge of the brown seaweed thallus, which experiences heavy light exposure (Lobban and Wynne 1981). Among the carotenoids from marine sources, Fxn contributes major (≥10%) of the total carotenoids in the environment. Fxn has numerous biological activities such as antioxidant, anti-obesity, anti-diabetic, hepatoprotective, neuroprotective, and anti-cancer activity (Naveen et al. 2019).

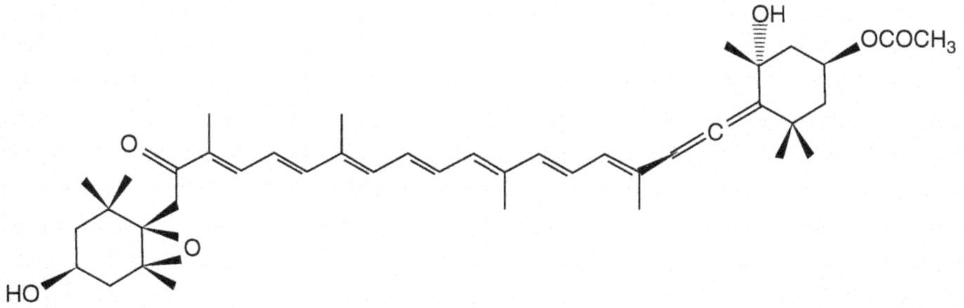

Fig. 32.2 Chemical structure of fucoxanthin, the major brown seaweed carotenoid

In a study conducted by Woo et al. (2010), Fxn lowered hepatic lipid content, increased faeces weight and faecal lipids in high fat diet (HFD) fed C57BL/6N mice. Similar finding was observed by Park et al. (2011) where Fxn supplementation showed a hepatoprotective effect in HFD fed mice by inhibiting the activity of fatty acid synthesis enzymes in the liver, normalized the hepatic glycogen content by regulating glucokinase. Tsukui et al. (2007) found increased levels of docosahexaenoic acid (DHA) in the liver of Fxn administered C57BL/6J mice. Further, experiments by the same research group demonstrated that administration of Fxn and fucoxanthinol improved the DHA and arachidonic acid levels in KK-Ay mice liver (Tsukui et al. 2009). The skin protective role of Fxn was studied by Heo and Jeon (2009). Fxn showed an anti-pigment effect in UV-B-induced melanogenesis by downregulating the melanogenic receptors and by the synthesis of prostaglandin E2 (Shimoda et al. 2010). Fxn treatment managed skin-wrinkle and skin photoaging formation in UV B-irradiated hairless mice (Urikura et al. 2011).

Fxn inhibited the progression of hyperglycemia and hypoinsulinemia by activating the insulin pathway (GLUT4 expression and translocation) in diabetic/obese KK-A(y) mice (Nishikawa et al. 2012). Ethanolic extract of Fxn from *Undaria pinnatifida* prevented insulin resistance in HFD induced obese C57BL/6J mice (Park et al. 2011). A study by Maeda et al. (2007) reported that Fxn treatment to diabetic/obese KK- A(y) mice in combination with fish oil attenuated the net body weight gain. Fxn exhibited an anti-proliferative effect by inhibiting the growth of epithelial cells (SRA 01/04) of the human (Moreau et al. 2006). Shiratori et al. (2005) demonstrated the anti-ocular inflammatory effect of Fxn in male Lewis rats where Fxn ameliorated the development of lipopolysaccharide-induced uveitis. In addition, Fxn showed an anti-angiogenic effect by inhibiting angiogenesis-related diseases like diabetic retinopathy, cancer, atherosclerosis, and psoriasis. Studies conducted by Sugawara et al. (2006) showed the anti-angiogenic effect of Fxn in endothelial cells of the human umbilical vein and rat aortic ring. In addition, Fxn displayed the suppression of new blood vessels development and inhibited differentiation of endothelial progenitor cells to endothelial cells.

Fxn was reported to induce apoptosis in breast cancer cells (Vijay et al. 2018). The results from the study by Zhu et al. (2018) showed Fxn treatment induced apoptosis and autophagy of human gastric cancer cells (SGC-7901) mediated by LC3, beclin-1, and cleaved caspase-3. Fucoxanthinol suppressed the integrin signals in human colorectal cancer cells (Terasaki et al. 2017). Liu et al. (2016) disclosed that Fxn treatment reduced the cell proliferation, invasion, migration, and increased activation by inhibiting the PI3K/Akt/mTOR expression. Fxn enhanced the cell cycle arrest and thereby increased the apoptosis in human bladder cancer cells (T24) by decreasing the mortalin, a multipotent chaperon regulation (Wang et al. 2014). Fxn extracted from the *Saccharin japonica* showed an anti-metastatic effect in melanoma cells (B16-F10) both *in vitro* and *in vivo* (Chung et al. 2013) and breast cancer cells (MCF-7 and MDA-MB-231), in addition, Fxn displayed cytotoxicity (De la Mare et al. 2013).

2.2 Polysaccharides

Seaweeds are known as a rich source of polysaccharides (Suleria et al. 2017) and are considered as sources of nutrients as well. Seaweed polysaccharides have a vast range of applications in the food and pharma industries (Renn 1997). Polysaccharides namely laminarin and fucoidan are found in brown seaweed (Menshova et al. 2014), ulvan in green seaweed and carrageenan in red seaweed (Mayakrishnan et al. 2013). The chemical structure of these major polysaccharides is displayed below (Fig. 32.3).

Fucoidan is an aqueous soluble sulphated heteropolysaccharide originated from the brown seaweed extracellular matrix (Wang et al. 2010). Fucoidan mainly consist of L-fucose and sulphate groups and L-fucose-4-sulfate as monosaccharide units (Fig. 32.3a). The skeleton of fucoidan is rich in fucose and sulphated depending on the position of fucose units (Zhao et al. 2018). The position of sulphated fucose unit and molecular mass contribute an important role in biological activity (Li et al.

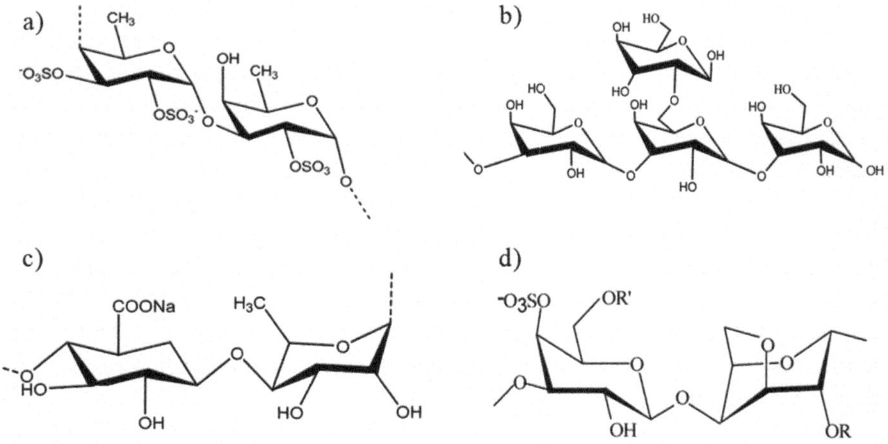

Fig. 32.3 Chemical structure of major seaweed polysaccharides. (**a**) Fucoidan; (**b**) Laminarin; (**c**) Ulvan; (**d**) Carrageenan

2008; Ale et al. 2011). Choi et al. (2007) found that the fucoidan extracted from *Sargassum fulvellum* showed a strong NO scavenger compared to α-tocopherol. Also, fucoidan showed more antioxidant activity than the sodium alginate extracted from the brown seaweed - *Cystoseira compressa* (Hentati et al. 2018). Recently, Ashayerizadeh et al. (2019) reported the antioxidant activity of fucoidan which was further correlated with the lower molecular weight and sulfate/fucose ratio.

Fucoidan is reported to exhibit anti-diabetic activity, increased insulin secretion, and inhibition of glycolytic enzyme activity (Havale and Pal 2009; Pozharitskaya et al. 2020). Dietary supplementation of fucoidan (1% and 2%) reduced the complications of obesity in the HFD induced obese mice (Kim et al. 2014). Research work by Synytsya et al. (2010) demonstrated that fucoidans extracted from the Korean brown algae - *Undaria pinnatifida* showed anti-tumor properties in several cancer cells including HepG2 (human hepatocellular liver carcinoma), Hela (human cervical), PC-3 (human prostate), and A549 (lung carcinoma). Yan et al. (2015) showed that fucoidan obtained from *Sargassum hemiphyllum* induced miR-29b expression mediated through inhibition of DNA methyltransferase 3B expression in human liver cancer cells. Lee et al. (2014) endorsed the anti-cancer potential of fucoidan, isolated from *Fucus vesiculosus* found to induce apoptosis in human mucoepidermoid carcinoma cells by activation of caspase-dependent apoptosis. Treatment of breast cancer cells (MCF-7) with fucoidans (50 μg/ml) potentiated apoptosis by caspase-8 activation (Palanisamy et al. 2017).

Laminarin is a polysaccharide extracted from brown seaweed mainly consist of β-(1–3)-glucan with β-(1–6)-linkages of 20–25 units (Nelson and Lewis 1974). Antioxidant activity of laminarin was compared with the lower molecular weight laminarin after γ-irradiation. Lower molecular weight laminarin showed higher antioxidant activity due to the presence of the carboxyl group (Choi et al. 2011).

Laminarin exhibited anti-inflammatory and immunostimulatory properties in RAW 264.7 cells by altering several inflammatory markers (Lee et al. 2012). Bobadilla et al. (2013) found the immunostimulatory activity of β-D-glucan of *Durvillaea antarctica* by increased activity (16.9%) of CD19+ B lymphocytes compared with the control. Sulfated laminarin reduced heparinase activity and hence considered as anti-inflammatory molecule (Kadam et al. 2015). Supplementation of laminarin significantly improved the antioxidant enzymes and immune response genes (IL-1β, TLR2, and IL-8) (Yin et al. 2014). The anti-cancer activity of laminarin was studied by Ji and Ji (2014). Recently, we studied the anti-obesity property of laminarin in HFD induced obese C57BL6 mice and reported reduced body weight, triglyceride, and cholesterol levels (Sharma and Baskaran 2021).

Carrageenan is a polysaccharide of red seaweeds (Rhodophyta) and the name carrageenan is an Irish name of seaweed "Carrageenan" means "little rock". Carrageenan typically occurs in various forms like ι-, κ-, λ- based on the source and the environmental condition. Antioxidant activity of ι-carrageenans, κ-carrageenans, and λ-carrageenans extracted from *Eucheuma spinosum*, *Eucheuma cottonii,* and *Gigartina acicularis/G. pistillata* were analysed *in vitro* and among the types, λ-carrageenan showed higher antioxidant potential (De Souza et al. 2007). Zhang et al. (2003) reported that the polysaccharide of *Porphyra haitanensis* and

Mastocarpus stellatus showed good antioxidant potential in animals. Anti-proliferative effect of κ-carrageenans extracted from *Hypnea musciformis* (Rhodophyceae) was determined in human cancer cells (MCF-7 and SH-SY5Y) (Souza et al. 2018). Carrageenan isolated from *Solieria chordalis* exhibited immunostimulatory activity, the treatment showed enhanced cytotoxicity, neutrophil phagocytosis, and lymphocyte proliferation stimulation (Stephanie et al. 2010).

Chen et al. (2007) found the λ-carrageenan oligosaccharides (150–300 μg/mL) inhibitory effect on tumor blood vessel endothelial cell differentiation by down-regulating the intracellular matrix metalloproteinase expression. Yuan et al. (2005) found the chemical alteration (acetylated, sulphated and phosphorylated) of κ-carrageenan oligosaccharides enhanced the antioxidant and immunomodulation activity. Carrageenan treatment for 6 days inhibited insulin resistance and glucose intolerance in C57BL/6J mice. The combination of HFD and carrageenan-induced the galectin-3 and galectin-3 binding to the insulin receptor, reduced tyrosine phosphorylation of the insulin receptor (Bhattacharyya et al. 2019).

2.3 Polyphenols

Seaweeds are also a rich source of polyphenols such as flavonoids, bromophenols, phlorotannins, phenolic terpenoids, and mycosporine-like amino acids. Polyphenols are highly hydrophilic secondary metabolites that are the defense mechanisms of seaweeds. Green and red seaweeds comprise of bromophenols, flavonoids, phenolics acids, phenolic terpenoids, and mycosporine-like amino acids (Heo et al. 2005; Corona et al. 2017; Gómez-Guzmán et al. 2018). While, phlorotannins are the unique class of polyphenols found exclusively in the brown seaweeds (Wijesekara et al. 2011).

Phlorotannins are polyphenolic compounds that are biosynthesized by acetate malonate pathway and typically formed by a group of complex polymers of phloroglucinol (1,3,5-trihydroxybenzene) (Fig. 32.4) (Wang et al. 2012; Li et al. 2017). Phlorotannins are reported for numerous pharmacological activities including antioxidant, anti-microbial, anti-diabetic, anti-inflammatory, anti-cancer, etc. (Javed et al. 2021). A study by Sathya et al. (2017) demonstrated the antioxidant activity of purified phlorotannins extracted from the brown seaweed *C. trinodis*. Numerous studies highlighted the neuroprotective effect of phlorotannins against oxidative stress and inflammation (Choi et al. 2015; Cui et al. 2019; Manandhar et al. 2019; Seong et al. 2019). In a study using xenograft mice model, oral administration of dieckol (300 mg/kg/week) significantly suppressed the tumor growth (Ahn et al. 2014). Naveen et al. (2021b) extracted total polyphenols and its profile revealed the presence of danshensu, luteolin, quercetin derivative, hydroxyl-ferulic acid, genistein, rosmarinic acid, acacetin derivative. Antioxidant potential of total phenolic was exhibited using DPPH (10.83 ± 2.81 μg/mL) and ABTS (58.85 ± 2.28 μg/mL) assays. Further, total phenolics revealed inhibition of glycolytic enzymes like α-amylase and α-glucosidase with IC50 value of 47.2 ± 2.9 μg and 28.8 ± 2.3 μg respectively.

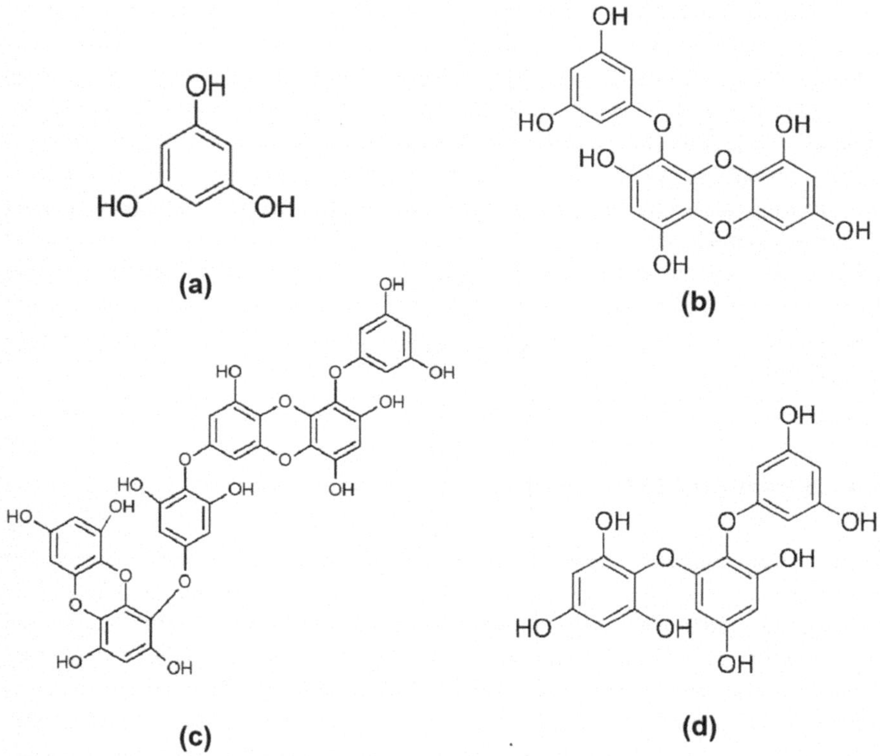

Fig. 32.4 Chemical structure of major phlorotannins; (**a**) Phloroglucinol; (**b**) Eckol; (**c**) Diekol; (**d**) Triphlorethol-A

Bromophenols, a kind of marine algae-derived polyphenolic compound comprised of one or more benzene rings, a varying degree of bromine, and hydroxyl-substituents. Bromophenols are isolated from vast marine algae including the red algae - *Rhodomela larix* (Katsui et al. 1967), brown algae (Chung et al. 2003), and green algae (Sun et al. 1983). Bromophenols are reported to possess a variety of biological activities including antioxidant, anti-microbial, anti-diabetic, anti-cancer, and anti-thrombotic effects. Bromophenols isolated from *Symphyocladia latiuscula* were reported for DPPH free radical scavenging activities (Duan et al. 2007). Bromophenol derivatives isolated from *Rhodomela confervoides* showed antibacterial activity against eight strains of gram-positive and gram-negative bacteria (Xu et al. 2003). Rawsonol, a novel brominated diphenyl methane derivative, isolated from the green algae - *Avrainvillea rawsoni* potentially inhibited HMG-CoA reductase activity (Carte et al. 1989). Bromophenols extracted from the brown algae - *Leathesia nana* exhibited cytotoxic effect against human cancer cell lines (Shi et al. 2009). Bromophenol derivatives 3-bromo-4,5-dihydroxy benzoic acid methyl ester and 3-bromo-4,5-dihydroxy-benzaldehyde isolated from *Rhodomela confervoides* inhibited cell viability of KB, Bel-7402, and A549 cells (Han et al. 2005).

Mycosporine-like amino acids are a group of secondary metabolites that are present exclusively in red seaweeds. Mycosporine-like amino acids majorly contribute towards photoprotection by absorbing UV radiation (Shick and Dunlap 2002; Navarro et al. 2018). A study by Gacesa et al. (2018) highlighted that porphyra-334 and shinorine induced Nrf2-regulated antioxidant response in primary dermal fibroblasts. In a study conducted in HaCaT keratinocytes, 4 different mycosporine-like amino acids showed increased proliferation and migration thereby exhibited wound healing ability (Orfanoudaki et al. 2020). Mycosporine-like amino acids from methanolic extracts from edible wild-harvested (*Chondrus crispus, Mastocarpus stellatus, Palmaria palmata*) and cultivated (*C. crispus*) marine red macroalgae showed an anti-proliferative effect in HeLa and U-937 cells (Athukorala et al. 2016).

2.4 Lipids and Fatty Acids

Seaweeds are generally recognized as low-energy food. The seaweeds comprised of low lipid content ranging from 0.4% to 5% dry weight with rich saturated fatty acids and palmitic acid content. ω-3 and ω-6 polyunsaturated fatty acids (PUFAs) form a major portion of seaweed lipids present majorly in brown seaweeds followed by green and red seaweeds (Mišurcová et al. 2011; Ganesan et al. 2019). Triacylglycerols constitute the major part of seaweed lipids and predominantly play a storage functions. Also, it is to be noted that the lipid and fatty acid composition of the seaweeds change depending on the environmental conditions such as light, salinity, level of nitrogen (Mishra et al. 1993; Floreto and Teshima 1998). A study by Abirami and Kowsalya (2012) reported the high occurrence of ω-3 fatty acids in *Ulva*, *Acanthophora*, and *Gracilaria* species. Seaweeds with high ω-3 fatty acids exhibited anti-hyperlipidemic, anti-inflammatory, anti-hypertensive, and inhibition of angiotensin I-converting enzyme (Fontenelle et al. 2018). Eicosapentaenoic acid (EPA) and DHA are the major ω-3 fatty acids observed in *Ulva* species which further contributed to its anti-inflammatory effects (McCauley et al. 2018). Administration of brown seaweed lipids extracted from *Undaria pinnatifida*, *Sargassum horneri,* and *Cystoseira hakodatensis* shown to lower lipid hydroperoxide levels in KK-Ay mice (Airanthi et al. 2011). Lipids from *Ulva armoricana* and *Solieria chordalis* were reported for anti-proliferative activity in human non-small cell bronchopulmonary carcinoma cells (NSCLC-N6) (Kendel et al. 2015).

3 Conclusion

Marine seaweeds are a great source of bioactives and have pharmacological importance in supporting human health. Even though seaweeds are of different kinds like brown, red, and green, they have their feature of bioactive contents. Although seaweed bioactives were reported for numerous health benefits, the integration of seaweeds in the human diet is limited to only a few Asian countries like Japan, China, and Korea. Hence, addition of seaweed bioactives in food, nutraceutical, pharmaceutical and industrial applications will further enhance the production and consumption of seaweed-based products. Also, owing to the high health benefits of seaweed bioactives, there is necessity to explore seaweed research which will help in identify highly effective bioactives for safe pharmaceutical application against health diseases and disorders.

Acknowledgement The first author Dr. J. Naveen greatly acknowledge Indian Council for Medical Research for Research Associateship (ICMR-RA)(F.No. 3/1/2/170/2020-Nut) and Dr. V. Baskaran acknowledge DST-SERB for the financial support (Award No. EMR/2017/002047; Dt. 14.11.2018).

References

Abirami RG, Kowsalya S (2012) Phytochemical screening, microbial load and antimicrobial activity of underexploited seaweeds. Int Res J Microbiol 3(10):328–332

Ahn JH, Yang YI, Lee KT, Choi JH (2014) Dieckol, isolated from the edible brown algae *Ecklonia cava*, induces apoptosis of ovarian cancer cells and inhibits tumor xenograft growth. J Cancer Res Clin Oncol 141:255–268. https://doi.org/10.1007/s00432-014-1819-8

Airanthi MK, Sasaki N, Iwasaki S, Baba N, Abe M, Hosokawa M, Miyashita K (2011) Effect of brown seaweed lipids on fatty acid composition and lipid hydroperoxide levels of mouse liver. J Agric Food Chem 59(8):4156–4163. https://doi.org/10.1021/jf104643b

Ale MT, Mikkelsen JD, Meyer AS (2011) Important determinants for fucoidan bioactivity: A critical review of structure-function relations and extraction methods for fucose-containing sulfated polysaccharides from brown seaweeds. Mar Drugs 9:2106–2130

Alghazeer R, Whida F, Abduelrhman E, Gammoudi F, Naili M (2013) In vitro antibacterial activity of alkaloid extracts from green, red and brown macroalgae from western coast of Libya. Afr J Biotechnol 12(51):7086–7091

Arivuselvan N, Radhiga M, Anantharaman P (2011) In vitro antioxidant and anticoagulant activities of sulphated polysaccharides from Brown seaweed (Turbinaria ornata) (Turner) J. Agardh. Asian J Pharm Biol Res 1:232–239

Ashayerizadeh O, Dastar B, Pourashouri P (2019) Study of antioxidant and antibacterial activities of depolymerized fucoidans extracted from Sargassum tenerrimum. Int J Biol Macromol 151(4). https://doi.org/10.1016/j.ijbiomac.2019.10.172

Athukorala Y, Trang S, Kwok C, Yuan YV (2016) Antiproliferative and antioxidant activities and mycosporine-like amino acid profiles of wild-harvested and cultivated edible canadian marine red macroalgae. Molecules 21(1):E119. https://doi.org/10.3390/molecules21010119

Bae JS (2011) Antithrombotic and profibrinolytic activities of phloroglucinol. Food Chem Toxicol 49(7):1572–1577

Bhattacharyya S, Feferman L, Tobacman JK (2019) Distinct effects of carrageenan and high-fat consumption on the mechanisms of insulin resistance in nonobese and obese models of type 2 diabetes. J Diabetes Res 2019:9582714, 14 pages

Bobadilla F, Rodriguez-Tirado C, Imarai M, Galotto MJ, Andersson R (2013) Soluble β-1,3/1,6-glucan in seaweed from the southern hemisphere and its immunomodulatory effect. Carbohydr Polym 92:241–248

Bourgougnon N, Lahaye M (1996) Annual variation in composition and in vitro anti-HIV-1 activity of the sulfated glucuronogalactan from Schizymenia dubyi (Rhodophyta, Gigartinales). J Appl Phycol 8(2):155–161

Carte BK, Troupe N, Chan JA, Westley JW, Faulkner DJ (1989) Rawsonol, an inhibitor of HMG-CoA reductase from the tropical green alga *Avrainvillea rawsoni*. Phytochemistry 28:2917–2919

Chen H, Yan X, Lin J, Wang F, Xu W (2007) Depolymerized products of λ-carrageenan as a potent angiogenesis inhibitor. J Agric Food Chem 55:6910–6917

Chevolot L, Foucault A (1999) Further data on the structure of brown seaweed fucans: Relationships with anticoagulant activity. Carbohydr Res 319(1):154–165

Cho EJ, Rhee SH (1997) Antimutagenic and cancer cell growth inhibitory effects of seaweeds. Prevent Nutrit Food Sci 2(4):348–353

Choi DS, Athukorala Y, Jeon YJ, Senevirathne M, Cho KR, Kim SH (2007) Antioxidant activity of sulfated polysaccharides isolated from *Sargassum fulvellum*. Prev Nutr Food Sci 12:65–73

Choi J, Kim H-J, Lee J-W (2011) Structural feature and antioxidant activity of low molecular weight laminarin degraded by gamma irradiation. Food Chem 129:520–523

Choi BW, Lee HS, Shin HC, Lee BH (2015) Multifunctional activity of polyphenolic compounds associated with a potential for Alzheimer's disease therapy from *Ecklonia cava*. Phytother Res 29:549–553. https://doi.org/10.1002/ptr.5282

Chung HY, Ma WCJ, Ang PO, Kim JS, Chen F (2003) Seasonal variations of bromophenols in brown algae (*Padina arborescens*, *Sargassum siliquastrum*, and *Lobophora variegata*) collected in Hong Kong. J Agric Food Chem 51:2619–2624

Chung T, Choi HJ, Lee J, Jeong H, Kim C, Joo M, Choi J, Han C, Kim S, Choi J, Ha K (2013) Marine algal fucoxanthin inhibits the metastatic potential of cancer cells. Biochem Biophys Res Commun 439:580–585

Collins KG, Fitzgerald GF, Stanton C, Ross RP (2016) Looking beyond the terrestrial: the potential of seaweed derived bioactives to treat non-communicable diseases. Mar Drugs 14:60. https://doi.org/10.3390/md14030060

Corona G, Coman MM, Guo Y, Hotchkiss S, Gill C, Yaqoob P, Spencer JPE, Rowland I (2017) Effect of simulated gastrointestinal digestion and fermentation on polyphenolic content and bioactivity of brown seaweed phlorotannin-rich extracts. Mol Nutr Food Res 61:1–10

Craft BD, Kerrihard AL, Amarowicz R, Pegg RB (2012) Phenol-based antioxidants and the methods used for their assessment. Compr Rev Food Sci Food Saf 11:148–173

Cui Y, Amarsanaa K, Lee JH, Rhim JK, Kwon JM, Kim SH, Park JM, Jung SC, Eun SY (2019) Neuroprotective mechanisms of dieckol against glutamate toxicity through reactive oxygen species scavenging and nuclear factor-like 2/heme oxygenase-1 pathway. Korean J Physiol Pharmacol 23:121–130

De la Mare JA, Sterrenberg JN, Sukhthankar MG (2013) Assessment of potential anti-cancer stem cell activity of marine algal compounds using an in vitro mammosphere assay. Cancer Cell Int 13:39

De Souza MCR, Marques CT, Dore CMG, Da Silva FRF, Rocha HAO, Leite EL (2007) Antioxidant activities of sulfated polysaccharides from brown and red seaweeds. J Appl Phycol 19:153–160

De Souza LM, Sassaki GL, Romanos MT, Barreto-Bergter E (2012) Structural characterization and anti-HSV-1 and HSV-2 activity of glycolipids from the marine algae *Osmundaria obtusiloba* isolated from Southeastern Brazilian coast. Mar Drugs 10:918–931

Duan XJ, Li XM, Wang BG (2007) Highly brominated mono- and bis-phenols from the marine red alga *Symphyocladia latiuscula* with radical-scavenging activity. J Nat Prod 70:1210–1213

Floreto EAT, Teshima S (1998) The fatty acid composition of seaweeds exposed to different levels of light intensity and salinity. Bot Mar 41:467–484

Fontenelle TPC, Lima GC, Mesquita JX, de Souza Lopes JL, de Brito TV, das Chagas Vieira Júnior F, Sales AB, Aragão KS, Souza MHLP, dos Reis Barbosa AL (2018) Lectin obtained from the red seaweed *Bryothamnion triquetrum*: secondary structure and anti-inflammatory activity in mice. Int J Biol Macromol 112:1122–1130

Gacesa R, Lawrence KP, Georgakopoulos ND (2018) The mycosporine-like amino acids porphyra-334 and shinorine are antioxidants and direct antagonists of Keap1-Nrf2 binding. Biochimie 154:35–44

Galasso C, Corinaldesi C, Sansone C (2017) Carotenoids from marine organisms: biological functions and industrial applications. Antioxidants 6:96

Ganesan AR, Tiwari U, Rajauria G (2019) Seaweed nutraceuticals and their therapeutic role in disease prevention. Food Sci Human Wellness 8:252–263

Gómez-Guzmán M, Rodríguez-Nogales A, Algieri F, Gálvez J (2018) Potential role of seaweed polyphenols in cardiovascular-associated disorders. Mar Drugs 16:250

Han LJ, Xu NJ, Shi JG, Yan XJ, Zeng CK (2005) Isolation and pharmacological activities of bromophenols from *Rhodomela confervoides*. Chin J Oceanol Limn 23:226–229

Havale SH, Pal M (2009) Medicinal chemistry approaches to the inhibition of dipeptidyl peptidase-4 for the treatment of type 2 diabetes. Bioorganic Med Chem 17:1783–1802

Hentati F, Delattre C, Ursu AV, Desbrières J, Le Cerf D, Gardarin C, Abdelkafi S, Michaud P, Pierre G (2018) Structural characterization and antioxidant activity of water-soluble polysaccharides from the Tunisian brown seaweed *Cystoseira compressa*. Carbohydr Polym 198:589–600

Heo SJ, Park EJ, Lee KW, Jeon YJ (2005) Antioxidant activities of enzymatic extracts from brown seaweeds. Bioresour Technol 96:1613–1623

Heo SJ, Jeon YJ (2009) Protective effect of fucoxanthin isolated from Sargassum siliquastrum on UV-B induced cell damage. J Photochem Photobiol B 95:101–107

Javed A, Hussain MB, Tahir A, Waheed M, Anwar A, Shariati MA, Plygun S, Laishevtcev A, Pasalar M (2021) Pharmacological applications of phlorotannins: a comprehensive review. Curr Drug Discov Technol 18(2):282–292

Ji CF, Ji YB (2014) Laminarin-induced apoptosis in human colon cancer LoVo cells. Oncol Lett 7:1728–1732

Jiao G, Yu G (2011) Chemical structures and bioactivities of sulfated polysaccharides from marine algae. Mar Drugs 9(2):196–223

Jurd KM, Rogers DJ (1995) Anticoagulant properties of sulphated polysaccharides and a proteoglycan from *Codium fragile* ssp. atlanticum. J Appl Phycol 7(4):339–345

Kadam SU, Tiwari BK, O'donnell CP (2015) Extraction, structure and biofunctional activities of laminarin from brown algae. Int J Food Sci Technol 50(1):24–31

Katsui N, Suzuki Y, Kitamura S, Irie T (1967) 5,6-dibromoprotocatechualdehyde and 2,3-dibromo-4,5-dihydroxybenzyl methyl ether: new dibromophenols from *Rhodomela larix*. Tetrahedron 23:1185–1188

Kazłowska K, Hsu T, Hou CC, Yang WC, Tsai GJ (2010) Anti-inflammatory properties of phenolic compounds and crude extract from *Porphyra dentata*. J Ethnopharmacol 128(1):123–130

Kendel M, Wielgosz-Collin G, Bertrand S, Roussakis C, Bourgougnon N, Bedoux G (2015) Lipid composition, fatty acids and sterols in the seaweeds *Ulva armoricana*, and *Solieria chordalis* from Brittany (France): an analysis from nutritional, chemotaxonomic, and antiproliferative activity perspectives. Mar Drugs 13(9):5606–5628. https://doi.org/10.3390/md13095606

Kim MH, Joo HG (2008) Immunostimulatory effects of fucoidan on bone marrow-derived dendritic cells. Immunol Lett 115:138–143

Kim M-J, Jeon J, Lee J-S (2014) Fucoidan prevents high-fat diet-induced obesity in animals by suppression of fat accumulation. Phytother Res 28:137–143. https://doi.org/10.1002/ptr.4965

Koishi AC, Zanello PR, Bianco EM, Bordignon J, Santos CNDD (2012) Screening of dengue virus antiviral activity of marine seaweeds by an in situ enzyme-linked immunosorbent assay. PLoS One 7(12):e51089. https://doi.org/10.1371/journal.pone.0051089

Kolender AA, Pujol CA, Damonte EB, Matulewicz MC, Cerezo AS (1997) The system of sulfated -linked D-mannans from the red seaweed *Nothogenia fastigiata*: Structures, antiherpetic and anticoagulant properties. Carbohydr Res 304(1):53–60

Lee JY, Kim YJ, Kim HJ, Kim YS, Park W (2012) Immunostimulatory effect of laminarin on RAW 264.7 mouse macrophages. Molecules 17(5):5404–5411

Lee HE, Choi ES, Shin J, Lee SO, Park KS, Cho NP, Cho SD (2014) Fucoidan induces caspase-dependent apoptosis in MC3 human mucoepidermoid carcinoma cells. Exp Ther Med 7:228–232

Li B, Lu F, Wei X, Zhao R (2008) Phlorotannins: structure and bioactivity. Molecules 13:1671–1695
Li Y, Fu X, Duan D, Liu X, Xu J, Gao X (2017) Extraction and identification of phlorotannins from the brown alga, *Sargassum fusiforme* (Harvey) Setchell. Mar Drugs 15(2):49
Liu Y, Zheng J, Zhang Y, Wang Z, Yang Y, Bai M, Dai Y (2016) Fucoxanthin activates apoptosis via inhibition of PI3K/Akt/mTOR pathway and suppresses invasion and migration by restriction of p38-MMP-2/9 pathway in human glioblastoma cells. Neurochem Res 41(10):2728–2751
Lobban CS, Wynne MJ (1981) The biology of seaweeds. University of California Press, Oakland
MacArtain P, Gill CI, Brooks M, Campbell R, Rowland IR (2007) Nutritional value of edible seaweeds. Nutr Rev 65:535–543
Maeda H, Hosokawa M, Sashima T, Miyashita K (2007) Dietary combination of fucoxanthin and fish oil attenuates the weight gain of white adipose tissue and decreases blood glucose in obese/diabetic KK-Ay mice. J Agric Food Chem 55:7701–7706
Magalhaes KD, Costa LS, Fidelis GP, Oliveira RM, Nobre LTDB, Dantas-Santos N, Camara RBG, Albuquerque IRL, Cordeiro SL, Sabry DA, Costa MSSP, Alves LG, Rocha HAO (2011) Anticoagulant, antioxidant and antitumor activities of heterofucans from the seaweed *Dictyopteris delicatula*. Int J Mol Sci 12:3352–3365
Manandhar B, Wagle A, Seong SH, Paudel P, Kim H-R, Jung HA, Choi JS (2019) Phlorotannins with potential anti-tyrosinase and antioxidant activity isolated from the marine seaweed *Ecklonia stolonifera*. Antioxidants 8:240
Marudhupandi T, Kumar TTA (2013) Antibacterial effect of fucoidan from *Sargassum wightii* against the chosen human bacterial pathogens. Int Curr Pharmaceut J 2(10):156–158
Mayakrishnan V, Kannappan P, Abdullah N, Ali Ahmed AB (2013) Cardioprotective activity of polysaccharides derived from marine algae: An overview. Trends Food Sci Technol 30(2):98–104
Mayer AM, Lehmann VK (2000) Marine pharmacology in 1998: Marine compounds with antibacterial, anticoagulant, antifungal, antiinflammatory, anthelmintic, antiplatelet, antiprotozoal, and antiviral activities; with actions on the cardiovascular, endocrine, immune, and nervous systems; and other miscellaneous mechanisms of action. The Pharmacologist 42(2):62–69
McCauley JI, Winberg PC, Meyer BJ, Skropeta D (2018) Effects of nutrients and processing on the nutritionally important metabolites of Ulva sp. (Chlorophyta). Algal Res 35:586–594
Menshova RV, Ermakova SP, Anastyuk SD, Isakov VV, Dubrovskaya YV, Kusaykin MI, Um B-H, Zvyagintseva TN (2014) Structure, enzymatic transformation and anticancer activity of branched high molecular weight laminaran from brown alga *Eisenia bicyclis*. Carbohydr Polym 99:101–109
Mishra VK, Temelli F, Ooraikul B, Shacklock PF, Craigie JS (1993) Lipids of the red alga, *Palmaria palmata*. Bot Mar 36:169–174
Mišurcová L, Ambrožová J, Samek D (2011) Seaweed lipids as nutraceuticals. Adv Food Nutr Res 64:339–355
Moreau D, Tomasoni C, Jacquot C, Kaas R, Le Guedes R, Cadoret JP, Muller-Feuga A, Kontiza I, Vagias C, Roussis V, Roussakis C (2006) Cultivated microalgae and the carotenoid fucoxanthin from *Odontella aurita* as potent anti-proliferative agents in broncho pulmonary and epithelial cell lines. Environ Toxicol Phar 22:97–103
Motshakeri M, Ebrahimi M, Goh YM, Othman HH, Hair-Bejo M, Mohamed S (2014) Effects of brown seaweed (Sargassum polycystum) extracts on kidney, liver, and pancreas of type 2 diabetic rat model. Evid Based Complement Alternat Med 2014:1
Navarro NP, Figueroa FL, Korbee N, Bonomi J, Álvarez-Gómez F, de la Coba F (2018) Mycosporine-like amino acids from red algae to develop natural UV sunscreens. In: Rastogi RP (ed) Sunscreens: source, formulation, efficacy and recommendations. biochemistry research trends. NOVA Science Publisher, New York, NY, pp 99–130
Naveen J, Baskaran V (2018) Antidiabetic plant-derived nutraceuticals: a critical review. Eur J Nutr 57:1275–1299
Naveen J, Madan KP, Baskaran V (2019) 22 biological activities and safety. In: Handbook of algal technologies and phytochemicals: two volume set, p 245

Naveen J, Madan Kumar P, Revathy B, Baskaran V (2021a) Nutritional facts of leafy vegetable *Lactuca sativa*. In: Advances in health and disease, p 33
Naveen J, Revathy B, Baskaran V (2021b) Profiling of bioactives and in vitro evaluation of antioxidant and antidiabetic property of polyphenols of marine *algae Padina tetrastromatica*. Algal Res 55(1):102250
Nelson TE, Lewis BA (1974) Separation and characterization of the soluble and insoluble components of insoluble laminaran. Carbohydr Res 33(1):63–74
Nishida Y, Kumagai Y, Michiba S, Yasui H, Kishimura H (2020) Efficient extraction and antioxidant capacity of mycosporine-like amino acids from red alga dulse *Palmaria palmata* in Japan. Mar Drugs 18:502
Nishikawa S, Hosokawa M, Miyashita K (2012) Fucoxanthin promotes translocation and induction of glucose transporter 4 in skeletal muscles of diabetic/obese KK-A(y) mice. Phytomedicine 19:389–394
Nishino T, Nagumo T (1991) The sulfate-content dependence of the anticoagulant activity of a fucan sulfate from the brown seaweed Ecklonia kurom. Carbohydr Res 214:193–197
Nwosu F, Morris J, Lund VA, Stewart D, Ross HA, McDougall GJ (2011) Anti-proliferative and potential anti-diabetic effects of phenolic-rich extracts from edible marine algae. Food Chem 126:1006–1012
Orfanoudaki M, Hartmann A, Alilou M, Gelbrich T, Planchenault P, Derbré S, Schinkovitz A, Richomme P, Hensel A, Ganzera M (2020) Absolute configuration of mycosporine-like amino acids, their wound healing properties and *in vitro* anti-aging effects. Mar Drugs 18:35
Palanisamy S, Vinosha M, Marudhupandi T, Rajasekar P, Prabhu NM (2017) Isolation of fucoidan from *Sargassum polycystum* brown algae: Structural characterization, *in vitro* antioxidant and anticancer activity. Int J Biol Macromol 102:405–412
Pangestuti R, Siahaan EA (2018) Seaweed-derived carotenoids. In: Bioactive seaweeds for food applications. Elsevier, Amsterdam, pp 95–107
Park HJ, Lee MK, Park YB, Shin YC, Choi MS (2011) Beneficial effects of *Undaria pinnatifida* ethanol extract on diet-induced-insulin resistance in C57BL/6J mice. Food Chem Toxicol 49:727–733
Pereira AG, Otero P, Echave J, Carreira-Casais A, Chamorro F, Collazo N, Jaboui A, Lourenço-Lopes C, Simal-Gandara J, Prieto MA (2021) Xanthophylls from the sea: algae as source of bioactive carotenoids. Mar Drugs 19:188
Pozharitskaya ON, Obluchinskaya ED, Shikov AN (2020) Mechanisms of bioactivities of fucoidan from the brown seaweed *Fucus vesiculosus*, L. of the barents sea. Mar Drugs 18:275
Ravishankar GA, Ranga Rao A (2019) Handbook of algal technolgies and phytochemicals: Volume-I food, health and nutraceutical applications. CRC Press. 322 p ISBN: 9780429054242
Rebours C, Marinho-Soriano E, Zertuche GJA et al (2014) Seaweeds: an opportunity for wealth and sustainable livelihood for coastal communities. J Appl Phycol 26:1939–1951
Renn D (1997) Biotechnology and the red seaweed polysaccharide industry: Status, needs and prospects. Trends Biotechnol 15(1):9–14
Saha S, Navid MH, Bandyopadhyay SS, Schnitzler P, Ray B (2012) Sulfated polysaccharides from *Laminaria angustata*: Structural features and in vitro antiviral activities. Carbohydr Polym 87:123–130
Sathya R, Kanaga N, Sankar P, Jeeva S (2017) Antioxidant properties of phlorotannins from brown seaweed *Cystoseira trinodis* (Forsskål) C. Agardh Arab J Chem 10:S2608–S2614
Seong SH, Paudel P, Choi JW, Ahn DH, Nam TJ, Jung HA, Choi JS (2019) Probing multi-target action of phlorotannins as new monoamine oxidase inhibitors and dopaminergic receptor modulators with the potential for treatment of neuronal disorders. Mar Drugs 17:377
Sharma PP, Baskaran V (2021) Polysaccharide (laminaran and fucoidan), fucoxanthin and lipids as functional components from brown algae (*Padina tetrastromatica*) modulates adipogenesis and thermogenesis in diet-induced obesity in C57BL6 mice. Algal Res 54:102187
Shi D, Li J, Guo S, Su H, Fan X (2009) The antitumor effect of bromophenol derivatives in vitro and Leathesia nana extract in vivo. Chin J Oceanol Limn 27:277–282

Shibata T, Fujimoto K, Nagayama K, Yamaguchi K, Nakamura T (2002) Inhibitory activity of brown algal phlorotannins against hyaluronidase. Int J Food Sci Technol 37(6):703–709

Shick JM, Dunlap WC (2002) Mycosporine-like amino acids and related gadusols: Biosynthesis, accumulation, and UV-protective functions in aquatic organisms. Annu Rev Physiol 64:223–262

Shimoda H, Tanaka J, Shan SJ, Maoka T (2010) Anti-pigmentary activity of fucoxanthin and its influence on skin mRNA expression of melanogenic molecules. J Pharm Pharmacol 62:1137–1145

Shiratori K, Ohgami K, Ilieva I, Jin XH, Koyama Y, Miyashita K,Yoshida K, Kase S, Ohno S (2005) Effects of fucoxanthin on lipopolysaccharide-induced inflammation in vitro and in vivo. Exp Eye Res 81:422–428

Silva TMA, Alves LG, de Queiroz KCS, Santos MGL, Marques CT, Chavante SF, Rocha HAO, Leite EL (2005) Partial characterization and anticoagulant activity of a heterofucan from the brown seaweed *Padina gymnospora*. Braz J Med Biol Res 38:523–533

Soares AR, Abrantes JL, Souza TML, Fontes CFL, Pereira RC, Frugulhetti ICPP, Teixeira VL (2007) In vitro antiviral effect of meroditerpenes isolated from the brazilian seawead *Stypopodium zonale* (dictyotales). Planta Med 73:1221–1224

Souza RB, Frota AF, Silva J, Alves C, Neugebauer AZ, Pinteus S, Rodrigues JAG, Cordeiro EMS, De Almeida AA, Pedrosa R et al (2018) *In vitro* activities of kappa-carrageenan isolated from red marine alga *Hypnea musciformis*: Antimicrobial, anticancer and neuroprotective potential. Int J Biol Macromol 112:1248–1256

Stephanie B, Eric D, Sophie FM, Christian B, Yu G (2010) Carrageenan from *Solieria chordalis* (Gigartinales): Structural analysis and immunological activities of the low molecular weight fractions. Carbohydr Polym 81:448–460

Sugawara T, Matsubara K, Akagi R, Mori M, Hirata T (2006) Antiangiogenic activity of brown algae fucoxanthin and its deacetylated product, fucoxanthinol. J Agric Food Chem 54:9805–9810

Suleria HAR, Masci PP, Zhao KN, Addepalli R, Chen W, Osborne SA, Gobe GC (2017) Anti-coagulant and anti-thrombotic properties of blacklip abalone (*Haliotis rubra*): *In vitro* and animal studies. Mar Drugs 15(8):240

Sun HH, Paul VJ, Fenical W (1983) Avrainvilleol, a brominated diphenylmethane derivative with feeding deterrent properties from the tropical green alga *Avrainvillea longicaulis*. Phytochemistry 22:743–745

Synytsya A, Kim WJ, Kim SM, Pohl R, Synytsya A, Kvasničkac F, Čopíková J, Park YI (2010) Structure and antitumour activity of fucoidan isolated from sporophyll of Korean brown seaweed *Undaria pinnatifida*. Carbohydr Polym 81:41–48

Tang J, Wang W, Chu W (2020) Antimicrobial and anti-quorum sensing activities of phlorotannins from seaweed (*Hizikia fusiforme*). Front Cell Infect Microbiol 10:586750

Terasaki M, Maeda H, Miyashita K, Tanaka T, Miyamoto S, Mutoh M (2017) A marine biofunctional lipid, fucoxanthinol, attenuates human colorectal cancer stem-like cell tumorigenicity and sphere formation. J Clin Biochem Nutr 61:25–32

Terme N, Boulho R, Kucma J-P, Bourgougnon N, Bedoux G (2018) Radical scavenging activity of lipids from seaweeds isolated by solid-liquid extraction and supercritical fluids. OCL 25(5):D505

Tsukui T, Konno K, Hosokawa M, Maeda H, Sashima T, Miyashita K (2007) Fucoxanthin and fucoxanthinol enhance the amount of docosahexaenoic acid in the liver of KKAy obese/diabetic mice. J Agric Food Chem 55(13):5025–5029

Tsukui T, Baba N, Hosokawa M, Sashima T, Miyashita K (2009) Enhancement of hepatic docosahexaenoic acid and arachidonic acid contents in C57BL/6J mice by dietary fucoxanthin. Fisheries Sci 75(1):261–263

Urikura I, Sugawara T, Hirata T (2011) Protective effect of fucoxanthin against UVB-induced skin photoaging in hairless mice. Biosci Biotechnol Biochem 75:757–760

Vani V, Venkateish VP, Nivya V, Baskaran V, Madan KP (2021) Nutritional and anti-cancer effects of carotenoids from *lactuca sativa*. In: Advances in health and disease, p 33

Vasanthi H, Rajamanickam G (2004) Tumoricidal effect of the red algae *Acanthophora spicifera* on Ehrlich ascites carcinoma in mice. Seaweed Res Utilizat 25:217–224

Venkateish VP, Vani V, Nivya V, Baskaran V, Madan KP (2021) Bioactives of *Lactuca sativa*: nutritional and clinical importance. In: Advances in health and disease, p 33
Vijay K, Sowmya PR, Arathi BP, Shilpa S, Shwetha HJ, Raju M, Baskaran V, Lakshminarayana R (2018) Low-dose doxorubicin with carotenoids selectively alters redox status and upregulates oxidative stress-mediated apoptosis in breast cancer cells. Food Chem Toxicol 118:675–690
Wang H, Ooi EV Jr, P.O.A. (2008) Antiviral activities of extracts from Hong Kong seaweeds. J Zhejiang Univ Sci B 9:969–976
Wang J, Zhang QB, Zhang ZS, Zhang H, Niu XZ (2010) Structural studies on a novel fucogalactan sulfate extracted from the brown seaweed *Laminaria japonica*. Int J Biol Macromol 47:126–131
Wang T, Jonsdottir R, Liu H, Gu L, Kristinsson HG, Raghavan S, Olafsdottir G (2012) Antioxidant capacities of phlorotannins extracted from the brown algae *Fucus vesiculosus*. J Agric Food Chem 60:5874–5883
Wang L, Zeng Y, Liu Y, Hu X, Li S, Wang Y, Li L, Lei Z, Zhang Z (2014) Fucoxanthin induces growth arrest and apoptosis in human bladder cancer T24 cells by up-regulation of p21 and down-regulation of mortalin. Acta Biochim Biophys Sin (Shanghai) Oct 46(10):877–884
Wijesekara I, Kim SK, Li Y, Li YX (2011) Phlorotannins as bioactive agents from brown algae. Process Biochem 46(12):2219–2224
Woo MN, Jeon SM, Kim HJ, Lee MK, Shin SK, Shin YC, Park YB, Choi MS (2010) Fucoxanthin supplementation improves plasma and hepatic lipid metabolism and blood glucose concentration in high-fat fed C57BL/6N mice. Chem Biol Interact 186:316–322
World Health Organisation (2018) WHO fact sheet. www.who.int/fact-sheet.com
Xu N, Fan X, Yan X, Li X, Niu R, Tseng CK (2003) Antibacterial bromophenols from the marine red alga *Rhodomela confervoides*. Phytochemistry 62:1221–1224
Yabuta Y, Fujimura H et al (2010) Antioxidant activity of the phycoerythrobilin compound formed from a dried Korean purple laver (*Porphyra* sp.) during in vitro digestion. Food Sci Technol Res 16(4):347–352
Yin G, Li W, Lin Q, Lin X, Lin J, Zhu Q, Jiang H, Huang Z (2014) Dietary administration of laminarin mproves the growth performance and immune responses in *Epinephelus coioides*. Fish Shellfish Immunol 41:402–406
Yan X, Chuda Y, et al. (1999) Fucoxanthin as the major antioxidant in *Hijikia fusiformis,* a common edible seaweed. Biosci Biotechnol Biochem 63(3):605–607
Yan MD, Yao CJ, Chow JM, Chang CL, Hwang PA, Chuang SE, Whang-Peng J, Lai GM (2015) Fucoidan elevates microRNA-29b to regulate DNMT3B-MTSS1 axis and inhibit EMT in human hepatocellular carcinoma cells. Mar Drugs 13:6099–6116
Yan T, Li HY, Wu JS, et al. (2017) Astaxanthin inhibits gemcitabine-resistant human pancreatic cancer progression through EMT inhibition and gemcitabine resensitization. Oncol Lett 14(5): 5400–5408
Yang L, Zhang LM (2009) Chemical structural and chain conformational characterization of some bioactive polysaccharides isolated from natural sources. Carbohydr Polym 76(3):349–361
Yuan H, Zhang W, Li X, Lü X, Li N, Gao X, Song J (2005) Preparation and in vitro antioxidant activity of κ-carrageenan oligosaccharides and their oversulfated, acetylated, and phosphorylated derivatives. Carbohydr Res 340:685–692
Zhang Q, Yu P, Li Z, Zhang H, Xu Z, Li P (2003) Antioxidant activities of sulfated polysaccharide fractions from *Porphyra haitanesis*. J Appl Phycol 15:305–310
Zhao Y, Zheng YZ, Wang J, Ma SY, Yu YM, White WL (2018) Fucoidan extracted from *Undaria pinnatifida*: Source for nutraceuticals/functional foods. Mar Drugs 16:321
Zhu Y, Cheng J, Min Z, Yin T, Zhang R, Zhang W, Hu L, Cui Z, Gao C, Xu S, Zhang C, Hu X (2018) Effects of fucoxanthin on autophagy and apoptosis in SGC-7901cells and the mechanism. J Cell Biochem 119(9):7274–7284

Chapter 33
Genetic and Genomic Approaches for Improved and Sustainable Brown Algal Cultivation

Ioannis Theodorou, Mallikarjuna Rao Kovi, Zhe Liang, and Hilde-Gunn Opsahl-Sorteberg

Abbreviation

ANN	Artificial neural network
°C	Degrees celsius
CESA/CesA	Cellulose synthase
COST	European Cooperation in Science and Technology
CRISPR	Clustered regularly interspaced short palindromic repeats
DEK1	Defective kernel1
DL	Deep learning
E. siliculosus	*Ectocarpus siliculosus*
FA	Filamentous actin
FAO	Food and Agricultural Organization of the United Nations
GE	Genome editing
GM	Genetic modification
GMO	Genetically modified organism

I. Theodorou
Department of Plant Sciences, Faculty of Biosciences, Norwegian University of Life Sciences, Ås, Norway

Morphogenesis of Macroalgae, LBI2M UMR8227, Station Biologique de Roscoff, CNRS-Sorbonne University, Paris, France
e-mail: ioannis.theodorou@nmbu.no

M. R. Kovi · H.-G. Opsahl-Sorteberg (✉)
Department of Plant Sciences, Faculty of Biosciences, Norwegian University of Life Sciences, Ås, Norway
e-mail: mallikarjuna.rao.kovi@nmbu.no; hilde-gunn.sorteberg@nmbu.no

Z. Liang
Biotechnology Research Institute, Chinese Academy of Agricultural Sciences, Beijing, China
e-mail: liangzhe@caas.cn

A. Ranga Rao, G. A. Ravishankar (eds.), *Sustainable Global Resources of Seaweeds Volume 2*, https://doi.org/10.1007/978-3-030-92174-3_33

H	Hour(s)
HD	Homeo domain
HSF	Heath schock factor
Mb	Megabases
RAC1	Ras-related C3 botulinum toxin substrate1
S. japonica	*Saccharina japonica*
S. latissima	*Saccharina latissima*
SNP	Single nucleotide polymorphism
TAL/TALE	Transcription activator-like effectors
TF	Transcription factor

1 Sustainable Exploitation for Ecological and Commercial Interests

Submarine kelp forests typically occupy 70% of cold and temperate marine coastal ecosystems, with correlated importance to our ecological and atmospheric balance as well as important resources for human interests with increasing attention worldwide (Bischof et al. 2006; Broch et al. 2019). In China and many other East-Asian countries, the production, processing, distribution and preparation of kelp (brown macroalgae in the order of Laminariales) for human food consumption is well-established. In 2018, a total of ten million tons of fresh *Saccharina japonica* was harvested from cultivation (FAO 2020). The annual increasing rate of production has been 6.7%. In Norway, the annual production of macroalgae could be increased up to 20 million tons by 2050 (Broch et al. 2019). The most commonly cultivated species along the Norwegian coast are *Saccharina latissima* and *Alaria esculenta* (Stévant et al. 2017). Potential biomass production of *S. latissima* alone was estimated to be 150–200 tons per hectare per year (Ye et al. 2015).

Harvesting by collection or trawling is still the main form of exploitation in Europe and Northern America; while East-Asia practices extensive cultivation filling whole bay areas. Harvesting the wild is still regarded safer to the ecological balance of the sea environment in much of Europe, despite the reduction in the wild population which may cause overexploitation and environmental factors. To offer preliminary protection some countries are allowing regrowth before new harvests. It has also been suggested to cut plants above the meristematic growth region to possibly allow established plants to live and produce for several rounds (Rolin et al. 2017). Cultivation may be another solution, since allowing increased production while protecting local wild populations from harvesting.

2 Genetic Know-How of Wild Seaweed Population and Evolutionary Position

Despite the extensive exploitation and economic interest in brown algae, a full understanding of the classification, evolution, and genetic relatedness across the seas are still not fully determined (Neiva et al. 2018; Nishitsuji et al. 2020). However, the phylogenetic and evolutionary placement of the brown algae should be clear as we get fossil findings complemented with genomic tools and genome sequences (Starko et al. 2019; Maloney et al. 2021). Full genome sequences have been done for microorganisms, land plants, and animals for improved biologic understanding. Such a knowledge base secures a better foundation to improve sustainability when it comes to a targeted framework for exploitation. It also allows for better plant protection and better security to keep ecological balances and even users health interests, for example, when we can evaluate possible breeding options to reduce iodine content in *S. latissima*. Macroalgae are still a new research field with few genomes sequenced and many biological characteristics still unknown. For instance, an isolation-to-distance model cannot fully explain the genetic relatedness of macroalgae, how they spread in our seas, meaning our ability to secure their future is still restricted (Breton et al. 2018). Interestingly, genome sequences of native and cultivated populations of *S. japonica* showed that the genetic variation is wider in wild populations than in the cultivated and that all cultivated individuals originate from one wild *S. japonica* accession (Ye et al. 2015). Evolutionary comparisons between brown algae genomes show that *Ectocarpus* and *S. japonica* are closely related, even though *Saccharina* has more and longer genes, and has an overall large genome expansion explaining partly, why the genome is three times larger (Liu et al. 2019; Ye et al. 2015).

2.1 Current Genetic Practices and Implications for Cultivation

String cultivation can use spores, gametophytes or sporophytes of local fertile populations from cross-breeding (Forbord et al. 2019; Goecke et al. 2020). Gametophytes can be kept under vegetative conditions for years before employing them in the cultivation process. A seeded string is left under optimal conditions for embryogenesis till the embryos reach the juvenile stage, subsequently, they are transferred to the sea or inland tanks. This practice secures the longevity of the wild populations without any genetic "pollution"(Hwang et al. 2019). Increased global warming is posing challenges to seaweeds as they prefer cooler temperatures. Germplasm collection and gene bank storage to secure future breeding for cultivation and conservation are receiving increased interest and attention. This is, however, not as trivial as storing dry seeds for land plants, but cryo-preservation of gametophytic material might be a preferred option (Wang et al. 2011; Barrento et al. 2016; Visch et al. 2019; Wade et al. 2020).

Many European countries have to meet precautionary principles such as cultivating specimens only originating from donor material collected from the near vicinity. It is assumed that the donor genotypes are only released into sea regions typically 10–20 km away. This is however, a rough estimate and might not be very efficient if releasing gametes only to individuals from the same populations. This because, if the concentration of cultivated plants becomes high, and these genetic crosses are on a few founder genotypes originally, this might lead to inbreeding depression, as if mating siblings, and this can be harmful for all individuals in the region. An additional issue is the spontaneous mutations happening constantly in nature as well as in "in vitro" cultivation, meaning this will add random and possibly selective genetic variation. This is part of all evolution and survival. Well-grounded breeding and cultivation of seaweed, rather than collecting, might be the best way of sustainable exploitation of our seas for nature and mankind.

2.2 Genetic Recombination and Spread of Gametes and Spores (Interbreeding)

Many brown seaweeds reach fertile maturity around winter (Fig. 33.1) like *S. latissima* (Lüning 1988). Mature sporophytes release spores that germinate to female or male gametophytes, producing gametes under optimal conditions. In most seaweeds and especially kelps, male gametes release is regulated by the female gametes' pheromones additionally promoting chemotactic movement (Maier et al. 2001; Kinoshita et al. 2017), where they stay alive for about a day. Fertilization of eggs generates the diploid life cycle with diploid sporophytes. Simultaneously fragments of sporophytes might generate gametes off-season, at least in lab conditions (Pang and Lüning 2004). This means restricting a cultivated strain in a sea area, keeping them from spreading to new regions is not trivial. In addition, it is known that in nature there are several kelp hybrids that are fertile (reviewed by Bartsch et al. 2008). In Asian countries, hybrids of *Saccharina angustata* with *Saccharina longissima* are planned to be employed to meet the warmer conditions to improve their issue with reduced stock cultures (Hwang et al. 2019).

Regarding migration of natural or cultivated populations, pieces of thalli from mature sporophytes can be the "vessels" responsible for transferring spores and germinated spores which can survive without optimal conditions for a long time up to years (Lüning 1980). To get a good understanding of the means and distances that spores might travel, comprehensive genetic studies and mapping need to be done along the world's long coastlines as initiated for some sea areas (Neiva et al. 2018). Before completing large-scale studies, we have limited knowledge of how far seaweed may spread. However, hydrodynamic models in combination with genetic resources from studied populations can assist in answering how they might be transported such long distances by currents, bulk water on ships, etc. (Mooney et al. 2018).

3 Genetic Resources and Possible Additions to Future Sustainable Seaweed Cultivation Options

Land plants have been adapted to meet human dietary needs for up to an estimated 10,000 years since we gathered and needed a more concentrated energy supply to meet the early settlers. Seaweed breeding is a common practice in Asia, and it is also getting increasing attention to meet seaweed interests in western waters linked to cultivation (Hwang et al. 2019; Goecke et al. 2020; Araújo et al. 2021).

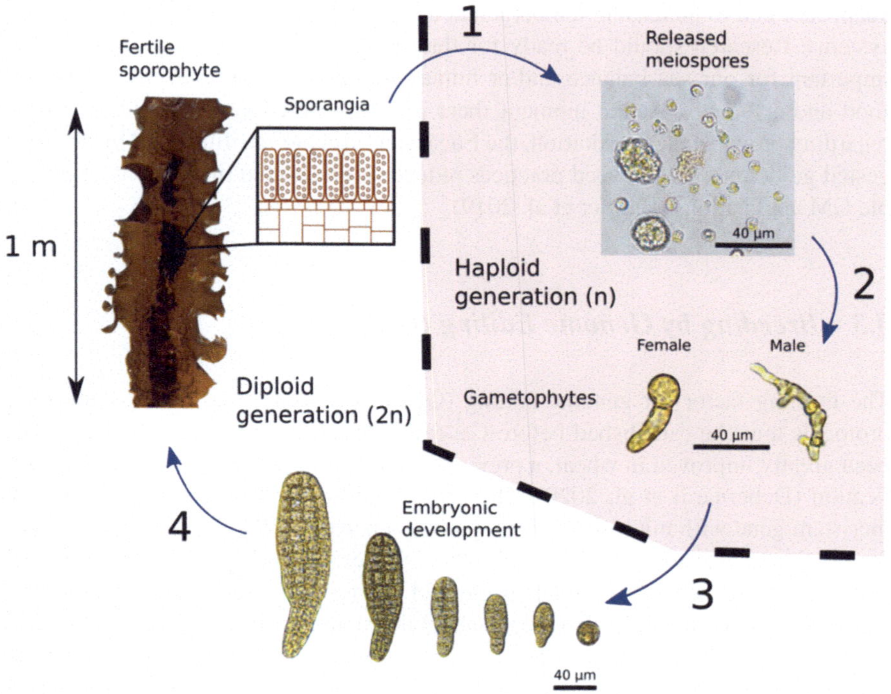

Fig. 33.1 The life cycle of *Saccharina latissima*. (1) Meiospores(n) are released from sporangia formed on the mature blade under specific light and temperature conditions (Lüning 1988). In nature, this is usually around late-autumn to winter. (2) Meiospores settle and germinate to dioicous gametophytes, 48H after the release (Lüning 1980). (3) Temperature, light conditions and biomass density are important parameters for the gametogenesis process (Lüning 1980; Ebbing et al. 2020), which usually takes around one week. The gametophytes can be kept in vegetative state for many years if kept under low red-light conditions. (4) Embryonic development (2n) is poorly understood including its exact time frame. Personal observations from our work on the development of *Saccharina* embryos, indicate a duration of about 1–2 months depending on the growth conditions. The earliest approximate time for a mature sporophyte to be fertile is 6 months (Pang and Lüning 2004)

3.1 Breeding by Intercrossing (Selective Breeding)

This might secure a wider, sound genetic background avoiding inbreeding depression from accumulating mutant allelic variants (Goecke et al. 2020).

3.2 Breeding by Classical Gene Modification (GM)

Gene modification (GM) by gene transfer can add tools to increase the gene bank for breeding, allowing adapting the cultivated cultivars to changing environment, user interests and increase the sustainable uses of our seas. This could allow us to adapt seaweed to increasing temperatures and their future survival for our sea ecosystems. Research should be ready for this solution if society should see this as important for our sea balance and/or human interests of seaweed production for food and/or energy. At the moment there are no adapted gene transfer methods regarding brown algae, in addition, the European "climate" as illustrated by the suggested guidelines for seaweed practices reduces interests and discussions of possible GM applications (Barbier et al. 2019).

3.3 *Breeding by Genome Editing (GE)*

The limiting factor for genome editing (GE) is still gene transfer, meaning GM protocols must be established before GE can be used. Recently this challenge has been greatly improved in wheat, a previously recalcitrant species for genetic modification (Debernardi et al. 2020). The progress came by combining the use of a meristem gene with micro-RNA for high frequencies of CRISPR/Cas9 gene transfer in all wheat genotypes tested and other species previously difficult to transfer genes to. This is promising for all species not having well-established gene modification protocols and might also help to adapt the method to macroalgae. This would be useful to do functional genomics to unravel gene function for high-resolution gene annotation of new genomes and applied interests possibly generate sterile plants for cultivation to avoid spreading bred genotypes to local native genotypes/populations.

4 Regulations

If CRISPR gene editing products will be regulated as GM, this might restrict the possible breeding options and results in part of the globe such as Europe (Callaway 2018; Asquer and Krachkovskaya 2021). As experiencing currently, genome editing is not classified in large parts of the world as GM. This means developing new varieties is less expensive and allows smaller breeders to use this promising tool. But for it not to be classified as GM, it must be one gene only as could happen in the wild. If piling genes in stacks, it will be GM and as such be of limited financial interests or be more expensive due to patent rights. Many turn to TILLING then to avoid this hurdle. This also illustrates how regulations affect options and solutions, and thereby the results for nature and humans. If the future increasingly demands tools to best manage our natural resources policies might adjust accordingly for the common good. Food resources are expected to meet new challenges as the human population increases in size and due to global heating (Fischer and Garnett 2016). Any regulations regarding GM and Genome edited organisms should take that into account when policies are made, to accommodate both the wellbeing of the ecosystem and live forms.

5 Genetic and Genomic Foundation

Prerequisites to secure wild populations, while harvesting seaweeds for human needs, include basic understandings of seaweed species and populations across our joint oceans. Achieving such basic knowledge takes genome sequences and mapping the genetic origin and variation within species. This will indicate the robustness of native populations regarding meeting gene flow from cultivated specimens as well as their possible adaptability to occurring climate changes such as increased sea temperature and changed salt concentrations.

However, how expected gene flow from cultivated individuals might affect native populations is still poorly documented. An exciting study showed that *Laminaria digitata* spores could spread longer than expected up to several kilometers (Brennan et al. 2014) and unpublished studies of *S. latissima* along Norwegian sea areas indicated genetic relatedness not necessarily connected to distance, but possibly the sea- current. Historic relatedness of cosmopolitan populations of *S. latissima* have also been linked to shifting glacial separation across the northern Pacific and Atlantic oceans and more generally an "allopatric speciation pump" has been suggested for widespread Amphi-oceanic organisms (Neiva et al. 2018).

Knowing the genetics, such as the population structure, evolution, relatedness, genetic variation and how the native populations migrate are of critical importance to set meaningful and effective regulatory framework/laws and develop a sustainable seaweed industry (Breton et al. 2018). In addition to possibly narrow genetic variation, environmental bottlenecks like increased sea temperatures will likely cause a strong selection pressure further affecting populations' robustness negatively. Therefore cultivation should be based on a sound management plan to secure both wild specimens and populations as well as the cultivated ones. Genome sequences and functional genomics will help meet related ocean management and sustainable uses of our global sea resources. This can predict effects of nutrient availability, reduced salt content, increased solar exposure and increased sea temperatures. One can even test possible genetic effects of cultivation, like effects of life up-side down if hanging from ropes where the holdfast is fixed growing downwards rather than in the wild where they grow from the bottom-up.

Currently, we are getting the first well-annotated full genome sequences, and historical detailed observations are complemented by high-resolution microscopy and transcriptomics to unravel the puzzle of early development and growth (work in progress at Roscoff marine station/Sorbonne and Norwegian University of Life Sciences). These are exciting times for marine biology and seaweed in particular, since the genetic and genomic part of the field is adding where there is still much to unravel. The European Commission is responsible for joint European marine resources, but we still lack an adopted regulatory framework for macroalgae exploitation, common laws, and international agreements for this emerging industrial development (Fischer and Garnett 2016; Lähteenmäki-Uutela et al. 2021). This was extensively covered and discussed in the Phycomorph COST action FA1406 (2015–2019), from many perspectives including genetic and ecological interests.

The resulting 200 pages guidelines named Pegasus were intended to help the authorities adopt mutual practices for our shared oceans' multiple interests (Barbier et al. 2019; Chopin and Tacon 2021). Safety issues of macroalgae such as harmful content of toxic components as arsenic (Yokoi and Konomi 2012; Wang et al. 2015) or high level of compounds that should be kept within a narrow window of minimal to maximal intake per day can also be met by genomics and possible breeding (Hwang et al. 2019; Goecke et al. 2020) to secure and reduce e.g. iodine content (*cf.* a review about the crucial role of seaweeds in iodine cycle Küpper 2015).

5.1 Genome and Pan-Genome Support

One of the most important "tools" in current genomics studies is to get access to a high-quality completely phased genome assembly. Without reference sequence availability, the scope of genomic research remains limited. For example, many sequencing applications rely on a reference assembly on which to map reads and genome-wide association studies rely on knowledge of the relative position of SNP markers. A reference genome also facilitates the development of molecular markers, efficient genetic diversity studies, as well as allowing comparisons of specific sequences such as genes, enabling the prediction of gene function across related species. There are about 1800 brown algae species, mostly living in cold sea waters. However, the harvested biomass comes from only a few species of the two orders Laminariales and Fucales (https://www.seaweed.ie/algae/phaeophyta.php). Only seven genomes of Phaeophyceae are published so far, *Ectocarpus siliculosus* (Cock et al. 2010) *Saccharina japonica* (Ye et al. 2015; Liu et al. 2019), *Macrocystis pyrifera* (unpublished, BioProject:), *Cladosiphon okamuranus* (Nishitsuji et al. 2016, 2020), *Nemacystus decipiens* (Nishitsuji et al. 2019), *Undaria pinnatifida* (Shan et al. 2020) and the fucale, *Sargassum fusiforme* (Wang et al. 2020). Sequencing has been done for another 40 seaweed genomes, yet not decoded or published (pers. Comm. Mark Cock). The most important brown algae *Saccharina japonica* was sequenced in 2015 (Ye et al. 2015)] and further annotated in 2019 (Liu et al. 2019) with an assembled draft genome of approx. 5805 Mb covering about 89% of the estimated genome. The genome size is approximately three times that of *Ectocarpus*, and of the 35,725 genes predicted and annotated many are related to halogen concentration, cell wall synthesis, development and defence systems. However, several other important brown algae species like *Saccharina latissima* and *Alaria esculenta* need to be sequenced to further accelerate the genomics studies in these species.

The emergence of the pan-genome concept, originally proposed for microbial species (Tettelin et al. 2005), has interesting implications for how highly heterozygous polyploid genomes will be presented in the future. Species pan-genomes have been extensively studied in prokaryotes, but evidence of species pan-genomes has also been demonstrated in eukaryotes such as plants and fungi. Many prokaryote and eukaryote functional and comparative genomics studies rely on the use of well-annotated reference genomes intended to be broadly representative of a given species. However, due to genetic and genomic variation between individuals within a species, reference genomes do not contain all the genetic information for that particular species (Parfrey et al. 2008). To capture the variation, it has become increasingly common to sequence multiple genotypes within species This 'pan-genome' of a species usually consists of two components. (1) The 'core' genome, containing genes conserved across all observed genomes from a species. These genes are usually, but not always, essential for the viability of an individual organism (Rouli et al. 2015). (2) The 'accessory' or 'dispensable' genome, containing genes specific to sets of isolate genomes or individual isolate genomes within a species. Thus, instead of relying on genome sequence information from single variety, pan-genomes concept can accelerate the breeding programs by capturing information from different genotypes within the species. In the future, pan-genomes are likely to be central for brown algae genomics and applications too.

5.2 *Transcriptome Studies*

Omics approaches in brown algae are still scarce and knowledge of their acclimation mechanisms to the changing climatic conditions can benefit from the application of RNA-sequencing. (Crépineau et al. 2000) firstly reported transcriptional analysis to *L. digitat*a with cell wall biosynthesis and halogen metabolism-related genes. The transcriptome studies were extensively performed in *S. japoni*ca, due to having a complete annotated genome sequence. Several biotic and abiotic stress responses in the transcriptome were well studied in these species. Recently, several RNA-seq. techniques were conducted to study the gene transcriptional patterns related to differences in temperature and light and the effects on different growing stages (Deng et al. 2012; Wang et al. 2013; Liu et al. 2014; Ding et al. 2019; Shao et al. 2019a, b). An interesting study conducted by Deng et al. (2012) on transcriptomic responses under blue light conditions showed 11,660 differentially expressed genes when compared to dark light. The temperature effect under 20 °C detected 947 up-or down-regulated genes (Liu et al. 2014). Several transcriptomic studies in *S. japonica* have greatly increased genome annotation. Transcriptome sequencing studies have further detected several key candidate genes involved in photosynthesis and heat resistance (Liu et al. 2019). In particular, the genes underlying the crucial alginate and mannitol biosynthesis during developmental were well studied (Shao et al. 2019b). However, more research is needed to be performed by integrating proteomics and metabolomics with transcriptomics to decipher the regulatory networks of *Saccharina*'s developmental biology. From a biogeographic approach, transcriptomics studies on different strains of *S. latissima* gametophytes grown in different temperatures (Monteiro et al. 2019b), explain the decrease of southern *S. latissima* populations from the Iberian peninsula and a move of the wild population towards northern and cooler waters. In addition, an insight is given to the warming climate effect on arctic *S. latissima* populations (Monteiro et al. 2019b; Li et al. 2020), where temperature increase till 15 C might positively affect production depending on the geographical latitude while reduced salt concentration might impair this effect depending on the cultivated strain. Interestingly, increased temperatures induced sex-biased gene expressional responses in gametophytes (Monteiro et al. 2019a), with the male gene expression pattern focusing on cell cycle and signalling while the female genes on increasing metabolic and energy production profiles.

5.3 Utilizing Machine Learning Methods for Big Data Analysis

With the rise of "big data", machine learning has become increasingly important to understand and translate massive information into biological functional meaning. Machine learning uses algorithms to parse data, learn from it, and make determinations without human intervention. Increasingly, new sources of data are being incorporated into breeding pipelines. Enormous amounts of data from field phenomics and genotyping technologies put data mining and analysis into a completely different level that is challenging from practical and theoretical standpoints. Predictions of the effects of changing environments on performance will help breeders to compare the results over multiple years to gain information about how candidate varieties will likely perform in a target environment. Improved methods of forecasting production can also be beneficial in making marketing decisions that could improve farm profitability (Johnson et al. 2016).

Deep learning (DL) is a recently developed machine-learning technique that provides good prediction capability with many advanced features, one of which is the deep multi-layered neural network architecture. A large number of neurons are used to capture complex, nonlinear relationships in big data (large datasets) (LeCun et al. 2015). DL has proven capable of improved prediction performance over traditional models for speech recognition, image identification, and natural language processing (LeCun et al. 2015). Most recently, however, DL has drawn the attention of systems biologists, who have successfully applied it to several prediction problems, e.g. the inference of gene expression (Chen et al. 2016), and the functional annotation of genetic variants (Zhou and Troyanskaya 2015). These successful applications in the fields of computational biology and systems biology have demonstrated that DL has a powerful capability of learning complex relationships from biological data, while we still need to verify and secure the results by independent experiments and human reasoning (Webb 2018). Moreover, there exist various tools and platforms allowing for DL. The most popular ones are Theano, TensorFlow, Keras, Caffe, PyTorch, TFLearn, Pylearn2 and the Deep Learning Matlab Toolbox (Bahrampour et al. 2016).

Multidisciplinary approaches are essential to advance the seaweed sector, where genomics and computation models for prediction and modeling play crucial roles. So far this is limited and poorly adapted to seaweed research when compared to other species, and this is expected to change in the near future. Mantri et al. (2020) described that artificial neural network (ANN) models were applied to predict the thermophysical properties of *S. latissima*, cadmium–zinc ions biosorption by the *Sargassum filipendula* and optimization of different physiochemical parameters for seedling production in *Gracilaria dura*. These studies demonstrated the efficiency and possible wide applications of machine learning approaches in the optimization of complex biological systems in seaweeds.

6 Cellular and Developmental Biology Insights of Brown Algae and Genomic Tools

Reexamination and annotation of the *Ectocarpus* genome, adding transcriptomics for functional studies indicate it has c. 25 chromosomes, 17,418 genes and an alternative splicing frequency of 1,6 similar to land plants and conserved micro-RNA suggesting regulatory roles (Billoud et al. 2014; Cormier et al. 2016). *S. japonica* has an estimated close to 36,000 genes, and their expanding genome is suggested important to allow a more complex body form than the *E. filamentous* one (Liu et al. 2019).

Regulatory promoter regions have not yet been identified for *Ectocarpus*, which is puzzling and also possibly why gene transfer protocols so far have failed. Other interesting genomic findings are that brown algae have more intron sequences, and differences in repetitive elements and an incorporated viral genome (Liu et al. 2019). There are reported up to 299 transcription factors (TF) in the sequenced brown algae genomes, including 22 heat-shock-factors (HSF) and 3 with homeoboxes suggesting some similarities to developmental set up in other species (Ye et al. 2015). From these homeobox genes, *SAMSARA* and *OUROBOROS* encode TALE homeodomain TFs which heterodimerise and important for the regulation of sporophytic development in *Ectocarpus*, similar to the KNOX/BEL heterodimers in land plants (Arun et al. 2019) showing a convergent evolution to the TALE HD TFs between brown algae and land plants.

Ectocarpus having the best annotated and understood genome among brown algae, has an interesting simple early filamentous development. It exhibits like land plants' root hairs and pollen tubes, tip growth. However, tip growth in *Ectocarpus* filaments follows an alternative strategy with a gradient of cell wall thickness along the dome of the tip, the apical cell conceding to a much thinner tip and increased cell wall stress (Rabillé et al. 2019). Land plants tip growth is mostly dependent on the chemical modification of the cell wall leading to loosening of the cell wall at the dome of the apical cell though the shanks are stiffer (Geitmann and Ortega 2009; Riquelme 2013). Further development and branching leads to architecture with some similarities to mosses (Coudert et al. 2017), with a particular pattern for the arising branches (Bail et al. 2008, 2011), which can genetically be disturbed. This pattern seems to be related with the different cell types characterized by their shape and position. Notably, while gametophytic and sporophytic generations in *Ectocarpales* are isomorphic (same overall morphology), ecologically the two forms occupy different niche where in nature often only one form can be present and at least for *E. siliculosus* both of them hold a parthenogenetic mode of propagation (Charrier et al. 2008; Couceiro et al. 2015).

Saccharina latissima 's life cycle is heteromorphic with a large conspicuous diploid sporophyte (kelp) and the microscopic gametophyte (Fig. 33.2). The diploid sporophyte develops its body plan during embryogenesis, not growing roots, branches, true vasculature or developing flowers as land plants do (Fritsch 1945). Early embryos develop by anticlinal division as a cell file, before the first periclinal divisions allows development of a flat one-cell layer thick undifferentiated sheet of cells. Later the lamina and stipe can be detected, with a transition zone between them. Now the divisions also give a three dimensional structure, several cell layers thick with an apical—basal axis and a superficial meristematic layer that covers the body of the kelp (meristoderm) (most recently covered in Theodorou and Charrier 2021). On a mature sporophyte, between the blade and the stipe, there is a transition zone where the meristoderm is particularly active and is responsible for the growth and the regeneration of the blade in case of removal or senescence (Rolin et al. 2017).

Mechanical forces are linked to cell shape during cell growth, setting land plant architecture (Du and Jiao 2020). A complex network of interactions between the cytoskeleton, transmembrane proteins and the cell wall is what keeps the integrity of the last, responding to the generated stresses via intracellular signaling. Especially, DEK1 is an essential transmembrane protein for cell wall integrity during the growth and division of land plants (Lid et al. 2005; Tran et al. 2017). DEK1 is also needed for three-dimensional orientation of embryo development beyond the heart stage, stem cell activity, and cell divisions in general (Perroud et al. 2014; Liang et al. 2015). This is a well-studied example of a transmembrane protein explained by genetics and functional studies. However, for brown algae there is a considerate amount of work on the cytoskeleton and cell wall connected to cell shape and ultrastructure (Katsaros et al. 2006; Bogaert et al. 2013; Charrier et al. 2019). It will be exciting to connect this to genomic tools to better understand how it is all related adding to our basic biological understanding of the organism. So far we know some differences between land plants and macro algae. The brown algal cell wall consists mostly of an amorphous matrix and only a small percentage of cellulose, while the origins of the different components show an evolutionary interest (Charrier et al. 2019). How the cytoskeleton of brown algae interacts with the cell wall is rather obscure, however, filamentous actin (AF) seems to be the main cytoskeletal element of the brown algal cells whereas the basic molecular toolbox for regulation of AF seems to be present, like RAC1 an essential protein related to intracellular signaling and AF ultrastructure (Muzzy and Hable 2008; Marston et al. 2019). Last but not least, it is known that CESAs positioning is mainly dependent on the microtubules dynamics and orientation, this role in brown algal cells seems to be played by the AF as has been shown by older works (Katsaros et al. 1996, 2002).

Regarding the physiology of the brown seaweeds, they perceive gravity and light differently than the land plants. Nevertheless, kelps exhibit their own unique

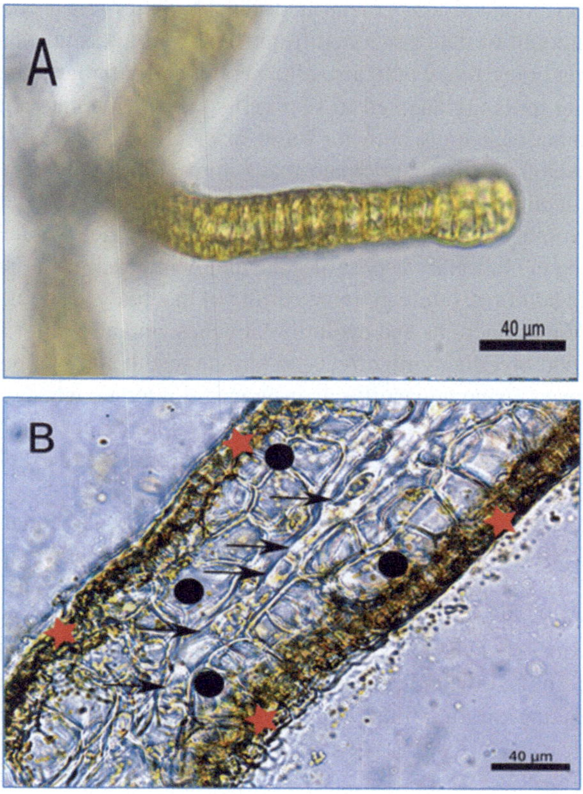

Fig. 33.2 Comparison of a 10 days embryo with a 2 months old juvenile of *Saccharina latissima*. (**a**) Side view of the one-celled monolayer of a 10 days embryo. (**b**) Cross-section of a juvenile blade consisting of different tissues. Red star: Meristoderm, black circle: Cortex and arrows: Medulla

complex vasculature system, raising the interest on studying the mechanisms related with long distance transport since it is an easy system to dissect and manipulate (Knoblauch et al. 2016). Wind is a major force impacting the growth and development of land plants (Gardiner et al. 2016), similarly in *S. latissima* it has been shown to affect the morphology and growth of the blade (Zhu et al. 2021), indicating though also the great plasticity that seaweeds have. Overall, we know little of the genetic background of brown seaweeds development, physiology, and growth. From what we know and with new tools being developed, we expect an exciting near future opening up a new world to us!

7 Conclusion and Future Perspectives

Compared to the many macroalgae species and brown algae in particular existing, only few macroalgae genomes are yet sequenced or published. Thus, pan-genomes and following potential resequencing projects are expected. That would provide valuable resources to identify genes that are central to understand macroalgae development and crucial additional tools to reach breeding goals. When the vast amount of data generated from transcriptomic studies are anchored to high quality genomes, that would also allow better and safer targeting efforts when genome editing tools are adapted to seaweed species. Genome editing would also allow good functional tools both for basic research and applied breeding. With editing tools established, we anticipate that future efforts will uncover the role and molecular functions of important genes in physiology and development of macroalgae. Notably, the special body plan of certain species like kelps, may allow us to explore the evolutionary aspect of gene and gene families to land plants and even animals. In addition, future genome editing of macroalgae should advance our understanding of the functions and evolution of genes, and also contribute to advances in feed and food breeding with effects on human health and wellbeing. The sugar kelp, experiences rising aquacultural/agronomical interest, which could be one key solution to meet increasing demands for food for future generations. Though still far from the streamlined practices of the eastern cultures, sugar kelp and seaweed aquaculture are also on the rise in the "west".

Acknowledgments: Thanks to co-workers and collaborators at NMBU, Roscoff/CNRS/Sorbonne and PGES/USDA/UC-Berkeley and especially Benedichte Charrier for all her support, patience and time, Jennifer C. Fletcher for our many long talks and Bernhard Billoud for making sense of and organizing the large numbers. Thanks also to all helping in daily work by guidance in the labs and making computation and communication flow. We look forward to seeing more and experiencing exciting times to come on and off shores above and below the sea surfaces.

References

Araújo R, Vázquez Calderón F, Sánchez López J et al (2021) Current status of the algae production industry in Europe: an emerging sector of the blue bioeconomy. Front Mar Sci 7:626389. https://doi.org/10.3389/fmars.2020.626389

Arun A, Coelho SM, Peters AF et al (2019) Convergent recruitment of TALE homeodomain life cycle regulators to direct sporophyte development in land plants and brown algae. elife 8:e43101. https://doi.org/10.7554/eLife.43101

Asquer A, Krachkovskaya I (2021) Uncertainty, institutions and regulatory responses to emerging technologies: CRISPR gene editing in the US and the EU (2012–2019). Regulat Govern. https://doi.org/10.1111/rego.12335

Bahrampour S, Ramakrishnan N, Schott L, Shah M (2016) Comparative study of deep learning software frameworks. arXiv preprint arXiv:151106435

Bail AL, Billoud B, Maisonneuve C et al (2008) Early development pattern of the brown alga *Ectocarpus Siliculosus* (Ectocarpales, Phaeophyceae) Sporophyte. J Phycol 44:1269–1281. https://doi.org/10.1111/j.1529-8817.2008.00582.x

Bail AL, Billoud B, Panse SL et al (2011) ETOILE regulates developmental patterning in the filamentous brown alga *Ectocarpus siliculosus*. Plant Cell 23:1666–1678. https://doi.org/10.1105/tpc.110.081919

Barbier M, Charrier B, Araujo R, et al (2019) PEGASUS -PHYCOMORPH European guidelines for a sustainable aquaculture of seaweeds. COST Action FA1406, Roscoff. https://doi.org/10.21411/2c3w-yc73

Barrento S, Camus C, Sousa-Pinto I, Buschmann AH (2016) Germplasm banking of the giant kelp: Our biological insurance in a changing environment. Algal Res 13:134–140. https://doi.org/10.1016/j.algal.2015.11.024

Bartsch I, Wiencke C, Bischof K et al (2008) The genus *Laminaria* sensu lato : recent insights and developments. Eur J Phycol 43:1–86. https://doi.org/10.1080/09670260701711376

Billoud B, Nehr Z, Le Bail A, Charrier B (2014) Computational prediction and experimental validation of microRNAs in the brown alga *Ectocarpus siliculosus*. Nucleic Acids Res 42:417–429. https://doi.org/10.1093/nar/gkt856

Bischof K, Gómez I, Molis M et al (2006) Ultraviolet radiation shapes seaweed communities. Rev Environ Sci Biotechnol 5:141–166. https://doi.org/10.1007/s11157-006-0002-3

Bogaert KA, Arun A, Coelho SM, De Clerck O (2013) Brown algae as a model for plant organogenesis. In: De Smet I (ed) Plant organogenesis: methods and protocols. Humana Press, Totowa, NJ, pp 97–125. https://doi.org/10.1007/978-1-62703-221-6_6

Brennan G, Kregting L, Beatty GE et al (2014) Understanding macroalgal dispersal in a complex hydrodynamic environment: a combined population genetic and physical modelling approach. J R Soc Interface 11:20140197. https://doi.org/10.1098/rsif.2014.0197

Breton TS, Nettleton JC, O'Connell B, Bertocci M (2018) Fine-scale population genetic structure of sugar kelp, *Saccharina latissima* (Laminariales, Phaeophyceae), in eastern Maine, USA. Phycologia 57:32–40. https://doi.org/10.2216/17-72.1

Broch OJ, Alver MO, Bekkby T et al (2019) The kelp cultivation potential in coastal and offshore regions of Norway. Front Mar Sci 5. https://doi.org/10.3389/fmars.2018.00529

Callaway E (2018) CRISPR plants now subject to tough GM laws in European Union. Nature 560:16–16. https://doi.org/10.1038/d41586-018-05814-6

Charrier B, Coelho SM, Bail AL et al (2008) Development and physiology of the brown alga *Ectocarpus siliculosus*: two centuries of research. New Phytol 177:319–332. https://doi.org/10.1111/j.1469-8137.2007.02304.x

Charrier B, Rabillé H, Billoud B (2019) Gazing at cell wall expansion under a golden light. Trends Plant Sci 24:130–141. https://doi.org/10.1016/j.tplants.2018.10.013

Chen Y, Li Y, Narayan R et al (2016) Gene expression inference with deep learning. Bioinformatics 32:1832–1839. https://doi.org/10.1093/bioinformatics/btw074

Chopin T, Tacon AGJ (2021) Importance of seaweeds and extractive species in global aquaculture production. Rev Fisher Sci Aquacult 29:139–148. https://doi.org/10.1080/23308249.2020.1810626

Cock JM, Sterck L, Rouz P et al (2010) The *Ectocarpus* genome and the independent evolution of multicellularity in brown algae. Nature 465:617–621. https://doi.org/10.1038/nature09016

Cormier A, Avia K, Sterck L et al (2016) Re-annotation, improved large-scale assembly and establishment of a catalogue of noncoding loci for the genome of the model brown alga *Ectocarpus*. New Phytol 214:219–232. https://doi.org/10.1111/nph.14321

Couceiro L, Gac ML, Hunsperger HM et al (2015) Evolution and maintenance of haploid–diploid life cycles in natural populations: The case of the marine brown alga Ectocarpus. Evolution 69:1808–1822. https://doi.org/10.1111/evo.12702

Coudert Y, Bell NE, Edelin C, Harrison CJ (2017) Multiple innovations underpinned branching form diversification in mosses. New Phytol 215:840–850. https://doi.org/10.1111/nph.14553

Crépineau F, Roscoe T, Kaas R et al (2000) Characterisation of complementary DNAs from the expressed sequence tag analysis of life cycle stages of *Laminaria digitata* (Phaeophyceae). Plant Mol Biol 43:503–513. https://doi.org/10.1023/A:1006489920808

Debernardi JM, Tricoli DM, Ercoli MF et al (2020) A GRF-GIF chimeric protein improves the regeneration efficiency of transgenic plants. Nat Biotechnol 38:1274–1279. https://doi.org/10.1038/s41587-020-0703-0

Deng Y, Yao J, Wang X et al (2012) Transcriptome sequencing and comparative analysis of *Saccharina japonica* (Laminariales, Phaeophyceae) under blue light induction. PLoS One 7:e39704. https://doi.org/10.1371/journal.pone.0039704

Ding H, Guo L, Li X, Yang G (2019) Transcriptome analysis of kelp *Saccharina japonica* unveils its weird transcripts and metabolite shift of main components at different sporophyte developmental stages. J Ocean Limnol 37:640–650. https://doi.org/10.1007/s00343-019-8019-y

Du F, Jiao Y (2020) Mechanical control of plant morphogenesis: concepts and progress. Curr Opin Plant Biol 57:16–23. https://doi.org/10.1016/j.pbi.2020.05.008

Ebbing A, Pierik R, Bouma T, et al (2020) How light and biomass density influence the reproduction of delayed Saccharina latissima gametophytes (Phaeophyceae). J Phycol 56:709–718. https://doi.org/10.1111/jpy.12976

FAO (2020) The state of world fisheries and aquaculture 2020. Sustain Act. https://doi.org/10.4060/ca9229en

Fischer CG, Garnett T (2016) Plates, pyramids, planet: developments in national healthy and sustainable dietary guidelines: a state of play assessment. FAO, Rome. http://www.fao.org/documents/card/en/c/d8dfeaf1-f859-4191-954f-e8e1388cd0b7/

Forbord S, Steinhovden KB, Solvang T et al (2019) Effect of seeding methods and hatchery periods on sea cultivation of *Saccharina latissima* (Phaeophyceae): a Norwegian case study. J Appl Phycol. https://doi.org/10.1007/s10811-019-01936-0

Fritsch FE (1945) The structure and reproduction of the algae. Cambridge University Press, Cambridge

Gardiner B, Berry P, Moulia B (2016) Review: wind impacts on plant growth, mechanics and damage. Plant Sci 245:94–118. https://doi.org/10.1016/j.plantsci.2016.01.006

Geitmann A, Ortega JKE (2009) Mechanics and modeling of plant cell growth. Trends Plant Sci 14:467–478. https://doi.org/10.1016/j.tplants.2009.07.006

Goecke F, Klemetsdal G, Ergon Å (2020) Cultivar development of kelps for commercial cultivation—Past lessons and future prospects. Front Mar Sci 8:110. https://doi.org/10.3389/fmars.2020.00110

Hwang EK, Yotsukura N, Pang SJ et al (2019) Seaweed breeding programs and progress in eastern Asian countries. Phycologia 58:484–495. https://doi.org/10.1080/00318884.2019.1639436

Johnson MD, Hsieh WW, Cannon AJ et al (2016) Crop yield forecasting on the Canadian Prairies by remotely sensed vegetation indices and machine learning methods. Agric For Meteorol 218–219:74–84. https://doi.org/10.1016/j.agrformet.2015.11.003

Katsaros C, Reiss H-D, Schnepf E (1996) Freeze-fracture studies in brown algae: putative cellulose-synthesizing complexes on the plasma membrane. Eur J Phycol 31:41–48. https://doi.org/10.1080/09670269600651171

Katsaros CI, Karyophyllis DA, Galatis BD (2002) Cortical F-actin underlies cellulose microfibril patterning in brown algal cells. Phycologia 41:178–183. https://doi.org/10.2216/i0031-8884-41-2-178.1

Katsaros C, Karyophyllis D, Galatis B (2006) Cytoskeleton and morphogenesis in brown algae. Ann Bot 97:679–693. https://doi.org/10.1093/aob/mcl023

Kinoshita N, Nagasato C, Motomura T (2017) Chemotactic movement in sperm of the oogamous brown algae, *Saccharina japonica* and *Fucus distichus*. Protoplasma 254:547–555. https://doi.org/10.1007/s00709-016-0974-y

Knoblauch J, Drobnitch ST, Peters WS, Knoblauch M (2016) In situ microscopy reveals reversible cell wall swelling in kelp sieve tubes: one mechanism for turgor generation and flow control? Plant Cell Environ 39:1727–1736. https://doi.org/10.1111/pce.12736

Küpper FC (2015) Iodine in seaweeds – Two centuries of research. In: Kim S-K (ed) Springer handbook of marine biotechnology. Springer, Berlin, Heidelberg, pp 591–596

Lähteenmäki-Uutela A, Rahikainen M, Lonkila A, Yang B (2021) Alternative proteins and EU food law. Food Control 130:108336. https://doi.org/10.1016/j.foodcont.2021.108336

LeCun Y, Bengio Y, Hinton G (2015) Deep learning. Nature 521:436–444. https://doi.org/10.1038/nature14539

Li H, Monteiro C, Heinrich S et al (2020) Responses of the kelp *Saccharina latissima* (Phaeophyceae) to the warming Arctic: from physiology to transcriptomics. Physiol Plant 168:5–26. https://doi.org/10.1111/ppl.13009

Liang Z, Brown RC, Fletcher JC, Opsahl-Sorteberg H-G (2015) Calpain-mediated positional information directs cell wall orientation to sustain plant stem cell activity, growth and development. Plant Cell Physiol 56:1855–1866. https://doi.org/10.1093/pcp/pcv110

Lid SE, Olsen L, Nestestog R et al (2005) Mutation in the *Arabidopisis thaliana* DEK1 calpain gene perturbs endosperm and embryo development while over-expression affects organ development globally. Planta 221:339–351. https://doi.org/10.1007/s00425-004-1448-6

Liu F, Wang W, Sun X et al (2014) RNA-Seq revealed complex response to heat stress on transcriptomic level in *Saccharina japonica* (Laminariales, Phaeophyta). J Appl Phycol 26:1585–1596. https://doi.org/10.1007/s10811-013-0188-z

Liu T, Wang X, Wang G et al (2019) Evolution of complex thallus alga: genome sequencing of *Saccharina japonica*. Front Genet 10:378. https://doi.org/10.3389/fgene.2019.00378

Lüning K (1980) Critical levels of light and temperature regulating the gametogenesis of three *Laminaria* species (Phaeophyceae). J Phycol 16:1–15. https://doi.org/10.1111/j.1529-8817.1980.tb02992.x

Lüning K (1988) Photoperiodic control of sorus formation in the brown alga *Laminaria saccharina*. Mar Ecol Prog Ser 45:137–144. https://doi.org/10.3354/meps045137

Maier I, Hertweck C, Boland W (2001) Stereochemical specificity of lamoxirene, the sperm-releasing pheromone in kelp (Laminariales, Phaeophyceae). Biol Bull 201:121–125. https://doi.org/10.2307/1543327

Maloney KM, Halverson GP, Schiffbauer JD et al (2021) New multicellular marine macroalgae from the early Tonian of northwestern Canada. Geology 49:743–747. https://doi.org/10.1130/G48508.1

Mantri VA, Kavale MG, Kazi MA (2020) Seaweed biodiversity of India: Reviewing current knowledge to identify gaps, challenges, and opportunities. Diversity 12:13. https://doi.org/10.3390/d12010013

Marston DJ, Anderson KL, Swift MF et al (2019) High Rac1 activity is functionally translated into cytosolic structures with unique nanoscale cytoskeletal architecture. PNAS 116:1267–1272. https://doi.org/10.1073/pnas.1808830116

Monteiro C, Heinrich S, Bartsch I et al (2019a) Temperature modulates sex-biased gene expression in the gametophytes of the kelp Saccharina latissima. Front Mar Sci 6:769. https://doi.org/10.3389/fmars.2019.00769

Monteiro C, Li H, Bischof K et al (2019b) Is geographical variation driving the transcriptomic responses to multiple stressors in the kelp Saccharina latissima? BMC Plant Biol 19:513. https://doi.org/10.1186/s12870-019-2124-0

Mooney KM, Beatty GE, Elsäßer B et al (2018) Hierarchical structuring of genetic variation at differing geographic scales in the cultivated sugar kelp *Saccharina latissima*. Mar Environ Res 142:108–115. https://doi.org/10.1016/j.marenvres.2018.09.029

Muzzy RA, Hable WE (2008) Rac1 function during fucoid development. Plant Signal Behav 3:717–719

Neiva J, Paulino C, Nielsen MM et al (2018) Glacial vicariance drives phylogeographic diversification in the amphi-boreal kelp *Saccharina latissima*. Sci Rep 8:1112. https://doi.org/10.1038/s41598-018-19620-7

Nishitsuji K, Arimoto A, Iwai K et al (2016) A draft genome of the brown alga, *Cladosiphon okamuranus*, S-strain: a platform for future studies of 'mozuku' biology. DNA Res 23:561–570. https://doi.org/10.1093/dnares/dsw039

Nishitsuji K, Arimoto A, Higa Y et al (2019) Draft genome of the brown alga, *Nemacystus decipiens*, Onna-1 strain: Fusion of genes involved in the sulfated fucan biosynthesis pathway. Sci Rep 9:4607. https://doi.org/10.1038/s41598-019-40955-2

Nishitsuji K, Arimoto A, Yonashiro Y et al (2020) Comparative genomics of four strains of the edible brown alga, *Cladosiphon okamuranus*. BMC Genomics 21:422. https://doi.org/10.1186/s12864-020-06792-8

Pang SJ, Lüning K (2004) Breaking seasonal limitation: year-round sporogenesis in the brown alga *Laminaria saccharina* by blocking the transport of putative sporulation inhibitors. Aquaculture 240:531–541. https://doi.org/10.1016/j.aquaculture.2004.06.034

Parfrey LW, Lahr DJ, Katz LA (2008) The dynamic nature of eukaryotic Genomes. Mol Biol Evol 25:787–794. https://doi.org/10.1093/molbev/msn032

Perroud P-F, Demko V, Johansen W et al (2014) Defective Kernel 1 (DEK1) is required for three-dimensional growth in *Physcomitrella patens*. New Phytol 203:794–804. https://doi.org/10.1111/nph.12844

Rabillé H, Billoud B, Tesson B et al (2019) The brown algal mode of tip growth: Keeping stress under control. PLoS Biol 17:e2005258. https://doi.org/10.1371/journal.pbio.2005258

Riquelme M (2013) Tip growth in filamentous fungi: a road trip to the apex. Annu Rev Microbiol 67:587–609. https://doi.org/10.1146/annurev-micro-092412-155652

Rolin C, Inkster R, Laing J, McEvoy L (2017) Regrowth and biofouling in two species of cultivated kelp in the Shetland Islands, UK. J Appl Phycol 29:2351–2361. https://doi.org/10.1007/s10811-017-1092-8

Rouli L, Merhej V, Fournier P-E, Raoult D (2015) The bacterial pangenome as a new tool for analysing pathogenic bacteria. New Microb New Infect 7:72–85. https://doi.org/10.1016/j.nmni.2015.06.005

Shan T, Yuan J, Su L et al (2020) First genome of the brown alga *Undaria pinnatifida*: Chromosome-level assembly using PacBio and Hi-C technologies. Front Genet 11:140. https://doi.org/10.3389/fgene.2020.00140

Shao Z, Wang W, Zhang P et al (2019a) Genome-wide identification of genes involved in carbon fixation in *Saccharina japonica* and responses of putative C4-related genes to bicarbonate concentration and light intensity. Plant Physiol Biochem 137:75–83. https://doi.org/10.1016/j.plaphy.2019.01.032

Shao Z, Zhang P, Lu C et al (2019b) Transcriptome sequencing of *Saccharina japonica* sporophytes during whole developmental periods reveals regulatory networks underlying alginate and mannitol biosynthesis. BMC Genomics 20:975. https://doi.org/10.1186/s12864-019-6366-x

Starko S, Soto Gomez M, Darby H et al (2019) A comprehensive kelp phylogeny sheds light on the evolution of an ecosystem. Mol Phylogenet Evol 136:138–150. https://doi.org/10.1016/j.ympev.2019.04.012

Stévant P, Rebours C, Chapman A (2017) Seaweed aquaculture in Norway: recent industrial developments and future perspectives. Aquacult Int 25:1373–1390. https://doi.org/10.1007/s10499-017-0120-7

Tettelin H, Masignani V, Cieslewicz MJ et al (2005) Genome analysis of multiple pathogenic isolates of *Streptococcus agalactiae*: Implications for the microbial "pan-genome.". Proc Natl Acad Sci 102:13950–13955. https://doi.org/10.1073/pnas.0506758102

Theodorou I, Charrier B (2021) Brown Algae: *Ectocarpus* and *Saccharina* as experimental models for developmental biology. In: Boutet A, Schierwater B (eds) Handbook of Marine Model Organisms in Experimental Biology. CRC Press, pp 27–46. https://doi.org/10.1201/9781003217503

Tran D, Galletti R, Neumann ED et al (2017) A mechanosensitive Ca 2+ channel activity is dependent on the developmental regulator DEK1. Nat Commun 8:1009. https://doi.org/10.1038/s41467-017-00878-w

Visch W, Rad-Menéndez C, Nylund GM et al (2019) Underpinning the development of seaweed biotechnology: Cryopreservation of brown algae (*Saccharina latissima*) gametophytes. Biopreserv Biobank 17:378–386. https://doi.org/10.1089/bio.2018.0147

Wade R, Augyte S, Harden M et al (2020) Macroalgal germplasm banking for conservation, food security, and industry. PLoS Biol 18:e3000641. https://doi.org/10.1371/journal.pbio.3000641

Wang B, Zhang E, Gu Y et al (2011) Cryopreservation of brown algae gametophytes of *Undaria pinnatifida* by encapsulation–vitrification. Aquaculture 317:89–93. https://doi.org/10.1016/j.aquaculture.2011.04.014

Wang W-J, Wang F-J, Sun X-T et al (2013) Comparison of transcriptome under red and blue light culture of *Saccharina japonica* (Phaeophyceae). Planta 237:1123–1133. https://doi.org/10.1007/s00425-012-1831-7

Wang D, Shimoda Y, Kurosawa H et al (2015) Excretion patterns of arsenic and its metabolites in human saliva and urine after ingestion of Chinese seaweed. Int J Environ Anal Chem 95:379–389. https://doi.org/10.1080/03067319.2015.1036860

Wang S, Lin L, Shi Y et al (2020) First draft genome assembly of the seaweed *Sargassum fusiforme*. Front Genet 11:590065. https://doi.org/10.3389/fgene.2020.590065

Webb S (2018) Deep learning for biology. Nature 554:555–557. https://doi.org/10.1038/d41586-018-02174-z

Ye N, Zhang X, Miao M et al (2015) *Saccharina* genomes provide novel insight into kelp biology. Nat Commun 6:6986. https://doi.org/10.1038/ncomms7986

Yokoi K, Konomi A (2012) Toxicity of so-called edible hijiki seaweed (*Sargassum fusiforme*) containing inorganic arsenic. Regul Toxicol Pharmacol 63:291–297. https://doi.org/10.1016/j.yrtph.2012.04.006

Zhou J, Troyanskaya OG (2015) Predicting effects of noncoding variants with deep learning–based sequence model. Nat Methods 12:931–934. https://doi.org/10.1038/nmeth.3547

Zhu G, Ebbing A, Bouma TJ, Timmermans KR (2021) Morphological and physiological plasticity of *Saccharina latissima* (Phaeophyceae) in response to different hydrodynamic conditions and nutrient availability. J Appl Phycol 33:2471–2483. https://doi.org/10.1007/s10811-021-02428-w

Index

A. Ranga Rao, G. A. Ravishankar (eds.), *Sustainable Global Resources of Seaweeds Volume 2*, https://doi.org/10.1007/978-3-030-92174-3

The manufacturer's authorised representative in the EU is Springer Nature Customer Service Centre GmbH, Europaplatz 3, 69115 Heidelberg, Germany. If you have any concerns regarding our products, please contact ProductSafety@springernature.com

Printed and bound by CPI Group (UK) Ltd, Croydon, CR0 4YY
30/09/2026
02225496-0017